Trace Metals in the Environment 1

Heavy Metals in the Environment

Trace Metals in the Environment

Series Editor: Jerome O. Nriagu
National Water Research Institute
Box 5050
Burlington
Ontario
Canada L7R 4AG

Trace Metals in the Environment 1

Heavy Metals in the Environment

Edited by

J.-P. Vernet

Institute F.-A. Forel, University of Geneva,
Versoix, Switzerland

Elsevier
Amsterdam – London – New York – Tokyo 1991

ELSEVIER SCIENCE PUBLISHERS B.V.
Molenwerf 1
P.O. Box 211, 1000 AE Amsterdam, The Netherlands

Distributors for the United States and Canada:

ELSEVIER SCIENCE PUBLISHING COMPANY INC.
655, Avenue of the Americas
New York, NY 10010, U.S.A.

ISBN 0-444-89064-5

This book is printed on acid-free paper.

Printed in the Netherlands

PREFACE TO THE SERIES

Environmental pollution with toxic metals has become a global phenomenon. Because of the growing apprehension about the potential effects of the metallic contaminants on various fauna and flora, the research on fundamental, applied and health aspects of trace metals in the environment has mushroomed to the point that it has become difficult to keep track of recent developments in particular topics. With the end of the Cold War, considerably more attention is expected to be paid to environmental degradation which will likely lead to more research on the consequences of local and global pollution with persistent toxic metals. The scientific literature on trace metals in the environment is therefore expected to continue to grow.

The series, *Trace Metals in the Environment*, is aimed at filling a need for a continuous forum to provide the reader with a charted route into the transdisciplinary studies in a rapidly evolving scientific field. Volumes in the series will offer critical evaluations and creative reviews covering all aspects of trace metals *and metalloids* in the environment, including their natural and anthropogenic sources, their distribution, behavior and fate in air, water and soil. Timely coverages are also expected on (but will not be restricted to) the local, regional and global cycles of trace metals and their disturbance by human activities; the speciation and dynamics of metallic contaminants; mathematical correlations and models; the accumulation of pollutant metals in animal and human food chains; the fundamental principles and mechanisms of toxicity; the biochemistry and symptoms of toxicity in the tissue, the organism and an ecosystem; control measures and technology; legislative agenda and protocols for setting standards; rehabilitation of contaminated ecosystems and the assessment of risks associated with increased metal pollution. Field and laboratory studies will be given equal prominence.

Volumes in the series thus will provide integrative and comprehensive accounts of current research on one of the major environmental problems of our time. They should be of interest to graduate students and practising scientists in the fields of toxicology, environmental science and engineering, public health and environmental control. It is my hope that they will be read by any scientist, engineer or medical practitioner who is concerned about the impacts of metallic contaminants on our health and our life support systems.

Since the study of trace metals and metalloids deals with a wide range of systems and principles, a flexible attitude has been adopted for the contents and the preparation of material for *Trace Metals in the Environment*. We welcome suggestions and encourage readers to submit proposals and manuscripts for volumes in the series. I take this opportunity to dedicate this and future volumes to scientists from all the disciplines who have contributed to the fund of knowledge on the biogeochemistry of toxic metals in our environment.

Jerome O. Nriagu
Series Editor

PREFACE TO VOLUME 1

The continuing interest in the role of heavy metals in the environment since the mid sixties has led to seven international conferences, the first in 1975 in Toronto, the 7th in 1989 in Geneva. At this last conference, over 1153 pages of extended abstracts, or more precisely condensed texts, were presented and these were published by CEP Consultants Ltd in Edinburgh. The present volume, a selection of the best papers from the Geneva conference, is a reference book of the actual state of research in heavy metals in the environment.

The originality of this volume is that it covers all fields related to the problems of heavy metals. Not only biologists and chemists will find it of interest, but also geologists, geochemists, sanitary engineers, physicians, ecotoxicologists and all scientists who deal with environmental problems. As the application of heavy metals is so large, this work attempts to give an overview as broad as possible of the problems in the various disciplines.

Through their effort and expertise the different authors have contributed to the success of this volume. The editor thanks all the authors for their important contribution and Elsevier Science Publishers B.V. in Amsterdam for their most appreciated editorial assistance.

Jean-Pierre Vernet

F.-A. Forel Institute
University of Geneva
Switzerland

May 1, 1991

LIST OF CONTRIBUTORS

F. AXENFELD	Norwegian Institute for Air Research, Lillestrom, Norway.
A. BOUDOU	Laboratoire d'Ecologie fondamentale et Ecotoxicologie, Université de Bordeaux, Talence, France.
P.G.C. CAMPBELL	Institut National de la Recherche Scientifique, Université du Québec, Sainte-Foy, Québec, Canada.
J.S. CHEN	Department of Geography, Peking University, Beijing, China.
C.Q. CHENG	Department of Geography, Peking University, Beijing, China.
H. DAHLGAARD	Department of Environmental Science and Technology, Riso National Laboratory, Roskilde, Denmark.
D.Z. DOLINAC	Institute of Navy Medicine, Split, Yugoslavia.
J. DOMINIK	Institute F.-A. Forel, University of Geneva, Versoix, Switzerland.
H. DÖRR	Institut für Umweltphysik der Universität Heidelberg, Heidelberg, Germany.
M.N. DUBRAVČIĆ	Institute of Navy Medicine, Split, Yugoslavia.
P.-Y. FAVARGER	Institute F.-A. Forel, University of Geneva, Versoix, Switzerland.
U. FÖRSTNER	Environment Engineering Section, Technical University Harburg, Hamburg, Germany.
E. GJENGEDAL	University of Trondheim, Department of Chemistry, Dragvoll, Norway.
S.K. GUPTA	Swiss Federal Research Station for Hygiene of Environment, Liebefeld-Bern, Switzerland.
G. HO	Department of Environmental Science, Murdoch University, Murdoch, Western-Australia.
H. HOFSTEDE	Department of Environmental Science, Murdoch University, Murdoch, Western-Australia.
B.D. HONEYMAN	Department of Marine Sciences, Texas A & M University, Galvestone, Texas, U.S.A.
C.P. HUANG	Department of Civil Engineering, University of Delaware, Newark, Delaware, U.S.A.
J.-P. HUANG	Graduate Institute of Environmental Engineering, National Chao Tung University, Shin Chu, Taiwan, R.O.C.
G. IRION	Environment Engineering Section, Technical University Harburg, Hamburg, Germany.

T.A. JACKSON	National Water Research Institute, Canada Centre for Inland Waters, Burlington, Ontario, Canada.
M. KERSTEN	Environment Engineering Section, Technical University Harburg, Hamburg, Germany.
L.K. KOOPAL	Department of Physical and Colloid Chemistry, Wageningen Agricultural University, Wageningen, The Netherlands.
P.I. KULIĆ	Institute of Navy Medicine, Split, Yugoslavia.
L.B. KURAJICA	Institute of Navy Medicine, Split, Yugoslavia.
E. LØBERSLI	Directorate for Nature Management, Trondheim, Norway.
J.-L. LOIZEAU	Institute F.-A. Forel, University of Geneva, Versoix, Switzerland.
C. MARTIN	Institute F.-A. Forel, University of Geneva, Versoix, Switzerland.
R. MAYER	Gesamthochschule Kassel, Landschaftökologie / Bodenkunde, Kassel, Germany.
A. MANGINI	Heidelberger Akademie der Wissenschaften, Heidelberg, Germany.
N.M. MLADLINIĆ	Institute of Navy Medicine, Split, Yugoslavia.
A. MOREHART	Department of Plant and Soil Sciences, University of Delaware, Newark, Delaware, U.S.A.
J. MÜNCH	Norwegian Institute for Air Research, Lillestrom, Norway.
K.O. MÜNNICH	Institut für Umweltphysik der Universität Heidelberg, Heidelberg, Germany.
M.M. NEDERLOF	Department of Soil Science and Plant Nutrition, Wageningen Agricultural University, Wageningen, The Netherlands.
M.M. ORLOV	Institute of Navy Medicine, Split, Yugoslavia.
J.M. PACYNA	Norwegian Institute for Air Research, Lillestrom, Norway.
J.J. PAGANO	State University of New York, Oswego, U.S.A.
F. RIBEYRE	Laboratoire d'Ecologie fondamentale et Ecotoxicologie, Université de Bordeaux, Talence, France.
P.H. SANTSCHI	Department of Marine Sciences, Texas A & M University, Galvestone, Texas, U.S.A.

E. SAOUTER	Laboratoire d'Ecologie fondamentale et Ecotoxicologie, Université de Bordeaux, Talence, France.
P.W. SCHINDLER	Department of Inorganic Chemistry, University of Bern, Switzerland.
W. SCHMITZ	Institut für Sedimentforschung der Universität Heidelberg, Heidelberg, Germany.
R.J. SCRUDATO	State University of New York, Oswego, U.S.A.
D. SPAN	Institute F.-A. Forel, University of Geneva, Versoix, Switzerland.
E. STEINNES	University of Trondheim, Department of Chemistry, Dragvoll, Norway.
A. TESSIER	Institut National de la Recherche Scientifique, Université du Québec, Sainte-Foy, Québec, Canada.
R.L. THOMAS	Geology Department, University of Windsor, Windsor, Ontario, Canada.
C.-T. TIEN	Maryland Department of the Environment, Baltimore, Maryland, U.S.A.
W.H. VAN RIEMSDIJK	Department of Soil Science and Plant Nutrition, Wageningen Agricultural University, Wageningen, The Netherlands.
J.-P. VERNET	Institute F.-A. Forel, University of Geneva, Versoix, Switzerland.
F.Y. WANG	Department of Geography, Peking University, Beijing, China.
F. WEBER	Institut für Umweltphysik der Universität Heidelberg, Heidelberg, Germany.

CONTENTS

Chapter 1
Atmospheric transport: large scale transport, models

EUROPEAN INVENTORY OF TRACE METAL EMISSIONS TO THE ATMOSPHERE

J.M. Pacyna, J. Münch and F. Axenfeld

1. INTRODUCTION

Emission inventories are the key to effective air quality management programs, serving as a basis for the planning of control strategies to achieve ambient air quality goals. In the past few decades, there has been growing concern about environmental contamination by trace elements. They are emitted during various high-temperature processes, such as coal and oil combustion in electric power stations and industrial plants, roasting and smelting of ores in non-ferreous metal smelters, melting operations in ferrous foundaries, refuse incineration, and kiln operations in cement plants. There are also natural sources of trace elements, such as windblown dust, volcano eruptions, forest wildfires, sea-salt emissions and vegetation.

First emission surveys were focusing on the assessment of releases of trace elements from single point sources, such as power plants or smelters. The emission data were based on source measurements and/or material balances for a given source (1, 2).

In the late 1970's long-range transport of trace elements was revealed and soon after it was realized that natural cycles of elements in the environment can be altered not only in the vicinity of emission sources but also a few thousand kilometers far from emission source regions.

Major scientific questions have been formulated: (1) to what extent anthropogenic emissions have already altered the biogeochemical cycles of trace elements, and (2) what is the scale of these perturbations. Information on anthropogenic emissions of trace elements became of primary importance in order to attempt the answers.

The objective of this work is to present emission survey for trace elements in Europe. Major source categories are discussed. Main technological parameters affecting atmospheric emissions of trace elements are presented. Recommendations are given for the further research within the field of trace element emission inventories.

2. EMISSIONS OF TRACE ELEMENTS FROM COMBUSTION OF FOSSIL FUELS

Combustion of fossil fuels to produce electricity and heat is the main source for anthropogenic emissions of Be, Co, Hg, Mo, Ni, Sb, Se, Sn and V. Combustion processes are also important sources of atmospheric As, Cr, Cu, Mn, and Zn in the atmosphere.

2.1 Electric Power Stations

Electric power stations emit a half of the total amounts of trace elements produced by the combustion of fossil fuels. Most of these emissions come from conventional thermal power plants and depend on: 1) the contamination of fossil fuels, 2) the physico-chemical properties of trace elements affecting their behaviour during combustion, 3) plant design defining the combustion conditions, and 4) the efficiency of control equipment.

2.1.1 Coal-fired Power Plants. There are large differences between element concentrations in coals from various producing fields in the world. For some elements, such as As, V or Zn these differences are up to three orders of magnitude. Thus, it is difficult to generalize on impurities in coal. However, it can be concluded that lignites and sub-bituminous coals are less contaminated by trace elements than bituminous coals (1).

The plant design, and particularly the burner configuration, influences the emissions of trace elements. Wet bottom boilers (cyclone) produce the highest emissions among the coal-fired utility boilers, because of the need to operate above the ash melting temperature. At a typical peak temperature of ca. $1550^{0}C$ the volatile trace elements in the coal ash evaporate (1). Later they condense as submicron aerosol particles, or on the surface of ash particles as the flue gas cools to 370^{0}-$450^{0}C$ in the convective heat transfer sections of the power plant. Details on the volatilization-condensation process are available from the literature (e.g. 4, 5, 6). The emission rates for other types of boilers, such as wall-fired and tangential units seem to be lower due to the lower temperature involved.

The other main parameter influencing the emission rates of trace elements from any source is the type and the efficiency of the control equipment. Coal-fired power plants are mostly equipped with either electrostatic precipitators (ESP), fabric filters, or wet scrubbers, with the ESP being most frequently used. The ESP's are particularly efficient in removing all types of particles <0.01 μm dia., and they can tolerate operating temperatures as high as 450^0 (7). Concerning the particles size distribution from the ESP's in coal-fired power plant, it was found that the particle mass containing trace elements is concentrated in two size ranges: 1) at ca. 0.15 μm dia., and 2) between 2 and 8 μm dia. (8). This information is particularly important when considering health effects, as well as the long-range transport of trace elements and their migration through the environment.

2.1.2 <u>Oil-fired Power Plants</u>. Oil-fired boilers do not require ash hoppers and ash pits. Fly ash removal equipment is not generally required, although such equipment is used at some oil-burning plants. Dust collectors are used during soot blowing. This equipment serves principally to collect particles >10 μm dia. The emission of particles containing trace elements from an oil-fired unit without stack gas cleaning is comparable to a coal-fired unit with better than 99% cleaning efficiency.

2.2 <u>Industrial, Commercial and Residential Boilers</u>

The other half of the trace element emissions from combustion of fossil fuels comes from industrial, commercial and residential boilers. Large industrial complexes generate electric power or process steam mainly in the pulverized and cyclone-type units. The units are similar in design to those discussed under electric utilities and the emission rates are the same. Some differences in the chemical composition and particle size of stack dust occur due to different control devices. High efficiency cyclones are mostly employed in industrial plants, but the ESP's are also used.

2.3 <u>Estimation of Emissions from European Power Plants</u>

More than 630 conventional power plants in Europe burning coal and oil were taken into account when presenting the modified version of the survey. Detailed information was obtained on the production capacity, the type of fuel burnt, and the geographical position of plants. In addition, information was available on the type of boilers used in a given country, and the type and efficiency of the control equipment. This information was used to estimate emission factors and the emisssions of trace elements from the European power plants and industrial, commercial and residential boilers. The results for As, Cd, Hg, Pb and Zn in 1982 are summarized in Table 1. The above elements have been selected due to their toxicity and ubiquitous appearance in the environment.

However, there were also some assumptions made resulting in increasing the inaccuracies of estimates. More than 70% of the total electricity production in the European conventional power plants comes from single-fired installations (e.g. hard coal power plants), ca. 25% from dual-fired installations (mainly hard coal/petroleum products and petroleum products/ natural gas) and the rest from triple-fired installations (mainly hard coal/petroleum products/natural gas). The use of multi-fired installations makes the emission calculations more difficult and results in some minor overestimation because very often hard coal emission factors are used to assess emissions from hard coal/natural gas-fired power plants.

Concerning the control equipment in coal power plants, a venturi wet scrubber system is more efficient in removing As, Cd and Pb from a flue gas stream than electrostatic precipitators (ESP) (9). The emission factors in Table 1 were calculated assuming only 99.5% efficiency of ESP in coal-fired power plants and this may cause an overestimation of the emission factors for some large power plants in Europe.

2.4 Control of Trace Element Emissions by Non-conventional Methods

Although the control of particle emissions is the most important device for trace element removal, flue gas desulphurization (FGD) is also efficient in retaining both particulate and vapour phase elements. These methods were recently reviewed by Smith (10), separately for wet and dry FGD systems. The use of FGD systems improve the removal efficiencies for As, Br, Cl, F, Hg, I and Se (11). The removal of halogens, mercuric chloride and selenium dioxide can be analogous to the SO_2 absorption.

Also some non-conventional methods of combustion such as fluidized bed combustion (FBC), may result in lower emissions of trace elements due to less volatilization at lower combustion temperatures.

Table 1: Emission factors and emissions of As, Cd, Hg, Pb, and Zn from fossil fuel combustion in electric utility, industrial, residential and commercial boilers in Europe in 1982.

Type of boiler	Emission factor		European emission in 1982, in t/y
	Unit	Value	
As			
A. Electric utility boiler			
- coal	mg/GJ	15-100	237
- oil	mg/GJ	8-60	90
B. Industrial, commercial and residential boilers	g/t fuel used	0.3-1.7	383
Cd			
A. Electric utility boiler			
- coal	mg/GJ	5-30	81
- oil	mg/GJ	4-30	43
B. Industrial, commercial and residential boilers	g/t fuel used	0.1-0.7	146
Hg			
A. Electric utility boiler			
- coal	mg/GJ	9-42	122
B. Industrial, commercial and residential boilers	g/t fuel used	0.1-0.6	133
Pb			
A. Electric utility boiler			
- coal	mg/GJ	50-300	844
- oil	mg/GJ	45-300	457
B. Industrial, commercial and residential boilers	g/t fuel used	1.2-7.7	1 606
Zn			
A. Electric utility boiler			
- coal	mg/GJ	75-500	1 190
- oil	mg/GJ	30-210	324
B. Industrial, commercial and residential boilers	g/t fuel used	1.7-11.5	1 777

3. EMISSIONS OF TRACE ELEMENTS FROM NON-FERROUS METAL INDUSTRY

The largest emissions of atmospheric As, Cd, Cu, In, and Zn come from the pyrometallurgical processes employed in the production of lead, copper and zinc. The type of technology employed in smelters, refineries, and other operations such as roasting, has a major influence on the trace element emission.

3.1 Primary Non-ferrous Metal Production

Generally, there are 3 steps in the non-ferrous metal production: 1) roasting of ores to remove sulphur, 2) smelting of roaster products, and 3) refining of smelting products. Roasting of ore concentrates and low grade ore enrichment are common in copper, nickel, lead, zinc and cadmium production. The processes occur at about 1200 ^{0}C, and volatile elements, such as arsenic, cadmium and mercury, are partly removed from the concentrates, and remain in the flue gases. As much as 95% of arsenic contained in copper ores may be driven off in the gases. Most of this is recycled. All the mercury in a smelter will vaporize early in the process and enter the gas stream, which is usually passed through an acid plant to remove mercury.

Smelting, can be either thermal (in horizontal or veritcal retort) or electrolytic extraction. The old, pyrometallurgic zinc plants employing thermal smelting is a major cadmium emissions source (12). The Cd emissions from primary copper and lead production are lower than the Cd emission from pyrometallurgic zinc plants.

Smelting and converting in copper-nickel smelters are important sources of As emissions. Behaviour of the metal depends primarily upon the process itself. Several smelting technologies are currently used in the copper-nickel industry, including reverberatory smelting, flush smelting, and the Noranda and Mitsubishi Continuous Copper Smelting Process. About 90% of copper is presently being produced by these processes. The rest comes from hydrometallurgic techniques. The portion of arsenic volatilized during the smelting operation ranges from 20% of the arsenic input for the reverberatory furnace to 85% for the Noranda reactor for making high grade matte (13). The effluent gases include also about 85% of the lead input in the copper-nickel ores and 80% of the cadmium input. Most of the metal is usually recycled in flue dust, but the remainder is emitted into the atmosphere.

Most of the airborne trace element emissions in lead ore processing are due to the roasting operations. Reduction in the blast furnaces is also a source of air pollution by trace elements. The emission rates of As, Cd, Pb and Hg from primary non-ferrous metal production are highly sensitive to differences in the efficiency of control devices. High-efficiency control devices are often employed in smelters. Emissions from roaster facilites, as well as from sintering processes, are mainly controlled with fabric filters and electrostatic precipitators.

3.2 Secondary Non-ferrous Metal Production

The secondary works produce ca. 20 per cent of the total Zn and Ni and ca. 40 per cent of the total Cu and Pb production.

The trace element emissions from secondary non-ferrous metal production are significantly lower than the emissions from primary works. Trace elements are introduced to the furnace of secondary works together with the scrap material. The chemical composition of input scrap and type and efficiency of control devices are two major factors that affect the trace element emissions from this source.

3.3 Estimation of Emissions from European Non-ferrous Metal Smelters

Atmospheric emissions of trace elements were estimated for 89 primary and 60 major secondary non-ferrous metal works in Europe in 1982. The emission factors were calculated separately for the production techniques presented above. Individual emission factors were calculated for some of these owrks, e.g. for the German plants. The emission factors were used together with the information on the production of copper, zinc and lead in the works to calculate the emissions presented in Table 2 for As, Cd, Hg, Pb and Zn.

4. OTHER INDUSTRIAL SOURCES OF ATMOSPHERIC TRACE ELEMENTS

Other major anthropogenic sources of atmospheric trace elements include high-temperature processes in steel and iron manufacturing, cement production and in a number of small plants where metals are applied in the production process. Emissions from 220 major steel and iron manufacturing plants in Europe in 1982 were calculated. Special emphasis was placed on the technology employed in the plants. The calculations for As, Cd, Pb and Zn are given in Table 3. Large differences can be observed between the emission factors for various production technologies due to the reuse of trace element containing scrap in the electric arc method. Much smaller differences exist between emissions from wet and dry types of kilns in cement production. As many as 227 cement plants were considered in the 1982 European survey.

Table 2: Emission factors (in g/t) and emissions of As, Cd, Hg, Pb and Zn (in t) from pyrometallurgical non-ferrous metal works in Europe in 1982 (14).

	Primary works						Secondary works			
	Cu production		Zn production		Pb production		Cu production		Pb production	
	Factor	Emission	Factor	Emission	Factor	Emission	Factor	Emission	Factor	Emission
As	1000-1500	3040	50-150	222	200-400	364				
Cd	10-400	257	50-1000	454	5-50	12	4	5	2.5	3.1
Mg			8-50	19	2-4	4				
Pb	1300-2600	3990	1200-2500	4030	3000-8000	3640	50-200	407	770	820
Zn	500-1000	2060	10000-18400	18400	50-120	134	500-1600	1780	300	272

Table 3: Emission factors (in g/t) and emissions of As, Cd, Pb, and Zn (in t) from iron and steel works in Europe in 1982 (14).

Operation	As		Cd		Pb		Zn	
	Factor	Emission	Factor	Emission	Factor	Emission	Factor	Emission
I Iron sintering	1.0	226	0.08	18	11.2	2550	27.0	6150
II Steel making:								
- Basic oxygen furnace (BO)			0.02	3	1.6	270	3.9	650
- Electric arc furnace (BO)			0.40	28	11.2	790	27.0	1910
- Open hearth furnace (OH)			0.03	3	2.7	290	6.5	700

5. EMISSIONS FROM REFUSE INCINERATION

Generally, waste incineration is gaining much interest, especially in Western Europe, due to emissions of Cd, Hg, Pb, Sb and other trace elements, and due to a need to incinerate an increasing amount of wastes.

Trace element emissions from municipal incinerators depend on the proportion of combustible and non-combustible material in the refuse input, the chemical composition of the input, the incinerator design, and the efficiency of control devices. As these factors vary from one country to another, the application of emissions factors is very limited. Thus, the national emissions from the literature were accepted for the 1982 European survey. Even these incomplete data show that incineration of wastes results in significant emissions of Cd, Hg, and Sb in Europe.

6. EMISSIONS OF Pb FROM GASOLINE COMBUSTION

Gasoline combustion is by far the largest source for the Pb emissions in Europe. This source contributed approximately 76% to the total Pb emissions, due to the large use of Pb additives in gasoline in 1982. For most of the European countries in 1982, the Pb content in gasoline was 0.4 g/l and low-lead and lead-free gasoline was only available in FRG, Norway, Sweden and Switzerland.

7. EMISSION INVENTORY FOR TRACE ELEMENTS EMITTED IN EUROPE IN 1982

Atmospheric emissions of trace elements from major anthropogenic sources in Europe were estimated on the basis of emission factors and statistical data on the consumption of raw materials and the production of industrial goods in 1982. Individual emission factors were used for major non-ferrous metal smelters and coal-, and oil-fired power plants in Europe. Collective emission factors were calculated for other source categories. The estimates are presented in Table 4. As many as 1500 point sources were considered in these calculations.

The country-by-country emission data for As, Cd, Hg, Pb and Zn in 1982 are presented in Table 5.

Table 4: European emissions of trace elements from various anthropogenic sources in 1982 (t/y).

Source	As	Be	Cd	Co	Cr	Cu	Hg	Mn	Mo	Ni	Pb	Sb	Se[1]	V	Zn	Zr
Conventional thermal power plants	327	21	124	787	1 196	1 377	122	1 011	352	4 580	1 301	122	155	12 600	1 514	732
Industrial, commercial, residential boilers	383	29	146	1 214	1 580	2 038	133	1 378	493	7 467	1 606	154	218	21 800	1 777	939
Fuel-wood combustion	30		15			1 500	13			375	420				599	
Gasoline combustion								92		1 330	68 337					
Mining	32					192		275		1 640	824				303	
Primary non-ferrous metal production																
- Copper	3 040[2]		257			7 850				140	3 992				2 056	
- Zinc	222[2]		454			440	19				4 029				18 401	
- Lead	364[2]		12			120	4				3 642				134	
Secondary non-ferrous metal production																
- Copper			5			61					407				1 781	
- Zinc															3 758	
- Lead			2								145				273	
Iron and steel manufacturing	227		53		15 400	1 710		10 770		340	3 900				9 409	
Refuse incineration	8		37		53	260	31	144		10	536	100	32	19	652	
Phosphate fertilizer production						77				77					193	
Cement production	205		12		663						613				3 744	
TOTAL	4 840	50	1 120	2 000	18 900	15 600	390[3]	13 600	850	16 000	89 750	380	420	34 400	44 600	1 670

1) Se emissions on particles.
2) Including emissions from secondary non-ferrous metal production.
3) Including the Hg emission of 71 t from chlor-alkali plants in Europe.

Table 5: Emissions of As, Cd, Hg, Pb and Zn from sources in various countries in Europe in 1982 (t/y).

Country	No. of point sources	Emission				
		As	Cd	Hg	Pb	Zn
ALBANIA	14	17.2	0.7	0.6	136	37
AUSTRIA	42	26.1	5.3	0.2	1122	235
BELGIUM	36	85.0	12.1	12.6	2097	694
BULGARIA	27	146.9	65.5	8.6	1569	1761
CSSR	55	93.7	21.6	14.9	1151	756
DENMARK	26	9.1	6.3	2.0	653	129
FINLAND	33	106.4	8.0	3.0	1122	217
FRANCE	94	143.7	31.8	16.7	8682	3637
GERMAN DEM.R	39	95.1	37.1	23.7	1749	819
GERMANY FED.R	233	351.0	81.1	64.2	5561	3699
GREAT BRITAIN	188	118.5	30.7	37.7	8615	2299
GREECE	14	14.8	3.2	1.5	1393	194
HUNGARY	49	16.1	4.7	2.9	596	199
ICELAND	2	0.1	.0	.0	39	1
IRELAND	10	4.1	0.8	0.2	437	43
ITALY	125	94.9	35.7	10.8	8591	1998
YUGOSLAVIA	35	272.0	85.8	6.4	1961	1958
LUXEMBURG	8	3.1	0.6	.0	165	93
NETHERLANDS	50	34.1	5.5	6.7	2205	294
NORWAY	20	41.2	2.1	1.3	727	117
POLAND	76	591.3	180.4	40.0	2956	4040
PORTUGAL	9	11.4	2.4	2.6	381	98
ROMANIA	51	116.2	43.4	16.1	1154	717
SPAIN	47	265.1	133.1	9.3	4227	3918
SWEDEN	49	180.9	16.4	7.3	1034	426
SWITZERLAND	30	3.7	0.9	0.1	450	62
SOVIET-UNION	137	2094.4	308.6	99.4	30924	13160
EUROPE	1499	~4840	~1120	~390	~89750	~44600

Total emissions of trace elements in Europe have also been spatially distributed within the EMEP grid of 150 km x 150 km. The emissions from point sources were distributed with the help of geographical coordinates of the sources. Information on population density within the EMEP grid has been used when distributing the emissions from area and mobile sources.

A spatial distribution of As, Cd, Pb and Zn emissions in 1982 is presented in Fig. 1-4, respectively.

8. ANNUAL CHANGES OF THE EUROPEAN EMISSIONS OF TRACE ELEMENTS

Any expectations of changes of emissions of trace elements from non-ferrous metal smelters in Europe are related primarily to changes of production technology and modernization of environmental control installations. The expansion plans of all major smelters in Europe have been reviewed and no indication of a major change of production technology was found for the next few years. As there is no significant change of the production of Pb, Cu, Zn and Ni in Europe and the modernization of environmental control installations concerns mainly the construction of new acid plants, it is suggested that the emissions of trace elements from European smelters in 1982 will continue to be valid for the next several years with only small changes.

Changes of the emission trends for coal and oil combustion in electric-power plants and industry are more difficult to predict, because they are related to the export-import movements of fossil fuels. An increase of the coal use in the European electric-power plants (ca. 2% per year) is related mainly to the increasing demand of electricity in Europe (ca. 1K% per year) and partly to the decrease of heat values of some European coals (e.g. 1% decrease per year for Polish coals). The expected increase of air pollution from coal combustion in the future due to the above factors, can be compensated by the tendency to extend the use of dual-fired installations with natural gas in Europe. Thus, the trace element emission from the European electric power plants does not seem to be changed in the near future.

Emissions of Pb from gasoline combustion decrease in Western Europe due to the lower use of Pb additives and the introduction of unleaded gasoline. In some Eastern European countries, however, there are quite opposite tendencies. The gasoline Pb emission in Europe decreased only from 68×10^3 t in 1982 to 64×10^3 t in 1985. However, many Western countries have decreased the Pb additives from 0.4 to 0.15 g/l beginning from 1986, so, the change of the Pb emissions between 1982 and 1987 or any later year will be more significant.

9. CHEMICAL SPECIFICTION OF TRACE ELEMENTS EMITTED FROM VARIOUS HIGH-TEMPERATURE PROCESSES

There is only limited information on the chemical speciation of trace elements emitted into the atmosphere, except for Hg.

The mercury from industries and power plants consists partly of elemental mercury and partly of oxidized forms. Rather limited information is available with respect to which volatile compounds dominate in the emission process, but it seems that reactive (or oxidized) forms are the more abundant in emissions from combustion processes. Mercury species other than elemental Hg (Hg(o)) and dimethyl mercury, $(CH_3)_2Hg$ can also be found. Mercury emitted as dimethyl mercury is a subject to relatively fast degradation to elemental mercury in the air (15). It was also concluded, that Hg(o) is mobilized to the atmosphere, where it is subjected to atmospheric oxidation processes to yield water soluble forms, and subsequently scavenged by wet or dry deposition (16).

The following classification of Hg compounds in air and natural water systems has been suggested (17):

1) Volatile: Hg(o), $(CH_3)_2Hg$;
2) Reactive: Hg^{2+}, HgX_2, HgX_3^- and HgX_4^{2-} with X = OH^-, Cl^- and Br^-; HgO on aerosol particles; Hg^{2+} complexes with organic acids;
3) Non-reactive: CH_3Hg^+, CH_3HgCl; CH_3HgOH and other organomercuric compounds, $Hg(CN)_2$; HgS and Hg^{2+} bound to sulphur in humic matter.

The major chemical species of As, Cd and Pb evolved during high-temperature processes, are presented in Table 6. Trivalent arsenic is the most common form of the element. Air samples from smelters or coal-fired power plants, contain mainly trivalent arsenic both in the vapour and particulate phases. This information is very important due to the fact that inorganic arsenic compounds are more toxic than organic arsenic

Table 6: Major chemical species of As, Cd and Pb emitted durin fossil fuel combustion and industrial processes

Process	As	Cd	Pb
Coal combustion	As(o), As_2O_3 As_2S_3	Cd(o), CdO CdS	$PbCl_2$, PbO, PbS, Pb
Oil combustion	As(o), As_2O_3 Organic arsines	Cd(o), CdO	PbO
Non-ferrous metal production	As_2O_3	CdO, CdS	PbO, $PbSO_4$, $PbO.PbSO_4$
Iron and steel manufacturing		CdO	PbO
Refuse incineration	As(o), As_2O_3, $AsCl_3$	Cd(o), CdO $CdCl_2$	Pb(o), PbO, $PbCl_2$

compounds, and the trivalent state more so than the other states (18). Methylated forms of arsenic are probably of minor significance.

Elemental cadmium and its oxide are the dominant chemical forms of this element emitted from all sources considered in this paper. These two forms seem to be the most toxic cadmium species, together with cadmium chloride (found in releases from refuse incineration).

The chemical forms of lead have also been studies (19). Lead halides are emitted in vehicle exhaust primarily as gases, condensing to form $PbCl_2$, PbClBr and $PbBr_2$ particles. Among the lead compounds arising from smelting operations, PbO and elemental Pb are identified in smelter emissions, but not in the ambient air outside the smelter. Concerning gaseous phases of lead, the tetra-alkyl lead compounds that are observed in the atmosphere, reflect their usage in gasoline, and their relative abundance changes during the atmospheric transport.

Considering the volatility of the As, Cd and Pb compounds in Table 6, As(o), As_2O_3, $AsCl_3$, Cd(o), and $CdCl_2$ are volatile, Pb(Cl,Br) is intermediately volatile, and CdO, Pb(o) and PbO are non-volatile at flue gas temperatures. The volatile species would be in the vapour state during high-temperature processes, whereas the non-volatile compounds would continue to be emitted largely as fly ash, even at higher temperatures.

10. RECOMMENDATIONS FOR FURTHER WORK ON TRACE ELEMENT EMISSIONS

The present experience and existing information on atmospheric emissions of trace elements entitle one to formulate the following tasks to be carried out in the near future.

1. A handbook of emission factors of trace elements from various source categories should be elaborated. The range of these factors shall illustrate the differences of various process technologies employed in the industry and the electric utilities.

2. An atlas of major point sources in Europe can be prepared on the basis of geographical location of individual sources.

3. A contribution of trace element emissions from natural sources needs to be evaluated for presenting the total emissions in Europe. This survey should take into account not

only sources located in Europe (e.g. volcanoes) but also long-range transport of pollutants from other continents and their deposition in Europe (e.g. a contribution of Mn and Cr transported within the Saharan dust).

4. Procedures for quality assurance and quality control need to be prepared for inventorying the trace element emissions in Europe.

5. An emission survey of trace elements from a variety of fugitive sources in Europe shall be approached.

6. Changes of trace element emission rates due to the introduction of new control technologies and new industrial processes shall be studied in a view of reduction strategies for air pollutants in Europe.

ACKNOWLEDGEMENTS

Most of the emission factors and emission values presented in this work was obtained during the projects granted by the German Environmental Protection Agency "Umweltbundesamt", Berlin and the Royal Norwegian Council for Scientific and Industrial Research (NTNF), Oslo.

Some parts of the work are based on discussions during the NATO Advanced Research Workshop on Control and Fate of Atmospheric Trace Metals, held in Oslo-Gardermoen, 12-16 September 1988.

REFERENCES

1 J.M. Pacyna, in: J.O. Nriagu and C.I. Davidson (Eds), Toxic Metals in the Atmosphere, Wiley, Chichester, 1986 33-52.

2 J.M. Pacyna, Coal-fired power plant as a source of environmental contamination by trace metals and radionuclides. Habilitation Monography No. 17-47. Technical University of Wroclaw, Wroclaw, Poland, 1980.

3 J.O. Nriagu and J.M. Pacyna, Nature, 333 (1988) 134-139.

4 R.G. Flagan and S.K. Friedlander, "Particle formation in pulverized coal combustion - a review". In: Proceedings of the Symposium on Aerosol Science and Technology of the 82nd National Meeting of the American Institute of Chemical Engineers, 1976, Atlantic City, NJ.

5 J.W. Kaakinen, Trace element study on a pulverized coal-fired power plant. Ph.D. dissertation, University of Colorado, Boulder, CO, 1979.

6 D.S. Ensor, S. Cowen, A. Shendrikar, G. Markowski and G. Waffinden, Kramer station fabric filter evaluation. EPRI Report CS-1669. Electric Power Research Institute, Palo Alto, Ca. 1981.

7 J.M. Pacyna, in: E.E. Pickett (Ed.), Atmospheric Pollution, Hemisphere Publishing Corporation, Washington, D.C., 1985.

8 D.S. Ensor, S. Cowen, R. Hooper and G. Markowski, Evaluation of the George Neal No. 3 Electrostatic Precipitator. EPRI Report FP-1145. Electric Power Research Institute, Palo Alto, Ca, 1979.

9 J.M. Ondov, R.C. Ragaini and A.H. Bierman, Environ.Sci. Technol., 13 (1979) 588-601.

10 I.M. Smith, Trace element from coal combustion: emissions. IEA Report CR-01, International Energy Agency, Coal Research, London, 1987.

11 P.-O. Moberg, M. Westermark and B. Noling, Migration of trace elements during flue gas desulfurization. KMH Report TR28, The Swedish State Power Board, Vällingby, Sweden, 1982.

12 M. Hutton, Cadmium in the European Community, MARC Report 26, Monitoring and Assessment Research Centre, London, 1982.

13 J.H. Reimers, S.R. Snelgrove and J.C. Taylor, Study on emission control technology for arsenic in the non-ferrous metallurgical industry. Environment Canada, Air Pollution Control Directorate, Mineral and Metal Processes Division, Ottava, 1977.

14 J.M. Pacyna, in: J.M. Pacyna and B. Ottar (Eds), Control and Fate of Atmospheric Trace Metals, Kluwer Academic Publishers, Dordrecht, 1989, 15-31.

15 F. Slemr, W. Slemr and G. Schuster, J. Geophys. Res., 86 (1981) 1159-1166.

16 C. Brosset, Water, Air and Soil Pollut., 16 (1985) 253-255.

17 O. Lindqvist and H. Rodhe, Tellus, 37B (1985) 136-159.

18 NRCC, Effects of arsenic in the Canadian environment. Ottawa, National Research Council Canada (NRCC No. 15391), 1978.

19 R.M. Harrison, Chemical speciation and reaction pathways of metals in the atmosphere. in: J.O. Nriagu and C.I. Davidson (Eds), Toxic Metals in the Atmosphere. John Wiley & Sons, New York, 1986, pp. 319-334.

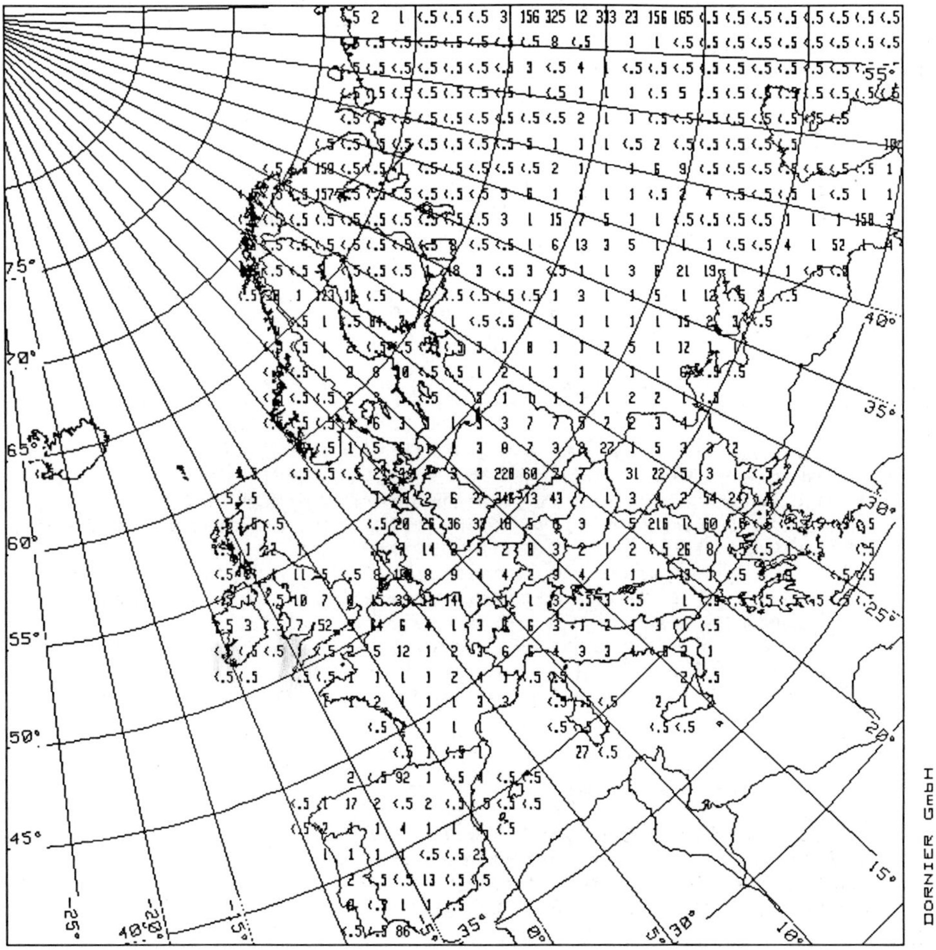

TOTAL: 4.9360×10^6 [KG/A]

SCALE: [10^3 KG/A]

Figure 1: Spatial distribution of As emissions (in t/y) in Europe in 1982 within the EMEP grid of 150 km x 150 km.

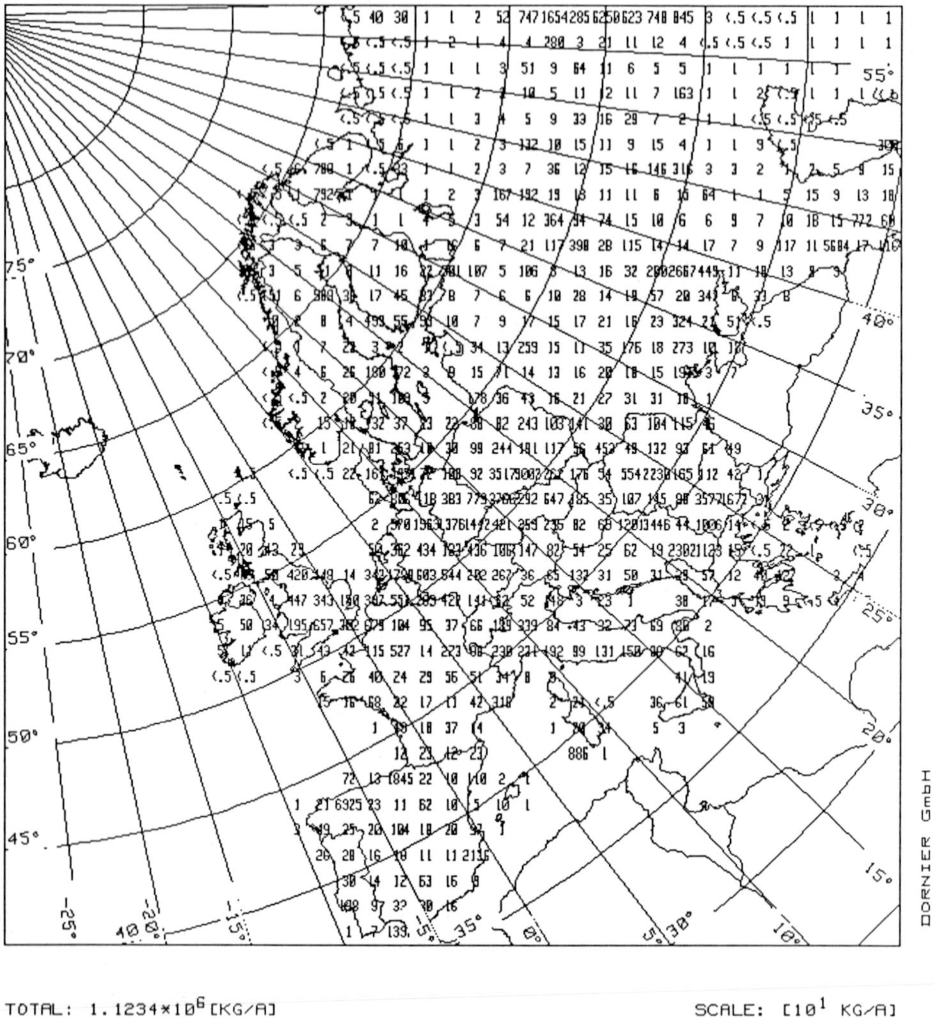

Figure 2: Spatial distribution of Cd emissions (in t/y) in Europe in 1982 within the EMEP grid of 150 km x 150 km.

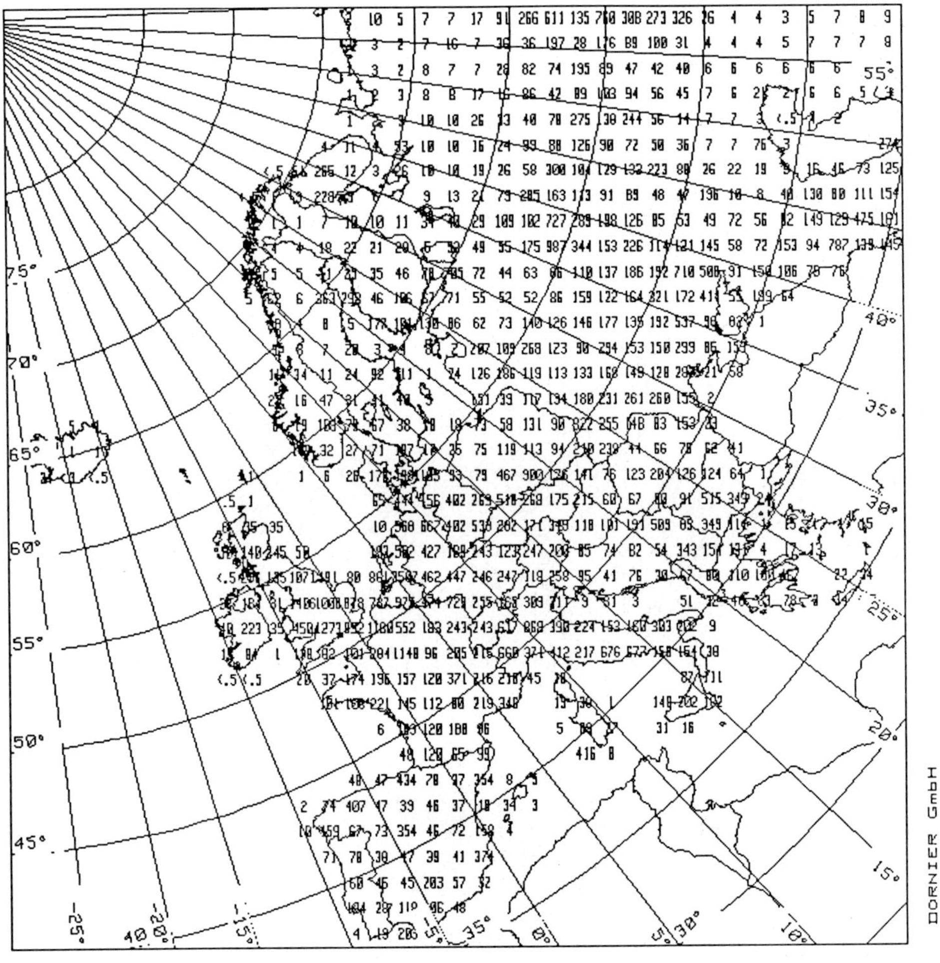

TOTAL: 8.9710×10^7 [KG/A]

SCALE: $[10^3$ KG/A]

Figure 3: Spatial distribution of Pb emissions (in t/y) in Europe in 1982 within the EMEP grid of 150 km x 150 km.

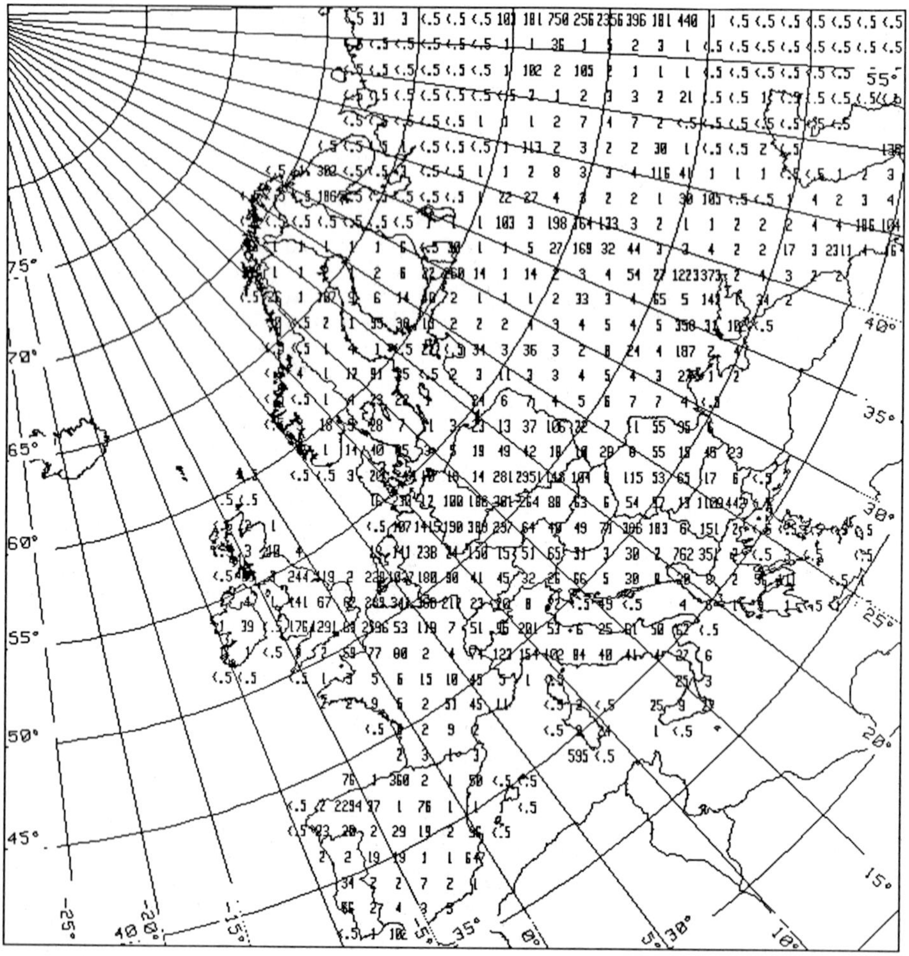

TOTAL: 4.1602×10^7 [KG/A]

SCALE: $[10^3$ KG/A]

Figure 4: Spatial distribution of Zn emissions (in t/y) in Europe in 1982 within the EMEP grid of 150 km x 150 km.

Chapter 2
Acid deposition and soil acidification

THE IMPACT OF ATMOSPHERIC ACID DEPOSITION ON SOIL AND VEGETATION

R. Mayer

ABSTRACT

Acid deposition from anthropogenic sources causes changes in terrestrial and in aquatic ecosystems. On land surfaces, soil is the most important link within the chain of acidifying processes influencing both vegetation and aquatic systems. The effects reach far beyond the easiliy visible changes in our environment such as forest decline and loss of freshwater fish.

Most important sources of acid deposition are the burning of fossil fuels and refuse, and intensive livestock farming. Soil **internal** acidifying processes which may be incduced by harvesting of biomass and other soil management practices, have to be distinguished from **external** acid input to soils. Impact on vegetation consists in direct damage to the leaf surface, leaching losses of nutrients, and increase in solubility and availability of toxic substances, including heavy metals, to leaf and root surfaces. Impact on soils can be subdivided into two complexes: loss of acid neutralizing capacity connected with the loss of nutrients like Ca, Mg, K from soil exchanger and soil minerals, and buildup of a base neutralizing capacity connected with the protonation of exchange sites and accumulation of cation acids such as Al, Mn, and Fe - often together with sulfate. In most soils the buffering rate, defined as the rate by which soils are able to consume protons **without** producing cation acids, cannot by far compete with the acid input in large parts of Europe and North America. As a consequence, acidity is - sooner or later - transported to the deeper underground and to subsequent aquatic systems, with an increase there in cationic acids (Al, heavy metals) and a complete loss of alkalinity. This comes up to a deterioration of the environmental conditions for most living organisms, including man who uses natural water bodies for drinking purposes.

1. INTRODUCTION

Deposition of acids from air pollution is a widespread phenomenon of global importance, with the most marked effects in those areas which are touched by air masses and precipitation from industrialized regions and urban agglomerations. Due to the long-range transport of pollutants, most surfaces in Central and Eastern Europe, and parts of Western, Southern and Northern Europe as well as eastern North America are strongly affected by acid deposition (1, 2). With world-wide progressive industrialization and industrial production the areas affected by acid deposition tend to increase and flow together.

After their deposition on land surfaces, acid substances interact with vegetation and soils in a way that a number of undesired effects may follow. One of these effects is the mobilization of heavy metals and aluminum from soils, being present there either as constituents of soil minerals, or from atmospheric deposition, fertilization, application of sewage sludge, refuse disposal, etc.

If we want to consider the impact of acid deposition, we have to distinguish natural processes of acidification within the ecosphere from man-made acid pollution.

2. SOURCES OF ACIDITY IN ATMOSPHERIC DEPOSITION

In order to estimate the acidifying effect of atmospheric deposition it is necessary to take into account all acid-/base-processes - natural as well as man-made - within soil and biosphere in which pollutants take part after deposition, i.e. all processes by which protons are produced, depending upon pH at the point of reaction, acid strength and concentration of all associated components in solution. Proton deposition with precipitation on any surface is not the only relevant parameter for judging the acid impact of pollutants. As main constituents of acidity in atmospheric input we have to consider the substances listed in Table 1 (3, 4):

TABLE 1

Main constituents of acidity in atmospheric deposition

		Effective pH-range
(P1)	CO_2 and carbonic acid ($CO_2.H_2O$)	pH > 5.0
(P2)	organic acids	pH > 4.5
(P3)	strong mineral acids and their precursors (SO_2/H_2SO_4; NO_x/HNO_3; Cl_2/HCl)	whole range
(P4)	metal cations forming weak hydroxides such as Al, Mn, Fe and heavy metals ("M_a-cations")	whole range
(P5)	ammonium ions NH_4^+ when transformed to ammonia ($NH_4^+ \rightarrow NH_3 + H^+$)	pH > 6,5
	or when incorporated into biomass ($NH_4^+ \rightarrow N_{org} + H^+$)	whole range

Process P1 - deposition of carbonic acid - is only relevant as a proton source when the pH at the point of reaction is above 5.0 . The limit below which most of our non-calcareous forest soils are ranging is the pH of 4.5, a level below which mobilization (solubility) of heavy metals is strongly increasing. In this context acid deposition with carbonic acid can therefore be left out of consideration as source of acidity.

Process P2 - deposition of organic acids - is in most areas of little importance especially if we are in a pH range below 4.5, where acid strength and concentration of most organic compounds occuring in atmospheric environment are not sufficient for them to serve as proton source comparable to other inputs. The same is true for M_a-cations - process P4 - acting as proton source by formation of hydroxo-complexes:

$$M_a^{n+} + H_2O \quad \text{---->} \quad M_aOH^{(n-1)+} + H^+$$

(these are the metals forming hydroxides of little solubility in neutral or moderately acid environments, like Al, Fe, Mn, heavy metals).

The most important input of acidity is usually linked to the deposition of strong mineral acids - process P3 - effective over the whole pH range of our soils.

Process P5 - transformation of ammonium to ammonia takes place only under neutral or alkaline conditions, not in acid soils. But we have to take into account the transformation of ammonium into organic binding. This process is a source of acidity wherever plant roots and/or microorganisms take up ammonium to build up biomass, i.e. in the upper soil layers.

When the input of acidity by atmospheric deposition is considered, it is necessary to take into account also the input of basicity, i. e. all processes introduced by atmospheric deposition through which protons are **consumed**. Most important here are:

- the input of dust particles, like carbonate particles, which are able to react with acids in the ecosphere fast enough to neutralize acids by consuming protons,
- the input of nitrate, if this nutrient ion is used after deposition by organisms (plants, microorganisms) to build up organic matter; the effect would be the deliberation of a bicarbonate ion and, in acid soils, the consumption of a proton.

Table 2 summarizes the main sources for acidity and basicity in atmosheric deposition.

3. IMPACT ON VEGETATION

We have to distinguish between direct effects of acidity and secondary effects by mobilization of toxic substances on plant surfaces. The most important direct effect is the leaching of nutrients from plant surfaces (Ca, Mg, K : "M_b cations" forming highly soluble hydroxides) (5-7). One aspect of this process is the development of nutrient deficiency, because the nutrient ion leached from the canopy must be restored via root uptake. The other aspect is the loss of alkalinity which leads to a neutralization of acidity in precipitation, but in the same time produces acidity in the soil horizons from which the supply of basic M_b cations takes place (rhizosphere) (5). In spite of the fact that acid precipitation is more or less buffered during canopy passage, the soil itself is subject to the acid load as a whole and soil

TABLE 2

Main components of acidity and basicity and their sources in atmospheric input

	Main source	Effective pH-range
ACIDITY		
(1) **Protons H^+**	SO_2 and NO_x from fossil fuel burning -> stong mineral acids	whole range
	CO_2 from atmosphere, respiration and fossil fuel burning	pH > 5.0
(2) **ammonium ions NH_4^+** when incorporated into biomass ($NH_4^+ \rightarrow N_{org} + H^+$)	agriculture, feedlots	whole range
or when transformed to ammonia ($NH_4^+ \rightarrow NH_3 + H^+$)	agriculture, feedlots	pH > 6.5
(3) **metal cations** forming weak hydroxides such as Al, Mn, Fe and heavy metals ("M_a-cations")	burning of fossil fuels and refuse; metal industry; soil dust	whole range
BASICITY		
(4) **bicarbonate HCO_3^-**	CO_2 from atmosphere, respiration and fossil fuel burning; carbonate particles from soil/fertilizer dust	pH < 5.0
(5) **nitrate NO_3^-** when incorporated into biomass ($NO_3^- + H^+ \rightarrow N_{org}$)	burning of fossil fuels	whole range

acidification is the consequence. Other direct effects of acid deposition on vegetation are the appearance of necrotic lesions on the surfaces of leaves, connected with the collapse of epidermis cells and slight chlorosis (1). This injury may be caused by acid precipition and/or by dry deposition of acidic hygroscopic aerosols. Also the surface structure of leaves and needles may be altered in such way that epicuticular waxes are destroyed and the cuticle is eroded, both layers acting as protective barriers towards toxic substances (8). Simultaneously, solubiltiy and mobility of such substances is increased by acidification of the leaf surfaces. Godt et al. (9) rinsed the twigs from spruce trees in the Solling area after a 3-weeks dry period with dilute acid (HNO_3; pH 4.0). The concentration in the first fraction of the wash-off solution is given in Table 3. It is compared with the average concentration in canopy drip measured over a period of several years at the same site (10). The concentrations at the beginning of a wetting cycle on the canopy surface are certainly high enough to reach toxic levels for the vegetation itself as well as for the organisms living in the canopy layer (insects, microorganisms, epiphytes).

TABLE 3

Heavy metal concentration in wash-off solution and canopy drip of spruce [after Godt et al. 1986 (9)]

	pH	Mn	Fe	Cu	Cd	Pb
wash-off after a 3-weeks dry period (initial fraction)	2.9	400	3030	71	20	935
average concentration in canopy drip after (10)	3.5	880	300	34	3	78

For a survey of the impact of air pollution on forest ecosystems and agricultural crops see (2, 8).

4. IMPACT ON SOILS

Emphasis is put here on acid soils, i.e. soils in a pH range below 5.0, because only under such conditions mobilization of heavy metals becomes a problem of actual relevance. In fact, most of the non-calcareous soils under forest in the regions subject to acid deposition are in that pH range.

The **external** sources of acidity (from deposition) have been discussed above. Soil **internal** sources of acidity have to be taken into account now. Again, the production of CO_2 by plant and soil organisms (soil respiration) acts as a source of acidity only at pH > 5.0, therefore soil respiration can not lead to a soil pH lower than 5. There remain two important internal sources of acidity in acid soils: 1. production of organic acids (podzolization) and 2. production of strong mineral acids, mainly nitric acid and - to a minor extent - sulfuric acid, by mineralization of organic bound nitrogen (nitrification) and sulfur, provided these compounds are not taken up by roots or soil organisms immediately after mineralization and are re-transformed into organic binding (6, 11).

Podzolization takes place predominantly in the upper humic soil horizons. Organic acids are infiltrating surface soils, but are normally not proceeding into the subsoil (in podsols, the depth of infiltration of organic acids is marked by the illuvial "spodic" Bh horizon). Under such conditions, this process can not be responsible for acidification of deeper subsoil and groundwater. It may though contribute considerably to the mobilization of heavy metals which is a prerequisite for the toxicity of these elements. There are situations where organic acids play a role in the acidification of deeper soil layers and groundwater/surface water systems (12, 13). Hereby, the production of organic acids may be a purely natural process, or may be induced by certain management practices. In these situations dissolved organic matter is found in seepage water, groundwater or surface water, as, for example, reported by Spitzy (14) from the Bavarian forest. The source of organic matter is usually located in the humic horizons of acid soils or bogs. It is quite difficult to estimate the contribution of theses processes to the acid load of soils or subsoils, because the nature of the dissolved organic matter is not well known. It can be estimated from the excess of strong acid anions over the sum of M_b-cations (15).

The process of nitrification can lead to the acidification of subsoils, since soils have little retention capacity for nitrate, and all nitrate not taken up by organisms or being transformed to N_2/NO_x by denitrification will finally be leached to deeper soil layers and groundwater under humid climatic conditions.

Moist soils, thanks to their mineral setup, provide alkalinity, but no acidity (except the rare cases where relevant amounts of sulfidic minerals are present). Therefore the soil is able to neutralize part of the acidity from external and internal sources by weathering of minerals. In this context, weathering can be considered as the transformation of bases of little solubility (carbonates, silicates) into bases of high solubility (bicarbonates, alkali and earth alkali hydroxides).

In considering the acid-/base processes in soil we have to take into account - as mentioned above - the consumption of acidity/protons by uptake of nitrate from atmospheric deposition or mineralization. This process may counteract the acidification originating from NH_4 uptake (from deposition and/or mineralization).

If we want to judge the long-term ability of a soil to neutralize acidity, we have to compare weathering rates (rates of alkalinity production) with the rate of acid input to soil plus net production of acidity within soil. Finally we have to account for the protons by formation and accumulation of biomass, either living (plants, soil organisms) or dead (humic substances), which takes place in all growing managed forest ecosystems not in steady state with respect to organic matter stores. This accumulation of biomass implies the accumulation of bases ("excess" M_b-cations) and an equivalent production of acidity in the soil compartment from which the cations were taken up.

Table 4 gives a list of all relevant sources and sinks of protons in the soil of two forest ecosystems in the Solling/Central Germany (16). The balance shows that only a very small fraction of the acidity in (non-calcareous) soil is buffered by soil internal alkalinity, i.e. weathering of silicates. Measurement of seepage flow below the root zone of soil show that, in fact, acidity is not transported to the deeper soil layers in the form of protons but rather as M_a cations ("cation acids"). The most prominent of these is Al as Al^{3+} and other ionic species, but Fe, Mn, and heavy metals play a role as well, depending upon type of minerals present in the weathering soil layer. These cations are carrying acidity into deeper soil layers and bedrock, and into groundwater and surface water, provided the pH in these systems is higher than in the soil so that the M_a cations react with water to form hydroxides and protons. M_a cations are toxic to soil and freshwater organisms.

TABLE 4

Proton input, output and turnover within soil in two forest ecosystems of the Solling/Central Germany (average rates 1971 - 1981, from Matzner 1984 [16])

	Beech	Spruce
	$keq.ha^{-1}.year^{-1}$	
Acid atmospheric deposition		
- **total proton deposition to soil**	**2.0**	4.0
- deposition of NO_3^- taken up by plants	-0.7	-0.1
- deposition of NH_4^+ taken up by plants	0.9	1.1
TOTAL ACID INPUT	**2.2**	**5.0**
Soil internal proton production		
- accumulation of excess M_b-cations		
- in wood increment	0.7	1.5
- in the humus layer	0.1	0.0
- leaching of organic acids	0.1	0.5
- H^+-buffering in the canopy (= acid input to the rhizosphere)	0.6	0.9
SUM OF INTERNAL ACID PRODUCTION	**1.5**	**2.9**
TOTAL ACID LOAD (EXTERNAL plus INTERNAL)	**3.7**	**7.9**
CONSUMPTION OF H^+ by weathering of silicates, discharge of M_b-cations into soil solution (estimate of soil internal alkalinity production)	**-0.2**	**-0.2**
BALANCE (net acid load of soil compartment)	**3.5**	**7.7**

In the case of calcareous bedrocks the buffering of acidity is followed by the mobilization of M_b cations (alkali and earth alkali metals) together with bicarbonate. Therefore, by this process acidity can not reach deeper layers or groundwater, provided the velocity of weathering is high enough (readily weatherable minerals, distribution of calcareous minerals in the fine fraction) to compete with input and source strength of acidity.

5. SOIL CHANGES BY ACIDIFICATION

The consequence of acid deposition to soil plus soil internal production of acidity is - as long as weatherable minerals are present - the consumption of protons connected with

- a decrease of acid neutralizing capacity (ANC) by loss and seepage of M_b-cations

and possibly

- an increase of base neutralizing capacity (BNK) by retention and accumulation of M_a-cations and protons.

The M_b-cations leached during the decrease of ANC can originate from the solid mineral phase (mainly silicates and carbonates) as well as from the exchangeable (adsorbed) fraction on weak acidic groups of minerals and organic matter. The buildup of BNK can take place by (exchangeable) adsorbtion of M_a-cations to acidic groups of minerals and organic matter, and by the immobilization of Al-ions together with sulfate (see below).From this we can define soil acidification as a decrease in ANC, and/or an increase in BNC, and/or an increase in acid strength of the acidic groups of the soil matrix (only the latter being linked to a decrease in soil pH!).

A large number of invenstigations in many countries with acid deposition have made evident the impact of acid air pollution on soils and subsequent aquatic systems (for a survey see 2, 17, 18), the most important facts being the following:

1. The average annual acid deposition in Germany is in the following range (19):

	$keq.ha^{-1}.year^{-1}$
deciduous forest, lowland and protected sites	1.2 - 2.6
decidous forest, mountains, exposed sites	2.0 - 3.5
coniferous forest, lowland and protected sites	2.0 - 2.7
coniferous forest, mountains, exposed sites	2.9 - 6.4

These figures can be considered as representative for Central Europe, forests in the southern part and agricultural crops ranging at the lower limits. The figures correspond well with the average annual emission of $S0_2$ and $N0_x$ in the F.R. of Germany (before 1989) of about 7 keq per hectare and year, part of which is neutralized during atmospheric transport. Higher rates of deposition than the ones given above are to be expected (and are reported) for countries with sulfur-rich coal burning, like (ex-)Eastern Germany, Poland and CSSR, and also for areas with high NH_3 emissions from livestock farming. Southern, Western and Northern Europe have, in general, a lower acid input if we exclude areas influenced by strong point sources.

2. The ability of non-calcareous soils to buffer acidity without being acidified (no decrease in pH) and without relvant mobilization of cation acids (such as Al) is in the order of 0.0 to 1.0 keq per ha and year, the magnitude being mainly dependet upon the rate of silicate weathering (2, 17, 20). This means that only a small fraction of acid input prevailing in Europe and parts of North America can be buffered by soils without generating secondary effects that are harmful to the ecosystem. Soil acidification is the inevitable consequence of an acid input of this magnitude.

3. Soil acidification during the last 20 to 30 years has been demonstrated for many forest areas by showing that several soil parameters (such as pH, stores of exchangeable cations, ANC, BNC) have changed during that time period. Nilsson (17) and Mayer (19) give surveys of such investigations which show that the pH drop in moderately acid soils amounts in many cases to more than one unit (from around 5.5 to < 4.5).

4. A very important mechanism by which acidity is retained and accumulated in soil is the storage of Al-ions together with sulfate, under the influence of a high sulfur input from the atmosphere (21). The Al-hydroxo-sulfate stored in large amounts in the soil has the potential to be mobilized and dissolved when pH or sulfate concentration in atmospheric input drop (for a review of case studies in which this process was observed see 15). Under such conditions, acidity can be carried down with seepage to the groundwater even if the input of atmospheric pollutants decreases, or if soil chemical conditions are changing due to climatic variations or changes in managemant practices (fertilization, clearfelling).

5. Acidification of groundwater and surface water systems

is a widerspread phenomenon in areas subject to acid deposition (2, 3, 15, 22). It is considered as a consequence of soil acidification enhanced and accelerated by acid atmospheric deposition (3, 23, 24). It is worthwhile to be mentioned once again that the presence of strong mineral acid anions like sulfate, nitrate and chloride is a prerequisite for the transfer of cationic acids to deeper subsoil and groundwater. For in naturally acid soils (podzols e. g., with a low atmospheric input of sulfates, nitrates etc.) there is no anion available that could be associated with the cationic acids to be translocated with seepage beyond the accumulation horizon (spodic B horizon). Therefore under the conditions of little input of strong acids, the soils are virtually "sealed" for acids.

Many of the areas in which water acidification is observed are also showing forest decline (3). It has been shown that forest decline for its part influences the quality of freshwater systems and hereby alters aquatic ecosystems (25). When the acid load of seepage water exceeds the rate by which weathering processes generate alkalinity, then acid is transported towards subsoil and groundwater. This takes place most often in the form of cation (M_a) acids such as Al-ions. As long as there are, during the passage of the water through deeper layers and bedrock, exchangeable basic M_b-cations or weatherable minerals, this may lead to a rise in pH and retention of M_a-cations. It seems that many cases the interaction between water flow and bedrock during saturated flow is not sufficient, due to high flow velocity and/or large pores and crevices, that complete buffering is not achieved. This is often true after heavy rainfall events and snowmelt, when soils are water saturated. In inclined surfaces we may have saturated subsurface flow where seepage water comes only into contact with acidified surface soil layers.

Hauhs & Wright (15) discuss two types of water acidification, both following each other under continuing acid input conditions. The first one is intensity controlled (typical for areas in Scandinavia with shallow soils, low rates of silicate weathering), the other one is capacity controlled, connected with a net decrease in base saturation (typical for areas with deeply weathered soils; the time scale of water acidification here is completely dependent upon the silicate weathering rate and soil depth).

Water which is influenced by acid deposition, and which is not well bufferd during passage of soil and bedrock, shows a chemical characteric which can be described as follows:

- The anionic composition is dominated by sulfate which originates from atmospheric deposition (provided the bedrock does not contain relevant amounts of sulfide or sulfate minerals).

- A high proportion of nitrate points at high atmospheric input and/or at an actual disintegration of organic matter (nitrification of organic bound nitrogen), a process which is often connected with forest decline.

- High concentrations of M_a-cations such as Al and heavy metals reflect, on the one hand, the geochemistry of the bedrock and the extent of acidification of the the soils and, on the other hand, the load of acid input from atmospheric pollution which the watershed has received during its history and is receiving actually (26).

This observation indicates the importance of acid atmospheric input and acidification of the soil compartment for the cycling of elements. The following chapter will therfore review the most important consequences for the heavy metal turnover in ecosystems.

6. IMPACT OF ACIDIFICATION ON HEAVY METAL CYCLING

It is a general characteristic of heavy metal compounds prevailing in the ecosphere that their solubility is highly dependent upon the pH of the solution, with a strong increase of solubility with decreasing pH (for a review see 27). The same is true for adsorptive binding and complexation of heavy metals on soil minerals and humus materials (28, 29). Since both translocation of heavy metals with the seepage water flow as well as the uptake by most soil organisms and plants require that the metal be in the solution phase, the influence of acidification on the cycling of heavy metals, their ecological relevance, and their transfer to aquatic systems is of primary importance. Usually the mobility of the metal, and by this the influence of acidification upon the behavior and relevance in the ecosphere, increase in the following order:

Pb, Cr < Cu < Ni < Zn < Cd << Co, Mn

but this order can be modified according to the presence of dissolved organic matter (and here again according to molecular size and type of complexing groups), to redox conditions and - with regard to biotic effects - to the toxicity of the metal.

Herms & Brümmer (30) discuss the influence of soil pH on the availability of heavy metals for plants and its consequences for agriculture. The effect of soil acidification on heavy metal cycling in forest ecosystems is reviewed for Germany by Mayer & Schultz (31), case studies from Scandinavia are reported by Bergkvist (32, 33).

There are indications that heavy metals and aluminium mobilized by soil acidification in fact disturb root growth of trees and forest vegetation as well as soil organisms, thus being a factor in forest decline and deterioration of forest ecosystems (2, 19, 34).

Before heavy metals appear, under the impact of acid deposition, in aquatic systems (groundwater, surface water bodies), usually a redistribution of the metals within the soil profile due to soil acidification takes place. Again this process is strongly dependent upon the buffer capacity of the soil solid phase, as demonstrated e. g. in a case study on different soil/bedrock types in the Harz mountains (35). In the same area, the concentration of heavy metals in springs and creeks from areas with various bedrocks as well as in lake sediments were investigated. Elements like Cd and Pb had concentrations 15 to 36 times higher in the water from creeks draining areas with rocks very poor in base minerals (quarzite) compared to creeks from areas with basaltic rocks (26). The lake sediments from this drainage basin showed a 4- to 24-fold enrichment in Zn, Pb, Cd, and Cu, compared to the natural background (36). Evidence was found that this enrichment is the consequence of atmospheric input of heavy metals together with acids.

7. CONCLUSIONS

As a result we may conclude that the impact of acid deposition has changed soils to an extent that changes in water chemistry are clearly visible. The large buffer capacity of most soils has so far prevented most groundwater and freshwater systems - exclusive those in the most sensitive areas with shallow soils on acid bedrock - from altering their chemistry in a way that, e. g., drinking water standards with respect to aluminum and heavy metals could no longer be met. But we have to keep in mind that the capability of soils to store large amounts of cation acids makes it an almost impossible task to get rid of and neutralize these acids once the capacity for retention is exhausted and all alkalinity is consumed. We therefore have to recognize the changes in water chemistry as an early warning to take measures for fast reduction of acid deposition.

REFERENCES

1 VDI-Kommission Reinhaltung der Luft 1987: Acid precipitation, Düsseldorf, 281 p.
2 P. Mathy ed. 1988: Air Pollution and Ecosystems. D. Reidel, Dordrecht, 981 p.
3 B. Ulrich 1984, Atmos. Environm. (Oxford), 18, 621-628.
4 B. Ulrich 1986, Z. Pflanzenern. Bodenk. 149, 702-717.
5 B. Ulrich 1983, in: B. Ulrich & J. Pankrath eds.: Effects of accumulation of air pollutants in forest ecosystems. D. Reidel, Dordrecht, 33-45.
6 E. Matzner & B.Ulrich 1984, Z. Pflanzenern. Bodenk. 147, 290-308.
7 S. Leonardi & W. Flückiger 1986, Allg. Forstz. 33, 825-828.
8 J. N. Cape & P. Mathy 1988: Scientific basis of forest decline symptomatology. Proc. Workshop Edinburgh 1988, Comm. of the Europ. Comm, Brussels, 410 p.
9 J. Godt, M. Schmidt & R. Mayer 1986. In: H. W. Georgii ed.: Atmospheric pollutants in forest ecosystems. D. Reidel, Dordrecht, 263-274.
10 R. Mayer 1981, Göttinger Bodenkundl. Ber. 70, 1 - 292.
11 M. Bredemeier 1987, Plant & Soil 101, 273-280.
12 D. F. Brakke, A. Henriksen & S. A. Norton 1987: The relative importance of acidity sources for humic lakes in Norway, Nature 329, 432-434.
13 J. O. Reuss, B. J. Cosby & R. F. Wright 1987, Nature 329, 27-32.
14 A. Spitzy 1984, Nationalpark Bayrischer Wald, Tagungsbericht "Wald und Wasser" Grafenau, Bd. II, 445-455.
15 M. Hauhs & R. F. Wright o.J.. Comm. Europ. Comm. Air Pollution Research Report 11, 42 p.
16 E. Matzner 1984 in: G. I. Agren ed., State and Change of Forest Ecosystems. Swed. Univ. Agric. Sci. Rep. 13, 303-311.
17 J. Nilsson ed. 1986: Critical loads for nitrogen and sulfur. Nordisk Ministerrad, Milijo rapport 1986/11, Kobenhavn,232 p.

18 J. Nilsson & P. Grennfelt eds. 1988: Critical loads for sulfur and nitrogen. Nordic council of ministers Rap. 15/1988, Kobenhavn, 417 p.
19 R. Mayer 1987. In VDI-Komm. Reinh. d. Luft, Düsseldorf, Schriftenreihe Vol. 5, 85-106.
20 H. Fölster 1985, in: J. I. Drever ed.: The chemistry of weathering. D. Reidel Dordrecht, 197-209.
21 J. Prenzel 1983 in: B. Ulrich & J. Pankrath eds.: Effects of accumulation of air pollutants in forest ecosystems, D. Reidel Dordrecht, 157-170.
22 R. F. Wright 1983, Water Qual. Bulletin, Burlington 8, 137 -143.
23 N. van Breemen, C. T. Driscoll, & J. Mulder 1984, Nature 307, 599-604.
24 J. O. Reuss & D. W. Johnson 1985, J. Environ. Qual. 14,26-31.
25 M. Hauhs 1985, Z. dt. Geol. Ges. 136, 585-597.
26 H. Heinrichs, B. Wachtendorf, K. H. Wedepohl, B. Rösner, & G. Schwedt 1986, N. Jahrb. Miner. Abh. Stuttgart, 156, 23-62.
27 W. L. Lindsay 1979: Chemical equilibria in soils. J. Wiley, New York, 449 p.
28 M. Schnitzer & S. U. Khan 1972: Humic substances in the environment. M. Dekker, New York, 327 p.
29 F. J. Stevenson & M. S. Ardakani 1972, in: J. J. Mortvedt, P. M. Giordano & W. L. Lindsay eds.:Micronutrients in agriculture, Madison Wisc., 79-114.
30 U. Herms & G. Brümmer 1980, Landwirtsch. Forsch. 33, 4, 408-423.
31 R. Mayer & R. Schultz 1987 in: S. E. Lindberg & T. C. Hutchinson eds.: Proc. Int. Conf. Heavy Metals in the Environment New Orleans 1988, Vol. 2, 402-406.
32 B. Bergkvist 1987, Water, Air and Soil Poll. 33, 131-154.
33 B. Bergkvist 1987, Forest Ecology and Managment (Elsevier, Amsterdam) 22, 29-56.
34 U. Avenhaus, U. Ebben, V. Glavac & R. Mayer 1988 in: J. P. Vernet ed.: Heavy Metals in the Envirnoment Proc. Int. Conf. Geneva 1989, 266-269.
35 H. Andreae & R. Mayer 1989 in: J. P. Vernet ed.: Heavy Metals in the Environment Proc. Int. Conf. Geneva 1989, 44-47.
36 J. Matschullat, H. Heinrichs, J. Schneider & M. Sturm 1987, Chemie d. Erde (Fischer, Jena) 47, 181-194.

IMPACT OF SOIL ACIDIFICATION ON THE MOBILITY OF METALS IN THE SOIL-PLANT SYSTEM

E. Løbersli, E. Gjengedal and E. Steinnes

SUMMARY

A survey of the present knowledge on metal mobility resulting from soil acidification and effects on plants is given.

Results from different studies on soil acidification and uptake of metals in terrestrial plants carried out in the authors' laboratory are discussed. The investigations are performed in natural ecosystems, and include both empirical studies and studies in manipulated catchments. Analyses of plants give a better estimate of metal bioavailability than a simple soil extraction. Our results provide strong evidence for increasing plant uptake of Rb due to increased acidity. The low base saturation found in surface soils in mesotrophic and eutrophic spruce forest ecosystems in southern Norway, compared with areas farther north, suggests a limited buffering capacity against further acidification. Results from this work provide evidence for increased soil leaching of base cations (Ca, Mg) and Mn from soil organic layers by soil acidification.The Cd content of vascular plants is substantially higher in Southern Norway than in areas farther north. A significant influence of soil acidity status on the plant avaiability of Cd is indicated from this work, but the higher plant uptake in the South may also simply reflect the higher load in the surface soils from atmospheric deposition, or even direct foliar uptake.

1. INTRODUCTION

The present degree of soil acidification is a result of the natural stress caused by the variability of the climate and by the aging and die-back of trees, the long-term utilization of biomass by man, which started 5000 years ago, and during the last century the phenomenon "acid deposition" from air pollution.

The term acid deposition may cover both air pollutants linked to the deposition of H^+ ions as well as those having the potential to produce acid in the soil immediately or later. This means that H^+, SO_4^{2-}, NO_3^- as well as SO_2, NO_x and NH_4^+ are all included.

The impact of increased atmospheric deposition on forest soils is being paid much attention, especially since damage to forest trees has become a widespread phenomenon.

There is a large variation in the potential effects of acid deposition on forest. This is owing to geographical variation in soil and vegetation properties and in the amount of pollutants deposited. Symptoms of forest damage or disfunction have become evident, but so far there seems to be no general consensus as to the relative importance of the possible causes of the reported anomalies. Probable contributing factors are usually closely interrelated to the emission pattern of pollutants, and additive or even synergistic effects might be expected. Among several hypotheses proposed as to the cause of tree anomalies are soil acidification leading to leaching losses of mineral nutrients and solubilization of toxic elements; disproportion of essential nutrients in soil and biomass due to surplus nitrogen deposition; acid leaching of essential nutrients from the canopy; damage due to gaseous pollutants, including oxidants and organics; infection by parasitic fungi, insects or viruses; and also climatic extremes.

The relationship between soil acidification, element cycling, biomass balance and ecosystem stability has been discussed by e.g. de Vries and Breeuwsma (1) and Ulrich (2).

The aim of this paper is to summarize present knowledge about the impact of soil acidification on the uptake of metal ions by vascular plants in natural ecosystems, with emphasis on boreal and alpine areas.

2. SOIL ACIDIFICATION, METAL MOBILITY AND EFFECTS ON PLANTS: SURVEY OF PRESENT KNOWLEDGE

2.1 Indirect effects of air pollutants on soil and vegatation

A recent accelerated pH decline accompanied by loss of cations in forest soils is mostly explained by influence of acid deposition from the atmosphere. The effects of acid precipitation on soil chemistry are strongly related to the mobility of the deposited anions in soil, and also to the cation exchange occurring in soil-plant systems. In periods with surplus water, replaced base cations originally adsorbed to soil particles can be leached from the soil, and both soil pH and base saturation percentage may decrease. Increased loss of base cations may gradually be compensated by base cations mobilized by the slow process of weathering. Increased deposition of N, which is likely to increase tree growth, may also increase soil acidity due to increased uptake of base cations and possibly also in the long term by increased nitrification.

As other elements are gradually leached from the soil during soil formation, there is a relative accumulation of Al. Particularily below pH 5 the solubility of Al increases sharply, and it has been suggested that continuous addition of acid by deposition from air, at a rate exceeding the buffer rate of silicate mineral weathering, may mobilize Al to an extent that forest ecosystems become unstable.

2.2 Plant availability of metals

Soil acidity is a main factor determining the bioavailability of metals (3). Soil acidification increases the release and leachability of many elements in soil, including toxic metals (4-7). An increase in soil acidity is therefore expeced to influence biological uptake of metals originating from indigenous minerals and those of anthropogenic origin. An increase in metal availability to plants does not imply, but may lead to, higher concentrations of these metals in plants. This might be an important link for aluminium and heavy metals to enter terrestrial food chains. Actual uptake of metals in plants is dependent not only on metal pools, however, but also on other soil properties, as e.g. Tyler documented (8) for *Anemone nemorosa*. Plants also regulate the uptake and translocate metals to a degree that varies among metals and plant species.

Metals exist in soils in different binding forms; including water soluble species; exchangeable species; metals bound in organic compounds, silicates, Fe-, Al-, and Mn oxides and hydroxides; and metals bound to phosphates. The soluble, the exchangeable and partly the organic bound fractions form the mobile and potentially bioavailable pool of metals. Heavy metals are known to form stable complexes or chelates with a variety of ligands, the average stability of which decreases with the electronegativity in the order Pb>Cu>Ni>Co>Zn>Cd>Mn (9, 10). Several factors including pH may influence the order (11). The chemical form of heavy metals in the soil solution is greatly dependent on the element considered, pH, fulvic acid content and presence of other compounds.

Metals may be mobilized due to ion exchange on soil colloids, release of complex bound metals or increased disintegration rates of minerals. Soil acidification is considered to raise the soil solution concentration of Al (6, 12-15). Other metals abundant in silicates, e.g. Fe, Mn, and alkali metals, are also expected to be mobilized due to increased mineral weathering. Potentially toxic concentrations of metal cations (Al^{3+}, Fe^{3+}, Mn^{2+}) in the soil solution can occur in acidic soils (16). Conditions of general nutrient deficiency are often associated with acidic soils, because of the tendency for soluble ions to be leached, or precipitated as metallic salts (17).

2.3 Effects on plants of increased availability of aluminium

The toxicity of dissolved Al to crop plants (18, 19), to tree roots (20), to freshwater biota (21-24) and under certain circumstances to man (25, 26) is considered to be a major deleterious effect of the acidification of the environment. Environmental effects of Al in terrestrial and aquatic ecosystems have recently been reviewed by Rosseland et al. (27).

Considerable interest has been paid to Al as an important growth-limiting factor in acid soils during the last century (28). Great interest has been focused on mobilization of Al in forest soils due to soil acidification (20, 29, 83). Al toxicity to tree roots has been suggested to be an important factor in forest decline in Europe (20). From reviews on impacts of Al on higher plants (27, 30, 31) it is evident that the mycorrhiza and fine root systems of terrestrial plants are adversely effected by high levels of inorganic monomeric Al. ("Labile Al"; $4<pH<6$: Al^{3+} + $AlOH^{2+}$ + $Al(OH)_2^+$ + $Al(OH)_3$ + $AlSO_4^+$). Mycorrhiza, the symbiosis between fungi and roots, is important for the water and nutrient uptake in plants.

Aluminium toxicity will primarily be expected in mineral soils low in organic matter. In the presence of organic matter and organic acids the availability of Al is reduced, due to complexation. The clay content and type of dominating clay mineral also influence the availability of Al. Soil rich in clay have a large Al fraction which can dissolve on acidification (30).

The most important effect of Al on plants is proposed to be reduced uptake of Ca, Mg, and P (18, 19). Aluminium may also induce K deficiency, and (less frequently) deficiency of Fe, Mn, Zn, and Cu (30). The Ca:Al balance in the soil solution is important, as increasing Ca concentrations in nutrient solution (32) or soil (33) counteract Al toxicity. Due to increasing solubility of Al and base cation losses by leaching under acid conditions, soil solutions may have a high level of Al and a low level of Ca. Critical conditions for spruce and beech arise as the Ca:Al ratio of the soil solution falls below 1.0, according to Rost-Siebert (34,35). In nutrient-rich soils, plants can manage better high concentrations of soluble Al. High levels of Ca, Mg, and P, but also of K and Si ameliorate Al toxicity, partly due to competitive interactions (30).

Experimental research has however also shown a wide range of tolerance to Al toxicity for various woody plants (36-38) and grass species (39-41). Plants thriving on acid soils may have developed mechanisms to tolerate high Al concentrations(30).

2.4 Metal leaching in soil

An increase in soil acidity may increase the leaching of metals downwards the soil profile. The leaching rate is also affected by amount of precipitation, presence of mobile anions and metal immobilization by plants and microorganisms. Leaching may transport metals, including essential nutrients, away from the main rooting zone which is the upper soil layer for most forest plants. An increased leaching rate may counteract the positive effect on solubility of increasing acidity with respect to bioavailability. The leachability as affected by an increase in acidity of the leaching solution differs greatly among metals. In Swedish forest soils the leaching of Ca, Cd, Mg, Mn, Na, Ni, Zn, and inorganic Al is known to increase with soil acidity, whereas for Cr, Cu, Fe, Pb, and organic Al the leaching is rather associated with solubility of organic matter in the soil solution (14). The influence of acidification on soil budgets of trace elements has not attracted as much concern as for the macronutrients. Strong acidification seems to reduce the litter decomposition rate, and the mobility of the organically bound metals appears to be regulated mainly by the formation of water-soluble organic compounds rather than directly by pH variations (6).

The influence of acid precipitation on the leaching of K, Ca, and Mg has attracted particular interest in the discussion of acid rain impact on terrestrial ecosystems. Already in the seventies Ulrich (42) claimed that there is a risk of Mg deficiency in trees due to increased Mg loss from acidified soils. Recently many studies have indicated that the "new type" forest damage in Europe is frequently associated with acute nutrient disturbances (43). Mg deficiency has been identified on acidic soils at high elevations in southern Germany (44-47), and fertilization has proven successful in reversing foliar symptoms (52). Spruce needle analyses have revealed nutritional disturbances for the elements Ca, Mg, K, Mn, and Zn (43). The deficiency is believed to be caused by ozone injury and foliar leaching (45) although a low soil supply also seems to be involved (47). Lower than normal levels of Ca and Mg have also been found in red spruce stands at high elevations in the Appalachian Mountains (48, 49). Nutrient deficiency seems to be a main factor in forest decline in parts of Central Europe (47, 50). Metals suggested to be in deficiency are mostly Mg, Ca, K, or Zn. P deficiency due to soil acidification has also been suggested (51).

2.5 Soil acidification in recent times

Increased soil acidity in some areas in Central Europe (53, 54) and Scandinavia (55-59) within the last decades is indicated by replications of earlier pH measurements. The pH reductions found are thought to be signs of a recent accelerated pH decline in forest soils. They are mostly explained by influence of acid deposition from the atmosphere. Regional comparisons of soil acidity status indicate higher soil acidity in the southern area compared to areas further north, which is less exposed to acid precipitation both in Sweden and Norway (60-62).

The recent pH reductions have been accompanied by loss of cations (63). Numerous studies have indicated that leaching of base cations from soil is enhanced by acid deposition (64-68).

Even though a considerable soil acidification has occurred in all forest types investigated in Europe, the dominating tree species has a strong influence on the acid-base properties of the soil and the metal cycling (7). The acidifying properties of spruce is greater than that of deciduous trees (72-75). This is partly due to the greater aerosol capturing capacity of conifers, but also to greater internal production of acid constituents in coniferous ecosystems (77, 84). The degree to which the acid deposition can increase soil acidity varies according to e.g. base cation reserves, mineral weathering rate and amount of and duration of acidic deposition (7). The influence of acid precipitation on weathering may vary strongly depending on the minerals concerned. As an example it has been indicated that weathering rate of feldspar is not influenced by soil acidity as long as the pH of the soil solution varies between 3 and 8 (85).

2.6 Heavy metal deposition and accumulation in soils

High concentrations of heavy metals can be toxic for forest ecosystems. Low input rates, however, may also affect terrestrial ecosystems by increased accumulation over long periods. Airborne heavy metal aerosols are filtered out of the air stream by the forest canopy or removed by wet deposition, and may subsequently be absorbed by the biomass. Dying tissue, epiphytic lichens and some mosses accumulate heavy metals particularly well. In the soil, e.g. humic acids, iron oxides, and clay minerals have a high fixation capacity for heavy metals. Most forest ecosystems are a sink for heavy metals. The solubility of heavy metal compounds however generally increases with decreasing pH, thus causing reduced fixation or increased mobilization in very acid soils. The latter may also occur during the decomposition phase in soil humus layers, e.g. after clear cutting.

Air quality measurements in southernmost Norway have revealed that considerable fluxes of certain metals are transported to southern Scandinavia from other European countries. Elements typical of these transboundary pollution aerosols include Pb, Zn, Cd, As, Sb, V, and Se (86). A study of heavy metals in ombrotrophic bogs in Norway indicates a substantial increase in the atmospheric supply of metals (Pb, Cd, As, Sb) in the southernmost part of Norway in recent times (78). Analyses of mosses have shown very distinct regional differences in the deposition patterns of metals (Pb, Cd, As, Sb, V, Zn) associated with long range atmospheric transport (79). The metals accumulate to a great extent in the organic surface soil where a corresponding geographic pattern is evident (80, 87, 88). Also higher plants have considerably higher contents of Cd, Pb, and Zn in South compared to Central Norway (81).

No causal relationship between long-range heavy metal immissions and forest decline is yet observed. In the long run, however, increasing heavy metal accumulation may pose a potential danger.

2.7 Sensitivity of heavy metals to vascular plants

Most heavy metals occur naturally in soils as rare elements. Plants require some of these elements in trace quantities but may be sensitive to higher concentrations. The release of heavy metals in biologically available forms, as a result of human activity, may damage or alter both natural and man-made ecosystems (11).

Soil acidification may, at least temporarily, reinforce the solubility and probably the availability of the heavy metal pools (11). Related to forest die-back, trees grown in acid soil showing symptoms of damages contained increased levels of heavy metals both in roots and aerial parts (89, 90).

In acid forest soils, Cu and Pb are chiefly present as fulvic acid complexes, whereas the free cationic species dominates for Zn and Cd (7). Heavy metals bound in fulvic acid and other humic complexes in the soil are considered less available than free metal ions and thus less toxic, at least to plant roots.

Toxicity of heavy metals (Zn, Cu, Cd, Pb) to vascular plants has recently been reviewed by Balsberg Påhlsson (92). She concludes that under unfavorable external plant conditions and a longer period of exposure, the possibility of a toxic effect arising on the vegetation from heavy metals cannot be excluded. Particularly tree seedlings and the fine root systems may be susceptible.

Besides time of exposure, the degree of toxicity is influenced by biological availability of the metal and interactions with other metals in the soil, nutritional status, age, and mycorrhizal infection of the plant. Although several plant species, including grasses, herbs and trees, are able to evolve tolerance (11), sensitive species or genotypes are supposed to be affected by heavy metals at relatively low concentrations in the soil solution (92).

The foliage of both conifers and deciduous trees growing in heavy metal polluted sites but lacking visible symptoms of damage was shown to have above-average concentrations of starch and total sugars (93). This is an early sign of physiological change, though the mechanisms are still obscure.

From solution culture experiments with vascular plants some general information on the relative toxicity of heavy metal ions may be obtained. On the average, at equimolar concentrations, the relative toxicity decreases in the order Cd>Cu>Pb>(=)Zn. Soil pH, availability of macronutrients and other environmental conditions may modify the sequence to some extent and greatly modify the concentrations necessary to bring about toxic reactions (11).

3. SOME EXAMPLES FROM STUDIES ON SOIL ACIDIFICATION AND METAL MOBILITY PERFORMED IN THE AUTHORS' LABORATORY

A thesis by E. Løbersli (94) focuses on the relationship between soil acidification and uptake of metals (including Al, Ca, Cd, Cu, Fe, K, Mg, Mn, Pb, Rb, Sr, Zn) in terrestrial plants. Plant metal uptake in relation to soil acidity and other soil factors is investigated by different approaches: a local acidification gradient study in the vicinity of a strong SO_2 source (95), a natural acidification gradient study comprising sites with different soil acidity status (96), a comparative study of corresponding vegetation types in areas exposed to different acid loads from the atmosphere (97), and a historical comparison of old and recent plant material from areas which have been exposed to considerable amounts of acidic deposition in recent times (98).

Manipulation experiments are being conducted in Norway to examine the effects of drastic changes in precipitation chemistry on soil and surface water acidification (99). In a clean area in western Norway two pristine catchments are being acidified by addition of H_2SO_4 and H_2SO_4+HNO_3, respectively. At an acidified site in southernmost Norway ambient acid precipitation is excluded by means of a roof, and clean precipitation is added beneath. A thesis in preparation by E. Gjengedal deals with the impact of soil acidification on the mobility of metals in the soil-plant system of these manipulated catchments. The mobility in soil and uptake by vegetation of major nutrients (K, Ca, Mg), and of metals which are released by weathering of minerals (Al, Mn, Rb), or are supplied by long-range atmospheric transport and acumulated in the upper layer of the soil (Pb, Cd, Zn), are studied (100).

From these different and quite independent approaches used to evaluate effects of soil acidification on metal uptake in plants some typical results will be discussed.

3.1 Rb - a sensitive indicator of acidification in terrestrial ecosystems

The uptake of Rb increases with decreasing base saturation of coniferous forest soils within a wide range of soils (97), and the uptake increases with decreasing distance from an acidification source (95). Several plant species from South Norway show higher levels of Rb in 1982 compared with around 1900 (98). The Rb concentrations in plants were found to be considerably higher in South Norway which has received a high deposition of acidifying substances in recent times, compared with Central Norway which is considerably less affected (97).

In an "acid addition experiment" in a clean area in Norway, in which metal uptake in a wide range of species is studied, significantly increased levels of Rb in vegetation following the acidification treatment are observed (100).

These results give significant evidence for increased plant availability of Rb with increasing acidity of soils, and they support the assumption that a recent increase in acidification of soils in southern Norway has increased the plant availability of Rb due to increased mineral weathering. These findings are in agreement with results presented elsewhere (103-105) which show increased biological uptake of this metal with increasing soil acidity.

From the present studies, and beech forest studies in Sweden (105) it seems that the acid/base controlled uptake of Rb is consistent for most organisms studied (vascular plants, fungi, insects and birds). This suggests biological uptake of Rb as a sensitive parameter of acidification of the terrestrial environment, since this element is not supplied in appreciable amounts through atmospheric deposition.
The chemicaly similar element Cs follows the same pattern as Rb. Higher concentrations in plants from South compared with Central Norway were observed (106). This indicates a corresponding response to soil acidification as observed for Rb. A similar regional pattern of Rb and Cs is suggested from analyses of surface water in Norway (107, 108).

Rb is not known to be essential to plants, but has been used as a chemical tracer for K in many studies (e.g. 101) and is suggested to substitute for K in some plant processes (102). The chemical properties of K are similar to those of Rb and Cs, but this element differs from the other two in being an essential nutrient to plants.

The metals Rb, Cs, and K appear to be intercorrelated in the plant materials analysed in this work (95, 97, 98). Extractable Rb and K are also positively correlated in surface soil (96). At increasing pH K is taken up preferentially to Rb (97). This is in accordance with results from Sweden (77, 103, 105). The uptake of these metals however varies in some respects. Some differences in the regulation of plant uptake of these metals are to be expected from the differences in the nutritional status.

3.2 Increasing leaching of Mn from soil by acidification

Mn appears to be easily leached from oligotrophic coniferous forest surface soils on acidification (94). The Mn content in plants increased with base saturation both in South and Central Norway in most plant species investigated. This contradicts results from Swedish studies from beech forests (105) showing significant decrease with increasing base saturation. Mn was however correlated with soil extractable Mn in all species analysed (97).

From the local acidification gradient study (95) it was evident that leaching had decreased plant available Mn at the most acidified sites, resulting in low plant contents.

In agreement with these results the "acid addition experiment" (100) has decreased the Mn uptake in *Betula pubescens*. Results reported from other experimental acidification studies also suggest decreased Mn availability to Picea abies (109) and to Pinus sylvestris (110).

In the "acid exclusion experiment" (100) at an acidified catchment in southernmost Norway, significant changes are observed as a function of time indicating that acid exclusion has initated reversal of ongoing soil acidification. Betula pubescens, Vaccinium uliginosum, Vaccinium myrtillus, and Molinia caerulea all show increasing uptake of Mn in the "restored" catchment compared to two control catchments.

In contrary to the experience from oligotrophic soils and soils exposed to a very strong acidification pressure, results from the historical study (98) (increase in Mn in Quercus spp.) and the regional comparison of mesotrophic / eutrophic vegetation types (97) (higher Mn contents in Anemone nemorosa from South Norway compared to Central Norway) suggest increased uptake of Mn at moderate acidification of more eutrophic soils, as also expected from other studies (8, 105, 111). The present results support the hypothesis that increased Mn availability with soil acidity occurs within certain soil pH limits (104). In the most acid raw humus soils a strong capacity for H^+ ions to replace Mn^{2+} probably leads to increased Mn leaching and lower plant availability (112).

3.3 Limited buffer capacity against further acidification at the most acid sites in South Norway

A very low base saturation at the most acid sites in South Norway suggests that further soil acidification may lead to nutrient deficiency on these soils (96). Extended soil acidification in the south is also evident from low ratios of soil extractable Ca:Al (96).

Several of the approaches used by Løbersli (94) indicate leaching of Ca and Mg from the surface soils by soil acidification and thereby reduced plant availability. This is evident from the decreased supply of Ca and Mg demonstrated in the most acid soils in the local acidification gradient study (95), and of Ca in the area affected by acid deposition as indicated from the historical study (98). The hypothesis that reduced availability of nutrients (Ca, Mg) may be one serious consequence of acid deposition (e.g. 66, 113-115) is also supported by relative differences in soil contents of exchangeable cations between edaphically similar sites in South and Central Norway (96), although these differences may also in part be explained by differences in bedrock geology.

In the "acid addition experiment" (100) the results indicate that the uptake of Ca in Betula pubescens is reduced following the acidification treatment. Reduced Ca uptake as a response to acid treatments has also been observed in other field experiments, e.g. in Pinus sylvestris (110) and Picea abies (67).

Mg is of special interest with regard to potential forest damage. Nutrient deficiency due to acidification has not been shown yet in the Norwegian forest monitoring programme, which is mainly concerned with spruce forests (116). Most of the sites in South and Central Norway studied in this work may be expected to be sufficiently supported by Mg due to sea salt influenced precipitation, containing relatively high levels of Mg (66, 117) and thus counteracting Mg deficiency. This assumption is supported by the historical study of plants indicating no decrease in Mg availability during the last century in South Norway (98).

3.4 Cadmium: Supplied by long range atmospheric transport and mobilized by soil acidification

The mobility of Cd in natural soils is enhanced by soil acidification (7, 104). Not only is Cd originating from local mineral material mobilized, but also most probably Cd supplied to the upper soil horizon by atmospheric deposition (80). In a study of metal contents in vascular plants growing in oligotrophic coniferous forest ecosystems in South and in Central Norway (81), samples from South Norway were consistently found to show levels several times higher. This observation corresponds to the similar regional difference observed in Cd deposition between South and Central Norway (118) which is presumably the major cause of the approximatly 5-10-fold higher Cd levels in the humus layer of the southern soils (87). The historical study (98) gives supporting evidence for an increase in the uptake of Cd (Anemone nemorosa, Quercus spp.) in South Norway during the last century. Cd is likely to have entered the plant predominantly by root uptake. Although at considerable deposition levels direct uptake from airborne surface contamination may become an important route of uptake even for higher plants (119), it is difficult to say whether the high plant uptake in the South depends on the more extensive soil acidification in this area (97) or if it simply reflects the higher load in the soil from atmospheric deposition, or even direct foliar uptake. At spruce forest localities in South Norway however Cd contents in several plant species showed a negative correlation with soil base saturation, both within eutrophic and oligotrophic sites, which indicates an influence of soil acidity status on the plant availability of Cd (97).

REFERENCES

1 W. de Vries and A. Breeuwsma, Water, Air and Soil Pollution, 28 (1986) 173-184.
2 B. Ulrich, in: E.D. Schulze and H. Zwölfer (Eds), Ecological studies, Springer Verlag, Berlin, 1987, pp. 11-49.
3 N.C. Brady, The Nature and Properties of Soils, Macmillan Publ. Comp., New York, 1984.
4 S.A. Norton, Water, Air and Soil Pollution, 7 (1977) 389-400.
5 E. Matzner and B. Ulrich, Zeitschrift für Pflanzenernährung und Bodenkunde., 144 (1981) 660-681.
6 B. Bergkvist, Water, Air and Soil Pollution, 31 (1986) 901-916.
7 B. Bergkvist, L. Folkeson and D. Berggren, Water, Air and Soil Pollution, 47 (1989) 217-286.
8 G. Tyler, Oikos, 27 (1976) 71-80.
9 D.P. Mellor and L. Maley, Nature, 159 (1947) 370.
10 H. Irwing and R.J.P. Williams,Nature, 162 (1948) 746-747.
11 G. Tyler, A.M. Balsberg Påhlsson, G. Bengtsson, E. Bååth and L. Tranvik, Water, Air and Soil Pollution, 47 (1989) 189-215.
12 D.W. Hanson, S.A. Norton and J.S. Williams, Water, Air and Soil Pollution, 18 (1982) 227-239.
13 B. Ulrich, in: B. Ulrich and J. Pankrath (Eds), Effects of Accumulation of Air Pollutants in Forest Ecosystems, D.Reidel Publ. Comp., Dordrecht, Netherlands, 1983, pp. 127-146.
14 B. Bergkvist, Water, Air and Soil Pollution, 33 (1987) 131-154.
15 E. Matzner, D. Murach and H. Fortmann, Water, Air and Soil Pollution, 31 (1986) 273-282.
16 H.W. Woolhouse, in: O.L. Lange, P.S. Nobel, C.B. Osmond and H. Ziegler (Eds), Physiological Plant Ecology. III. Responses to the Chemical and Biological Environment. Encyclopedia of Plant Physiology, New series, Springer Verlag, Berlin, 12C (1983) 245-300.
17 I.H. Rorison and D. Robinson, Plant, Cell and Environment, 7 (1984) 381-390.
18 C.D. Foy, in: E.W. Carson (Eds), The plant root and its environment,University Press of Virginia, Charlottesville, 1974, pp. 601-642.
19 C.D. Foy and A.L. Fleming, The physiology of plant tolerance to excess available aluminium and manganese in acid soils. American Society of Agronomy, Special publication, Symposium in crop tolerance, 1987, pp.301-327.
20 B. Ulrich, R.T. Mayer and K. Khanna, Soil Science, 130 (1980) 193-199.
21 C. Cronan and C.L. Schofield, Science, 204 (1979) 306.
22 I. Muniz and H. Leivestad, in: D. Drabløs and A. Tollan, Ecological impacts of acid precipitation. Proc. Int. Conf., Sandefjord, Norway, 1980, pp. 320-321.
23 C.M. Wood and D. McDonald, Annales Societé Royale Zoologique de Belgique, 117 (1987) 399-410.
24 J. Hermann, in: J. Landner (Eds), Speciation of metals in water, sediment and soil systems. Lecture Notes in Earth Sciences, Springer Verlag, Berlin, 11 (1987) 157-161.
25 M. Wills and J. Savory, Environ. Health Perspect., 63 (1985) 141-147.
26 T. Vogt, Water quality and health. Study of a possible relation between aluminium in drinking water and dementia, Sosiale og økonomiske studier 61, Statistisk sentralbyrå,

Oslo, 1986.
27 B.O. Rosseland, T.D. Eldhuset and M. Staurnes, Environmental Geochemistry and Health, 12 (1990) 17-27.
28 T.C. Hutchinson, Heavy Metals in the Environment, Proc. Int. Conf., Vol. 1 (1983) pp. 17-26.
29 B.W. Bache, J. Geol. Soc., London, 143 (1986) 699-706.
30 M. Andersson, Water, Air and Soil Pollution, 39 (1988) 439-462.
31 T.D. Eldhuset, Communications of the Norwegian Forest Research Institute, Ås, Norway, 40 (1988) 1-19 (in Norwegian, English summary).
32 K. Rost-Siebert, Allgemeine Forstzeitschrift, 38 (1983) 686-689.
33 G. Abrahamsen, Phil. Trans. R. Soc. Lond. B, 305 (1984) 369-382.
34 K. Rost-Siebert, Allgemeine Forstzeitschrift, 26/27 (1983) 686-689.
35 K. Rost-Siebert, Aluminium Toxicity in Seedlings of Norway Spruce (Picea abies Karst.) and Beech (Fagus sylvatica L.), Contribution to the Workshop on Aluminium Toxicity to Trees, Uppsala May 14-16, 1984, Stencil.
36 L.H. McCormick and K.C. Steiner, Forest Science, 24 (1978) 565-568.
37 T.C. Hutchinson, L. Bozic and G. Muno-Vega, Water, Air and Soil Pollution, 31 (1986) 283-294.
38 P.J. Ryan and S.P. Gessel, Plant and Soil, 96 (1986) 239-257.
39 C. Hackett, J. Ecol., 53 (1965) 315-333.
40 D.T. Clarkson, Journal of Ecology, 54 (1966) 167-178.
41 M.S. Davies and R.W. Snaydon, Journal of Applied Ecology, 10 (1973) 47-55.
42 B. Ulrich, Forstwiss. Centralbl., 94 (1975) 280-287.
43 R. Hüttl, "Neuartige" Waldschäden und NäHrelementversorgung von Fichtenbeständen (Picea abies Karst. in Südwestdeutchland Freiburger Bodenkundliche Abhandlungen Heft 16, Freiburg im Breisgau, 1985.
44 W. Von Zech and E. Popp, Forstwiss. Centralbl., 102 (1983) 50-55.
45 K. E. Rehfuess, Allgemeine Forstzeitschrift, 38 (1983) 601-610.
46 W. Von Zech, TH. Suttner and E. Popp, Water, Air, and Soil Pollution, 25 (1985) 175-183.
47 H.W. Zöttl and R.F. Hüttl, Water, Air and Soil Pollution, 31 (1986) 449-462.
48 A.J. Friedland, G.J. Hawley and R.A. Gregory, Plant and Soil, 105 (1988) 189-193.
49 W.P. Robarge, J.M. Pye and R.I. Bruck, Plant and Soil, 114 (1989) 19-34.
50 H.W. Zöttl,Experentia, 41 (1985) 1104-1113.
51 D. Paré and B. Bernier, Canadian Journal of Forest Research 19 (1989) 132-134.
52 H.W. Zöttl, R.F. Hüttl, S. Fink, C.H. Tomlinson and J. Wisniewski, Water, Air and Soil Pollution, 48 (1989) 87-109.
53 H. Butzke, Forst- und Holzwirt, 21 (1981) 542-548.
54 K.E. Rehfuess, Water, Air and Soil Pollution, 48 (1989) 1-20.
55 L. Hallbäcken and C.O. Tamm, Scandinavian Journal of Forest Research,1 (1986) 219-232.
56 U. Falkengren-Grerup, Environmental Pollution, 43 (1987) 79-90.
57 U. Falkengren-Grerup, Soil acidification and vegetation changes in South Swedish forests, Thesis, Department of ecology, Lund University, Sweden, 1989.
58 M. Berden, S.I. Nilsson, K. Rosen and G. Tyler, Soil acidification - extent; causes and consequencies. An evaluation of literature information and current research

National Swedish Environmental Protection Board, SNV, Report 3292, Sweden,1987.
59 E. Dahl, Acidification of soils in the Rondane mountains, South Norway, due to acid precipitation, Økoforsk Report, Ås, Norway, 1 (1988) 1-53.
60 E. Steinnes, Jord og Myr, 9 (1985) 248-254 (in Norwegian).
61 C.O. Tamm and L. Hallbäcken, Ambio, 17 (1988) 56-61.
62 E.I. Aune, E. Dahl and A.K. Løes, Communications of the Norwegian Forest Research Institute, Ås, Norway, 42 (1989) 133-146.
63 U. Falkengren-Grerup, N. Linnermark and G. Tyler, Chemosphere 16 (1987) 2239-2248.
64 G.W. Rutherford, J. vanLoon, S.F. Mortensen and J.A. Hern, Canadian Journal of Forest Research, 15 (1985) 848-854.
65 G. Abrahamsen and A.O. Stuanes, Water, Air and Soil Pollution, 31 (1986) 865-878.
66 G. Abrahamsen, H.M. Seip and A. Semb, in: D.C. Adriano and M. Havas (Eds.), Acidic Precipitation. Vol. I.Case Studies, Springer Verlag, New York, 1989, pp. 137-177.
67 A.O. Stuanes, G. Abrahamsen and B. Tveite, in: P. Mathy (Ed.), Air Pollution and Ecosystems. Proc. Int. Symp., Grenoble, France, 18-22 May 1987, pp. 248-253.
68 T.R. Moore, Canadian Journal of Forest Research, 17 (1987) 370-378.
69 M. Hornung, Soil use and management, 21 (1986) 24-28.
70 J.O. Reuss, B.J. Cosby and R.F. Wright, Nature, 329 (1987) 27-32.
71 E.C. Krug and C.R. Frink, Science (Washington D.C.), 221 (1983) 520-525.
72 B. Nihlgård, Oikos, 22 (1971) 302-314.
73 B. Ulrich, E. Ahrens and M. Ulrich, Ecological studies, 2 (1971) 171-190.
74 A.S. Messenger, Soil Science Society of America Proc., 39 (1975) 698-702.
75 A.S. Messenger, Forest Ecology and Management, 3 (1980) 113-125.
76 N.V. Hue, G.R. Craddock and F. Adams, Soil Science society of America. Journal, 50 (1986) 28-34.
77 G.Tyler, D. Berggren, B. Bergkvist, U. Falkengren-Grerup, L. Folkeson and Å. Rühling, in: T.C. Hutchinson and K.M. Meema (Eds.), Effects of Atmospheric Pollutants on forests, wetlands and agricultural ecosystems. NATO ASI Series, Vol. G16, Springer-Verlag Berlin-Heidelberg, 1987, pp. 347-359.
78 O.Ø. Hvatum, E. Steinnes and B. Bølviken, in: S.E. Lindberg and T.C. Hutchinson (Eds.), Heavy Metals in the Environment, Proc. Int. Conf., Vol 1 (1987) pp. 201-203.
79 Å. Rühling, L. Rasmussen, K. Pilegård, A. Mäkinen and E. Steinnes, Survey of atmospheric heavy metal deposition - monitored by moss analysis, Nordic Council of Ministers NORD, 21 (1987) 44 p.
80 E. Steinnes, W. Solberg, H.M. Pettersen and C.D. Wren, Water, Air and Soil Pollution, 45 (1989) 207-218.
81 W. Solberg and E. Steinnes, Heavy Metals in the Environment, Proc. Int. Conf., Heidelberg, Vol. 1 (1983) pp. 170-173.
82 I.T. Rosenqvist, Science of the Total Environment, 10 (1978) 39-49.
83 A.S. Gee and H. Stoner, in: M.B. Usher and D.B.A. Thompson (Eds.),Ecological change in the uplands, Blackwell Sc. Publ., Oxford, 1988, pp. 273-387.
84 S. Woodin, Journal of Applied Ecology, 26 (1989) 749-761.
85 H.C. Helgeson, W.M. Murphy and P. Aagaard, Geochim. Cosmochim. Acta, 48 (1984)2405-2432.
86 C.E. Amundsen, J.E. Hanssen, A. Semb and E. Steinnes, Long range atmospheric transport of trace elements to Southern

Norway, Atmospheric Environment, submitted 1991.
87 R.O. Allen and E. Steinnes, in: D. Drabløs and A. Tollan (Eds.), Ecological Impact of Acid Precipitation, Proc. Int. Conf. Sandefjord, Norway, 1980, pp. 102-103.
88 E. Steinnes, in: B. Yaron, G. Dagan and J. Goldschmid (Eds.), Pollutants in Unsaturated Zone, Springer Verlag, Heidelberg, 1984, pp. 115-122.
89 R. Mayer and H. Heinrichs, Zeitschrift für Pflanzenernährung und Bodenkunde, 144 (1981) 637.
90 B. Von Bredow, A. Buggert, A. Eckhoff, B. Hollstein, M. Neumann, R. Schindel, A. Weber, S. Zech and V. Glavac, Allgemeine Forstzeitschrift, 22 (1986) 551.
92 A.M. Balsberg Påhlsson, Water, Air and Soil Pollution, 47 (1989) 287-319.
93 A.M. Balsberg-Påhlsson, Canadian Journal of Botany, 67 (1989) 2106-2113.
94 E. Løbersli, Soil acidification and metal uptake in plants, Thesis in preparation, Department of botany, University of Trondheim, AVH, Norway, 1991.
95 E. Løbersli and E. Steinnes, Water, Air, and Soil Pollution, 37 (1988) 25-39.
96 E. Løbersli and E. Steinnes, Chemical properties of coniferous forest soils from South and Central Norway, in preparation, 1991.
97 E. Løbersli and E. Steinnes, Metal uptake in coniferous forest plants growing on soils of different acidity, in preparation, 1991.
98 E. Løbersli and E. Steinnes, Environmental Monitoring and Assessment, 15 (1990) 111-129.
99 R.F. Wright, E. Lotse and A. Semb, Nature, 334 (1988) 670-675.
100 E. Gjengedal and E. Steinnes, in: J.P. Vernet (Ed.), Heavy Metals in the Environment, Proc. Int. Conf., Vol. 2 (1989) pp. 40-43.
101 M. Fried, G. Hawkes and W.Z. Mackie, Soil Science Society of America Proc., 23 (1959) 360-362.
102 A. Kabata-Pendias and H. Pendias, Trace Elements in Soils and Plants, CRS Press Inc., Florida, 1984.
103 G. Tyler, Communications in Soil Science and Plant Analysis, 14 (1983) 1075-1089.
104 G. Tyler, E. Steinnes, L. Folkeson, E. Løbersli and E. Nyholm, Försurningens inverkan på metallupptagning i växter och djur, Slutrapport MIL 3 prosjektet: Försurningens inverkan på metalltransport i miljön, Nordisk Ministerråd (Nordic Council of Ministers), miljörapport 3 (1986) 39 p (in Swedish).
105 L. Folkeson, N.E.I. Nyholm and G. Tyler, The Science of the Total Environment, 96 (1990) 211-233.
106 J. Caplan, E. Løbersli, R. Næumann and E. Steinnes, Journal of Radioanalytical and Nuclear Chemistry, Articles, 114 (1987) 13-19.
107 R.O. Allen and E. Steinnes, Geological Survey of Norway Bulletin, 409 (1986) 35-48.
108 E. Steinnes, in: J. Låg (Ed.), Geomedical consequences of chemical composition of freshwater, Norwegian University Press, Oslo, 1987, pp.35-41.
109 G. Ogner and O. Teigen, Plant and Soil, 57 (1980) 305-321.
110 G. Abrahamsen, B. Tveite and A.O. Stuanes, in: Lavender (Ed.) Proc. IUFRO Conference on Woody Plant Growth in a Changing Physical and Chemical Environment, Vancouver, 27-31 July, 1987.
111 G.H. Godo and H.M. Reisenauer, Soil Science Society of America Journal, 44 (1980) 993-995.
112 H. Neite, Zeitschrift für Pflanzenernährung und Bodenkunde,

152 (1989) 441-445.
113 C.W. Cronan, W.A. Reiners, R.L. Reynolds and G.E. Lang, Science (Washington, D.C.) 200 (1978) 309-311.
114 G. Abrahamsen, in: D.S. Shriner, C.R. Richmond and S.E. Lindberg,Atmospheric Sulphur: Environmental Impact and Health effects, Ann Arbor Science Publication, Ann Arbor, MI, 1980, pp. 397-416.
115 D.W. Johnson, D.D. Richter, G.M. Lovett and S.E. Lindberg, Canadian Journal of Forest Research, 15 (1985) 773-782.
116 OPS (Overvåkingsprogram for skogskader), Overvåkingsprogram for skogskader, Årsrapport 1988, Norsk institutt for skogforskning (Norwegian Forest Research Institute), Ås, Norway, 1989, 52 p. (in Norwegian, English summary).
117 J. Låg, Acta Agric. Scand., 18 (1968) 148-152.
118 J.P. Rambæk and E.Steinnes, in: Nuclear Methods in Environmental and Energy Research, (CONF-800433), U.S. Dept. of Energy, 1980, pp. 175-180.
119 M.F. Hovmand, J.C. Tjell and H. Mosbaek, Environmental Pollution (Series A), 30 (1983) 27-38.

Chapter 3
Soil interaction

ASSESSMENT OF ECOTOXICOLOGICAL RISK OF ACCUMULATED METALS IN SOILS WITH THE HELP OF CHEMICAL METHODS STANDARDIZED THROUGH BIOLOGICAL TESTS

S.K. Gupta

1. INTRODUCTION

During last few years problems releated to soil contamination have been discussed not only among common men but also among politicians. In this paper, only the inorganic pollutants especially heavy metals would be discussed. There are two following prerequisite for proper implementation of soil protection:

1. Soils already low in contaminants must be protected from future contamination through stringent control on emissions sources.
2. The polluted soils either need to be cleaned or atleast cleaned to the levels which are safe for good human health, good plant growth (quantitiy and quality) and optimum growth of soil micro-and macroorganisms.

The concentration of metals in a given soil, if above the normal concentration range then one has reasons to suspect an additional input of metals in soils through following three important routes:

- agricultural inputs
- air (emission through traffic, metal industry, waste incineration ect.)
- parent rocks specially rich in metal or metals (e.g. Serpentine)

Accumulated pollutants in soils can be transported to different important environmental components e.g. soil, water and plant:

- as gas or as component of dust particles via air
- as volatile substances via air
- as dissolved substances in the rain water or in soil leachate in deep layers of soil profiles and finally to the ground water.

Besides the direct transport of pollutants via air water and soil to other components of the enivronments, there are following indirect routes as well:

- Soil- Plant- Men
- Soil-Plant-Animal

Several pot experiments in greenhouse and in laboratory have been used to estabilish the ecotoxicological (plants and soil microorganisms) concentration in soils through chemical methods. However, most of these experiments were carried out on artificially contaminated soils. These concentrations have been used to fix the limiting values of metals in soils. This paper present the results of a Offsite-Greenhouse and Onsite-Miniplot Field experiments carried out on anthropogenic (in situ) contaminated soils to assess the eco toxicological effects of soil metals. Further, the results of microorganisms growth experiments carried out mainly with artificially contaminated soils are discussed to evaluate the ecotoxicological reelevant concentration in soils. The estimated concentrations are compared with the present Swiss guide values which were estabilished with the help of pot experiments in green house using mainly artificially contaminated soils.

2. HOW TO ASSESS ECOLOGICAL RISK OF ACCUMULATED METALS IN SOILS

The best way to assess the ecological risk of heavy metals in soils is by ascertaining the critical concentration in soils. The critical concentrations are the metal concentrations either in soils, or in soil solution or in nutrient solutions or in drinking water, which can trigger either

- phytotoxic or zootoxic concentrations in plants and or
- causes detrimental effects on the growth and activities (Fig. 1) of soils microorganisms
- causes adverse effects on human health

Mainly, metals in soil are distributed in two phases i.e. in solid and solution. From the practical point of view, there can only be two possible analytical criteria assess the ecotoxicolog effects of accumulated metals in soils.

2.1 Total Metal Content
2.2 Metal in Soil Solution

2.1 TOTAL METAL CONTENT (Pedogenic and Anthropogenic)

Total metal (Metal in Solid + Solution phases of soil) content of a soil can be determined either by fusion analysis or perchloric acid digestion. Although, the fact remains that both of these methods are laborious and time consuming.

An approximation of total metal should also be done by carrying out fractionation of soil, which means succesive extraction. A Total sum of metal extracted , is usually close to total and give good idea about the binding strength of metals, but is time consuming and laborious. In contrast, strong acids (such as HNO_3) ot chelationg agents (buffered in acidic range) gives an good approximation for total content and are suitable for routine analysis.

Total metal contents are excellent criteria to judge the extent of metal accumulation in soil:

$$\text{Degree of metal accumulation} = \frac{\text{Total metal (meq.)}}{\text{Metal binding capacity}} \times 100$$

Fig. 1

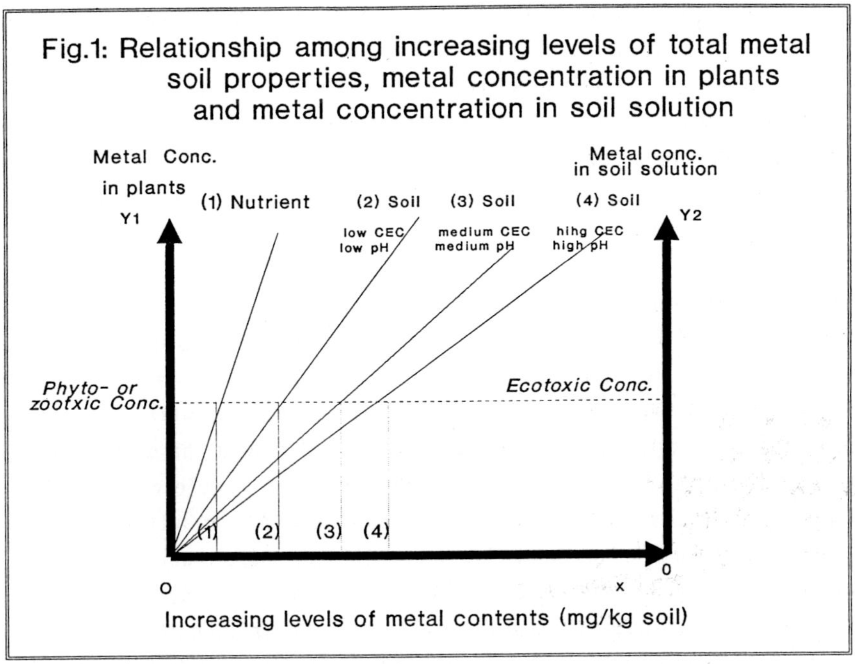

In order to induce phyto- or zootoxic metal conc. in a plant, which is grown on different soils, different amounts of total metal are needed to be added. These amounts depend upon the pH and CEC of growth media. The Fig. 1 schematically shows the regulatory effects of CEC, pH and total metal content of soils on phyto- or zootoxic conc. of a given plant (Ref 4). This fact has been substantiated by using results of two growth experiments presented in Table 1. Graded doses of Zn and Cd were applied in the form of salt and results are presented in Table 1 (Ref 4).

Table 1
Amounts of Total Metal to be Added in different Soils to Induce Phyto- or Zootoxic Theshold Values in Plants

Soil Properties			Calculated total Metal* to be added (mg/kg dry soil)	
Nr.	pH	CEC meq/100g	Zn clover	Cd grass
1.	5.6	8	60	0.9
2.	5.7	13	144	0.9
3.	5.9	57.4	210	2.5
4.	5.1	4.5	-	0.8
5.	5.6	16.6	-	1.2
6.	6.9	15.2	-	4.0

* These values are calculated from response curve

It can be seen that in soil 1, phytotoxic threshold Zn concentration i.e. 200 mg/kg dry plant in red clover leaves was induced by addition of a total of 60 mg Zn/kg dry soil (calculated from conc. curves), whereas max. tolerable Zn load proposed by Kloke is 300 mg/kg. In contrast, in soil 3 one requires to apply 210 mg Zn/kg dry soil so as to induce the phytotoxic threshold value in plants. Similary, in order to induce zootoxic threshold Cd concentration of 1 mg Cd/kg plant in rye grass grown on 6 different soils require varying amout of total Cd (Table 1).

2.2 METAL CONCENTRATION IN SOIL SOLUTION

The distribution of metals in soil-soild (capacity), soil-solution (intensity) on one hand and on the other hand the concentration in the plant grown on them (ecological significant) and growth of microorganisms are affected by several factors. The Soil-solution also designated as the liquid-phase is composed of water and dissolved ions or charged and wncharged complexes (Ref 1). The composition of the soil solution tells about the net result of variety of rections goingon in the soil but fails to tell the contribution of each individual reaction. Soil solution bathes the plant roots and soil microorganisms obtain their water, inorganic and partial organic nutrition (Ref 1).

It means that if soil solution is simulated to as it is found in nature then the metal concentrations in simulated in soil solution, it would be possible to assess ecotoxic metal concentration for soil microorganisms and plants. Further, there is expected to be consitstent and uniform in different soils irrespective of their properties. It means, soil solution metal concentration in conjunction with total metal content would be the best criteria to predict ecological danger of accumulated metals.

Attemps have been made to simulate the soil solution with the help of chemical solvents such as acetic acid, ammonium acetate + EDTA etc. These solvents shift the solid-solution eqilibria to solution side so that metal concentration is high enough to measure with simple analytical techniques. These historic solvents usually fail to simulate the metal concentration of soil solutions and thus bioavailable fraction of metal.

In contrast, neutral salt solutions such as $NaNO_3$, or Electroultrafiltration (EUF) are suitable extractants to simulate soil solution because neither it does affect the soil pH nor exchange the cations from soil solids. EUF method is time consuming and is not suitable for routine analysis.

Fig 2: Soil solution : A Reflection of Soil Properties and Properties of Plant Roots

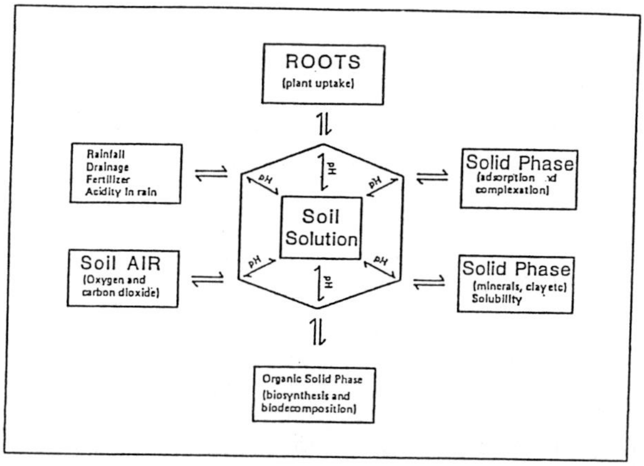

Metal concentration in soils

The results of two growth and one laboratory experiment carried out using contaminated soils (anthropogenic or artificially) are described in following paragraphs, are used to find the validity of above mentioned **hypothesis**.

2.2.1 OFFSITE-GREENHOUSE EXPERIMENT (In situ: anthropogenic contaminated soils)

2.2.2 ONSITE MINIPLOT FIELD EXPERIMENT (In situ: anthropogenic contaminated soils)

2.2.3 LABORATORY GROWTH EXPERIMENTS WITH SOIL MICROORGANSIMS (artficially contaminated soils)

2.2.4 COMPARISION OF ECOTOXIC CONCENTRATIONS IN SOILS (estabilished with soils either artificially contaminated or in situ contaminated)

2.2.1 OFFSITE GREENHOUSE EXPERIMENT

A pot experiment in green house was conducted using 10 soils which are exposed in the field to different metal emission sources such as metal coating plant, highway, refuge incineration plant, compost ans sludge. Redish and ryegrass were used as test plants. Three replications were used for each soil. In order to achieve optimum growth, each pot received a basal doses of nitrogen, phosphorus, potassium and magnesium. Soil and plant samples of both experiments were analysed for the following parameters:

Soils samples which were air dried and passed through 2 mm sieve were analysed for pH, CEC, HNO_3-extractable metals, $NaNO_3$-extractable metals. Plants samples were dried and analysed for metals such Cd, Zn, Ni and Pb. Atomic Absorption spectrophotometer was used for metal analysis.
In this growth experiment, the Zn and Cd concentration in soil and in test plants were found to exceed the normal range, whereas Pb and Ni concentrations were in normal range. The Zn concentration in radish tubers grown in 4 soils representing four different polution sources are presented in Fig 3.

Fig. 3 Effect of soil Zn on Zn content in radish tubers (in 4 anthropogenic contaminated soils).

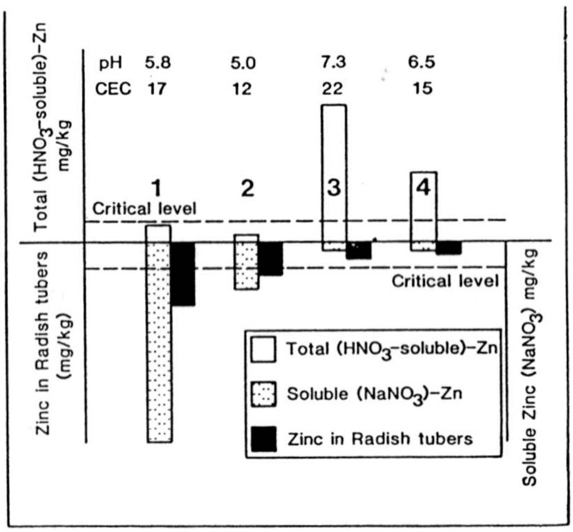

The total Zn content (HNO_3-souble) in all four soils execeeded or reached to the limiting value for total zinc in soil e.g. 200 mg Zn/kg dry soil. Further the phytotoxische threshold of Zn (200 mg Zn/kg plant) exceeded in only tow soils (Ref 5 and 7, Soil Nr. 1 & 2). The Zn-concentration in radish turbers grown in 4 representative soils are presented in Fig. 1. The low concentration of Zn in radish tubers in soils 3 & 4 were due to high pH (>6) and high cation exchange capacity. in contrast, $NaNO_3$ extractable soil Zn concentration in soil Nr.1 and 2 exceeded the critical Zn concentration of 1 mg Zn/kg dry soil and one observe phytotoxic concentration of zinc in radish tubers.

2.2.2 ONSITE-MINIPLOT FIELD EXPERIMENT

A miniplot ($1m^2$) Field Experiment was carried out on 25 fields, which are contaminated through different metal emission sources ans distributed in different parts of Switzerland. The miniplot locations are scattered through out Switzerland. These experiments was carried out for two years e.g. 1987 and 1988. Lettuce and ryegrass were used as test plants. At each location, both crops were seperately replicated for three times.

Soil and plant samples of both experiments were analysed for the following parameters:

Soils samples which were air dried and passed through 2 mm sieve were analysed for pH, CEC, HNO_3-extractable metals, $NaNO_3$-extractable metals. Plants samples were dried and analysed for metals such Cd, Zn, Ni and Pb. Atomic Absorption spectrophotometer was used for metal analysis (Ref 3).

The relationship between HNO_3-soluble, $NaNO_3$-soluble metal concentration in soils with metal concentration in plants are presented in Table 2. It is evident that neutral salt soluble metal concentration can assess the eco-toxicological metal concentration in soil better as compared to HNO_3.

Table 2: Relationship (correlation coefficients= r) zwischen Soil-(HNO_3-and $NaNO_3$-soluble) und metal concentration in lettuce leaves and rye grass (log)

Soil-extrakts	Test plants: Lettuce leaves			Rye grass		
	Zn	Cu	Cd	Zn	Cu	Cd
HNO_3	0,08	0,84	0,04	0.47	0,77	0,33
$NaNO_3$	0,84	0,86	0,69	0,83	0,81	0,66

Primary purpose of the mini plot experiment was to test the guide values of heavy metals of Swiss Ordinance on Pollutants (VSBo Ref 2) in soils, which were mainly estabilished with the help of greenhouse experiments, under field conditions and in situ contaminated soils.

In fact, the values which were obtained from pot experiments were halved to create a safty factor fix the VSBo values. It means that the plants growing in soils containing double the VSBo values should not exceed the phyto- or zootoxic concentration in them.

Table 3 presents the results of mini plot experiments considering the critical (phyto or zootoxic) plant concentration of 20 , 200 and 1 mg/kg dry matter for copper, zinc and cadmium, respectively. Lead and nickel are not considered because investigated plots either showed very low contamination or in very few locations exceeded the limiting values.

Table 3: Number of locations, where the $NaNO_3$-soluble metal contents lies double the VSBo-Guide values (Ref 3) and the plant contents are below the critical concentration . Number of total locations given in bracket.

Metal	Rye grass	Lettuce leaves
Copper	1 (34)	3 (32)
Zinc	2 (35)	2 (33)
Cadmium	0 (43)	*

*The cadmium content in lettuce leaves poses special problem because even by 0,004 mg $NaNO_3$-soluble Cd/kg soil the zootoxic concentration in plants are observed. These concentrations are supposed to be background concentratin in soils. So the problems of elevated concentration of Cd in sensitive plants can not be solved with the present guide values.

The Fig. 4 shows the relationship between $NaNO_3$ and HNO_3 soluble Copper concentration in soils and the critical phtotoxic concentration of copper in rye grass. The results show that HNO_3- soluble metal concentration of a given soil is not sufficient to predict bioimpact of metal or metals present in soil. Under field condtions, one can predict the possible bioimpact of given metals on plants by measuring metal conc. in salt extract of soils (Ref 3).

Fig. 4

Fig. 4 Effect of HNO3 and NaNO3 soluble copper concentration in different soils an the copper contents of rye grass

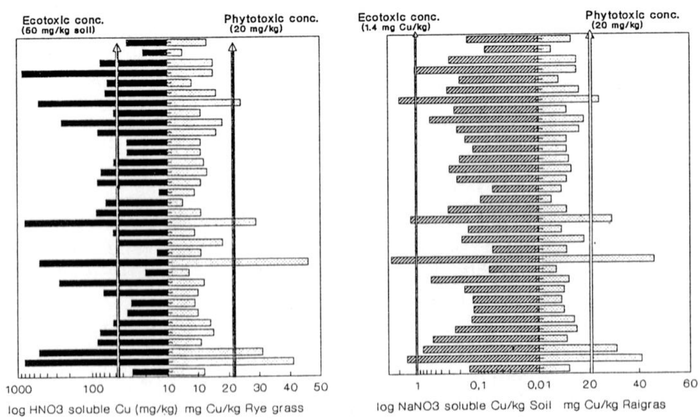

2.2.3. LABORATORY GROWTH EXPERIMENTS WITH SOIL MICROORGANISMS

Effect of Total- and $NaNO_3$-Soluble Cd on Soil Cd & Rel. % Carbon mineralisation:

At constant Cd-Application rate of 16 mg/kg soil (Fig. 5) the rel. % carbon mineralization values for sludge treated and without sludge treated acid Sandy soil (Steinhof) are singificantly different.

Fig. 5: Effect of Relationship among HNO_3-(total) & $NaNO_3$-soluble Soil Cd and rel. % Carbon mineralizaton.

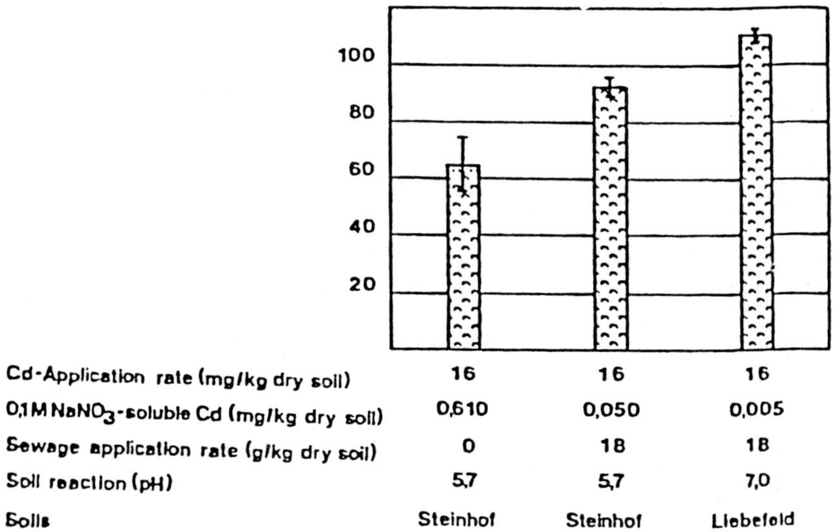

These differences can not be explained with help of total metal content of both soil and treatments because the total Cd contents were similar (Ref8). However, these differences in rel. % carbon mineralisation could easily be explained through differences in $NaNO_3$-solubles Cd contents (Steinhof: sludge untreated e.g. 0.61 and for sludge treated 0.05 mg Cd/kg dry soil).

2.2.4 COMPARISION OF ECOTOXICOLOGICAL CONCENTRATION IN SOILS (STADDARIZED WITH PLANT AND SOIL MICROORGANISMS TESTS)

In conclusion, it is to be noted that even such good agreement between the ecotoxic values estimated with pot experinments and mini plot field experiments, there are indications that the the $NaNO_3$ soluble values for copper should be reduced. In three locations though the copper concentrations in $NaNO_3$ extracts are below the critical values but the the growth of the rye grass was significantly affected which could be attributed to growth of soil microorganisms.

Tabel 4: Comparision of estimated ecotoxic metal concentrations in soils

Metal	Estimated Ecotoxic metal concentrations (in soil solution simulated with $NaNO_3$ (mg/kg Boden)		
	Artificaially contaminated		In situ (anthropogen) contaminated
	pot experiment (VSBo, 1986) Ref 2	Soil microorganisms (Laboratory experiment) Ref 9	(Miniplot field experiment) Ref 3
Zn	0.5	0.6	1.0
Cu	0.7	0.5	0.7
Cd	0.03	-	0.02
Ni	0.2	0.4	0.2

3. SUMMARY

During last few years problems releated to soil contamination have been discussed not only among common men but also among politicians. There are two following prerequisite for proper implementation of soil protection:

1. Soils already low in contaminants must be protected from future contamination through stringent control on emissions sources.
2. The polluted soils either need to be cleaned or atleast cleaned to the safe levels for good human health , good plant growth (quantitiy and quality) and optimum growth of soil micro-and macroorganisms.

Accumulated pollutants in soils can be transported to different important environmental components e.g. soil, water and plant:

- as gas or as component of dust particles via air
- as volatile substances via air
- as dissolved substances in the rain water or in soil leachate in deep layers of soil profiles and finally to the ground water

This paper present the results of a Offsite-Greenhouse and Onsite-Miniplot Field experiments carried out on anthropogenic contaminated soils to assess the ecotoxicological effects of soil metals. Further, the results of microorganisms growth experiments carried out mainly with artificial contaminated soils are discussed to evaluate the ecotoxicological reelevant concentration in soils. The estimated concentrations are comparable with the present Swiss guide values which were estabilished with the help of pot experiments in green house usingf mainly artificially contaminated soils.

4. REFERENCES

1. G. Sposito, Thermodynamics of soil solutions. Oxford Claredon Press, 1-223.
2. VSBO (Swiss Ordinance on "Heavy metal content in soils") 1986: Swiss Bundesrat. 1ST Sept. 1986 Nr. SRR 814.01.
3. S. Gupta, H. Häni and A. Blatter, Methodik zur Bestimmung biologisch relevanter Schwermetallkonzentrationen im Boden und Ueberprüfung der Auswirkungen auf Testpflanzen sowie Mikroorganismen in belasteten Gebieten. Schrifenreihe der FAC Nr. 2. Swiss Federal Research Station for Agricultural chemistry and Hygiene of Environment, 3097 Liebefeld-Bern, Switzerland, 1-54.
4. S. Gupta, 1984: Schweiz. Landw. Fo. 23 (3), 209-225
5. H. Häni and S. Gupta 1985: Proc. of Seminar organized by the CEC, Brussels in Rome (Italy). pp 157-167.
6. S. Gupta, Stadelmann, F.X., H. Häni, and A. Rudaz, Toxicological and Environmental Chemistry (1984),8(2/3), 173-184
7. S. Gupta, Proc. of Int. Conf. on the Heavy metals in the Environment, Genf (1989), Publ. CEP-Consultants Ltd. 197-200.
8. S.Gupta, H. Häni, and A. Rudaz, Proc. of Int. Conf. on Heavy metals in the Environment, Athens (1985).
9. F.X. Stadelmann und E. Santschi-fuhrimann, Publ. Swiss Fed. Res. Station for Agril. Chemistry and Hygiene of Environment, Liebefeld-Bern (1987).

THE EFFECT OF ADDITION OF BAUXITE REFINING RESIDUE (RED MUD) ON THE BEHAVIOUR OF HEAVY METALS IN COMPOST

Harrie Hofstede and Goen Ho

1. Introduction.

Within the Perth Metropolitan area of Western Australia the councils of Perth city, Stirling and Wanneroo are planning to process the organic fraction of their municipal waste through composting. The end product of the composting process, compost, generally has a high organic matter content. The functional groups of the humic substances and the high specific surface area give compost a high cation exchange capacity, and compost contains nutrients which are gradually released during the mineralisation process. These properties make compost a highly desirable soil amender for the poor Spearwood and Bassendean sands of the Swan Coastal Plain on which Perth is located.

However, a representative compost sample from a municipal compost plant in Auckland, New Zealand, considered to produce similar refuse to Perth, was analysed for heavy metals. Each metal was present at a considerable level (Table 1).

The sandy soils of the Swan Coastal Plain have a low pH and a low content of organic matter (< 1 %) and clay. In such an environment the mobility of heavy metals can be high. The potential high mobility of metals in the soils will limit the safe application volume of compost in order to comply with the guidelines for soil application of metal containing wastes. This will subsequently be a drawback on the extent of the compost market and the marketing of compost.

In order to address the problem of heavy metal mobility in soils, it was hypothesised that bauxite refining residue ('red mud') which is produced in abundance locally, could be a suitable amendment to compost. Red mud is a sodium saturated loam with a high alkalinity (pH 9 - 12) and a high cation exchange capacity (15-45 meq/100 g). It could reduce the mobility of heavy metals through metal hydroxide and carbonate precipitation and adsorption. An additional advantage is that the adsorption complex of red mud can retain applied fertiliser, preventing the nutrients from leaching out.

In order to improve the dispersed structure of the clay particles as a result of sodium saturation and to neutralise the alkalinity, the effect of addition of gypsum

($CaSO_4.2H_2O$), which is a locally produced waste product from the superphosphate industry, was included in the study.

Table 1
Results of Auckland municipal compost analysis and heavy metal application guidelines applied in the EEC and Australia; heavy metals in mg/kg dry matter.

	Auckland	EEC*, (recomm.)	max. load* (kg/ha.yr)	Australia (1) (sew. sludge)
cadmium	6	5	0.25	8 - 20
copper	380	300	10	1200
nickel	36	50	2	100
lead	608	750	10	500
zinc	1460	1000	25	1800
chromium	225	150	10	500
iron	1.6 %			
manganese	560			500
pH	7.2			
ash %	40			

*Proposal for EEC Council Directive for the use of sludge in agriculture (25-5-84).

The objective of this study was to gain insight into the behaviour of the extractable and leachable heavy metals in compost and the effect that addition of red mud or red mud - gypsum might have on reducing the mobility of heavy metals. The ultimate aim is to reduce the mobility of the heavy metals such that the dispersion of the heavy metals through plant uptake and groundwater leaching is within acceptable levels with regard to public health.

2. Soluble organic matter fraction and its interaction with heavy metals

2.1 Fractionation of soil organic matter

Roughly organic matter can be divided into a humic and a non - humic fraction (Figure 1). The non - humic matter fraction consists of products such as fats, waxes, oils, resins, water soluble polysaccharides, hemicellulose, cellulose and proteins (2). The humic matter on the other hand is a relatively stable by-product of the microbial transformation of a great variety of biochemical compounds derived from dead cells and extracellular products of various organisms (3).

The humic matter fraction can be separated into a fraction which is soluble in an alkaline solution and in a fraction which is not. According to Mortensen (4), the average age of the soluble organic matter fraction is 50 - 250 years whereas the non - extractable fraction, called the humin fraction, has an average age of 2000 years. The alkaline solution soluble humic matter is better known as humic and fulvic acids.

At low pH the humic acid molecules are aggregated into colloidal particles with very little or no charge which causes them to precipitate (5). In this way the fulvic and humic acids can be separated.

Because humic substances can form stable complexes with metal ions, minerals, non - humic compounds and certain anions (4), the soluble humic matter fraction can play an important role in the transportation of these products through the soil. The transportation of heavy metals by humic and fulvic acids will be discussed in more detail.

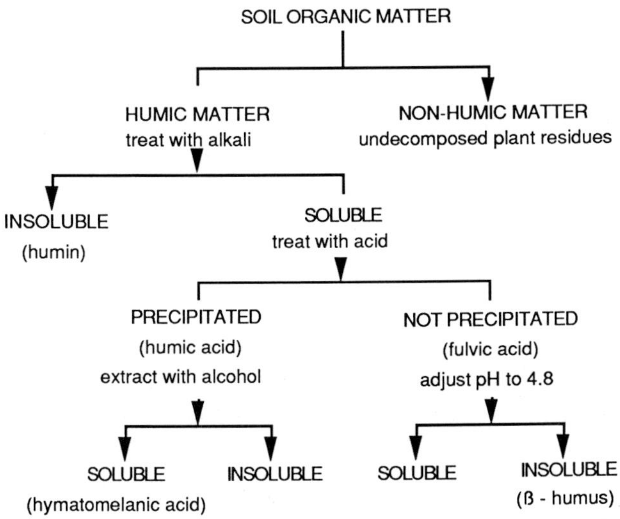

Figure 1 *Fractionation of humus based on extraction of soil organic matter and further subdivision of the extracted material (2).*

2.2 Nature and properties of humic and fulvic acids

It is important to note that no sharp division exists between the three humic matter fractions, humins, fulvic acids and humic acids. They are all part of an extremely heterogeneous polymer system and the differences between the subdivisions are due to variations in elemental composition, acidity, degree of polymerisation and molecular weight (6). Molecular weight ranges from a few hundred for fulvic acids to tens of thousands for the humic acids and the humin fractions.

The principal reason for the chemical differences is probably the difference in source material. Environmental factors may impose additional diversification (3). But also the degree of maturity of the humic matter causes variations. When humic matter becomes more mature, the aromatic compounds become more highly condensed, oxygen and nitrogen bearing functional groups are split off, the aromaticity of the humus increases and unsaturated aliphatic chains may condense with each other.

2.2.1 **Basic structure and functional groups**

A brief discussion of the structural nature and functional groups of humic and fulvic acids is given here. Jackson *et al.* (6) presented a detailed review on this topic.

Humic acids

Basically humic acids all have aromatic nuclei in the form of flat lattices of cyclic polymerized carbon. The basic structure consists largely of condensation products of compounds such as catechol, syringaldehyde and 3,5-dehydroxybenzoic acid (Figure 2). Side chains are connected which carry various functional groups such as carboxyl, phenolic and alcoholic hydroxyl, methoxyl, carbonyl and quinone groups.

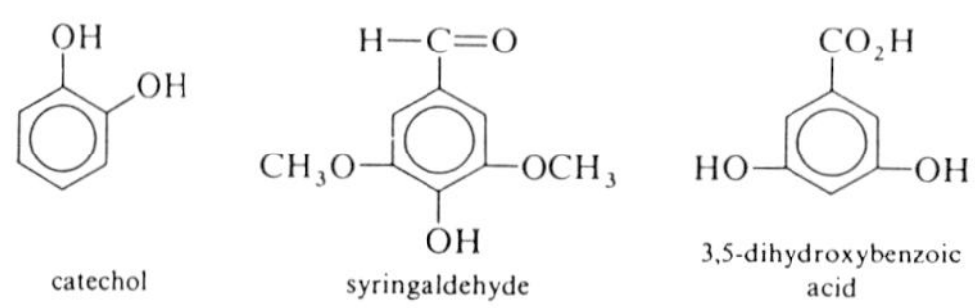

Figure 2 *Some of the typical compounds found among the products of chemically decomposed humic acids (7).*

Fulvic acids

Fulvic acids contain aromatic rings which form an open structure in a two- or three-dimensional array (Figure 3). The open structure is held together at least partly by hydrogen bonds among the functional groups. The functional groups identified in humic acids have also been identified in fulvic acids.

Functional groups content

More important than the structural networks of humic and fulvic acids are their functional groups content. It has been concluded that fulvic acids contain more functional groups, particularly carboxyl, per molecule than do humic acids (Table 2). There are no methoxyl and quinone functional groups to be found in fulvic acid as alcoholic hydroxyl in humic acid. Functional groups are also unevenly distributed among the molecular weight fractions (8).

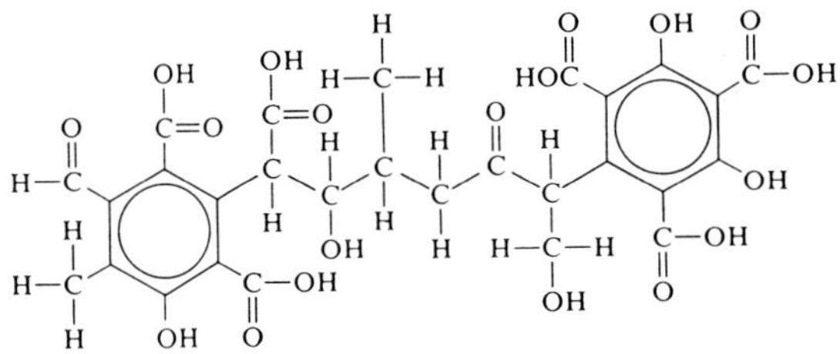

Figure 3 *Typical type of compound composing fulvic acid (7).*

Table 2
Number of functional groups per molecule of soil humic acid according to Manskaya and Drozdova and soil fulvic acids according to Schnitzer and Skinner, Gamble and Schnitzer (6).

Functional group	Humic acid	Fulvic acid	Structure
carboxyl	3 - 4	6 - 8	- COOH
phenolic hydroxyl	3 - 4	2 - 4	ø - OH
alcoholic hydroxyl	-	3 - 4	C - OH
methoxyl	± 1 - 2% by weight	-	C - O - C
carbonyl	-	2 - 3	C = O
quinones	3 - 4	-	O = ø = O

ø = benzene ring

2.3 **Metal - humic matter complexes**

Humic and fulvic acids behave as negatively charged species due to the ionization of their acidic carboxyl and hydroxyl groups. That is why they can attach to positively charged particles like metal ions and form stable complexes. Shapiro (9) found that the yellow organic acids in coloured waters can keep considerable quantities of ferric iron in solution even though the pH and redox potential are high. After esterification of the -COOH and -OH groups the iron - binding capacity disappeared. Stevenson (10) found evidence that COOH groups play the prominent role in complex formation. Because fulvic acids have more carboxyl and hydroxyl functional groups, they probably can complex more heavy metals than humic acids.

It is suggested that humic compounds of different origins differ in their metal-binding behaviour because of different functional group contents (3). The lower molecular weight fractions seem to have a greater metal holding capacity than the higher molecular weight fractions because of a relatively higher functional group content.

The ions or molecules that are bonded or co-ordinated with the metal are termed ligands. In structure the ligands are bonded to the metal through two or more different ligand atoms (Figure 4). They can form part of heterocyclic rings. Rings formed in this way are termed chelate rings. A metal complex which contains one or more chelate rings is called a metal chelate and the ligands involved are termed chelating ligands. Detailed information about this type of complex formation has been given by Houghton (11).

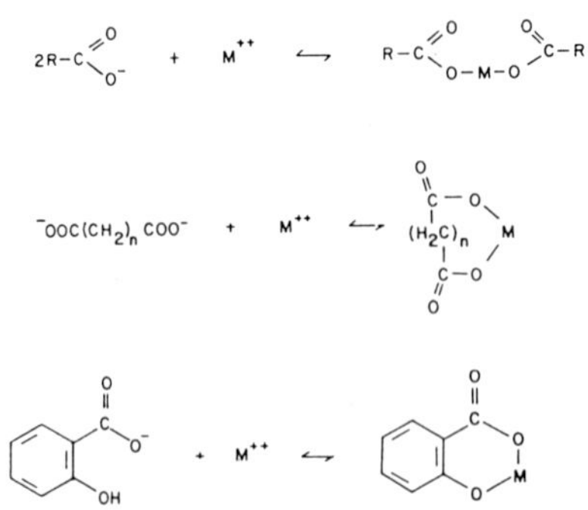

Figure 4 *Different types of metal-ligand complexation (12).*

2.3.1 **Stability constants.**

To evaluate the potential hazard of heavy metal - organic matter complexes transported through the soil, the main factors that control the extent of metal-soluble organic matter interactions should be investigated. These main factors according to Mantoura *et al* (13) are:

(i) The value of the stability constant for the metal-humic matter complex.

(ii) The pH which provides a measure of the competition from carbonate and hydroxy ligands.

(iii) The major ion concentrations, which control the competition for humic matter.

The last two factors can be measured easily. The determination of the stability constant of the humic matter - metal complex is not easy however, due to the complecated nature of the binding of metal ions by humic substances.

Complex formation can be regarded as a competitive reaction between the metal ion and H^+ for reactive sites on the ligand. For a two-step process, the reactions are:

$$HA + M^{2+} \xrightarrow{k_1} MA^+ + H^+ \quad [1]$$

and

$$HA + MA^+ \xrightarrow{k_2} MA_2 + H^+ \quad [2]$$

where the successive constants k_1 and k_2, called stepwise formation constants (7) are given by:

$$k_1 = \frac{(MA+)*(H+)}{(HA)*(M^{2+})} \quad [3]$$

$$k_2 = \frac{(MA_2)*(H^+)}{(HA)*(MA^+)} \quad [4]$$

The overall formation constant, K, is given by:

$$K = k_1 * k_2 = \frac{(MA_2)*(H^+)^2}{(HA)^2*(M^{2+})} \quad [5]$$

There are different methods to measure this constant such as ion exchange equilibria and the method of continuous variation (14), potentiometric titration (10), gel filtration (15) and the paper chromatography (16).

In general, the stabilities of complexes of the various metals seem to follow the Irving - Williams order of stabilities of chelates formed by metal ions with ligands (13). According to this order, the stability of Mg < Ca < Cd = Mn < Co < Zn = Ni < Cu < Hg. Takamatsu *et al.* (16) found with a paper chromatography test with metal-humic acids a slightly different order : Mn < Co = Cd < Zn < Ni < Pb < Hg < Cu. The stability constants for five heavy metal species presented in Table 3 do not confirm this order though. Although humic acids have less metal binding capacity, the stability constants are substantially higher than those of fulvic acids according to Table 3.

The influence of pH on the stability constant

From the data presented in Table 3, it can be concluded that the pH of the soil solution has a significant effect on the stability constant, which increases with increasing pH. This is probably caused by two different processes.

Firstly, the pH is important in determining the solubility and the state of hydrolysis of the metal cations and therefore their ability to complex with humic or fulvic acids. Secondly the charge of the humic matter depends on the pH. At low pH values, humic acid molecules for instance are aggregated into colloidal particles with very little or no charge. If the pH is increased the negative charge on the molecule increases as dissociation occurs. At the same time the aggregates disintegrate, as the molecules become more hydrated, until they are all completely dispersed in solution (5).

Table 3
Average stability constants (as log K) for fulvic (FA) and humic acid (HA) - heavy metal complexes at varying pH and ionic strength (I).

	Cadmium	Copper	Nickel	Lead	Zinc	Reference
FA from B_h	-@	5.78	3.47	3.09	1.73	(17, 18)
horizon in	-@@	8.69	4.14	6.13	2.34	derived by ion-
podzol profile						exchange method.
FA from B_h	-*	3.3	3.1	2.6	2.4	(14)
horizon in	-**	4.0	4.2	4.1	3.7	
podzol profile	-***	4.7	4.5	3.6	3.2	
	-****	2.6	2.4	2.1	2.0	
3 humic acids from diverse sources	6.9	8.9	-	8.7	-	(14), derived by potentiometric titration method. I = 0.0.
peat FA	4.57	7.85	4.98	-	4.83	(13)
peat HA	-	8.29	-	-	-	derived by gel
lakes	4.64	9.23	5.21	-	5.19	complexometry.
rivers	-	9.54	-	-	5.39	pH=8.0, I=0.02,
lochs	4.91	9.55	5.25	-	-	
sediments	-	10.46	-	-	5.43	
seawater	4.69	9.30	5.46	-	5.29	

*pH=3.0, I = 0.1 ***pH = 3.0, I = 0.00 @pH=3.5
**pH=5.0, I = 0.1 ****pH = 3.0, I = 0.15 @@pH=5.0

Kiekens, (19) showed that in fact there are two stability maxima. One occurring at pH 6 and another at pH 9. This might be attributed to the dissociation of functional carboxyl (-COOH) and hydroxyl (-OH) groups.

The pH also influences the order of stability. Schnitzer and Skinner (17, 18) reported an order change from Mg > Mn > Zn > Ca > Co > Pb > Ni > Fe > Cu at pH 3.5 to Mg > Zn > Ca > Co = Mn > Ni > Fe > Pb > Cu at pH 5.0.

The influence of ionic strength on the stability constant.

Schnitzer and Hansen (17) noted a marked effect of ionic strength on the stability constants of metal-fulvic acid complexes (Figure 5). With increasing ionic strength (I), the stability constants decrease. This is probably due to the increase in competition between the different ions in solution to complex with organic matter when the ionic strength of the solution increases. Because the ionic strength of the soil solution is normally low, log K values obtained at very low ionic strength are more likely to represent conditions existing in the soil solution than those at high ionic strength (11).

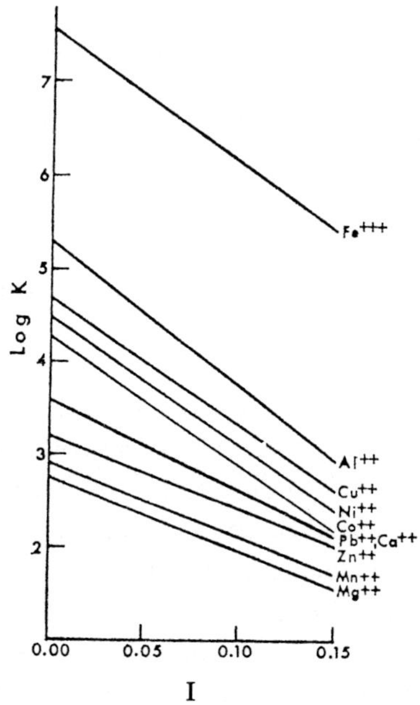

Figure 5 *Effect of ionic strength on log-K values of metal-fulvic acid complexes at pH 3 (14).*

2.3.2 Valency change of metal ions by humic matter

Humic matter not only forms complexes with metal ions, they seem to be able to reduce these cations as well. This is apart from soil environmental factors and micro-organisms which are known to have reducing capacities too. Ghosh *et al* (20) found after sterilizing the soil in order to eliminate microbial activity, that humic substances are independently able to change the valence state of metal ions such as iron, copper and manganese. Takamatsu *et al* (16) found a reduction of Hg^{2+} by humic acids during a chromatography test.

2.4 Clay particle-humic matter interactions

Humic substances are also able to form complexes with minerals like clay particles. Roy (21) found that small amounts of humic matter could disperse kaolinite particles in suspension. On the other hand humic matter could conceivably cement clay particles together to form stable aggregates (3).

Tan (22) made an attempt to investigate the complex formation between clay and humic and fulvic acids to obtain more detailed information. After studying infrared spectra from the clay particles and the humic and fulvic acids he concluded that humic acids might have attacked the clay mineral, liberating and complexing Si and Al ions from tetra- and octra-hedral positions.

2.5 Soil transportation of metal - humic matter complexes

From computer based models it has been shown that a complicated web of interactions can be expected in any natural water that contains hazardous concentrations of heavy metals along with reactive organic compounds (23). The extent of solubilization of heavy metals depends upon a number of factors, including the stability of the complex; the concentration of the complexing agent; pH; and the nature of the metal deposit. The situation is even more complicated in the soil solution.

The pH of the soil solution plays a major role in determining whether heavy metal - humic matter complexes become soluble and can be transported through the soil. Tan and Binger (24) studied the effect of soil humic acid on aluminium toxicity in corn plants in a low pH sand culture. Addition of humic acids prevented a sharp reduction in dry matter production and aluminium toxicity symptoms disappeared. The high amount of aluminium had complexed with the insoluble humic acids at low pH. Hardgrove (25) found at pH values greater than 5 that Al-organic matter complexes could become solubilized and then adsorbed by plant roots.
Lagas and Loch (26) found that iron, zinc, nickel and lead are mobilized in synthetic landfill leachate with fatty acids.

3. Bauxite refining residue

3.1 Bauxite refining residue production

Virtually all alumina [Al_2O_3] is refined from bauxite via the Bayer process developed by Kart J. Bayer in 1887 (27). The Bayer process takes advantage of the fact that gibbsite [$Al(OH)_3$] and boehmite [AlOOH] are soluble in sodium hydroxide [NaOH]. In Western Australia, the bauxite is high in gibbsite. Crushed bauxite is digested in strong caustic soda at a temperature of 143°C and a pressure of at 250 kPa (28). The reactions are:

$$Al(OH)_3 + Na^+ + OH^- \text{-------}> Al(OH)_4^- + Na^+ \qquad [6]$$

$$AlOOH + OH^- + Na^+ + H_2O \text{-------}> Al(OH)_4^- + Na^+ \qquad [7]$$

Other constituents of bauxite, with the exception of clays and organic matter, do not dissolve and generally pass through the process unchanged or only slightly changed (29). The clay minerals, that is reactive silica, mainly as kaolinite, in Darling Range bauxite dissolve and recrystalise under high temperature (>200°C) and pressure (> 34 x 10^6 Nm^{-2}) as relatively insoluble sodium-aluminium-silicates known as the "Desilication Product (DSP)". DSP is not a single compound but a series of zeolites (30). The reaction of kaolinite in the process illustrates desilication:

$$Al_2O_3{\cdot}2SiO_2{\cdot}2H_2O + 6NaOH \text{-----}> 2NaAlO_2 + 2Na_2SiO_3 + 5H_2O \qquad [8]$$

$$2Na_2SiO_3 + 2NaAlO_2 + (2+x)H_2O \text{-----}> Na_2O{\cdot}Al_2O_3{\cdot}2SiO_2{\cdot}xH_2O + 4NaOH \qquad [9]$$

With the Al_2O_3 in solution, the next step is to remove the solid residue by sedimentation and filtration. This separation must be as complete as possible because any remaining solids will contaminate the product. The coarse fraction of the residue larger than 100µm can be removed easily by liquid - solid cyclones or more primitive settling chambers. The desanded slurry is then pumped to a thickener which removes the remaining residue by sedimentation.

The residue, both sand and mud, are separately washed with residue pond water by countercurrent decantation. The aim of washing the residue is to minimize the amount of soluble salt in the residue and the cost of evaporation that the dilution introduced. After washing, the residue is discarded to tailing ponds (27).

The major environmental concern of alumina production is disposal of the bauxite residue. The production of residue averages 1 tonne per tonne alumina and varies according to the type of bauxite. In Western Australia, the total bauxite residue produced was about 12 million tonnes in 1986 (31).

3.2 **Bauxite residue soil classification**

Bauxite residue consists of a coarse fraction of size > 150µm called red sand and a fine fraction of size < 150µm called red mud. The sandy fraction, which is predominantly quartz, is about 30 to 50% of the total residue solids (31).

Fine bauxite residues mainly consist of very fine particles with a diameter smaller than 0.02 mm. Fine bauxite residue samples from the Kwinana Refinery of

Alcoa consisted of 40% by weight out of particles less than 2 μm in diameter (Table 4). This particle distribution classification gives red mud a very high specific surface on which both anions and cations can be adsorbed.

Table 4

Particle size distribution of bauxite residue.

Fraction (mm)	Point Comfort (32)	Kwinana (33)	Mobile (34)
		Weight (%)	
Sand (2-0.02)	8	30	5
Silt (0.02-0.002)	66	30	25
Clay (<0.002)	26	40	50

3.3 **Metal binding ability of fine bauxite residue**

Cation exchange

The exchange of cations on red mud is determined by two groups of minerals, the zeolites (DSP) formed during the refining process and the non - zeolites such as haematite, Al - goethite, muscovite and illite (Table 5).

Table 5

Mineralogical composition of bauxite residues from various sources.

Mineral	Kwinana (35)	Jamaica (36)	Brazil (36)	Greece (36)
		Composition (weight %)		
Quartz	16.3	0	2.2	0
Haematite	13.9	57.5	18.7	47.0
Goethite	15.4	1.0	22.7	Trace
Zeolite (DSP)	10.6	5.5	5.1	7.0
Aluminia goethite	5	0	7.0	0
Muscovite	10.1	-	-	-
Boehmite	6	Trace	2.1	1.2
Gibbsite	9	-	-	-
Calcite	4.6	-	-	-
Anatase/Rutile	1.6	1	0	2.4
Tricalcium aluminate	3.8	1	0	2.4

The cation exchange sites can be classified as surface exchange sites and internal exchange sites. The DSP frame work consists of a network of very fine channels that only allow passage of monovalent cations. It has been found that 70 % of the sodium in red mud is adsorbed on the internal sites. The internal exchange capacity was found to be 50 meq/100 g. These sites are however not accessable for the much larger divalent or trivalent heavy metals.

The surface exchange sites of the non - zeolitic minerals and the DSP have a combined cation exchange capacity of 25 meq/100 g. These sites are available for adsorption of heavy metals.

Precipitation

The total alkalinity in red mud is 5 - 6 g/l, resulting in a pH of 11- 12. The high alkalinity is due to the presence of NaOH, Na_2CO_3, $NaHCO_3$ and $Na_2(AlO)_3$ in the entrained liquor. As a result of atmospheric exposure in the red mud impoundments, the soluble hydroxide reacts with carbon dioxide to form bicarbonate and subsequently carbonate depending on the pH. Both hydroxide and carbonate precipitates of heavy metal have a very low solubility and are thus useful as a tool to control the release of immobilised metals, in conjunction with pH control.

3.4 **Gypsum as an amendment of red mud**

In general, addition of acidic ameliorants to a sodic soil can initiate permeability increases due to both electrolyte concentration and cation exchange effects (37, 38, 39, 40). Increases in the electrolyte concentration in soil solutions reduce the dispersion of clay in soil which results in a higher soil permeability. The exchange of sodium by other cations can reduce the exchange sodium percentage (ESP) in soil, which in turn enhances the water permeability of soil. In strongly alkali soils, the cation exchange effect seems to be more important for permanent improvement in soil properties than just increasing the hydraulic conductivity. It permanently reduces the ESP in soil (41).

Different ameliorants including $FeSO_4$, FeS, S and $CaSO_4$ have been used to neutralize the alkalinity and reduce the sodicity of bauxite residue (42, 43, 44, 45, 46, 47,48, 32, 49). In Western Australia, waste gypsum from a fertilizer plant in Kwinana, located near Alcoa's refinery is readily available as product for neutralisation of bauxite residue.

The chemical reaction of red mud and entrained liquor with gypsum is shown below:

$$Na_2X + CaSO_4 \text{ ------> } CaX + Na_2SO_4 \qquad [10]$$

Na_2X in weathered bauxite residue exists mainly as carbonate. Hence $CaCO_3$ forms as the end products of the above reactions. Since carbonate is insoluble , the reaction equilibria moves to the right. This facilitates further release of sodium from the residue, and the removal of soluble carbonates. The soluble sodium salts can then be removed by leaching.

Gypsum is a commonly used amendment for sodic and alkaline soil reclamation because of its low cost and general availability. Through the precipitating of CO_3^{2-} to $CaCO_3$, gypsum dissolution can reduce soil pH to a level similar to a calcareous soil (32, 44, 45, 49). The theoretical pH of a soil buffered with calcium carbonate in equilibrium with atmospheric carbon dioxide is about 8.3 (50). Moreover, Ca^{2+} dissolved from gypsum increases soil electrolyte concentration and can exchange with sodium in the exchangeable sites. The replacement of sodium from the exchange sites has a permanent effect in reducing the soil sodium status. This will reduce soil ESP and increase the rate of water infiltration.

Gypsum was found to be an effective ameliorant for red mud. Five percent gypsum was found to be sufficient to neutralize the alkalinity of red mud (44. 47). Excess gypsum must be applied due to the slow release of the potential alkalinity from red mud which may result in a gradual increase in pH, sodium concentration and sodium adsoption ratio (SAR), (48, 49). Excess gypsum can provide red mud neutralization capacity to counteract further release of alkalinity.

The reduction in alkalinity is also of importance to the stability of organic matter mixed with red mud. A high alkalinity reduces the stability of organic matter through oxidation and thus increases solubilisation and leaching. This may have as a consequence leaching of metals that have formed stable complexes with the organic matter.

4. Methodology

The study had a three way approach. The first step was to assess the effect of red mud on the extractability of a selection of metals, deemed relevant, using $CaCl_2$, $BaCl_2$, DTPA and HCl as extractants. In the second part the interaction of heavy metals with organic acids was studied through fractionation of the soluble fulvic and humic acids. In the third and final part percolation columns were used to determine metal leaching from compost - red mud mixtures.

Two series of mixtures were prepared from compost and 0, 5, 10 and 15% red mud (RM) or red mud neutralized with gypsum (RMG). Each sample weighed 5 gram. The mixtures were extracted by mechanical shaking for two hours in a 10:1 liquid to solid ratio with 50 ml 0.5 M $BaCl_2$ and with 0.1 M HCl, to determine the exchangeable adsorbed, easily soluble (0.5 M $BaCl_2$) and the exchangeable adsorbed, organically complexed and metals associated with Fe- and Mn-oxides (0.1 M HCl) amounts of lead and zinc (26). After extraction, the samples were centrifuged in a

Sorvall ultracentrifuge at 2000 rpm for 15 minutes and filtered over a 0.45 μm membrane filter.

Two other series of mixtures, 0, 10, 20, 30, 40 and 50 % RM or RMG, were extracted in a identical manner with 0.005 M DPTA, at pH 7.3 to determine plant available metals (51), and with 0.01 M $CaCl_2$ to determine soil solution soluble amounts of cadmium, copper and nickel.

The fulvic and humic acids were extracted by adding 200 ml of a 0.1 M sodium hydroxide solution to 50 gram air dried and ground compost with 0, 5, 10 and 15 gram red mud added. The samples were extracted for two hours on a Coulter mixer. Then the samples were centrifuged at 5000 rpm for 20 minutes on a Sorvall RC-5B superspeed centrifuge to remove particulate matter. The supernatants were filtered over a 0.45 μm membrane filter.

The humic acids were separated from the extracts by reducing the pH to 1 with 6 M HCl, allowing the humic acids to precipitate for one hour, and subsequent centrifugation at 15,000 rpm for 20 minutes (52, 53). The humic acids were washed with 0.1 M HCl and redissolved in 0.1 M NaOH.

The UV-visible spectra (200 - 800 nm) of the fulvic and humic acids extracts were measured. In order to determine the maturity of the compost by calculation of the Δlog K value and the E4/E6 ratio ('Welt'-index), the absorbances at wavelengths 400, 465, 600 and 665 nm were measured.

The extracted fulvic and humic acids were fractionated over a Sephadex G 10 gel filtration column. The different fractions were analysed for UV-absorbance in-line at a wavelength of 254 nm with a LKB 8300 Unicord II UV-analyser and recorded on a Rikadenki pen recorder. The organic acid fractions were collected by a LKB fraction collector and stored for analysis.

The extracts and fractions were analysed for cadmium, copper, nickel, lead and zinc on a GBC atomic absorption spectrophotometer. A heavy metal balance could thus be made.

Two percolation columns (50 mm x 1000 mm)were constructed and filled (500 mm) with a mixture of compost and red mud or red mud-gypsum in a ratio of 60 : 40, and with a porosity of 75 % (96.6 cm^3). The columns were leached with 0.01 M $CaCl_2$ through feeding using a Marriotte bottle set-up. The leachate was collected in a LKB fraction collector and each fraction was analysed for cadmium, copper, nickel and zinc.

5. Results and discussion

5.1 Batch extractions

5.1.1 *Exchangeable and weak- acid extractable cations*

$BaCl_2$ was able to extract 2.8 % of the total zinc content of compost without red mud addition. Increasing amounts of red mud resulted in a decrease (41 -18 mg/kg) in zinc extraction down to 1.2 % of total zinc content (Table 6). The decrease could be caused by zinc adsorption onto red mud. Addition of red mud increases the adsorption capacity by providing more adsorption sites.

In the hydrochloric acid extract, the zinc concentration decreased from 55 to 8 mg/kg (85 %) as a result of red mud addition. A considerable part of this reduction may be attributed to some neutralisation of the hydrochloric acid by the alkalinity in the added red mud, resulting in a reduced strength of acid solution and thus the ability of the solution to extract or dissolve cations.

Lead levels extracted by $BaCl_2$ were low, only a maximum of 1.9 % of the total lead content. This was expected as lead forms very stable complexes with organic matter, particularly with humic acids (54). Increasing red mud addition did reduce the extractable lead significantly, but no lead was detected with 15 % red mud (Table 6).

Table 6
Extractable concentrations of zinc and lead (mg/kg) with increasing red mud additions to the compost.

% red mud	$BaCl_2$		HCl	
	zinc	lead	zinc	lead
0	41	2.4	55	4.4
5	29	1.6	24	bd
10	22	1.9	8	2.6
15	18	bd	8	bd

bd = below detection limit.

During extraction with $BaCl_2$ it is likely that barium ions will prefer to exchange with sodium adsorbed to red mud. A combination of high concentrations of sodium and barium in solution can then extract lead from its stable organic matter complex. Increased red mud addition will increase the cationic strength of the cation exchange extractant with sodium. Increased red mud addition also increases the available cation exchange capacity.

Acid extractable lead was nearly double the exchangeable lead level. This shows that some organically complexed lead was extracted, something that $BaCl_2$ is incapable of extracting. The reduced acid extractability of lead as a result of red mud addition can partly be attibuted to adsorption of lead to red mud, though acid neutralisation would have affected the lead extraction as well. Thus in both cases red mud addition reduced the amount of extractable zinc and lead in the compost.

5.1.2 *Plant available and soil solution soluble cations*

DTPA extraction showed that red mud addition might have different effects on the plant availability of heavy metals (Table 7). Cadmium decreased significantly, copper increased slightly, whereas nickel did not appear to be affected. The degree of complexation of heavy metals with organic matter, adsorption to red mud and preference for complexation by DTPA interact to determine the extractable amount from compost. The slight increase in plant available copper is likely to be associated with the increased solubilisation of the organic fraction as a result of the red mud addition. Copper is known to be very strongly bound by organic matter (55). Plant availability is high for nickel, 9 mg/kg (25 %) and cadmium, 2.1 mg/kg (46 %). This suggests a relatively low stability constant for nickel and cadmium complexes with the organic matter fraction in compost.

Table 7

The effect of red mud on plant available and soil solution soluble cadmium, copper and nickel concentrations (mg/kg).

% red mud	plant- available (DTPA)			leachable ($CaCl_2$)		
	cadmium	copper	nickel	cadmium	copper	nickel
0	2.2	8.6	8.9	-	1.7	1.8
10	1.9	8.9	8.1	-	1.6	1.8
20	1.7	8.9	8.7	-	1.5	1.9
30	1.5	8.9	8.7	-	1.5	1.9
40	1.3	9.7	8.8	-	1.7	1.9
50	1.0	10.3	8.8	-	1.7	2.0

The $CaCl_2$ extraction showed that addition of red mud did not affect heavy metal release from compost. Precipitation could be ruled out as concentrations did not exceed the solubility product. The extractable nickel was 3.5% of total content in compost, copper 0.13% and cadmium very low. The results of the batch extraction tests suggest that most heavy metals leached from compost are bound to organic matter and not in free cationic form, otherwise red mud addition would have a more significant effect on heavy metal removal from the leachate.

5.2 Percolation columns

The percolation experiment shows that gypsum amended red mud reduces heavy metal leaching (Figures 6, 7, 8). The reduction is in the order of Zn > Cu > Ni. The reduction was not caused by pH, because the pH of the leachates did not appear to be affected by gypsum addition. The red mud - gypsum amended compost column showed a lighter coloured effluent than red mud amended compost column, indicating less soluble organic matter in the leachate. As a result, the leached metals must be adsorbed or complexed onto this soluble organic matter, as any free copper, nickel or zinc would precipitate as their hydroxide or carbonate or adsorb to red mud clay particles.

An explanation for lower organic matter solubilisation can be the higher ratio divalent/monovalent cations in the mixture through addition of calcium as gypsum. A high calcium to sodium ratio prevents the dispersion of organic acids. Additionally the acidic character of gypsum has a considerable capacity to reduce the alkalinity of red mud, as a 5 % gypsum amendment reduces the pH of red mud from 12 to 8.3 (47). A reduction in the alkalinity of the red mud will reduce the solubilisation of organic acids.

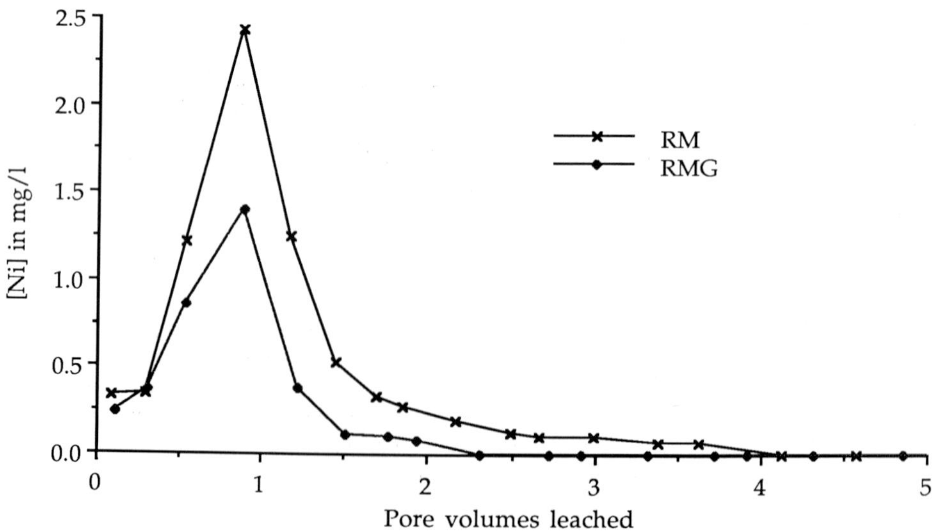

Figure 6 *Nickel leaching from red mud (RM) and red mud - gypsum (RMG) amended compost.*

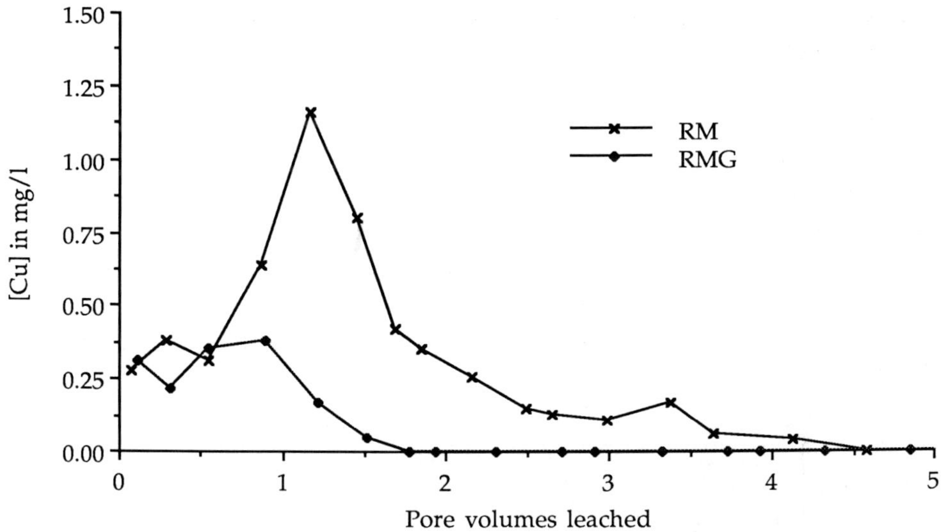

Figure 7 *Copper leaching from red mud (RM) and red mud - gypsum (RMG) amended compost.*

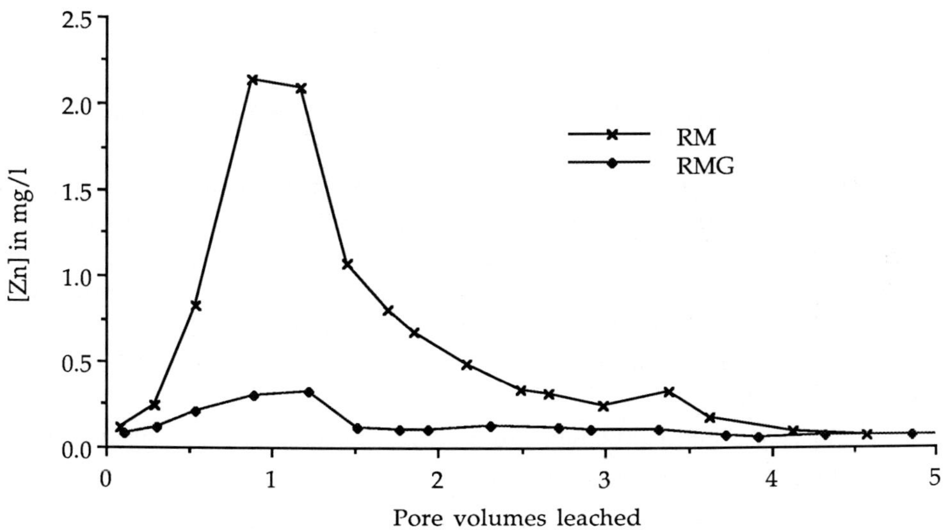

Figure 8 *Zinc leaching from red mud (RM) and red mud - gypsum (RMG) amended compost.*

5.3 Interaction of heavy metals with the soluble organic fraction

5.3.1 *Characterisation of humic and fulvic acids*

The UV-scan over the range 200-800 nm shows a similar UV transmission pattern for both humic and fulvic acids (Figure 9).

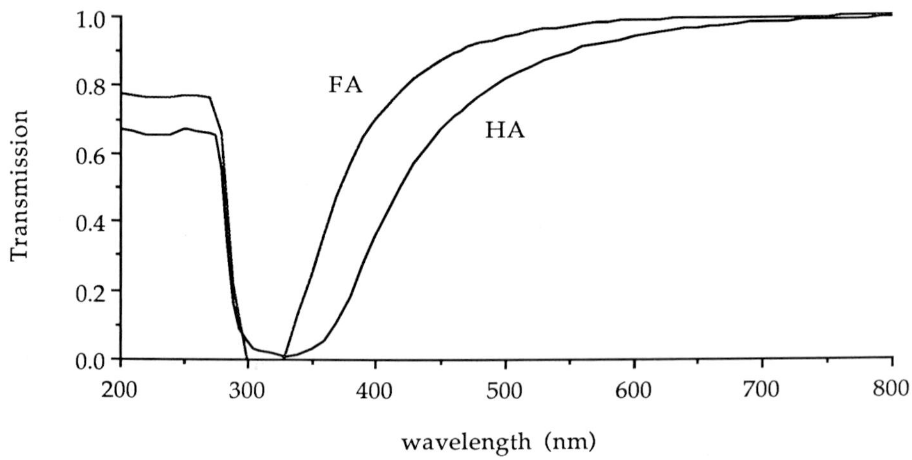

Figure 9 *The UV and visible spectra of fulvic and humic acids in compost.*

Transmission remains constant in the range 200-270 nm and is higher for fulvic acids (0.78) than for humic acids (0.68). From 270 to 320 nm transmission dropped to almost zero. From 350 nm transmission increased again to a maximum (1.00) at 800 nm. Experimentally, it was confirmed that the high adsorption in the 270-320 nm range was not caused by the soluble anionic zincate.

In Table 8, the E4/E6 ratio's and Δlog K values are given for the fulvic and humic acids.

Both Δlog K and E4/E6 ratio are used as indicators for the maturity or state of humification of organic material (56,57). A decreasing Δlog K and an increasing E4/E6 ratio indicate an increasing state of stabilisation. During stabilisation the higher molecular weight humic acids (>10,000) are broken down into the lower molecular weight fulvic acids. The threefold higher E4/E6 ratio for fulvic acids than for humic acids indicate a well matured compost.

Table 8
UV absorbance at four wavelengths and the E4/E6 ratios and ΔlogK values of fulvic (FA) and humic (HA) acids in compost.

Compound	Wavelength (nm)				Δlog K	E4/E6
	400	465	600	665		
Fulvic acids	0.90	0.29	0.04	0.01	1.35	29.0
Humic acids	1.94	0.83	0.17	0.08	10.40	10.4

5.3.2 *Fractionation of soluble organic acids*

5.3.2.1 *Fulvic acids*

The fulvic acid extract was separated in three UV visible fractions at 254 nm (Figures 10, 11). The distribution coefficients (K_{av}) are given for each peak (Table 9). The K_{av} for the first peak approaches zero. This indicates that the molecular weight of the largest fulvic acid fractions exceed 700 g/mol. This is in agreement with the results of Giovanozzi-Sermanni (58), who separated compost humic and fulvic acids on a Sephadex G-100 column. The graphs show that red mud reduces the level of heavy metals in solution. The levels of cadmium, nickel and lead were below their detectable limits on AAS.

Behaviour of zinc in extracts

The zinc peaks indicate either cationic zinc adsorption to organic matter with a higher molecular weight or that zinc is in solution as $Zn(OH)_6^{4-}$ or $Zn(OH)_4^{2-}$ (Figure 10). The pH of the sample applied to the column is 10.2, but the elution liquid has a pH of 13 (0.1 M NaOH) resulting in a pH value of the eluted sample approaching 13. Zinc in anionic form could be adsorped onto red mud by Al and Fe oxides in a similar manner than PO_4 (59).

In a percolation column experiment with a compost/RMG ratio of 70 : 30, it was found that zinc concentration peaked at the same elution volume as the absorbance at 465 nm (64). This indicates zinc adsorption onto this molecular weight fraction. However, the pH of the eluent in this case is 8 ($\pm$ 0.1) ($Ks = 1.2 * 10^{-11}$). This exceeds the solubility product of $Zn(OH)_2$ at pH 8 so most, if not all, zinc ions present in the eluent were adsorbed in this study to the organic fraction with UV absorbance peaking at wavelength 465 nm. Stability constants of metals adsorbed to humic substances increase with increasing pH (6).

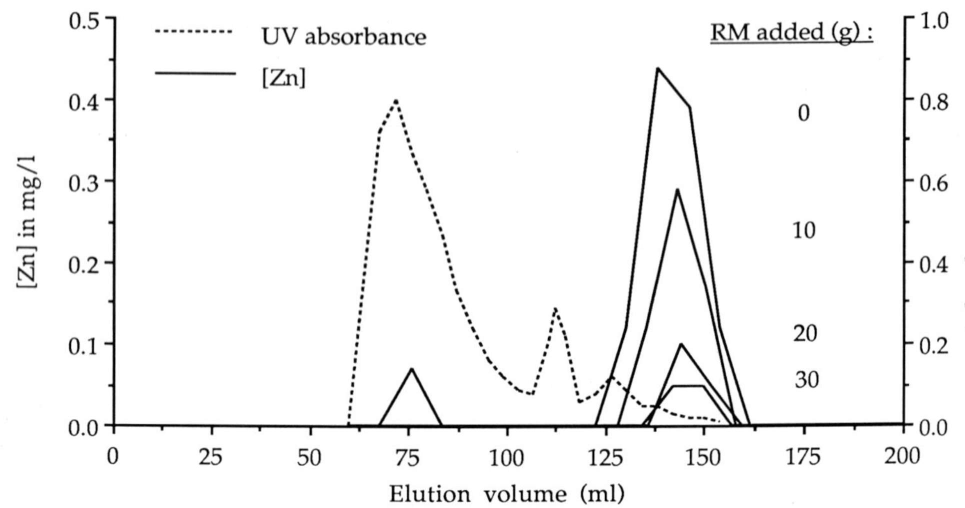

Figure 10 *The effect of red mud on the distribution of zinc over the different molecular fractions of a compost fulvic acid extract.*

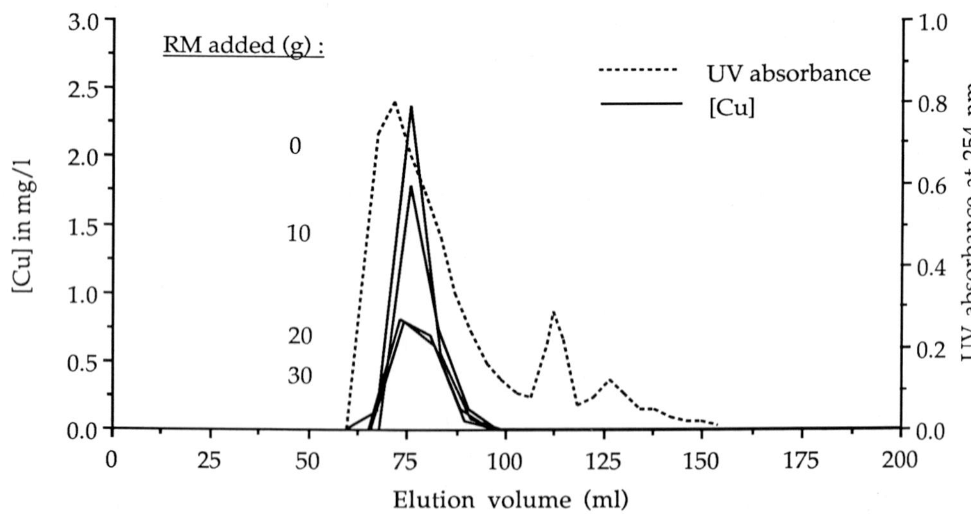

Figure 11 *The effect of red mud on the distribution of copper over the different molecular fractions of a compost fulvic acid extract.*

Zinc was not detected in the humic acid extract. Addition of red mud shows a consistent decrease in zinc concentration from 1.45 to 0.1 mg/l (93 %). Reduction can be due to adsorption of $Zn(OH)_4^{2-}$ onto aluminous goethite in red mud. Zinc removal through precipitation of $Zn(OH)_2$ is unlikely as the reduction has no correlation with a decreasing pH. Residual zinc will stay soluble during the fractionation with 0.1 M NaOH.

Behaviour of copper in extracts

The elution diagrams (Figure 11) show that copper elutes at the same time as soluble organic matter at wavelength 254 nm. All copper must be adsorbed to organic matter. Any copper in solution would precipitate as Cu(OH)2 at the elution pH of approximately 13. Copper must be strongly adsorbed, as sodium will compete strongly for adsorption sites on the functional groups of the organic acids. Increasing red mud addition resulted in a reduction in copper concentration from 2.37 to 0.8 mg/l or 66 %. During a column experiment where soil organic matter in the leachate was measured at 265 nm, copper peaked at the same elution volume as did the electrical conductivity (salt leaching) at pH 8 (Ks = $3.6 * 10^{-11}$)(57). It can be concluded that all copper leached from the G - 10 column is chelated to the fulvic acids.

Table 9 shows that the addition of red mud had no influence on the distribution coefficient and the adsorbance pattern of the organic matter (maximum and 0.5 x peak width). The red mud thus did not change molecular weight distribution nor the concentration of fulvic acids.

Table 9

Kav values of fulvic acids obtained from compost/red mud mixtures separated on a Sephadex G-10 column.

red mud (g)	KavI	KavII	KavIII
0	0.02	0.37	0.49
5	0.0	0.34	0.45
10	-0.01	0.34	0.47
15	0.0	0.34	0.47

mean stand. dev. : 0.00(± 0.01), 0.34 (± 0.01), 0.46 (± 0.02).

5.3.2.2. *Humic acids*

The separation of the humic acid extract on a G - 10 column was insufficient to give more information about the molecular weight distribution. The humic acids had a molecular weight exceeding 700 g/mol. The tailing of the peak showed that there is a range of fractions with a molecular weight of around 700 g/mol. In table 10 the copper and zinc concentrations are given.

Red mud appears to remove metals through adsorption of free cations, eg zinc, and seems also capable of reducing the amount of copper adsorbed to the fulvic acids, without an increase in free ionic copper. This shows that red mud is able to immobilise cationic as well as metals complexed to the soluble organic fraction.

Table 10

The distribution of copper and zinc (mg/kg) in the soluble organic fraction of compost.

red mud (g)	Copper			Zinc		
	fulvic	humic	ionic	fulvic	humic	ionic
0	24.2	-	bd	bd	-	8.2
5	17.4	0.65	bd	bd	0.03	4.5
10	10.0	0.38	bd	bd	bd	2.8
15	10.7	-	bd	bd	-	1.6

bd = below detection limit.

The Figures show that only a small amount of metal is adsorbed onto the humic and fulvic acids, compared to the total amount in the compost. Zinc is in solution as zincate, due to the high pH. Copper is not present in free ionic form, as it will precipitate. Most metals are complexed to the insoluble organic matter fraction (humins).

6. Conclusions

The use of red mud as an amendment to compost to reduce mobility of heavy metals is centred around its ability to adsorb free cations strongly. Most heavy metals however are adsorbed or complexed onto organic matter. Red mud seems able to adsorb heavy metals complexed to the soluble organic matter, which is mainly fulvic acids. Because fulvic acids are soluble over a wide pH range (1-13) the use of pH as a tool to reduce metal leaching is irrelevant.

The sodium and alkalinity in red mud contributed to the solubilisation of organic matter. Through this process red mud addition increased the leaching of heavy metals complexed to soluble organic matter. Free metal ions precipitate as hydroxides, carbonates or are adsorped to red mud minerals.

Red mud neutralised by gypsum reduced metal leaching (>50 %) considerably, probably due to a reduction in organic matter solubilisation. This combination leaches the lowest amounts of metals.

Red mud does not reduce plant availability of heavy metals. It is possible that DTPA forms a stronger complex with metals than red mud does.

Adsorption isotherms of several heavy metals onto red mud is currently being investigated as is the addition of red mud during the composting process to immobolise metals before organic matter complexation.

Acknowledgement

Perth City Council and Alcoa of Australia are acknowledged for their generous financial sponsorship.
Gerard de Jong and Wim Beltman are acknowledged for assisting in the research.

References

1 A. S. Awad, A. D. Ross and R. A. Lawrie, Guidelines for the use of sewage sludge on agricultural land, New South Wales Department of Agriculture and Fisheries, Australia, 1989.

2 F. J. Stevenson, in: C. A. Black *et al.* (Ed.), Methods of soil analysis, Part 2 : Chemical and microbio-logical properties, American Society of Agronomy, Madison , U.S.A., 1965. pp. 1409 - 1421.

3 T. A. Jackson, .Soil Sci. 119 (1975) 56 - 64.

4 J. L. Mortensen, In : C. A. Black *et al* (Ed.) Methods of soil analysis, Part 2 : Chemical and micro-biological properties, American Society of Agronomy, Madison, U.S.A. 1965, pp. 1140 - 1408.

5 G. S. P. Ritchie and A. M. Posner, J. Soil Sci. 33 (1982) 233 - 247.

6 K. S. Jackson, I. R. Jonasson and G. B. Skippen, Earth Sci. Reviews, 14 (1978) 97 - 146.

7 S. E. Manahan, in: Environmental Chemistry Third edition, Willard Grant Press, Boston, Massachusetts, 1979, pp 55 - 80.

8 D. S. Gamble, C. H. Langford and A. W. Underdown, in: Complexation of trace metals in natural waters. (Ed. C. J. M Kramer and J. C. Duinker). Proceedings of the international symposium, Texel, The Netherlands, May 2 - 6, 1983, Martinus, pp. 349 - 356.

9 J. Shapiro, J. Am. Wat. Works Assoc. (1964) 1062 - 1082.

10 F. J. Stevenson, Soil Sci. Soc. Am. J., 40 (1976) 665 - 672.

11 R. P. Houghton, Metal complexes in organic chemistry, Cambridge University Press, London, 1979.

12 A. L. Page, A. C. Chang, G. Sposito and S. Mattigod, in: I. K. Iskandar (Ed.), Modeling wastewater renovation, John Wiley and sons, New York, 1981.

13 R. F. C. Mantoura, A. Dickson and J. P. Riley, Estuarine and Coastal Marine Sci., 6 (1978) 387 - 408.

14 M. Schnitzer, and E. H. Hansen, Soil Sci. 109 (1970) 333 - 340.

15 J. Baham, N. B. Ball and G. Sposito, J. Environ. Qual. 7 (1978) 181 - 188.

16 T. Takamatsu, R. Kusakabe and T. Yoshida, Soil Sci. 136 (1983) 371 - 381.

17 M. Schnitzer, and S. I. M. Skinner, Soil Sci., 102 (1966) 361 - 365.

18 M. Schnitzer and S. I. M. Skinner, Soil Sci., 103 (1967) 247 -252.

19 L. Kiekens, In: Utilization of sewage sludge on land: rates of application and long term effects of metals. pp. 126 - 134. Proceedings of a seminar held in Uppsala June 7-9, 1983.

20 K. Ghosh, A. Chattopadhyay and C. Varadachari, Soil Sci., 135 (1983) 193 - 196.

21 H. Roy, Indian Ceramic Soc. Transactions, 30 (1971) 149 - 151.

22 K. H. Tan, Soil Sci., 120 (1975) 188 - 194.

23 G. Sposito, F. T. Bingham, S. S. Yadav and C. A. Inouye, Soil Sci. Soc. Am. J. 46 (1982) 51 - 56.

24 K. H. Tan and A. Binger, Soil Sci. 141 (1986) 20 - 25.

25 W. L. Hargrove , Soil Sci., 142 (1986) 179 - 181.

26 Lagas, P. and J. P. G. Loch, RID report 81-21, 1981, National Institute for Water Supply, The Netherlands.

27 L.K. Hudson,in: A.R. Burkin (Ed.), Production of Aluminium and Alumina. Critical Reports on Applied Chemistry Vol. 20, John Wiley & Son, 1987, pp.11-46.

28 B.J. Robson, Use of Bayer Process solid residues (red mud and process sand) for the renovation of wastewater - in particular the removal of ammonium, M.Sc. Thesis, Murdoch University, 1982, 114 pp.

29 K. Solymar, T. Ferenczi and J. Zoldi, The effects of technology on the properties of red mud. Travaux ICSPBA, 15 (1979) 287-305.

30 M.G. Leiteizen, L.A. Pashkevich, I.B. Firfarova and D.I. Tsekhovolskaya, in: Research Inst. Non-ferrous Metal (Ed), Proc. 2nd Conf. VAMI and FKI Leningard, 1975, pp. 167-179.

31 D.J. Glenister, in: N. Stockton (Ed.), Tailings Disposal and Management Symposium, Murdoch University, Western Australia, 1987, pp.50-70.

32 L.R. Hossner, R.H. Loeppert and J.K. Bush, Reclamation and vegetation of bauxite residue, Alcoa Report No 1. Soil and Crop Sci. Dept., Texas A & M Univ., Coll. Station, 1984, 100 pp.

33 M.R. Thornber, J.H. Taplin, and C.A. Huges, A mineralogical and chemical investigation of various Alcoa red mud waste materials. CSIRO Report No. 1592R, 1985, 36 pp.

34 B.K. Parekh and W.M. Goldberger, in: Proceedings of The Sixth Mineral Waste Utilization Symposium, U.S. Bur. of Mines and IIT Research Institute. Chicago, Illinois, May 2-3, 1978, pp. 122-132.

35 ALUTERV-FKI, Report on chemical, mineralogical and organic compound analysis of sodium aluminate liquor and bauxite samples, coarse sand and fine mud residues from the Kwinana and Pinjarra plants of Alcoa of Australia Ltd., Budapest, 1983.
36 J. Zambo, Red mud: its properties, handling, storage and utilization. Travaux ICSOBA. 15(1979) 265-276.
37 S. Miyamoto, R.J. Prathen and J.L. Stroehlen, Plant and Soil, 43 (1975) 573-585.
38 J. Loveday, Aust. J. Soil Res., 14 (1976) 361-371.
39 I. Shainberg, R. Keren and H. Frenkel, Soil Sci. Soc. Am. J., 46 (1982) 113-117.
40 R.L. Cates Jr., V.A. Haby, E.O. Skogley and H. Ferguson, J. Environ. Qual., 13 (1984) 252-256.
41 R.K. Gupta and R.R. Singh, Soil Sci. 146(4) (1989) 277-286.
42 J.R. Meecham and L.C. Bell, Aust. J. Exp. Agric. Husb. 17 (1977), 679 - 688.
43 R.D. Fuller, E.D.P. Nelson and C.J. Richardson, J. Envir. Qual. 11 (3) (1982) 533 - 539.
44 N.J. Barrow, Aust. J. Agric. Res., 33 (1982) 275-285.
45 S.C. Ward, Reclaim. Reveg. Res., 2 (1983) 177-190.
46 S.C. Ward, The use of the fine residue from bauxite refining as a soil amendment. Ph.D. Thesis. Murdoch Univ. 1986, pp. 318.
47 G.E. Ho, P. Newman, K. Mathew, and B. Parker, Red Mud Report No. 2. School of Environmental Sci., Murdoch Univ., Western Aust. 1984, 136 pp.
48 G.E. Ho, P.W.G. Newman, K. Mathew, and H. De Potter, CHEMECA, 85 (1985), 102-108.
49 L.R. Hossner, R.H. Loeppert, H.J. Woodard, T.J. Moore, and T.L. Thompson, Alcoa Report No, 3. Soil and Crop Sci. Dept., Texas A & M Univ., Coll. Station, 1986, 128 pp.
50 G.H. Bolt and M. G. M. Bruggenwert, Soil Chemistry. A. Basic Elements. Elsevier, Amsterdam,1978, 281 pp.
51 W. L. Lindsay, Norvell W. L., Soil Sc. Soc. Am. J., 42 (1978) 421.
52 K. M. Holtzclaw, G. Sposito and G. R. Bradford, Soil Sci. Soc. Am. J., 40 (1976) 255 - 258.
53 K. M. Holtzclaw, D. A. Keech, A. L. Page, G. Sposito, T. J. Ganje and N. B. Ball, J. Env. Qual., 7 (1978) 124 - 127.
54 H. T. Hofstede, in preparation.
55 W. E. Emmerich, L.J. Lund, A. L. Page, A. C. Chang, J. Env. Q., 11 (1982) 178 - 181.
56 G. Almendros, A. Polo, E. Dorado, Agrochemica, 27 (1983) 310 - 325.
57 Y. Chen, N. Senesi, M. Schnitzer, J. Soil Sc. Soc. Am., 41 (1977) 352 - 358.
58 G. Giovannozzi-Sermanni, In: M. de Bertoldi, et al. (Ed.), Compost : production, quality and use, Elsevier, 1986, 190 - 197.
59 M. Kayaalp, G. Ho, K. Mathew and P. Newman, Water (Melbourne), 15 (1) (1988) 26-29.

60 H. T. Hofstede, Beltman W., Behaviour of heavy metals in red mud amended compost, Environmental Sciences, Murdoch University, Western - Australia, 1988, unpublished.

Chapter 4
Regional studies: freshwater and marine environment

THE REGULATION OF HEAVY METAL CONCENTRATIONS IN NATURAL AQUATIC SYSTEMS

Paul W. Schindler

INTRODUCTION

Heavy metals belong to the group of elements whose hydrogeochemical cycles have greatly been accelerated by man. Anthropogenic metal emissions into the atmosphere of metals such as Pb, Hg, Zn, Cd and Cu are one to three orders of magnitude higher than natural fluxes (Sposito and Page, 1984; Baccini, 1990). Similarly, direct anthropogenic inputs into the hydrosphere by waste disposal or by processes such as the corrosion of zinc coated tubings have increased. As a consequence these elements are expected to become increasingly accumulated in natural reservoirs. An increase in heavy metal concentrations during the last century is well documented (Fig. 1, Table 1) for lake sediments (e.g. Nriagu et al., 1979; Imboden et al., 1980; Rippey et al., 1982; Müller, 1982). An increase in trace metals concentrations in sea water and fresh waters is, however, not obvious since earlier data on the trace metal concentrations in these systems suffer from unadequacy of sampling technique as well as from a lack of reliable analytical tools. Similarly, an accumulation of heavy metal in soils can be recognized from heavy metal profiles in undisturbed forest soil systems (Fig. 2) but we have little information on a related increase of heavy metal concentrations in the soil solution. We are thus challenged by the question of how a given aquatic systems reacts to a change in the input parameters. The question can only be answered if the pertinent transport processes in which metal ions participate are known. Among these processes, the reactions that follow the scheme

$$\text{dissolved matter} = \text{particulate matter}$$

are of special interest since they modify the mobility of the metal. Examples for such transformations are

$$Mn(II) + \tfrac{1}{2}\ O_2 = MnO_2$$

$$2\ Fe(III) + O_2 = Fe_2O_3$$

metals in solution = metals adsorbed at suspended matter.

The suspended matter mentioned in the last equation consists of inorganic particles (oxides such as Fe_2O_3, clay minerals) as well as of biota and biological debris. It is the purpose of this article to review possible mechanisms operative in transforming dissolved metals ions into particulate form. Specifically it is aimed

- to offer evidence for a control of trace metal concentrations by adsorption desorption reactions;
- to introduce the surface complexation model as an useful tool for describing the adsorption of metal ions;
- to discuss the nature of the adsorbing surfaces and to describe cycling and ultimate removal of trace metals in selected aquatic systems.

REGULATION BY SOLUBILITY EQUILIBRIA: SILLÉNS OCEAN MODEL.

Lars Gunnar Sillén's (1961) work on the physical chemistry of sea water is doubtless the beginning of modern chemical oceanography. On the basis of Goldschmidt's estimates of global chemical mass balances, Sillén discussed the composition of sea water in terms of an equilibrium model. Sea water was assumed to be a saturated solutions of the underlying marine sediments. Specifically the concentrations of trace metals were assumed to be controlled by solubility equilibria involving oxides and/or carbonates: In the case of a divalent metal the controlling equilibria could thus be

$$MO_{(s)} + 2\ H^+ = M^{2+} + H_2O \qquad {}^*K_{s0} = [M^{2+}][H^+]^{-2} \qquad (1)$$

or

$$MCO_{3(s)} + 2\ H^+ = M^{2+} + CO_{2(g)} + H_2O \qquad {}^*K_{ps0} = [M^{2+}]P_{CO2}[H^+]^{-2} \qquad (2)$$

In the absence of ligands that form complexes with M^{2+} the solubility S is equal to $[M^{2+}]$ and we have either

$$S = [M^{2+}] = {}^*K_{s0}[H^+]^{-2} \qquad (1a)$$

or

$$S = [M^{2+}] = {}^*K_{ps0}P_{co2}[H^+]^{-2} \qquad (2a)$$

In the presence of a ligand L that forms complexes $ML_n{}^{2+}$ with the stability constant

$$\beta_n = [ML_n{}^{2+}]/[M^{2+}][L]^n \qquad (3)$$

the solubility is given by

$$S = [M^{2+}] + \sum^{n}[ML_n{}^{2+}] = [M^{2+}](1+\sum^{n}\beta_n[L]^n)$$

and thus by

$$S = {}^*K_{s0}(1+\sum^{n}\beta_n[L]^n)[H^+]^{-2} \qquad (1b)$$

or

$$S = {}^*K_{ps0}(1+\beta_n[L]^n)P_{co2}[H^+]^{-2} \qquad (2b)$$

It should be noted, that within the scope of Silléns model, the trace metals concentrations in the sea water only depend on pH, P_{co2}, the concentrations of complexing ligands and related parameters. It would , however, not be affected by increasing inputs since any increase would be balanced by precipitation. The increase in input would, however result in a change in the composition of the sediments. Hence a sedimentary record such as the one presented in Figure 1 is in accordance with Sillén's model. Although Sillén (in contrast to many of his followers) was fully aware of the crudeness of his approach, he was " stroke by the resemblance of the equilibrium model and the real system". Sillén's provocation initiated a broad activity and numerous efforts to prove, improve or disprove his conception (e.g. Kramer,

1967; Stumm and Morgan, 1970; , Broecker, 1974). The equilibrium concept was extended to all kind of natural aquatic systems including soil solutions (Lindsay, 1979; Sposito, 1981). Progress in analytical chemistry of sea water and extended work on the solubility of oxides and carbonates revealed that for most of the heavy metals sea water is undersaturated with respect to any thinkable pure solid. The development is illustrated in Table 2. It was then assumed, that heavy metals might by minor constituents of solid solutions although the nature of such solid solutions remained obscure. The details of Sillén's pioneering work were to be revised, the concept of equilibrium models remained an essential part of modern aquatic chemistry.

MODELS BASED ON ADSORPTION/DESORPTION EQUILIBRIA

Fundamentals

The idea of solubility equilibria as controlling processes (and thus precipitation of pure solids as mechanisms to transfer trace metals from the water column to the water-sediment interface) had to be abandoned. In search of other processes that might transform dissolved heavy metals into settling particles adsorption at inorganic particulate matter (such as clay minerals and iron(III)hydroxide) or incorporation into biota were considered (Schindler, 1975 a). These processes are also designated as "scavenging". Similarly, the partitioning of trace metals between solid soil constituents and the soil solution ware assumed to be controlled by adsorption processes (Figure 2). Assuming such transformations

$$\text{M (dissolved)} = \text{M (particulate)} \qquad (4)$$

to be at equilibrium, the ratio of the concentrations of dissolved and particulate matter can be described by the distribution coefficient

$$D = \{M_p\}/[M] \qquad [dm^3/kg] \qquad (5)$$

where

$\{M_p\}$ = Concentration of particulate metal in the adsorbing or incorporating particle [mol/kg]

$[M]$ = Concentration of dissolved metal [mol/dm^3]

The distribution coefficient D for a given metal ion and a given adsorbent is a conditional constant. Its values depends on temperature and pressure, on the concentration of the metal ion in the solution. It is further affected by the solution parameters pH and pϵ as well by the presence of other metal ions and ligands. It will be shown, that the connection between D and the above mentioned solution parameters can conveniently be formulated on the basis of the surface complexation model. For the case of τ'_M being a constant for a given aquatic system, eq. (6) states that $[M]_{(t)}$ is a liner function of $[M_i]_{(t)}$. Hence any change in the composition of the inflowing water is directly reflected in the composition of the system. Moreover, since the concentrations in the water column and the underlying sediments are related to each other by the distribution coefficient D, any increase in trace metal concentration in the water column leads to a corresponding increase in the trace metal content of the sediments.

A 1-box steady state lake model

The equation (2) has been used as a basis for a simple one box (Fig. 3) steady state mode (Schindler, 1975 b ; Imboden and Lermann, 1984). The steady state condition implies that processes that add dissolved and particulate metal to the lake system are balanced by processes that remove metals from the lake system either by the outflowing rivers or by transfer into the lake sediments. The lake model connects the steady state concentration in the lake with the concentration in the inflow as follows:

$$[M]_{(t)} = \tau'_M [M_i]_{(t)}, \text{ where} \qquad (6)$$

$$\tau'_M = \frac{(\Phi(1 + A_i D\tau'_p) + DB\tau'_p)}{(\Phi(1 + A_i D) + DB)} \qquad (7)$$

$$[M_i]_{(t)} \geq [M]_{(t)} \geq \tau'_p[M_i]_{(t)} \qquad (8)$$

Where

$[M]_{(t)}$ = Total concentration of M in the lake [mol/dm^3]

$[M_i]_{(t)}$ = Total concentration of M in the inflow [mol/dm^3]

t

τ'_M = Relative residence time of M in the lake. $\tau'_M = \tau_M / \tau_{H2O}$

τ'_p = relative residence time of particulate matter in the lake. $\tau'_p = \tau_p / \tau_{H2O}$

Φ = Flux of water (in and outflow) [dm^3s^{-1}]

B = in situ production of particulate matter (SiO_2, MnO_2, FeOOH, organic debris, biota) in the lake [kgs^{-1}]

A comparison of the prediction of equation (6) with field data is presented in Table 3.
The important parameters that actually control the concentrations $[M]_{(t)}$ in an aquatic system are thus the concentrations of M in the inflowing rivers and the relative residence time of the metal in the system. This relative residence time τ'_M is in turn dependent on the distribution coefficient D, (Figure 4) the amount of allochtonous ($A_i\Phi$) and autochthonous (B) material imported or produced per unit time and the settling velocity of this particulate matter as expressed by its relative residence time τ'_p. The latter quantity is largely dependent on the size and the form of the aquatic system and thus a characteristics of the individual reservoir.

A 1-box steady state ocean model

If one neglects inputs by precipitation and output by formations of marine aerosols the eq. (3) applies for the ocean as well as for lake systems.(Fig. 5, Schindler, 1975b). The relative residence time is then given by

$$\tau'_M = \frac{\Phi + \tau'_p(\Phi A_i + B)D}{(\Phi A_i + B)D} \approx \frac{\Phi}{(\Phi A_i + B)D} \qquad (9)$$

Again D is the master variable that controls the relative residence time. As is the case of lake systems the lover limit of τ'_M is given by τ'_p. In contrast to a lake system, there is apparently no upper limit for τ'_M and thus $[M]_{(t)}$. This result is a consequence of the assumption that removal of M is restricted to scavenging. In reality, upper limits would be imposed by the above mentioned solubility equilibria.

Heavy metal transport in soils

Adsorption equilibria are assumed to control metal concentrations in soil solutions. The distribution coefficient D is thus an important parameter in the equations used to model the transport of heavy metals in soils. The average penetration depth z_p of a metal M can for instance be estimated from the equation (Bolt, 1978)

$$z_p = \frac{z_0}{\varphi + (1-\varphi)dD} \qquad (10)$$

where z_0 is the total amount of M containing rain (m), φ is the porosity of the soil and d is the bulk density of the soil. Within the scope of the classical convection - dispersion model, the one dimensional transport is described by

$$R(\delta c/dt) = ql(\delta^2 c/\delta z^2) - q(\delta c/\delta z) \qquad (11)$$

where q is the Darcy flux (cm s^{-1}) and l is the dispersitivity (cm). R is the retardation factor defined by

$$R = 1 + dD/\varphi \qquad (12)$$

UNDERSTANDING TRACE METAL ADSORPTION: THE SURFACE COMPLEXATION MODEL.

Fundamentals

In this section we shall preponderantly be concerned with the adsorption of metal ions from aqueous solutions at (hydr)oxide surfaces. The concepts and models introduced can, however, be extended to include the adsorption at clay minerals and even at biological membranes. The pH value of the solution is the master variable that governs the adsorption of metal ions at (hydr)oxide surfaces: High pH values promote adsorption. With a given oxide, it is observed, that each metal ions has a characteristic range of 1 to 2 pH units where raises from zero to almost hundred percent. For metal ions such as Ca^{2+} and Mg^{2+} the pH range of adsorption was usually found to lie well above the pH of the zero point of charge (pH_{pzc}) of the adsorbing (hydr)oxide. It was thus concluded that adsorption of metal ions is based on electrostatic interaction with the negatively charged surface and that in can thus conveniently be described by double layer models. The observation of Co^{2+} ions being adsorbed by positively charged TiO_2 surfaces (James and Healy, 1972) led to a change in this perception. Since earlier IR investigations had already revealed the presence of surface hydroxyls at oxide and hydroxide surfaces, adsorption now assumed to consist of complex formation of the adsorbed metal with the surface hydroxyls (Hohl and Stumm, 1976; Schindler et al., 1976). The development if this idea eventually lead to the surface complexation model (Schindler. 1981; Schindler and Stumm, 1987; Dzombak and Morel, 1990). The most important reactions of surface hydroxyl groups are collected in Table 4. Surface hydroxyls are both weakly acidic (reaction a) and weakly basic (reaction b). A noticeable exception is the surface silanol group ≡Si-OH. It basicity is extremely low and $\equiv Si\text{-}OH_2{}^+$ - if existing - must be a very strong acid. The reaction c describes the formation of a binary surface complex and explains the fact that adsorption of metal ions is connected with a release of hydrogen ions. The average number of protons released per adsorbed metal ion was found to range between 1 and 2. This can be explained by the simultaneous occurrence of the reactions c and d. Alternatively the observed stoichiometry can as well be explained by the formation of a ternary surface complex S-O-M-OH. The surface hydroxyl can be exchanged against dissolved ligands L (reaction e). This ligand exchange, that is favoured by low pH values is the principal mechanism for anion adsorption. Metal-ligand complexes ML can be adsorbed by formation of ternary surface complexes. Two different types of ternary surface complexes have so far been identified. Reaction f describes the formation of a type A ternary surface complex where the metal ion M is coordinated to both deprotonated surface hydroxyls and to ligands L. The formation of type A ternary surface complexes can be expected for the case that L is a monodentate or bidentate ligand (i.e. a ligand that caries one or two ligand atoms and occupies one or two sites in the coordination sphere of the metal ion). For the case of L being a polydentate ligand, the formation of a type B surface complex (reaction g), is anticipated: Part of the ligand atoms are used to replace surface hydroxyls and to thus anchor the ligand to the surface and the remaining ligand are utilized to coordinate the metal ion M.

Thermodynamics

In this section we will briefly summarize the conventions adapted for the description of the thermodynamic stability of surface complexes and outline the experimental methods used for evaluating the pertinent stability constants.

Conditional and intrinsic constants. Let's consider the equilibrium (equation c in Table 4)

$$\equiv S\text{-}OH + M^{z+} = \equiv S\text{-}OM^{(z-1)+} + H^{+}$$

The Intrinsic microscopic equilibrium constant (King, 1965) is defined by the equation

$$^{*}K^{s}_{(int)} = \frac{a_{\equiv SOM}\ a_{H}}{a_{\equiv SOH}\ a_{M}} \qquad (13)$$

where a_i denotes the activity of a surface species i. The surface activity of i can be expressed as the product of its surface concentration {i} (in mole per kg of adsorbing solid) and its surface activity coefficient y_i (in kg mole^{-1}):

$$^{*}K^{s}_{(int)} = \frac{\{\equiv SOM\}\ y_{\equiv SOM}\ a_H}{\{\equiv SOH\}\ y_{\equiv SOH}\ a_M} \qquad (14)$$

The surface concentrations {≡SOH} and {≡SOM} are experimentally accessible quantities. It is sometimes convenient to express the concentrations of the surface species in the unites as the concentrations of dissolved species (i.e. in mole dm^{-3}). The conversion can be performed with the aid of an equation such as

$$[\equiv SOH] = \frac{A}{V}\ \{\equiv SOH\} \qquad (15)$$

where A is the amount of solid adsorbent (kg) and V is the volume of the aqueous solution (dm^3). It is, unfortunately, in most cases not possible to experimentally evaluate the surface activities a_H and a_M. For the case, that surface and solution are at electrochemical equilibrium, it is possible to evaluate a_H and a_M from the activites of H and M respectively in the aqueous phase as follows:

The equilibrium condition is

$$\mu^{*}_{(H,aq)} = \mu^{*}_{(H,surf)}\ ;\quad \mu^{*}_{(M,aq)} = \mu^{*}_{(M,surf)} \qquad (16)$$

The electrochemical potential μ^* of a species i in a phase j is given by

$$\mu^{*}_{(i,j)} = \mu^{*0}_{(i,j)} + RT\ ln\ a_{(i,j)} + z_iF\ _{(j)} \qquad (17)$$

where $_{(j)}$ is the potential at the phase j and z_i is the charge number of the species i. In the presence of an ionic medium of constant ionic strength the activities in the aqueous solution can - without any loss of rigor - be replaced by concentrations. By choosing the standard states to be the same in both phases

$$\mu^{*0}_{(H,aq)} = \mu^{*0}_{(H.surf)}\ ;\ \mu^{*0}_{(M,aq)} = \mu^{*0}_{(M.surf)}$$

and assigning a zero potential to the aqueous solution one obtains from (14) and (17)

$$^{*}K^{s}_{(int)} = \frac{\{\equiv SOM\}[H^{+}]}{\{\equiv SOH\}[M^{z+}]} \frac{y_{\equiv SOM}}{y_{\equiv SOH}} \exp((z-1)F/RT) = {}^{*}K^{s}\Gamma \qquad (18)$$

or

$$^{*}K^{s} = {}^{*}K^{s}_{(int)}(\Gamma)^{-1} \qquad (19)$$

where

$$^{*}K^{s} = \frac{\{\equiv SOM\}[H^{+}]}{\{\equiv SOH\}[M^{z+}]} \qquad (20)$$

is the experimentally accessible conditional constant and

$$\Gamma = \frac{y_{\equiv SOH}}{y_{\equiv SOM}} \exp((z-1)F/RT) \qquad (21)$$

collects terms that contribute to deviation from ideality. Generally, the equilibrium

$$m \equiv S\text{-}OH + M^{z+} = (\equiv S\text{-}O)_{m}M^{(z-m)+} + m\,H^{+}$$

will be decsdibed the equilibrium constant

$$^{*}\beta_{m}{}^{s}_{(int)} = \frac{\{(\equiv SO)_{m}M\}[H^{+}]^{m}}{\{\equiv SOH\}^{m}[M^{z+}]} \frac{y_{(\equiv SO)mM}}{(y_{\equiv SOH})^{m}} \exp((z-m)F/RT) \qquad (22)$$

<u>Evaluation of intrinsic stability constants; The validity range of conditional constants.</u> The forgoing paragraph leaves us with the problems i) to evaluate intrinsic stability constants from the experimentally accessible conditional constants, and ii)

to trace the validity range of the conditional constants. The former problems requires the evaluation of conditions where Γ approaches unity, the latter problem requires the evaluation of conditions where Γ remains approximately constant. Any solution of these problems requires consideration of the physical nature of the factors that are responsible for non-ideal behavior. According to Schindler and Stumm (1987) these factors are:

a) the above mentioned surface potentials that originate from the adsorption of charged species. We first look for conditions where the surface potentials vanish. For the simple case that only one charged species is adsorbed, the surface potential will vanish at zero surface coverage. For the case of simultaneous adsorption of several charged species, the surface potential will again vanish at zero surface coverage or at zero charge conditions where the charges of the various adsorbed species compensate each other. Next we evaluate conditions where the surface potentials are (approximately) constant. Obviously, for constant temperature and pressure, the surface potential at a given surface in a solution of given ionic strength remains constant as long as the concentration of adsorbable ions in the solution remains constant. Under conditions, where the surface charge is dominated by $\equiv S\text{-}O^-$ and $\equiv S\text{-}OH_2^+$ groups, the surface potential is controlled by the pH of the solution, and it will be constant as long as the pH of the aquatic systems remains constant.

b) lateral interactions (both attraction and repulsion) based on coulombic and van der Waals forces. Again these lateral interactions will vanish at zero surface coverage and they will be constant at constant surface coverage and consequently at constant composition of the aqueous solution.

c) surface heterogeneity i.e. the presence of $\equiv S\text{-}OH$ groups with different binding properties that are in turn reflected by different $K^s_{(int)}$ values. At low surface coverage, adsorption will be restricted to involve $\equiv S\text{-}OH$ groups with maximum $K^s_{(int)}$ values. The contribution of surface heterogeneity to Γ will remain constant as long as the nature of the reacting $\equiv S\text{-}OH$ groups is constant.

The above considerations indicate that the influence of both surface potentials and lateral interactions decrease with decreasing surface coverage so that the difference between K^s and $K^s_{(int)}$ diminishes and may eventually escape experimental observation. Hence, intrinsic stability constants can be obtained from observations at very low surface coverage or by extrapolation conditional constants obtained at various surface coverage to zero coverage conditions. Alternatively, the surface potential may be calculated with the aid of double layer models. For details see Schindler and Stumm (1987) and Schindler (1990). On the other hand, it can be seen that conditional constants remain valid as long as the composition of the aqueous phase remains constant. One would the value of a conditional constant expect to vary if an important paraameter such as the pH value of the aquatic system is changed. It was, however, observed, that at low surface coverage the conditional constants describing the adsorption of trace metals at SiO_2 (Schindler et al., 1976), Al_2O_3 (Hohl and Stumm, 1976) and TiO_2 (Fürst, 1976) are nor dependend on the pH of the solution.

Stabilities of surface complexes

This section discusses the factors that control the stabilities of binary surface complexes. Stabilities of type A ternary surface complexes are discussed by Schindler (1990). Stability constants of some binary surface complexes are collected in Table 5. A comprehensive compilation of the stability constants of binary surface complexes on amorphous iron(III)hydroxide has recently been published by Dzombak and Morel (1990). For a given surface the data in Table 5 suggest the following series of increasing stability

$$Ba^{2+} < Ca^{2+} < Mg^{2+} < Cd^{2+} < Ni^{2+} < Co^{2+} < Cu^{2+} < Pb^{2+} < Fe^{3+}.$$

A similar series

$$Cd^{2+} < Cu^{2+} \approx Pb^{2+}$$

has been reported for bacterial cell walls as adsorbents (Gonçalves et al., 1987). The same series is observed for the stabilities of hydroxo complexes in solution. Moreover, it has been reported that for amorphous silica (Schindler et al., 1976), Al_2O_3 (Hachiya et al., 1984), α-FeOOH and amorphous $Fe(OH)_3$

(Balistrieri et al., 1981, Dzombak and Morel, 1990) the stability constants of the binary surface complexes $\equiv S\text{-}OM$ and $(\equiv S\text{-}O)_2M$ correlate very well with the stability constants of the corresponding hydroxo complexes MOH and $M(OH)_2$ respectively. For a given metal ion, the stability of the binary surface complex increases in the series

$$\equiv Si\text{-}OH < \equiv Al\text{-}OH \approx \equiv Fe\text{-}OH < \equiv Ti\text{-}OH$$

Calculating the distribution coefficient D; comparison with field data.

The obtained values for the stability constants of surface complexes permit the calculation of D values. Under the assumption that the distribution of a metal M between particles and solution is based on adsorption/desorption equilibria, $\{M\}_p$ the concentration of M in the solid phase is the sum of the concentrations of the individual surface complexes $(\equiv S\text{-}O)_m M^{(z-m)+}$

$$\{M\}_p = \sum^{m} \{(\equiv SO)_m M = \Sigma[[M^{z+}]^* \beta_m{}^s \{\equiv SOH\}^m [H^+]^{-m} \qquad (22)$$

In the presence of dissolved ligands $L_1, L_2, .., L_z$, the total concentration of dissolved metal is given by

$$[M] = [M^{z+}] + \Sigma\Sigma[M(L_z)_n = [M^{z+}](1 + \Sigma\Sigma[M^{z+}]\beta_{z,n}[L_z]^n) \qquad (23)$$

where $\qquad \beta_{z,n} = [M(L_z)_n]/[M^{z+}][L_z]^n$

is the stability constant of the complexed $M(L_z)_n$. Combining the equations (22) and (23) we obtain

$$D = \frac{\sum^{m} {}^*\beta_m{}^s \{\equiv SOH\}^m [H^+]^{-m}}{1 + \sum\sum^{zn} \beta_{z,n} [L_z]^n} \qquad (24)$$

Eq.(24) predicts an increase of D with increasing pH (Fig.6) and a decrease of D with either increasing concentrations of dissolved ligands L or increasing stability constants $\beta_{z.n}$ of the dissolved metal-ligand complexes. Both predictions are in agreement with field observations. Kramer (1975) has observed that D values of various trace metals in Canadian lakes were decreasing in cases of lake acidification by acid rain and related decrease in pH. There a numerous reports on trace metal mobilization in soils upon soil acidification. As to the effect of dissolved ligands, Santschi (1975) reported a seasonal change of D(Cu) in L.Biel that was attributed to a seasonal change of DOC and thus concentration of organic ligands. Quantitative treatment (Table 6, Fig.6) shows, however, that d values calculated with pure oxides as model sediments differ by one to three order of magnitudes from field data. This disagreement essentially reflects our ignorance of the nature of the adsorbing surfaces. In addition, laboratory studies have revealed that D-values are dependent on the concentrations on suspended particles (Galicia, 1989) and even changing with time.

THE NATURE OF THE SCAVENGING PARTICLES

Allochtonous particulate matter is predominantly the result of mineral weathering and soil erosion and it therefore mainly composed of silicates and oxides. Autothonous particulate matter contains predominantly biota and biological debris. The nature of the scavenging particles is thus dependant on the ration of the fluxes of allochtonous matter (ΦA_i) and autothonous matter (B). The water flux Φ depends largely on the area S_d of the drainage basin, and B depends on the lakes surface ares S_l. L.Biel has a S_d/S_l ration of 200 and a $\Phi A_i/B$ ration of 1.4 (Schindler, 1975b). The particulate matter in this lake is dominated by inorganic constituents. On the other hand, L.Constance has a S_d/S_l ration of approximately 20. Its particulate matter is thus dominated by biota and by organic debris. Sigg (1987) has found, that sedimentation of heavy metals in L.Constance is largely controlled by adsorption and/or uptake by biota. This biological control is even more pronounced in the open ocean that is well isolated from allochtonous sources. The striking correlations of nutrients and

(some) heavy metals in North Pacific (Bruland, 1980) profiles suggest that aquatic microorganisms control the concentrations of many trace metals in the surface ocean. Morel and Hudson (1985) suggest that Redfields model of mutual control of biota and nutrients may be extended to trace metals such as Mn,Fe,Ni,Cu Zn and Cd.
Transport by biota and biological debris is important for the upper part of the water column. The settling biota are, however, reoxidized and the adsorbed or incorporated trace metals are given back to the aqueous solution and transported back to the surface by upswelling (Li, 1981). The ultimate removal is based on adsorption and incorporation into clays or oxides.In lake systems it is often MnO_2 that takes over in regions near the water sediment interface. This is especially true for aquatic systems that become temporarily anaerobic. Under such conditions Mn(II) accumulates at the sediment water interface and is reoxidized upon upwards diffusion. the freshly formed MnO_2 is known to be an efficient trace metal scavenger.

SCAVENGING KINETICS

The steady state models described by the eqs.(6) and (9) assume establishment of the adsorption-desorption equilibria. Nyffeler et al. (1984, 1986) have reported that uptake of radioisotopes by suspended particles is often slow and that the time scale for scavenging may be in the same other of magnitude as the residencae time of the radionuclide in the water column. Using a simple two step model for adsorption kinetics, they were able to successfully model the rate of trace model uptake by scavenging particles in mesocosms and lacustrine limnocoralls. More recently, Honeyman and Santschi (1989) presented a modified two step model of scavenging kinetics i.e. for the transfer

dissolved (non-filtrable) metal = particulate (filtrable) metal.

They assume a rapid adsorption of trace metals on the surface of colloidal (non-filtrable) particles followed by a slower aggregation of these colloidal particles to form larger, filtrable particles. For details the reader is referred to the contribution by Santschi and Honeyman to this volume (p. 229).

CONCLUSIONS

The steady state model together with the surface complexation model gives an outline of the factors that control the fate of trace metals in natural aquatic systems and in soil solutions. It permits to predict in a given system the effect of changing environments upon trace metal concentrations. As to be expected from the set up of these models, these predictions are qualitatively correct but numerically unsatisfactory. The main reasons for this failure in quantitative predictions lies in our ignorance of the nature of the scavenging particles and in neglecting scavenging kinetics.

Acknowledgements

The work was financially supported by the Swiss National Foundation

REFERENCES

Baccini, P. (1989), The control of heavy metal fluxes from the anthroposphere to the environments, In Heavy metals in the environment, J.-P. Vernet, ed., Vol. 1 , Geneva.

Broecker, W. S. (1974), Chemical Oceanography, Harcourt Brace Jovanovich, New York.

Bruland, K. W. (1980), Oceanographic distribution of cadmium, zinc, nickel and copper in the North Pacific, Earth Plant. Sci. Lett. 47, 176.

Bruland, K. W. (1983), Trace Elements in the Sea, In Chemical Oceanography, Vol. 8, p.157, Academic Press, London.

Dzomback, D. A. and Morel, F. M. M. (1990), Surface Complexation Modeling, John Wiley & Sons, New York.

Fürst, B. (1976), Das koordinationschemische Adsorptionsmodell: Oberflächenkomplexbildung von Cu(II), Cd(II) und Pb(II) an SiO_2 (Aerosil) und TiO_2 (Rutil), Ph.D. Thesis, University of Bern.

Gonçalves, M. L. S., Sigg, L., Reutlinger, M. and Stumm, W. (1987), Metal ion binding by biological surfaces: Voltametric studies in the presence of bacteria, Sci. Tot. Environ., 60, 105.

Huang, C. P. and Stumm, W. (1973), Specific adsorption of cations on hydrous -Al_2O_3, J. Colloid Interface Sci., 43, 409.

Hohl, H. and Stumm, W. (1976), Interaction of Pb2+ with hydrous -Al_2O_3, J. Coll. Interface Sci., 55, 281.

Honeyman, B. D. and Santschi, P. H. (1989), A Brownian pumping model for oceanic trace metal scavenging: Evidence from Th isotopes , J. Marine Res., 47, 951.

Imboden, D. M., Tschopp, J. and Stumm, W. (1980), Die Rekonstruktion früherer Stofffrachten in einem See, Schweiz. Z. Hydrol., 42, 1.

James, R. O. and Healy, T. W.(1972), Adsorption of hydrolyzable metal ions at the oxide-water interface, J. Colloid Interface Sci., 40, 82.

Kramer, J. R. (1967), Equilibrium models and composition of the Great Lakes, In Equilibrium concepts in natural water systems, R. F. Gould,ed., The American Chemical Society, Washington D.C.

Kramer, J. R. (1975), Personal communication.

Li, Y.-H. (1981), Ultimate removal mechanisms of elements from the ocean, Geochim.. et Cosmochim.. Acta, 45, 1659.

Lindsay, W. L. (1979), Chemical Equilibria in Soils, John Wiley & Sons, New York.

Morel, F. M. M. and Hudson, R. J. M. (1985), The geobiological cycle of trace elements in aquatic systems: Redfield revisited, In Chemical processes in lakes, W. Stumm (ed), John Wiley & Sons, New York.

Müller, K. (1982), Die Bielerseesedimente 1878-1978, Ph.D. Thesis, University of Bern.

Njiagu, J. O., Kemp, A. L. W., Wong, H. K. T. and Harper, N. (1979), Sedimentary record of heavy metal pollution in Lake Erie, Geochim. Cosmochim. Acta, 43, 247.

Nyffeler, U. P., Li, Y.-H. and Santschi, P. H. (1984), A kinetic approach to describe trace element distribution between particles and solution in natural systems, Geochim. Cosmochim. Acta, 48, 1513.

Nyffeler, U. P., Santschi, P. H. and Li, Y.-H. (1986), The relevance of scavenging kinetics to modeling of sediment-water interactions in natural waters, Limnol. Oceanogr., 31, 277.

Rippey, B.,Murphy, R. J. and Kyle, S.W. (1982), Anthropogenically derived changes in the sedimentary flux of metals in Luogh Neagh, North Ireland, Environ. Sci. Technol., 16, 23.

Santschi, P. H. (1975), Chemische Prozesse im Bielersee, Ph. D.

Thesis, University of Bern.

Schindler, P. W. (1967), Heterogeneous equilibria involving oxides, hydroxides, carbonates and hydroxide carbonates, In Equilibrium concepts in natural water systems, R. F. Gould, ed., The American Chemical Society, Washington D.C.

Schindler, P. W. (1975 a) Removal of trace metals from the ocean: A zero order model. Thal. Juglosl. 11, 101.

Schindler, P. W. (1975 b) The regulation of trace metal concentrations in natural water systems: a chemical approach, Proc. First Speciality Symposium on Atmospheric Contribution to the Chemistry of Lake Waters. Int. Assoc. Great Lakes Res., 1975, 132.

Schindler, P. W., Fürst, B., Dick, R. and Wolf, P. U. (1976), Ligand properties of surface silanol groups, I. Surface complex formation with Fe3+, Cu2+, Cd2+ and Pb2+, J. Colloid Interface Sci., 55, 469.

Schindler, P. W. (1981) , Surface complexes at oxide/water interfaces, In Adsorption of inorganics at the solid/liquid interface, M. A. Anderson and A. J. Rubin , eds., Anna Arbor, Michigan.

Schindler, P. W. and Stumm, W. (1987), in Aquatic surface chemistry, W. Stumm, ed., John Wiley & Sons, New York.

Schindler, P. W. (1990), Co-adsorption of metal ions and organic ligands: formation of ternary surface complexes, In Reviews in mineralogy, Vol. 22, Mineral-water interface geochemistry, M. Hochella and A. White, eds., Mineralogical Society of America, Washington D.C.

Sigg, L. (1987), Surface chemical aspects of the distribution and fate of metal ions in lakes, In Aquatic surface chemistry, W. Stumm, ed., John Wiley & Sons, New York.

Sposito, G. and Page, A. L. (1984), Cycling of metal ions in the soil environment, In Metal ions in biological systems, H. Sigel, ed., Vol. 18, Circulation of metals in the environment, Marcel Dekker, New York.

Sposito, G. (1984), The surface chemistry of soils, Oxford University Press, New York.

Stumm, W. and Morgan J. J. (1970), Aquatic Chemistry, 1st edition, Wiley Interscience, New York.

TABLE 1: Anthropogenic heavy metal inpiuts into L.Biel sediments in tons per year (Müller, 1982). The data were obtained from analysis of 17 dated sedimentary cores.

Period	Zn	Cu	Pb
1878-1889	4	1.5	1.1
1890-1899	6	1.6	1.7
1900-1909	11	3.0	2.4
1910-1919	14	3.4	2.7
1920-1929	18	4.2	2.6
1930-1938	19	4.9	2.5
1939-1947	24	4.5	2.5
1948-1954	31	5.5	3.4
1954-1958	39	7.1	5.0
1959-1963	37	7.7	6.3
1964-1968	29	7.0	6.0
1969-1973	28	7.2	6.2
1974-1977	21	5.4	5.8

TABLE 2: Copper(II) in sea water

Schindler (1967) calculated the solubility of CuO in an aqueous solution at equilibrium with the atmosphere assuming the solubility equilibrium

$$CuO_{(s)} + CO_{2(g)} = CuCO_3^{\,0}$$

$$\log K_s = \log [CuCO_3](P_{CO2})^{-1} = -3.73$$

$$\log [CuCO_3] = -3.73 + \log P_{CO2}$$

Assuming a value of log P_{CO2} = - 3.52 he thus obtained a value of log [$CuCO_3$] = -7.25 whitch was close to the that time assumed copper(II) - concentration in the sea of log [Cu(II)] = -7.3. According to Bruland (1983) the total copper concentration in sea water now appears to range between 0.5 and 6 nmol kg^{-1}.

TABLE 3: Steady state scavenging model: Comparison of field data (Santschi, 1975) and model predictions (Schindler, 1975 b). Concentration are given in ppb.

Metal	$[M_i]_{(t)}$	$[M]_{(t)}$ (calc)	$[M]_{(t)}$ (obs)
Zn	9.0	3.8	3.7
Cu	3.8	2.4	2.9

The calculation were performed with the following parameters:

Φ: 2.1 E5 dm^3s^{-1} ; A_i: 1.7 E-5 kg/dm^{-3} ; B: 2.56 kgs^{-1} ;

τ'_p: 0.303 ; $D_{(Zn)}$: 1.7 E5 dm^3kg^{-1} ; $D_{(Cu)}$: 3.9 E4 dm^3kg^{-1};

TABLE 4. Reactions of surface hydroxyl groups

Acid base reactions:

$\equiv S\text{-}OH = \equiv S\text{-}O^- + H^+$ (a)

$\equiv S\text{-}OH + H^+ = \equiv S\text{-}OH_2^+$ (b)

Formation of binary surface complexes

$\equiv S\text{-}OH + M^{z+} = \equiv S\text{-}OM^{(z-1)+} + H^+$ (c)

$2 \equiv S\text{-}OH + M^{z+} = (\equiv S\text{-}O)M^{(z-2)+} + 2 H^+$ (d)

Ligand exchange

$\equiv S\text{-}OH + L = \equiv S\text{-}L^+ + OH^-$ (e)

Formation of ternary surface complexes

$\equiv S\text{-}OH + ML^{z+} = \equiv S\text{-}OML^{(z-1)+} + H^+$ (f)

$\equiv S\text{-}OH + ML^{z+} = \equiv S\text{-}LM^{(z+1)+} + OH^-$ (g)

TABLE 5. Stability constants of binary surface complexes (25^0)

Solid oxide	M^{z+}	$\log {}^*K_1{}^s$	$\log {}^*\beta_2{}^s$	I	Ref.
$-Al_2O_3$	Mg^{2+}	-5.4	-	0.1 M $NaNO_3$	a
	Ca^{2+}	-6.1	-	0.1 M $NaNO_3$	a
	Ba^{2+}	-6.6	-	0.1 M $NaNO_3$	a
	Cu^{2+}	-2.1	-7.0	0.1 M $NaClO_4$	b
	Pb^{2+}	-2.2	-8.1	0.1 M $NaClO_4$	b
$SiO_{2(am)}$	Fe^{3+}	-1.77	-4.22	3.0 M $NaClO_4$	c
	Co^{2+}	-6.2	-11.7	1.0 M $NaClO_4$	d
	Ni^{2+}	-6.3	-11.9	1.0 M $NaClO_4$	d
	Cu^{2+}	-5.52	-11.19	1.0 M $NaClO_4$	c
	Zn^{2+}	-5.75	-11.43	1.0 M $NaClO_4$	d
	Cd^{2+}	-6.09	-14.20	1.0 M $NaClO_4$	c
	Pb^{2+}	-5.09	-10.68	1.0 M $NaClO_4$	c
TiO_2	Cu^{2+}	-1.52	-4.96	1.0 M $NaClO_4$	e
(Rutile)	Cd^{2+}	-3.17	-10.48	1.0 M $NaClO_4$	e
	Pb^{2+}	0.15	-1.96	1.0 M $NaClO_4$	e

a) Huang and Stumm (1973); b) Hohl and Stumm (1976) ; c) Schindler et al., (1976); d) Schindler (1975 b); e) Fürst (1976).

TABLE 6. Comparison of D values for a fresh water model system with field data.

a) Example: Calculating D of Cu for a fresh water model system

Model system: $Cu^{2+} - SiO_2 - CO_3^{2-} - H_2O$

Model parameters:
- pH = 8.0
- $[CO_3^{2-}]$ = 2.5 E-6 mole dm^{-3}
- {≡Si-OH} = 1.28 mole kg^{-1}
- Temperature: 25^0 C
- Pressure: 1 bar
- Ionic strength: 0

Equilibria:

*) ≡Si-OH + Cu^{2+}	=	≡Si-OCu$^+$ + H^+	$\log {}^*K^s_1 = -5.52$
*) 2 ≡Si-OH + Cu^{2+}	=	$(\equiv Si\text{-}O)_2Cu^0$ + 2 H^+	$\log {}^*\beta^s_2 = -11.19$
Cu^{2+} + CO_3^{2-}	=	$CuCO_3^0$	$\log K_1 = 6.75$
Cu^{2+} + 2 CO_3^{2-}	=	$Cu(CO_3)^{2-}$	$\log \beta_2 = 9.92$
Cu^{2+} + OH^-	=	$CuOH^+$	$\log K_1 = 6.3$
Cu^{2+} + 2 OH^-	=	$Cu(OH)_2^0$	$\log \beta_2 = 12.8$

$$\sum_m K^s_m\{\equiv SiOH\}^m[H^+]^{-m} = 1.06\ E5 \qquad \sum_z\sum_n \beta_{z,n}[L_z]^n = 22.4$$

$$D = 4.53\ E3$$

*) These values where evaluated at ionic strength 1.0 and assumed to be valid at ionic strength 0. The error introduced by this simplification is hard to estimate.

b) Comparison field data

Metal	log D (model)	log D (field)		
		L.Biel	L.Constance **)	L.Zürich**)
Cu	3.66	4.6	4.5	4.0 - 5.0
Zn	4.71	5.2	5.0	4.6 - 5.1
Cd	2.31		4.7	5.0 - 5.7
Pb	4.97		6.0	5.7 - 6.5

**) Data from Sigg (1987)

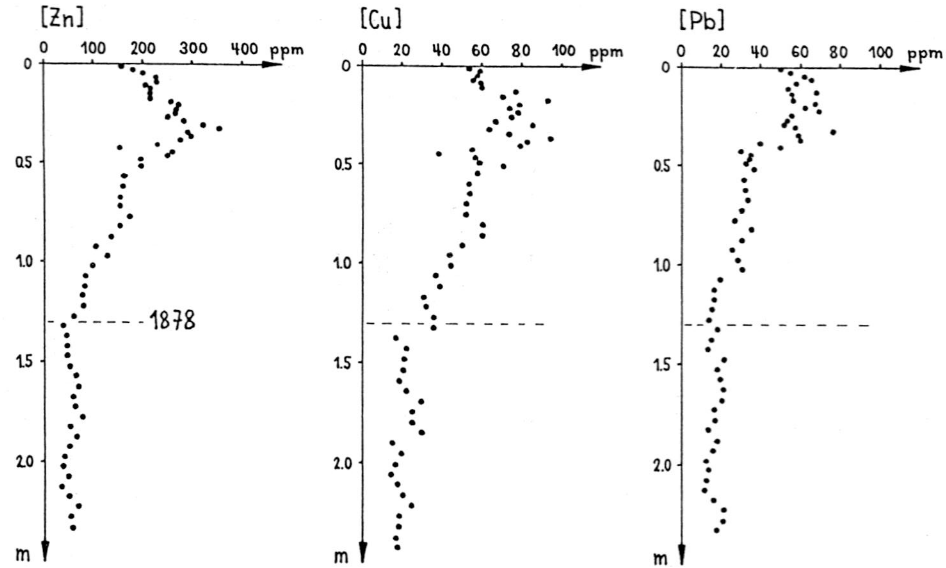

Fig. 1. Trace metals concentrations in a dated sedimentary core from L. Biel (Switzerland). From Müller (1982). The background values for Zn, Cu and Pb are approximately 40,20 and 20 ppm respectively.

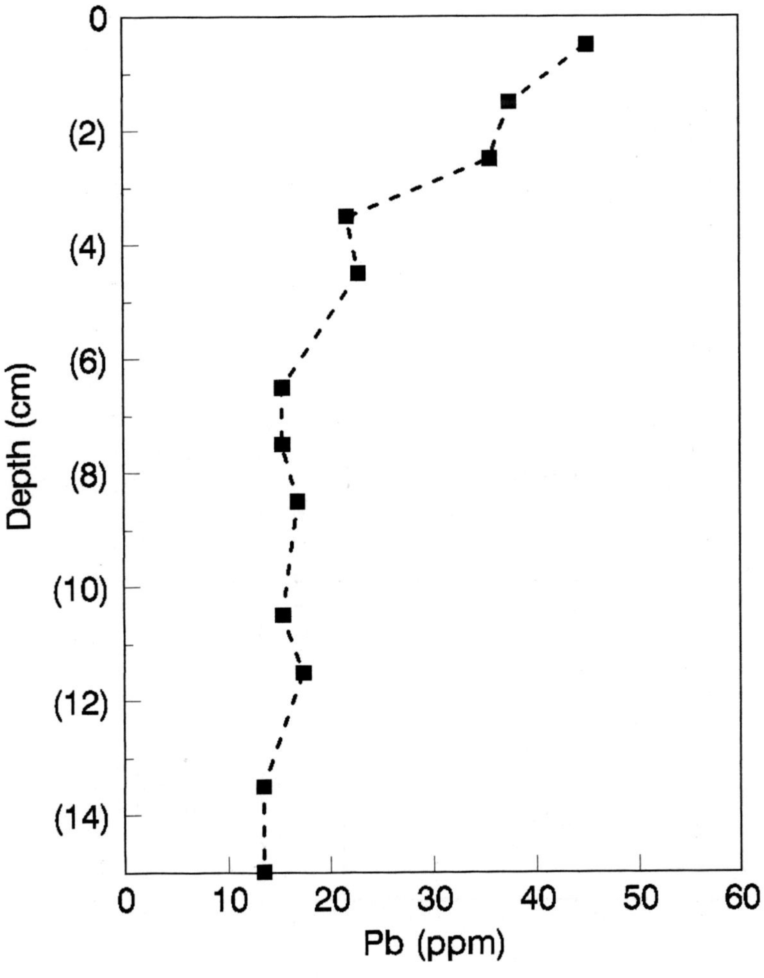

Fig. 2. Lead profile of a Swiss forest soil from the drainage basin of L. Biel. The natural background is at approximately 15 ppm. From Nyffeler (1984, personal communication). Note that the Pb content of the top layer is approximately the same as on the top layer of L. Biel sediments (Fig. 1).

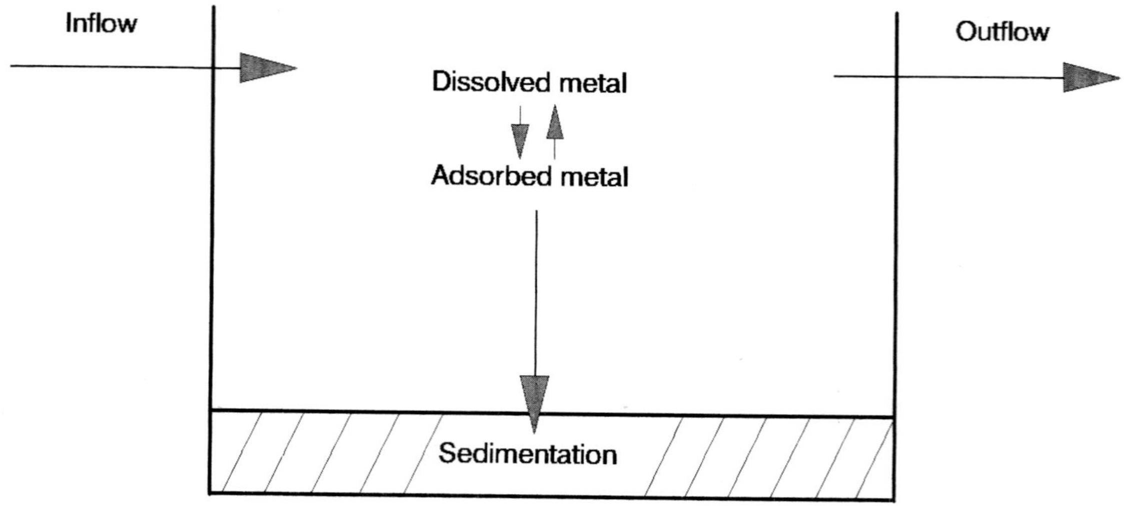

Fig. 3. A one-box steady state lake model. The adsorption desorption equilibrium between the water column and the scavenging particles is established. The adsorbed metals are carried down into the sediments by the settling particles. Adapted from Schindler (1975 b).

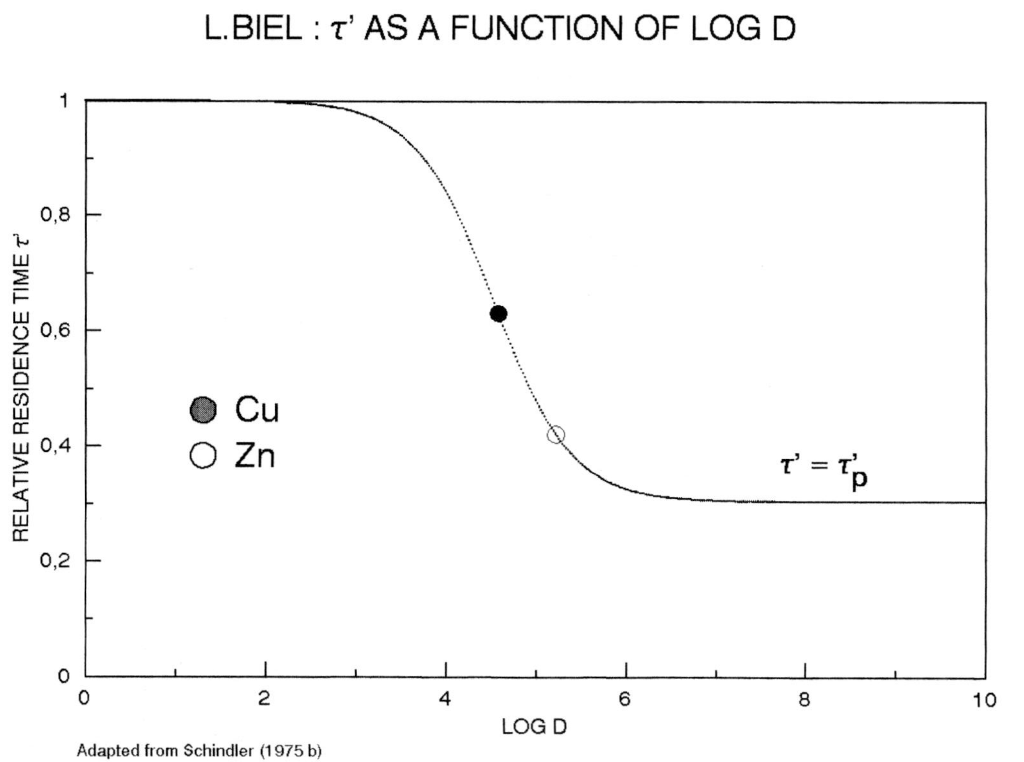

Fig. 4. Relative residence time for metals in L. Biel as a function of the distribution coefficient D. The solid line has been calculated with the aid of eq. (7) using the parameters given in Table 3.

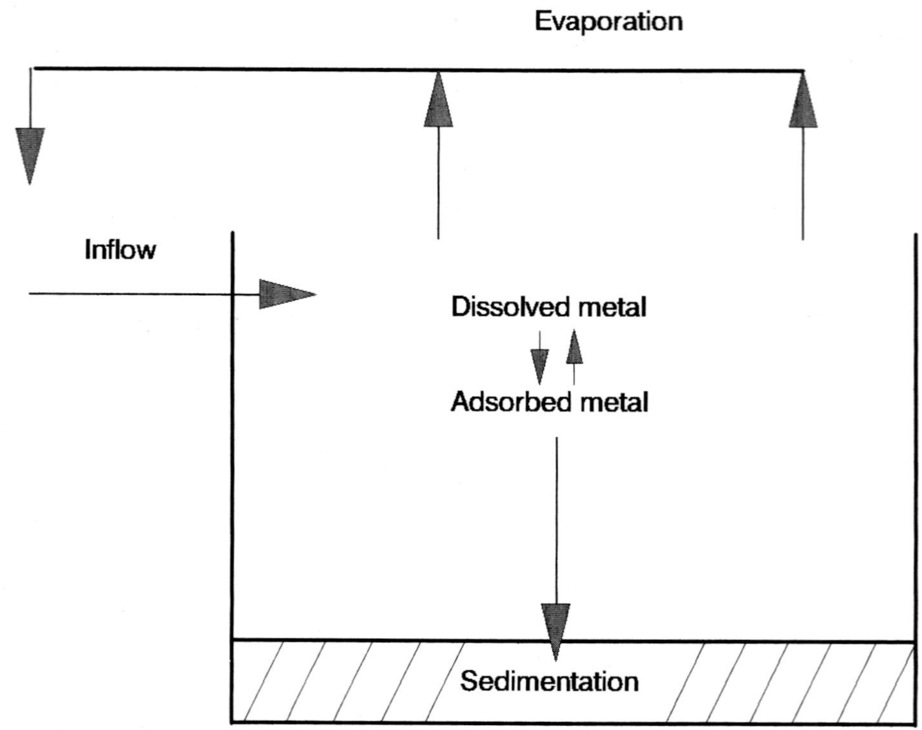

Fig. 5. A one-box steady state ocean model. Adapted from Schindler (1975 b).

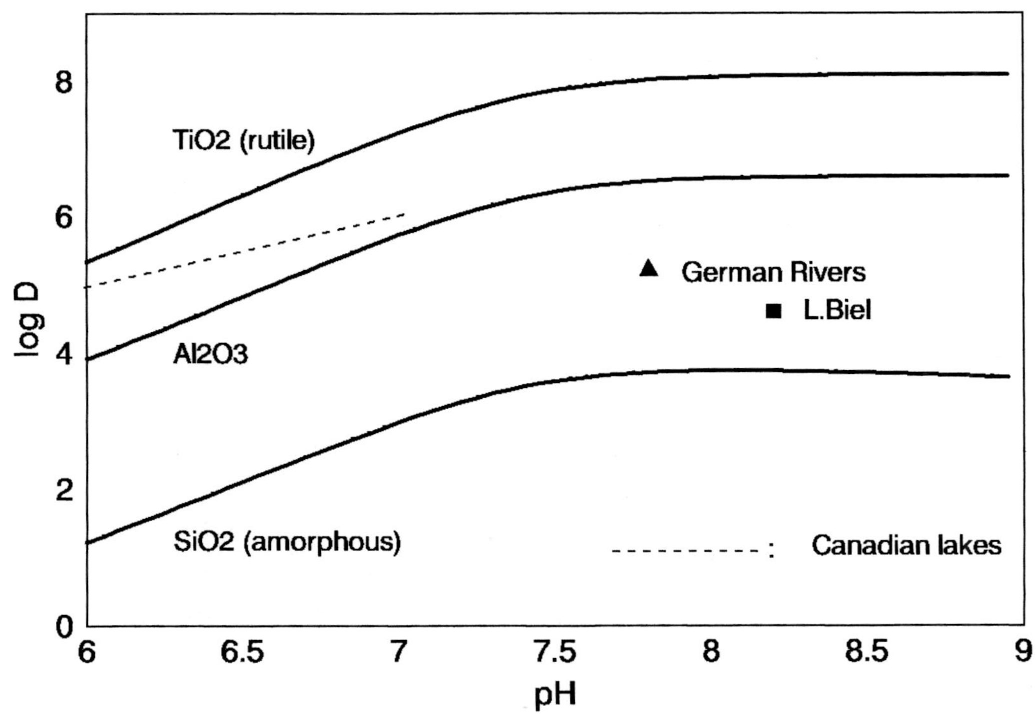

Fig. 6. Log D as a function of pH. The solid lines have been calculated with the aid of eq. 24. The following species where considered: $\equiv S\text{-}OCu^+$, $(\equiv S\text{-}O)_2Cu^0$, $CuCO_3{}^0$, $Cu(CO_3)_2{}^{2-}$, $CuOH^+$, $Cu(OH)_2{}^0$. The pertinent equilibrium constants where taken from the Tables (5) and (6). Adapted from Schindler (1975 b).

GEOGRAPHICAL TENDENCY OF GEOCHEMICAL CHARACTERISTICS AND BINDING CAPACITY OF SUSPENDED MATTER FROM THE FIVE MAIN RIVERS OF EAST CHINA FOR HEAVY METALS

J.S. Chen, C.Q. Cheng and F.Y. Wang

ABSTRACT

Samples of suspended particulate matter were collected from the five main rivers of East China. the Amur River. the Songhua River. the Yellow River. the Yangtze River and the Pearl River. The samples were analysed for particle size distribution. composition of clay minerals. contents of Fe-Mn Oxides. components of humin substances and cation exchange capacity. The Cadmium and Copper adsorption isotherms on the samples of suspended matter were extablished using adsorption experiments to determine the cation exchange capacity. The results show that. controlled by the geographical and geological conditions in basins of the rivers mentioned above. the geochemical characteristics and binding capacity of suspended matter present the following rules: the Montmorillnite/Kaolinite ratio (M/K) decreases gradually from the north rivers to the south rivers; the contents of clay particle. total organic matter and the ratio of Humic acid/Fulvic acid (H/F) decrease too. exceptly the Yellow River; Contrary. the contents of $Fe(OH)_3$. Humin substances increase from the north to the south. The CEC decreases gradually from the north rivers to the south ones. present the latitudinal zonality. The characteristic values of adsorption isotherms show that the adsorption capacities of suspended matter for the heavy metals from the north rivers are larger than those from the south rivers. exceptly lowest in the Yellow River.

INTRODUCTION

Suspended matter plays an extremely important role in the transport and geochemical cycling of trace metals in aquatic system. Its physical and geochemical characteristics influence its binding capacity for trace metals and the river's self-purification (Hunt and Fitzgerld 1983. Laxen 1983). Unfortunately. basic data and information on suspended matter from Chinese rivers are very lacking.

The Amur River. the Songhua River. the Yellow River. the Yangtze River and the Pearl Rivr are the five largest rivers in China. account for upwards of 90 percent of all the area covered by Chinese exterior drainage basins. The Yangtze River and Yellow River are also two of the largest rivers in the world. These rivers flow in a very different physical geographical zones from Northeast China with a cold snowy forest climate and dark brown earth, to South China with a moist subtropical climate and red earth (Table 1). The suspended particulate matter from these five rivers have different geochemical chracteristics and different binding capacity for heavy metals. The purpose of this paper is to quantify the geographical differences in geochemical characteristics and binding capacity of suspended matter from these five rivers.

Table 1. Physical conditions of the five main river basins

River	Average latitude	Annual mean temperature (℃)	Annual mean precipitation (mm)	Soil type
Amur River	49.0° N			dark brown earth
Songhua River	46.5° N	1.0	600	black earth, dark brown earth
Yellow River	37.5° N	8.5	375	drab soils, grey drab soils
Yangtze River	31.0° N	14.5	1300	yellow brown earth
Pearl River	24.0° N	20.2	1650	lateritic red earth, red earth

OUTLINE OF STUDY AND METHODS

SAMPLE COLLECTION

Using a portable high-pressure filtration system (Fig. 1), 20-50 liters of river water were filtered through a 0.45 m membrance filter of celulose acetate. The samples of suspended particulate matter were collected in July and August, 1986 during flood discharge at several sites from each of the above rivers, far from urban and industrial areas. The sampling location are shown in Fig. 2.

The samples collected from the north rivers, e.g., the Amur River and Songhua River, were taken near the shores, while those from the south rivers were taken near the main flow.

The samples were dried at less than 60℃ by evaporation methods and kept at 4℃.

ANALYTICAL METHODS

The grain-size differentiation of suspended matter samples were conducted by using a Micron Photo Sizer, Mode SKC-2000, according to Stokes principles. The clay minerals of suspended matter samples were determined with X-ray diffraction. The content of $Fe(OH)_3$ determined using the Tessier-Campbell sequential extraction procedure (Tessier et al. 1979, Harrison et al. 1981). The components of humin substances of suspended matter samples were separated by NaOH and HCl and determined by Chromic acid using the volumetric analysis methods (Nanjing Soil Institute, China, 1978).

The cation exchange capacity was measured by shaking 1 to 5g of suspended matter sample with 250 ml of a 2M solution of NH_4^+ until no change in these concentrations occurs. The amount taken up was measured and expressed as mmol's per 100g dry material (Grim, 1962).

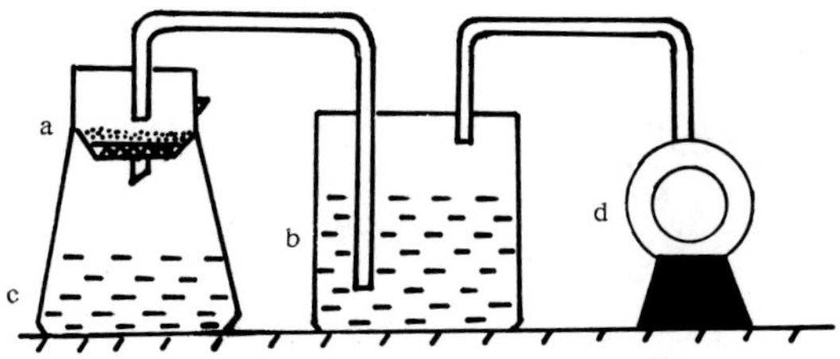

a 0. 45 m membrane filtration equipment hermeticallysealed

b chamber filled with unfiltrated water sample

c chamber filles with filtrated water sample

d pump

Fig. 1 A portalbe high-pressure filtration system

ADSORPTION EXPERIMENTAL METHODS

The prepared suspended particulate matter samples were dispersed in distilled water. The suspensions were then dispensed volumetrically into a flask, and pH was adjusted at 5. 5±0. 2 with microliter additions of NaOH or HCl. Known volumes of Cadmium (or Copper) solution wer added to produce the final concentraction of Cadmium in the range of 0. 1-1. 0 m. e. l^{-1}. The flasks were shaken at 25±1℃ for 20h, and the solid phase separated by filtration through a 0. 45 m membrane filter of celulose acetate. The quantity of metal adsorbed on the suspended matter was calculated by difference between the amount of metal added and the final soluble concentration. Metal ions adsorbed on the walls of the flasks was found to be negligible for these concentrations using Hitachi Model 180-80 Atomic Absorption Spectrophotometer. All laboratory glassware was precleaned by soaking in 10% HNO_3, and reagents were anala R (analytical) grade or

Fig. 2 Sampling location

better.

RESULTS AND DISCUSSION

GEOCHEMICL CHARACTERISTICS OF SUPENDED MATTER FROM THE RIVERS

The grain size distribution. the relative contents of clay minerals. the contents of Fe -Mn Oxides. the components of humin substances and the cation exchange capacity (CEC) are some important geochemical Parameters of suspended matter. which are controlled by the physical geographical conditions. weathering processes and soil formations of the corresponding river basins. and influence the binding capacity of the suspended matter for heavy metals. These characteristics of suspended matter from the five main rivers of East China are given in Talbe 2. Table 3. Table 4 and Table 5. respectively.

Granin size distribution

As shown in Table 2. the sequence of clay particle (<0. 005mm) content in suspended matter is: the Amur River > the Pearl River > the Songhua River > the Yangtze River > the Yellow River. The tendency of clay particle content in suspended matter from the north rivers to the south ones corresponds well with the geographical tendency of clay particle content in soils from the corresponding river drainage basins in East China (Nanjing Soil Institute. China. 1978).

Clay minerals

Determined by X-ray difraction. within the clay fraction the clay minerals occupy a dominant position. Table 3 shows the relative contents of Montmorillonite and Kaolinite. From the north to the south. the ratio of Montmorillonite/Kaolinite (M/K) decrease gradually. The relative contents of clay minerals in suspended matter also correspond with those in soils from the concerned river drainage basins in East China (Nanjing Soil Institute. China. 1978).

Fe-Mn Oxides

The contents of $Fe(OH)_3$ in suspended matter are shown in Table 4. The contents of $Fe(OH)_3$ increase gradually from the north rivers to the south rivers. mainly owing to the increasing contents of $Fe(OH)_3$ in soils with increasing fersiallitization from the north to the south in East China.

Organic matter

Suspended matter contain small amounts of organic compounds. either derived from the land by water or wind erosion. or produced in the water itself. The organic compounds are mainly bituments. lipids. and humin substances. which offer extensive sites for both physical and chemical sorption (Golterman. 1983). The contents of TOM

Table 2. The grain-xize distribution of suspended matter (%)

River	Site	<0. 05mm	<0. 01mm	0. 005mm	<0. 002mm
Amur River	1	100	78. 9	71. 2	53. 6
	2	100	74. 1	68. 5	58. 2
Songhua River	3	100	68. 9	53. 2	49. 1
	4	100	76. 1	66. 5	56. 6
Yellow River	5	100	51. 3	31. 9	21. 5
	6	100	21. 1	17. 3	13. 5
Yangtze River	7	100	57. 8	48. 5	39. 6
	8	100	67. 6	54. 5	41. 9
Pearl River	9	100	70. 1	57. 3	51. 2
	10	100	78. 0	67. 0	55. 3

Table 3. The relative contents of clay minerals of suspended matter

River	Site	Montmori-llonite(%)	Illite (%)	Kaolinite (%)	M/K
Amur River	1	52. 0	32. 0	16. 0	3. 25
Songhua River	3	52. 0	33. 0	14. 0	3. 75
	4	60. 0	25. 0	15. 0	4. 00
Yellow River	5	42. 0	28. 0	30. 0	1. 40
	6	44. 0	20. 0	36. 0	1. 20
Yangtze River	7	16. 0	41. 0	43. 0	0. 37
	8	14. 0	38. 0	48. 0	0. 30
Pearl River	9	14. 0	33. 0	53. 0	0. 26
	10	13. 0	24. 0	63. 0	0. 20

(total organic matter) decrease from the Amur.

river to the Yellow River. then increase from the Yellow River to the Pearl River. The contents of Humin (Humic acid + Fulvic acid) increase gradually from the north rivers to the south rivers. while the ratio of Humic acid/Fulvic acid (H/F) decraeses except the Yellow River (Table 4).

Table 4. The contents of $Fe(OH)_3$ of suspended matter (%)

River	Site	$Fe(OH)_3$
Amur River	1	3. 93
Songhua River	3	4. 46
Yellow River	5	4. 66
Yangtze River	7	4. 72
Pearl River	9	5. 24

Table 5. The components of humic substances of suspended matter

River	TOM in SM (%)	Relative content in TOM (%)			
		Humic acid	Fulvic acid	Humin	H/F
Amur River	5. 23	14. 46	15. 69	70. 15	0. 92
Songhua River	3. 93	10. 22	13. 38	76. 40	0. 76
Yellow River	0. 61	3. 44	17. 45	79. 11	0. 20
Pearl River	2. 61	3. 65	9. 36	87. 01	0. 44

Table 6 shows that CEC of suspended matter are higher in the north rivers than in the south rivers. lowest in the Yellow River. Its corresponding with the content sequences of clay particle and Humin (Table 1. Table 5) suggests that. as well as clay particle and humin contents. CEC^{es} was controlled by their composition and properties. Therefore. we define CEC^{es} (CEC of effective solbents) as

$$CEC^{es}\ (m.\ e/100g) = \frac{CEC(m.\ e/100g)}{\text{contents of clay particle + organic matter (\%)}}$$

The CEC^{es} of suspended matter from the five main rivers of East China relate well to the clay particle composition. silica-alumina ratio. H/F of the representive soils in the corresponding river basins (Table 6). and present the latitudinal zonality rule (Fig. 3).

CADMIUM AND COPPER ADSORPTION ISOTHERMS ON SUSPENDED MATTER SAMPLES FROM THE FIVE RIVERS

Copper and Cadmium are found in most Chinese rivers. especially in the south rivers.

duing to the wide distribution of metalliferous ores and non-berrous industries in South China. They are also two kind of very toxic and relatively accessible metals (Wood. 1974). So we chose Copper and Cadmium ions as solbents in the adsorption experiments. Copper and Cadmium adsorption isotherms on suspended particulate matter were all of the Langmuir type in which saturation usually occurred. The adsorption isotherms can also be described in Freundlich type. The Langmuir type isotherm used was where G° is

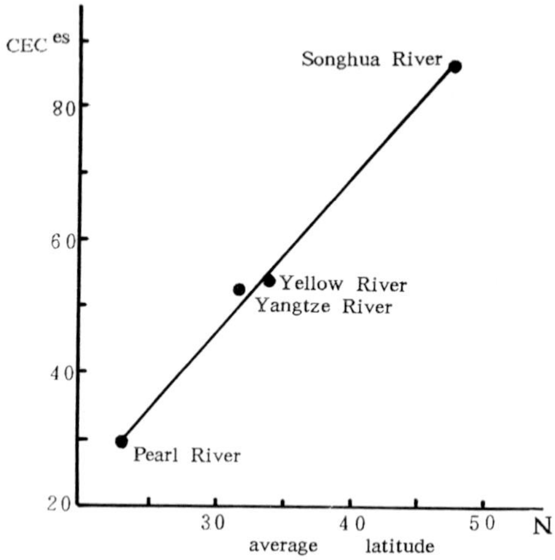

Fig. 3 CEC^{es} and average latitudes of the rivers

$$G = G^{\circ}\ C/(A+B) \qquad (1)$$

or

$$\frac{1}{G} = \frac{1}{G^{\circ}} + \frac{A}{CG} \qquad (2)$$

the adsorption value. C is the corresponding equilibrium concentrations of Cadmium (or Copper). G° is the saturated adsorption value and A is the adsorption coefficient.

The Freundlich type isotherm used was

$$Q = KC^{1/n} \qquad (3)$$

where k. n are empirical factors. $n > 1$.

Fig. 4 shows the Cadmium adsorption isotherms with Cadmium for each of the suspended matter from the five main rivers of East China. The Copper adsorption isotherms are shown in Fig. 5. The characteristic values of Cadmium adsorption isotherms. G and other parameters are presented in Table 7.

Table 7 and Fig. 4. Fig. 5 show that the binding capacities of suspended matter for

Table 6. Comparison of CEC. CEC and some geochemical characteristics of soils in corresponding river basins

River	CEC	CEC^{es}	Clay minerals*	Si/Al	H/F
Songhua River	0. 52	85. 84	H. M	3. 54	2. 17
Yellow River	0. 24	56. 11	H. V	3. 22	2. 51
Yantze River	0. 17	53. 12	H. V. K	2. 70	0. 44
Pearl River	0. 08	30. 05	K	1. 89	0. 19

* H: Hydromicas

M: Montmorillonite

V: Vermiculite

K: Kaolinite

Table 7. The characteristic Values of Cd adsorption isotherms of suspended matter

		Freundlich type			Langmuir type		
River	Site	1/n	K	R	G°	A	R
Amur River	1	0. 356	0. 300	0. 930	0. 254	0. 066	0. 983
Songhua River	3	0. 375	0. 297	0. 984	0. 254	0. 075	0. 981
Yellow River	5	0. 375	0. 178	0. 994	0. 164	0. 092	0. 994
Yangtze River	7	0. 340	0. 158	0. 999	0. 158	0. 110	0. 998
Pearl River	9	0. 334	0. 097	0. 999	0. 100	0. 128	0. 991

Copper and Cadmium are higher in northern China rivers than in southern China rivers.

CONCLUSION

The five main rivers of East China. the Amur River. the Songhua River. the Yellow River. the Yangtze River and the Pearl River. flow in a very different physical geographical zones. controlled by the physical geographical conditions. weathering processes and soil formations of the corresponding river basins. The main geochemical characteristics of suspended matter. such as the contents of clay particle. the composition of clay minerals. the contents of Fe-Mn oxides and the components of humin substances present the geographical differentiation from the north rivers to the south rivers. Corresponding with the physical geographical conditions of these river drainage basins and the geochemical characteristics of suspended matter. CEC. CEC^{es} and the binding capacity of suspended matter from these five rivers for heavy metals have the same geographical distribution tendency.

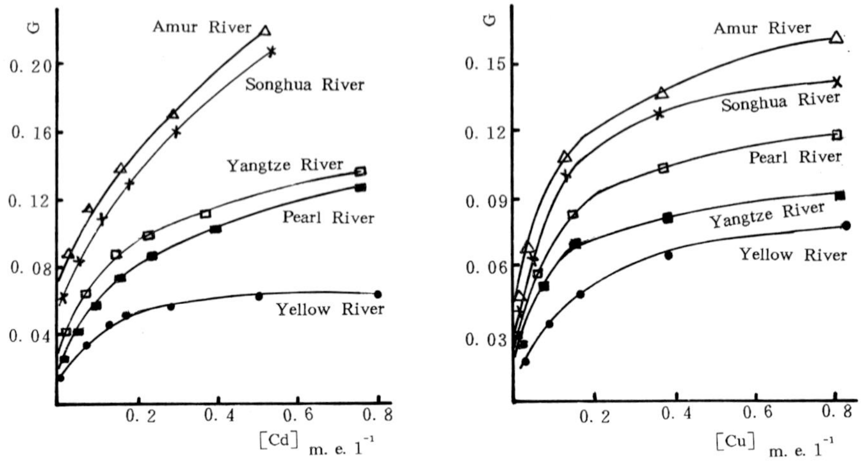

(a) (b)

Fig. 4 Adsorption isotherms of suspended matter in rivers of East China
a: Cd b: Cu

ACKNOWLEDGEMENT

We thank Dr S. Tao. B. S. Den and F. Tang who helped us collect the samples. We acknowledge the analytical support provided by Dr G. S. Gao and L. Z. Yang. We are also grateful to Professor E. D. Ongley in Rivers Research Branch National Water Research Institute. Canada. for text processing the manuscript.

REFERENCES

1 C. Hunt and W. Fitzegerald. Marine Chemistry. 12(1983). 255-280.
2 D. P. H. Laxen. The Science of the Total Environment. 30(1983). 129-146.
3 A. Tessier. P. G. C. Campbell and M. Bisson. Anal. Chem. . 51(1979). 84.
4 R. M. Harrison. D. P. H. Laxen and S. J. Wilson. Envir. Sci. Technol. . 15 (1981). 1378-1382.
5 Nanjing Soil Institute. Physical and Chemical Analysis for Soil. Shanghai Science Press. Shanghai. 1978. pp. 136-142 (in Chinese).
6 R. E. Grim. Clay mineralogy. McGraw-Hill Publ. . New York. 1962. pp. 596.
7 Nanjing Soil Institute. Soils of China. Science Press. Beijing. 1978 (in Chinese).
8 H. L. Golterman. P. G. Sly and R. L. Thomas. Study of the Relationship between

Water Quality and Sediment Transport, Unesco, 1983, pp. 18.
9 J. S. Chen, L. Den and B. S. Den, Hydrobiologia, 176/177(1984), 159-170.
10 U. Forstner and G. T. W. Wittmann, Metal Pollution in the Aquatic Environment, Springer-Verlag, Berlin, 1979.
11 J. S. Chen, C. J. Zhang, G. S. Gao and C. Q. Chen, Scientia Geographica Sinica, 6 (1986), 322-332 (in Chinese).

PARTICULATE TRACE METALS IN SURFACE WATERS OF THE NORTH SEA

M. Kersten, G. Irion and U. Förstner

1. INTRODUCTION

Particles in seawater play a major role in regulating the transport and fate of trace metals discharged into marine waters as a result of natural weathering processes and human activities. This is particularly true in coastal waters, where particulate matter from river discharge interact with seawater, but also in areas where fronts of nutrient-rich water lead to enhanced primary productivity. All the processes of transfer of trace elements from terrestrial to aquatic environments are in fact more or less constrained by particle-water interactions (1). These processes include the supply of inorganic rock weathering products and organic substances from river runoff, resuspension of previously deposited sediments, aeolian fallout, production and degradation of pelagic biota, but also geochemical adsorption-desorption, precipitation-dissolution, and complexation reactions at the solid-solution interface. While in a conceptual sense these processes are generally recognized, the present state of knowledge is such that detailed information on the mechanisms that result in trace metal enrichment of particulate matter is still lacking (2).

In general, the concentration of a heavy metal in the particulate phase, expressed in terms of weight of metal per unit weight of dry solid material (µg/g), is much higher than that of the same metal in the dissolved phase (µg/l). Suspended particulate matter (SPM) tend to accumulate trace elements. It is therefore possible to exploit the relative enrichment of the particulate phase in order to evaluate the level of contamination in an aquatic system. The capacity of suspended particulate matter (SPM) to scavenge trace elements, as well as some of their spatial and temporal trace element variability, has several causes. The most frequently cited and investigated control factors of particulate trace element accumulation include the grain size distribution, surface area and the concentration or quality of geochemical components.

Physical grain size fractionation and subsequent chemical analyses can be used for both spatial and temporal trend identification and accurate determination of trace element enrichment factors (3). One procedure entails normalization of bulk chemical data to the percent of a selected size range, and the other requires an actual physical separation of a selected size range prior to chemical analysis, typically <63 or <20 µm. If these techniques are not used, intersample comparisons can be improper because intersample grain-size differences can superimpose metal concentration gradients which do not reflect actual conditions. Normalization of bulk metal concentrations to Al contents representing the clay fraction have been discussed recently as an alternative by the working group on "Marine Sediments in Relation to Pollution" and of the "Advisory Committee on Marine Pollution" of the International Council for the Exploration of the Sea (ICES). Compositional studies of the quality of trace metal-binding geochemical components, on the other hand, are more ambiguous than grain-size normalization techniques because they generally are based on chemical extractions. The interpretational difficulties with these methods have been described previously (4). Only a very few reports exist on application of these methods to SPM (5).

Surface area is another factor important in controlling particulate trace element concentrations and variability, because many of the processes involved in sediment-trace element interactions fall into the general category of surface reactions or surface chemistry. Only three papers that directly link sediment-trace element chemistry and measured surface area were located by an extensive literature review (6). The apparent lack of interest in surface area in relation to particulate trace element studies is surprising, but may result from the perception that surface area and grain size are interchangeable. This may occur because grain-size measurement (distributions or separations) are relatively simple to perform. However, grain-size measurements are also relatively ambiguous after removal of the organic matter. Such components as Fe/Mn oxihydroxides and organic matter can be deposited as thin surficial coatings on clay minerals, and then these coatings rather than the underlying substrate acts as trace element scavengers. As surface area increases, the concentration of these geochemical phases tend also to increase, but the surface area can also be much smaller as that of the clay substrate due to agglomeration effects induced by that coatings (6).

Element concentration distributions have been recently documented for sediments (7,8) and suspended particulate matter (9,10) of the North Sea. Though these reports include data on a relatively wide range of heavy metals relevant for the marine environment, no published data exist for the metalloids Hg and As. The above cited observations on distribution, dispersal and deposition of heavy metals, however, evidenced increased concentrations in the central and northern part of the North Sea far remote from any direct input source. It is therefore of interest whether this observation can be confirmed for all the toxic elements of the CEC priority pollutants list, i.e. Hg, Cd, and Pb. Moreover, the question arised whether the variety found in regional distribution patterns could be explained by anthropogenic contamination or by natural processes. The present paper summarizes results of a field investigation on particulate trace metals in North Sea water. The purpose is to apply the above cited methods and evaluate the particulate element distribution patterns gained for the entire North Sea. We follow thereby the approach which was exemplified by Horowitz and Elrick for stream bed sediments from the entire U.S.A. (6). Moreover, we attempt to explain the composition of the SPM in relation to source and in-situ particle production by use of interrelations between the SPM characteristics and trace element concentrations.

2. MATERIALS AND METHODS

In January to March 1987, during a sampling cruise in the North Sea, SPM samples were collected for major and trace element analysis. The entire North Sea was covered by a total amount of 145 sampling stations, but only from 115 locations SPM was collected during appropriate weather conditions as depicted in Fig. 1. Additionally, on 21 transects between stations distributed over the entire area large-volume samples were collected to enable a more detailed geochemical analysis. This effort was part of a multidisciplinary experiment (ZISCH) designed to evaluate the physical, chemical and biological oceanography of the North Sea during winter, when meteorological conditions should prevent patchiness by biological productivity.

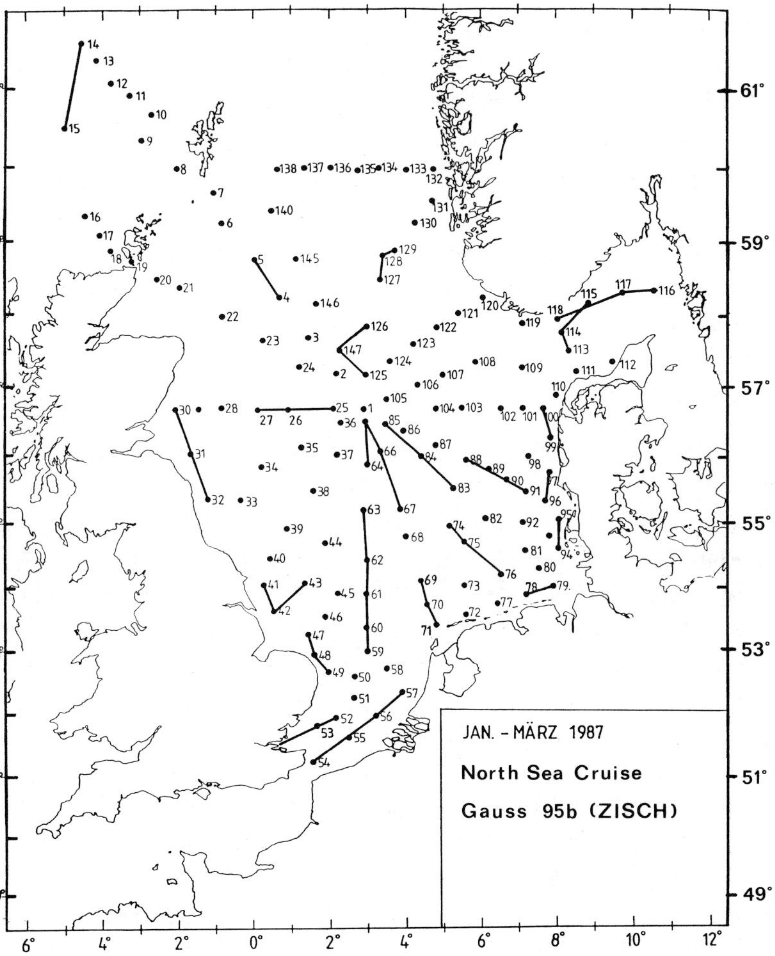

Fig. 1. Sampling locations and transects in North Sea surface waters (10 m) where SPM was collected during January - March 1987

For the various analytical methods to be applied enough SPM material had to be collected at each station. Thus, 200-l samples of surficial water were collected from 10 m depths using a PVC tube and a portable impeller pump equipped with a plastic material head. SPM samples were separated within 4 h of water collection into the HDPE tanks by use of flow-through centrifugation (Fig. 2). Comparisons involving both field and laboratory trials have shown that continuous-flow centrifugation is a viable alternative to in-line filtration for the separation of SPM for subsequent trace metal analysis (11-13). The centrifuges (HERAEUS Labofuge 15000) mounted on gimbals operate at a maximum speed of 10,000 Upm with a flow rate of 0.8 l/min and a mean water residence time inside the rotor of about 20 sec. This rate is well below that recommended for low (< 30 mg/l) SPM concentrations and very fine mean grain size (< 10 μm) of the SPM (13). Large-volume samples (up to 10 m^3 of water) were separated directly on-line during transects between stations using a Padberg centrifuge which operate at a maximum speed of 20.000 Upm with a flow rate of 10 l/min and a particle residence time of about 30 sec.

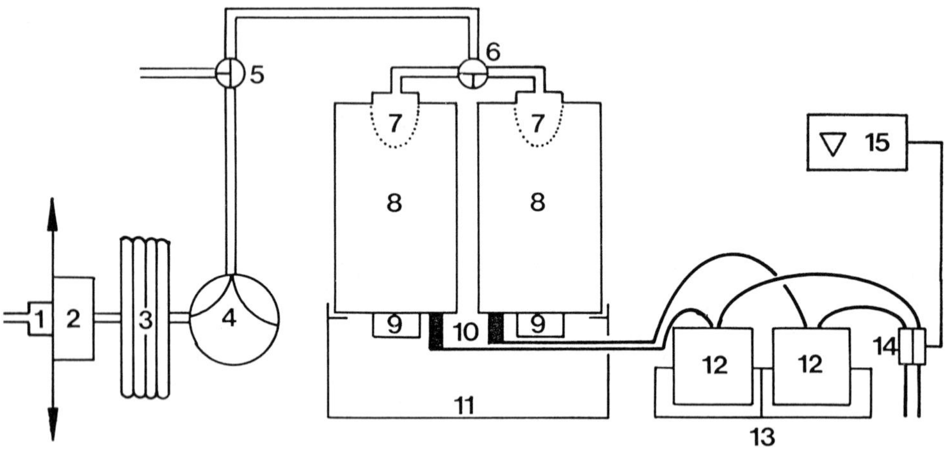

Fig. 2. Scheme of the SPM separation equipment installed onboard RV "Gauss": (1) inlet tube with reflux valve, mounted at (2) a flow sail with hydrowire brackets, and connected to (3) a PVC spiral-mantle tube (100 m, 3/4"); (4) flexible impeller pump with epoxy head (ITT Jabsco E80, rate of application 15 l/min); (5) by-pass and (6) distribution valve (both PVC); (7) plankton net (100 μm); (8) HDPE barrels (200 l); (9) magnetic stirrers with hanging bars (Ikamag Maxi M1); (10) magnetic inflow valves for the zentrifuges; (11) bearing frame; (12) continuous-flow zentrifuges (Heraeus Labofuge 15000); (13) gimbals; (14) flow rate indicator; (15) control unit.

The minimum size of particles (with assumed density of 1.8 g/cm^3) that can be separated under the experimental conditions is about 0.5 μm (12), which is equivalent to the nominal pore size of 0.45 μm membrane filters most commonly used in filtration of natural water samples. A check for recovery using in-line filtration (0.45 μm) in parallel at some of the stations showed that the percentage of loss changes according to SPM density. In waters of high turbidity (> 10 mg/l), weight loss by centrifugation as compared to filtration is less than 1 %, while in waters of low turbidity (< 1 mg/l), a loss of top 10 % was observed. However, even though we know that this very fine fraction not recovered is much richer in particulate trace metals (12), the induced error will not go beyond 20 %. The Heraeus centrifuge rotor can hold 300 ml of water. The walls of that rotor, on which the SPM material was collected, were covered with Teflon supports that could be easily removed and replaced between sampling. These Teflon supports were frozen for transport to our institute. In the laboratory, the material was washed from the supports and rinsed twice by Milli-Q water to remove any remaining seasalts using an usual laboratory centrifuge, and freeze-dried. After back-weighing, the material was subjected to digestion and analysis. A significant release of metals from the SPM samples during resuspension was not expected, because other workers have reported minimal metals loss during dispering and sieving procedures (14). Depending on the amount of suspended matter in the water samples and the centrifuges used, the yield ranged between 15 mg and top 40 g by dry-weight.

For dissolution of the SPM samples, they were first digested with hydrofluoric acid in open Teflon vessels till dryness to degrade the silicate matrix. The residue was subsequently dissolved in a HNO_3-H_2O_2 admixture to yield clear solutions for ICP-AES analysis of major and trace elements (15). Sample solutions were additionally analyzed for the trace elements As, Cd, and Pb using a graphite furnace AAS with Zeeman background correction (P.E. 3030). Hg was determined with a HGAAS technique (P.E. 3030 equipped with the hydride generation system MHS-20), but on subsamples which have not been treated with HF before HNO_3-H_2O_2 digestion. All samples were analyzed in dublicate. Precision and bias (usually better than ± 10 %) were monitored by the concomitant digestion and analysis of reference sediments. Quality control of the analysis was checked by an intercomparison exercise with satisfactory results (16). Particulate organic carbon (POC) was determined on subsamples using a CHN-Analyzer (17). Viable plankton was virtually absent during the sampling period (except in the Skagerrak which was sampled at a beginning phytoplankton bloom) so that only the contents of detrital (dead) organic matter and/or of bacteria were determined. The concentrations are expressed in *%* (major elements) or *μg/g* (minor and trace elements) of the SPM dry weight, while the SPM concentration is given as *mg/l* of seawater.

Grain-size separations were made following the procedure described by Müller (18). Freeze-dried subsamples were separated into the nominal grain-size fractions < 2, 2-6.3, 6.3-20, 20-63, 63-90 and > 90 μm by repeated sedimentation in Atterberg cylinders. After drying at 80 °C the clay fraction was digested with conc. nitric acid and metal determination was carried out as described above. Surface area was determined for appropriate-sized sample aliqouts using the single-point BET method. Freeze-dried samples were initially outgassed for 1 h at 150 °C under argon. Surface area was automatically calculated using N_2 after adsorption under liquid nitrogen and desorption at room temperature using a Ströhlein Areameter (19). It should be noted that this procedure measures "external" surface area of the dried, bulk material which may not agree with that of wet material as measured, e.g., by the methylene blue method. Moreover, considering the fractal dimension of the adsorbents surface (20) this method produces just operationally defined but comparable results. Precision and accuracy were checked by the BCR reference material CRM 172 (Quartz grains, 2.50 m^2/g). Precision was better than ± 20 %.

TABLE 1
Physical and chemical data for the large-volume samples from the 26 North Sea transects

Transect No.	SPM mg/l	S.A. m^2/g	POC %	Al %	Ca %	Fe %	Mn	Cu	Pb	Cd µg/g	As	Hg
4- 5	0.35	8.3	3.54	3.38	7.18	2.17	381	50	82	2.50	<0.5	0.27
14-15	0.44	1.9	5.55	1.30	1.70	0.77	150	15	62	0.22	0.7	0.09
25-27	0.63	4.8	2.66	2.00	3.00	1.11	210	13	35	0.26	-	-
30-32	0.52	10.3	5.58	-	-	2.00	644	29	67	0.35	-	-
41-43	1.86	5.5	4.43	4.02	3.43	2.33	596	20	73	0.26	7.2	0.34
47-49	2.92	12.2	3.42	4.73	4.58	2.72	617	24	70	0.22	-	0.32
52-Thames	7.32	15.3	1.82	5.80	4.31	2.77	358	27	58	0.15	15.1	0.34
54-56	2.54	8.6	3.16	2.70	7.60	1.62	250	17	39	0.24	-	-
56-57	1.58	7.4	5.62	2.23	6.34	1.56	305	9	45	0.33	3.1	0.26
59-61	5.74	12.3	2.54	4.38	4.75	2.77	461	22	45	0.25	10.6	0.36
61-63	1.22	9.2	5.40	3.57	4.07	2.35	355	19	47	0.34	1.7	0.18
64-67	0.36	-	8.84	3.15	3.04	2.28	455	5	56	0.92	1.1	0.20
69-71	1.42	10.4	3.92	3.60	5.04	2.32	398	14	55	0.23	<0.5	0.42
71-72	11.51	11.9	3.28	6.78	6.41	3.05	792	17	51	0.15	16.8	0.40
74-76	2.50	11.3	2.66	4.36	4.33	3.32	578	14	62	0.19	2.8	0.27
78-79	3.24	11.9	3.60	4.45	6.07	2.66	620	18	75	0.37	4.1	0.51
83-85	0.52	8.5	8.13	4.39	3.89	2.75	444	13	73	1.45	<0.5	0.22
88-91	0.81	9.8	6.18	4.56	5.08	2.83	542	15	72	0.39	<0.5	0.23
94-95	6.46	14.6	3.51	5.29	5.34	2.67	905	18	61	0.33	13.8	0.39
96-97	2.29	11.5	5.88	4.54	5.21	2.51	822	13	82	0.44	-	-
99-100	2.25	6.4	6.99	3.73	5.12	1.91	497	10	54	0.30	-	-
113-115	1.08	3.1	25.40	0.64	1.05	0.34	178	6	13	0.56	-	-
116-117	1.72	11.3	23.92	0.56	1.06	0.30	152	7	12	0.60	<0.5	0.14
117-118	0.97	3.7	25.48	0.61	1.17	0.37	200	6	14	0.83	2.2	0.09
125-126	0.53	7.0	6.97	3.02	3.02	2.17	339	3	55	0.50	2.1	0.18
127-129	0.34	-	5.83	3.50	6.73	2.09	414	2	50	1.24	<0.5	0.33
Median	1.07	8.9	5.55	3.60	4.58	2.32	414	14	57	0.35	(5.8)	0.28

Operationally defined geochemical phase determinations were carried out on the transect samples using the procedure developed by Kersten and Förstner (5). This entailed a five-step extraction sequence as follows: (1) a digestion in a solution of 1 M ammonium acetate adjusted to pH 7 and shaken for 2 h at room temperature (adsorbates); (2) an 1 M sodium acetate-acetic acid buffer adjusted to pH 5 for 5 h at room temperature (adsorbates and carbonates); (3) a 0.1 M ammonium oxalate-oxalic acid buffer adjusted to pH 3 and shaken for 24 h at room temperature in a dark box (amorphous Fe/Mn-oxides); (4) a solution of 30 % H_2O_2 adjusted to pH 2 with

HNO_3, heated at 85 °C for 5 h and evaporated till dryness, followed by digestion in 1 M ammonium acetate in 6 % HNO_3 (pH 2) overnight at room temperature (organic matter). The residue was dissolved by mineral acids as described above for comparison with the bulk concentrations. Two 0.5 g aliquots for each of the 21 selected samples were treated in separate sealed 50 ml centrifuge tubes. After each extraction, the aliquots were centrifuged for 20 min at 3000 r.p.m. The supernates collected for trace metal analysis as described above. Matrix matched standards were used. The data for all steps were thereby adjusted for procedural blank concentrations, which were except for Zn in the acetate extractants relatively low.

Precision and bias were evaluated by an intercomparison exercise using BCR reference sediments (21), which gave satisfactory results. Precision significantly worsened, however, with step 1 and 4 where concentrations of some metals were near detection limits.

TABLE 2

Selected correlation coefficients for chemistry, grain size and surface area of the transect samples ($n \leq 26$, $p = 0.05$).

	surface	POC	Al	Fe_{red}[a]
%<2µm	.69	n.s.	n.s.	n.s.
%<6.3µm	.74	n.s.	n.s.	n.s.
%<20µm	.69	n.s.	n.s.	n.s.
Fe	.58	-.81	.94	.98
Fe_{ex}[b]	.59	.98	-.82	-.83
Cd	n.s.	.47	n.s.	-.50
Cd_{ex}[b]	-.56	.87	-.66	-.68
Pb	.37	-.72	.87	.94
Pb_{ex}[b]	.40	.91	-.68	-.71
Cu	.39	-.59	.42	.83
Cu_{ex}[b]	-.63	.83	-.70	-.78
As	.67	-.58	.58	n.s.
Surface		-.50	.64	.66
POC			-.76	-.83
Al				.97

[a]reducible Fe fraction
[b]exchangeable metal fractions

3. RESULTS AND DISCUSSION

3.1 General physical and chemical characteristics of the SPM

In order to simplify the discussion and to help clarify the various interpretations, the bulk chemical, grain size, surface area, and geochemical association data for the 26 large-volume samples from the North Sea transects have been tabulated together (Table 1). In addition, to provide a numerical value for the strength of the various interrelations, Pearson correlation coefficients were calculated for the various parameters after all the data had been log-transformed (Table 2). Only correlations significant at the 95 % or better confidence limit are presented.

3.2 Distributions of metal concentrations in the North Sea

Geographical distributions of Cd, Pb, As and Hg have been given in Figs. 3a-d. Distribution patterns of 10 other major and trace elements, together with that of SPM, POC, TPP and chlorophyll-a equivalents, have been given elsewhere (22). The total concentrations of these four elements in the SPM samples show the following distributions in the surface water:

- Cd (Fig 3a). The Cd-content in the SPM samples was high in the central and northern North Sea and in the Skagerrak (mostly > 2 µg/g, maximum 14.2 µg/g). In the southern North Sea the Cd-content was generally < 1 µg/g, reaching minimum values of 0.12 µg/g. A significantly positive correlation was found between Cd and POC, while a negative was found between Cd and Al (22).

- Pb (Fig. 3b). The Pb-content was lowest in the central North Sea and in the Skagerrak (< 50 µg/g). In coastal areas around the North Sea, especially off the British Island, it was generally higher (> 50 µg/g, reaching nearly 200 µg/g). A significantly positive correlation was found between Pb and Mn (22).

- As (Fig. 3c). The As-content was generally below 25 µg/g (median value 7.4 µg/g), but higher in the southern part than in the northern part of the North Sea. As-enriched SPM was found especially in the German Bight and in the eastern North Sea, with values around 20 µg/g. A significantly positive correlation was found between As and Fe, and a negative between As and TOC (22).

- Hg (Fig. 3d). The distribution pattern of this metalloid is similar to that of Cd, with higher concentrations in the northern part and lower in the southern part of the North Sea, where the Hg-content was generally < 0.35 µg/g, reaching minimum values of 0.09 µg/g. A striking fact is that the Humber and Elbe plumes are recognizable by their increased Hg concentrations.

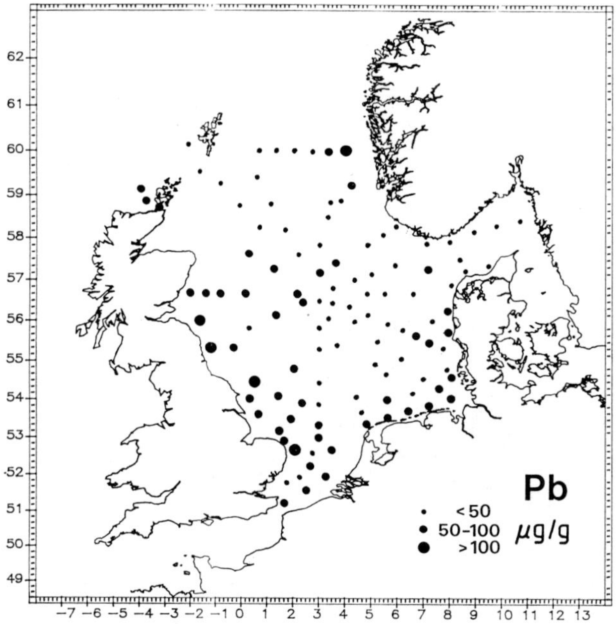

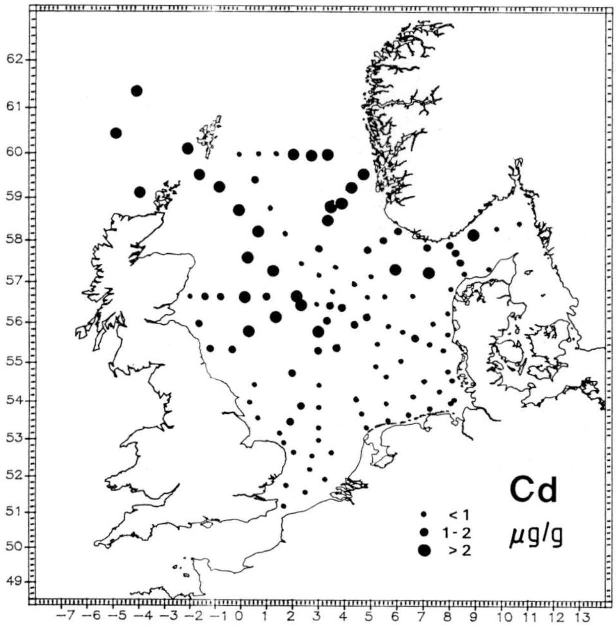

Fig. 3a,b. Concentration (µg/g) distribution maps for particulate Cd and Pb in the North Sea based on SPM samples from surficial waters (10 m depths) in winter 1987.

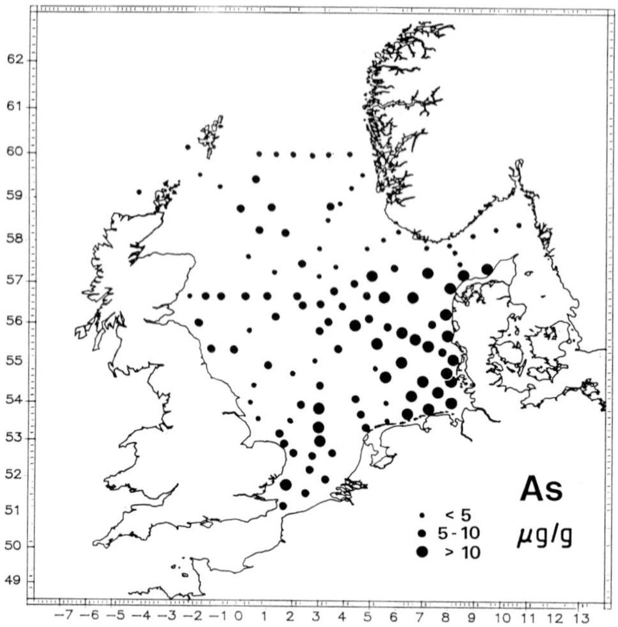

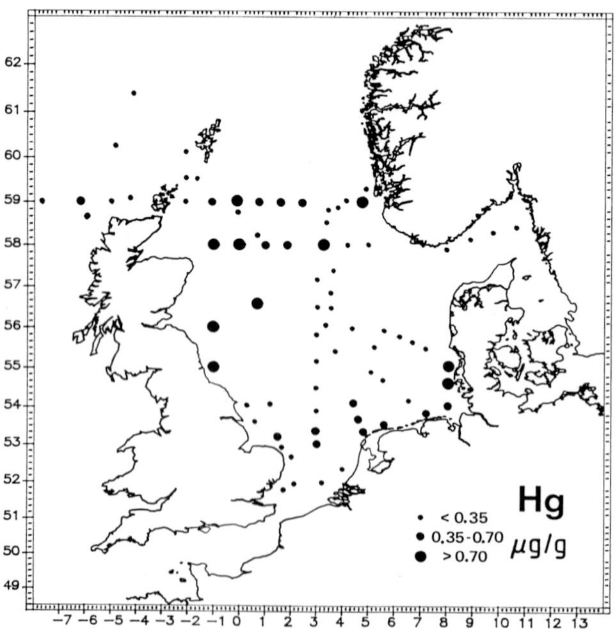

Fig. 3c,d. Concentration (µg/g) distribution maps for particulate As and Hg in the North Sea based on SPM samples from surficial waters (10 m depths) in winter 1987.

Before going on to discuss briefly the distributions of that elements, it seems necessary to obtain some background information on their SPM carrier. To decide whether the elements were associated with the inorganic or organic fraction, they were correlated with Al and POC, respectively. Moreover, this distinction can also be obtained from the relation between the element contents and total SPM concentrations in the water column. Duinker (23) has shown that most element contents are related to the SPM concentrations: Al, e.g., tend to increase with suspended matter concentration, whereas POC increases strongly at low SPM content. This inverse relation is reflected in the negative correlation coefficient between Al and POC (Table 2). It is simply due to the fact that the resuspended mineral particles predominates at high SPM content in thewater column, while at low SPM concentrations biogenic matter is the predominant component of in the permanently suspended fraction. Pb and As showed about the same relation with SPM concentrations as Al, while for Cd, Cu and Hg about the same hyperbolic relation as for POC was found (Fig. 4a-g). From this similarity it is obvious that Pb and As fall into the mineral particle-bound group, whereas Cd, Cu and Hg seem to be bound predominantly to organic matter. At higher SPM concentrations, the element concentrations tend to approach a baseline at those levels reported for non-industrial background sediment composition, except for Hg (24). This threshold value was 53 µg/g for Pb, 0.16 µg/g for Cd, and 17 µg/g for As. The value for Hg (0.35 µg/g) is somewhat obscured by the fact that the Elbe and Humber plumes have both high SPM and particulate Hg concentrations, the latter probably due to anthropogenic contaminations. It is now possible to make allowances for varying SPM concentrations of the samples. Readings which are considerably above the metal vs. SPM concentration curve can be termed anomalous. Anthropogenic impact by an increase in contaminated SPM input will be readily detected in areas with a high SPM loading for all the four elements. It is obvious that Pb and As concentrations higher than that values are indicative for anomalous element enrichment. For the elements Hg and Cd, however, it seems to be difficult to differentiate natural and anthropogenic enrichments simply from its total particulate element concentration at a low SPM loading.

The above brief considerations give a sufficient background for interpretation of the distributions of particulate Cd, Pb, As and Hg concentrations in the surficial North Sea waters as shown in Figs. 3a-d. In spite of the low density of sampling sites, certain areas of high element contents in the SPM samples indicate that the distributions are not even, and that mixing of suspended matter in the North Sea is far from being complete. A clear division is obvious in the composition of suspended matter between a northern part including the northern and central North Sea and most of the Skagerrak and Norwegian Trench, and a southern part including the coastal areas. Cd contents vary between 0.1 and > 14 µg/g (median 1.57 µg/g) which agree well with the ranges reported earlier (9,10). While higher Cd and Hg contents were found in the northern than in the southern part of the North Sea, quite an opposite distribution was found with As contents.

The reason for that inverse pattern for Cd and As must be seen mainly in the different SPM concentrations found in these areas (22,25). SPM concentrations vary considerably between

North Sea areas showing a drastic change from the shallow (15-30 m) Southern and Eastern Bights to the deeper (70-100 m) central North Sea. The central and northern North Sea are strongly influenced by the inflow from the North Atlantic with SPM concentrations of less than 0.1 mg/l because of the large water depth. In the shallow southern areas, wind- and tidal current-induced stresses on the floor resuspend bottom sediments into the water column (25). This resuspended matter determines significantly the composition of SPM. Run-off by the rivers and authigenic Fe/Mn-oxides precipitated from epibenthic fluxes are another significant source of inorganic components to SPM in the costal environments of the southern North Sea (26). In the deeper central and northern North Sea resuspension of bottom sediments much less affects the composition of the surface water suspended matter. Here organic matter derived from biological productivity is the major component of SPM (25). The Skagerrak, which receives suspended matter from the Baltic Sea in its surface waters, take up a separate position. The inverse distribution pattern of Cd and As can thus be understood from their predominance for different scavenging substrates as discussed above.

Though atmospheric inputs are suspected to be the major source of Cd into the North Sea, recent studies evidenced that Cd from contaminated aerosols is leached to nearly 100 % upon contact with seawater (27). Hence, enhanced particulate Cd and Hg concentrations in that areas may be the result of low particulate loads, biological scavenging processes, and remineralization of biogenic detritus at the sediment-water interface rather than simply being an input of contaminated particles. These processes lead to the rapid decrease of Cd enrichment in particles following the most likely pathway from waste combustion as a potential source to the North Sea sediments as the intermediate or final sink (8). The mixing zone of both nutrient-rich Atlantic and metal-enriched North Sea water may be seen therefore to represent a last "biogeochemical barrier" to prevent dissolved toxic metals such as Cd and Hg from escaping into the ocean.

Enhanced particulate Pb concentrations were found along coastal areas, especially in the western parts of the North Sea. The large inputs delivered by the estuaries (28) seem not to reach far beyond the coastal zone. Particle scavenging and sedimentation in the plumes is in fact very effective in diminishing the Pb impact to the marine environment (2). Atmospheric input, however, is suspected to be the major source of anthropogenic Pb in the North Sea (22). From atmospheric deposition modeling, a significant decrease of the total Pb deposition is found from land to sea, which is still in good agreement with the particulate Pb distribution in surface waters (Fig. 3a). High Pb contents have also been found in the northwestern North Sea. Since Pb, in contrast to Cd, is not predominantly enriched by organic matter but bound to the mineral and oxide components of SPM, other than natural sources might be responsible for the excess Pb load in these areas. Though about half of the Pb load is leached from aerosols upon contact with sea-water (27), in the shallower southern part of the open North Sea, this Pb fraction is scavenged by a higher particulate load which increases the removal of anthropogenic Pb by relatively short particle residence times.

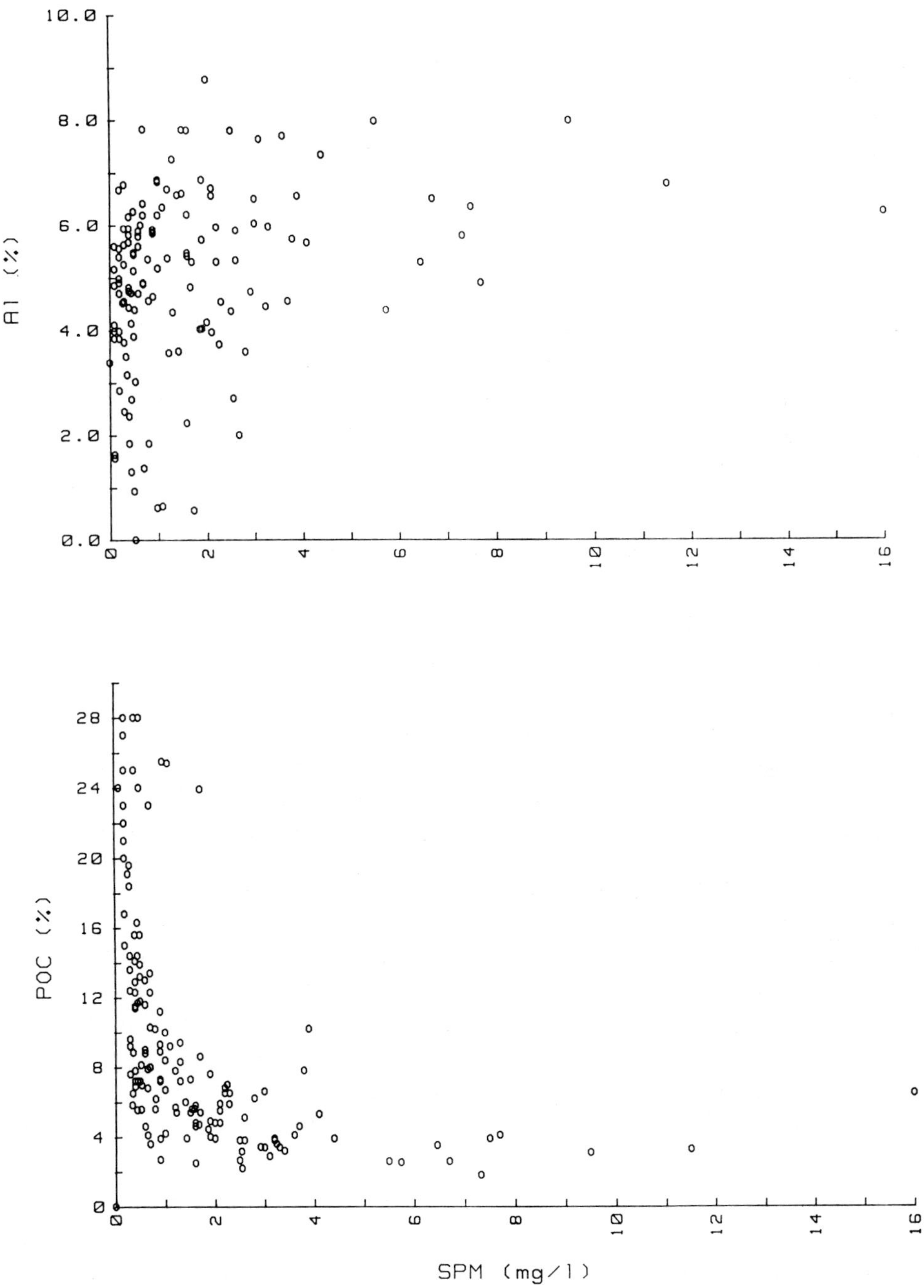

Fig. 4a-b. Diagrams showing the relationship between Al and POC contents in SPM (μg/g), and SPM concentrations (mg/l) in North Sea water.

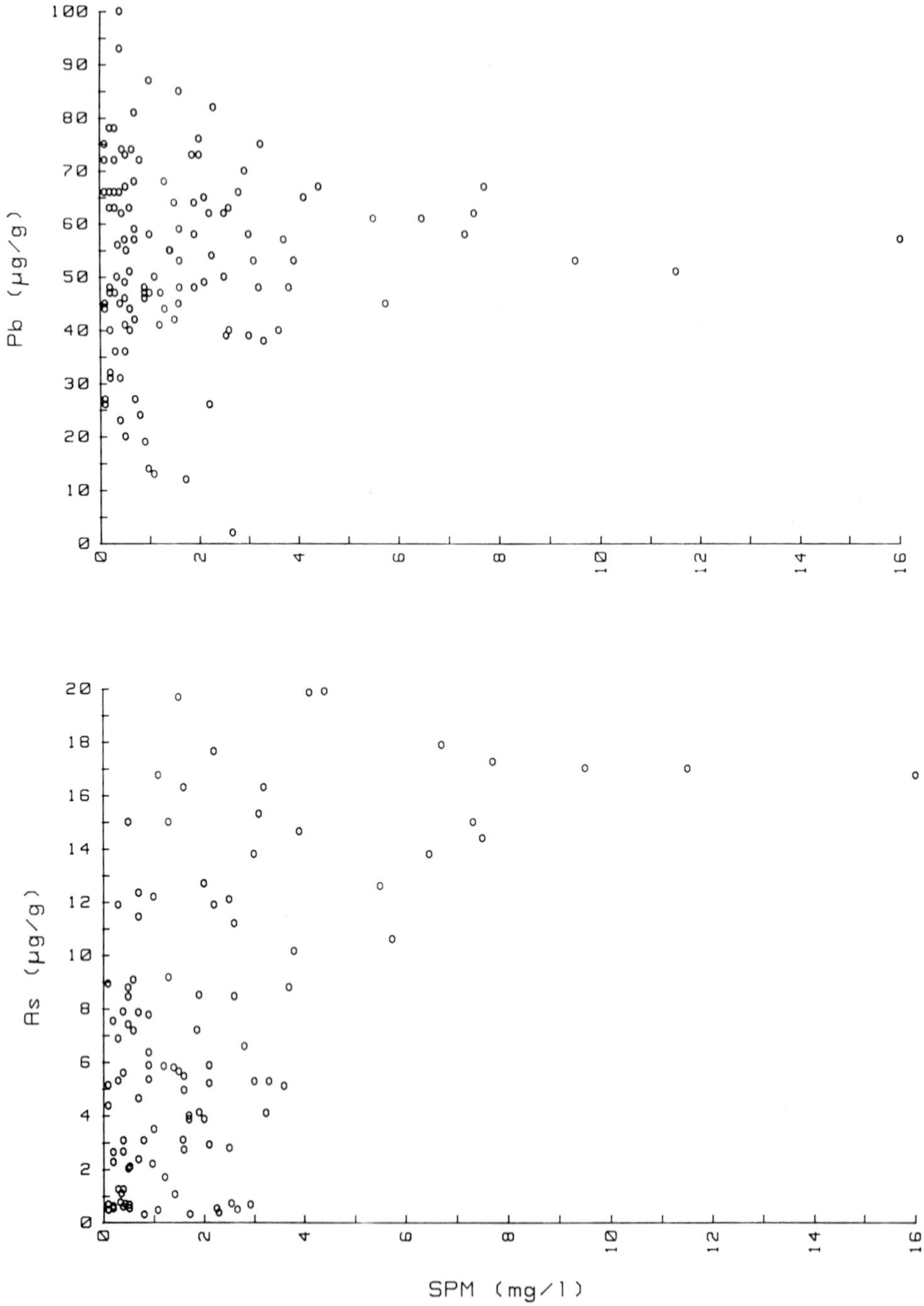

Fig. 4c,d. Diagrams showing the relationship beteen the Pb and As contents in SPM (μg/g), and SPM concentrations (mg/l) in North Sea water.

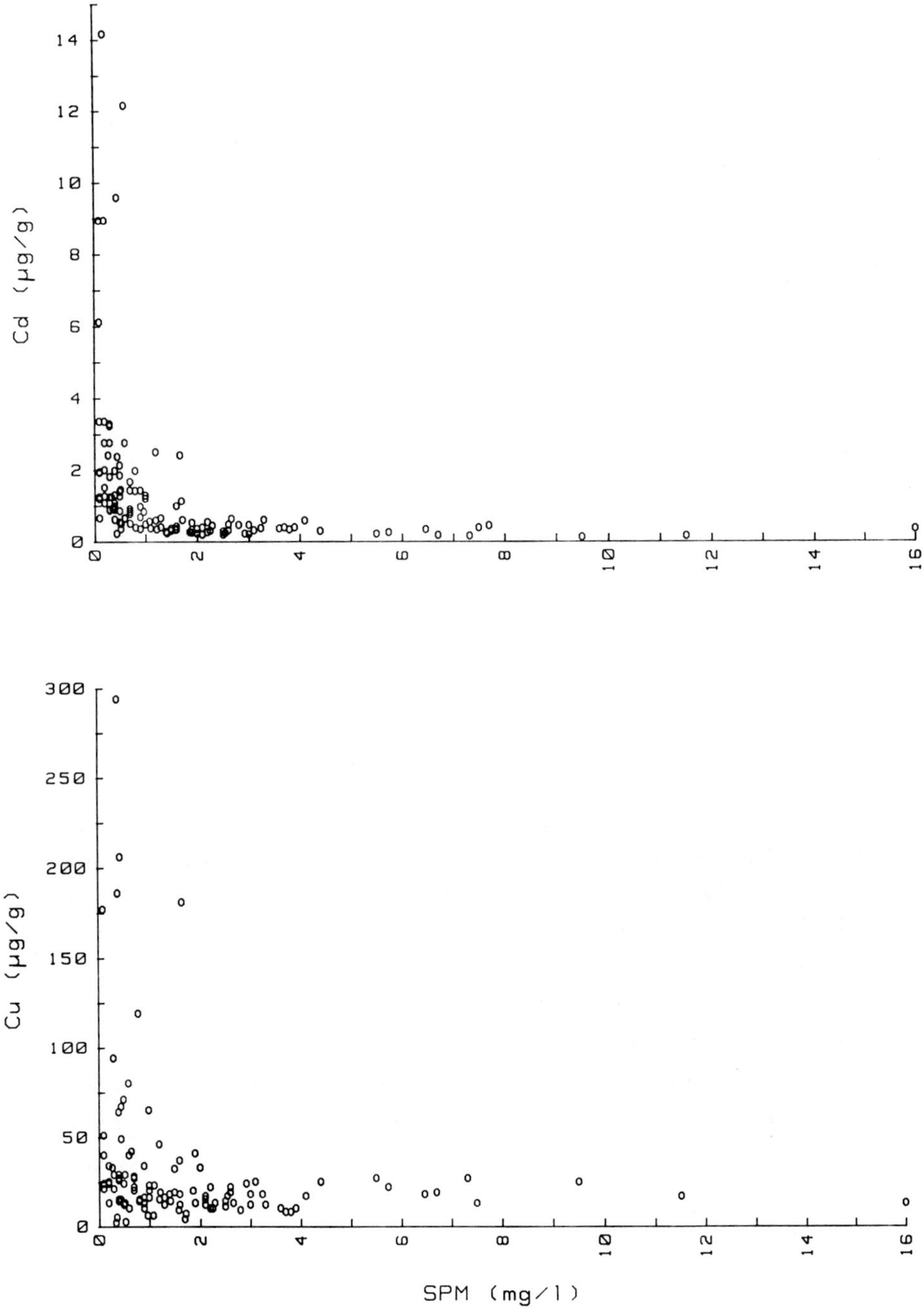

Fig. 4e,f. Diagrams showing the relationship beteen the Cd and Cu contents in SPM (µg/g), and SPM concentrations (mg/l) in North Sea water.

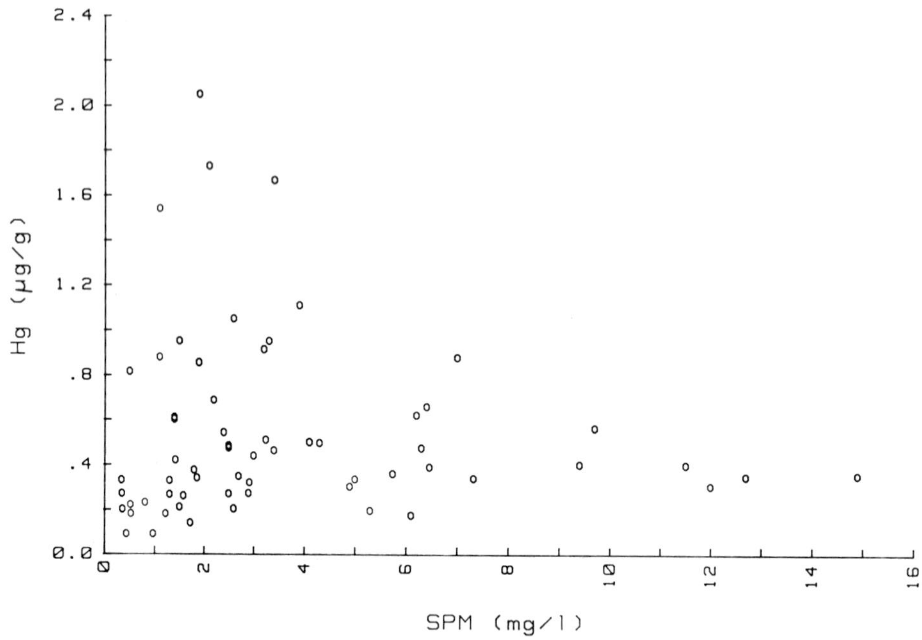

Fig. 4g. Relationship between Hg content in SPM, and SPM concentration in the North Sea.

The above considerations suggest that normalization of the SPM metal concentrations to Al contents for most of the trace metals considered would lead to an improvement in evaluation of the spatial variations of all elements. However, if the concentration of trace elements in SPM of the North Sea shelf waters results from the relative proportions of inorganic sediment particles and organic components as in case of Cd, a normalization method can be tried using both the Al and POC contents (29). The contribution of the inorganic particles can be calculated from the more or less constant metal-to-Al ratio (R_n) found at high SPM concentrations. If organic particles result from in-situ primary production they presumably would have a metal-to-POC ratio (R_c) similar to that of phytodetritus. This assumption is corroborated by the particulate C/P ratios which were similar to the Redfield ratio (22). Thus, if these two types of SPM compounds are the major contributors to the observed total SPM metal concentration, their contribution should be approximated by:

$$C_{obs} = (R_n \times Al_{obs}) + (R_c \times POC_{obs}) \qquad [1]$$

where C_{obs}, Al_{obs} and POC_{obs} are the measured concentrations of trace metals, Al and POC in the

SPM sample to be evaluated. For Pb and As, a more significant contribution of Al-bearing components can be expected ($R_n > R_c$), but quite the opposite for Cd, Cu and Hg ($R_n < R_c$). Unfortunately, no data on trace metal contents of phytoplankton exists for the open North Sea to prove this hypothesis.

3.3 Surface area/grain size/trace element relations

Mean grain sizes of the 21 selected samples were determined graphically from cumulative curves following the procedures of Müller (18). They fall into a quite narrow range (Fig. 5), while surface area shows a fairly wide range from as low as 1.9 to 15.3 m^2/g (mean 8.8 m^2/g, Table 1). Chemical analyses of the separated size fractions from each sample display the expected trends: as grain size decreases, trace element levels increase (Table 3). The rationale for this interrelation is discussed in the introduction. Correlation coefficients calculated for bulk sample chemistry vs. the percent <2 µm, <6.3 µm, and <20 µm size fractions are not as significant as expected, because of the relatively small variation in grain-size distribution between the SPM samples. Thus an improvement in intersample comparison and clarification of the significant spatial trends in element distributions was not obtained from grain-size fractionation.

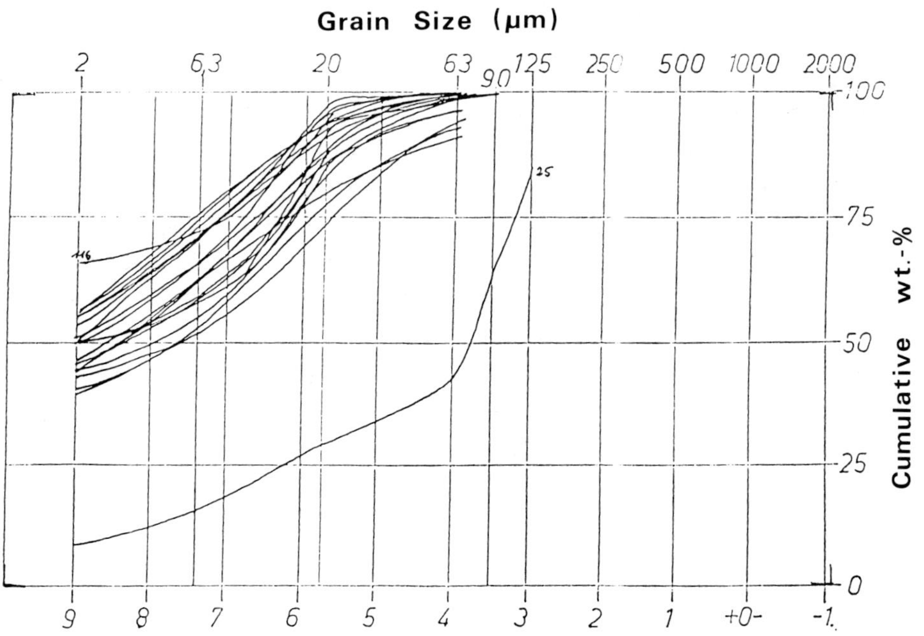

Fig. 5. Diagram showing cumulative grain-size curves for SPM from 21 North Sea transects.

Correlation coefficients calculated for bulk sample surface area vs. the various grain-size fractions are very strong, but do not increase with decreasing size as expected. In fact, the correlations between surface and the percent <20 µm, <6.3 µm and <2 µm size fractions are about the same (Table 2). This result seem to indicate that surface area is affected by more than just grain size of the individual particles. However, as claimed already in the methods section, the procedure used for surface area determination measures external surface area as the interlaminar spaces of many materials, including clay mineral aggregates. Though correlation coefficients calculated for bulk sample chemistry vs. bulk sample surface area are stronger than those for bulk sample chemistry vs. grain-size measurements (in fact, no significant correlation was found with the latter), the grain-size distribution found with the 21 SPM samples from the entire North Sea do not vary to such extent that a clearcut correlation could be drawn from the results. Moreover, a significant increase of trace metal content with increasing surface area (Table 3) was found only for the mineral-bound elements as discussed above. For POC, and the trace elements Cu and Cd bound predominantly to organic matter, correlations between surface area and sample chemistry were less significant.

TABLE 3

Mean trace metal contents of selected grain-size fractions of SPM samples from 21 North Sea transects

Element	<2µm	2-6.3µm	6.3-20µm
Zn	198	124	118
Cr	94	51	49
Ni	39	32	32
Co	23	14	12
Pb	96	41	36
Cu	32	19	19
Cd	0.28	0.18	0.17

3.4 Geochemical phase associations of trace metals in SPM

The above results suggest that for bulk trace metal concentrations SPM geochemical phase composition could have a more significant effect than any one single size fraction or surface area. The results of the various operationally defined geochemical phases investigated in the large-volume SPM samples are summarized in Table 4. The oxalate buffer agent should liberate metals bound in non- or partially crystalline iron and manganese oxide or hydrous oxide phases which should be present predominantly as coatings on particle surfaces or as small particulates. The fractions containing or related to these reducible Fe phases are in fact most important for the non-residual SPM-trace element (except Cd) chemistry. Since there is a positive correlation between this phase and surface area, it is not surprising to find also a positive correlation between surface area and bulk metal concentrations (Table 2). A strong negative correlation between bulk POC contents and the reducible Fe fraction is correborated by the strong negative correlation coefficients between bulk trace metal and POC contents (except Cd, for which the reducible fraction is of minor importance). A strong interrelation between amorphous Fe-oxides and their possible substrates (clay minerals) is corroborated by the strong correlation between bulk Al contents (which represent the amount of clay), surface area, and the reducible Fe fraction.

This result is also supported by previous work with size fractions of North Sea sediment. Förstner et al. (3) used in that work a five-step sequential extraction to operationally quantify carbonates, Mn oxide, amorphous Fe oxide and oranic matter in the various size fractions. They found that larger sediment particles (i.e. > 20 µm) are preferentially coated by (easily reducible) amorphous Fe oxihydroxides and Mn oxides. Moreover, the geochemical fractionation results indicate that there is no significant correlation between the reducible Fe fraction and the <2 µm size fraction. Reducible Fe fractions tend to be concentrated in the coarser size fractions, lending support to the hypothesis that Fe/Mn-oxide coatings are enriched in coarser fractions. The exact mechanism of this coarse-fraction enrichment is unclear; however, the data support the suggestion that particle size plays in fact a dual role in controlling the concentrations of trace metals in sediments and suspended matter. Increase in surface area/volume tend to increase concentrations, however, because coarser particles may spend more time at the sediment-water interface, where the chemical environment is conducive to the formation of these oxide coatings (26). Thus the metal concentrations in coarser SPM fractions are higher than predicted (Table 3).

TABLE 4

Geochemical fractions of Fe, Mn, Zn, Pb, Cu, and Cd in the large-volume SPM samples (mean ± S.D. in % of total extracted)

Fraction		Fe	Mn	Zn	Cu	Cd
exchangeable	(pH 7)	0.4±0.9	20.5±18.5	5.5± 5.4	5.1±7.1	30.6±14.6
carbonatic	(pH 5)	0.3±0.8	26.3± 5.0	18.2± 9.2	2.1±2.0	35.5± 7.0
reducible	(pH 3)	35.1±5.5	37.3±14.1	38.1± 6.0	36.4±3.8	13.1± 6.4
oxidizable	(pH 2)	< 0.1	2.0± 0.6	< 0.1	< 0.1	9.3± 7.0
residual	(HF/HNO_3)	64.1±5.6	13.9± 5.7	38.1±10.2	56.4±9.9	11.4± 5.6

The oxidizable organic fraction is for all investigated metals including Cd and Cu surprisingly of minor importance. According to the bulk chemistry data correlation analysis this fraction should be dominating at least for Cd. However, there are strong positive correlation coefficients between bulk POC contents and metal concentrations in the exchangeable fraction, but negative between the reducible Fe fraction and the exchangeable trace metal fractions, which indicate that organic matter rather than Fe oxihydroxides provides for the desorbable trace element binding sites (Table 2). In fact, the best correlation was found between exchangeable Fe and POC content with a coefficient of 0.98. However, the exchangeable fraction is only important for Cd binding (Table 4). Since there is a negative correlation between POC and surface area, it is not surprising to find no significant positive correlation between total Cd concentration and surface area.

These results add additional support to the view that the interrelation between surface area and the concentration and retention of particle-associated trace metals is more complex than predicted from the surface area/volume trends for the size classes of particles considered here. The sequential extraction data also re-emphasize the significance of amorphous Fe-oxides and organic substrates for trace metal accumulation. However, they indicate also that these high surface area geochemical phases are not concentrated in the finest grain-size fractions which explains the weak correlations between these size fractions and bulk metal chemistry. Horowitz and Elrick (6) have shown that the variable nature of these geochemical phases may be due to two factors: (1) overall concentration of the geochemical phase, and (2) its form or composition. Both organic matter and Fe oxides tend to exist in two major forms, as coatings on particles and as discrete particles. Though they represent relatively small proportions of the overall sample weight, coatings may cement small suspended particles together forming water-stable aggregates ("macroflocs") which increases their mean grain size and, thus, decrease its surface area. Removal of the "cement" breaks down the agglomerates to their original, smaller, individual component particles which have larger surface areas than their aggregates (25).

With respect to grain size/surface area interrelations, the geochemical phases discussed above can act therefore as either a surface area inhibitor (its removal produces a surface area increase) or a surface area contributor (its removal produces a surface area decrease). Organic matter (more common in the finest grain-size fractions of SPM) may act as a surface area inhibitor, while the high surface-area oxide phases (more common in the coarse-grained SPM fractions) act as contributors. This hypothesis would also explain the importance of the coarser size fractions, if significant amount of metal-accumulating agglomerates fall into the > 6.3 μm fraction.

4 CONCLUSIONS AND IMPLICATIONS

This study demonstrate the interrelation among, and the interdependence of, grain size, surface area and the composition of SPM, as well as their combined effect on particulate trace metal concentrations in the North Sea. The conclusions which can be drawn from this work with respect to pollution reconnaissance studies have implications beyond that case study.

4.1 Normalization of particulate metal concentrations to SPM grain size or surface area

A good deal of the literature concerned with grain size/trace element relations has dealt with selecting an appropriate size fraction or range for eliminating the grain-size effect, so that spatial or temporal trends of associated trace elements can be determined. However, as indicated above, the presence of water-stable coatings would tend to cast doubt on the use of other than bulk SPM chemistry. In fact, disaggregation of that coatings for grain-size determinations gives a false indication of grain-size distributions emphasizing the presence of fine-grained particles which do not exist as individual trace metal accumulating grains in the aquatic environment. Variable results may be due to differences in methods used to accomplish the grain-size separation. The more vigorous the separation, the less likely it is to retain aggregates. Trace metals released by sample pretreatment and chemical dispersants used in grain-size determinations would redistribute onto the newly exposed external surfaces of fine (clay) particles. Though normalization of particulate metal concentrations to grain size is a prerequisite for numerical quality criteria such as the "Geoaccumulation Index" (30), the difficulties associated with selecting an appropriate grain-size fraction for that purpose tend also to cast doubt on the value of that quality criteria for SPM monitoring.

If surface area were substituted for grain size, the problem of selecting the most appropriate size range to deal with the grain-size effect might be eliminated. In fact, surface area seems to be more important as grain size in affecting and controlling trace metal accumulation in SPM. It may be the parameter of choice when dealing with suspended matter samples having small masses because the BET method requires small sample aliquots, and it describes external surface which may not depend on the actual grain size distribution of the individual particles as indicated above. Surface area is especially in suspended matter affected by geochemical composition rather than by grain size. Here, the composition and concentration of surface coatings have a greater effect on surface area than grain size. However, coatings may change the trace metal sorption capacity within the same surface area range depending on the element considered. Most trace elements seem to be accumulated on reducible amorphous Fe-oxide coatings. If to evaluate only the spatial distributions of these elements, normalization to surface area may be helpful. However, toxic elements such as Cd are bound predominantly to organic matter which is less efficiently substituted by surface area due to its variable nature. Surface area seems therefore not be helpful to clarify both spatial and temporal trends of these elements.

4.2 Evaluation of the spatial particulate metal distribution by normalization with respect to SPM composition

Composition has the most significant effect on both grain size and surface area. It has also a significant effect on trace element levels. Of the various geochemical components investigated, those associated with reducible Fe-oxides appear to be the most important in affecting surface area and concentrations of the trace elements Pb, As and Hg. Organic matter is more important in controlling the trace element concentrations of Cd and Cu. The anomalies found in the interrelations between trace element concentrations and grain-size or surface area could be explained by compositional effects. In conclusion, it appears to be most suitable to use composition for substituting and normalizing any grain-size or surface area effect. However, it is not necessary to use an amorphous Fe-oxide extraction as has been suggested by Horowitz and Elrick (6). Although this study has similarly indicated that reducible Fe could be used as a normalizing factor in the same way as surface area, the strong correlation ($r = 0.97$) between amorphous Fe and bulk Al concentrations suggests that both elements can be used interchangeable. However, due to the divergent behaviour of such elements as As and Cd it is necessary also to include POC contents as suggested by equation [1]. Grain size, surface area and composition are so interrelated that this equation seems to be best suited to substitute for all these effects on spatial or temporal trace element concentration distributions. In fact, most of the particulate trace metal enrichments in the open North Sea can be attributed to extreme SPM composition characteristics. Areas with significant anthropogenic (i.e., anomalous) particulate trace metal enrichments in the open North Sea, which have not been already assimilated in the natural trace metal cycle, cannot be clearly recognized. This result is not surprising with regard to the high reactivity of metals in atmospheric aerosols (27), which is the only relevant anthopogenic impact in that area. Contamination of SPM by toxic elements like Hg is still confined to estuarine plumes such as that of the Elbe and Humber rivers.

5 ACKNOWLEDGEMENTS

This research was funded through grants from German Federal Ministry for Science and Technology. Dr. H. Albrecht, German Hydrographic Institute, Hamburg, provided the large-volume samples and POC data. Geochemical analysis and surface area measurement were carried out at the Central Laboratory for Analytical Chemistry, University of Technology, Hamburg, by Mrs. S. Koelling, Mrs. M. Rieck, and Mrs. H. Sonnenberg. The grain-size separation was carried out by Mrs. B. Krause. W. Kienz assisted in the preparation of some of the figures.

REFERENCES

1 K.K. Turekian, Geochim. Cosmochim. Acta 41 (1977) 1139-1144.
2 B.D. Honeyman and P.H. Santschi, Environ. Sci. Technol. 22 (1988) 862-871.
3 U. Förstner, W. Calmano and J. Schoer, Thalassia Jugosl. 18 (1982) 97-122.
4 M. Kersten and U. Förstner, in: G.E. Batley (Ed), Trace Element Speciation: Analytical Methods and Problems, CRC Press, Boca Raton, 1989, pp. 245-317.
5 M. Kersten and U. Förstner, Mitt. Geol.-Paläontol. Inst. Univ. Hamburg 58 (1985) 631-645.
6 A.J. Horowitz and K.A. Elrick, Appl. Geochem. 2 (1987) 437-451.
7 G. Irion and G. Müller, Proc. Int. Conf. Heavy Metals in the Environment, New Orleans, USA, September 1987, CEP Consultants, Edinburgh, 1987, pp. 38-41.
8 M. Kersten, M. Dicke, M. Kriews, K. Naumann, D. Schmidt, M. Schulz, M. Schwikowski and M. Steiger, in: W. Salomons, et al. (Eds), Pollution of the North Sea, Springer, Berlin, pp. 300-347.
9 L. Brügmann, Beitr. Meereskde. 55 (1986) 3-18.
10 R.F. Nolting and D. Eisma, Neth. J. Sea Res. 22 (1988) 219-236.
11 H. Etcheber and J.M. Jouanneau, Estuar. Coastal Shelf Sci. 11 (1980) 701-707.
12 H.A. van der Sloot and J. C. Duinker, Environ. Technol. Lett. 2 (1981) 511-520.
13 A.J. Horowitz, K.A. Elrick and R.C. Hooper, Hydrol. Processes 2 (1989) 163-184.

14 A. Dong, G.V. Simsiman and G. Chesters, Soil Sci. 139 (1985) 97-99.
15 S. Koelling, J. Kunze and C. Tauber, Fresenius Z. Anal. Chem. 332 (1988) 163-184.
16 D.H. Loring and R.T.T. Rantala, Mar. Chem. 24 (1988) 13-28.
17 K. Grasshoff, M. Erhardt and K. Kremling, Methods of Seawater Analysis, Verlag Chemie, Weinheim, 1983.
18 G. Müller, Methods in Sedimentary Petrology, Hafner, New York, 1967.
19 R.S. Mikhail and E. Robens, Microstructure and Thermal Analysis of Solid Surfaces, Wiley, Chichester, 1983.
20 D. Avnir, D. Farin and P. Pfeifer, J. Chem. Phys. 79 (1983) 3566-3571.
21 D. Robbe, Proc. Int. Conf. on Heavy Metals in the Environment, Vol. 1, New Orleans, September 1987, CEP Consultants, Edinburgh, pp. 360-362.
22 M. Kersten, W. Kienz, S. Koelling, M. Schröder and U. Förstner, Vom Wasser 75 (1990) 245-272.
23 J.C. Duinker, Spec. Publ. Int. Ass. Sed. 5 (1981) 451-459.
24 U. Förstner, Chemiker Z. 105 (1981) 165-174.
25 D. Eisma and J. Kalf, Neth. J. Sea Res. 21 (1987) 265-285.
26 F. Dehairs, W. Baeyens and D. Van Gansbeke, Estuar. Coastal Shelf Sci. 29 (1989) 457-471.
27 M. Kersten, M. Kriews and U. Förstner, Mar. Chem. (submitted in 1991).
28 R. Hupkes, Pollution of the North Sea imposed by West European Rivers, ICWS-Report 90.03, Amsterdam, 1990.
29 H.L. Windom, R.G. Smith and C. Rawlinson, Mar. Chem. 27 (1989) 283-297.
30 G. Müller, Umschau 79 (1979) 778-783.

Chapter 5
Bio-accumulation

BIOLOGICAL AVAILABILITY OF METALS IN SEDIMENTS: ANALYTICAL APPROACHES

Peter G.C. Campbell and Andre Tessier

1. INTRODUCTION

Freshwater sediments act as an important reservoir for metals of anthropogenic origin. The fluxes of metals to sediments may be "gravitational", i.e. settling of particulate material with its associated metals (1-2), or "diffusive", in those cases where dissolved metal concentrations in the water column exceed those in the sediment interstitial waters, notably in acid lakes (3-4). In both cases the overall result is a marked enhancement of metal concentrations in the upper strata of lake sediments (2) - concentrations can attain levels 1000 - 5000 times higher (on a $\mu g/g$ fresh weight basis) than those in the overlying water column.

The contamination of surficial sediments by metals of anthropogenic origin is of potential ecological importance. In effect, the sediment-water interface serves as the habitat for an often abundant and diverse community (5) that can include epipelic microalgae, rooted aquatic plants, bacteria, detritivorous, filter-feeding and predatory invertebrates, and bottom-dwelling fish. These benthic organisms have been shown to exert a great influence on lake metabolism (6). In addition, toxic effects on benthic organisms could alter benthic community structure and thus in turn affect other organisms, e.g. predatory fish.

Given the superposition of high metal concentrations and an abundance of benthic organisms, an obvious question arises: to what extent are these sediment-bound metals bioavailable? This question is not simply of academic interest - the response will have important management implications. In designing programmes to restore polluted environments, resource managers are almost invariably confronted with the problem of "contaminated sediments" - as defined by geochemists. Before deciding on costly remedial measures (e.g.: dredging, capping, ploughing, *in situ* treatment) managers would clearly like to know whether the elevated metal levels in the offending sediments are indeed causing deleterious effects on the benthic community. Note in this context that controlled laboratory feeding studies (often involving metal-doped synthetic sediments) have indicated that the bioavailability of a given metal may differ by as much as 1000-fold among different sediment types (7-8). Clearly it would be of considerable interest to be able to determine metal partitioning in a natural sediment, and then relate this partitioning to metal bioavailability.

This chapter deals with the partitioning of metals in sediments and the possible links between this partitioning and the bioavailability of the metals. Emphasis has been placed on oxic sediments in freshwater environments, i.e. those at the sediment-water or sediment-root interface and thus of biological importance to rooted aquatic plants and to deposit- and suspension-feeders. This is the most important zone of contact between the detritus-based food web and metals associated with bottom sediments (9). Treatment of this complex subject will necessarily be brief; interested readers are referred to several recent reviews for greater detail (10-12).

2. PARTITIONING OF METALS IN SEDIMENTS

Metals associated with freshwater sediments may occur in dissolved or colloidal form in the sediment interstitial water, as constituent elements present in the essentially insoluble products of physical weathering (i.e. in detrital or residual minerals) or in a variety of secondary forms:

e.g.
- adsorbed at particle surfaces (Fe/Mn oxides; clays; humic flocs)
- present in lattice positions in secondary minerals or occluded in amorphous Fe/Mn oxides
- associated with organic matter (living micro-organisms; freshly deposited detritus; refractory organic material).

In the present paper, we emphasize these secondary forms and the metals in the sediment interstitial water, i.e., those that are more reactive and thus more likely to be bioavailable.

Current approaches available for determining metal partitioning in sediments are summarized in Table 1. **Direct determination** of metal-solid associations in a sediment matrix has rarely proved feasible. The great variety of possible metal forms, their tendency to exist in complex organic-mineral associations,

TABLE 1
Possible approaches for determining metal partitioning in aquatic sediments.

Approach	Principle	Assumptions	Examples
Direct Determination	Identify individual sediment components according to their known properties (e.g. by X-ray diffraction)	-	(15)
Mathematical Modelling	Calculate equilibrium distribution of metal among various sorbents	Sorptive equilibrium attained; discrete binding constants; independent behaviour of sorbents	(4) (16) (17) (18) (19)
Partial Extraction	Extract specific metal forms from sediment matrix	Selective partial extraction feasible; negligible readsorption of solubilized metals	(10) (20) (21) (22)

the amorphous nature of the most important potential sinks (e.g. Fe/Mn oxides; humic acids), and the relatively low metal concentrations involved (normally < 1000 $\mu g/g$) have all conspired to limit this approach. A possible means of circumventing this problem has recently been tested in the laboratory and in the field (13-14). When inert (Teflon^R) sheets are inserted vertically in lake sediments and left there, diagenetic Fe and Mn oxides are deposited in distinct bands at the oxic/anoxic interface (Fig. 1). Careful retrieval of the sheets affords a virtually unperturbed sample of the Fe or Mn oxide, which can then be characterized with conventional analytical instruments and techniques.

Equilibrium partitioning models are based on the premise that metals attain sorption equilibrium between the sediment interstitial water and the various solid surfaces available for sorption. These models are analogous to those that have been used with success to calculate the speciation of metals in solution. Necessary input data for this approach include estimates of the sorbed metal concentrations, the binding intensities and binding capacities for each sediment sorbent, and the relative abundance of each sorbent. Binding capacities and intensities are available

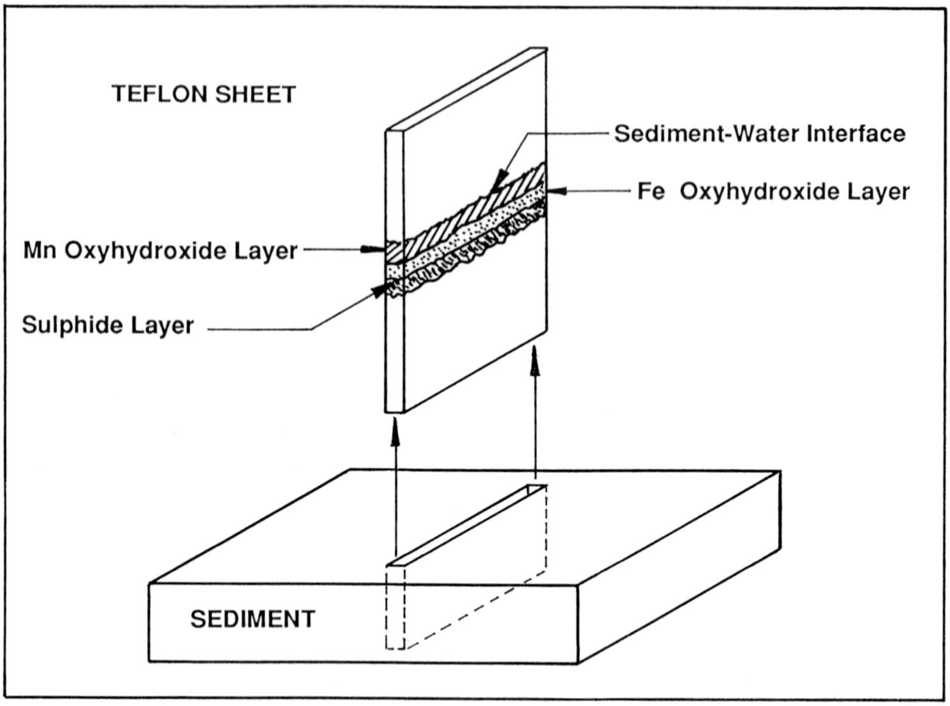

Fig. 1: Use of Teflon^{R} sheets to collect diagenetic Fe and Mn oxides in distinct bands at the oxic/anoxic interface - see (14) for details.

for individual sediment components, as studied in the laboratory, but only recently have attempts been made to extend this approach to complex multi-component natural systems (4,17-19). This approach shows considerable promise for estimating metal partitioning in sediments, but applications are currently limited by the virtual absence of field-derived data, and by difficulties in experimentally verifying the model predictions (cf. the limitations of the extraction schemes, as mentioned below) (10).

Partial extraction procedures offer a third approach for determining metal partitioning in sediments. Designed for use with oxic sediments, these methods make use of relatively mild reagents selected to attack particular non-detrital solid phases. These reagents may be used singly or in sequence (weak -> strong) to evaluate the distribution of metals among the various fractions described above and/or to determine the concentrations of various sorbing phases (9).

The uses and limits of the partial extraction approach have been reviewed recently (10,11,23). Potential difficulties are conveniently grouped in four categories:

- sensitivity to procedural variables
- post-extraction readsorption
- selectivity
- sample preservation.

(i) Sensitivity to procedural variables. Metal partitioning as determined by partial extractions is sensitive to such factors as the extraction sequence, the choice of the reagents used for the various extraction steps, the contact time between the sediment and the extractant, and the sediment : extractant ratio (10,23). It follows that the observed distribution of a metal among various fractions does not necessarily reflect the relative contributions of discrete sediment phases, but rather should be considered as operationally defined by the methods of extraction.

(ii) Post-extraction readsorption. After extraction (and destruction of certain easily solubilized phases), partial readsorption of the initially liberated metals might be anticipated. If extensive readsorption did occur, then the partitioning observed would be determined by the concentration and type of surface sites available after partial extraction, and it would bear little or no relationship to the metal partitioning that existed in the original sediment. Indeed, several authors have suggested that such post-extraction readsorption may render meaningless the results of partial extraction schemes (24-27). Recent work (28) suggests however that the problem may be less important than previously feared.

Oxic lake sediments were subjected to a five-step sequential extraction procedure involving the use of neutral $MgCl_2$ (F1), an acetic acid/Na acetate pH 5 buffer (F2), hydroxylamine hydrochloride (F3A,B) and finally an acidic solution of hydrogen peroxide (F4). The importance of post-extraction readsorption of As, Cd, Cu, Ni, Pb and Zn was evaluated by measuring the recovery of small amounts of trace elements added each extraction step. For the five extraction steps and six elements studied, significant losses ($P < 0.05$) of the added metal were noted in only one case; in extractions with the pH 5 acetate buffer (i.e., F2), > 90% of the spiked lead was lost, presumably by sorption onto the residual sediment. For all other cases the spikes were recovered within the limits given by the analytical precision (28).

(iii) Selectivity. The reagents available for the extraction of sediments (e.g., concentrated inert electrolytes; weak acids; reducing agents; oxidizing agents; chelating agents; strong alkalies; strong mineral acids) are inherently unselective - any sequential extraction procedure will unavoidably suffer from a certain lack of selectivity. Specific limitations of the extraction approaches currently in use include the possible coextraction of specifically adsorbed and carbonate-bound metals, as well as the coextraction of sulphide-bound and organic-bound metals. Despite these limitations, tests to date (20,21,29) suggest that selectivity is reasonably satisfactory for low-sulphide (oxic) sediments. For sulphide-rich (anoxic) sediments, however, several lines of evidence indicate that the various extraction procedures currently in use are compromised by the tendency of metal-containing sulphide phases to be leached progressively rather than selectively from the host sediment (22,29). The development of partial extraction procedures for anoxic sediments would be a fruitful area for future research, given the (paleolimnological) interest in determining metal partitioning in the lower (anoxic) strata of sediment cores (2).

(iv) Sample preservation. Unavoidable delays between sediment collection and analysis usually necessitate some form of sample storage and preservation. Various preservation techniques have been examined (22); wet storage (1-4 ^{o}C); freezing; freeze-drying; drying under a nitrogen stream (20 ^{o}C); drying under air in a convection oven (105 ^{o}C). No storage method tested completely preserved the metal distribution patterns observed in the fresh sediments. Drying of the sediment (air-drying; N_2-drying; freeze-drying) proved particularly harmful and clearly should be avoided at all costs - acceptable preservation techniques include wet storage (short-term, dark, 1-4 ^{o}C) or freezing.

3. GEOCHEMICAL ESTIMATES OF METAL BIOAVAILABILITY

How can the analytical approaches described above be used to estimate metal bioavailability in sediments? A general framework has been developed to relate metal partitioning in sediments to metal bioavailability (10-11), both for organisms living in intimate contact with the sediments but incapable of ingesting particulate material (Type A: benthic algae; rooted aquatic plants) and for organisms capable of ingesting sediments and taking up metals from the particulate phases (Type B: benthic invertebrates).

3.1 General considerations

For Type A organisms, the obvious geochemical phases to consider are the sediment interstitial water and those solid phases with bound metals in equilibrium with the interstitial water. The definition of "interstitial" water will differ among benthic species and may include, in different cases (10):

- anoxic waters for organisms in contact with deep sediments;
- oxic waters towards the sediment surface for organisms living within the sediment, but whose primary environmental contact is above the redox interface;
- waters at the sediment-water interface for suspension feeders.

For most metals, the key factor influencing metal bioavailability in the interstitial water will be the free aquo-ion concentration, $[M^{z+}]$, a difficult variable to measure. An alternative to the direct measurement of $[M^{z+}]$ is the determination of the concentrations of metals associated with a particular sediment constituent, $\{S_n\text{-OM}\}$, together with the concentrations of the solid phase, $\{S_n\text{-OH}_x\}$, responsible for binding the metal in this form. If it is assumed that the free metal ion concentration is controlled by adsorption (4), then the following relation can be derived (11,31):

$$[M^{z+}] = K * \frac{\{S_n\text{-OM}\}\ [H^+]^x}{\{S_n\text{-OH}_x\}} \qquad [1]$$

In this approach, the expression $\{S_n\text{-OM}\} * [H^+]^x / \{S_n\text{-OH}_x\}$ represents a surrogate measure of $[M^{z+}]$ and should thus be a useful predictor of metal bioaccumulation and toxicity.

The reasoning described above also applies to Type B organisms, but to these considerations must be added an evaluation of the possible uptake of metals during the digestion of the sediments in the gut (10). Since even in the gut metals must normally pass into solution before they are assimilated, the obvious geochemical phases to be considered are those that are likely to be affected by passage through the digestive tract. Digestive chemistry thus assumes considerable importance. In addition, given the known tendency of Type B organisms to be selective in their feeding/ingestion habits, sediment particle size must also be considered.

This combined theoretical construct for Type A and B organisms provides a specific basis for the development of relevant analytical approaches to studying sediments. Examples are given below.

3.2 Type A (macrophytes)

Despite the known capacity of rooted aquatic plants to concentrate various metals from their immediate environment (concentration factors as high as 10^3 to 10^4 have been observed with respect to the ambient water), relatively few studies have considered the **prediction** of metal accumulation by aquatic macrophytes (10). If metal uptake from sediment interstitial water were important, one might anticipate correlations between the free metal ion concentration in the interstitial water ($[M^{z+}]$ or its surrogate) and metal levels in the underground parts of the plants (for submerged rooted species in a field setting, the least ambiguous indicator of metal bioavailability in the sediments will be the metal concentrations in the underground parts of the plants - (31)). To test this hypothesis, surficial sediment samples and five species of rooted aquatic plants were collected at sites located along a gradient of decreasing metal contamination, downstream from a mining and smelting complex at Rouyn-Noranda in northwestern Québec, Canada. Five plants were collected at each site and dissected; the underground parts were cleaned, dried, digested (HNO_3/H_2O_2) and analyzed for Cu, Pb and Zn. The sediment samples from the same sites were wet-sieved (<850 μm) and subjected to a sequential extraction procedure to determine metal partitioning (20,32).

Regression analysis was used to examine the relationship between metal concentrations in the plant roots/rhizome ($[M]_r$) and in various metal fractions extracted from the sediment. Since the pH varied little over the limited study region, $[H^+]^x$ was dropped from equation [1] and relationships were sought between the metal levels in the plant (dependent variable) and the ratio {Fe-OM} / {Fe-OH_x} in the host sediment. The usefulness of this ratio as a predictor varied among metals (Cu > Pb, Zn) and among plants (*Potamogeton richardsonii* > *Eleocharis smallii*, *Nuphar variegatum*, *Sparganium americanum* > *Glyceria borealis*) - see Fig. 2 and Table 2. In this connection, it should be recalled that two pathways exist for metal accumulation in the underground parts of rooted aquatic plants: direct uptake from the sediment interstitial water; foliar uptake from the water column followed by basipetal translocation from the leaves and stems to the rhizome and/or roots (33). The absence of a relation between $[M]_r$ and the ratio {S_n-OM} / {S_n-OH_x} for certain plants (e.g., *G. borealis*) or a particular metal may well reflect an important foliar contribution to metal bioaccumulation. These preliminary results suggest that the pondweed *P. richardsonii* may prove useful as a biomonitor species for sediment-associated metals. To confirm the potential

TABLE 2
Prediction of metal levels in the underground parts of five species of rooted aquatic plants on the basis of the ratio {Fe-OM/FeOH_x} in the host sediments.

Plant	Correlation coefficient (r)[a] Copper		Lead		Zinc	
Potamogeton richardsonii	0.85	(10)**	0.79	(10)**	0.94	(10)**
Sparganium americanum	0.92	(6)**	(0.14)	(7)	0.91	(6)**
Nuphar variegatum	0.74	(12)**	(0.52)	(5)	(-0.30)	(13)
Eleocharis smallii	0.81	(7)*	0.80	(7)*	(0.01)	(6)
Glyceria borealis	(-0.02)	(9)	(-0.13)	(9)	(-0.11)	(9)

[a]An asterisk (*) indicates $P < 0.05$; (**) indicates $P < 0.01$. The values in parentheses correspond to the number of field sites.

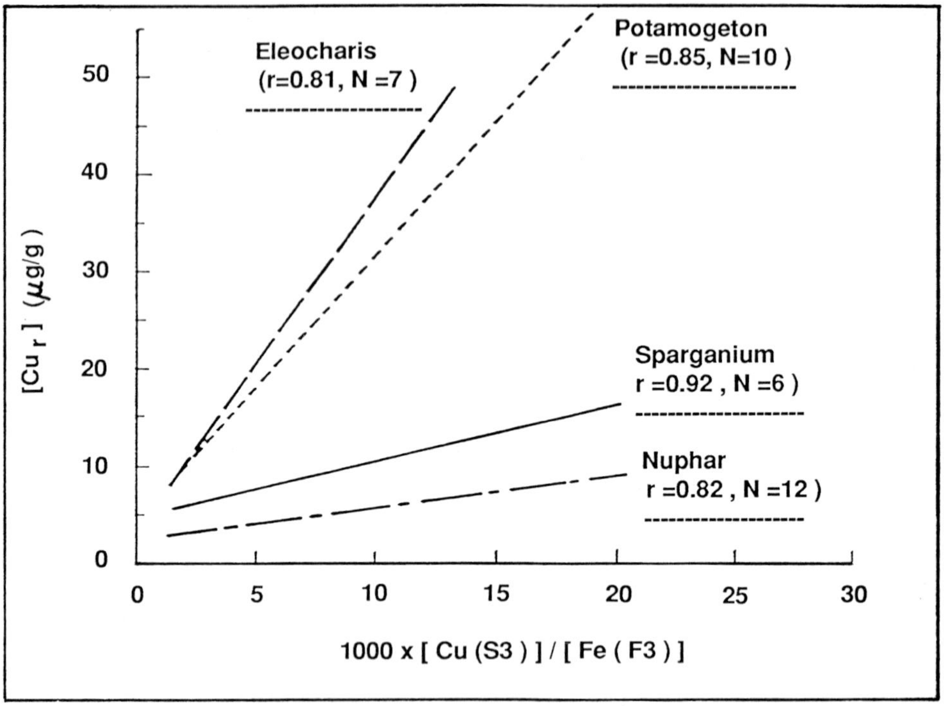

Fig. 2: Relationships between the concentration of copper, $[Cu]_r$, in the underground parts of various rooted aquatic plants and the ratio [Cu(S3)]/[Fe(S3)] in the host sediments. Note that [Cu(S3)] corresponds to the Cu concentration removed from the sediment by sequential extraction with $MgCl_2$, NaOAc/HOAc (pH 5) and $NH_2OH{\cdot}HCl$, whereas [Fe(F3) corresponds to the amorphous iron oxide fraction extracted by $NH_2OH{\cdot}HCl$. As explained in the text, the ratio [Cu(S3)]/[Fe(F3)] is taken as a surrogate measure of the free Cu^{2+} concentration in the sediment interstitial water.

of this species and the usefulness of the expression $\{S_n\text{-}OM\} * [H^+]^x / \{S_n\text{-}OH_x\}$ as a surrogate for $[M^{z+}]$, it will however be necessary to extend the study to a wider geological setting.

3.3 Type B (macroinvertebrates)

Benthic invertebrates are directly exposed to sediment-bound metals and are capable of accumulating metals from interstitial waters and/or from ingested sediment (8,10). If water is the more important exposure vector, then one would anticipate correlations between the free-metal ion concentration in the interstitial water ($[M^{z+}]$ or its surrogate) and metal levels in the organisms ($[M]_{benthos}$). This hypothesis is currently being tested for filter- and deposit-feeding invertebrates; preliminary results for filter-feeders, organisms whose nutritional strategy exposes them to large quantities of water from above the sediment-water interface, are discussed briefly below.

The experimental approach involves field studies at lacustrine sites located along spatial geochemical gradients (pH, [M]). Measured variables include metal body burdens in indigenous clams (Elliptio complanata; Anodonta grandis), ambient pH, metals in adsorptive equilibrium with the sediment constituents $\{S_n\text{-OM}\}$, together with the concentrations of the solid phase(s), $\{S_n\text{-OH}\}$, responsible for binding the metal in these forms. Regression analysis is used to examine the relationship between metal concentrations in the clams (= $[M]_{benthos}$) and the ratio $\{Fe\text{-OM}\} * [H^+]^x / \{Fe\text{-OH}_x\}$ in the host sediment. Preliminary results for M = Cd and E. complanata are encouraging (Table 3).

TABLE 3
Prediction of Cd accumulation by the freshwater clam Elliptio complanata (30).

Sediment geochemical variable	N	Correlation
Total Cd	13	0.43
[Fe-OCd]	13	0.37
$[Fe\text{-OCd}]/[Fe\text{-OH}_x]$	13	0.46
$[Fe\text{-OCd}] * [H^+]^{1.2}/[Fe\text{-OH}_x]$	13	0.82

4. CONCLUSIONS

Trace metals of anthropogenic origin tend to accumulate in lacustrine sediments, where they pose a potential danger to benthic organisms - both those that ingest sediments (Type B) and those that simply live in close association with the sediments (Type A). Despite the complexity of the environmental and biological factors involved, results obtained to date suggest that the coupling of metal partitioning models with relatively simple geochemical measurements will provide useful predictions of metal concentrations in sediment-dwelling organisms.

ACKNOWLEDGEMENTS

The overall approach to the determination of biologically available metals in sediments, as presented in this paper, has been developed in collaboration with our colleagues J.C. Auclair, R. Carignan and R.L. Hare at INRS-Eau. In addition, many of the concepts were refined while one of us (PGCC) worked on a review document for the National Research Council of Canada (10). Financial support from the Natural Sciences and Engineering Research Council of Canada, from the Wildlife Toxicology Fund, and from the Québec Fonds pour la Formation de Chercheurs et l'Aide à la Recherche is gratefully acknowledged.

REFERENCES

1 L. Sigg, in: W. Stumm (Ed.), Chemical Processes in Lakes, J. Wiley & Sons, New York, NY, 1985, pp. 283-310.
2 E.A. Livett, Adv. Ecol. Res., 18 (1988) 65-177.
3 R. Carignan and A. Tessier, Science, 228 (1985) 1524-1526.
4 A. Tessier, R. Carignan, B. Dubreuil and F. Rapin, Geochim. Cosmochim. Acta, 53 (1989) 1511-1522.
5 R. Wetzel, Limnology, W.B. Saunders, Philadelphia, PA, 1983, Chapters 15 and 16.
6 D.W. Schindler, in: W. Stumm (Ed.), Chemical Processes in Lakes, J. Wiley & Sons, New York, NY, 1985, pp. 225-250.
7 S.N. Luoma and E.A. Jenne, in: R.E. Wildung and H. Drucker (Eds), Biological Implications of Metals in the Environment, ERDA Symp.Ser., 42 (1977) 213-230.
8 S.N. Luoma, S.N., Sci. Tot. Environ., 28 (1983) 1-22.
9 S.N. Luoma and J.A. Davis, Marine Chem., 12 (1983) 159-181.
10 P.G.C. Campbell, A.G. Lewis, P.M. Chapman, A.A. Crowder, W.K. Fletcher, B. Imber, S.N. Luoma, P.M. Stokes and M. Winfrey, Biologically Available Metals in Sediments, National Research Council Canada, NRCC Report No. 27694, 1988, 298 p.
11 P.G.C. Campbell and A. Tessier, in: A. Boudou and R.Ribeyre (Eds), Aquatic Ecotoxicology, CRC Press, Boca Raton, FL, 1989, pp. 125-148.
12 S.N. Luoma, Hydrobiologia, 176 (1989) 379-396.
13 R.R. DeVitre, N. Belzile, G.G. Leppard and A. Tessier, in: Proc. Int. Conf. Heavy Metals in the Environment, Geneva, CEP Consultants Ltd., Edinburgh, U.K., 1987, pp. 217-220.
14 N. Belzile, R.R. DeVitre and A. Tessier, Nature, 340 (1989) 376-377.
15 E.S. Pilkington and L.J. Warren, Environ. Sci. Technol., 14 (1979) 295-299.
16 E.A. Jenne, D.M. Ditoro, H.E. Allen and C.S. Zarba, in: Proc. Int. Conf. Chemicals in the Environment, Lisbon, J.N. Lester, R. Perry and R.M. Sterritt (Eds), CEP Consultants Ltd., Edinburgh, U.K., 1986, pp. 560-568.
17 S.N. Luoma, Marine Chem., 20 (1986) 45-59.
18 A. Tessier, F. Rapin and R. Carignan, Geochim. Cosmochim. Acta, 49 (1985) 183-194.
19 S.M. Oakley, P.O. Nelson and K.J. Wilkinson, Environ. Sci. Technol., 15 (1981) 474-480.
20 A. Tessier, P.G.C. Campbell and M. Bisson, Anal. Chem., 51 (1979) 844-851.

21 A. Tessier, P.G.C. Campbell and M. Bisson, J. Geochem. Explor., 16 (1982) 77-104.
22 F. Rapin, A. Tessier, P.G.C. Campbell and R. Carignan, Environ. Sci. Technol., 20 (1986) 836-840.
23 U. Förstner, Hydrobiologia, 91 (1982) 269-84.
24 P.S. Rendell, G.E. Batley and A.J. Cameron, Environ. Sci. Technol., 14 (1980) 314-318.
25 C. Kheboian and C.F. Bauer, Anal.Chem., 59 (1987) 1417-23.
26 P.M.V. Nirel and F.M.M. Morel, Water Res., 24 (1990) 1055-1056.
27 A. Tessier and P.G.C. Campbell, Water Res., 25 (1991) 115-117.
28 N. Belzile, P. Lecomte and A. Tessier, Environ. Sci. Technol., 23 (1989) 1015-1020.
29 F. Rapin and U. Förstner, in: Proc. Int. Conf. Heavy Metals in the Environment, Heidelberg, CEP Consultants Ltd., Edinburgh, U.K., 1983. pp. 1074-1077.
30 A. Tessier and P.G.C. Campbell, in: Metal Speciation in the Environment, S. Guçer, F.F. Adams, E. Izdar and D. Klockow (Eds), Springer-Verlag, Berlin, 1991 (in press).
31 H.H. Schierup and V.J. Larsen, Aquat. Bot., 11 (1981) 197-210.
32 P.G.C. Campbell, A. Tessier, M. Bisson and R. Bougie, Can. J. Fish. Aquat. Sci., 42 (1985) 23-32.
33 P. Denny, Biol.Rev., 55 (1980) 65-92.

SYNTHESIS OF MERCURY CONTAMINATION MECHANISMS OF A BURROWING MAYFLY *(HEXAGENIA RIGIDA)*: METHODOLOGICAL BASES AND PRINCIPAL RESULTS

E. Saouter, F. Ribeyre and A. Boudou

SUMMARY

Experimental studies using a three compartment ecotoxicological model "water, natural sediment and *Hexagenia rigida* nymphs" revealed very high bioaccumulation capacities of this burrowing mayfly for two mercury compounds, $HgCl_2$ and CH_3HgCl. Contamination processes are strongly dependent on exposure conditions : contamination sources, Hg concentrations in the biotopes, length of exposure, etc.. By studying the metal organotropism between two principal target organs - gills and gut -, it was possible to link specific contamination routes (direct or trophic) with the chemical forms of mercury and the initial contamination sources (water column or sediment). Mercury partitioning in the water (dissolved and particulate phases) and in the sediment (interstitial water and geochemical phases) was studied, and transformations of the two Hg chemical forms were quantified. All these experimental approaches, taking into account the three fundamental poles of Ecotoxicology (abiotic, biotic and contamination factors) led to a better understanding of the range and kinetics of the contamination mechanisms for this intra-sedimentary species.

1. INTRODUCTION

Studies on the contamination of freshwater systems by trace metals clearly show the importance of sediments as reservoirs for storage and chemical transformations of these elements and also as a secondary contamination source, through release processes (1-3).

Metal fluxes to sediments are strongly dependent on inputs of anthropogenic origin. They result from gravitational and diffusive processes : the former generally predominate, owing to the strong affinity of metals to suspended matter ; the latter are the result of transfers from dissolved metal species in the aquatic phase (4-5). Metal partitioning in sediments results from actions and interactions of a great number of factors : physico-chemical properties of the metal considered, contamination modalities, relative importance of sediment sorbents, abiotic parameters, etc.. Many factors are able to induce, either directly or indirectly, metal transfers between the upper layers of sediments and the water column and/or the trophic networks. Among these factors are mechanical stirring (currents, dredging, navigation,...), bioturbations (burrowing species) and changes in the physico-chemical characteristics of biotopes (e.g. lake acidification) (6). In the same way, abundance and diversity of benthic and intra-sedimentary communities - bacteria, microalgae, rooted macrophytes, annelids, molluscs, insects,... - represent a potential trophic contamination source for the aquatic networks and several terrestrial predators, such as piscivorous birds or man. In all cases, metal bioavailability is closely dependent on partitioning in the sediments and also on chemical transformations occuring in this compartment.

Our research objectives are based on an experimental approach to mercury bioaccumulation, transfers and chemical transformations within freshwater systems. All the experiments take into account the three fundamentals of ecotoxicology : abiotic, biotic and contamination factors. This approach is based on devising and using multispecies and interactive ecotoxicological models (water, natural sediments and species representative of different trophic levels) (7-8). According to the criteria of complexity and representativity in relation to natural contaminated systems, these models represent, from a methodological point of view, an intermediate position between field studies on the one hand and monospecific tests or bioassays on the other.

The first step of this new methodological approach has consisted essentially on the quantification of the actions and interactions of different abiotic factors (pH, temperature, photoperiod, light intensity) and contamination factors (chemical forms of mercury, contamination sources, ...) on mercury bioaccumulation by various species of rooted macrophytes (8-10).

More recently, similar studies have been developed using a benthic species : a burrowing mayfly, *Hexagenia rigida* (11-13). Several authors have described the biological and ecological characteristics of this species (14-17). The choice of this species is based on several biological and ecological properties. The nymphs are detritivores and live in burrows which they construct in the upper layers of freshwater lentic sediments (silty substrates). The nymphal stage is the longest period in the life cycle, between 10 and 24 months in natural conditions, depending on the climate, and has a high tolerance for fluctuations in abiotic factors, such as temperature, pH or dissolved oxygen (18-19).

The nymphal stages of *Hexagenia rigida* thus provide a very good means of studying at the laboratory level mechanisms of mercury bioaccumulation and transfer in relation to quantitative and qualitative modifications in several ecotoxicological factors. Contamination processes of this benthic species are fairly complex, however, being based on the transfer not only of mercury linked to sediment particles ingested by the nymphs - trophic route - but also of metal present in the water (interstitial water in the sediments and water column) - direct route -. These two routes are always more or less concomitant, but with one or other route predominating, according to whether contamination of the experimental systems is initially via water column or via the sediment compartments.

In this paper we shall present a synthesis of the main results from our experimental approach to the bioaccumulation, transfer and chemical transformations of two mercury compounds ($HgCl_2$ and CH_3HgCl). These results were obtained in the context of our research programme using a three compartment system "water, natural sediment and *Hexagenia rigida* nymphs", contamination of the experimental units (EU) being based either via the sediment or the water column (Fig. 1). Complementary studies of the metal partitioning in the biotopes were set up, in order to obtain a better understanding of the differences in distribution and bioavailability between the two compounds. Finally, chemical transformations of inorganic or organic mercury initially introduced in the water or in the sediment - methylation and demethylation reactions - were quantified.

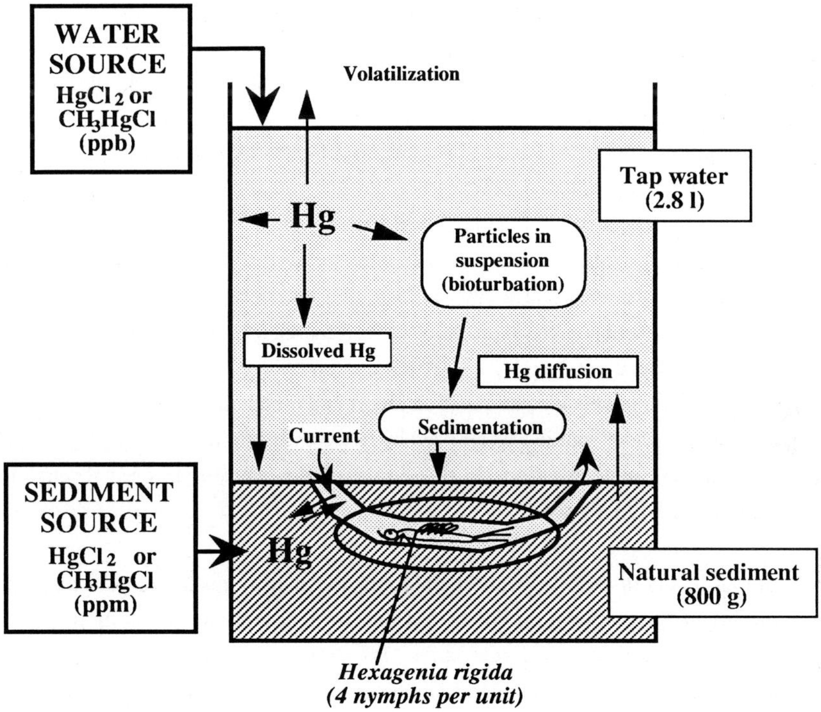

Fig. 1. Basic structure of the experimental model "water, natural sediment, *Hexagenia rigida* nymphs" : principal mercury exchanges between the different compartments, after initial contamination via the water or the sediment source.

2. MATERIAL AND METHODS

2.1 Culture technique for *Hexagenia rigida nymphs*

Mass culture of nymph stages was initiated in our laboratory from eggs collected each summer from Lake Winnipeg (Freshwater Institute, Winnipeg, Canada) and stored at +4°C. The culture technique (20-21) has been adapted in our laboratory (12-13). Hatching was carried out in dechlorinated tap water, saturated with oxygen and raising the temperature from 4°C to 24°C in 4°C stages, every 48 hours. In these conditions, percentage hatched was over 95%. Newly hatched nymphs were gently transferred to glass tanks (25x25x30 cm), using a Pasteur pipette. Each tank contained a natural sediment layer (depth : 5 cm), identical to sediment used for the experiments. Nymphs selected for the experiments were 15-25 mm long and 25-80 mg fresh weight. Growth heterogeneity was important inside each culture tank and it was necessary to collect a large number of individuals in order to obtain, after a screening process, sufficient nymph batches for the experiments.

2.2 Experimental design

The basic structure used in the experiments was the experimental unit (EU) : a glass tank (12x12x30 cm - lined with a plastic bag) containing sediment (depth : 5 cm), 2.8 l of dechlorinated tap water (depth : 20 cm) and *Hexagenia rigida* nymphs (4 nymphs/EU) (Fig. 1). Sediment was taken from the banks of the Garonne River, upstream from Bordeaux. It was a very homogeneous silt, rich in clays (75-80%), with a low organic carbon particle content (2% on average). The natural level of total mercury was 0.124 ± 0.012 $mgHg.kg^{-1}$ (fresh weight), with a fresh weight/dry weight ratio of about 2.0. A large quantity of sediment was homogenized by mechanical mixing and two batches were made up, one for the control units and for the EUs contaminated by the water source, and the second for the sediment source, to be enriched with mercury by means of aqueous solutions of CH_3HgCl or $HgCl_2$ (Merck - 500 $mgHg.l^{-1}$). After a second homogenization, sediment samples were taken from each batch to check contamination levels and metal distribution in the substratum. The contamination by the water source was realized by twice daily additions of CH_3HgCl or $HgCl_2$ in the water column ; Hg concentrations in this compartment were monitored throughout the experimental period, 2 or 3 times a week.

Water column levels in the EUs were maintained constant by means of periodic additions during the experiment, to compensate for losses due to evaporation and sampling (mercury determinations). EUs were placed in larger tanks (140x65x30 cm), which were themselves in enclosed containers. Each tank had thermoregulation equipment, which was very efficient, due to the large volume of water constantly stirred by submerged pumps. Light was produced by two neon tubes (Sylvania F36W/GRO) positioned at 45 cm from the surface of the EUs and operated by a timer switch. Average light intensity at the surface was 2000 lux (Quantum Sensor, LI190SB). Discontinuous aeration in EU was produced by air pumps (RENA 301), the diffuser being placed in the upper layers of the water column, in order not to disturb the sediment-water interface too much.

Four nymphs were introduced into each EU. In order to minimize weight heterogeneity between organisms, nymph selection was based on four weight classes and a similar biomass was then achieved in each EU, by introducing one nymph of each class into each unit. The nymphs were added 10 days after water and sediment had been introduced in the EUs. This delay allowed the physico-chemical conditions of the water phase to stabilize (22-23). No food supply was added during the experiment.

All the experiments were developed using complete experimental designs, which permit to quantify the actions of the different factors studied and also their interactions, using multiple regression method (24-26). Complete experimental designs and orthogonal polynomials simplified interpretation of the effects of each regressor because of the independence of the regression coefficients. The regressor coding was giving by Snedecor and Cochran tables (27). 99% confidence limits were adopted to select the terms of the regression models. In this paper, most of the results presented will be based on theoretical models, representative of all the observations made during our research programme (12).

2.3 Total, organic and inorganic mercury determination

For many benthic species, the determination of contaminants accumulated in the whole organism is often difficult because of the sediment inside the digestive tract (28-29). To clear the intestine of its contents after the contamination phase, nymphs were collected from the EUs and introduced in a new sediment without mercury. The time required for clearing the gut - 6 hours at 20°C - was determined from preliminary experiments (12) ; this was in agreement with results published by Zimmerman and Wissing (30).

Total mercury determination was carried out by flameless atomic absorption spectrometry (VARIAN AA475). Nymphs and sediment samples were first mineralized by nitric acid attack (pure HNO_3) in a pressurized medium (borosilicate glass tubes), at 95°C for three hours. A bromine salts treatment was applied before the addition of stannous chloride (31). The detection limit was 5 ngHg. The validity of the analytical method was checked periodically by means of intercalibration exercises and reference standards (NBS, Washington ; IEAE, Monaco ; KFA, Jülich).

Total mercury determination in the gills and the gut of the nymphs was determined using radioactive isotope ($CH_3{}^{203}HgCl$ - Amersham, UK, and $^{203}HgCl_2$ - Dupont, USA). Radioactivity was measured in an "LKB Wallac 1282 Compugamma" gamma particle counter. Correction was made for radioactive decay and for background radiation. Measurements were carried out at the "Institut National de la Recherche Scientifique, INRS-Eau", University of Quebec, Ste Foy (Canada).

Organic and inorganic mercury determinations were carried out by flameless atomic absorption spectrometry, the metal being preconcentrated on gold wire and the organic and inorganic forms separated by anion exchange chromatography. All these measurements were achieved at the Institute of Applied Physical Chemistry, KFA, Jülich (Germany), according to Horvat's method (32-33).

Mercury accumulation in *Hexagenia rigida* (whole organism) was expressed by the concentration criterion ($\mu gHg.g^{-1}$, fresh weight) ; Hg bioaccumulation in the two organs selected - gills and gut - was expressed as relative burden (%). The natural level of total mercury in the nymphs was 130 ± 12 $ngHg.mg^{-1}$ (fresh weight).

3. RESULTS AND DISCUSSION

The bioaccumulation of the two mercury compounds - $HgCl_2$ and CH_3HgCl - in the nymphs is strongly influenced by the contamination source - water column or sediment. When the EUs were initially contaminated via the sediment source (identical initial concentrations of inorganic mercury and methylmercury), average Hg concentrations in the nymphs differed by between 40 and 60 fold, more in certain cases, in favour of sediments enriched with organic mercury (Fig. 2). Similar studies on the accumulation of organic and inorganic mercury by Shrimps (*Crangon crangon*) via the trophic route also showed very high contamination differences (a factor close to 40, in favour of methylmercury) (34).

When Hg is introduced in the water column, differences between total Hg concentrations in the nymphs tended to be very small, between 1.2 and 1.8, yet always in favour of EUs contaminated by methylmercury.

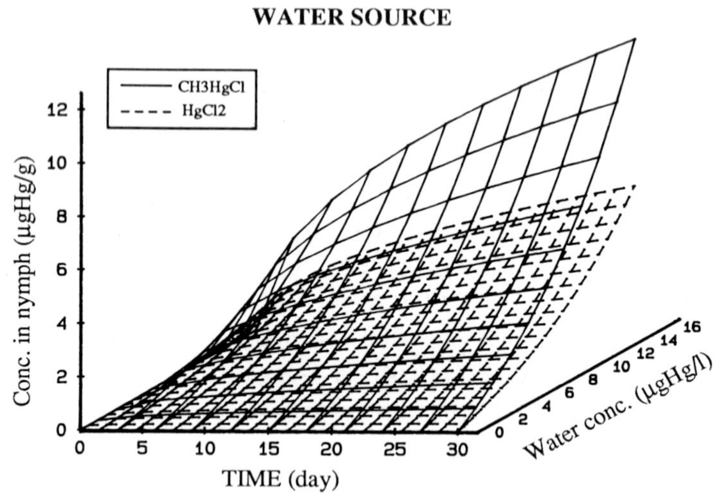

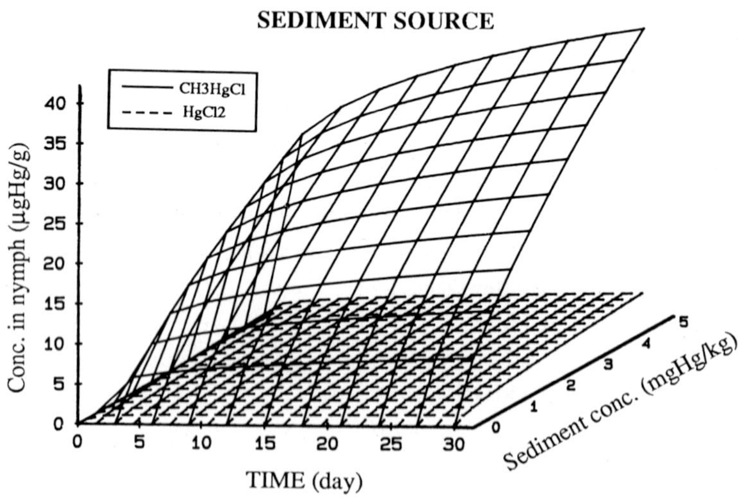

Fig. 2. Synthetic representation of total mercury concentrations accumulated in the nymphs ($\mu gHg.g^{-1}$, fresh weight), as a function of the contamination sources (water column or sediment), the chemical forms of mercury ($HgCl_2$ and CH_3HgCl), the metal concentration in the water column or in the sediment and the exposure length.

Although it is very difficult to achieve a strict comparison between the two contamination sources, it is clear that contamination via the water source does give rise to much greater transfers of mercury into the organisms than contamination via the sediment source. Daily additions of mercury into the water column of the EUs produced concentrations in this compartment of about $\mu gHg.l^{-1}$, after 28 days' exposure, whereas the initial contamination levels in the sediment were about $mgHg.kg^{-1}$. Corresponding average Hg concentrations measured in the nymphs were in the same level order of magnitude, of about $mgHg.kg^{-1}$. In terms of bioconcentration factors (BCF = $C_{organism}/C_{biotope}$), it can be seen that accumulation capacities are greater from the water source by a factor of almost 1000. In fact, these differences could be linked to the degree of bioavailability of the metal in the water phase and in the sediment (2-3, 35).

The kinetics of accumulation also varied greatly depending on the contamination source. Via the sediment source, mercury bioaccumulation, asymptotic in nature, was very rapid, the maximum concentrations in the nymphs being observed after 10-15 days' exposure ; after this period, an "equilibrium" between the ad-absorption of the metal and its elimination gives rise to a well-defined plateau tendency. We should note that for exposures of over 30 days, we observed some fluctuations in this plateau tendency, with sharp decreases in the metal concentrations accumulated in the organisms, followed by a resumption of the bioaccumulation kinetics. These phenomena may be associated with the specific type of growth of *Hexagenia rigida* nymphs, which involves a series of moults. The rapid turn-over of the biological structures during the phase that precedes the moult may be a very efficient decontamination mechanism for the nymphs. We have no precise informations available on this phenomenon, but it is interesting to note that a chronological approach to the bioaccumulation of cadmium, zinc and lead by this mayfly species, under similar experimental conditions, gives rise to similar observations (36).

When the EUs were contaminated via the water source, mercury accumulation by the nymphs was also asymptotic, but the period of equilibrium between the metal's entering and leaving the organisms occurred very much later (Fig. 2). The differences between the two contamination sources could be associated with the specificities of the contamination modalities : when mercury compounds were added twice daily to the water column, the contamination pressure tended to remain constant throughout the experiment, and logically this exposure conditions contribute to maintain the metal transfers between the biotope and the organisms during the experiment. When the EUs were contaminated via the sediment source, however, the two mercury forms were added to the systems in a single addition, at the beginning of the experiment and the bioavailability of the metal may therefore decrease over time, in relation to the high complexation capacity of the particulate phase of the sediment.

Differences in levels of bioaccumulation between the two mercury forms tended also to vary according to the metal concentration in the biotopes :

- for the sediment source, these differences were not proportional to the amounts of mercury initially introduced into this compartment. They increased as mercury concentrations in the sediment increased, the relationships between bioaccumulation and sediment contamination level being close to linear for methylmercury, but highly asymptotic for the inorganic compound. We should note that whatever the metal concentrations initially introduced into the sediments, the accumulation kinetics were always similar, with a plateau tendency appearing after several days' exposure ;

- for the water source, the relationships between Hg burdens in the nymphs and metal concentrations in the water column were close to linear or even slightly exponential, for the two mercury compounds.

Mercury distribution in the main organs of *Hexagenia rigida* nymphs revealed some considerable variations, depending on the chemical form of the metal and the contamination source considered (Fig. 3). By looking at the mercury organotropism, we should be able to estimate the respective importance of the direct and trophic contamination routes.

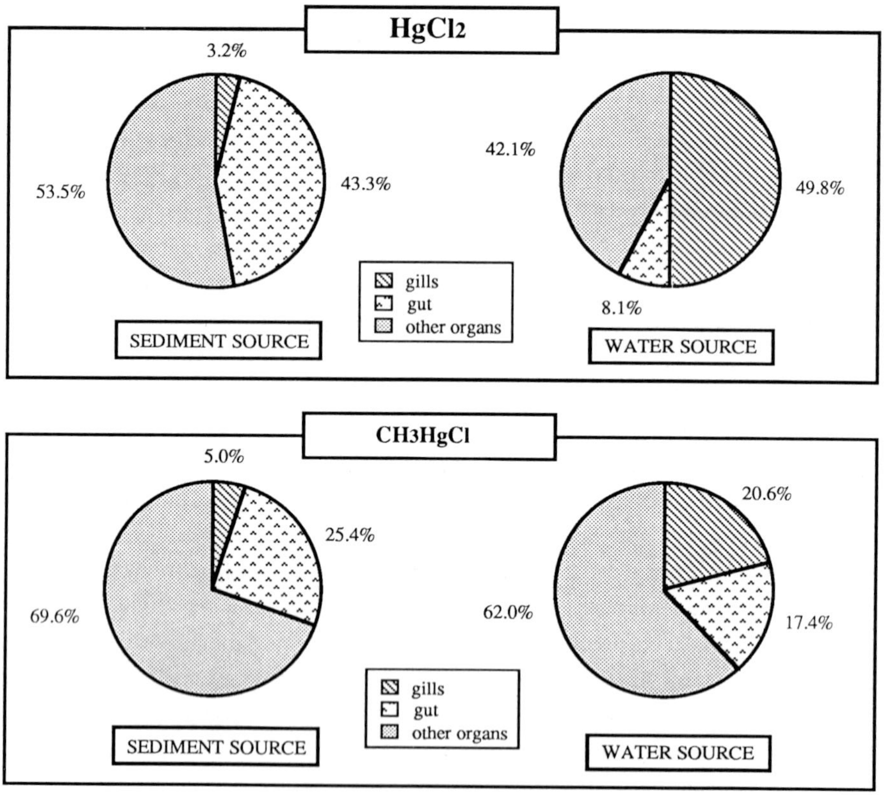

RELATIVE WEIGHT OF THE ORGANS (% / body weight, dry weight)
gills : 6.3%
gut : 6.7%
other organs : 87%

Fig. 3. Relative total mercury burdens (%) in the gills, the gut and the other organs of *Hexagenia rigida* nymphs, after contamination of the EUs via the sediment or the water source, by $HgCl_2$ or CH_3HgCl .

After contamination by inorganic mercury, via the sediment source, the relative mercury burdens were, on average, 43% in the gut and 3% in the gills. Via the water source, however, only 8% of the metal was accumulated in the gut, as opposed to approximately 50% in the gills. This mercury distribution would lead us to suppose that for the sediment source, the trophic contamination route, via the sediment ingested, was more important ; for the water source, the direct route, via the water column and/or the water in the burrows was more predominant. Nevertheless, in both instances, a high proportion of the metal was located in the rest of the body (53% and 42% respectively) ; this includes the cutaneous layer around the organism and all the organs and tissues not sampled (muscle mass, circulatory system, nervous system, haemolymph, etc.).

When the EUs were contaminated with methylmercury, via the sediment source, 25% of the metal was accumulated in the digestive tract, as opposed to 5% in the gills. As before, this distribution tends to suggest the predominance of the trophic route, by the intermediary of ingested sediment. With the organic form of mercury, the contrast between the metal burdens of these two organs was much less marked than when the inorganic mercury was used, and almost 70% of the accumulated metal was in the rest of the body. This distribution can be interpreted by attributing to the gut a very high capacity for the absorption of organic mercury, which is then stored in the different organs and tissues within the nymphs. In the case of the water source, mercury burdens in the gills and the gut were 20% and 17% respectively. The difference between these readings is not great: this may be due either to a contamination predominantly via the direct route, but with inter-organ transfers bringing large amounts of the metal into the gut by means of the haemolymph ; or it may be due to a mixed "direct + trophic" contamination. In both instances, more than 60% of the metal was nevertheless accumulated in the rest of the body ; so the same hypotheses apply as were put forward above in relation to the "sediment - CH_3HgCl" source, i.e. when there is a high rate of transfer across the biological barriers or a high level of fixation on the cutaneous interface.

Results obtained with other aquatic biological models (36-40) are in agreement with these observations. Metal burdens in the gut and the gills differ widely, with one or other being greater, depending on whether organisms are contaminated by the trophic or the direct route. Also, when contamination is via the trophic route, large quantities of inorganic mercury appear to bind on the gut wall, while absorption rates remain relatively low, thus explaining the low contamination levels measured in the whole organisms (Fig. 2). Methylmercury, on the other hand, crossed the digestive barrier of the nymphs much more easily, with bioaccumulation capacities, at the organism level, 40 to 60 times greater than those observed after contamination of the sediment by the inorganic compound.

In fact, mercury accumulated in the gills or the gut of *Hexagenia rigida* nymphs does not derive only from the water column or ingested sediment respectively. From the life style of the nymphs and the mercury exchanges established between water column and sediment compartments of the EUs, we can suppose that contamination is via the two routes, direct and trophic. The respiratory activity of the nymphs inside their burrows, with a current of water running through permanently (gill fluttering), does indeed contribute towards an increase in inter-compartment exchanges.

It is important to stress the role that the nymph's cutaneous coating may play in relation to mercury fixation and/or absorption. This barrier consists of a large surface area, in direct contact with the surrounding medium, and it is therefore able to participate in mercury accumulation by means of complexation reactions at the interface. With the micro-dissection procedure that we have adopted we are not able to estimate the quantities of mercury fixed on this barrier ; future experiments, using autoradiography (^{203}Hg), will permit to evaluate the role played by this barrier towards Hg fixation.

Lastly, it should also be noted that the relative Hg burdens in the gills and the gut sometimes represent very high concentrations, considering the low weight of these two organs relative to the rest of the body (6.3% and 6.7% respectively). Our estimates for the gut, for example, give concentrations of over 100 $\mu gHg.g^{-1}$ (fresh weight), for an initial concentration of 10 $mgHg.kg^{-1}$ ($HgCl_2$) in the sediment.

In this synthesis of results relating to the bioaccumulation of mercury compounds by *Hexagenia rigida*, at whole organism and main organ level, we are able to show the complexity of the phenomena involved and how closely dependent they are on initial contamination sources, chemical forms of mercury, contamination levels of the sources, length of exposure, ... For a better understanding of these processes, complementary studies were undertaken, based for example on the determination of mercury partitioning in the sediment, using chemical extractions (1,41-44) or on the quantification of the transformations of the two chemical forms of mercury initially introduced into the EUs.

When mercury is initially introduced into the sediments, its distribution between the dissolved and particulate phases differs, according to the chemical form considered - $HgCl_2$ or CH_3HgCl -. Our comparative study on the Garonne sediment in suspension in water shows that in the case of mercuric chloride, only 0.2% of the total mercury introduced was found in the dissolved phase ; for methylmercury, the percentages were between 2 and 5%. Moreover, with identical extraction conditions (1), the quantities of metal recovered from the oxic layers of the sediment after extraction with HCl (HCl 1N - iron and manganese oxide-bound Hg), ranged from 2 to 9% for the inorganic mercury and from 50 to 60% for the methylmercury. For mercury obtained after extraction with NaOH (NaOH 0.1N - humic acid fraction), the corresponding percentages are 13-17% and 15-20%. It was not possible to achieve satisfactorily experimental conditions for the determination of the metal extracted with H_2O_2 (H_2O_2 30% v/v - oxidisable organic matter) ; an important amount of mercury loss by volatilization was observed when samples were treated with peroxide (12,45).

Quantification of Hg dissolved and particulate forms in samples taken from the water column (filtration over nylon membrane - 0.45 µm - Nalgene 25 mm) in EUs contaminated by the sediment source shows that the metal concentrations are always greater when the sediment is first enriched with methylmercury. The presence of the metal in the aquatic phase is due to release processes from the sediments, strongly influenced by the bioturbation activity of the nymphs. The various results obtained demonstrate that the organic form of mercury has a smaller overall capacity to bind on the sediment particles. When mercury is introduced into the water column, average Hg concentrations in this compartment remain similar for the two compounds, but the amount of dissolved metal is always greater when the EUs are contaminated with the organic form. Differences between inorganic and organic mercury are very closely dependent on the extent of the particle phase (turbidity of the water column). This phase is itself dependent on the degree of activity of the nymphs, the exposure period and the physico-chemical characteristics of the medium (e.g. temperature or pH). In any case, when mercury is initially added to the water column, the chemical form of the metal influences its distribution in this compartment much less than in the sediment source.

The fact that only small differences were observed between the two compounds in relation to mercury accumulation in the nymphs when contamination was via the water source could be linked with these results. In terms of bioavailability, there would seem to be little differences between $HgCl_2$ and CH_3HgCl as far as contamination conditions for the organisms in the water source were concerned.

Similarly, the very wide differences observed between the quantities of mercury accumulated in the nymphs, depending on the form of the metal introduced into the sediment - $C_{CH3HgCl}/C_{HgCl2} > 40$ - could be linked to the differences observed in metal concentrations in the interstitial water in the sediments or in the water column (release processes) : mercury is more abundant in these two compartments, and thus comparatively more bioavailable, when the sediment is initially enriched with methylmercury.

Based on these differences in the distribution of the two chemical forms of mercury throughout the biotopes, according to the contamination source, other arguments can be put forward to explain the differences observed between the amounts of Hg bioaccumulated in *Hexagenia rigida* nymphs.

First, let us review the physico-chemical properties of the two mercury compounds and how they could be linked with the transport processes across biological membranes. Many authors have established a close link between the high accumulation capacity of methylmercury and its liposolubility. In fact, the partition coefficient "n-octanol/water" (P) is 2.54 for methylmercury and 0.61 for $HgCl_2$ (46). These values are very low compared to highly liposoluble compounds, such as chlorinated hydrocarbons (47). Similar measurements on a "phospholipid (DPPC)/water" system gave similar P values for methylmercury (48). Studies on trans-membrane fluxes, using phospholipidic membrane models (planar bilayer membranes or BLM) show that methylmercury has a low affinity for the hydrophobic core of the bilayers, although its capacity for diffusion through the membrane is very great, and this explains how it is transported so efficiently (48-49). Moreover, the trans-membrane fluxes of inorganic mercury and methylmercury can be modified to a great extent, according to the physico-chemical conditions of the medium, especially pH and even more particularly the concentration in chloride ions (pCl). These processes are directly linked to the chemical speciation of dissolved mercury, where neutral species ($HgCl_2$ and CH_3HgCl for example) are better able to diffuse through the membranes than negatively charged ($HgCl_3^-$ or $HgCl_4^{2-}$)(50-51). Biochemical and biophysical studies of the binding of mercury compounds on membrane ligands have attributed a predominant role to the SH protein groups (thioloprive property). Recent studies in our Laboratory, in collaboration with the Paul Pascal Research Centre (CNRS, Bordeaux) show that phosphatidylserine and phosphatidylethanolamine represent a new class of binding sites in cell membranes, the specific interaction between inorganic Hg and the primary amine groups on their polar heads giving rise to large-scale disturbances in the membrane micro-fluidity (52).

At the level of the biological barriers - gill epithelium and walls of the digestive tract -, structural and ultrastructural approach were realized in order to distinguish certain specific features of mercury accumulation and absorption. While the exchange surfaces of these two organs are very extensive - microvillosities of the middle gut and secondary lamella in the gills -, conditions for contact with the external environment and with the sediment ingested are quite different (12). Because of the presence of the peritrophic membrane which envelopes ingested sedimentary material, contamination via the trophic route requires that mercury pass through the dissolved phase in order to cross this first barrier and to reach after the apical face of the enterocytes. Thus, if the metal concentrations in the interstitial water are greater, as has been observed in the case of methylmercury, this may improve the metal accessibility to the interface zone between the lumen of the gut and the digestive tract wall. Mercury bound with the organic matter in the sediments can take advantage of different stages in the digestive process (proteolysis, for example, and liberation of small polypeptides or amino-acids) in order to be transported through the peritrophic membrane and then absorbed at the apical face of the enterocytes.

The ad- and absorption processes at the gut barrier level, in relation with the metal bioavailability, are minimized in the case of the gills, which are in constant contact with the surrounding environment. This obviously favours mercury transfers, especially of the dissolved chemical species. Moreover, our structural analysis of the gills reveals the presence of large muscle masses, especially at the base of the gill lamellae, which are in close contact with the haemolymph. This, in turn, represents an ideal site for mercury accumulation in this organ (12).

Many reactions of mercury complexation take place amongst particulate and dissolved ligands, which may contribute to strong ecotoxicological modifications within our experimental systems, especially in relation to mercury partitioning, physico-chemical properties and bioavailability. Running parallel with these processes, the inorganic and organic forms of the metal initially introduced into the water column or the sediment compartment may also be undergoing chemical transformations : methylation reactions, which produce from the inorganic mercury (HgII), methylmercury (CH_3HgX) or dimethylmercury ($(CH_3)_2Hg$, a volatile form) ; demethylation reactions, which transform the organo-mercurials into inorganic mercury (HgII) and in a second phase into elementary mercury (Hg^0, a volatile form)(3, 53, 54).

In our experimental conditions, we have recently begun to analyse these chemical transformation processes, by quantifying organic and inorganic forms of the metal accumulated in the sediment and in the *Hexagenia rigida* nymphs (whole organism level). The first results obtained refer to the mercury chemical form initially introduced into the water column or sediment compartments of the EUs ($HgCl_2$ or CH_3HgCl) and to the natural levels in the biotopes and in the organisms, in order to determine whether or not chemical transformations had taken place. Note that with the determination method used, it was not possible to separate and quantify the different organic forms of the metal (methylmercury, dimethylmercury, etc.).

When mercuric chloride was used to contaminate the sediment compartment of the EUs, analysis of samples collected after 28 days'exposure shows that a significant amount of organic mercury was produced in the substrate. Concentrations were fairly low, between 2 and 11 $ngHg.g^{-1}$, which correspond to 0.09 and 0.8% respectively of the total mercury concentrations present in the sediment. These levels of methylation are in agreement with the results found in natural environments (1, 55-58).

When initial contamination of the EUs was with methylmercury, however, via the sediment source again, results show very high demethylation rates, which varied according to the depth of the substrate. Hence, depending on the abiotic conditions (temperature and pH), 4 to 33% of the organic compound initially introduced into the upper strates (oxic layer) was transformed into inorganic mercury ; in the lower layers (anoxic zone), these percentages were higher and more homogeneous at around 70% on average. We should mention that in several studies, rapid methylmercury decomposition has been observed, whether the compound was naturally occurring or artificially introduced into the sediments (56, 59-62).

For the water source, no determinations were made on samples collected from the water column ; on the other hand, we have some results for the upper layers of the sediment, at the water-sediment interface. When the EUs were contaminated with mercuric chloride, organic Hg was not detected in the surface layers of the sediment ; when contamination was with CH_3HgCl, however, very high levels of inorganic Hg were measured, corresponding to about 95% of the total mercury concentrations accumulated in this compartment ; these results clearly indicate very high rates of demethylation.

Inorganic and organic Hg determinations carried out on *Hexagenia rigida* nymphs produced some similar qualitative conclusions :

- when EUs were contaminated with mercuric chloride, organisms contained quantities of organic mercury at the end of the experiment (28 days for example) which were significantly greater than the natural background in the control nymphs, with relative burdens reaching a maximum of 3% in relation to the average total mercury accumulated in the nymphs ;
- when EUs were contaminated with methylmercury, a considerable proportion of the metal present in the nymphs after 28 days'exposure was in the inorganic form, with relative burdens between 25 and 55%, depending on the initial contamination source and the physico-chemical conditions of the environment.

At the stage we have currently reached in our experimental approach, it is not yet possible to establish any direct links between all the data collected, nor to define the mechanisms responsible for the chemical transformations. Nevertheless, several comments can be made :

- when the nymphs were contaminated via the sediment source, there were wide differences in mercury bioaccumulation, depending on the chemical form of the metal initially introduced. These differences could be as great as a factor of 40, in favour of methylmercury. Inorganic and organic Hg determinations in the sediment at the end of the exposure period show that a high proportion of methylmercury had been demethylated, suggesting that the differences in the bioaccumulation capacities for the two compounds have, in fact, been underestimated when based on the initial concentrations of $HgCl_2$ and CH_3HgCl in the sediment ;

- in relation to the contamination of the EUs via the water source, we have no direct information on the extent of the chemical transformation reactions of the two compounds in the water column. Analysis of samples collected in the upper sediment layers, however, when methylmercury was added to the water column, shows that the metal present in this interface zone was mostly inorganic, as a result of direct water-sediment transfers or/and deposits of suspended particles. If this result is extrapolated throughout the whole experimental system, this phenomenon could provide an explanation for the very small differences observed between the accumulation capacities of the nymphs for the two experimental conditions "water source - $HgCl_2$" and "water source - CH_3HgCl" ;

- the presence of significant amounts of organic Hg in the nymphs after contamination of the experimental systems with mercuric chloride may be linked with the production of this organic form observed in the sediment ; it may also be due to methylation reactions within the organisms, especially in the digestive tract. Ultra-structural studies have indeed revealed the presence of a great number of bacteria in the gut lumen (12).

The presence of inorganic mercury in *Hexagenia rigida* nymphs after contamination of the EUs by methylmercury (water or sediment source) can also be correlated with the demethylation reactions observed in the biotopes. As has already been shown, inorganic mercury can accumulate in the nymphs when this compound is present in the surrounding environment, its physico-chemical properties and bioavailability being the reason for differences noted between processes of binding to and crossing the biological barriers. As well as methylation reactions, we can also suppose that demethylation reactions take place in the nymph's digestive tract, as a result of bacterial activity. Abiotic processes, such as UV radiations, could also participate to the production of Hg(II) and Hg^0 (52) ; experiments are actually under current work in order to compare the effects of different light intensities and photoperiods on these processes.

4. CONCLUSION

Our ecotoxicological approach to the processes of mercury bioaccumulation, transfers and chemical transformations within freshwater systems, based on a three compartment experimental model "water, natural sediment, *Hexagenia rigida* nymphs", provides a wide range of results. They reveal very strong effects of the different abiotic and contamination factors taken into account : temperature, pH, chemical forms of the metal, initial contamination sources (water column or sediment), length of exposure, ... The preliminary results on the quantification of the chemical transformations of inorganic and organic Hg, in the sediment and in the nymphs, after initial contamination of the EUs by the water column or the sediment, show the high potentialities of this experimental model to reveal and quantify these essential processes, and to analyse the actions and interactions of many ecotoxicological factors.

Our research programme is actually directed towards a more detailed and mechanistic approach of these processes, in order to better understand the complex links occurring in natural conditions between mercury dispersion and partitioning in the biotopes, chemical transformations (methylation and demethylation reactions), biodisponibility, bioaccumulation and trophic transfers along the food chains and networks.

Acknowledgements

We would like to thank P.G.C. Campbell, L. Hare and A. Tessier (INRS eau, Québec) for their assistance in this work and M.K. Friesen for eggs collection. We would like also to express our appreciation to the different members of our laboratory. The research program was supported by the French Ministry of Environment and the Commission of the European Communities (DG XII, Brussels), the CNRS and Bordeaux I university.

REFERENCES

1 W.J. Langston, J. Mar. Biol. Ass. U.K., 62 (1982) 667-684.
2 W. Salomons and U. Forstner, Springer-Verlag, Berlin, 1984.
3 P.G.C Campbell, A.G. Lewis, P.M. Chapman, A.A Crowder, W.K. Fletcher, B. Imber, S.N. Luoma, P.M. Stokes and M. Winfrey, CNRC, Ottawa, Canada, K1A OR6, 1988.
4 A. Bourg, in: Boudou A. and Ribeyre F. (Eds), Aquatic Ecotoxicology : fundamental concepts and methodologies. CRC Press, Boca-Raton, USA, vol I, 1989, pp. 107-123.
5 P.G.C. Campbell and A. Tessier, in: A. Boudou and F. Ribeyre (Eds), Aquatic Ecotoxicology : fundamental concepts and methodologies. CRC Press, Boca Raton, USA, vol 1, 1989, pp. 125-148.
6 D.W. Schindler, Sci. 239 (1988) 149-157.
7 F.Ribeyre, Ecotox. Environ. Safety, 9 (1985) 346-363.
8 F.Ribeyre and A. Boudou, in: A. Boudou and F. Ribeyre (Eds), Aquatic Ecotoxicology : fundamental concepts and methodologies. CRC Press, Boca Raton, USA, vol 2, 1989, pp. 3-45
9 R. Maury, A. Boudou, F. Ribeyre and P. Engrand, Aquat.Toxicol. 12 (1988) 213-228.
10 F.Ribeyre and A. Boudou, in: W. Wang, J.W. Gorsuch and W.R. Lower (Eds), Use of plants for Toxicity Assessment. STP 1019, ASTM, Philadelphia, 1990, pp. 97-113
11 E. Saouter, F. Ribeyre, A. Boudou, in: J.P. Vernet (Ed.), Heavy metals in the environment. CEP Consultants, Edinburgh, 1989, pp.378-381.
12 E. Saouter, Bordeaux I University, Ph.D., n° 544, 1990.
13 E. Saouter, F. Ribeyre, A. Boudou and R. Maury-Brachet, Environ. Poll. 69 (1991) 51-67.
14 F. Neave, Contrib. Can. Fish. N.S., 7 (1932) 177-201.
15 C.R. Fremling, Trans. Amer. Fish. Soc., 97 (1968) 278-280.
16 W.P. McCafferty, Trans. Am. Entomol. Society., 101 (1975) 447-504
17 M.C. Zimmerman, Miami University, Ph.D., 1977.
18 R.E Craven, and B.E. Brown, Am. Midl. Nat., 82 (1969) 346-358.
19 R.H. Peterson, Can. Tech. Rep. Fish. Aquat. Sci., 1523 (1987) 1-7.
20 M.K. Friesen, J.F. Flannagan and J.F. Layrence, Can. Entomolo., 108 (1979) 665-673.
21 M.K. Friesen, in: S.G. Laurence (Ed.), Manual for the culture of selected freshwater invertebrates. Publ. Spec. Can. Sci. Halieut. Aquat.,1982.
22 R. Maury and P. Engrand, Bordeaux I University, Ph.D., n°667, 1986.
23 F.Ribeyre, Bordeaux University, Ph.D., n°957, 1988.
24 G.E.P. Box, W.G. Hunter and J.S. Hunter, Wiley, New-York, 1978.
25 R. Tomassone E. Lesquoye and C. Millier, I.N.R.A., Masson, Paris, 1983.
26 P. Dagnelie, Presses agronomiques de Gembloux, Belgique, 1984.
27 G.W. Snedecor and W.G. Cochran, Ass. Coordin. Techn. Agri, Paris, 1971.

28 C. Amiard-Triquet, C. Metayer and J.C. Amiard, Rev. Int. Oceanogr. Med., 73-74 (1984) 27-34.
29 L. Hare, P.G.C. Campbell, A. Tessier and N. Belzile, Can. Aquat. Fish. Sci., 46 (1989) 451-456
30 M.C. Zimmerman and T.E. Wissing, Freshwat. Biol., 8 (1978) 269-277.
31 B.J. Farey, L.A. Nelson and M.G. Rolph, Analyst, 103 (1978) 656-659.
32 M. Horvat, K. May, M. Stoeppler and A.R. Byrne, Appl. Organometallic Chem., 2 (1988) 515-524.
33 K. May, M. Stoeppler and K. Reisinger, Toxicol. Environ. Chem., 13 (1987) 153-159.
34 H.Y. Riisgard and P. Famme, Mar. Poll. Bull., 17 (1986) 255-257.
35 S.N. Luoma, Sci. Total Environ., 2 (1983) 73-82.
36 L. Hare, E. Saouter, P.G.C. Campbell, A. Tessier, F. Ribeyre, and A. Boudou, Can. J. Fish Aquat. Sci., 47 (1990) 39-47.
37 A. Boudou, and F. Ribeyre, W.A.S.P. 26,(1985) 137-148.
38 D.G. King, and I.M. Davies, Mar. Poll. Bull., 18 (1987) 40-47.
39 L. Albanus, L. Frankerberg, C. Grant, U. Haartman, A. Jernelov, G. Nornberg, M. Rudalv, A.Schutz, and S. Skerfuing, Environ. Res., 5 (1972) 425-442.
40 P.A. Cunningham and M.R. Tripp, Mar. Biol., 31 (1975) 321-334.
41 R.P. Gambrell, R.A Khalid, V.R. Collard, C.N. Reddy and W.H.Jr. Pattrick, in Dredging: Environmental Effects and Technology. Proceeding of Wodcom VII, World Dredging Conference, San Pedro, Calif., 1976, pp. 579-604.
42 A. Tessier, P.G.C. Campbell and M. Bisson, Anal. chem., 51 (1979) 844-851.
43 S.N. Luoma and G.W. Bryan, Sci. Total Environ., 17 (1981) 165-196.
44 S. Karlsson, K. Hakansson and B. Allard, J. Environ. Sci. Health., 6 (1987) 549-562.
45 P.G.C. Campbell and A. Tessier, in: J.P. Vernet (Ed), Heavy metals in the environment. CEP Cons., Edinburgh, UK, 1989, pp. 516-525.
46 S. Halbach, Arch. Toxicol., 57 (1985) 139-141.
47 D.R. Nimmo, in: G.M. Rand and S.R. Petrocelli (Eds), Fundamentals of aquatic toxicology. Hem. Publ. Corp., Washington, 1985, pp. 335-373.
48 J.R. Lakowicz and C.J. Anderson, Chem. Biol. Interact., 30 (1980) 309-323.
49 E. Bienvenue, A. Boudou, J.P. Desmazes, C. Gavach, D. Georgescauld, J. Sandeaux and P. Seta, Chem. Biol. Interact., 48 (1984) 91-101.
50 J. Gutknecht, J. Memb. Biol., 61 (1981) 61-66.
51 A. Boudou, D. Georgescauld and J.P. Desmazès, in : J.O. Nriagu (Ed), Aquatic Toxicology. Wiley, New York, 1983, pp. 117-136.
52 M. Delnomdedieu, A. Boudou, J.P. Desmazès and D. Georgescauld, Biochim. Biophys. Acta, 986 (1989) 191-199.
53 J.M. Wood, in: J.P. Vernet (Ed), Heavy metals in the environment. CEP Cons., Edinburgh, UK, vol 1, 1989, pp. 1-12.
54 T. Barkay, Appl. Environ. Microbiol., 53 (1987) 2725-2732.
55 P.D. Bartlett and P.J. Craig, Water Res., 15 (1981) 37-47.
56 A. Furutani and J.W.M. Rudd, Appl. Environ. Microbiol., 40 (1980) 770-776.
57 P.S. Ramlal, J.W.M. Rudd, A. Furutani and L. Xun, Can. J. Fish. Aquat. Sci., 42 (1985) 685-692.
58 S.M. Callister and M.R. Winfrey, W.A.S.P., 29 (1986) pp.453-465.
59 M.R. Winfrey and J.W. Rudd, Environ. Tox. Chem., 9 (1990) 853-869.
60 A. Kudo, D.R. Miller, H. Akagi, D.C. Mortimer, A.S. De Freitas, H. Nagase, D.R. Towsen, and R.G. Warnock, Prog. Water Technol., 10 (1978) 329-339.
61 P.S. Ramlal, J.W.M. Rudd and R.E. Hecky, Applied Environ. Microbiol., 51 (1986) 110-114.
62 F. Baldi, W.A.S.P., 38 (1987) 111-119.

Chapter 6
Microbial adaptation and microbial interactions

EFFECTS OF HEAVY METALS AND SELENIUM ON MERCURY METHYLATION AND OTHER MICROBIAL ACTIVITIES IN FRESHWATER SEDIMENTS

T.A. Jackson

1. INTRODUCTION

Bacteria in natural waters and sediments mediate the chemical transformations and biogeochemical cycling of nutrients and toxic substances (1, 2), besides being important links in aquatic food chains (3); therefore, inhibition or stimulation of these organisms by pollutants may have far-reaching consequences for aquatic ecosystems (4, 5, 6). As the species which comprise a microbial community interact with each other in various ways ranging from mutualism to antagonism (1, 7) and differ widely in their functions, requirements, and limits of tolerance, pollution may have a number of direct and indirect effects on the species composition, activities, and population size of the microflora. The effects vary with environmental conditions, which control the chemical speciation, bio-availability, and toxicity of pollutants as well as microbial growth and activities (8). The complexity of these processes is compounded if there is a mixture of pollutants, which may interact synergistically or antagonistically, or if a pollutant can act as a nutrient as well as a toxic agent, as is the case with copper and zinc.

This report, a brief preliminary version of which has been published elsewhere (9), deals with effects of copper (Cu), cadmium (Cd), zinc (Zn), mercury (Hg), and selenium (Se) on the microbial transformation of inorganic bivalent Hg into methyl Hg (CH_3Hg^+), and on heterotrophic microbial activity in general, in lake sediments. Methylation of Hg by free-living bacteria and fungi is a widespread phenomenon in aquatic and sedimentary environments (10) and is not limited to bodies of water affected by point-sources of Hg pollution (11-13). It is a cause for serious concern because CH_3Hg^+, though far less abundant than inorganic Hg, is much more bio-available and toxic (10). Unlike inorganic Hg, CH_3Hg^+ undergoes biomagnification up the food chains and is preferentially accumulated by fish, whose flesh may thus be rendered unsafe for human consumption (13, 14). Many different kinds of micro-organisms that vary widely in their ecological requirements are responsible for the synthesis and subsequent decomposition (demethylation) of CH_3Hg^+ (10, 15, 16); therefore, these reactions occur over a wide range of environmental conditions, although anaerobic environments rich in organic nutrients but poor in sulfides tend to maximise CH_3Hg^+ production (10-12, 15, 17). A shift in some environmental variable or exposure of the microbial community to pollutants could change the net rate of CH_3Hg^+ production (the balance between CH_3Hg^+ formation and breakdown) by inhibition or stimulation of microbes (selective effects on particular species or general effects on the microflora as a whole), or by alteration of the forms and availability of the inorganic Hg, or by a combination of these factors (9-12, 15, 17-19).

Little is known about the influence of heavy metals and Se on Hg methylation and other activities of natural microbial communities (5, 6), but both inhibitory and stimulatory effects have been reported (5, 9). Besides being of fundamental ecological interest, such phenomena have a bearing on problems of environmental protection involving metal-polluted waters. Se has attracted special attention because it counteracts the toxicity of Hg, Cd, and other heavy metals (20), and it has been noted that both Se and Zn may interfere with the accumulation of Hg by fish (17, 21). Indeed, treatment of Hg-polluted natural waters with Se has been tentatively proposed as an expedient for suppressing the bio-accumulation of Hg (21). However, Se itself may be toxic and is all the more hazardous because it undergoes biomagnification (22). Consequently, there is a need to know more about the broader ecological significance of Se and metal-Se interactions as well as the effects of heavy metals and Se on the activities of microbial communities in natural waters.

2. METHODS AND MATERIALS

Effects of heavy metals and Se, singly and in mixtures, on Hg methylation and other microbial activities in sediments were examined by means of laboratory experiments and analyses performed on samples of mud collected from different lakes in Canada. The mud employed in the experiments was dredged from Tyrell Lake, Saskatchewan (near Flin Flon, Manitoba) (23); East Mynarski Lake, Manitoba (12); the east basin of Clay Lake, Ontario (24-28); and the eastern part of Pasqua Lake, Saskatchewan (18, 29). Samples were also collected for chemical analysis from several other lakes clustered around Flin Flon (Schist Lake, Phantom Lake, West Nesootao Lake, and Hamell Lake) (30). Clay Lake and Pasqua Lake are contaminated with Hg and other pollutants from industrial and urban sources, Tyrell Lake and the other lakes near Flin Flon are polluted with metals - notably Cu, Cd, and Zn - from a mining and smelting complex, and East Mynarski Lake is unpolluted (not counting the ubiquitous trace pollutants due to atmospheric fallout). Pasqua Lake is situated in the semi-arid prairie region of the Great Plains, but all the other lakes are in the Boreal forest zone of the Canadian Shield.

The mud samples were analysed for total Cu, Cd, and Zn by atomic absorption spectrophotometry (AAS) following digestion with hot $HNO_3/HClO_4$ (30), and total Hg was determined by flameless AAS after digestion with hot H_2SO_4/HNO_3 or HCl/HNO_3 (12, 18, 25). CH_3Hg^+ was extracted from the sediments with $CuSO_4$, NaBr, H_2SO_4, and toluene, and then dissolved in a mixed solution of aqueous $Na_2S_2O_3$ and ethanol, purified by rinsing with benzene, converted to CH_3HgI by treatment with KI, taken up in benzene, and determined by gas chromatography (GC) (12, 18, 19).

The experimental techniques have been described in full elsewhere (12, 19, 30) but are summarised briefly as follows. The reactions were carried out in stoppered flasks fitted with valves for sampling head gas. The flasks and all solid experimental materials except the mud samples used as sources of micro-organisms were autoclaved beforehand, but solutions were sterilised by filtration through autoclaved 0.45 μm membranes. Replicate portions of homogenised mud were enriched with natural organic nutrient substrates (plant fragments) and suspended in equal volumes of water containing <u>either</u> a fixed concentration of dissolved inorganic Hg(II) in the form of $HgCl_2$ with or without different concentrations of dissolved $CuCl_2$, $CdCl_2$, $ZnCl_2$, or Na_2SeO_3 (or a mixture of two of these compounds), <u>or</u>, alternatively, different concentrations of dissolved $HgCl_2$. The slurries were unbuffered or else were buffered with $CaCO_3$ or Tris to maintain a constant pH close to 7. All slurries were purged with N_2 gas and incubated in the dark at 20-25°C under an atmosphere of N_2 except in one experiment, in which air was used in order to create aerobic conditions. Each experiment conformed to one of two basic designs: either (a) the incubation time was varied from 0 to 14 days but the quantity of metal or Se supplement was kept constant, and batches of replicate experimental slurries were compared with control slurries containing no added test substance; <u>or</u> (b) the incubation time was the same for all systems (7 or 8 days), but the quantity of metal or Se was varied and the starting conditions were assessed by means of slurries analysed at the outset of the term of incubation.

After incubation, head-space gas was analysed for carbon dioxide (CO_2) and methane (CH_4) by GC, and the pH and oxidation-reduction potential (Eh) of each slurry were measured with a pH meter. CH_3Hg^+ was extracted from the slurries by the methods outlined above, and was usually determined by GC. In the experiments employing Clay Lake mud, however, CH_3Hg^+ was estimated as undifferentiated "organic Hg," defined as the total Hg content of the ethanolic $Na_2S_2O_3$ solution, determined by flameless AAS after digestion of the solution with HNO_3/K_2CrO_7 and hot $H_2SO_4/KMnO_4/K_2S_2O_8/H_2O_2$ (28). Analysis of a suite of extracts by both GC and AAS revealed a highly significant (though nonlinear) correlation between CH_3Hg^+ as such and organic Hg ($r = 0.964$; $P < 0.001$), indicating that organic Hg was equivalent to CH_3Hg^+.

3. RESULTS AND DISCUSSION

3.1 Inhibition and stimulation of microbial activities by different metal and selenium concentrations

3.1.1 Effects of metals. Experiments on the effects of increasing Cu, Cd, and Zn concentrations on CH_3Hg^+ production in sediments demonstrated that microbial communities in different lakes may differ radically in their responses to heavy metal contamination (Fig. 1). The nature of the response depended on the quantities of Hg and Zn present in the sediment before the addition of metal supplements (Table 1), apparently indicating that adaptation of microbial populations to high metal concentrations through the evolution of metal-tolerant strains under field conditions was of key importance in the severely metal-polluted lakes but not in the mildly polluted or unpolluted lakes (see below).

In mud from the "unpolluted" East Mynarski Lake, whose metals are at background levels (Table 1), CH_3Hg^+ production barely changed with increasing metal concentration up to a critical threshold level, which was lowest in the case of Cd, and then declined sharply and continuously (Fig. 1A). Apart from a very slight stimulation of Hg methylation by Cu at concentrations below the toxic threshold (in the range 0.01-10 μM Cu), all three trends are essentially smooth and

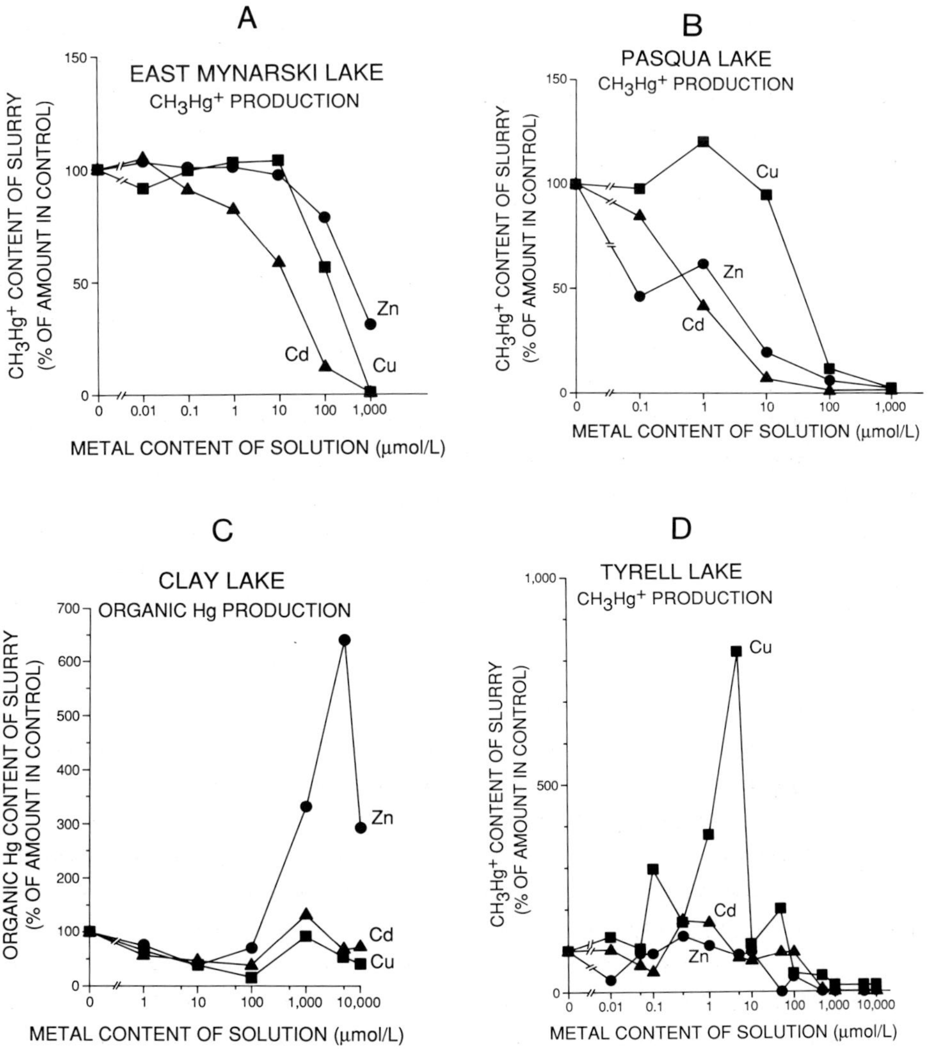

Fig. 1. Effects of different quantities of Cu, Cd, and Zn on CH_3Hg^+ production in sediments from (A) East Mynarski Lake, (B) Pasqua Lake, (C) Clay Lake, and (D) Tyrell Lake on incubation for 7 days under N_2. (<u>Note</u>: "organic Hg" is equivalent to CH_3Hg^+.)

featureless, indicating that the only major effect of the added metals was the expected inhibition of methylation at higher metal concentrations. The sensitivity of methylators to metal poisoning increased in the order Zn < Cu < Cd, but the effects of all three metals were qualitatively similar. Pasqua Lake mud, which is moderately polluted with Hg (Table 1), gave comparable results, but

TABLE 1

The mean total Hg, Zn, Cu, and Cd concentrations in the sediments used in the experiments, together with a qualitative assessment of the importance of metal-tolerant microbial species in these sediments based on inspection of experimental results.

Sampling site	Development of metal tolerance strong (+) or weak to absent (0)	Total metal concentration (nmol/g dry wt.)			
		Hg	Zn	Cu	Cd
Clay Lake (east basin)	+	33.9	6,690	667	85
Tyrell Lake	+	3.8	1,810	220	4.0
Pasqua Lake (eastern part)	0	0.57	1,700	548	4.4
East Mynarski Lake	0	0.44	1,660	567	1.9

there was a somewhat more pronounced stimulation of CH_3Hg^+ production - by both Cu and Zn, but not by Cd - at the lower metal levels (0.1-10 μM) (Fig. 1B). However, the dominant effect, again, was inhibition at the higher metal levels. Toxicity in this case increased in the order Cu < Zn < Cd.

In contrast to the muds from East Mynarski Lake and Pasqua Lake, the severely polluted muds from Clay Lake and Tyrell Lake, which have the highest Hg and Zn concentrations (Table 1), showed major nonrandom oscillations between inhibition and stimulation of CH_3Hg^+ production, resulting in alternation between CH_3Hg^+ maxima and minima, with increasing Cu, Cd, or Zn content (Figs. 1C, 1D) up to a threshold above which the metals were consistently inhibitory (Fig. 1D). All three metals gave similar patterns of variation, although the absolute values of the maxima and minima differed. In Clay Lake mud the abundance of CH_3Hg^+ (organic Hg) declined progressively with increasing metal concentration, reached a minimum, then rose steeply to a maximum, equaling or even far exceeding the amount in the control (to which no metals other than Hg had been added), and then declined again (Fig. 1C). The three metals did not differ much in their inhibitory effects but diverged widely in their stimulatory effects, the degree of stimulation increasing in the order Cu < Cd << Zn. In Tyrell Lake mud there was a succession of maxima and minima in CH_3Hg^+ yield (the peak values, again, commonly equaling or greatly exceeding the control values) until, at the highest metal levels (> 100 μM), the oscillations were damped down owing to unrelieved suppression of Hg methylating activity (Fig. 1D). Cu far exceeded Zn and Cd in its stimulatory effects.

The patterns of variation displayed by Clay Lake and Tyrell Lake sediments were essentially the same for buffered and unbuffered systems, even though the CH_3Hg^+ yields tended to be higher in the buffered slurries (pH ~6.8-8.0) than in the unbuffered ones, which were acidic (pH 4.2-6.1) (cf. 19, 31). The experiment employing Tyrell Lake mud amended with Cd was repeated once with the addition of a buffer ($CaCO_3$) and then a second time without a buffer but with neutralisation of the Cd stock solution, whose pH was acidic (5.04) owing to metal hydrolysis. The patterns of variation in CH_3Hg^+ production with increasing Cd concentration were nearly the same in all three experiments (Figs. 2A, 3A, 3B), demonstrating that the complex effects of the metal were remarkably reproducible, besides being insensitive to considerable variations in ambient pH.

Variations in the rate of CH_3Hg^+ production in Tyrell Lake mud with increasing Cu, Cd, or Zn concentration corresponded closely with variations in other microbial functions, signifying that the changes in microbial Hg transforming activities were accompanied by more general changes in the nature and activities of the microflora. Thus, the peaks in CH_3Hg^+ abundance coincided with CO_2 maxima and usually, though not invariably, with pH maxima and corresponding Eh minima (Figs. 2, 3), suggesting that the ranges of heavy metal abundance which fostered accelerated CH_3Hg^+ production did so at least partly because they promoted heterotrophic microbial growth and activity in general (cf. 11, 12, 15). Nonetheless, the variations in the abundance of CH_3Hg^+ could not have been caused by the changes in pH and Eh, as the suppression of pH (and Eh) fluctuations by the addition of a buffer did not alter the pattern of variation in CH_3Hg^+ production (Fig. 3A). (At Cd concentrations > 500 μM the pH and Eh did fluctuate despite the presence of $CaCO_3$, but methylation was totally blocked by the Cd). Presumably the CH_3Hg^+ and CO_2 levels and pH and Eh represented in Figure 2 are correlated because the values for all of these parameters depend on the intensity of heterotrophic microbial activity.

Since CH_3Hg^+ is formed from inorganic Hg and is therefore especially abundant in Hg-polluted bodies of water, the quantity of

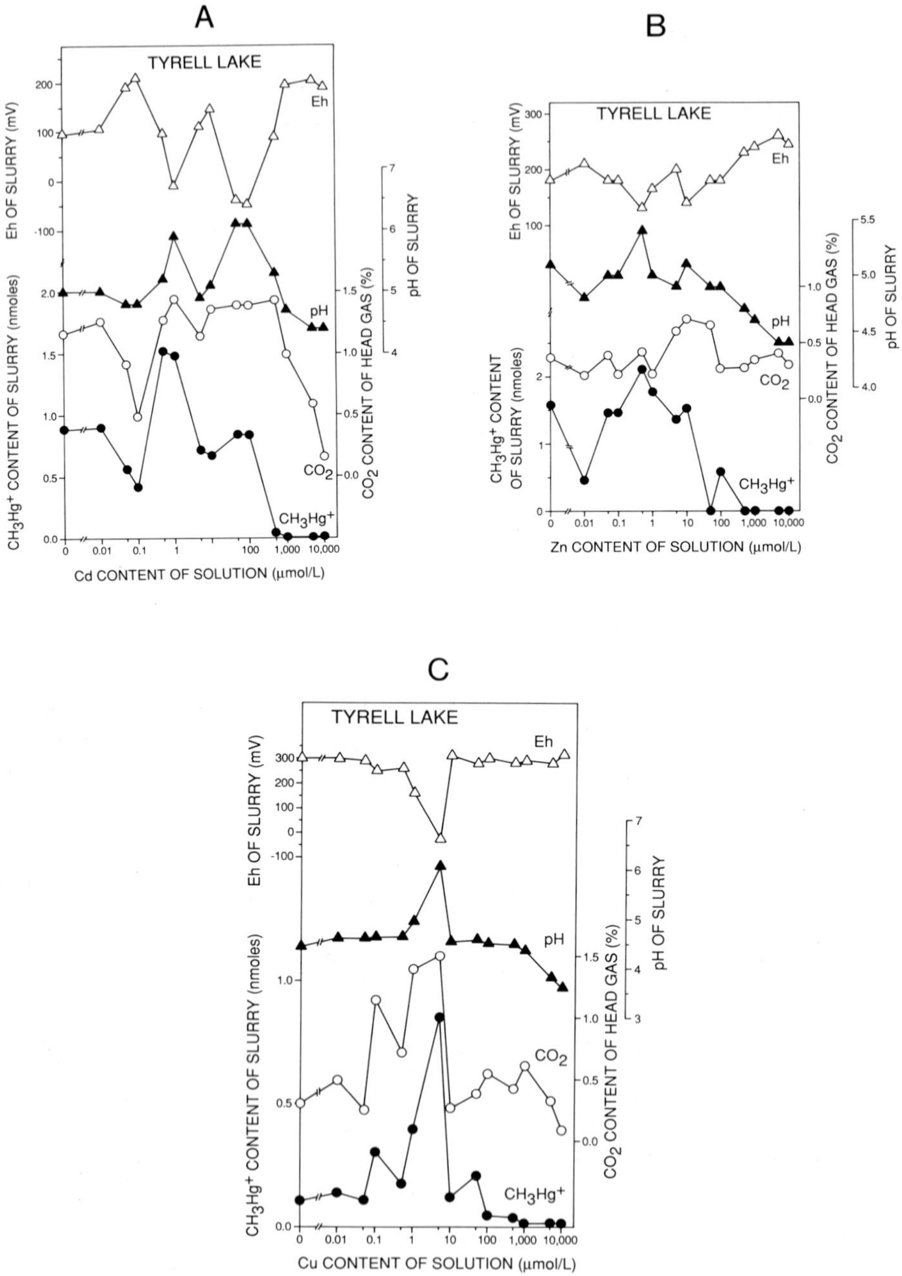

Fig. 2. Effects of different quantities of (A) Cd, (B) Zn, and (C) Cu on CH_3Hg^+ and CO_2 production, Eh, and pH in Tyrell Lake sediment on incubation for 7 days under N_2.

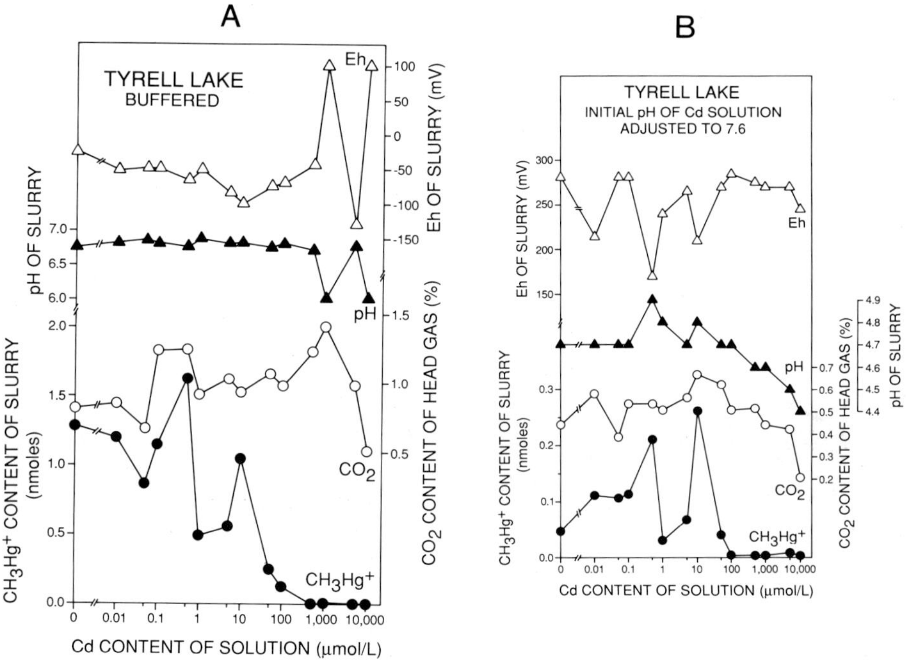

Fig.3. Effects of different quantities of Cd on CH_3Hg^+ and CO_2 production, Eh, and pH in Tyrell Lake sediment on incubation for 7 days under N_2, with (A) a buffer ($CaCO_3$) added to the sediment and (B) the acidity of the Cd solution neutralised prior to start of experiment.

CH_3Hg^+ produced by a mass of sediment under constant ambient conditions would be expected to rise continuously with increasing inorganic Hg concentration, finally peaking and then declining as the microflora succumbed to Hg poisoning at sufficiently high Hg levels. Although experimental data reported in the literature have substantiated this conception (31, 32), the results of an experiment employing Clay Lake mud presented a more complex picture. CH_3Hg^+ (organic Hg) tended to increase, as expected, with increasing inorganic Hg, but superimposed on this trend was a succession of maxima and minima like those induced by Cu, Cd, and Zn (Fig. 4A). Although this pattern of variation is clearly discernible in a log-log plot of organic Hg produced against inorganic Hg added (Fig. 4A), substitution of a linear scale for the organic Hg shows that only one of the maxima is of major significance (Fig. 4B). Very little organic Hg was generated until the inorganic Hg concentration exceeded 1,000 μM, when the quantity of organic Hg rose abruptly to

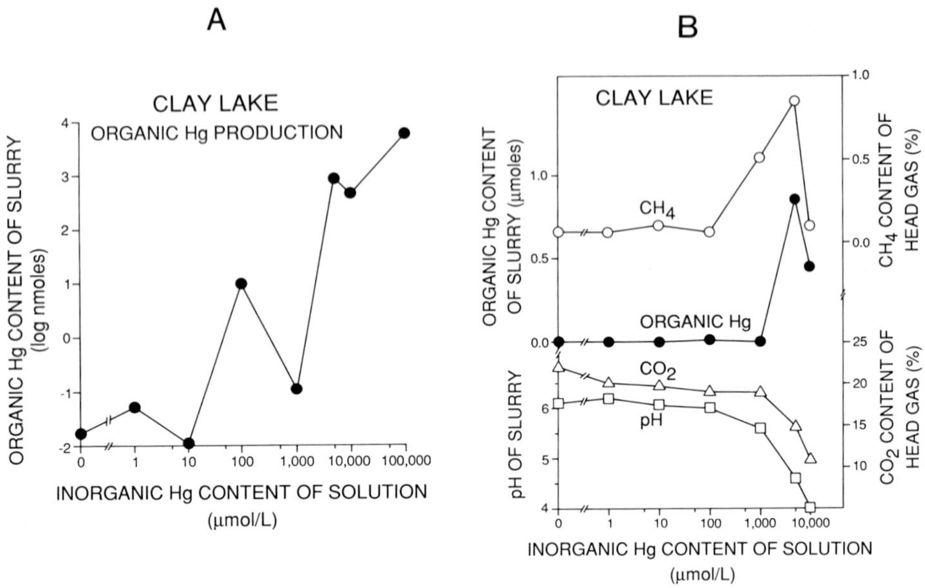

Fig. 4. Effects of different quantities of inorganic Hg on organic Hg (CH_3Hg^+), CH_4, and CO_2 production and pH in Clay Lake sediment on incubation for 7 days under N_2, with organic Hg plotted on (A) a log-log scale and (B) a linear-log scale.

form a large peak coinciding with a peak in CH_4 concentration and a sharp decline in both pH and CO_2 concentration (Fig. 4B). This suggests a major upsurge in methylation rate owing to activation of Hg-methylating CH_4 bacteria. The organic Hg peak was probably not caused by the drop in pH, as organic Hg or CH_3Hg^+ production was generally depressed, not enhanced, at low ambient pH values. The efficiency of Hg methylation apparently depended, above all, on the methylating efficiency or overall activity or abundance of the particular species of methylating microbe that were best adapted to a certain narrow range of inorganic Hg concentrations and thus were the dominant methylators in that range.

The phenomena described thus far occurred under experimental conditions, but are they of any importance under natural conditions? Analytical data for sediment samples from different lakes in the vicinity of Flin Flon strongly suggest that they are. The CH_3Hg^+ content of the samples alternately decreased and increased as a function of total Cd content, producing a regular zigzag pattern similar to the patterns produced experimentally by sediments from Clay Lake and Tyrell Lake (Fig. 5; cf. Figs. 1C, 1D, 2, 3) (cf. 19,

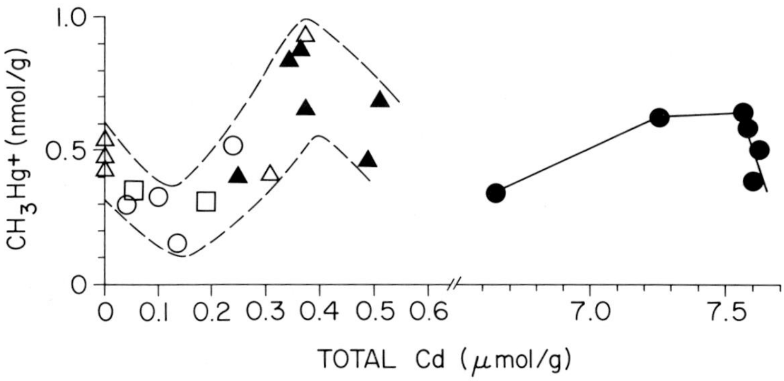

Fig. 5. Relationship between CH_3Hg^+ and Cd in sediment samples from lakes near Flin Flon (Northwest Arm of Schist Lake, ●; Inlet Arm of Schist Lake, ○ ; Phantom Lake, ▲ ; Hamell Lake, △ ; West Nesootao Lake, □). Originally published elsewhere (30).

Fig. 22A), probably for the same reasons. In contrast, inorganic (unmethylated) Hg content simply correlated positively with Cu, Cd, and Zn content (reflecting the fact that all the heavy metals came from the same source of pollution) and showed no trace of a zigzag pattern (30, Fig. 14.4C).

What general principles can be evoked to explain the various observed effects of heavy metals on microbial activities? Specifically, why did severely Hg and Zn contaminated sediments show large alternating downswings and upswings in the intensity of CH_3Hg^+ production and associated microbial activities with increasing metal concentration, whereas unpolluted sediment showed little more than a continuous decline in methylating activity? The probable reason is that microbial communities in metal-polluted sediments included a number of more or less metal-tolerant species or strains which had evolved in response to their metal-rich environment, whilst the microbial community in the unpolluted sediment consisted mostly of metal-sensitive species (33-36). The series of CH_3Hg^+ maxima and minima observed in Clay Lake and Tyrell Lake mud are attributable to ecological succession in the microbial community - specifically alternate suppression of metal-sensitive methylating microbes and stimulation of tolerant ones. With a rise in metal concentration, the more sensitive species were increasingly inhibited, causing a fall in methylation rate. Beyond a critical threshold concentration, their activities had diminished so drastically that certain relatively tolerant species, which had been kept in check by competition or other antagonistic relations with the sensitive ones, were able to grow, initiating a resurgence of methylating activity. With a further rise in metal concentration, these tolerant species, in turn, were inhibited by the metal and finally supplanted by still more tolerant species, and the process continued until the metal concentration was so high that even the most tolerant species were suppressed. The microbes contributing to the observed effects may well have included not only methylators but also demethylators and species which did not mediate Hg transformations but influenced the activities of those that did. The apparent absence or scarcity of tolerant species in the unpolluted East Mynarski Lake mud is readily explained by the fact that organisms in an unpolluted environment are not under pressure to evolve adaptations to toxic pollutants. Note that the intermittent stimulation of Hg methylation was probably not due to any nutritional function of the element added: Major stimulation occurred only in sediments that had high metal concentrations to begin with and were therefore unlikely to be deficient in nutrient trace elements (although the bio-availability of the element as well as its abundance in the sediment must be taken into consideration). Besides, the commonly observed inhibition at lower levels of metal supplement followed by stimulation at higher levels is the opposite to the expected result of supplying a deficient micronutrient.

3.1.2 Effects of selenium. Se, when added to Tyrell Lake sediment, had effects similar to those of Cu, Cd, and Zn (Fig. 6; cf. Figs. 2, 3). Overall, CH_3Hg^+ tended to decline with increasing Se, but superimposed on this trend were CH_3Hg^+ peaks coinciding with CO_2 peaks and, with one exception, coinciding with pH and Eh fluctuations as in Figures 2 and 3B. At all Se concentrations $\geq$ 50 μM, CH_3Hg^+ production was totally suppressed, but CO_2 continued to be generated in abundance and even formed a large peak at 1,000 μM Se, implying the existence of an active microflora. These observations are best explained by selective inhibition of methylation at the highest Se levels (probably owing to strong fixation of Hg by selenides), not by general inhibition of microbial activity.

Although it would not be right to derive sweeping generalisations from such limited data or to assume that experimental data are necessarily pertinent to processes occurring under natural conditions, the results of the experiment raise questions about the advisability of treating Hg-polluted natural waters with Se in order to make the Hg less toxic or less bio-available (21). Effects of Se on the accumulation of Hg by the aquatic biota are beyond the scope

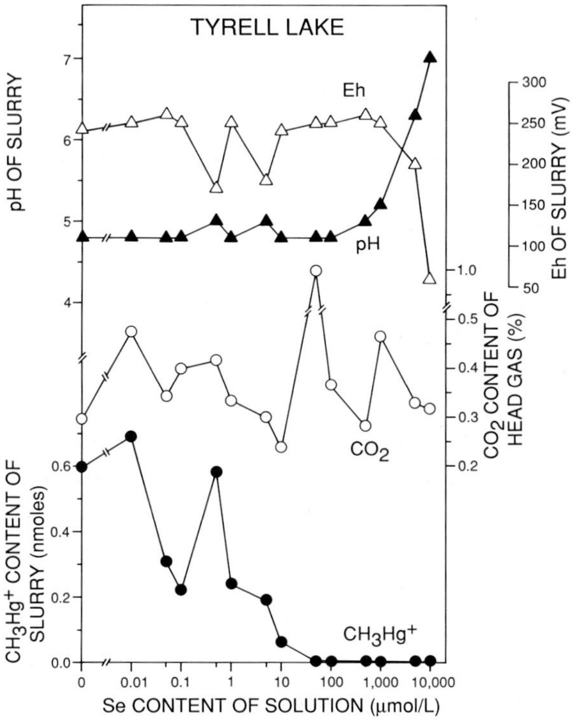

Fig. 6. Effects of different quantities of Se on CH_3Hg^+ and CO_2 production, Eh, and pH in Tyrell Lake sediment on incubation for 7 days under N2.

of this paper; but the effects of Se on Hg methylation are evidently complex, inconsistent, and thus unreliable, as they may alternate between inhibition and stimulation of CH_3Hg^+ production over a wide range of Se concentrations. Above all, unacceptably high concentrations of Se, which is hazardous in its own right, would seem to be needed to block methylation effectively. Thus, a concentration of 3,950 μg Se/L (50 μM) or more, which is orders of magnitude higher than what is considered safe, was required to achieve total suppression of Hg methylation. In natural waters, Se concentrations as low as 10 μg/L, the legal upper limit for drinking water in Canada (21), can be harmful to fish (22), and Se at concentrations exceeding 2-5 μg/L is believed to be capable of undergoing biomagnification in food chains, resulting in toxic effects (22). The massive doses of Se which would apparently be needed to suppress methylation would be out of the question, and even far lower doses could have lethal or sublethal effects on fish. In general, the alteration of complex, delicately balanced natural systems by the introduction of a potentially toxic chemical whose ecological effects are not fully understood could well have unforeseen harmful consequences. Using one environmental poison to counteract another could turn out to be like the folly of the legendary British king Vortigern, who hired Saxon mercenaries to beat back invading Picts and Scots, thereby unwittingly inviting the Saxon conquest of Britain.

3.2 Effects of different metal and selenium concentrations on microbial activities over time

3.2.1 Stimulatory effects of metals under anaerobic and aerobic conditions. At a concentration of only 10 μM, Cd had no significant effect on CH_3Hg^+ or CO_2 production or on pH or Eh in Tyrell Lake mud over a span of 14 days, but the same molar concentration of Zn caused a sudden major upsurge in both CH_3Hg^+ and CO_2, accompanied by a drop in Eh and a rise in pH, after day 7 (Fig. 7, A-D). The increase in CH_3Hg^+ caused by Zn was nearly six-fold. As with Cd, 10 μM Cu did not affect Hg methylation appreciably, as revealed by a separate experiment. Evidently the Zn promoted heterotrophic microbial activity in general as well as Hg methylating activity in particular (cf. Fig. 2B). However, when the experiment was repeated under aerobic conditions, the results were altogether different: Zn had only a trifling, temporary positive effect on Hg methylation, whilst Cd had a small, though consistent, positive effect after day 3 (Fig. 8). Similarly, the metals had only small and not entirely consistent effects on CO_2 production, Eh, and pH: There was no dramatic upsurge of activity as was caused by Zn under anaerobic conditions.

The reasons for the profound difference between the effects of Zn under aerobic and anaerobic conditions are not known, but they could involve differences in the forms and bio-availability of Zn or differences in the Zn requirements of the aerobic and anaerobic organisms (Zn being an essential micronutrient for all organisms). The data could be interpreted as meaning that under anaerobic conditions the Zn promoted the growth of certain microbes by making good a micronutrient deficiency. This seems unlikely in view of the large amount of Zn already present in the sediment and the small amounts presumably required for microbial growth, but conceivably most of that Zn might have been in biologically unavailable forms. In any case, these striking results indicate that with alternation between reducing and oxidising conditions on a lake bottom the role

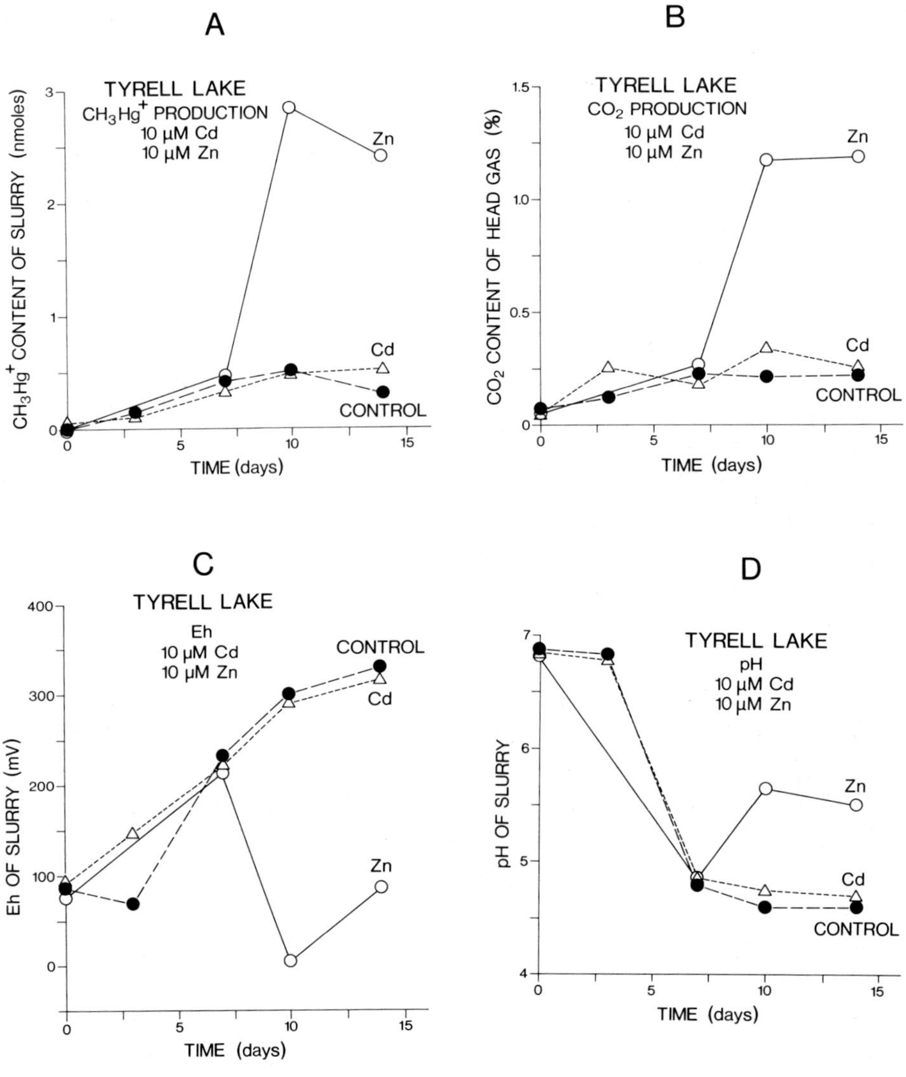

Fig. 7. Variations in (A) CH_3Hg^+ production, (B) CO_2 production, (C) Eh, and (D) pH in Tyrell Lake sediment as functions of incubation time in the presence and absence of 10 μM Cd and 10 μM Zn when incubated under N_2.

of a metal in microbial ecology may be drastically altered, even if there is no change in the total concentration of the metal.

3.2.2 <u>Effects of metal-metal and metal-selenium mixtures as compared with effects of the individual elements</u>. At metal concentrations of 100 μM, both Zn and Cd depressed Hg methylation, Cd being much more toxic than Zn (Fig. 9A). Nevertheless, a mixed

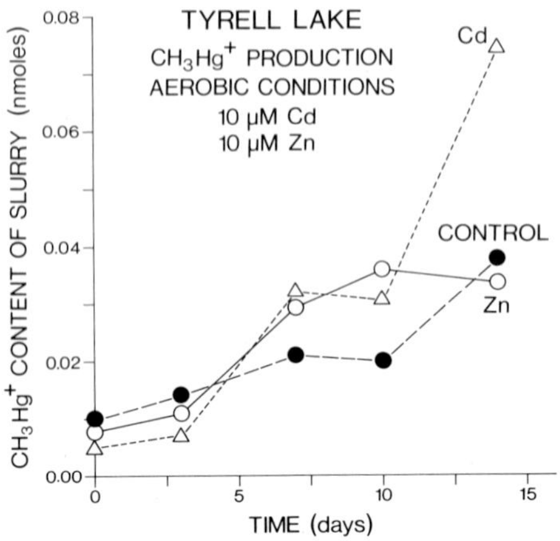

Fig. 8. Variations in CH_3Hg^+ production in Tyrell Lake sediment as functions of incubation time in the presence and absence of 10 μM Cd and 10 μM Zn when incubated under air.

solution containing 100 μM Cd + 100 μM Zn had practically the same effect as Cd alone (Fig. 9A). Thus, Cd was antagonistic to the action of Zn, completely cancelling the toxic effect of the Zn without its own toxic effect being modified. A possible explanation is that Cd combined preferentially with the binding sites or enzymes of the microbes, excluding the Zn.

In a similar experiment employing 100 μM each of Cd and Cu, and 100 μM Cd + 100 μM Cu in mixed solution, Cu stimulated methylation slightly and, when mixed with Cd, ameliorated or offset the inhibitory effect of the Cd to about the same extent (Fig. 9B). The results suggest that the two metals, when mixed, exerted separate, independent effects without interacting with each other.

In a third experiment involving 100 μM each of Cd and Se, and 100 μM Cd + 100 μM Se, Se almost completely suppressed Hg methylation, being far more effective in that regard than Cd (Fig. 9C). Although both elements were highly toxic, the mixture of Se and Cd had the same impact as Se alone (Fig. 9C). As Se protects organisms against the toxicity of Cd as well as Hg (20), it may be supposed that the Se, while suppressing Hg methylation, also prevented Cd from exerting an inhibitory effect.

In none of these experiments could the impact of the metals or Se on Hg methylation be attributed to inhibition of heterotrophic microbial activity in general. Consequently, effects on particular microbial species or on particular types of biochemical activity would seem to be probable.

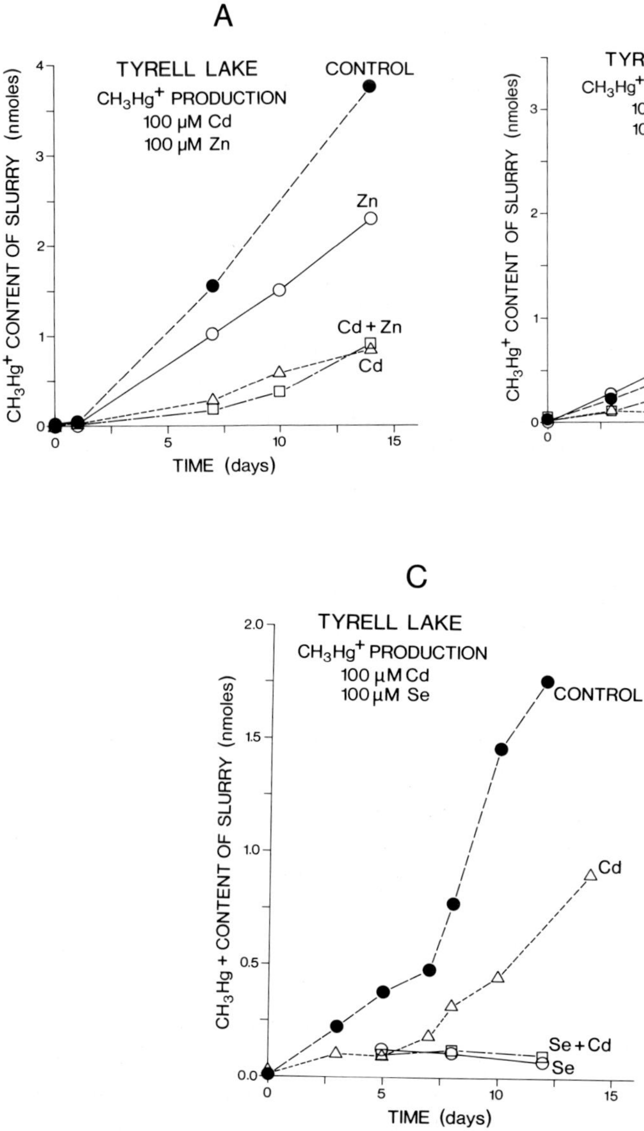

Fig. 9. Variations in CH_3Hg^+ production in Tyrell Lake sediment as functions of incubation time in the presence and absence of 100 μM Cd, Zn, Cu, and Se, individually and in binary mixtures, when incubated under N_2:
(A) Cd, Zn, and Cd + Zn;
(B) Cd, Cu, and Cd + Cu;
(C) Cd, Se, and Cd + Se.

3.3 Correlations between biological effects of metals and various properties of the metals

The effects of Cu, Cd, and Zn on microbial Hg methylating activity varied systematically and significantly with certain properties of the metals as represented by measurements gleaned from the chemical literature. Such relationships provide a basis for interpreting and predicting the behaviour of different metallic elements in various natural environments with reference to fundamental laws of chemistry and biochemistry (37-40).

In East Mynarski Lake mud, metal toxicity, expressed as the metal concentration required to achieve a 50% reduction in CH_3Hg^+ production (ED_{50}), was inversely correlated with the standard entropy of the metal sulfide (41) (Fig. 10A). In other words, the greater

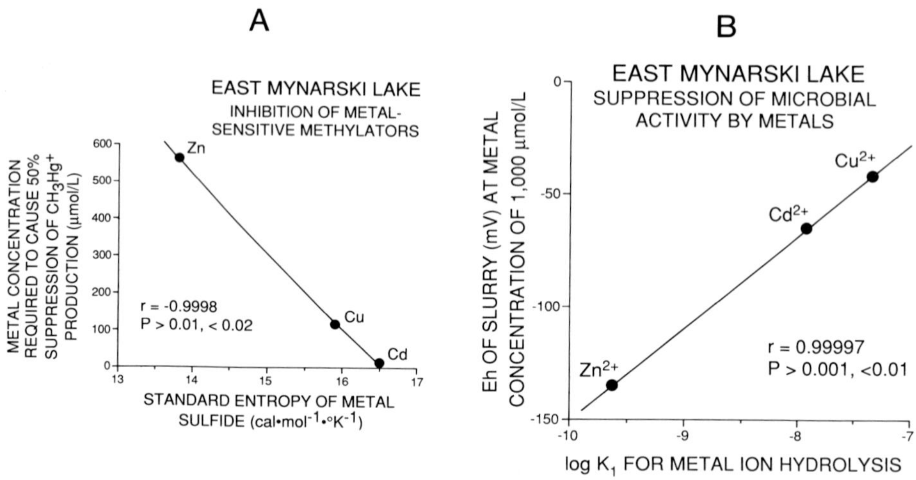

Fig. 10. Relationship between effects of Cu, Cd, and Zn on microbial activities and properties of the metals in East Mynarski Lake sediment: (A) ED_{50} for CH_3Hg^+ production plotted against the standard entropy of the metal sulfide; (B) Eh at metal concentration of 1,000 μM plotted against the equilibrium constant (log K_1) for metal ion hydrolysis.

the stability of the metal-sulfide bond as measured by the entropy (but not the enthalpy) term of the Gibbs free energy equation, the more effective was the metal in suppressing Hg methylation, the toxicity increasing in the order Zn << Cu < Cd. However, the toxicity of the metals to microbial activity in general, as measured by Eh at a metal concentration of 1,000 μM (high Eh being symptomatic of poor microbial growth), was positively correlated with the equilibrium constant (log K_1) for metal ion hydrolysis (42)

(Fig. 10B); in this case the degree of toxicity increased in the order Zn << Cd < Cu. The correlation shown in Figure 10A could mean that inhibition of CH_3Hg^+ production was caused by the binding and inactivation of sulfhydryl (-SH) groups of enzymes by the metals, and dependence of bond stability on entropy rather than enthalpy hints at the involvement of chelation (39, 43). The correlation in Figure 10B also suggests inhibition of enzymes by the binding of metals by active sites, as the polarisability of the metal ion, and therefore the covalence of metal-ligand bonds, tend to increase with log K_1 for metal ion hydrolysis (44). Thus, both of these relationships suggest inhibition of enzymes, but the difference in the order of increasing toxicity of the metals and in the elemental properties controlling the degree of toxicity implies that the enzymes and organisms involved and the nature of the metal-enzyme reactions are not the same in the two cases.

In Pasqua Lake mud, the ED_{50} values for CH_3Hg^+ production correlated positively with the second ionisation potential of the metal (45) (Fig. 11). Thus, the greater the covalent character of

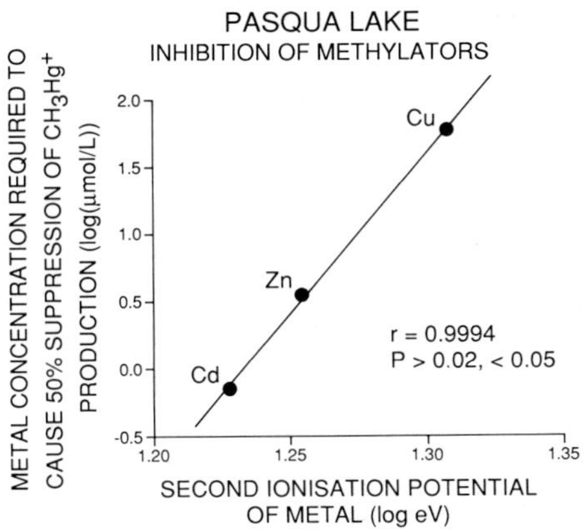

Fig. 11. Relationship between effects of Cu, Cd, and Zn on CH_3Hg^+ production (as represented by ED_{50} values) and the second ionisation potentials of the metals in Pasqua Lake sediment.

metal-ligand bonds (insofar as ionisation potential contributes to covalence), the less toxic the metal, the toxicity increasing in the order Cu < Zn < Cd. Essentially the same result was obtained for Tyrell Lake mud using the net quantity of CH_3Hg^+ generated by the end of a 10-day incubation period as the measure of relative toxicity (Fig. 12A). Furthermore, metal toxicity increased with increasing solubility of the metal sulfide (46) (Fig. 12B). In contrast, then,

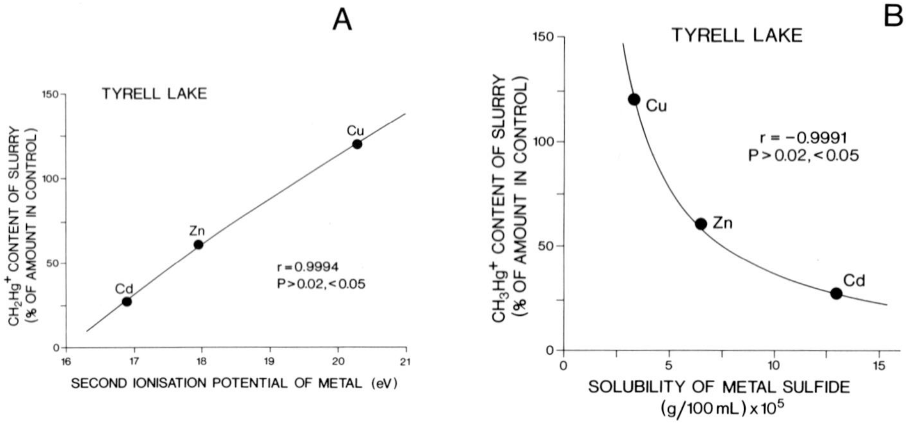

Fig. 12. Relationship between effects of Cu, Cd, and Zn on CH_3Hg^+ production (as represented by percent of CH_3Hg^+ in control slurry) and (A) the second ionisation potentials of the metals and (B) the solubilities of the metal sulfides in Tyrell Lake sediment.

to the East Mynarski Lake sediment, in which toxicity increased with increasing strength of metal-sulfide bonding, Pasqua Lake and Tyrell Lake sediments showed a decrease in toxicity with increasing strength of metal-sulfide bonding. A likely explanation for these results is that metal poisoning in Pasqua Lake and Tyrell Lake was limited primarily by protective metal-scavenging substances such as sulfides in the external environment of the microbes. Indeed, metal-binding agents may have been excreted by metal-resistant microbes as a means of rendering the metals less bio-available and less toxic.

In Clay Lake mud the relationship between biological response to metals and particular properties of the metals varied as the biological effect shifted from net inhibition to net stimulation of organic Hg production with rising metal concentration (see Fig. 1C). The ED_{50} values for the inhibitory phase (at metal concentrations in the range 0-100 μM) gave a negative correlation with the "excess polarising strength" of the metal cation (47) (Fig. 13A), meaning that metal toxicity increased as metal-ligand bonds became more highly covalent (insofar as excess polarising strength contributes to covalence). This may reflect the binding of metals by active sites, though not necessarily -SH groups (cf. Fig. 10A), of microbial enzymes, inhibition of the enzymes becoming increasingly severe with increasing bond strength and covalence. Relative toxicity increased in the order Zn << Cu ~ Cd. Nevertheless, toxicity expressed as the maximum percent suppression of organic Hg production (with respect to the control), which occurred at metal concentrations of 10-100 μM,

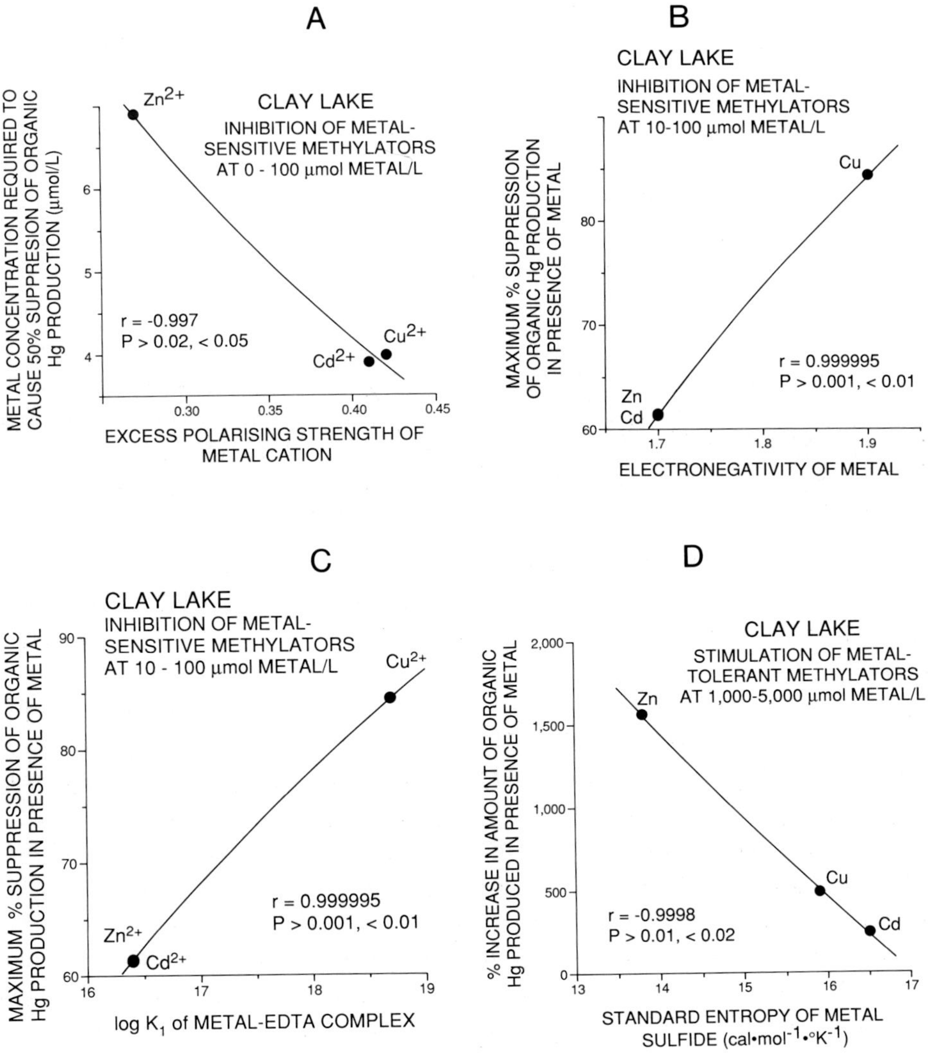

Fig. 13. Relationship between effects of Cu, Cd, and Zn on organic Hg (CH_3Hg^+) production and properties of the metals in Clay Lake sediment: Plots of (A) ED_{50} for production of organic Hg against excess polarising strength of metal cation; maximum percent suppression of organic Hg production against the (B) electronegativity of the metal and (C) stability constant (log K_1) of the metal-EDTA complex; and (D) percent increase in organic Hg production during the stimulatory phase against the standard entropy of the metal sulfide.

correlated positively with the electronegativity (44) of the metal (Fig. 13B) and the stability constant of the EDTA complex of the metal (42) (Fig. 13C). Thus, toxicity increased in the order Cd = Zn << Cu. Again, this suggests increasingly strong inhibition of enzymes with increasing strength and covalent character of bonds between metals and ligands at the active sites of enzymes, and chelation may be involved. In this case, however, electronegativity rather than excess polarising strength is the decisive factor contributing to the degree of covalence, possibly signifying that the enzymes and microbial species involved are not the same in the two cases. Note that ED_{50} simply measures inhibition of the most sensitive microbial species, whereas maximum percent suppression may reflect a steady state between toxicity and detoxification, or the existence of a mixed population of sensitive and tolerant species, near the critical metal concentration above which inhibition gave way to stimulation (D. Scott, personal communication). In the range of metal levels at which stimulation occurred (1,000-5,000 μM), the degree of stimulation (percent increase in organic Hg production) showed an inverse relationship with the standard entropy of the metal sulfide (Fig. 13D), decreasing in the order Zn > Cu > Cd. This could reflect partial inhibition owing to the inactivation of the -SH groups of enzymes by metals, or it could mean that the more readily a metal is immobilised by sulfides the less available it is to suppress metal-sensitive Hg methylating species which keep the metal tolerant methylators in check.

To summarise, the unifying principle underlying the diverse relationships illustrated in Figures 10-13 seems to be that the toxicity of a metal to a particular microbial population inhabiting a particular sedimentary environment is determined by the strength and stability of covalent bonding between the metals and certain key ligands - either active sites of microbial enzymes, in which case bonding <u>causes</u> inhibition, or complexing agents that happen to be present in the environment or were produced by the microbes for protection against metal poisoning, in which case bonding helps to <u>prevent</u> inhibition. This hypothesis accounts for the apparent paradox that toxicity may either increase or decrease with increasing affinity of the metals for sulfides.

As the nature and abundance of the ligands vary from one site to another, or with changing conditions at a single site, owing to differences in the species composition of the microflora and the physicochemical characteristics of the sedimentary matrix, so, too, do the metal properties which contribute to covalence and bond strength vary in their relative importance, and hence the relative toxicity of the metal varies accordingly. For instance, in one set of circumstances Cd may be more toxic than Cu (Figs. 10A, 11, 12A, 12B, 13D), but in another the reverse may be true (Figs. 10B, 13B, 13C) or the two elements may be nearly equal in toxicity (Fig. 13A).

4. CONCLUSIONS

A number of tentative conclusions and generalisations have been drawn from the research results. Although they are based on sound empirical data comprising both experimental and field observations and are consistent with a wider body of well established biological and chemical knowledge, they are to be regarded more as reasonable inferences and working hypotheses than as proven facts. Further work is needed to confirm and extend them and to determine how generally they apply to the natural world. With this cautionary note borne in mind, the principal interpretations may be summarised as follows:

- Contamination of lakes with Cu, Cd, Zn, Hg, and Se may have complex, variable effects involving both inhibition and promotion of Hg methylation and other microbial activities in sediments. A progressive increase in the amount of heavy metal or Se introduced into a freshwater environment already contaminated with metals is apt to cause alternating decreases and increases in CH_3Hg^+ production and corresponding fluctuations in other indicators of microbial activity, such as Eh, pH, and the production of CO_2 and CH_4. Such complex, inconsistent patterns of variation are attributable to ecological succession arising from suppression of relatively metal-sensitive microbial species followed by opportunistic flourishing of metal-tolerant species which had previously been kept in check by metal-sensitive competitors and antagonists. The episodic upsurges in CH_3Hg^+ production may be manifestations of nonspecific stimulation of heterotrophic microbial activity or selective stimulation of particular species such as CH_4 bacteria. Metal toleration in the microbial community presumably evolved as an adaptation to the high metal concentrations in the sediments.

- In contrast, introduction of progressively increasing quantities of heavy metals into a previously unpolluted lake is likely to result simply in a steady decline in Hg methylating activity with little or none of the episodic resurgence of activity seen in polluted sediments. This is ascribed to a lack of metal-tolerant species.

- Insofar as it would affect Hg methylation, the treatment of Hg-polluted lakes with Se to ameliorate the biological effects of the

Hg is of doubtful value and desirability as a method of environmental management, although evidence and arguments in its favour have been advanced by other workers. Judged within the scope of this report, Se is unreliable because its effects may alternate between inhibition and promotion of CH_3Hg^+ production over a wide range of Se concentrations, and the Se concentrations required to ensure total suppression of methylation are so high that the Se itself would probably become an ecological menace.

- The ecological effects of heavy metals may change drastically with fluctuations in the value of an environmental parameter such as dissolved O_2 concentration, as demonstrated by Zn, which, in one set of experiments, strongly stimulated CH_3Hg^+ production under anaerobic conditions but not in the presence of O_2.

- At concentrations of metals and Se that depress Hg methylation, certain element mixtures may interact antagonistically. Cd may cancel the toxicity of Zn, and Se may cancel the toxicity of Cd, without there being any change in their own toxicity. However, Cd and Cu seem to act independently of each other.

- The toxicity of heavy metals to sedimentary micro-organisms varies systematically as a function of certain metal properties, and these relationships may provide a basis for interpretation and prediction of metal behaviour in natural environments with reference to fundamental laws of chemistry and biochemistry. The relative toxicity of Cu, Cd, and Zn to microbes and the particular elemental property that controls toxicity vary with environmental conditions, probably owing to variations in the nature of the microflora (and its metal-sensitive enzymes) and the nature and abundance of environmental ligands. Thus, Cd may be more toxic than Cu in one set of circumstances, but the reverse may be true in another, and toxicity may either increase or decrease with increasing affinity of metals for sulfides. Although the relationships are diverse and not altogether consistent, there is a unifying principle that can explain them all: Relative toxicity apparently depends on various factors which contribute to the formation of strong, stable covalent bonds with ligands. Increasing strength of covalent bonding presumably results in an <u>increase</u> in toxicity when it involves the bonding of microbial enzymes but causes a <u>decrease</u> in toxicity if the metal is bonded to ligands in the external environment of the microbes.

ACKNOWLEDGMENTS

The technical work was done by K. Supeene, A. Adams, C. Baron, K. Parkkari, E. Thomson, and R. N. Woychuk, and the research was financed by the Government of Canada (Department of the Environment). I thank D. Scott for helpful comments.

REFERENCES

1 M. Alexander, Microbial Ecology, John Wiley and Sons, New York, 1971.

2 M. Alexander, in: R.R. Colwell and J. Foster (Eds.), Aquatic Microbial Ecology. Proceedings of conference sponsored by American Society for Microbiology, Clearwater Beach, Florida, 7-10 February, 1979, a Maryland Sea Grant Publication, University of Maryland, College Park, 1979, pp. 110-123.

3 L.R. Pomeroy, in: R.R. Colwell and J. Foster (Eds.), Aquatic Microbial Ecology. Proceedings of conference sponsored by American Society for Microbiology, Clearwater Beach, Florida, 7-10 February, 1979, a Maryland Sea Grant Publication, University of Maryland, College Park, 1979, pp. 85-109.

4 M.D. Baker, W.E. Inniss, C.I. Mayfield, and P.T.S. Wong, Water Res., 17 (1983) 925-930.

5 D.G. Capone, D.D. Reese, and R.P. Kiene, Appl. Environ. Microbiol., 45 (1983) 1586-1591.

6 C.E. Nugent, G.J. Atchison, D.W. Nelson, and A.W. McIntosh, Hydrobiol., 70 (1980) 69-73.

7 J.H. Slater, in: R.R. Colwell and J. Foster (Eds.), Aquatic Microbial Ecology. Proceedings of conference sponsored by American Society for Microbiology, Clearwater Beach, Florida, 7-10 February, 1979, a Maryland Sea Grant Publication, University of Maryland, College Park, 1979, pp. 15-24.

8 H. Babich and G. Stotzky, Ecol. Bull. (Stockholm) 35 (1983) 315-323.

9 T.A. Jackson, in: J.-P. Vernet (Ed.), Heavy Metals in the Environment, vol. 1. Proceedings of International Conference, Geneva, September, 1989, CEP Consultants Ltd., Edinburgh, 1989, pp. 65-68.

10 K. Beijer and A. Jernelöv, in: J.O. Nriagu (Ed.), The Biogeochemistry of Mercury in the Environment, Elsevier, Amsterdam, 1979, pp. 203-210.

11 R.A. Bodaly, R.E. Hecky, and R.J.P. Fudge, Can. J. Fish. Aquat. Sci. 41 (1984) 682-691.

12 T.A. Jackson, Can. J. Fish. Aquat. Sci. 45 (1988) 97-121.
13 J.W. Huckabee, J.W. Elwood, and S.G. Hildebrand, in: J.O. Nriagu (Ed.), The Biogeochemistry of Mercury in the Environment, Elsevier, Amsterdam, 1979, pp. 277-302.
14 H.L. Windom and D.R. Kendall, in: J.O. Nriagu (Ed.), The Biogeochemistry of Mercury in the Environment, Elsevier, Amsterdam, 1979, pp. 303-323.
15 T. Fagerström and A. Jernelöv, Water Res. 6 (1972) 1193-1202.
16 M. Shariat, A.C. Anderson, and J.W. Mason, Bull. Environ. Contam. Toxicol. 21 (1979) 255-261.
17 A. Björnberg, L. Håkanson, and K. Lundbergh, Environ. Pollut. 49 (1988) 53-61.
18 T.A. Jackson, Can. J. Fish. Aquat. Sci. 43 (1986) 1873-1887.
19 T.A. Jackson, Appl. Organometallic Chem. 3 (1989) 1-30.
20 R.A. Rimerman, D.R. Buhler, and P.D. Whanger, in: S.D. Lee (Ed.), Biochemical Effects of Environmental Pollutants, Ann Arbor Science Publishers, Ann Arbor, Michigan, 1977, pp. 377-396.
21 M.A. Turner and J.W.M. Rudd, Can. J. Fish. Aquat. Sci., 40 (1983) 2228-2240.
22 A.D. Lemly and G.J. Smith, Fish and Wildlife Leaflet 12, U.S. Department of the Interior, Fish and Wildlife Service, Washington, D.C., 1987.
23 W.G. Franzin, G.A. McFarlane, and A. Lutz, Environ. Sci. Technol., 13 (1979) 1513-1522.
24 F.A.J. Armstrong and A.L. Hamilton, in: P.C. Singer (Ed.), Trace Metals and Metal-Organic Interactions in Natural Waters, Ann Arbor Science Publishers, Ann Arbor, Michigan, 1973, pp. 131-156.
25 T.A. Jackson and R.N. Woychuk, in: U.S. Environmental Protection Agency (Eds.), Restoration of Lakes and Inland Waters. Proceedings of International Symposium on Inland Waters and Lake Restoration, Portland, Maine, 8-12 September, 1980, U.S. EPA, Washington, D.C., 1980, pp. 93-101.
26 T.A. Jackson, J.W. Parks, P.D. Jones, R.N. Woychuk, J.A. Sutton, and J.D. Hollinger, Hydrobiol. 92 (1982) 473-487.
27 J.W.M. Rudd, M.A. Turner, A. Furutani, A. Swick, and B.E. Townsend, Can. J. Fish. Aquat. Sci. 40 (1983) 2206-2217.
28 J.W. Parks and A.L. Hamilton, Hydrobiol. 149 (1987) 159-188.

29 W.F. Warwick, in: D.A. Murray (Ed.), Chironomidae: Ecology, Systematics, Cytology, and Physiology, Pergamon Press, Oxford, 1980, pp. 255-267.
30 T.A. Jackson, in: J.O. Nriagu (Ed.), Environmental Impacts of Smelters, John Wiley and Sons, New York, 1984, pp. 551-578.
31 T.A. Jackson, in: Governments of Canada and Manitoba, Technical Appendices to the Summary Report, Canada-Manitoba Agreement on the Study and Monitoring of Mercury in the Churchill River Diversion, vol. 2, chapt. 8.
32 S. Jensen and A. Jernelöv, Nature, 223 (1969) 753-754.
33 T.C. Hutchinson and P.M. Stokes, Water Qual. Parameters, AST STP 573, American Society for Testing and Materials (1975) 320-343.
34 J.D. Nelson, Jr. and R.R. Colwell, Microbial Ecol. 1 (1975) 191-218.
35 C. Houba and J. Remacle, Microbial Ecol. 6 (1980) 55-69.
36 P.L. Foster, Freshwater Biol. 12 (1982) 41-61.
37 E. Somers, Nature 187 (1960) 427-428.
38 T.A. Jackson, Environ. Geol. 2 (1978) 173-189.
39 T.A. Jackson, in: Proceedings of International Conference on Management and Control of Heavy Metals in the Environment, London, September, 1979, CEP Consultants Ltd., Edinburgh, 1979, pp. 457-460.
40 T.A. Jackson, G. Kipphut, R.H. Hesslein, and D.W. Schindler, Can. J. Fish. Aquat. Sci. 37 (1980) 387-402.
41 K.C. Mills, Thermodynamic Data for Inorganic Sulphides, Selenides, and Tellurides, Butterworths, London, 1974.
42 L.G. Sillén and A.E. Martell, Stability Constants of Metal-Ion Complexes: Special Publication no. 17, 1964, and Supplement no. 1, Special Publication no. 25, 1971, The Chemical Society, London.
43 F. Basolo and R.C. Johnson, Coordination Chemistry, W.A. Benjamin, New York, 1964.
44 B.E. Douglas and D.H. McDaniel, Concepts and Models of Inorganic Chemistry, Blaisdell (Ginn and Co.), Waltham, Massachusetts, 1965.
45 C.S.G. Phillips and R.J.P. Williams, Inorganic Chemistry, vol. 1, Oxford University Press, Oxford, 1965.
46 R.C. Weast (Ed.), CRC Handbook of Chemistry and Physics, 63rd Edition, CRC Press, Boca Raton, Florida, 1982.
47 R. Rich, Periodic Correlations, W.A. Benjamin Inc., New York, 1965.

Chapter 7
Health effects of metals

HEALTH EFFECTS OF THE WORKERS EXPOSED TO AIRBORNE LEAD

M.M. Orlov, N.M. Mladinić, P.I. Kulić, D.Z. Dolinac,
M.N. Dubravčić and L.B. Kurajica

SUMMARY

Health effects of 56 workers exposed to high concentration of lead in the air (up to 23.0 mg/cu.m. in PVC industry were investigated. The following parameters were examined periodically in all workers during 2-3 years: Blood Pb (Pb-B), Dehydratase Delta ALA (ALAD) and Free Eritrocyte Porphiryn (FEP).

The results confirm good correlation between Pb-exposure and biochemical parameters. Moreover, the reduction in motor nerve conduction velocity and some tooth abnormalities in exposed workers were found.

1. INTRODUCTION

In Split's PVC industry one group of the workers was exposed to airborne lead (aerosols and dust) in the process of weighting and pouring (WAP) stabilizers. The WAP process cannot be a source of dangerous lead poisoning in ordinary industry production. However, in this case we have two main characteristics which caused these conditions:

- last years Pb-stabilizers were delivered in big sacks and workers had to weight and pour them in the fixed portion in a small and bad-ventilated workshop;
- many of the workers did not use all health protective fittings during the WAP operation.

2. SUBJECT AND METHODS

A total of 56 Pb-exposed workers and 20 non-exposed persons were observed in this study. The exposed workers were divided in two groups formed on the basis of the different concentration of lead exposure:

- I group covers the workers who are directly involved in the WAP operation.
- II group covers the workers who have others jobs close to WAP operation place.

From the first group, 7 workers were separated in I(a) group, because of existing of the big lead-poisoning effect before the period of investigation (FEP over 3.0 μmol/l).

For all workers we investigated blood Pb (Pb-B), blood ALA dehydratase (ALAD) and free eritrocyte porphiryn (FEP), at the intervals which depended of the degree of lead absorption. Pb-B was measured with graphite furnace technic on the AAS Perkin-Elmer model 2380 using "addition" method /1,2/, ALAD according to European standardized method /3/, FEP with microphotofluorometric determination /4/.

For the workers of the first group reticulocyte (RET) were measured periodically.

Nerve conduction velocity of several workers from each group were examined, too. The results are obtained on the electro-myograf "Alvar-Myodine II P" by using direct stimulation with bipolar not insulated electrode.

Teeth control of the workers was done according to DMF values. Furthermore, we tried to find out daily Pb-exposition due to Pb-concentration in a human saliva. The determination was done directly with the graphite furnace technic on the ASS.

3. RESULTS AND DISCUSSION

3.1. Airborne lead contamination

As we previously mentioned, the WAP workers made their activities in a small and bad-ventilation workshop. They spent 2-3 hours a day in the contaminated area. According to the results of the Security Institute of Zagreb, the average concentration of dust and aerosols of Pb was 12,8 mg/m^3 in the vicinity of WAP operation place. In the nearest rooms the concentration was from zero up to 1.0 mg/m^3. We found out that the average concentration in these rooms was 0.86 mg/m^3.

3.2. Biochemical parameters

As the most interesting results of all investigations we are presenting the values of Pb-B, ALAD, FEP and RET for I(a) group of workers (Table 1).

TABLE 1
Biochemical parameters of I(a) group

Worker	WAP operation (in years)	Pb-B µmol/l	ALAD U/lE	FEP µmol/l	RET No %
1 AM	18	2.76(16)*	11.4(15)	17.86(12)	23(3)
2 NS	11	1.92 (7)	6.9 (7)	10.67 (7)	20(3)
3 JL	14	1.96 (8)	9.9 (6)	3.62 (4)	17(2)
4 AG	12	2.18(11)	13.0(11)	3.39 (7)	19(2)
5 BP	9	1.08 (6)	14.8 (6)	5.18 (6)	10(2)
6 ZS	6	1.89 (5)	11.5 (5)	3.35 (5)	18(2)
7 NP	9	0.95 (8)	25.5 (9)	3.21 (8)	19(2)

* = in brackets: number of effected measurements

TABLE 2

Biochemical parameters of exposed workers (average values)

Group	Lead (mg/m^3)	Number work.	meas.	Pb-B (µmol/l)	ALAD (U/IE)	FEP (µmol/l)	RET (‰)
I	12.80	41	138	1.45*(0.43)	14.9°(4.75)	3.64*(5.10)	18.6(7.0)
Ia		7		1.80(0.59)	13.0(6.0)	6.94(5.60)	18.0(5.8)
Ib		34		1.32(0.34)	16.1(3.8)	1.76(0.99)	18,8(8.6)
II	0.86	15	34	0.88 (0.20)	24.2 (3.8)	0.92 (0.42)	
III (C.G.)	Ø	20	20	0.76 (0.17)	35.1 (3.2)	0.87 (0.30)	

* Statistically significant; $p < 0.05$ I gr. vs II and III gr.

° Statistically significant; $p \leq 0.05$ I gr. vs II and III gr., II vs III gr.

In the same manner we examined the other 34 workers who were directly involved in the WAP operation, without significant intoxication before the period of investigation. The achieved results for this main group are very equalized and confirm the permanent intoxication of the workers engaged in such WAP activities. In the group II, 15 workers who have other duties in the vicinity of WAP operation place were observed as well as 20 patients from the control group III.

The results of all groups are presented in Table 2, as average values with standard deviation and statistical treatment. The values of Pb-B, ALAD and FEP indicate a general trend of a higher level of lead absorption and effects on exposed workers. High concentration of lead in the air (12.8 mg/m^3) has obviously strong effects on all three parameters. In some cases we proved out that in these conditions, workers could be intoxicated on the level close to saturnism per year.

For example, you may see the incident of worker IA (woman):

TABLE 3
Biochemical parameters for IA - patient

WAP operation (in months)	Pb-B µmol/l	ALAD U/lE	FEP µmol/l
12	1.22	4.2	6.97
13	3.44	2.9	7.55
Sick leave			
15	1.15	18.6	4.12
17	1.13	18.3	2.75
21	0.58	16.9	1.33

In 13 months IA was intoxicated in the level close to saturnism with high content of Pb in blood, depressed of delta ALA and high concentration of free eritrocyte porphiryn. During this period there were no visible clinical effects. After the sick leave and change of work place, IA eliminated effects of Pb-intoxication during the eight month period. However, the value of ALAD still indicates the contact with Pb.

Low concentration of lead in the air was without significant effects on Pb-B and FEP parameters on the exposed workers. But, we discovered an important change in values of ALAD, which proved us that this parameter is the most sensitive index of lead absorption. Big difference among the values of FEP between I(a) and I(b) group shows that the disturbance of this parameter lasts much longer than the others.

The correlation between all three parameters is shown in Table 4 and Figure 1, where the values of Pb-B, ALAD and FEP covering the period of investigation for the worker 1 AM. According to /5/ this is an ordinary example of saturnism form.

TABLE 4
Biochemical parameters for 1 AM

Time	Pb-B µmol/l	ALAD U/1E	FEP µmol/l
1	2	3	4
20.08.1987.	3.09	8.3	
17.11.1987.	2.95		26.05
10.05.1988.	3.09	7.1	
15.06.1988.	2.90	7.6	
21.07.1988.	3.45		30.00
21-23.07.1988. EDTA TREATMENT			
06.09.1988.	2.41	2.5	16.00
13.09.1988.	4.40	4.5	18.40
27.09.1988.	3.73	9.5	15.55
29-09.-01.10.1988. EDTA TREATMENT			
04.10.1988.	2.91	15.2	14.00
08.11.1988.	3.16	13.6	17.00
15.12.1988.	2.15	8.0	23.90
19.01.1989.	2.12	19.1	20.90
11.04.1989.	1.79	15.2	12.10
09.05.1989.	2.02	16.6	11.50
07.09.1989.	1.59	23.9	
22.02.1990.	2.15	13.2	8.90
14.-16.03.1990. EDTA TREATMENT			
22.02.1991.	1.65		9.50

Generally, in this case we have a gradual decrease of Pb-B and FEP and increase of ALAD in some sinusoidal form. Besides the effect of EDTA treatment a normal blood-lead fluctuation, influenced by many different reasons has to be included.

After the first treatment with EDTA we noticed a great diminish of Pb-B and FEP, and increase of ALAD. For 1-2 weeks the oposite effects begins due to the disturbance of equilibrium lead in blood in comparison with the whole body. In the second EDTA treatment, changes in the biochemical parameters have the same direction, but the changes are buffered because of the previous equilibrium effect,

The number of reticulocytes for the first group of the workers is slightly higher than the normal value (5-15 %), what points out the small effect of Pb-absorption on this blood parameter. Nevertheless, the correlation between Pb-absorption and the number of reticulocytes not exists.

3.3. Lead effects on the teeth

During the period of investigation we carried out the teeth control of the exposed workers of group I. Changes on the gingive were discovered only in

Notice: We are grateful to the Clinical and Toxicological laboratory of the Institute for Medical Research and Occupational Health - Zagreb for the results during EDTA treatment.

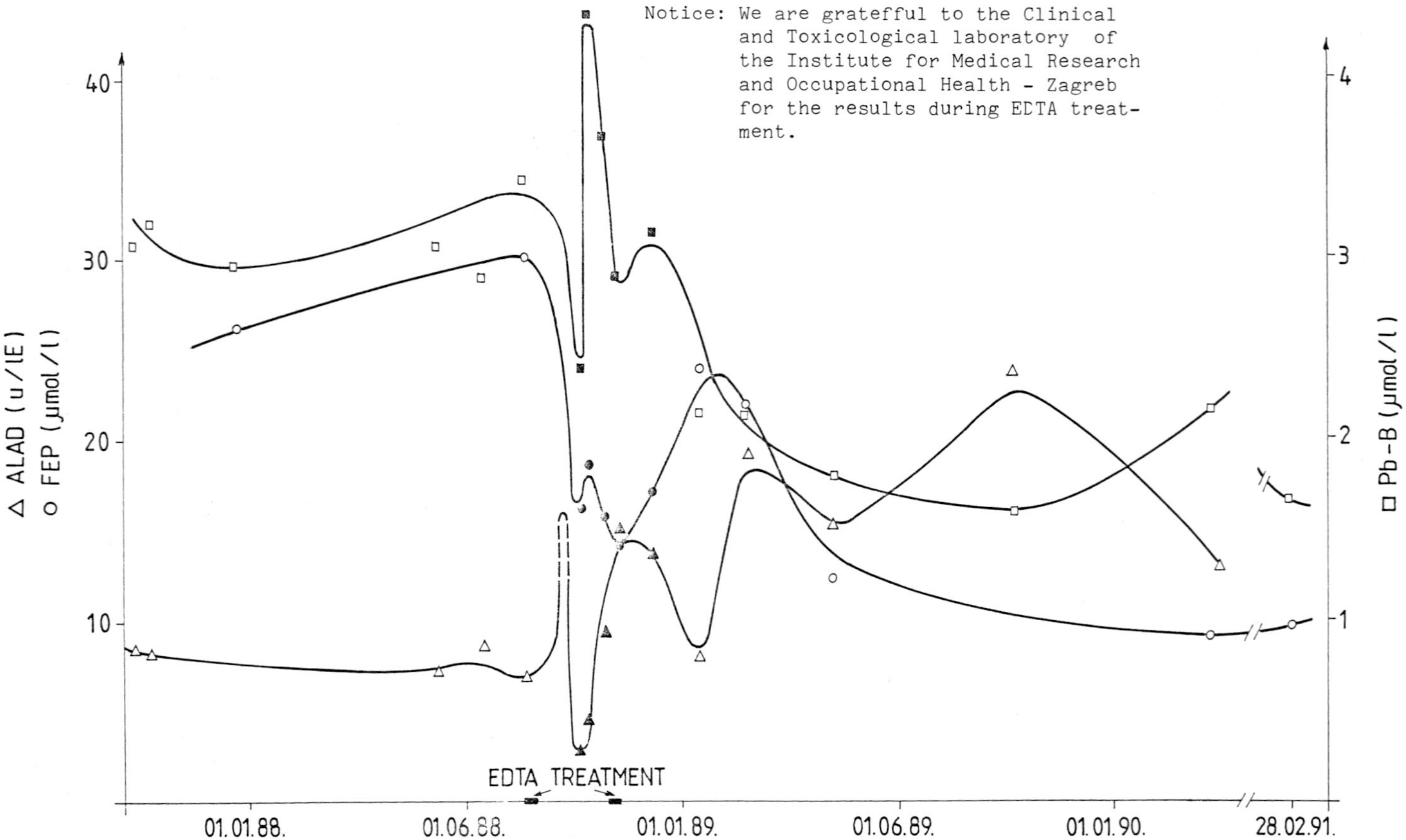

Figure 1. CORELATION BETWEEN BIOCHEMICAL PARAMETERS

workers 1 AM and 4 AG. Besides, in the extracted teeth, we found 166 mg/kg and 57.5 mg/kg Pb-T respectively. The average of Pb-T amounting to 6.2 mg/kg is made from the five teeth examination of the patient from the control group.

The average DMF value for all workers of group I is 16.8 what is significant increase in comparison with the control group (DMF of control group is 12.6).

Preliminary results of Pb in saliva indicate that this parameter can be daily indicator of Pb-absorption. We found out the average concentration of 0.068 mg/1 Pb in 25 exposed workers of group I.

3.4. Nerve conduction velocity of exposed workers

Conduction of radial motor (RMCV) and sensory nerve (RSCV) like conduction of deep peroneal motor nerve (DMPCV) and superficial peroneal sensory nerve (SPSCV) were measured.

TABLE 5
Nerve conduction velocity of exposed workers

Parameters	Average conduction velocity value (m/s)		
	I(a) group (7)	I(b) group (7)	Control group (7)
RMCV	42	54	62
RSCV	53	56	66
DPMCV	40	47	53
SPSCV	44	48	55

From Table 5 the motor and sensory conduction of deep peroneal nerve and sensory conduction of superficial peroneal nerve were significantly decreased in group I(a). The moderately decreased values were obtained in group I(b).

The selective motor neuropathy with prominent wrist drop is very often described in the literature. We found the reduction of motor and sensory conduction velocity in the lower and upper extremities, according to the demielination figure, which shows us the close correlation with lead exposure.

3.5. Statement of working ability of the workers exposed to lead (SWA)

Although final purpose of our experiment was not to find biochemical parameters as criterion for SWA, we tried to incorporate it in this problem.

Even from 1967, Beritić /5/ gave very simple and usable directions for SWA purpose. He used three exposition zones due to dynamics of biological reactions caused by lead:

I - zone of harmless exposition

II - zone of significant absorption (presaturnism form) and

III - zone of manifest poisoning.

Workers were distributed to one of that three groups on the basis of the different values of the following biological parameters: Blood Pb (Pb-B), urine lead (Pb-U), urine coproporphiryn (Cp-U), delta ALA in urine, protoporphiryn in eritrocytes, reticulocytes and basophilic stippled RBC counts.

In the later publications, Beritić /6/, Telišman /7/, Skender /8/ and Buljat /9/ use other biochemical parameters as criteria for SWA: Dehydratase delta ALA in blood (ALAD), free eritrocyte porphiryn (FEP) and zinc protoporphiryn in blood (ZPP). In general, authors retain the earlier zones for SWA, but they changed their expressions:

I - zone of normal exposition (worker capable for working with lead but with necessary health control upon expiration of 12 months period),

II - zone of abnormal but acceptable exposition (worker capable for working with lead but with necessary health control upon expiration of 3 months period),

III - zone of abnormal nonacceptable exposition (work is forbidden with necessary health control upon expiration of 3 weeks period).

Later experience proved out that these three above mentioned zones are not enough to express all relevant possibilities in criteria for SWA. Namely, regarding to these suggestions, the main part of exposed workers very easy entered in the zone III (ALAD < 5 or ZPP > 1) what have strong effect on their production efficiency. For this reason, Kurajica suggests to insert a new zone between II and III zone as abnormal one but still acceptable. In this II(a) zone in the exposed workers has found just one of biochemical parameters with abnormal and nonacceptable value and according to this fact, they can work but with very often controls. For that reason III zone is changed and both existing parameters are nonacceptable (ALAD < 5 and ZPP > 1).

Considering all mentioned above, such as our results, we may proposed four zones to be determined for SWA regarding to the parameters: ALAD, FEP and Pb-B, as it shown in Table 6 /10/.

Table 6
Statement of working ability of the workers exposed to lead

Biochemical parameters	ZONE I Normal	ZONE II (abnormal but acceptable)	ZONE III (abnormal but still acceptable)	ZONE IV (abnormal and nonacceptable)
ALAD (U/1E)	> 26	26 to 5	< 5 and/or	< 5 and/or
FEP (μmol/1E)	< 1.26	1.26 to 5	> 5 and/or	> 5 and/or
Pb (μmol/1 blood)	< 1.48	1.48 to 3	> 3	> 3

Remark:

Observed worker belongs to IV zone in the case when at least two (2) of relevant parameters have abnormal nonacceptable values.

Observed worker belongs to III zone in the case when the most one (1) of relevant parameters has abnormal nonacceptable value.

3.5.1. I - Zone of normal exposition - All three relevant parameters are in the area of normal values. Workers are capable for working with lead but with necessary health control upon the expiration of 12 months period.

3.5.2. II - Zone of abnormal but acceptable exposition - All parameters have abnormal but acceptable values. Workers are capable for working, but with necessary health control upon the expiration of 3 months period.

3.5.3. III - Zone of abnormal but still acceptable exposition - Only one of three parameters have abnormal nonacceptable value while the others have abnormal but still acceptable ones. Workers are still capable for working with lead, but with necessary health control upon the expiration of 3 weeks period.

3.5.4. IV - Zone of abnormal nonacceptable exposition - At least two od three relevant parameters have abnormal nonacceptable values. Workers uncapable for working with lead, but with necessary health control upon the expiration of 3 weeks period.

For the workers who belong to IV zone is permisible to work with lead when at least two relevant parameters "returned" in III zone, but with the condition that visible clinical indicators not exist.

If the worker during two years falls at least twice in IV zone, he becomes permanent uncapable for working with lead.

REFERENCES:

1. *M. Stoppler, K. Brandt and T.C. Rains, Analyst, 103 (1978) 714-722.*
2. *M. Stoppler, K. Brandt, Z. Anal. Chem. 300 (1980) 372-380.*
3. *A. Berlin and K.H. Schaller, Z. Klin. Biochem. 12 (1974) 389-390.*
4. *J.J. Chilsolm and D.H. Brown, Clin. Chem., 21 (1975) 1669-1682.*
5. *M. Buljat, S. Kovač, V. Karačić, E. Keršanc, D. Dimov and T. Baretić, Archives of Industrial Hygiene and Toxicology, 34 (1983) 15-31.*
6. *T. Beritić, Archives of Industrial Hygiene and Toxicology, 18 (1967) 1-9.*
7. *T. Beritić, D. Prpić-Majić, V. Karačić and co, J. Occup. Med., 19 (1977) 551.*
8. *S. Telišman, Archives of Industrial Hygiene and Toxicology, 30 (1979) 49-72.*
9. *Lj. Skender, D. Prpić-Majić, V. Karačić, Archives of Industrial Hygiene and Toxicology, 33 (1982) 227-238.*
10. *L.B. Kurajica, M. Orlov, Archives of Industrial Hygiene and Toxicology, in press for 1991.*

Chapter 8
Radionuclides as chronometers and tracers

RADIOISOTOPES AS TRACERS FOR THE INERACTIONS BETWEEN TRACE ELEMENTS, COLLOIDS AND PARTICLES IN NATURAL WATERS

Peter H. Santschi and Bruce D. Honeyman

ABSTRACT

The self-cleaning capacity of natural waters for "particle-reactive" elements depends on the concentration of particle surface sites and on the rates of renewal of these sites. Particles of all sizes are introduced to the water by primary production, from rivers, through the atmosphere and by resuspension of surface sediments, while particle aggregation and break-up maintain the particle size spectrum and, thus, available surface sites for adsorption throughout the particle-size classes. Because of the conservative (i.e. soluble) nature of some of the parent nuclides of the U/Th decay series (e.g. U, Ra), particle-reactive daughter nuclides are introduced into aquatic systems with well-defined source functions. Some of these radionuclides, particularly certain species of Th, are among the most powerful tracers for studying the processes controlling the self-cleaning capacity of natural waters. In this paper, recent experimental and theoretical evidence elucidating the applications of Th nuclides as tracers for particle-particle interactions and the role colloids play in trace metal scavenging are discussed.

1. INTRODUCTION

Due to their strong "particle-affinity" and their defined production rate in the ocean, natural radionuclides (i.e., ^{234}Th, ^{230}Th, ^{228}Th, ^{210}Pb, ^{210}Po, ^{7}Be, ^{10}Be, ^{231}Pa) have been used in the last several decades to define the time scales of oceanic scavenging. Classically, the unspecific term "scavenging" has been applied to include a suite of uptake mechanisms which lead to the removal of highly particle-reactive elements from the water column. Scavenging processes include sorption (adsorption onto particle surfaces, absorption by particles, and surface precipitation; ref 1), "coprecipitation" by iron and manganese oxides, coagulation of colloids into, or to produce, larger particle aggregates, surface diffusion of ions or molecules through biological membranes, films or coatings, repackaging of particles into fecal pellets by swimming filter-feeding organisms, or impaction of sinking particles onto marine snow or surface-"fluff" overlying consolidated sediments.

Characteristic times for the transfer of radionuclides from an operationally-defined solution phase to particles large enough to be retained by filters, and for the sedimentation or resuspension of these particles, can be calculated for shallow and deep water bodies from a knowledge of the daughter production rate and from the concentration of the soluble mother (e.g., ^{238}U, ^{234}U, ^{228}Ra) and particle-reactive daughter nuclide (e.g., ^{234}Th, ^{230}Th, ^{228}Th, respectively) activities, in the water and on particles. Despite more than a decade of detailed measurements (by counting the decay products or the atoms) of Th and other radionuclides in the ocean, we still do not understand the exact (physical, chemical or biological) meaning of the rate measurements calculated from oceanic surveys of Th profiles.

For example, it has been found that 1) partition coefficients between solution and particle phases and 2) rate constants highly correlate with suspended particle concentration (e.g., refs 2-5), with particle flux (refs 6-7), or particle residence time (ref 6). Such observations confound an interpretation in the sense of equilibrium surface complexation. Furthermore, since oceanic residence times calculated from U/Th series disequilibrium are too long to be interpreted on adsorption time scales (days to months to years instead of milliseconds to seconds), other processes for controlling the removal of Th and other isotopes from the solution phase must be invoked.

It is critical to note that while the overall scavenging process is generally considered to describe the transfer of radionuclides, produced in the solution phase, to scavenging particles, the initial physico-chemical state of the daughter radionuclides is rarely, if ever, determined. Instead, operational definitions of "dissolved", such as by filtration through 0.45μm filters, are employed.

The only hypothesis which, so far, can consistently explain all the observations to date of the behavior of Th nuclides and other particle-reactive elements in natural waters is the involvement of colloids in the uptake process as intermediate carriers of the adsorbed isotopes from the truly-dissolved phase to filter-retained particles (ref 8). Examples of colloidal material include natural organic macromolecules such as humic matter, iron oxyhydroxide, aluminous and siliceous (clay mineral) phases.

Most radionuclides in natural waters exist in ultratrace, sub-femtomolar concentrations. The only exceptions are only ^{238}U (10^{-8} -10^{-9} M) and ^{40}K (10^{-6} M - 10^{-9} M) in sea and fresh waters, respectively (ref 9). Because of the low concentration of trace metals, high-energy surface sites of small amounts of colloidal particles in natural waters can effectively compete against solution complexes for cationic radioactive and stable trace metals in solution. Colloidal carrier phases, or host phases, therefore become much more important in natural waters for ultratrace elements such as selected radionuclides and stable trace elements (e.g., Cu, Zn, Cd, Hg, Pb, Sn, which often occur at pM levels) (e.g., see review in ref 10). For example, at 1 mg/l of suspended particle and 1 mg/l DOC concentration (with 10% at MW 10^4 or larger), typical for surface waters, one would expect about 10^6 particles/cm^3, and 10^{13} colloidal molecules/cm^3. For trace metals at pM levels, (~10^9 atoms/cm^3) and selected radionuclides at 10^{-15} to 10^{-20} M (~10 to 10^6 atoms/cm^3), the number of colloids and filter-retained particles becomes comparable or larger than the number of dissolved ions in a given volume of water. It is understandable then that the particle size distribution and composition below 1 μm, which, as of yet, cannot be unambiguously measured, becomes important for metal ions at ultratrace concentrations.

2. EVIDENCE FOR THE IMPORTANCE OF COLLOIDS

The role of submicron-sized particles in affecting the behavior of particle-reactive solutes has long been recognized, although the evidence in many instances has been indirect. Colloid transport through porous media has been reviewed by McDowell-Boyer et al. (ref 11), their importance in surface waters by Buffle (ref 12), and the literature on the formation and properties of radiocolloidal dispersions has been summarized by Kepak (ref 13).

Buddemeier and Hunt (ref 14) studied radionuclide transport in groundwaters pumped from a nuclear bomb test cavity and found that the transition and lanthanide radionuclides, which moved considerably further than expected from ^{3}H and other conservative species, were essentially all associated with the colloidal (<0.2 μm) fraction. Sheppard et al. (ref 15) concluded that much of the mobility of radionuclides in soil waters is probably due to soil particles in the 2-5 nm range. Moore and co-workers (refs 16, 17) examined the partitioning of added ^{234}Th among dissolved, colloidal and particulate fractions and found that up to 64 percent of the total ^{234}Th added to the water was associated with the colloidal (1 nm-0.2 μm) size fraction. Baskaran and co-workers (ref 18) measured the distribution of natural ^{234}Th in the colloidal fraction in the Gulf of Mexico, and found that 15-85 % of the "dissolved" Th is associated with colloids. Nash and Choppin (ref 19) estimated that in natural waters at 0.1 mg l^{-1} DOC (humate [Hu] = 4×10^{-7} eq l^{-1}) the ratio of [ThHu]/[Thdiss] would be about 4×10^{10}. Therefore, we expect that most of Th "dissolved" in ocean water is associated with colloidal humic acids. Sequential filtration and ultrafiltration experiments by Santschi et al. (ref 20, 21) have shown that radioisotopes of particle reactive elements are involved in dynamic cycles of coagulation, aggregation and disaggregation of particles and colloids in the water column of coastal marine ecosystems; and Buchholtz et al. (ref 22) demonstrated the aggregation effect of colloidal radionuclides in laboratory batch experiments with marine sediments.

All equations described in the following text can be found in Table 1. A glossary of symbols follows the list of reference citations.

3. COLLOIDAL-PUMPING

The "colloidal-pumping" model for trace element transfer from the dissolved phase to relatively large particles via a colloial intermediate (ref 8) consists of two steps: 1) Dissolved species, e.g., Me^{z+}, rapidly form complexes with surface sites on both colloidal and filter-retained particles (Fig. 1); this step most likely has characteristic times on the scale of milliseconds to minutes. Consequently, chemical equilibrium is quickly reached and may be described by association constants for large and colloidal particles, K_f and K_c, respectively. 2) "Slow" particle aggregation. Colloidal particles, operationally defined as part of the "dissolved" phase, slowly aggregate with larger, filter-retained particles becoming part of the filter-retained particle size class. Particles aggregate and disaggregate at rates B and D, respectively (Fig. 1b,c). Most likely, the particles that constitute the filter-retained particle-size fraction consist of clusters of large and small particles, of varied history and composition, aggregated together (Fig. 1c) and subject to continual reworking. As the solid-phase components of particle clusters gradually make their way to the ocean floor, elements associated with colloids may periodically cycle through the "dissolved" pool as particle clusters aggregate (B), disaggregate (D) and are remineralized (R_m) as the consequence of hydrodynamic forces, changing chemical conditions and biological activity.

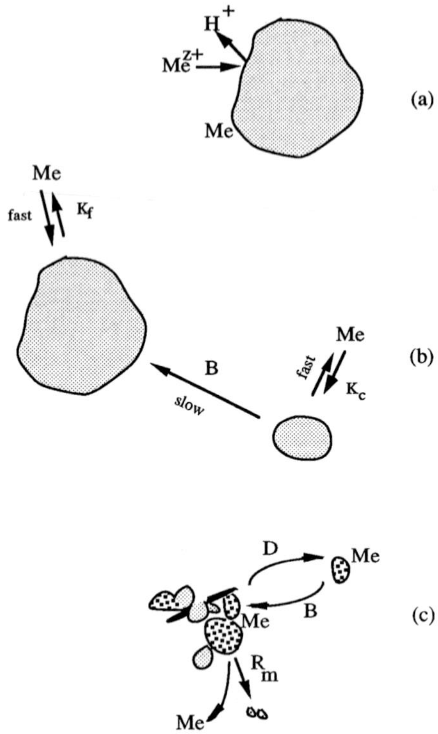

Figure 1. Phenomenological model of adsorption coupled to coagulation consists of two steps: a) Rapid formation of trace element (Me^{z+})/particle surface site complexes. Characteristic complex formation times are assumed to be relatively rapid, perhaps on the order of 10^{-3} to 10^{-2} sec; b) Slower aggregation of colloids onto larger, filter-retained particles. K_f and K_c are the conditional solution/particle distribution coefficients for the trace element with filter-retained particles and colloids, respectively; c) Natural particle cluster. Clusters form and disaggregate at rates B and D, respectively, transferring sorbed radionuclides and trace elements to and from the submicron pool. Clusters also remineralize, R_m, releasing Me to the dissolved pool (from ref 8).

The "Brownian-pumping" model of Honeyman and Santschi, (ref 8) relates the coagulation/sedimentation equations of Farley and Morel (ref 23), with the measured particle concentration (C_p) as the master variable, to the empirical scavenging rate equations derived from U/Th series nuclide disequilibrium measurements in the water column (see Table 1).

The rate of transfer of a trace element or radionuclide from the truly dissolved state to the filter-retained particle size fraction via the colloidal particle, is dependent on the rate of aggregation of colloids into particle clusters of sedimentation size (and which are assumed to be synonymous to particles which can be filter-retained, as described by equations 1-3, and f_{cd}, the fraction of the element of interest not retained by a filter but which are associated with colloids (Fig. 2, Eqn. 4, 8d).

Table 1. Equations cited in text. (Refer to the glossary for symbol definitions).

$$-\frac{dC_p}{dt} = B_{ds}C_p^{2.3} + B_{sh}C_p^{1.9} + B_bC_p^{1.3} = B_b'C_p \qquad (1)$$

$$-\frac{dC_p}{dt} = \frac{dC_c}{dt} = B_b'C_p = B_bC_p^{0.3}C_p \qquad (2)$$

$$-\frac{dM_f}{dt} = \Gamma_C\frac{dC_c}{dt} = B_b'C_p\,\Gamma_C \qquad (3)$$

$$k_f = (B_b' + B_{sh}')\,f_{cd}\,C_p \qquad (4a)$$

$$B_b' = B_bC_p^{0.3} \qquad (4b)$$

$$B_{sh}' = B_{sh}C_p^{0.9} \qquad (4c)$$

$$\frac{d[Th_{part}]}{dt} = R_f'\,[Th_{diss}] - R_r\,[Th_{part}] \pm J \qquad (5)$$

$$R_f' = \frac{[Th_{part}]}{[Th_{diss}]}\,(R_r + \lambda_{Th} + \lambda_p) = R_f\,C_p^{\Theta} \qquad (6)$$

$$K_d = \frac{R_f'}{R_r\,C_p} = \frac{R_f\,C_p^{\Theta}}{R_r\,C_p} = \frac{R_f\,C_p^{\Theta-1}}{R_r} \qquad (7a)$$

$$K_d = K_f\,/\,(1 + K_c\,C_c) \qquad (7b)$$

$$f_f = [Me_f]/[Me_T] = K_f\,C_f/\Sigma \qquad (8a)$$

$$f_c = [Me_c]/[Me_T] = K_c\,C_c/\Sigma \qquad (8b)$$

$$f_d = [Me_d]/[Me_T] = 1/\Sigma \qquad (8c)$$

$$f_{cd} = f_c/(f_c + f_d) = K_c\,C_c/(1 + K_c\,C_c) \qquad (8d)$$

$$\Sigma = 1 + K_f\,C_f + K_c\,C_c \qquad (8e)$$

$$G = \sqrt{0.3\,\varepsilon/\nu_f} \qquad (9)$$

$$\varepsilon = C_d\,U^3/h \qquad (10)$$

$$\varepsilon = u_*^4/\nu_f \qquad (11)$$

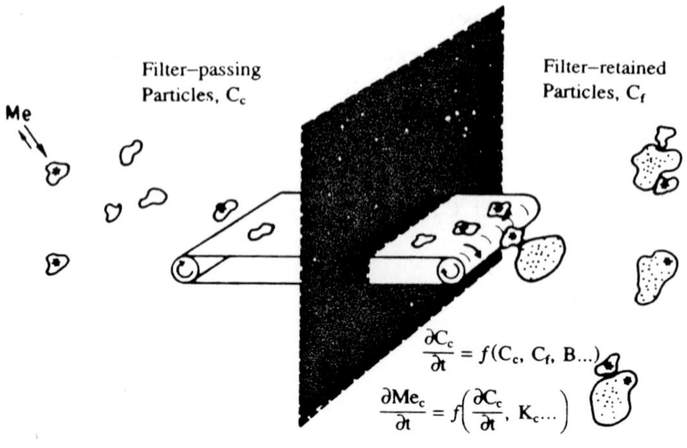

Figure 2. Colloidal-pumping: the transfer of dissolved trace elements to filter-retained particles via a colloidal intermediate. The rate of coagulation of colloidal particles with filter-retained particles is a function of particle numbers, which is assumed to be proportional to the surrogate, C_p. The rate of transfer of a trace element or radionuclide, Me, from the truly dissolved state to the filter-retained-particle fraction is a function of the solution/colloid distribution coefficient, K_c, in addition to the coagulation rate. "*" represents particles "tagged" with radioactive metal species. (after ref 8)

The coupling of coagulation and sorption as a mechanism for scavenging can be verified in laboratory experiments (ref 24). In our laboratory studies, the question of colloid identification (a problem central to field studies) was circumvented by producing hematite (α- Fe_2O_3) with ^{59}Fe as a structural part of the iron-oxide lattice. This allowed us to directly account for the pool of filter-passing particles and to assess the influence of these particles on the kinetics of metal appearance in the filter-retained particle size class and on the equilibrium distribution of trace elements between filter-retained and filter-passing particle-size fractions.

For example, Figure 3 shows the relative filter-retained and filter-passing ^{59}Fe activity for an initially homodisperse batch system containing 4 ppm of 250 nm ^{59}Fe-hematite particles. Filter-retained ^{59}Fe activity at $t = 0$ represents hematite particles coagulated onto the filter surface. Loss of ^{59}Fe activity from the filter-passing fraction is exactly compensated by its appearance in the filter-retained fraction. Figure 4a shows results from a similar system but with ^{113}Sn added. A two stage filtration system was used to measure the ^{113}Sn residing within three "pools": >400 nm, 30 to 400 nm and <30 nm. Since no detectable ^{59}Fe label passed the 30 nm filter, ^{113}Sn activity in the <30 nm fraction was considered representative of truly dissolved ^{113}Sn. Note that dissolved ^{113}Sn remained constant but that particulate ^{113}Sn was transferred from the <400 nm particle pool to the >400 nm pool. If we had not specifically accounted for all of the hematite particles, e.g., if only a 400 nm filter had been used, then "sorption" would have appeared to be slow (solid symbol, Fig. 4b). In reality, sorptive equilibrium is rapidly achieved (open circles) and the observed kinetic effect is due solely to the aggregation of colloids into particles which can be filter-retained. A more detailed description of these and other laboratory experiments which study the coupling of surface adsorption and coagulation is published elsewhere (ref 24).

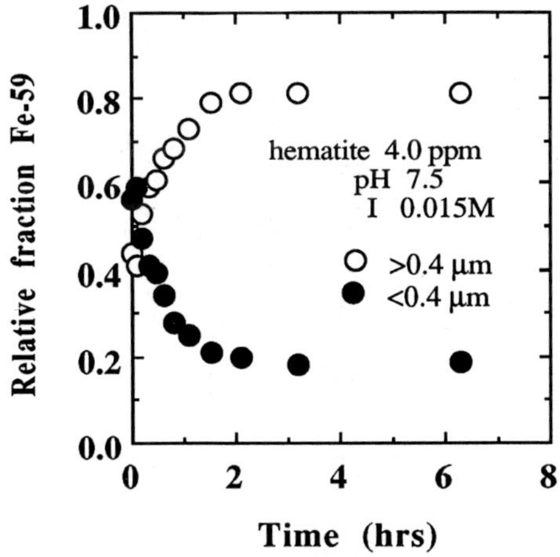

Figure 3. Hematite is transferred from the filter-passing fraction to the filter-retained fraction by coagulation. Approximately one third of the hematite coagulates with the filter surface. The symmetry of the curves indicates that particles are not lost to beaker walls; as such, the shift in ^{59}Fe distribution is due only to colloidal aggregation (from ref 24).

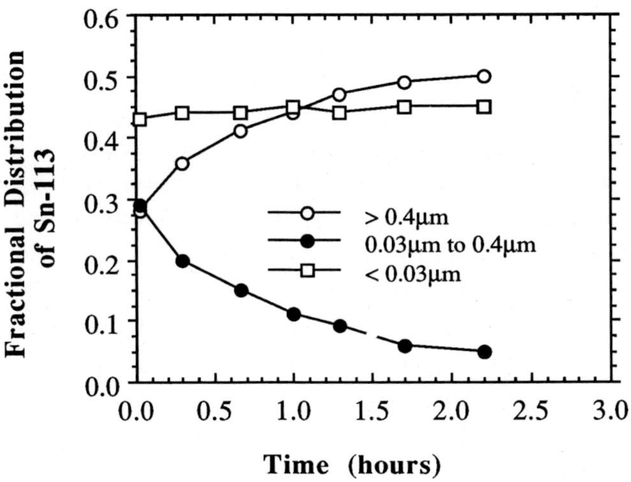

Figure 4a. Distribution of ^{113}Sn among >400 nm, 30-400 nm and dissolve pools. The 120 nm hematite particles were initially homodisperse and coagulated at the rate shown in Fig. 3. Adsorption of Sn is essentially instantaneous and invariant with time as shown by the constant dissolved fraction. The kinetic effect in the transfer of Sn(IV) between the particle pools is due solely to coagulation (from ref 24).

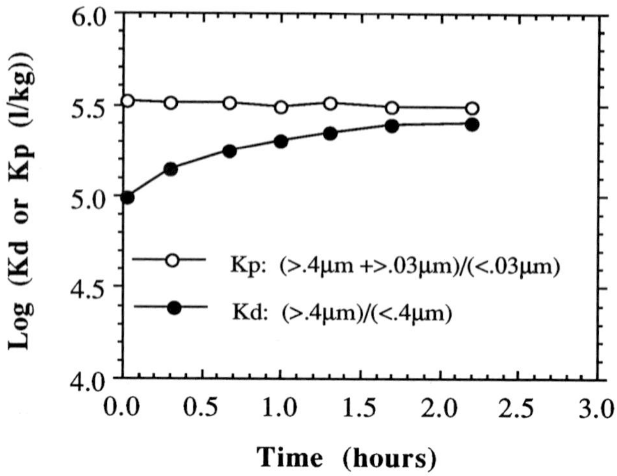

Figure 4b. Comparison of distribution coefficients. Closed symbols: an operationally-defined "dissolved" fraction. Open symbols: ratio of sorbed to truly-dissolved ^{113}Sn. Particle aggregation gives the appearance that sorption is slow when the 0.4μm particle size fraction is used as a reference frame (from ref 24).

4. COMPARISON OF OCEANIC SCAVENGING WITH COAGULATION RATES

A general expression for the rate of "scavenging" of dissolved radiogenic metals by settling particles in a water column (e.g., refs 26, 5) is given by the sum of production and removal rates (Eqn. 5, table 1). At steady state, both sides of Eqn. 5 (Table 1) are equal to zero and the equation may be rearranged to represent R_f' in terms of measurable values (Eqn. 6, Table 1). Figure 5 is a plot of the log of R_f' as a function of log particle concentration. R_f' is not constant but increases with C_p. Characteristic sorption times range from 10^3 days in the deep sea to less than 1 day in a high particle concentration (C_p) estuarine environment. The solid lines in Fig. 5 represents model estimates for coagulation/sedimentation rate constants, using the parameter values of Table 2 and a mixing depth of h = 6 m. Line 1 is for sedimentation driven by Brownian coagulation only (Table 2), and line 2 includes the influence of shear coagulation (Table 3), using a typical value of G of 1 sec^{-1} for the oceanic mixing rate at moderate energy dissipation rates. Particle aggregation by shear is usually important only at Cp > 1 mg/l. The lines indicate the conditions such that $R_f' = B_b' + B_{sh}'$: that is, the limiting scavenging rate driven by colloidal pumping. The line has initially a slope of 0.3 as expected from Eqn. 3 ($B_b' = B_b\ C_p^{0.3}$) and increases as the region of the shear coagulation mechanism at moderate energy dissipation rates is reached (line 2). The values of 0.5 (e.g., (refs 27,28), 2.5 g/cm^3 (ref 29) and 6m for α_b, ρ_e and h, respectively, were used in the calculation of Lines 1 and 2 (Fig. 4a). Since none of the parameters are, themselves, functions of C_p, any change in their values will shift the coagulation/sedimentation rate line in a direction parallel to the solid line. As it is likely that the values of α_b, ρ_e and h are maximum values, this line represents an upper limit for "scavenging". The reciprocal of R_f', for values corresponding to the coagulation line, approximately represents the time for 63% of the colloidal particles to be transferred to aggregates and be removed by sedimentation (τ_c). Thus, τ_c ranges from about 25 days at $C_p = 10^{-8}$ kg/l to approximately 1 day at 10^{-4} kg/l.

5. THORIUM REMOVAL IN HIGH ENERGY DISSIPATION ENVIRONMENTS

Coastal regions, with their high energy dissipation rates and their relatively high particle concentrations and fluxes, are areas where turbulent mixing energies could significantly affect colloidal pumping rates onto particles. Using data sets on ^{234}Th removal rates from coastal waters with different energy dissipation rates, the predictions by the scavenging model can be tested for areas with similar particle concentrations (i.e. 10^1 mg/l) but different energy dissipation rates (10^{-1} to 10^1 erg g^{-1} s^{-1}). Included are data from Texas (ref 30), the Irish Sea (refs 31,32), and Narragansett Bay (refs 33, 34). The high energy dissipation rates, ε, in the first two areas cause small particle coagulation rates to be dominated by turbulent shear, and to a lesser extent, by Brownian action, while in the third area, where energy dissipation rates are lower, coagulation rates are mainly given by the Brownian mechanism (with kernel B_b). Shear coagulation rates (with kernel B_{sh}), which depend on the rate of energy dissipation (refs 35, 36), can be estimated from first principles as shown below.

The rate constant for coagulation/sedimentation, k_f, is given by equation 4 in Table 1. At lower particle concentrations and sizes (ref 23), only the Brownian and shear coagulation terms are important. The shear coagulation kernel, B_{sh}, contains the current shear factor, G, which is a function of the energy dissipation rate, ε (equation 10, Table 1).

For the purpose of calculating shear coagulation rates, it is crucial how ε is calculated: a) average value over the water column, calculated from energy loss through small scale turbulence in the water, b) value in the viscous boundary layer, where most energy is dissipated by friction with the bottom. There are orders of magnitude difference between the two estimates.

a) Water column averaged values of ε can be calculated using equations given by Garrett (ref 37), i.e. equation 11 in Table 1.

b) The energy dissipation rate, ε, in the viscous sublayer (ref 35) can be estimated from equation 12, Table 1.

Energy is mostly dissipated near the sediment-water interface, thereby imparting momentum to the particles by impaction and resuspension. It is, therefore, likely that in shallow environments, shear coagulation can be higher than calculated from the water column averaged energy dissipation rate. For example, there is some recent evidence that interfacial shear coagulation mechanisms are dominating coagulation rates in an embayment on the coast of Massachusetts, USA (ref 36). Water column averaged values for ε can provide, however, lower limits for the calculation of shear coagulation rates. Average ε values, taken from the literature, range from 6 erg g^{-1} s^{-1} for Salem Sound, Massachusetts (ref 36), 3 erg g^{-1} s^{-1} for the Great Bay Estuary in New Hampshire (ref 38) to 0.25 (ref 39) and 0.002 - 1.0 erg g^{-1} s^{-1} (ref 40) for Narragansett Bay, Rhode Island. For Galveston Bay, we estimate values for ε of 1-2 erg g^{-1} s^{-1}, depending on wind and tidal conditions.

Table 2. Parameter values (ref 24) used to calculate the Brownian coagulation coefficient, B_b (after ref 23).

$$B_b = 1.33\ (S/h)^{0.6} \rho_e^{-0.3} \alpha_b^{0.4} K_b^{0.4} \qquad (mg^{-0.3} L^{0.3} d^{-1})$$

Stokes settling:

$$S/h = (1/6\pi^2)^{0.33}\ (g/3\nu h)\ (\rho_e - \rho_f)/\rho_f \qquad (cm^{-2} s^{-1})$$

Brownian coagulation kernal:

$$K_b = 2kT/3\mu \qquad (cm^3 s^{-1})$$

Parameter	Description	Dimension	Value*
g	grav.accel.	$cm\ s^{-2}$	980
ν	kin. visc.	$cm^{-2} s^{-1}$	0.0096
μ	dyn. visc.	$cm^2\ s^{-1}$	0.01
k	Boltzman	$erg\ ^\circ K^{-1}$	1.4×10^{-16}
T	absolute temp.	$^\circ K$	283
ρ_e	particle den.	$g\ cm^{-3}$	$f(d_p)$
ρ_f	fluid den.	$g\ cm^{-3}$	1.05
α_b	collision eff. factor	(-)	0.1-0.5
h	depth of vert. homogeneous water column	cm	variable

* Values of ν, μ, ρ_e, ρ_f and T reflect typical oceanic conditions.

Table 3. Parameters used to calculate shear and differential settling kernals (K_i) and coefficients (B_i) (after ref 23).

Shear:

$$K_{sh} = G/\pi \qquad (s^{-1})$$

$$B_{sh} = 10.0\ (S/h)^{0.15} \rho_e^{-0.9} \alpha_b^{0.1} K_b^{0.1} \alpha_{sh}^{0.75} K_{sh}^{0.75} \qquad (mg^{-0.9} L^{0.9} s^{-1})$$

Differential settling:

$$K_{ds} = (6/\pi)^{0.33}\ (g/12\nu)\ (\rho_e - \rho_f)/\rho_f \qquad (cm^{-1} s^{-1})$$

$$B_{ds} = 3.12\ (S/h)^{-0.32} \rho_e^{-1.3} \alpha_b^{0.17} K_b^{0.17} \alpha_{ds}^{1.15} K_{ds}^{1.15} \qquad (mg^{-1.3} L^{1.3} s^{-1})$$

G has the dimension s^{-1}. Other parameters are defined in Table 2. The subscript "b" indicates a parameter for Brownian coagulation.

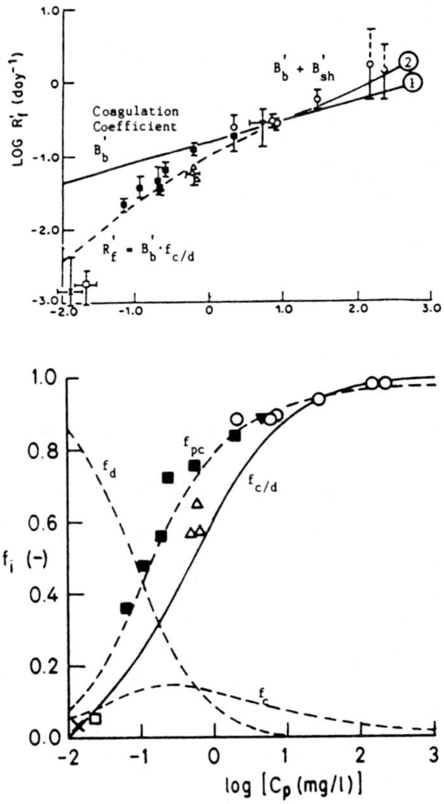

Figure 5. a) Comparison of the pseudo-first order sorption rate constant, R_f' determined from isotopic disequilibrium (after ref. 5) with calculated Brownian pumping sorption rate constants, as a function of particle concentration. Parameter values as in Table 1 except: $\alpha_b = 0.5$, $\rho_e = 2.5$ g/cm^3 and h = 6m. a) Solid lines: 1) Brownian coagulation rate constant (B_b'); 2) Brownian and shear coagulation. Broken line: B_b' f_{cd} with $K_c = K_f = 10^{7.0}$; log C_c = 0.7 log Cp - 2.6. b) Postulated distribution of ^{230,234}Th for the conditions represented by the broken line above. Symbols are as defined in Table 1. The fraction of Th in the filter-retained-particle fraction, f_f, was derived from the sources listed above. The solid line is for the case where $f_{cd} = 1$ at all particle concentrations. At particle concentrations greater than approximately 1 mg/l, metal sorption is controlled by particle coagulation. (from ref 8).

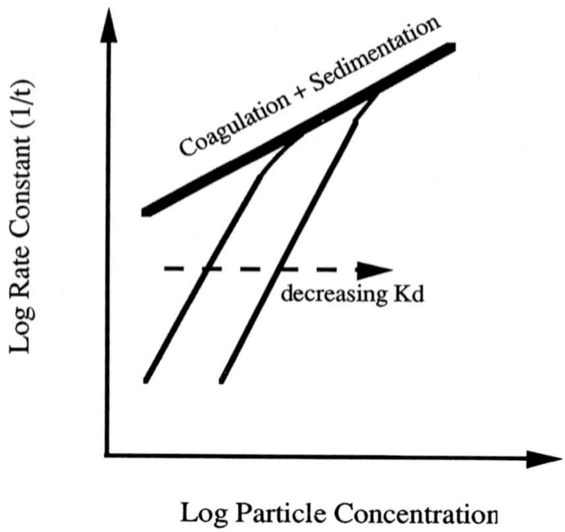

Figure 5c. Schematic illustration of the relationship among master variables R_f, C_p and K_d. At a constant C_p, trace elements with progressively-smaller K_ds will exhibit systematically-smaller scavenging rate constants.

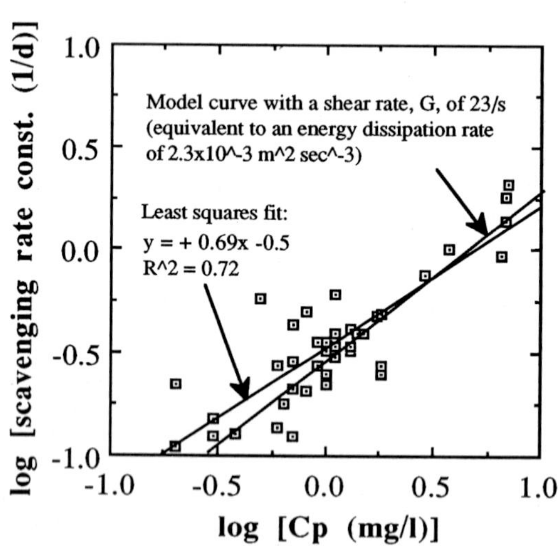

Figure 6. Model simulations of particle uptake constants, k_f, which calculated from $^{234}Th/^{238}U$ disequilibrium for high shear environments, as a function of particle concentration.

Assuming reasonable (instead of maximum) estimates of the coagulation parameters, i.e., a collision efficiency factor, α, of 0.1 (ref 28), and particle density in the ocean of 1.3 g cm^{-3}, the particle uptake (i.e. removal) rate constants reported from the measurement of the disequilibrium between ^{234}Th and ^{238}U in the water column of the Irish Sea (refs 31,32), Galveston Bay (ref 30) and Narragansett Bay (refs 33.34) can be reasonably-well simulated using our model (Fig. 6. Table 2).

The higher than average value of ε of 20 erg $g^{-1}s^{-1}$, needed for model fitting the Irish Sea data (Figure 6) is not unreasonable in light of the high average overall energy dissipation rate for this area (refs 37, 41), and considering the importance of interfacial shear coagulation (ref 36).

Table 4. Comparison of predicted to measured values of particle uptake rate const., k_f

Bay	ε (erg $g^{-1}s^{-1}$)	C_p (mg/l)	model predicted k_f (d^{-1})	k_f from measured $^{238}U/^{234}Th$ ratio (d^{-1})
Narragansett (Refs. 33,34)	0.2-1	1-10	0.1 - 0.6	0.1 - 1
Galveston (ref 30)	~ 1-2	30	~ 0.6 - 2*	1.4

*) assuming f_{cd} = 0.25 to 0.7 (ref 18)

6. THORIUM PARTITIONING BETWEEN PARTICLES OF COLLOIDAL AND FILTER-RETAINED DIMENSIONS

The overall transfer of Th from the dissolved phase to the filter-retained particle fraction must include the partitioning of Th between solution and colloids; more specifically, the fraction of Me associated with coloids, f_c, must be determined. Unfortunately, the mass (C_c) and adsorption characteristics of colloids (K_c) in the oceanic water column are not well known. However, it is possible to use equilibrium K_d data to estimate the relative magnitude of these parameter's values.

Figure 7 shows log K_d as a function of log C_p for $^{230,234}Th$ and ^7Be (ref 42). An assumption is that $C_p = C_f$. At low values of C_p, log K_d for Th appears to be independent of C_p. As C_p increases, log K_d decreases monotonically. Equation 7b (Table 1) describes the relationship between K_f, K_c, C_c and K_d for a system containing two particle-size classes. For conditions such

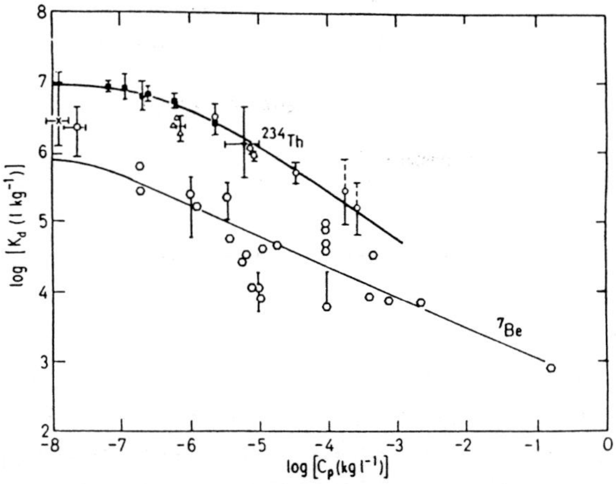

Figure 7a. Log K_d versus log C_p for $^{230,234}Th$ (from ref 8) and ^7Be.

that the product of K_c and C_c is small compared to 1, (< 0.05), K_d is constant and equal to K_f. Therefore, Fig. 7 indicates that $K_f = 10^{6.8}$ to $10^{7.0}$. As C_p increases, C_c will also increase and the term K_cC_c becomes significant relative to 1 (i.e., > 0.05). At this point, K_d will have a reciprocal dependence on C_p, the exact relationship depending on the functional relationship between C_c and C_p. Since K_c and K_f are assumed to be constant, the slope of the log K_d versus log C_p plot (for C_ps > 10^{-6} kg/L) indicates that $C_c = f(Cp^{0.7})$. This is also the slope suggested by Morel and Gschwend (ref 43) for the relationship between the "microparticles" and "macroparticles" in Gschwend and Wus (ref 44) data. If it is assumed that $K_c = K_f$, then it follows that log C_c = 0.7 log C_p - 2.6. Such a relationship between C_c and C_p includes the particle-concentration ratio needed to model Nyffeler et al (ref 45) data for marine sediments and is in the proximity of that for the lake and stream sediments described by Gschwend and Wu (ref 41). The alternative description follows from the kinetic approach of Nyffeler et al (ref 45) in which K_d is expressed in terms of a ratio of rate constants (Eqn. 7a, Table 1) where $R_f' = f(Cp^{\Theta})$. At high C_p, Θ approaches 0.3, i.e., the exponential coefficient for the dependency of B_b' on C_p. At low C_p, Θ approaches 1.0. Thus, log K_d is proportional to $(\Theta - 1)\cdot\log C_p$. At low C_p, K_d is independent of C_p (the slope of log K_d versus log C_p is zero). As the

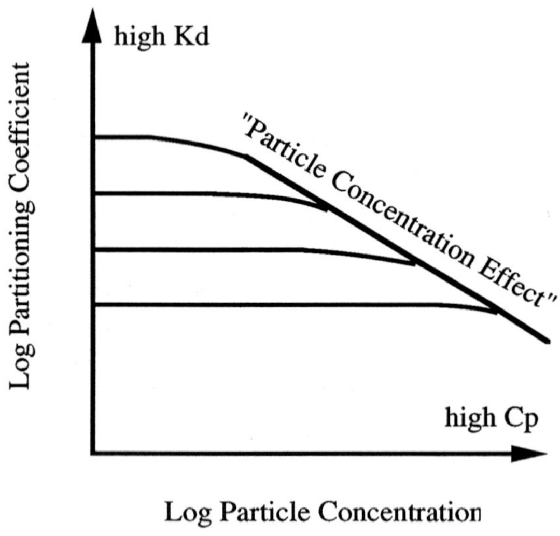

Figure 7b. Hypothetical relationships among master variables. The lower the relative value for the trace element's K_d the higher the value of C_p required for the onset of the "particle-concentration effect."

sorption rate constant approaches B_b', Θ approaches 0.3 and log K_d versus log C_p takes on a slope of (0.3 - 1) or -0.7. The model calculations were made with the assumption that R_r is independent of C_p; the value of R_r required to fit the calculations to the Th data was 0.04 d^{-1}. This value is in agreement with values from Narragansett Bay (P. Santschi, unpublished data: R_r = 0.004 - 0.7 d^{-1}), marine particles from sediment traps (ref 44: R_r = 0.04 to 0.14 d^{-1}), the English Channel (ref 46: R_r = 0.05 d^{-1}) and Puget Sound (ref 47: R_r = 0.3 d^{-1}). It is also not much higher than the values calculated for deep-sea suspended particles (ref 26: R_r = 0.004 to 0.017 d^{-1}; refs 48-50: $R_r \approx 0.018$ d^{-1}).

7. CONCLUSIONS

Over the last five years or so, colloids have been increasingly implicated in affecting trace-element behavior, for example in producing slow "sorption" kinetics and the particle-concentration effect. However, much of the evidence for the role of colloids in affecting the fate of trace elements and radionuclides has been circumstantial because of the difficulty in identifying submicron particles and in tracing their movement in natural waters. The emerging evidence suggests that colloids may play a pivotal part in the biogeochemical pathways of particle-reactive elements.

Finally, the model calculations by Honeyman and Santschi (ref 8), suggest that, in an inversion of the colloid "problem", we may be able to use Th isotopes as natural "coagulometers" for particle aggregation and "rate meters" for the disaggregation process (Fig. 8). Both are processes which, so far, have not been experimentally well determined in oceanographic settings.

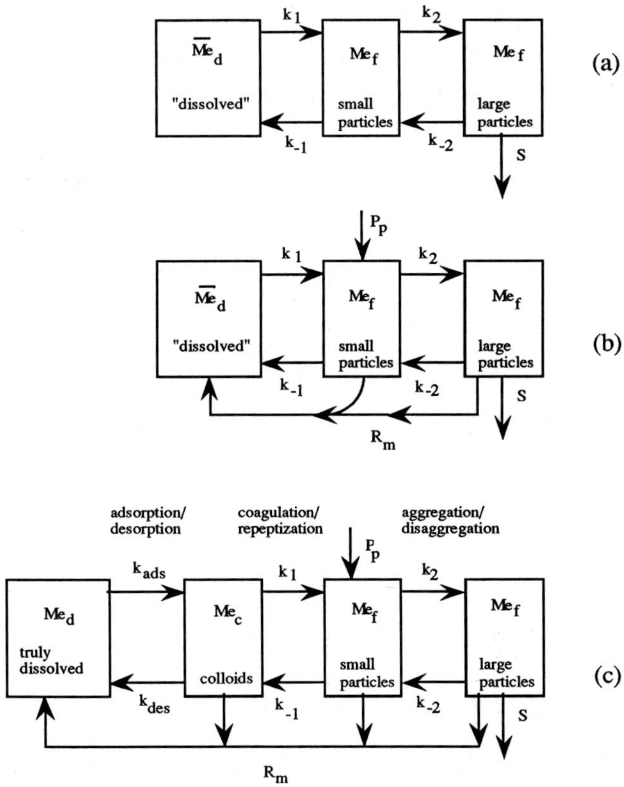

Figure 8. This figure depicts the current view on particle dynamics in the ocean. The interactions between the different boxes can be modelled using Th isotopes. a) Bacon and co-workers as well as Nozaki and co-workers have calculated rate constants for the interactions shown. $\overline{Me}_d$ represents operationally-defined "dissolved" species of Me and includes both truly-dissolved Me as well as colloid-associated species. Note, also, that the "small" particles are still relatively large: 10μm and greater. b) Sarmiento, Clegg, Whitfield and others have added a term for a remineralization reaction, R_m, which can originate from all particle pools and a primary-production term, P_p. c) Honeyman, Santschi and others are proposing a colloidal phase which can take up Th rapidly through sorption/desorption reactions but which coagulate relatively slowly with the small particle pool. Many seemingly anomalous observations of Th behavior in the ocean can be satisfactorily explained by the inclusion in the model of a colloidal pool of particles.

8. ACKNOWLEDGEMENTS

This material is based in part upon work supported by the Texas Advanced Research Program under Grant No. 4697, the Texas Institute of Oceanography, and the National Science Foundation (Grant OCE-9012103).

9. LITERATURE CITED

1 G. Sposito, in: J.A. Davies and K.F. Hayes (Eds.), Geochemical Processes at Mineral Surfaces, American Chemical Society Symposium Series No. 323, Washington, D.C., 1986, Chapter 11.
2 S. Krishnaswami, D. Lal, and B.L.K. Somayajulu, Earth Planet. Sci. Lett. 32 (1976) 403-419.
3 Y.-H. Li, L. Burkhardt, M. Buchholtz, P. O'Hara, and P.H. Santschi, Geochim. et Cosmochim. Acta, 48 (1984) 2011-2019.
4 B.A. McKee, D.J. DeMasterand C.A. Nittrouer, Cont. Shelf Res., 6 (1986) 87-106.
5 B. Honeyman, L.S. Balistrieri and J.W. Murray, Deep Sea Res., 35 (1988) 227-246.
6 P.H. Santschi, Limnol. and Oceanogr. 29/5, (1984) 1100-1108.
7 K.H. Coale and K.W. Bruland, Limnol. Oceanogr., 30 (1985) 22-33.
8 B.D. Honeyman and P.H. Santschi, J. Mar. Res., 47/4 (1989) 950-992.
9 P.H. Santschi and B.D. Honeyman, Radiat. Chem. Phys. 34(2), (1989) 213-240.
10 K.W. Bruland, Chem. Oceanography., Vol. 8, Academic Press, New York, 1983, pp. 157-221.
11 L.M. McDowell-Boyer, J.R. Hunt and N. Sitar, Water Res. Res., 22 (1986) 1901-1921.
12 J. Buffle, Complexation Reactions in Aquatic Systems. An Analytical Approach, Ellis Horwood, Chichester , 1987 .
13 F. Kepak,. in: A. Evans and M. Muramatsu (Eds.), Radiotracer Techniques and Applications, Marcel Dekker, Inc., New York, 1977, Chapter 8.
14 R.W. Buddemeier and J.R. Hunt, Appl. Geochem., 3 (1988) 535-548.
15 J.C. Sheppard, M.J. Campbell, T. Cheng and J.A. Kittrick, Environ. Sci. Technol., 11 (1980) 1349-1353.
16 S.E.H. Niven and R.M. Moore, in: J.C. Guary, P. Gueguenìat and R.J. Pentreath (Eds.), Radionuclides: A Tool for Oceanography, Elsevier Applied Science, New York, 1989, pp. 111-120.
17 S.B. Moran, and R.M. Moore. Geochim. et Cosmochim. Acta, 53 (1989) 2513-2524.
18 M. Baskaran, G. Benoit, and P.H. Santschi, The role of sub-micron particles in the removal of throium isotopes from the oceanic water column, in preparation
19 K.L. Nash and G.R. Choppin., J. Inorg. Nucl. Chem., 42 (1980) 1045-1050.
20 P.H. Santschi, Y.-H. Li, J. Bell, D. Adler, M. Amdurer and U.P. Nyffeler, Geochim. Cosmochim. Acta, 47 (1983) 201-310.
21 P.H. Santschi, Y.-H. Li, P. O'Hara, M. Amdurer, D. Adler and P. Doering, J. Mar. Res., 45 (1987) 1007-1048.
22 M. Buchholtz, P.H. Santschi and W.S. Broecker,. In: T.H. Sibley and C. Myttenaire (Eds.), Application of Distribution Coefficients to Radiological Assessment Models, Elsevier Applied Science Publ., New York, 1986, pp. 192-205.
23 K.J. Farley and F.M.M. Morel, Environ. Sci. Technol., 20 (1986) 187-195.
24 B.D. Honeyman and P.H. Santschi, Coupling adsorption and particle aggregation: kinetic and equilibrium studies using ^{59}Fe-labeled hematite. Environ. Sci. Technol. (1991), in press.
25 L. Liang, Effects of surface chemistry on kinetics of coagulation of submicron iron oxide particle: (α-Fe_2O_3) in water. Ph.D. Thesis, California Institute of Technology, 1988.
26 M. Bacon and R. Anderson, J. Geophy. Res., 87 (1982) 2045-2056.
27 D. Lal, Earth and Planet. Sci. Lett., 49 (1980) 520-527.
28 J.K. Edzwald, J.C. Upchurch and C.R. O'Melia, Environ. Sci. Tech., 8 (1974) 58-63.
29 I.N. McCave, Deep Sea Res., 31 (1984) 329-352.
30 M. Baskaran, Ch. Coleman, G. Benoit and P.H. Santschi,. EOS, 71/2, (1990) 71.
31 P.J. Kershaw, and A. Young, J. Environ. Radioactivity, 6 (1988) 1-23.
32 P.J. Kershaw, P.A. Gurbutt, A.K. Young and D.J. Allington, in: J.C. Guary, P. Gueguenìat and R.J. Pentreath (Eds.), Radionuclides: A Tool for Oceanography, Elsevier Applied Science, New York, 1989, pp. 131--142.

33 P.H. Santschi, Y.-H. Li and J. Bell, Earth Planet. Sci. Lett. 45 (1979) 201-213.
34 P.H. Santschi, D. Adler, M. Amdurer, Y.-H. Li and J. Bell. Earth Planet. Sci. Lett. 47 (1980), 327-335.
35 H.B. Fisher, J.E. List, R.C. Koh, R.C. Imberger and N.H. Brooks, Mixing in Inland and Coastal Waters, Academic Press, New York, 1979.
36 K. Newman, S.L. Frankel and K.D. Stolzenbach, Environ. Sci. Technol., 24, (1990) 513-518.
37 C.J.R. Garrett, J.R. Keeley and D.A. Greenberg, Atmosphere-Ocean, 16 (1978) 403-423.
38 W.S. Brown and R.P. Trask, J. Phys. Oceanogr., 10 (1980) 1742-1754.
39 E.R. Levine and K.E. Kenyon, J. Geophys. Res., 80 (1975) 1683-1888.
40 S.W. Nixon, D. Alonso, M.E.Q. Pilson and B.A. Buckley, in: J.P. Giesy (Ed.), Microcosms in Ecological Research, DOE Symposium Series 52, USADE TIC.CONF-781101, 1980, pp. 818-849.
41 G.I. Taylor, Phil. Trans. Roy. Soc. London, A220 (1919) 1-93.
42 B.D. Honeyman and P.H. Santschi, Environ. Sci. Technol., 22 (1988) 862-871.
43 F.M.M. Morel and P.M. Gschwend, in: W. Stumm (Ed.), Aquatic Surface Chemistry: Chemical Processes at the Particle-Water Interface", John Wiley, New York, 1987,. Chapter 15.
44 P.M. Gschwend and S.-C. Wu, Environ. Sci. Technol., 19 (1985) 90 -96.
45 U.P. Nyffeler, Y.-H. Li, and P.H. Santschi, Geochim. et Cosmochim. Acta, 48 (1984) 1513-1522.
46 R.M. Moore and G.E. Millward, Geochim. et Cosmochim. Acta, 52 (1988) 113-118.
47 H.W. Jannasch, B.D. Honeyman, L.S. Balistrieri, and J.W. Murray, Geochim. et Cosmochim. Acta, 52 (1988) 567-577.
48 Y. Nozaki, Y. Horibe and H. Tsubota, Earth Planet. Sci. Lett., 54 (1981) 203-216.
49 M. Bacon and R. Anderson, J. Geophys. Res., 87 (1982) 2045-2056 .
50 S.L. Clegg, and M. Whitfield, Deep-Sea Res., 37 (1990) 809-832 .

GLOSSARY OF SYMBOLS

B_{ds} = rate coefficient for differential settling ($mg^{-1.3}l^{1.3}d^{-1}$)

B_{sh} = rate coefficient for fluid shear ($mg^{-0.9}l^{0.9}d^{-1}$)

B_b = rate coefficient for Brownian motion ($mg^{-0.3}l^{0.3}d^{-1}$)

B_b' = pseudo first order rate coefficient for Brownian motion (d^{-1})

C_c = colloid mass concentration ($mg\ l^{-1}$)

C_f = filter–retained particle concentration ($mg\ l^{-1}$)

C_p = total particle concentration ($mg\ l^{-1}$) = $C_f + C_c$

Me_f, Me_c, Me_d, Me_T = trace–element concentration on filter–retained particles, colloid–sized particles, truly dissolved and total ($M\ l^{-1}$)

Γ_C = moles of trace element adsorbed per mass of colloids ($M\ g^{-1}$)

Th_{part}, Th_{diss} = "particulate" and "dissolved" Th concentration ($M\ l^{-1}$)

λ_{Th} = radioactive decat constant (d^{-1})

λ_p = removal rate constant for particles (d^{-1})

R_r = pseudo–first–order rate constant for reverse reaction (d^{-1})

R_f'= pseudo–first–order rate constant for forward reaction (d^{-1})

J = other production or sink terms for the metal (e.g., production by another nuclide, removal by radioactive decay or particle sinking)

$K_{d'}$ = partition coefficient as particulate/filter–passing activity ($l\ kg^{-1}$) for equilibrium (K_d) and non–equilibrium (Q_d) conditions

K_f = conditional equilibrium association constant for filter–retained particles ($l\ kg^{-1}$)

K_c = conditional equilibrium association constant for colloidal particles ($l\ kg^{-1}$)

f_a = fraction of total trace element which is sorbed = $f_c + f_f$

f_c = fraction of total trace element which is associated with colloids

f_f = fraction of total trace element which is associated with filter–retained particles

f_{cd} = fraction of filter–passing trace element associated with colloids

α_b = collision efficiency factor

ρ_e = effective particle density

h = depth of homogeneous water column

μ = dynamic viscosity

ν_f = kinematic viscosity ($cm^2\ s^{-1}$)

C_d = drag coefficient ($2.5\ 10^{-3}$)

U = average current velocity ($cm\ s^{-1}$)

$u_* = U_{100}\sqrt{f/8}$ = friction velocity ($cm\ s^{-1}$)

f = 0.02 = friction or drag coefficient

U_{100} = current velocity at 100 cm above seabed ($cm\ s^{-1}$)

STABLE LEAD AND ^{210}Pb IN GERMAN LAKE SEDIMENTS AND SOILS: A TRACER FOR ANTHROPOGENIC LEAD EMISSION

H. Dörr, A. Mangini, W. Schmitz, F. Weber and K.O. Münnich

1. INTRODUCTION

Anthropogenic lead emission increased since about 1750 primarily due to non-ferrous and ferrous metal production. Since 1950 the dominant anthropogenic lead source in the Federal Republic of Germany is burning of leaded gasoline. While 'industrial' lead is emitted from point sources, 'gasoline' lead emission is assumed to be distributed rather homogeneously over the whole country. The natural, radioactive lead isotope ^{210}Pb (half-life 22.4 years) is produced by α decay of ^{222}Rn . Atmospheric ^{210}Pb , like gasoline lead is incorporated into the aerosol, and thus being subject to atmospheric mixing and transport. After an average lifetime of about 5 - 10 days, both ^{210}Pb and lead exhaust aerosol particles are removed from the atmosphere by dry and wet deposition. The atmospheric ^{210}Pb component in lake sediments and soils ($^{210}Pb_{exc}$) is distinguished from the in-situ produced ^{210}Pb by simultaneous measurement of ^{226}Ra , the radioactive mother of ^{222}Rn . In sediments we assume the in situ produced ^{210}Pb to be in radioactive equilibrium with ^{226}Ra , while in soils we have to correct for loss of ^{210}Pb due to the escape of ^{222}Rn through the soil surface. In this paper we compare anthropogenic lead profiles in lake sediments and soils with the reconstructed atmospheric input function for gasoline lead, and discuss the use of the gasoline lead maximum as a time marker for sediment dating and sedimentation rate determination.

2. MATERIALS AND METHODS

Sediment cores LA5, SM5, and SM6 from the central and eastern part of Lake Constance were collected in September 1979. Cores CS6, CS7, CS8, and CS10 were sampled on a north–south transect from Friedrichshafen to Romanshorn in 1986. Coring, sampling procedures and radioisotope analyses are described by (1–3). Sediment Cores from Black Forest lakes Wildsee (S2WIL), Herrenwieser See (S2HER) and Feldsee (S2FEL) were taken in 1983. The sampling sites of the sediment cores are shown in Fig. 1 as full squares. A detailed description of the sampling procedure, heavy metal analyses and geochemical parameters are given by (4). The geogenic lead content in the sediments is estimated from the measured lead concentrations in the deepest (oldest) samples (see Table 1). The geogenic lead contribution is subtracted from the total lead concentration to obtain the anthropogenic lead content in the sediments.

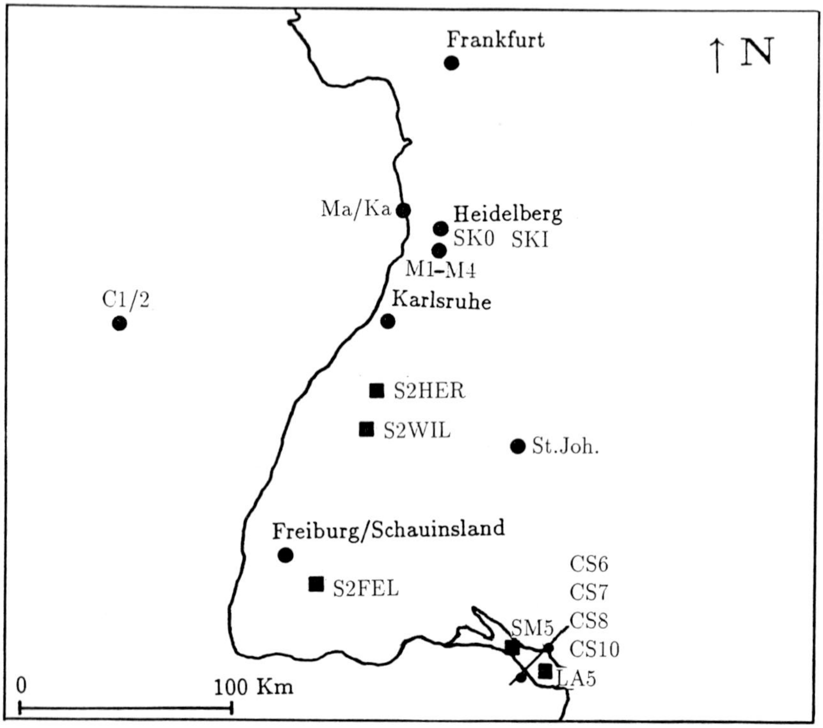

Fig. 1. Map of sample locations (south–west Germany). Sediment samples are indicated by full squares, soil samples by full circles.

Soil cores were taken in stainless steel cylinders, 12 cm long, 72 cm^2 cross section, and cut into slices 1 – 2 cm thick. All samples were taken in forested sites to ensure sampling of undisturbed soil cores. The samples M1 – M4 were taken about 15 km south of Heidelberg in 1987. All other samples were taken in 1988. SK0 and SKI were collected about 15 km east of Heidelberg. The core Ma/Ka (Mannheim/Karlstern) was taken in a small forested area north of Mannheim, surrounded by highly frequented roads and highways (BAB6, BAB659). The sample Berl. was taken in a forested area in Berlin (Grunewald). The sample St.Joh (St.Johann) was taken ca 50 km south of Stuttgart and 6 km east of Reutlingen. C1/2 was sampled near the French-German boarder, ca. 20 km north east of Nancy and ca. 50 km south west of Saarbrücken. The location of the soil sampling sites are shown in Fig. 1 as full circles.

The sites were classified with respect to population and road/traffic density as urban, semi–urban, and rural. The classification of the sampling sites is given in Table 1 together with soil type and average soil pH in the 0 – 10 cm layer.

TABLE 1
Classification with respect to population and road/traffic density of the sampling sites, soil type, soil pH, and geogenic lead content c_{geo}.

site		c_{geo} [ppm]	soil type	soil pH
LA5	semi – urban	14		
SM5	semi – urban	22		
S2FEL	rural	75		
S2WIL	rural	20		
S2HER	rural	17		
Ma/Ka	urban	0	sand	3.3
Berl.	urban	10	loam	4.2
M1	semi – urban	0	sand	7.7
M2	semi – urban	60	loam	8.0
M3	semi – urban	0	sand	3.5
M4	semi – urban	20	clay/loam	5.0
St.Joh.	semi – urban	0	clay/loam	4.4
SK0	semi – urban	20	loam	3.2
SKI	semi – urban	20	loam	6.8
C1/2	rural	10	clay	4.4

The sites Ma/Ka and St.Joh. are part of the 'Bodenmeßnetz Baden-Württemberg' of the Environment Ministry of Baden-Württemberg, F.R.G. Heavy metal analyses were carried out under the response of LFU-Karlsruhe (Landesanstalt für Umweltschutz). ^{210}Pb, ^{226}Ra , ^{134}Cs , and ^{137}Cs in soil samples were measured by γ spectrometry (5). Chernobyl cesium is distinguished from bomb cesium by the known ^{134}Cs /^{137}Cs ratio of Chernobyl cesium (6). Stable lead was measured by atomic absorption. To obtain the anthropogenic lead content $c_{anthr.}$ in soil samples, the total lead concentration c_{tot} is corrected for the geogenic lead contribution using

$$c_{anthr.} = c_{tot} - (1 - c_{org}) \cdot c_{geo} \qquad [1]$$

c_{geo} is determined from the measured lead concentration in depths where the anthropogenic lead content can be neglected. The organic matter concentration c_{org} (in weight %) is measured by heat combustion. The factor (1 - c_{org}) in equation [1] considers that soil organic material is free of geogenic lead. The geogenic lead concentrations c_{geo} at the sampling sites are given in Table 1.

3. ANTHROPOGENIC LEAD IN LAKE SEDIMENTS

Fig. 2 shows stable lead concentration profiles in the sediment of Lake Constance (SM5) and of Herrenwieser See (S2HER) in the Black Forest. Lead concentrations are given in ppm (mg Pb per kg dry weight). The depth is given as accumulated weight to correct for sediment compaction. The time marks on the right hand ordinate are calculated from the average sediment accumulation rate (see Table 2). Sedimentation rates are determined by ^{210}Pb dating (1, 4).

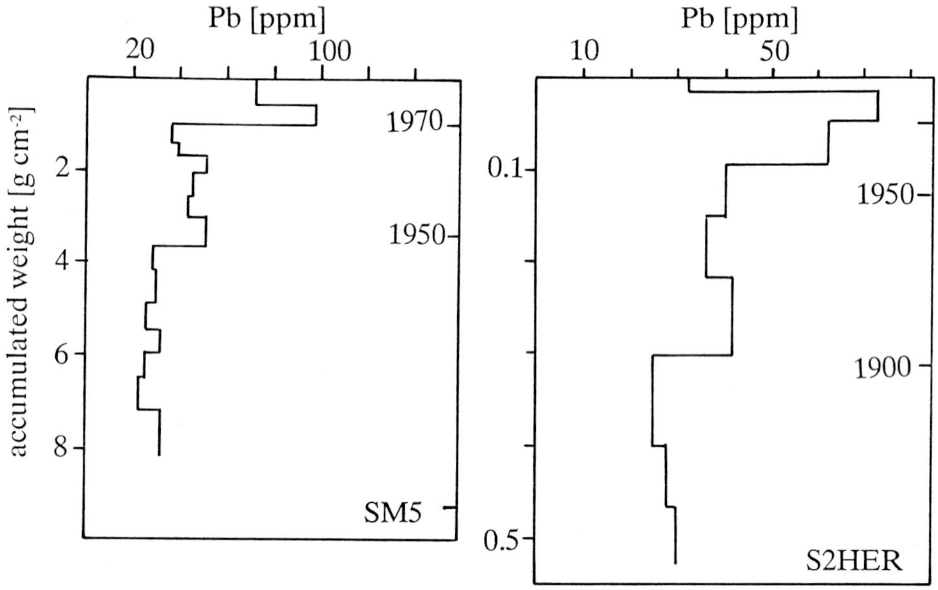

Fig. 2. Stable lead in sediments of Lake Constance (SM5) and Herrenwieser See, Black Forest (S2HER). The time marks on the right hand ordinate are calculated from the average sediment accumulation rate (see Table 2). Locations are shown in Fig. 1.

The sediments show a pronounced lead concentration maximum in the upper part of the profile (2 – 5 cm depth). ^{210}Pb dating shows that maximum lead concentrations occured in the early seventies. This is observed in lakes with sedimentation rates between 0.0037 g cm^{-2} a^{-1} (S2HER) and 0.118 g cm^{-2} a^{-1} (SM5), and pH values between pH = 5.5 in the Black Forest lakes and more than pH = 7 in Lake Constance. In general, high lead concentrations in a sediment are caused by high concentrations in the sediment material, by a temporarily low sedimentation rate, and by an enrichment of lead due to changing chemical conditions in the sediment profile. These mechanisms can not be distinguished from stable lead data alone. Temporal variations of the sediment accumulation rate and enrichment of lead in specific sediment layers can be identified from additional ^{210}Pb data:

The flux density of stable lead P_z, and of the radioactive lead isotope J_z for any depth z is given by:

$$P_z = c_z \cdot s_z \quad [2]$$

$$J_z = a_z \cdot s_z \quad [3]$$

where s_z is the sedimentation rate of layer z, c_z is the stable lead concentration, and a_z is the $^{210}Pb_{exc}$ activity concentration in layer z. Assuming the ^{210}Pb flux $J_{0,z}$ (corrected for radioactive decay) to the sediment to be constant with time (2,7,8) the ratio c_z/a_z is a measure of the stable lead flux:

$$c_z/a_z = P_z/J_z \quad [4]$$

a_z decreases exponentially with time due to radioactive decay, and thus the ratio c_z/a_z increases with sediment depth even in layers where c_z remains at a constant (geogenic) level. Note that enrichment of lead and changes of the sedimentation rate act on both stable lead and ^{210}Pb concentrations. Thus, these processes do not influence the lead/^{210}Pb ratio.

Fig. 3 shows the depth distribution of the stable lead /$^{210}Pb_{exc}$ ratio ($Pb/^{210}Pb_{exc}$) in two Lake Constance sediment cores (LA5, SM5) and in two Black Forest lakes (S2WIL, S2FEL). The error bars are calculated from the statistical counting error of the ^{210}Pb measurement and a 10 % error of the stable lead concentrations. The error bars in y direction indicate the depth intervall of the individual samples. The plotted curves are cubic spline fits to the data.

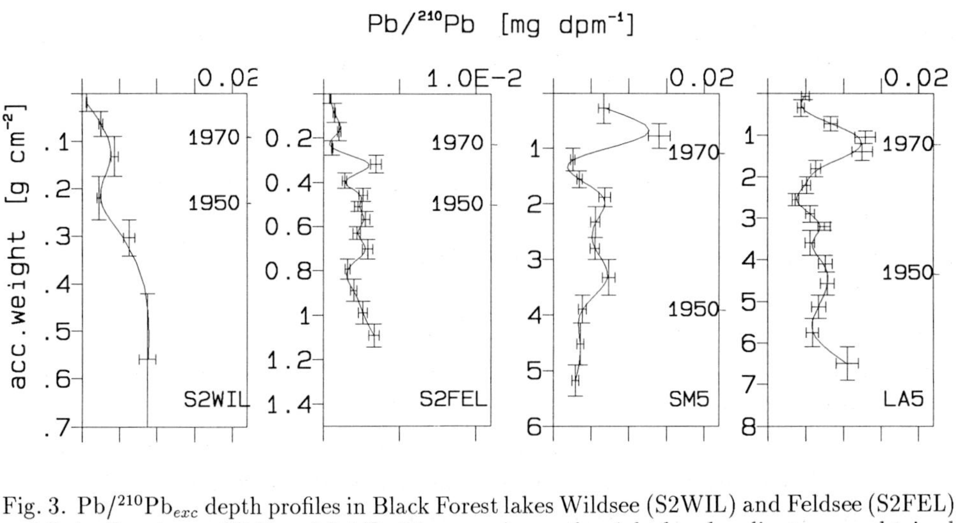

Fig. 3. $Pb/^{210}Pb_{exc}$ depth profiles in Black Forest lakes Wildsee (S2WIL) and Feldsee (S2FEL) and Lake Constance (SM5 and LA5). Time marks on the right hand ordinates are obtained from ^{210}Pb dating.

In all sediments the $Pb/^{210}Pb_{exc}$ profiles show a similar trend. A pronounced maximum is observed in layers originating from about 1970. At greater depth the increase of the $Pb/^{210}Pb_{exc}$ ratio is due to radioactive decay of ^{210}Pb at a constant stable lead level (see S2FEL and LA5). Assuming a time–constant geogenic lead concentration of the sediment material the $Pb/^{210}Pb_{exc}$ depth profiles reflect the atmospheric input of anthropogenic lead to the sediment showing maximum lead concentrations in the atmosphere in the early seventies.

3.1 Atmospheric lead and ^{210}Pb input function

We reconstructed the atmospheric gasoline lead input function for the Federal Republic of Germany from annual fuel consumption (9) and average lead content in gasoline, because the major source of anthropogenic lead in the atmosphere is burning of leaded gasoline (10). Fig. 4 shows the annual gasoline lead exhaust in the Federal Republic of Germany since 1950. The error bars indicate a 10% uncertainty of the gasoline lead content and the gasoline consumption data. The plotted curve is a cubic spline fit to the lead exhaust data. Gasoline lead exhaust in the FRG steadily increased since 1950 as a consequence of rising fuel consumption and reached maximum values between 1970 and 1975. After reduction of the gasoline lead content in 1972 and 1976 to 0.4 gl^{-1} and 0.15 gl^{-1}, respectively, emission then abruptly decreased to about half of the maximum level. From 1986 to 1990 gasoline lead exhaust again decreases due to an increase of the contribution of unleaded gasoline from about 10 % in 1986 to about 70 % in 1990 at a nearly constant gasoline consumption.

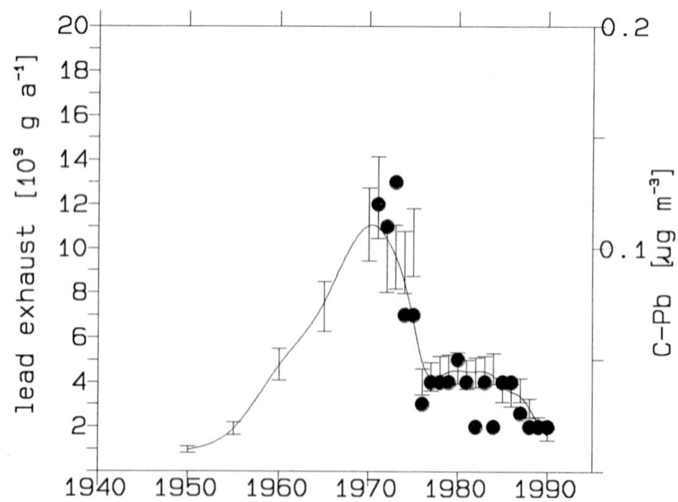

Fig. 4. Gasoline lead exhaust in the FRG since 1950. Full circles indicate annual mean atmospheric lead concentrations at the Schauinsland (right ordinate). The curve is a cubic spline fit to the lead exhaust data (vertical bars).

The total integrated gasoline lead exhaust in the FRG since 1950 is about $250 \cdot 10^9$ g Pb. If we simply assume the total gasoline lead to be deposited homogeneously and the territory of the FRG to be a closed system, i.e. neglecting lead import and export, the total lead deposition in the FRG is 1 g m^{-2}. Lead deposited in the close vicinity of highways and major roads is of minor importance for the total gasoline lead balance (11) Measurements of atmospheric lead concentrations at several urban and rural sites in the F.R.G since 1972 (12) are in good agreement with the lead exhaust data of Fig. 4. As an example, Fig. 4 shows the average annual lead content in the atmosphere at the Schauinsland/Freiburg (full circles). The right ordinate is obtained from the linear regression of atmospheric lead content versus yearly gasoline lead exhaust. A good correlation is observed between the annual mean of the local atmospheric lead content at the 'UBA stations' (run by the Umweltbundesamt, Berlin) and average gasoline lead exhaust in the FRG in the period 1972 – 1986. From this correlation we estimate the local gasoline lead source density in urban and semi–urban regions to be about 10 and 4 times higher than the average lead source density in rural regions (11).

The atmospheric ^{210}Pb content shows strong seasonal but practically no year to year variations due to a comparably short atmospheric residence time of about 5–10 days (13–15) The long–term average of the atmospheric ^{210}Pb content in Fribourg/CH (measured between 1973 and 1982) is 26.4±0.9 dpm 10^{-3} m^{-3} (16).

3.2 Anthropogenic lead inventories in lake sediments

Comparison of the atmospheric gasoline lead input function (see Fig. 4) with Pb/^{210}Pb profiles in lake sediments (see Fig. 3) indicates that stable lead concentrations in sediments reflect the anthropogenic gasoline lead emissions into the atmosphere. To confirm the assumption that the source of anthropogenic lead in sediments is the burning of leaded gasoline, we compare the anthropogenic lead inventories in the examined lake sediments with the total gasoline lead deposition.

TABLE 2
Sedimentation rate s obtained from ^{210}Pb dating, sedimentation rate s_{Pb} obtained from the position of the gasoline lead maximum (see section 3.3), atmospheric ^{210}Pb flux j_0, and anthropogenic lead inventory Q in sediment cores. c_{max}/a_{max} is the lead/^{210}Pb ratio in the layer showing maximum lead concentration.

	s [gcm^{-2}a^{-1}]	j_0 [dpm cm^{-2}a^{-1}]	Q [g m^{-2}]	c_{max}/a_{max} [μg dpm^{-1}]	s_{Pb} [g cm^{-2}a^{-1}]
S2FEL	.015	1.4	0.75	3.5	.027±.003
S2WIL	.007	.14	0.23 (0.90)	4.3	.011±.001
S2HER	.0037	.055	0.11 (1.1)	6.9	.008±.001
LA5	.133	.89	2.09	12.9	.151±.019
SM5	.118	.71	1.75	14.0	.097±.012
CS6		.96	4.37		.317±.040
CS7			3.91		.283±.035
CS8			4.68		.167±.021
CS10			2.59		.117±.015

Table 2 gives a summary of the anthropogenic lead inventories in the sediments of Lake Constance and of the Black Forest lakes together with the average sediment accumulation rate and the average atmospheric ^{210}Pb flux to the sediments. The average anthropogenic lead inventory in the sediments of Lake Constance of 3.2 g Pb m^{-2} is in good agreement with the inventory expected from gasoline lead exhaust in semi–urban regions. The comparatively low anthropogenic lead inventories and ^{210}Pb fluxes in the Black Forest lakes Wildsee (S2WIL) and Herrenwieser See (S2HER) are due to incomplete retention of lead and ^{210}Pb , because of the very low pH of the water. Calculations based on ion and heavy metal concentrations in precipitation and in– and outflowing water from these lakes show that only 19 % and 30 % of the input is retained in the lakes (4). Assuming an average ^{210}Pb flux from the atmosphere for the South German region of 0.55 dpm cm^{-2} a^{-1} (5) we correct the anthropogenic lead inventory by the ratio of the expected and observed ^{210}Pb flux (10 % and 27 % retention at S2HER and S2WIL, resp.). The lead inventories corrected for incomplete retention are given in brackets in Table 2. The average anthropogenic lead inventory in the sediments of the Black Forest lakes of 0.9 g Pb m^{-2} is also in good agreement with the inventory expected from the gasoline lead exhaust in rural regions.

Another confirmation of the assumption that the stable lead profiles in lake sediments reflect the gasoline lead emmission is obtained from the maximum Pb/^{210}Pb ratio in the sediments. With a maximum atmospheric lead content in rural and semi–urban regions of about 0.1 and 0.4 μg m^{-3}, respectively (see section 3.1) and an average atmospheric ^{210}Pb concentration of about 25 dpm 10^{-3} m^{-3} (measurements from Fribourg/CH, (16)) the expected maximum Pb/^{210}Pb ratio is about 4 μg dpm^{-1} and 16 μg dpm^{-1} in rural and semi–urban regions, respectively, being in good agreement with the average Pb/$^{210}Pb_{exc}$ ratios in the sediments of the Black Forest lakes (5 μg dpm^{-1}) and of Lake Constance (14 μg dpm^{-1}) (see Table 2).

3.3 Determination of sedimentation rates from the gasoline lead maximum

If we assume the sediment layer with maximum lead concentration to be deposited in 1971 (see Fig. 3) we determine the sedimentation rate s_{Pb} from the time elapsed since 1971 and the accumulated weight above that layer:

$$m_z = \Delta z \cdot \rho_z \qquad [5]$$

$$s_{Pb} = \frac{\sum_{z=0}^{Pb-max} m_z}{(samplingtime - 1971)} \qquad [6]$$

Δz is the depth intervall, and ρ_z is the dry bulk density of layer z. In addition to the Lake Constance cores LA5 and SM5 we determined the sedimentation rates of cores taken on a north–south transect over Lake Constance (CS6, CS7, CS8, and CS10). The stable lead profiles are shown in Fig. 5. The sedimentation rates obtained from equation [6] are given in

Table 2. The errors are determined from an assumed ± 1 year uncertainty of the gasoline lead maximum in the atmosphere, and from the depth intervall of the sample with maximum lead concentration. Sedimentation rates obtained from equation [6] are slightly higher than those obtained from ^{210}Pb dating (see Table 2). This is supposedly due to the fact that ^{210}Pb dating gives an average sediment accumulation rate over the past 30 – 100 years, while 'gasoline lead dating' gives a more recent sedimentation rate. An increased sedimentation rate in the past 10–20 years is indeed indicated by comparatively low $^{210}Pb_{exc}$ activities in the upper sediment layers (1, 2).

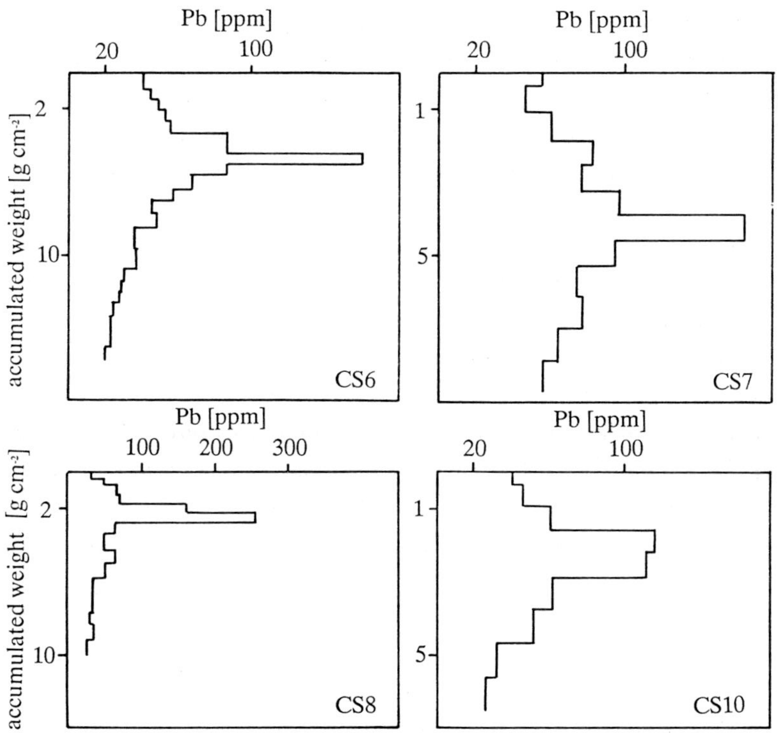

Fig. 5. Stable lead profiles from the north–south transect 'Friedrichshafen–Romanshorn' over Lake Constance.

4. ANTHROPOGENIC LEAD AND ^{210}Pb IN SOILS

4.1 Anthropogenic lead inventories in soils

Investigations of the displacement of atmospheric ^{210}Pb and solid soil organic material, using bomb produced radiocarbon as a tracer, showed that lead is 'displaced' together with the soil organic matter only (17). The velocity of downward migration of lead w_{Pb} in forest soils was found to be identical with the 'sedimentation rate' $w_{org} = J_{org}/\rho_b$ of organic material (J_{org} is the annual flux of organic matter to the soil as litter fall and ρ_b is the dry bulk density)

(5). The average downward velocity of lead in soils is about 1 mm a^{-1}. If we assume the geochemical behaviour of stable lead and ^{210}Pb in soils to be identical, the anthropogenic gasoline lead is expected to be contained in the topmost organic rich soil layers. Fig. 6 shows anthropogenic lead profiles in forest soils at a rural, semi–urban, and urban site. The lead content is given per unit of soil volume to correct for increasing soil bulk density with depth. Note that this representation of lead concentration profiles in soils is equivalent to that of lead profiles in lake sediments, where the lead content per unit of soil mass is plotted versus the accumulated weight. More anthropogenic lead profiles in soils are shown in (11).

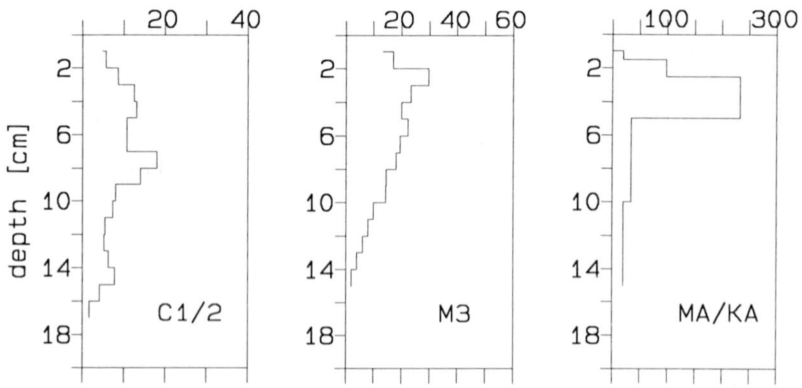

Fig. 6. Anthropogenic lead profiles in undisturbed forest soils at a rural (C1/2), semi–urban (M3), and urban (Ma/Ka) site.

The anthropogenic lead content is determined from equation [1] and from the geogenic lead concentrations shown in Table 1. Migration velocities obtained from ^{210}Pb data (5) and anthropogenic lead inventories in the investigated soils are shown in Table 3.

TABLE 3
^{210}Pb downward velocity v_{Pb}, average annual ^{210}Pb flux to the soil j, and anthropogenic lead inventory I in undisturbed forest soils.

	v_{Pb} [mm a^{-1}]	j [dpm cm^{-2} a^{-1}]	I [g m^{-2}]
C1/2	1.3	.71	1.4
M1	1.6	.55	3.6
M2	1.2	.82	4.6
M3	.6	.47	2.2
M4	.8	.60	2.6
SK0	.4	.94	4.4
SKI	.6	.75	6.2
St.Joh	.8	.56	4.1
MA/Ka	.9	.72	10.8
Berl.	.8	.43	5.3

Fig. 6 shows that practically all anthropogenic lead is contained in the first 20 cm of the soil. The average anthropogenic lead inventories at the rural site C1/2 is 1.4 g m^{-2}, at the semi–urban sites 4.0 $\pm$ 1.3 g m^{-2}, and at the urban sites 8.0 $\pm$ 2.8 g m^{-2}. The anthropogenic lead inventories in soils are in good agreement with the inventories predicted from total gasoline lead exhaust (see section 3.1). The observed inventories, however, depend on the local deposition rate, and, thus, represent an upper limit of the average anthropogenic lead content in soils due to an increased deposition in forest ecosystems. Assuming a space–constant ^{210}Pb concentration in the atmosphere the average ^{210}Pb flux obtained from soil profiles is a direct measure of the local deposition rate. A representative average lead inventory $\bar{I}$ is, thus, obtained by normalizing the observed inventory I to an average ^{210}Pb flux $\bar{j}$ to the soil:

$$\bar{I} = I \cdot \bar{j}/j \qquad [7]$$

With the lead inventories and ^{210}Pb fluxes from Table 3 and an average ^{210}Pb flux $\bar{j}$ = 0.55 dpm cm^{-2} a^{-1} (5) the average lead inventory is about 1 g m^{-2} in rural regions, about 3.1 $\pm$ 0.6 g m^{-2} in semi–urban regions, and 7.5 $\pm$ 0.7 g m^{-2} in urban regions.

4.2 Depth distribution of anthropogenic lead

The anthropogenic lead profiles show a concentration maximum at depths between 3 and 7 cm. This can, however, not simply be interpreted like in lake sediments as the gasoline lead concentration maximum. In contrast to lake sediments, where the anthropogenic elements are deposited together with the sediment itself, the depth distribution of atmospheric trace–elements in soils depends on the depth of infiltration ('wash–in' by precipitation), on mixing processes (bioturbation), and on the mobility (velocity of downward migration) of the considered elements in the soil. Trace–elements are attached to aerosols and are removed from the atmosphere by dry and wet deposition. The depth of infiltration into the soil, thus, depends on the intensity of precipitation and the contribution of wet and dry deposition, as well as on the size distribution and chemical composition of the aerosol particles. The lead exhaust aerosol particles and atmospheric ^{210}Pb , which is attached to aerosol particles of different size and chemical composition, are, thus, not expected to infiltrate into the same depth. We simulate infiltration, mixing, and downward migration with a simple, chromatographic box model. The model is based on the equation of continuity

$$\frac{\partial c(z,t)}{\partial t} = -div j - \lambda \cdot c + Q \qquad [8]$$

where c is the tracer concentration, j is the flux density, and Q is the input from the atmosphere. $\lambda = 0$ for stable elements and $\lambda = \ln 2/t_{1/2}$ for radioactive isotopes ($t_{1/2}$ is the radioactive halflife). After a subdivision of the soil into N boxes of height H (assuming homogeneity in x and y direction) equation (8) reads

$$\frac{\Delta c_i}{\Delta t} = \frac{j_i - j_{i+1}}{H} - \lambda \cdot c_i + f_i \cdot \frac{Q}{H} \qquad [9]$$

for each box i (i = 1 to N). Δt is the time–step of calculation. The flux density out of box i is $j_i = w \cdot c_i$, where w is the velocity of downward migration. f_i is the relative fraction of the annual input from the atmosphere that directly enters box i ($\sum_{i=1}^{N} f_i = 1$), and thus represents the infiltration depth. The box height H, simulating statistical fluctuations in the downward movement ('numerical dispersion' (18)), is the mixing parameter. A 'mixing coefficient' D can be defined by $D = H^2/\Delta t$.

The *infiltration depth* of gasoline lead by wet deposition is estimated from the depth distribution of Chernobyl cesium. Due to the relatively small migration velocity of cesium in soils (1 – 2 mm a^{-1}) and particularly the short time since deposition in 1986 the influence of transport and mixing on the Chernobyl cesium in the soil can be neglected (5). Measurements of the depth distribution of Chernobyl cesium in European forest soils showed that the infiltration depth is between 1 to 5 cm. The *mixing parameter* H is determined by model simulation of the bomb cesium depth profile. After a short–period input in the early sixties, the broadening of the bomb cesium peak during downward movement in the soil is a good tracer for mixing processes. The bomb cesium maximum in undisturbed forest soils is in a depth of 3 to 6 cm. The width of the initially 'sharp' bomb cesium peak is between about 2 and 5 cm. Modelling of bomb cesium depth profiles gives mixing parameters of H = 0.5 to 2 cm ($D = 8 \cdot 10^{-9}$ to $1.3 \cdot 10^{-7}$ $cm^2\ s^{-1}$). The bomb cesium input function is reconstructed from data of (19). The model calculation of bomb cesium profiles starts in 1950. The *migration velocity of lead* is determined from model simulation of the atmospheric ^{210}Pb profile. We assume the atmospheric ^{210}Pb input to the soil as constant with time during the past 200 years. The model is run for 200 years to establish steady state conditions. The migration velocity of ^{210}Pb ranges between 0.35 mm a^{-1} and 1.6 mm a^{-1}. Applying these parameters and the gasoline lead input function shown in Fig. 4 to the model we simulate the depth distribution of anthropogenic lead in soils. The curves in Fig. 7 show the model results of bomb cesium, atmospheric ^{210}Pb , and anthropogenic lead in the soil at site SKI. The model curves are normalized to the maximum concentrations of the ^{210}Pb , ^{137}Cs , and anthropogenic lead profiles. The model parameters are Δ t = 1 year, H = 1 cm ($D = 3 \cdot 10^{-8}$ $cm^2\ s^{-1}$), infiltration depth = 3 cm (homogenously distributed), $w_{Cs} = 1.0$ mm a^{-1}, and $w_{Pb} = 0.6$ mm a^{-1}. Deviations of the model curve from the concentrations below 8 cm depth might be due to an incorrect correction for geogenic lead, or to an additional contribution of non–gasoline lead to anthropogenic lead before 1950. Sensitivity tests of the model show that the lead concentration increase from the soil surface to the depth of the concentration maximum depends on the shape of the gasoline lead input function, while the depth position of the lead concentration maximum is mainly determined by the infiltration depth. Anthropogenic lead profiles at other sites are also satisfactorily reproduced with the gasoline lead input function and the model parameters obtained from radioisotope data.

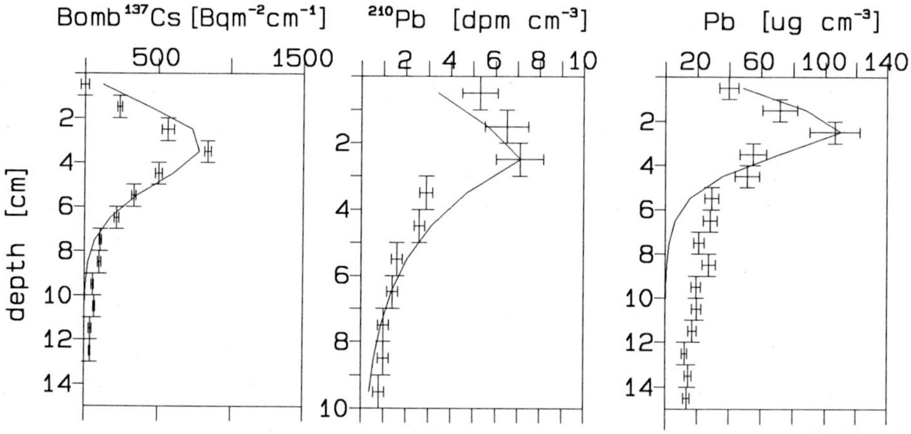

Fig. 7. Model simulation of the bomb cesium, atmospheric ^{210}Pb and anthropogenic lead profiles in the soil at site SKI.

5. CONCLUSIONS

Lead profiles in lake sediments and soils are found to reflect gasoline lead emission since 1950. A pronounced lead concentration maximum is found in various lakes where sediment accumulation rates differ by more than an order of magnitude. Comparison of sediment accumulation rates determined by the ^{210}Pb method and by the displacement of the gasoline lead maximum shows that the mobility of lead in sediments is sufficiently small to use the gasoline lead maximum as a time marker in sediment dating. Anthropogenic lead inventories in undisturbed, organic rich forest soils are in agreement with total gasoline lead deposition. Modeling of the lead profiles in soils confirms that the origin of anthropogenic lead in soils is burning of leaded gasoline. Atmospheric ^{210}Pb proofs to be a suitable tracer to evaluate representative regional lead levels and and to study lead transport in the soil. Moreover, atmospheric ^{210}Pb , Chernobyl and bomb cesium are useful tracers to describe the behaviour of anthropogenic substances in the soil.

REFERENCES

1 J. Dominik, A. Mangini and F. Prosi, Environ. Geol. 5, 3 (1984) 151–157.
2 J. Dominik, A. Mangini and G. Müller, Sedimentology, 28 (1981) 653–677
3 U. Christian, Diplomarbeit, Inst. für Umweltphysik, Universität Heidelberg 1987.
4 K. H. Feger, Freiburger Bodenkundliche Abhandlungen, 17 (1986) 1–253
5 H. Dörr and K. O. Münnich, (Water, Air, and Soil Pollution), in press.
6 Dörr and K. O. Münnich, Naturwissenschaften, 74 (1987) 249–251.
7 P. G. Appleby, F. Oldfield, R. Thompson and P. Huttunen, Nature, 280 (1979) 53–55.
8 F. Oldfield, P.G. Appleby and R.W. Battabee, Nature, 271 (1978) 339–342.

9 Mineralölwirtschaftsverband e.V. Hamburg, 1986.
10 J. M. Pacyna, A. Semp and J. E. Hanssen, Tellus, 36B (1984) 163–178
11 H. Dörr, A.Mangini, W.Schmitz and K. O. Münnich, Naturwissenschaften, 77 (1990) 428–430.
12 Umweltbundesamt Berlin, Jahresbericht 1986, Dr.Fricke, personal communication.
13 K. Fukuda and S. Tsunogai, Tellus, 5 (1975) 514 - 521.
14 W. C. Graustein and K. K. Turekian, Journ. Geophys. Res., 91, D13 (1986) 14355 –14366.
15 S. Tsunogai, T. Kurata, T. Suzuki and K. Yokata, Journ. Atmos. Chem. 7 (1988) 384 - 407.
16 Kommission zur Überwachung der Radioaktivität, Fribourg/CH, 1982.
17 H. Dörr and K. O. Münnich, Radiocarbon, 31, 3 (1989) 655–663.
18 J. Bear and A. Verruijt, Modeling Groundwater Flow and Pollution, D. Reidel Publishing Company, Dordrecht, 1987.
19 H. Bachhuber, K. Bunzl, W. Schimmack and I. Gans, Nuclear Technology, 59 (1982) 291–301.

VARIATION IN RADIONUCLIDE LOSS RATES FROM BALTIC *MYTILUS EDULIS*

Henning Dahlgaard

ABSTRACT

Two long-term *in situ* loss studies with 12 trace elements in Baltic *Mytilus edulis* showed large differences between biological half lives and between retentions. The results indicate that it is important to be aware of variations in bioindicator response, when using *Mytilus edulis* as a bioindicator for metals and radionuclides. Variation in growth rates are also considered important, since they have a direct effect on the concentrations of metals and radionuclides. Furthermore, energy metabolism may have an effect on the loss rates of the elements from mussels. It is suggested that "starving" mussels will show a prolonged retention of trace elements; i.e. when energy metabolism is low, trace element turnover also becomes low. Combined with the lack of growth dilution in such animals, this may lead to much higher metal and radionuclide concentrations in starving mussels than in fast growing animals.

INTRODUCTION

The mussel, *Mytilus edulis*, is probably the most widely sampled bioindicator organism in marine monitoring programmes for various pollutants, mainly radionuclides and metals (see e.g. 1-4). One of the reasons for its popularity is its wide distribution. It is found from temperate to arctic waters, from fully marine areas to estuaries showing permanent salinities as low as 3 - 5 ‰. It is furthermore quantitatively dominating in several low-water environments, e.g. in large parts of the Baltic sea (5). Another interesting characteristics of Mytilus is its filter feeding. A medium sized mussel with 1 g fresh weight of the soft parts clears $\sim 50\ell$ of seawater per day for particles above $\sim 4\ \mu m$ (3).

There are several advantages of including biological indicator organisms like mussels in monitoring programmes: biomagnification improves detection limits, time integration due to long term retention smooths fluctuating environmental levels of pollutants and the sensitivity is greatest for the most bioavailable elements and chemical species. These advantages are, however, not without costs. By introducing biological samples in a monitoring programme, one inevitably introduces a variation with several environmental parameters such as e.g. season and salinity. It is important to know this variation in bioindicator response, as insignificant differences between data sets could otherwise be misinterpreted. Nevertheless, the variation does not prohibit a meaningful use of the bioindicators.

In addition to the "applied" aspects of improving the use of mussels in bioindicator programmes, the study of trace element cycling in marine and estuarine environments, including its variation, forms an important part of our general understanding of nature's "household". Even if such basic ecological aspects are not presently "fundable", they may be the most interesting in the long run.

This paper refers to two long-term loss studies performed *in situ* in the Baltic Sea. Due to its low salinity and its low rate of water exchange, the Baltic is a sensitive environment deserving special attention.

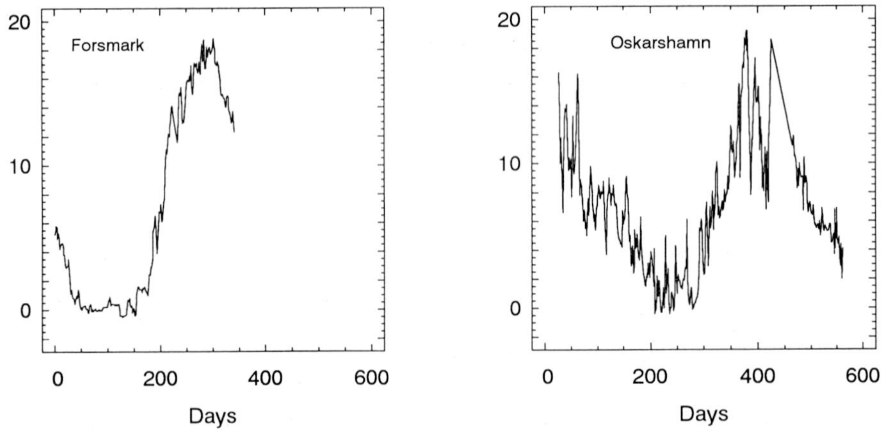

Fig. 1. Water temperatures (°C) during the two experiments at Forsmark October 1983 - August 1984 and Oskarshamn June 1985 - December 1986.

MATERIALS AND METHODS

Mussels, *Mytilus edulis*, were allowed to accumulate 13 different radionuclides in the laboratory for 4-5 days before they were brought back to the environment from which they had been collected. Twelve different elements were examined - one of them in two oxidation states: ^{54}Mn, ^{60}Co, ^{65}Zn, ^{99m}Tc, ^{110m}Ag, ^{134}Cs, ^{144}Ce, ^{155}Eu, ^{237}Np, ^{238}Pu, ^{239}Pu, ^{241}Am and ^{244}Cm. The amount of carrier added with the radionuclides was kept as low as possible in order to keep the metal concentrations at a natural low level (6). After being contaminated, the mussels were placed in small net bags and suspended in the water column in the cooling water intake area off two Swedish nuclear power plants. The animals were thus away from their natural habitat for only ~1 week during contamination. Measurements were made by multielement gamma and alpha spectrometry on individual mussel soft parts sampled at different intervals after being contaminated.

The chosen sites resulted in a steady high rate of water exchange and they kept the mussels from freezing in. The first experiment was performed in the Forsmark Biotest facility on the Bothnian coast 120 km north of Stockholm from October 1983 to August 1984. The second was performed at Oskarshamn 250 km south of Stockholm from June 1985 to December 1986. Details of the Forsmark experiment are given elsewhere (6). The Oskarshamn experiment differed from the Forsmark experiment in a number of ways: Temperature curves for the two sites are given in Fig. 1 showing that the Forsmark experiment started shortly before a long winter, whereas the Oskarshamn experiment started at mid summer followed by a relatively mild winter. The Oskarshamn site shows a stable salinity of ~7 ‰, whereas the Forsmark salinity is only 5 - 6 ‰. This probably results in slower growth rates and smaller maximal sizes at Forsmark (5). In accordance with this, no growth could be demonstrated in the Forsmark animals (6), whereas the Oskarshamn mussels clearly gained soft parts as well as shell weight during the first 5 months of the experiment (Fig. 2). Trace element levels at the chosen sites monitored by mussels were similar to other Baltic locations (7-8)

Fig. 2 gives details of the animal growth during the Oskarshamn experiment. As the Forsmark experiment had indicated that the largest animals might have been selectively chosen for sampling, leaving smaller animals for the last samples (6), the Oskarshamn animals were randomly split up in separate small cages, each containing 10 animals, at the beginning of the experiment. Due to animal deaths, some of the samplings consisted of two containers. Totally, 187 animals were sampled and analyzed individually out of 249 initially contaminated. At the start of the Oskarshamn experiment, the average length of the 249 animals was 27.0 ± 2.4 (SD) mm. Fig. 2 shows a fast growth during the first 5 months of the experiment. During the following winter, growths stopped, and apparently soft parts weights stagnated the following summer, in spite of shell growth. As a result, the condition index, given as the ratio between dry weights of soft parts and shells, decreased after the first 5 months of the experiment. In spite of the random sample splitting at the beginning of the experiment, shell weights and shell lengths apparently decreased during the winter. This could be explained by size dependent mortality. Maybe a higher mortality of the largest animals is also the explanation of the decreasing sizes towards the end of the Forsmark experiment, rather than the suspected selective sampling mentioned above.

Biological half lives ($T_{½b}$), excluding growth dilution, were calculated by linear regression of log (radionuclide content) versus time, where the slopes (λ) of the regression lines relates to biological half lives as $-\lambda = \log(2) / T_{½b}$. The calculation basis for each element was the total soft

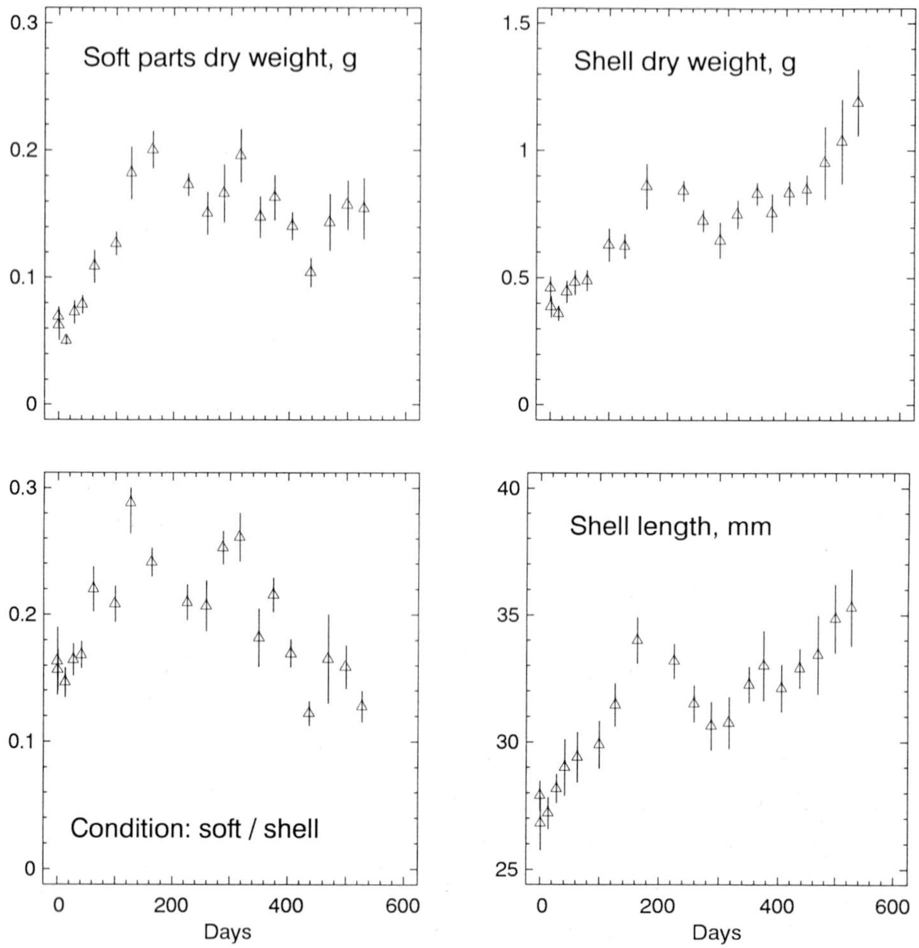

Fig. 2. *Mytilus edulis*, weight, length and "condition" during Oskarshamn experiment. Error term: ±SE

parts radionuclide contents, decay-corrected to the same time. The dilution due to growth is thus not included in the present biological half lives. Normal probability plots showed that the data were log-normally distributed (Fig. 3). The plotted loss curves (Figs. 4 - 5) thus give geometric means and SE of means of log (values).

RESULTS

Loss curves for 8 elements, comparing the two experiments, are given in Figs. 4 - 5. As Cm and Am were seen to behave as analogues in an earlier *Mytilus* experiment, Am from the

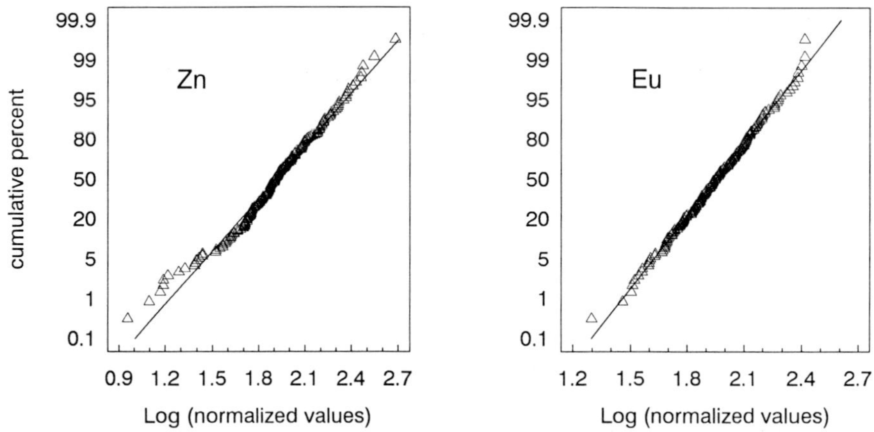

Fig. 3. Oskarshamn experiment. Logarithmic normal probability plots for the 187 Zn and Eu values (total soft parts contents) normalized to a geometric mean of 100 for each sampling time.

Forsmark experiment has been compared with Cm from the Oskarshamn experiment in Fig. 5. Biological half lives, excluding growth dilution, and retention percentages for the slow compartments are given in Table 1. Data from the first 2 months of loss have been excluded from the analysis of regression in order to avoid interference with fast compartments. For slopes being not significantly different from zero, the biological half lives have been given as ∞.

For most elements, it is seen that there exists a slow compartment with a substantial half life. However, some interesting differences appear between the two experiments. In general, the half lives seem to be longer and the retentions smaller in the Oskarshamn experiment. Another difference is that for the Forsmark experiment loss rates were clearly correlated with season (6), which was not the case at Oskarshamn.

Differences in accumulation of the different elements, expressed as concentration ratios of the radionuclide tracers (Bq g^{-1} soft parts / Bq ml^{-1} water) measured before the loss study, are given in Table 2. The concentration ratios were furthermore combined with the retentions in Table 1 and the length of the accumulations, to give the long-term retained concentration ratio increments obtained per day of accumulation (Table 2). These data show that even when the initial accumulations were not very different, several elements (Ag, Pu, Mn, Co, Ce) were accumulated much better in the slow compartment in the Forsmark experiment than at Oskarshamn, whereas Zn, and to a lesser extent Eu and its analogues Am and Cm, seems to be taken up at comparable rates in the two experiments. Silver showed the largest differences.

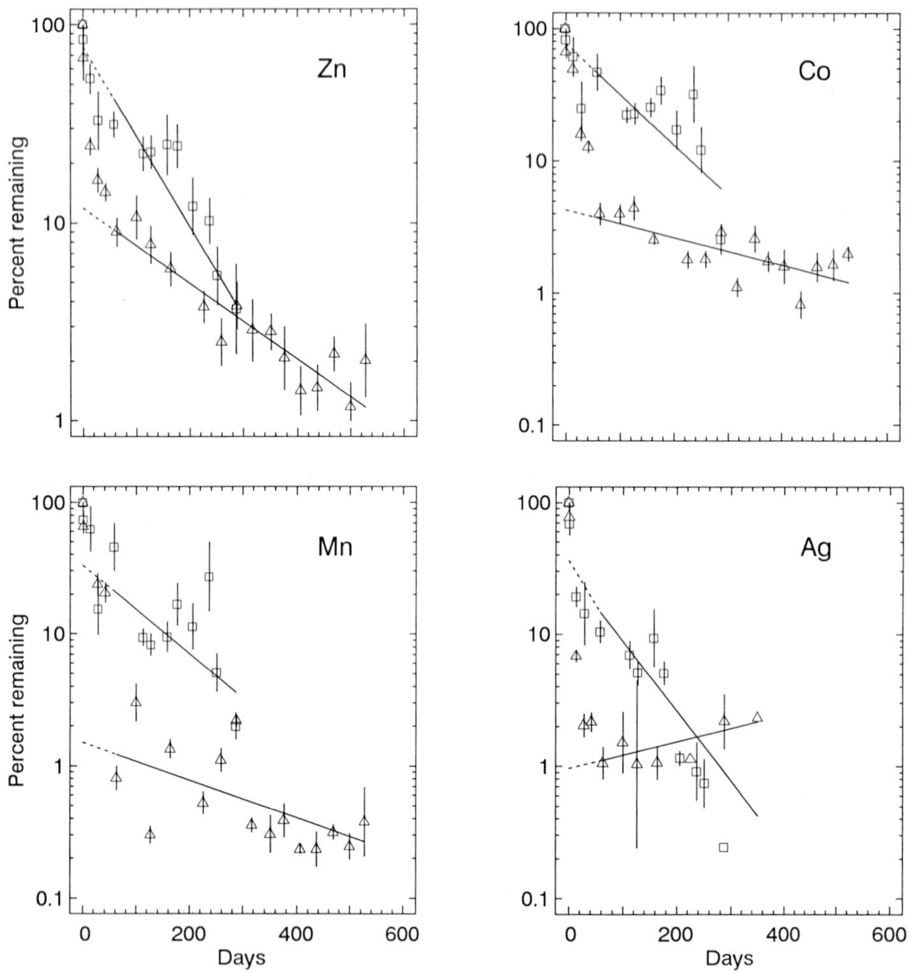

Fig. 4. Loss of radionuclide tracers of Zn, Co, Mn and Ag from *Mytilus edulis* soft parts. Forsmark (□) and Oskarshamn (▵) experiments. Logarithmic means and SE.

DISCUSSION

The causes of the variation between the two loss experiments are not obviously clear. The possibility of experimental artefacts, e.g. caused by differences in water chemistry during contamination of the mussels, cannot be ruled out completely. However, these experiments are less vulnerable to artefacts than pure laboratory experiments, as the animals were only out of their natural environment for ~ 1 week during contamination. As the experimental animals were taken from the loss sites, effects of any genetic or physiological differences between different mussel populations, which could result in reduced viability due to "stress", can be ruled out. Genetic

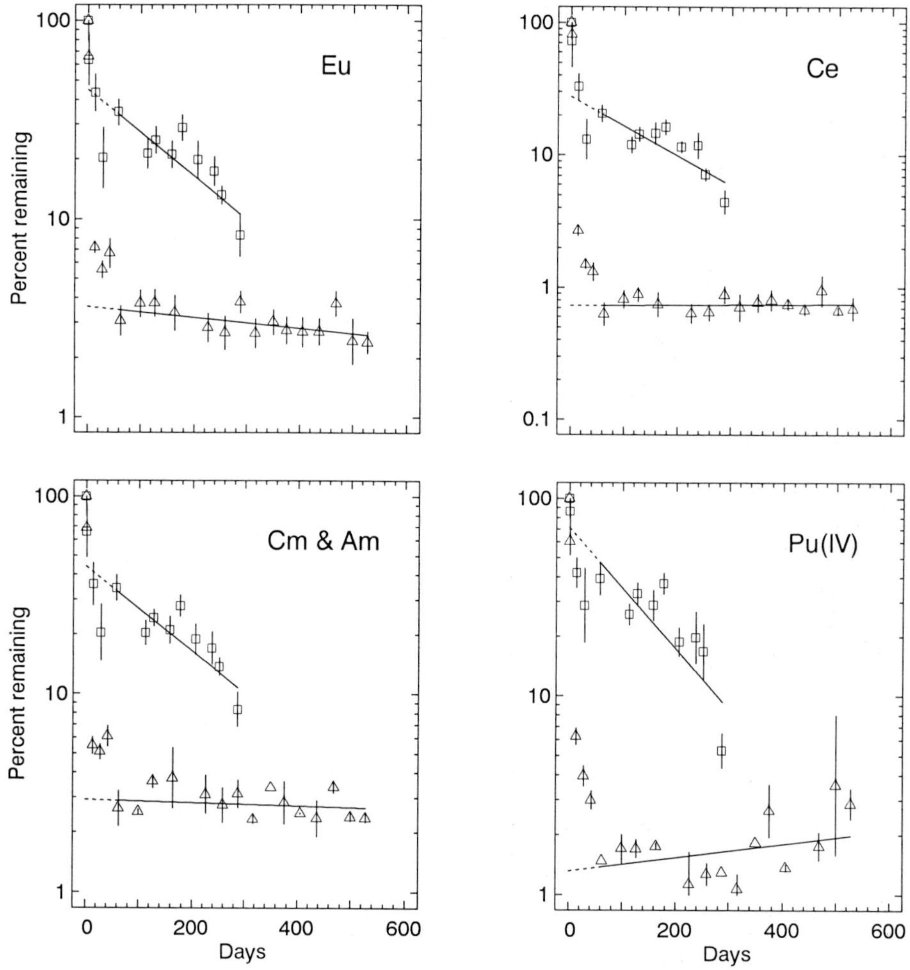

Fig. 5. Loss of radionuclide tracers of Eu, Ce, the analogues Cm and Am and Pu(IV) from *Mytilus edulis* soft parts. Forsmark (□) and Oskarshamn (△) experiments. Logarithmic means and SE.

differences between the two mussel populations may well exist (9-10), but it is not considered likely, that they could contribute significantly to the observed differences in trace element metabolism.

Salinity affects the accumulation of several elements (3), but the observed salinity differences are too small (cf. above) to explain the observed metabolic differences as a direct salinity effect. However, indirectly, salinity may probably be one of the factors determining the difference in growth between the two populations (5). It has previously been stated that the energy metabolism is an important factor determining trace element metabolism in mussels (6, 11-12). In the Forsmark

Table 1: *Mytilus edulis* soft parts. Biological half lives (excluding growth dilution) and retentions of slow pools in two Baltic loss experiments.

	Forsmark Nov. 83 - Aug. 84		Oskarshamn June 85 - Dec. 86	
Element	Halflife days	retention %	Halflife days	retention %
Manganese	90	33	212	1.5
Cobalt	79	77	287	4.3
Zinc	67	77	158	11.9
Technetium	∞	12	-	-
Silver	45	36	∞	0.97
Caesium	-	-	78	0.074
Cerium	134	28	∞	0.73
Europium	137	45	∞	3.6
Neptunium	133	82	-	-
Plutonium(IV)	98	71	∞	1.3
Plutonium(VI)	-	-	∞	4.9
Americium	141	44	-	-
Curium	-	-	∞	2.9

Table 2 Concentration ratios (cf.text) obtained before the loss experiments and long-term retained concentration ratio increments per day of accumulation for the two *Mytilus* experiments.

	Forsmark Nov. 83 - Aug. 84		Oskarshamn June 85 - Dec. 86	
Element	CR 136 h	Retained CR/d	CR 101 h	Retained CR/d
Manganese	350	20	250	0.89
Cobalt	900	122	850	8.7
Zinc	950	129	3000	85
Technetium	12	0.28	-	-
Silver	6500	413	2650	6.1
Caesium	-	-	146	0.026
Cerium	2000	99	5630	9.7
Europium	350	28	950	8.2
Neptunium	10	1.5	-	-
Plutonium(IV)	400	50	720	2.2
Plutonium(VI)	-	-	265	3.1
Americium	400	31	-	-
Curium	-	-	1160	8.0

experiment, this was supported by the similar retentions of all the trace elements, studied during the winter period under "cold" natural temperatures, and those studied during the winter in the artificially heated part of the experiment. Since the distinct seasonal effect thus could not be explained as a simple temperature effect, a seasonal variation in available food (suspended phytoplankton) and the resulting variation in the mussels energy metabolism was suggested as the key factor (6).

Why, then, is this seasonal effect not seen in the Oskarshamn experiment? Differences in season at the beginning of the two experiments could be of importance. I will put up the following hypothesis to explain what has happened, assuming that the previous hypothesis on a close relation between energy metabolism and trace element metabolism is still valid: During the first 5 months of the Oskarshamn experiment, i.e. from June till December, the energy metabolism was high and there was a significant growth of both soft parts and shell (Fig. 2). Under these circumstances, considered ideal for trace element turnover, the fast pools of radionuclides, accumulated during the contamination period, were lost during the first 2 months of the experiment (Figs. 4 - 5). In the Forsmark experiment, starting in late autumn, the fast pools were apparently also exhausted after 2 months (6), but that might have been a misinterpretation, triggered by the Bothnian Sea winter conditions (ice cover for 4 months) with so low levels of suspended food that the mussels lost weight in spite of the low temperatures; at the same time, loss of the radiotracers was halted. What was believed to be increased loss rates from slow pools next spring, might have been the resumed loss of activity from faster pools that had not been sufficiently cleared before the winter conditions stopped further loss. If this is correct, it also explains why the retention in the "slow" pools appeared larger and the biological half lives during the next summer period of the Forsmark experiment were taken to be faster (6) than for the slow pools of the Oskarshamn experiment; the slow pools in the Forsmark mussels might still have been overlapped by relatively fast pools during all the 8 months of the experiment, because of the generally low energy turnover. Finally, the totally lacking seasonal effect on loss rates in the Oskarshamn experiment might not only be due to the complete elimination of the fast pools during the first 2 months of loss. It could also partly be an effect of low energy metabolism during the last part of the experiment, as growth of unknown reasons actually stagnated after the first 5 months (Figure 2).

When sampling mussels in pollution monitoring programmes, the total soft part concentration of metals or radionuclides is normally used. The retained concentration ratios (Table 2) will be important for the total concentrations measured in animals in a monitoring programme, as they are a measure of the amount of metal that virtually accumulates day by day. Another parameter determining the long-term concentrations is growth, as it dilutes accumulated material in the slow pools. Growth rates do not have to be very fast before they are more important in lowering the concentrations than the biological half lives of the slow pools. Growth dilution can be described by a "half life due to growth", $T_{½g}$, calculated by exponential regression like the biological half life, but based on inverse soft parts dry weights. For the first 5 months period of the Oskarshamn experiment, where growths was fast (Fig. 2), $T_{½g}$ was estimated at 85 days, whereas a regression over the total period gave a half life of 450 days. This implies, that in the Oskarshamn experiment growth dilution would have been important during the first 5 months of the experiment, but not later when the growth stagnated.

Unless the differences observed between the two experiments are caused by experimental artefacts, the consequences must be that similar variations are to be expected whenever *Mytilus* is used as a bioindicator organism. Even if some of the observed differences may be explained by "artefacts" due to prolonged retention of slow pools during periods with low levels of available food, as discussed above, such differences obviously also appear under natural conditions, and they are therefore relevant when mussels are used from different locations in bioindicator monitoring programmes. Taking the above hypothesis on a close coupling between energy metabolism and trace element metabolism as true, it should be expected that metal and radionuclide concentrations would be much higher in a starving mussel population and in non-growing individuals than in fast growing animals. This may explain why higher concentrations of several metals have been observed in the Baltic than in the Kattegat, in spite of similar water concentrations (7-8), and it may give an alternative explanation of differences observed (2,13) in the response of *Mytilus* and another bioindicator, the brown algae *Fucus vesiculosus*, between Kattegat and the Baltic.

CONCLUSIONS

- Long term loss rates and retentions of several radiotracers appeared very different in two Baltic *Mytilus edulis* experiments. At least part of these differences may be explained by low food availability.
- It is suggested that "starving" mussels, not having access to sufficient food to sustain growth, show a prolonged retention of trace elements, i.e. when the energy metabolism is reduced, the rate of trace element turnover is also reduced. Such a coupling between energy metabolism and trace element metabolism includes not only biologically active elements like Co, Zn, Mn and Ag, but also elements with no known biological function like Pu, Am, Cm, Ce and Eu.
- Under normal food and growth conditions, a dilution of the accumulated trace elements due to new growths may be more important in determining trace element concentrations than the actual loss of metals or radionuclides from the animals.
- It should therefore be expected that metal and radionuclide concentrations may be much higher in a starving mussel population than in a fast growing one. This is very important when

interpreting mussel data in bioindicator monitoring programmes.

ACKNOWLEDGMENTS

Logistic support from the Swedish Environmental Protection Board, especially J. Anderson in Oskarshamn and B. Nymann and E. Krogh in Forsmark, is acknowledged. This project was partly funded by the Radiation Protection Research Programme of the European Economic Community 1980-1989 and by the Radioecology Project of the Nordic Liaison Committee for Atomic Energy under the Nordic Council of Ministers 1978-1985.

REFERENCES

1 E.D. Goldberg, V.T. Bowen, J.W. Farrington, G. Harvey, J.H. Martin, P.L. Parker, R.W. Risebrough, W. Robertson, E. Schneider, E. Gamble, Environmental Conservation, 5 (1978) 101-125.
2 D.J.H. Phillips, Quantitative aquatic biological indicators, Applied Science Publishers, London, 1980.
3 H. Dahlgaard, Bioindicators for Monitoring Radioactive Pollution of the Marine Environment, Risø-R-443, Risø National Laboratory, 1981.
4 E.D. Goldberg, M. Koide, V. Hodge, A.R. Flegal, J. Martin, Estuarine, Coastal & Shelf Science, 16 (1983) 69-93.
5 N. Kautsky, Mar. Biol., 68 (1982) 117-133.
6 H. Dahlgaard, Mar.Ecol.Prog.Ser., 33 (1986) 157-165.
7 D.J.H. Phillips, Mar. Biol., 43 (1977) 283-291.
8 D.J.H. Phillips, Mar. Biol., 46 (1978) 147-156.
9 B.F. Theisen, Ophelia, 17 (1978) 135-142.
10 H.-P. Bulnheim, E. Gosling, Helgoländer Meeresunters., 42 (1988) 113-129.
11 G.W. Bryan, J. mar. biol. Ass. U.K., 53 (1973) 145-166.
12 T. Borchardt, Mar. Biol., 76 (1983) 67-76.
13 D.J.H. Phillips, Environ. Pollut. 18 (1979) 31-43.

HISTORY OF MERCURY CONTAMINATION RECONSTRUCTED FROM HIGH-RESOLUTION RADIO-ISOTOPIC DATING OF SEDIMENT CORES IN LAKE GENEVA

J. Dominik, J.-L. Loizeau, P.-Y. Favarger, J.-P. Vernet
and R.L. Thomas

1. INTRODUCTION

Mercury contamination of the alpine Rhone sediments has been monitored yearly since 1970 (1). This monitoring revealed an important pollution event occurring in 1971-1972 (2). Concentrations in the river sediments at the mouth of the Rhone increased from 0.17 in 1964 to nearly 1.5 μg/g, with a mean content of the watershed sediments close to 0.7 μg/g, which is more than one order of magnitude higher than the natural background level.
Four surveys (1970/71, 1978, 1983 and 1988) of the mercury concentration in surficial sediments of Lake Geneva yielded the following mean concentrations : 0.7, 0.75, 0.43 and 0.47 μg/g, respectively (3). These values are, however, not readily comparable because of the differences in sampling strategy.
Mercury concentrations in older sediments were obtained by Vernet and Favarger (4) from sediment cores. They noticed an important contamination in sediments deposited between 1940 and 1950 in the "Petit Lac".
Mercury contents were also measured in a number of fishes from Swiss lakes (5, 6) and were found to be higher in Lake Geneva (0.38 μg/g wet weight) than in the other lakes (0.03-0.10 μg/g).
The aims of this study are to: 1) reconstruct the history of mercury pollution in Lake Geneva using all available sediment cores dated with radioisotopic methods and; 2) define input sources and estimate the mercury discharge to the lake. Further, we report evidence of perturbations of sedimentary record. A number of factors, such as slumps, turbidites, erosion, biological and physical mixing etc., complicate the interpretation of these records. In this paper we show how the local heterogeneities of sediments can modify the record of Hg pollution and how different radionuclides can be used to detect and interpret perturbations in the sediment profiles.

2. SETTING AND METHODS

Fourteen gravity cores of varying types (Rheineck modified, Ambühl, Benthos and modified Pfleger corers) have been recovered between 1976 and 1988 from Lake Geneva sediments (figure 1).

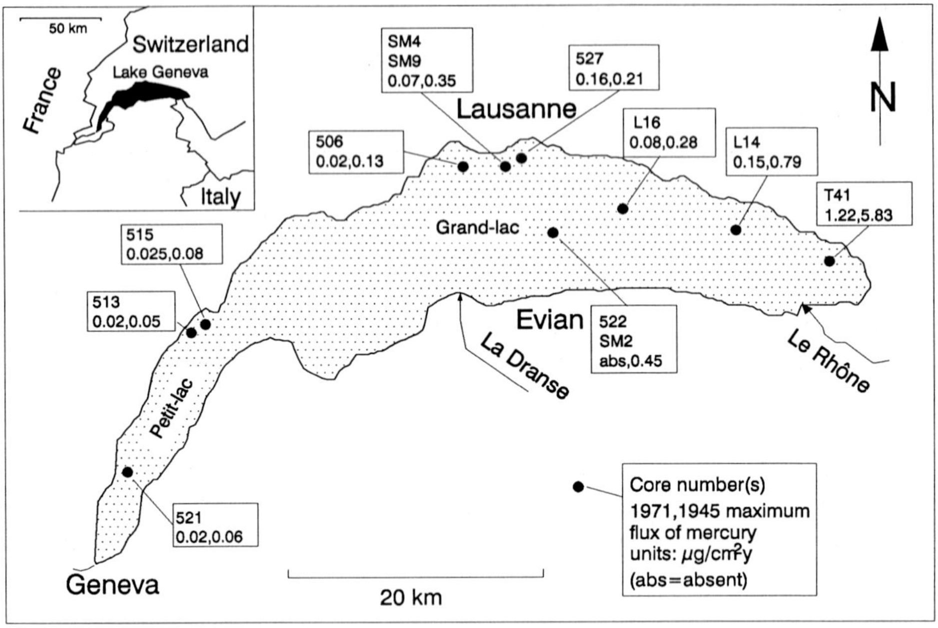

Figure 1: Location of dated sediment cores from which maximum mercury fluxes are calculated.

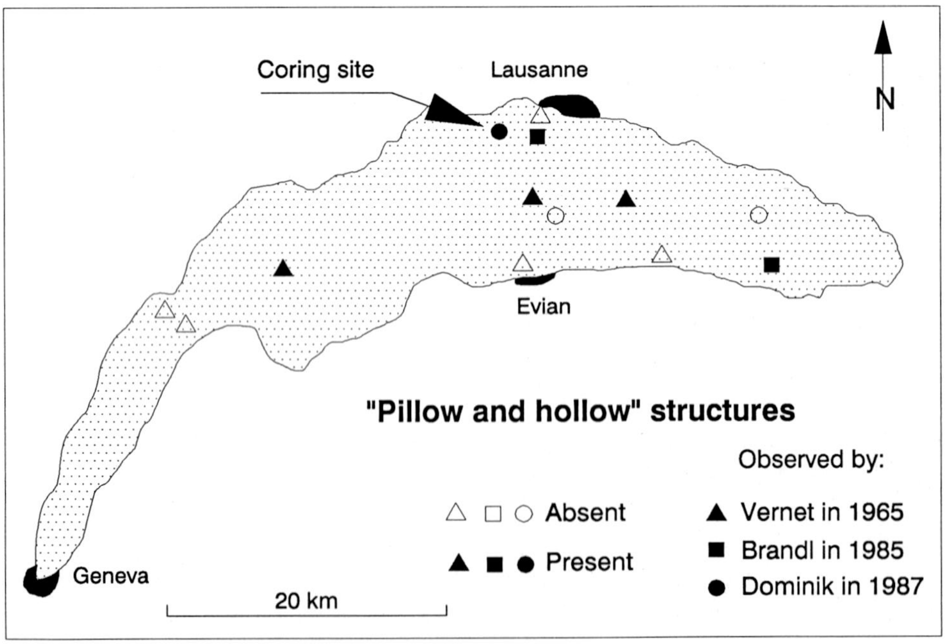

Figure 2: Reported observations of the pillow-hollow structures (7, 22) from Lake Geneva bottom sediments and the location of the coring site.

Several cores (SM4 - SM10) were taken in 1987 with the submersible "F.-A. Forel" at 132 m water depth, from a small area located south-west from Lausanne (fig. 2), in order to examine the influence of a peculiar bottom morphology on the contamination record. A large, but poorly defined area of Lake Geneva sediments shows undulated morphology, which we call a "pillow-hollow" structure. It consists of "pillows", often forming rounded hexagons, 40 to 60 cm in diameter, surrounded by the smaller, triangular "hollows" (fig. 3). The hollows are usually 5 to 20 cm deep. This type of morphology was first described from Lake Geneva by Vernet (7), who suggested a number of processes leading to the formation of pillows and hollows: two different current regimes or convection current in sediments. He also noticed that the bottom dwelling fish - burbot - (*lota lota*) may use the existing depressions as niches (fig. 4). Sturm (8) examined sediment cores from a pillow-hollow pair and found a deficit of ^{137}Cs in the hollow, which could suggest bio-erosion by *lota lota* at these sites.

We studied the sediment cores taken selectively from the pillows (cores SM4, SM9) and adjacent hollows (SM5, SM10). Three additional cores (SM6, SM7, SM8) located within a distance of about 15 cm one from each other (fig. 5) were taken to obtain lithological profiles.

Repetitive sampling was also performed in the central part of the lake at 309 meter depth, within the area of a recent turbidite deposition.

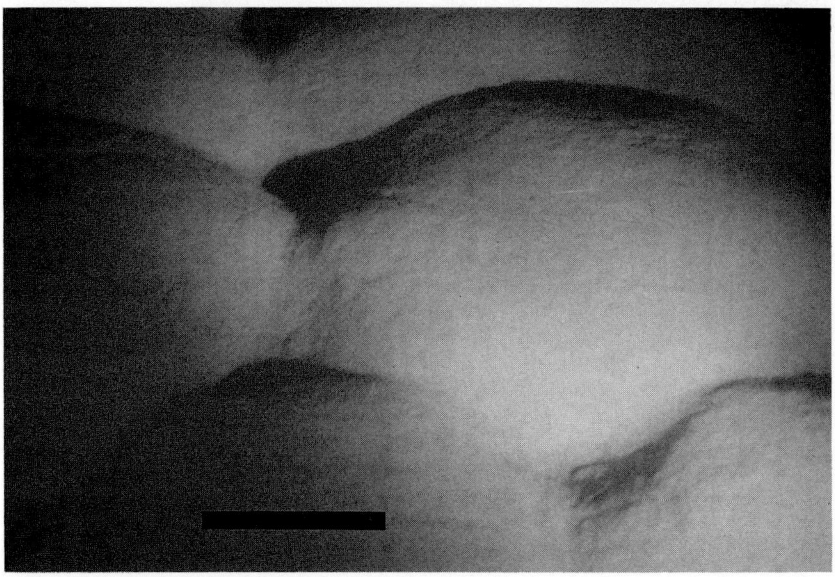

Figure 3: Pillow - hollow morphology of bottom sediments south-west from Lausanne at 132 m water depth. Horizontal bar represents about 20 cm.

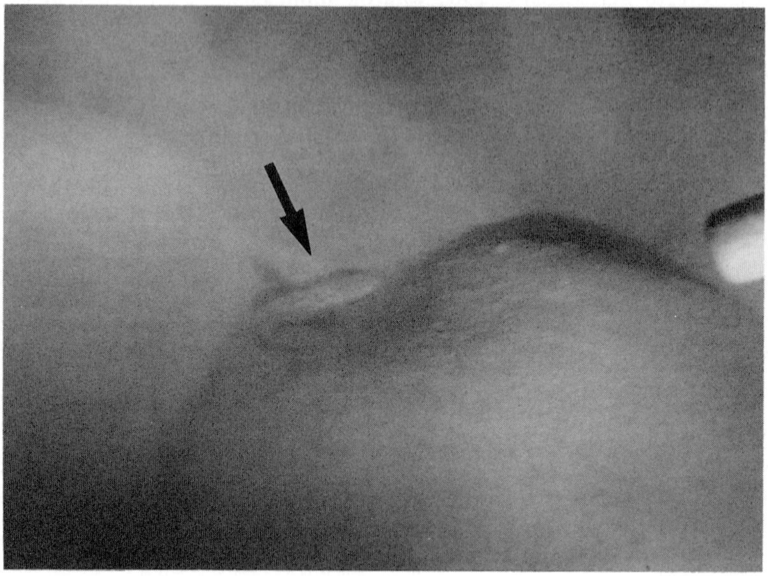

Figure 4: Young burbot (*lota lota*) dwelling in a hollow.

Figure 5: Closely spaced coring with the submersible "F.-A. Forel" (cores SM6 - SM8).

The absolute time scale was established in selected cores using ^{210}Pb and/or ^{137}Cs methods. The ^{210}Pb method relies on the decay of the excess of this radionuclide over ^{226}Ra in the sediments and allows dating for the last 100 years (9, 10).
The ^{137}Cs can be used as a time marker in the sediments due to its known input history. The maximum fallout from nuclear tests in the atmosphere is marked as the peak activity in the sediment layer deposited in 1963/1964 (the date of January 1st 1964 is attributed to ^{137}Cs maximum in this paper). In more recent cores a strong maximum of ^{137}Cs, accompanied by ^{134}Cs, close to the sediment surface, marks the date of fallout from the Chernobyl accident (May 1st 1986). As the ratio of $^{134}Cs/^{137}Cs$ at the time of fallout is known, the ^{137}Cs from Chernobyl accident can be distinguished from residual activity of ^{137}Cs originating from nuclear tests and still delivered to the lake via soil derived particles. For dating of Hg contamination events the positions of the ^{137}Cs and Hg peaks were determined by a polynomial of 2nd degree, involving usually five points scattered on both sides of the peak.

^{210}Pb was obtained via alpha-counting of its daughter ^{210}Po under the assumption of secular equilibrium (11). ^{226}Ra was gamma-counted with a low background Ge well-detector via ^{214}Pb and ^{214}Bi, after allowing for ingrowth into equilibrium in glass vials during 14 days. In most cases ^{210}Pb-excess was calculated as difference between ^{210}Pb and ^{226}Ra activities in the same sample. As, in general, ^{226}Ra activities varies little with depth in the same core, the average or interpolated values were adopted in the absence individual measurements. The ^{134}Cs, ^{137}Cs, as well as ^{234}Th and ^{7}Be were counted directly with Ge well-detector. The last two short-lived radionuclides were used in a number of cores to check for the completeness of surface sediment recovery or the effects of bioturbation. All results are shown with 1σ counting uncertainty.

Mercury analyses were performed by cold vapor atomic absorption technique after a nitric-hydrochloric acid digestion. For samples collected before 1983 only the fraction inferior to 63 μm was analyzed, while for the other samples analyses were performed on the total sediment. As the sediments involved are fine lacustrine deposits containing generally very few sand-size particles (except for turbidites) the differences between the two analyses are considered negligible for the purpose of this study.

3. RESULTS

3.1. Mercury contamination chronology

The longest profile giving the best overview of mercury contamination history in Lake Geneva sediments was obtained from core 521, recovered in the "Petit Lac" (fig. 6A). A smooth and exponential decrease of ^{210}Pb-excess with depth testifies to a quiet and relatively constant sedimentation rate in the last century of 0.123 g/cm^2y. These conditions of a uniform sedimentation are confirmed by a smooth decrease of porosity with depth and by a very good agreement of the mean sedimentation rate on the time scale of 100 years with the position of ^{137}Cs peak at 7 cm depth, which is related to nuclear bomb test maximum in 1963/64. Small deviations of ^{210}Pb activities from the ideal exponential decrease with depth may suggest minor variations of the sedimentation rate especially in the uppermost 10 cm and from 27 to 37 cm depth. A separate fitting of these intervals, however, will not improve markedly the quality of dating of the Hg maxima, as the uncertainty of their age is essentially determined by the sampling step in this core. The profile of mercury concentrations shows a slight increase starting at the end of the last century and rising to the most prominent maximum dated in this core to about 1950. In more recent sediments a slow decrease is observed with a secondary maximum at about 1973. These two maxima are related to specific mercury contamination, because other heavy metals do not follow the same evolution.

Similar Hg profiles, sometimes with better resolution, have been obtained from many cores but only a few of them were dated with ^{210}Pb method and the only reference point is the maximum of ^{137}Cs activity in 1964. The maximum mercury concentration in the sediments varies from 0.8 $\mu g/g$ near Geneva (core 521) to 5.6 $\mu g/g$ close to the Rhone mouth (core T41 in fig. 6B). This maximum value is higher than these observed in some other contaminated lakes (12), such as Lake Ontario (3.95 $\mu g/g$) and Lake Erie (4.8 $\mu g/g$), and half of the maximum observed in the heavily polluted Lake St-Clair (10.3 $\mu g/g$).

Results of dating with ^{210}Pb methods from 4 cores are given in table 1 and figure 7. The date of the maximum mercury pollution is approximately 1945-1946, while the minor peak is related to the 1971 Rhone sediment pollution. This latter result is confirmed by the ^{137}Cs time scale from which the mean date of 1971.6 (± 1.7) is inferred.
The dating of the older pollution with ^{137}Cs is not straightforward because the sedimentation rate seems to decrease between the forties and present, particularly in the vicinity of the Rhone mouth (13). This is probably related to the construction of major dams in the watershed in 1960's and 1970's.

3.2. Small scale inhomogeneities and perturbated sedimentary record

Lithological profiles from cores SM6, SM7 and SM8 taken selectively from the pillow-hollow structures from the site shown in figure 2 can be correlated, notably due to a distinct dark layer and a fine lamination below it (fig. 8). Above the dark layer the correlation is difficult because of black iron sulfide tint, but it is clear that both the oxidized surface and dark reduced layers are much thinner in the core from the hollow than in these from the pillow.

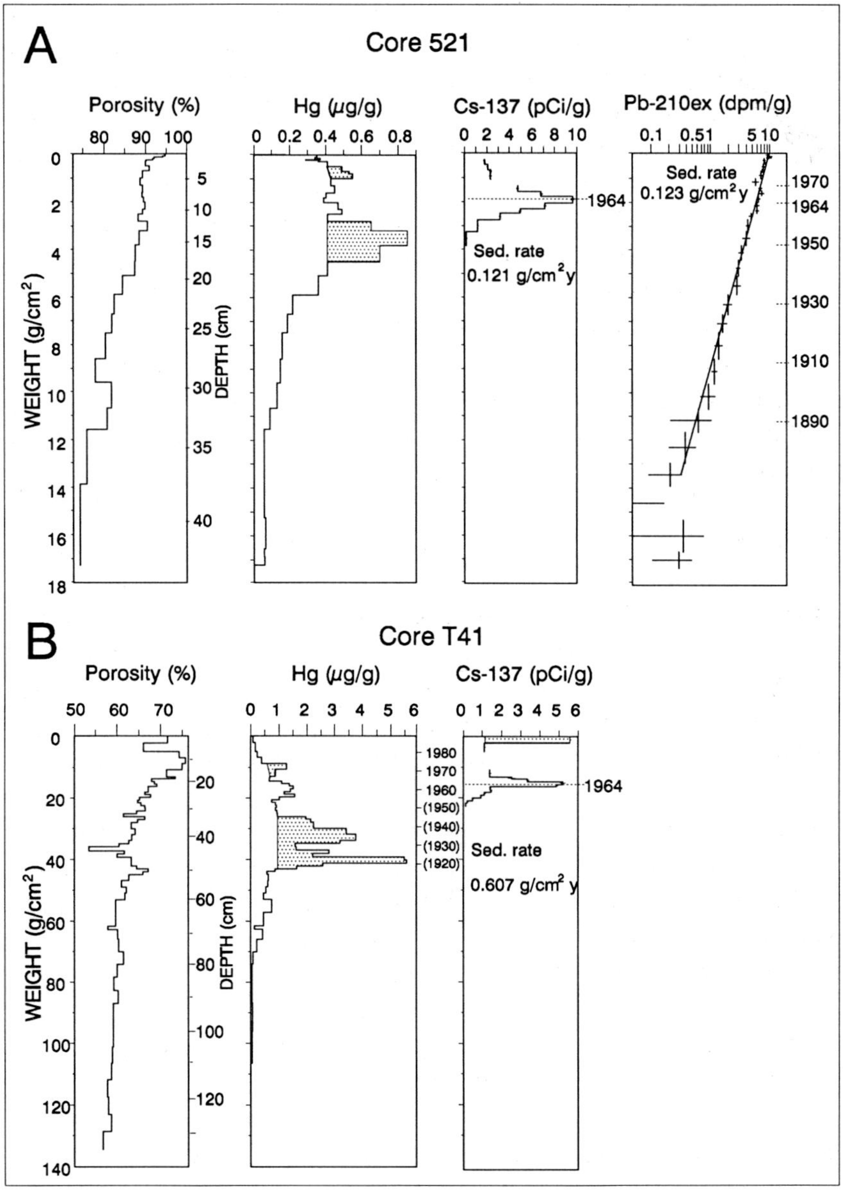

Figure 6: A) ^{210}Pb and ^{137}Cs dated core from the "Petit Lac" showing full history of anthropogenic mercury deposition in Lake Geneva with two characteristic peaks (shadowed). In this and following diagrams, the depth in sediments is expressed as the cumulative weight of dry sediment per 1 cm^2. A non-linear depth scale in cm is shown on the right.
B) High resolution Hg profile from the region close to the Rhone delta showing one additional peak dated at 1960 with ^{137}Cs. The major contamination peak show a local minimum related to a turbidite layer. ^{137}Cs originating from Chernobyl is shown as shadowed area. The time scale (in parentheses) extrapolated below the ^{137}Cs peak in probably not correct (see explanations in text).

Table 1.
Dating of Hg pollution peaks and sedimentation rates.

^{210}Pb dating				^{137}Cs dating			
Core	1st pollution	2nd pollution	Sedimentation rate g/cm^{2}y	Core	1st pollution	2nd pollution	Sedimentation rate g/cm^{2}y
SM2	1946	absent[a]	0.101	T41	(1922)	1972	0.606
SM4	1946	1969	0.120, 0.139[b]	L14	(1926)	1974	0.158
SM9	1942	1970	0.120, 0.149[b]	L16	(1925)	1972	0.069
521	1950	1973	0.123	522	(1946)	absent[a]	0.101
				SM2	(1946)	absent[a]	0.106
				527	(1949)	1972	0.199
				SM4	(1948)	1969	0.120
				SM9	(1944)	1970	0.134
				506	(1936)	1971	0.131
				515	(1946)	1971	0.141
				513	(1936)	1972	0.073
				521	(1951)	1973	0.121

[a] eroded by a turbidite
[b] variation of rate with time

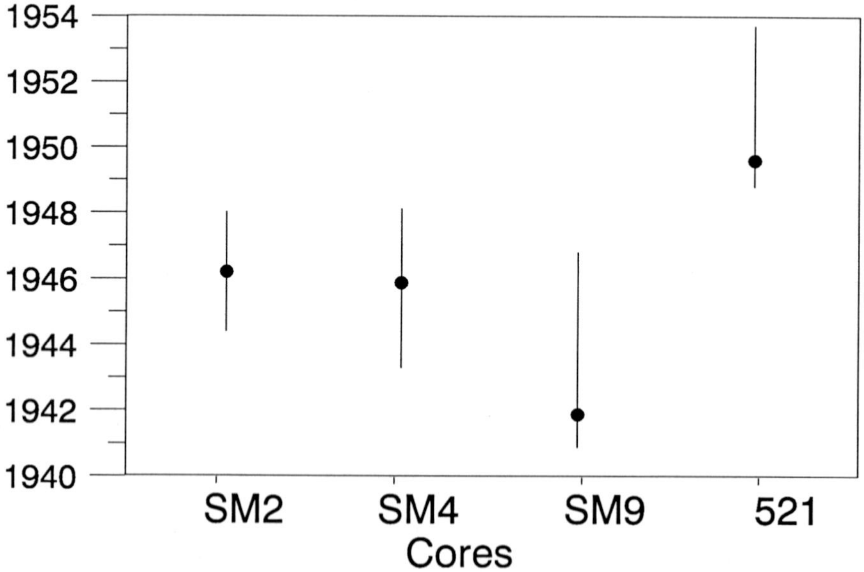

Figure 7: The date of the older pollution peak estimated with the ^{210}Pb method. Points mark the date of the maximum Hg concentration determined with a polynomial of 2nd degree. Lines indicate the error resulting from the sampling interval alone.

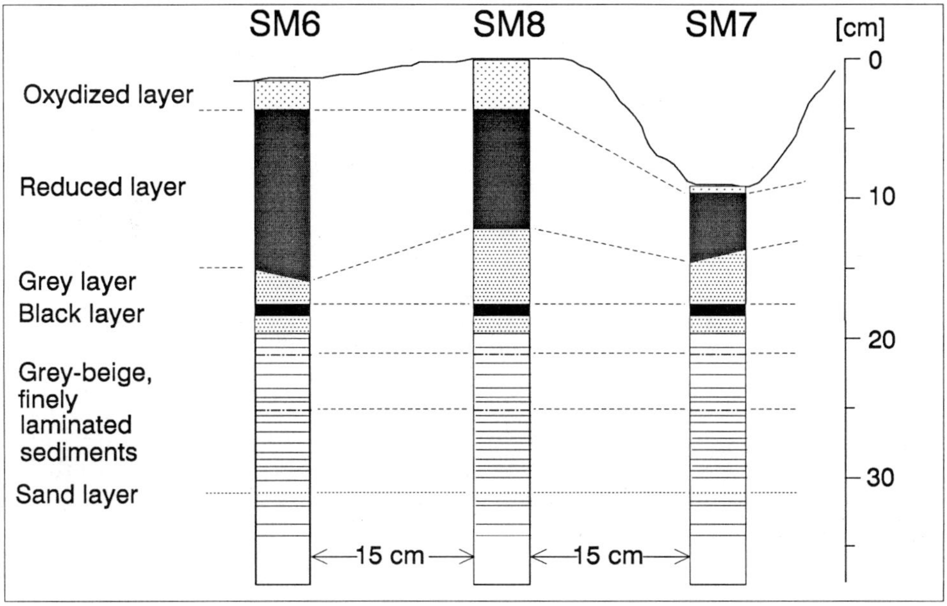

Figure 8: Lithological profiles of cores from a pillow (SM6, SM8) and an adjacent hollow (SM7). Note an excellent correlation below the black layer.

The profiles of porosity, ^{210}Pb, ^{137}Cs and Hg from the core pairs SM4 - SM5 and SM9 - SM10 are shown on figures 9A and 9B, respectively. The profiles of these parameters and the rate of sedimentation obtained from the radionuclides are very similar in the cores recovered from pillow sites. The decrease of ^{210}Pb is fitted in two separate intervals as there seems to be a small difference of slopes between the uppermost 14 cm and underlying sediment (14 - 20 cm). With the activities of ^{210}Pb-excess falling below 1 dpm/g in the deeper part of the cores, the uncertainty of the individual data is too large to attempt a determination of sedimentation rate.

One core from the same area taken on the pillow has also been dated by Schuler et al. (14) at EAWAG in Zurich. They obtained a sedimentation rate of 0.36 cm/y with ^{210}Pb method, equivalent to about 0.13 g/cm^{2}y, in excellent agreement with our results.

Porosity appears to be a very sensitive parameter for detection of changes in sediment character. The porosity profiles in all four cores (fig. 9A and 9B) show a characteristic minima (marked with broken lines on figures) which correspond to a sediment layer located immediately below the dark horizon shown on the lithological profiles (fig. 8). The low porosity layers are generally associated with the coarser-grained sediments. In this layer the ^{210}Pb and Hg also show local minima.

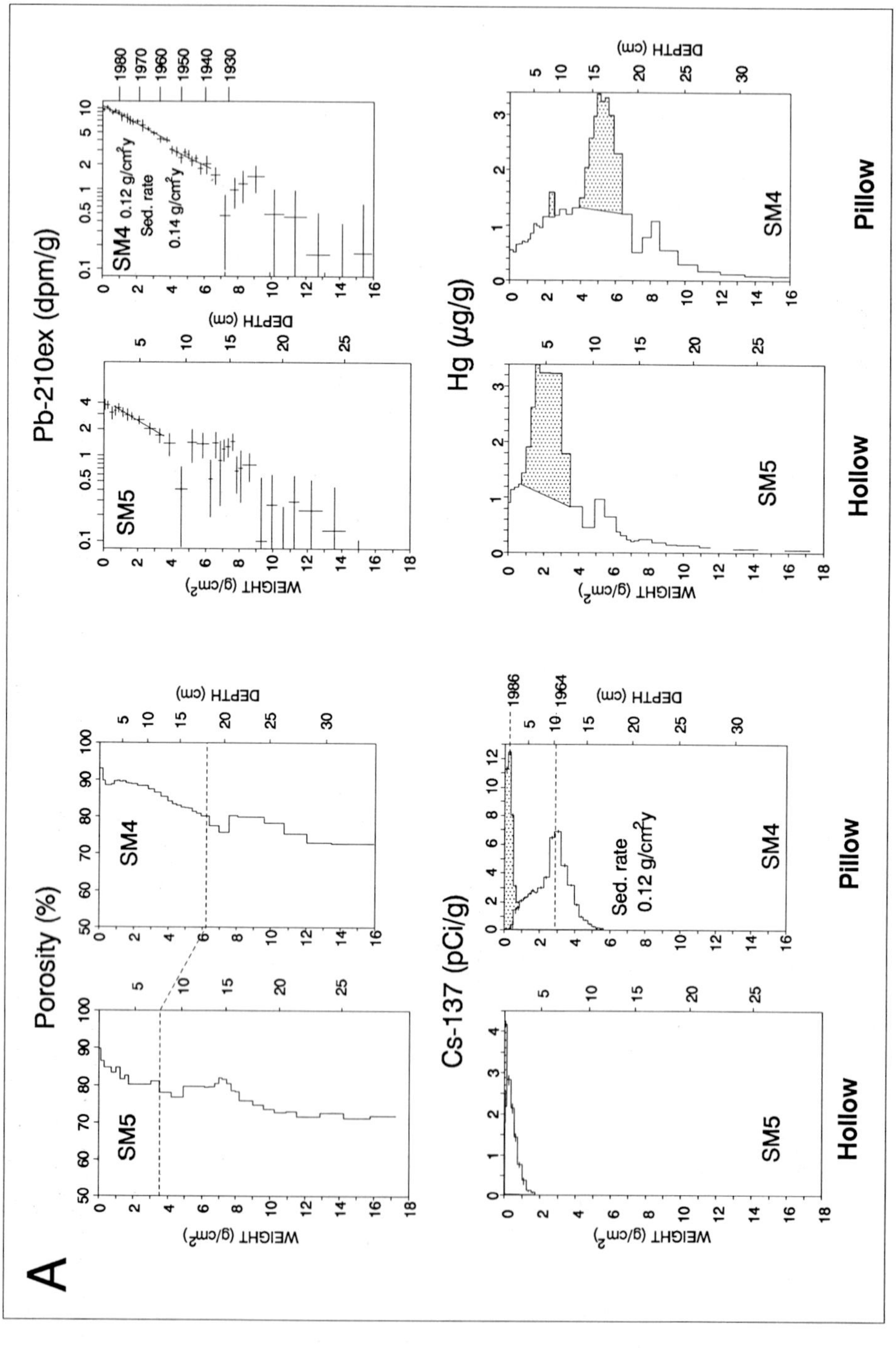
A
Porosity (%)
Pb-210ex (dpm/g)
Cs-137 (pCi/g)
Hg (µg/g)
SM5
SM4
SM4 0.12 g/cm²y
Sed. rate
0.14 g/cm²y
Sed. rate
0.12 g/cm²y
1980
1970
1960
1950
1940
1930
1986
1964
WEIGHT (g/cm²)
DEPTH (cm)
Hollow
Pillow

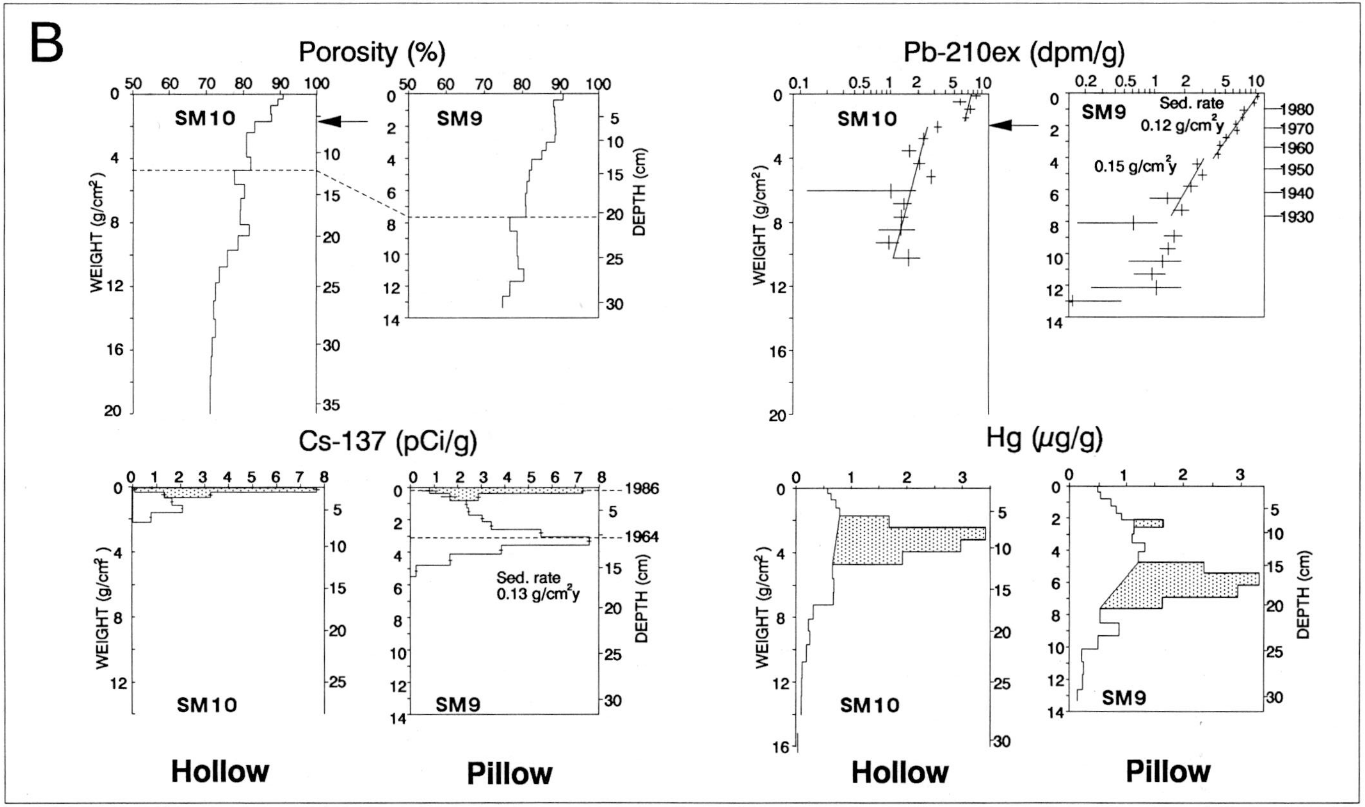

Figure 9: Porosity, ^{210}Pb-excess, ^{137}Cs and Hg profiles from two pairs of cores recovered from pillows and hollows. Time scale is obtained from ^{210}Pb-excess data. ^{137}Cs from Chernobyl is shown as shadowed surface.
A) SM4 - pillow and SM5 - hollow,
B) SM9 - pillow and SM10 - hollow, arrows indicate the sharp decrease of porosity and ^{210}Pb-excess in core SM 10.

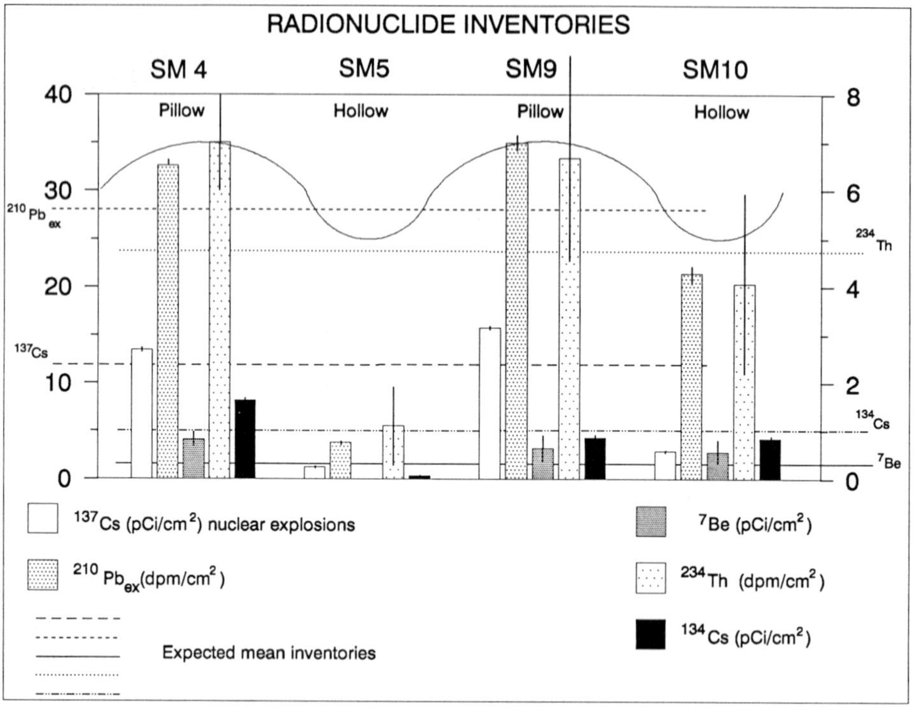

Figure 10: Radionuclide inventories in cores from pillows and hollows. The left scale refers to the inventories of ^{137}Cs and ^{210}Pb, the right one to the inventories of ^{7}Be, ^{234}Th and ^{134}Cs. Vertical bars represent the measured inventories and horizontal lines show the expected mean inventories inferred from a sediment trap experiment carried on at the similar water depth, in the central part of the lake (our unpublished data). The inventory of ^{137}Cs is obtained from fallout data corrected for decay to the date of sediment sampling.

The inventories of long-lived radionuclides from the pillow sites are higher than the inventories expected from measured flux into the sediment traps in lake Geneva (^{210}Pb) or known from fallout monitoring (^{137}Cs from nuclear tests). This strongly suggests that material resuspended from the hollow sites is not transported far away but resettles in the same region, increasing the radionuclide inventories in the pillow sites. As the surface area of hollows is much smaller than those of pillows the excess of radionuclide inventories on pillow sites is only 20 - 25 percent. Slightly higher excess of inventories of about 40-50 percent was found by Schuler et al. (13) in one core from the same area. As this increase is strongly time dependent, it may vary considerably from place to place.

There are distinct differences in the radionuclide and Hg profiles between the pillow and hollow sites. The activity of ^{210}Pb-excess in the surface layer in core SM5 (hollow) is much lower than those in core SM4 (pillow). Assuming that, initially, the ^{210}Pb activities were the same in these two cores, about 3.6 g/cm^2 of surface sediments is missing in the hollow. At the sedimentation rate of about 0.12 g/cm^{2}year in the upper section of the profile, this represents about 30 years of normal sedimentation.
The activity of ^{137}Cs is present in core SM5 at the surface only, and the total of cesium activity, from both the nuclear tests and Chernobyl accident, is much reduced in comparison to the familiar profile obtained from core SM4.
A different pattern is observed in another hollow site (SM10). While the surface activity of ^{210}Pb-excess is not very different from those in the nearby pillow site (SM9), a sharp decrease of activity occurs at about 6 cm depth. Accordingly, the surface ^{137}Cs activities originating from Chernobyl are very similar in cores SM9 and SM10, but the peak from the nuclear tests is largely missing in the latter core. These observations strongly suggest that erosional episodes occurred in hollow sites, but they were not synchronous at different hollows.
The Hg profiles from both pillow sites are similar to those from the core 521 and other undisturbed cores. They exhibit the two contamination events, dated at 1946 and 1969 in core SM4 (or 1942 and 1970 in core SM9). Both profiles seems to show a small peak prior to the strongest one. This is due to a lower concentration in the low porosity layer composed of coarse-grained sediments. In sediments recovered from hollows only the peaks of the older contamination are present and are found at a lesser depth than in sediments from pillow sites. The younger part of record, including 1971 contamination, is missing.

The sediment perturbations which occurred in hollows can conveniently be analyzed by a comparison of the radionuclide inventories. The total inventories of ^{210}Pb-excess, ^{137}Cs originating from the nuclear tests, ^{134}Cs from the Chernobyl accident and short-lived radionuclides, ^{7}Be (half-life 52.3 days) and ^{234}Th (24.1 days) are shown in figure 10.
In the undisturbed pillow sites the radionuclide inventories are very similar with the one exception of Chernobyl ^{134}Cs which is higher in core SM4 than in core SM9. Considering the pillow-hollow pair SM4 - SM5, one observes a strong deficit in the inventories of all radionuclides in the hollow suggesting that the last erosional event took place quite recently, as ^{7}Be is missing and ^{234}Th-excess (over its parent ^{238}U) and Chernobyl ^{134}Cs inventories are very low. In the core pair SM9 - SM10 short-lived radionuclides (^{7}Be and ^{234}Th) and Chernobyl cesium inventories do not differ significantly between the pillow-hollow sites, indicating undisturbed recent sedimentation at both sites. This is in agreement with the profile of ^{210}Pb-excess in core SM10 showing a perturbation at 6 cm depth, but not at the surface. Consequently, the inventory of ^{210}Pb-excess in core SM10, being lower than in the nearby pillow site, is higher than in the hollow site SM5, where surface sediments with the high ^{210}Pb activities are not preserved.

3.3. Large scale perturbations

One of the most common perturbations in sedimentary record in lakes and ocean sediments are induced by turbidity currents. They can remove a portion of previously deposited "pelagic" sediments and/or introduce a discontinuity in the form of rapidly deposited sand and silt horizons. While the latter effect is relatively easy to identify by a careful examination of sediment lithology, porosity and grain-size, the former may be difficult to detect. In the following paragraph we illustrate just one example of perturbation due to the deposition of a turbidite in the central plain of Lake Geneva.

The turbidite horizon 6 - 9 cm thick, covering about 25 km^2 of lake bottom (15) was deposited most probably in 1977 or 1978 (13). This turbidite layer is clearly delimited as a low porosity interval in figure 11, corresponding also to a very low ^{210}Pb-excess. Activity of this nuclide does show an exponential decrease in the sediment sections above and below the thick turbidite. In these two sections the rates of sedimentation were found to be 0.106 and 0.101 g/cm^2y, respectively. These mean rates are in fairly good agreement with the mean value of 0.124 g/cm^2y obtained by Gandais (16) from sediment trap measurements in 1986 and 1987 at the same station.
The determination of sedimentation rates is necessary but not sufficient to establish a time scale in this core because we ignore the erosive effect of turbidite deposition. Deposition of turbidite can be considered as an instantaneous event, thus the ^{210}Pb activity above and below the turbidite layer should be the same, provided no underlaying sediment was removed by the turbidity current. As can be seen in figure 11 the ^{210}Pb activity is distinctly lower below the turbidite layer than above it. From the difference we can calculate a loss of 9.3 years of sedimentation, which corresponds to the erosion of 2.8 cm of sediment below the turbidite. This reasoning might seem quite speculative, but it is supported by independent evidence. From the position of ^{137}Cs peak, just below the turbidite we calculate the loss of 3.1 cm of sediment, corresponding to 10.3 years of sedimentation. From the above considerations we conclude that "pelagic" sediments deposited between 1967 and 1977 were eroded by the turbidity current.
Using the sedimentation rate determined with ^{210}Pb method in the lower part of the core and the ^{137}Cs peak activity as the reference horizon marking the year 1964, a reliable time scale can be established to interpret the mercury profile in this site.

The major contamination event at 23 cm depth is dated at about 1946, in good agreement with the results from other cores. The contamination dated in a number of sediment cores at 1971 is not preserved here due to the sediment erosion by the turbidity current. Local minima of Hg concentration, porosity, ^{210}Pb and ^{137}Cs at 19 cm depth are related to a thin silty layer, possibly a distal turbidite.

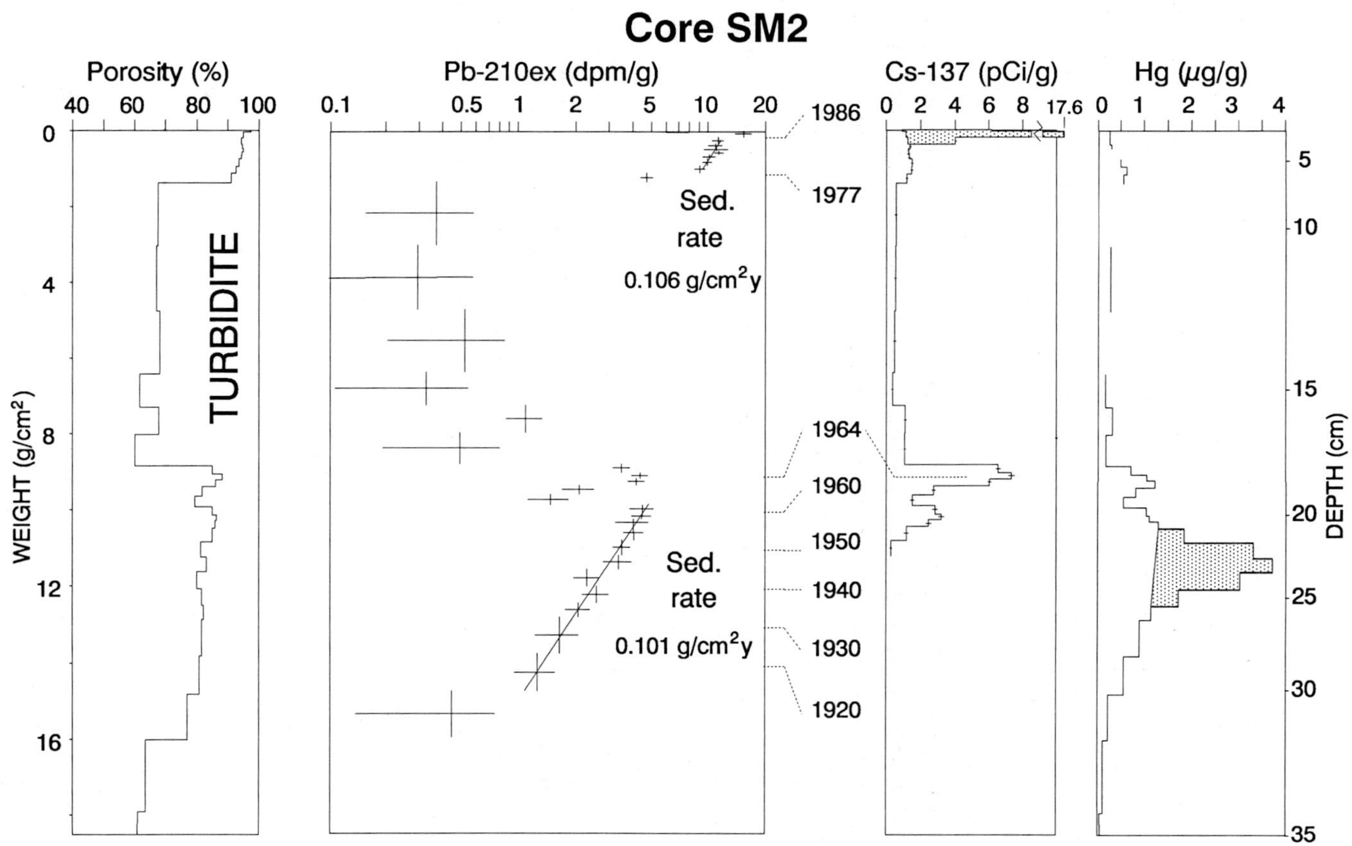

Figure 11: Dating of Hg profile in the core perturbated by the turbidite deposition. Time scale is established from combined ^{210}Pb and ^{137}Cs methods.

4. DISCUSSION

The determination of sedimentation rate in numerous cores enables the computation of mercury fluxes to the lake sediments. Maximum fluxes of the 1945 and 1971 events in excess of the general pollution level are given in figure 1. The Hg inventories of both episodes decrease with distance from the Rhone mouth (fig. 12). The trend for both events is similar, almost parallel. It has been demonstrated (2) that the 1971 pollution source was the Rhone River, we can thus conclude that the 1945 pollution came from the same source.

The estimation of mercury inventories and their origin in different parts of the lake is shown in figure 13. The inventories and their different fractions were obtained as an average of several cores in each part of the lake and weighted by the corresponding surface area.

The estimations of the total amount of Hg delivered by the two specific pollution episodes (fig. 14) give an input of 2.7 and 33 metric tons (with a maximal flux of 0.8 and 3.6 tons/year) for the 1971 and 1945 events, respectively.

The sources of the 1971 pollution in the watershed were known from a survey of the Rhone sediments which traced the contamination upstream, pointing to the emissions from the chemical industries (3). In contrast nothing is known about the 1945 pollution sources. The sole information we have is that mercury importation in Switzerland was at its highest during the period 1940-1945 (fig. 15).

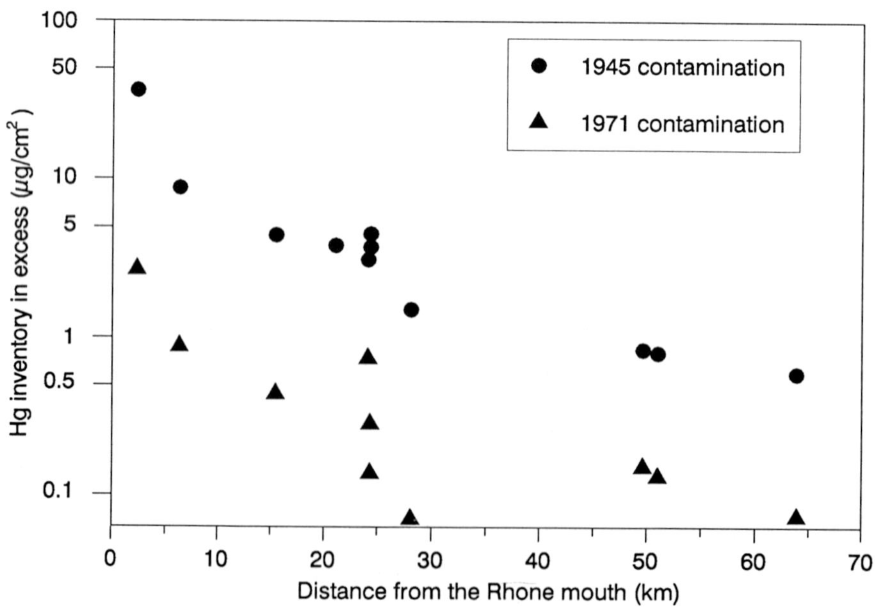

Figure 12: Plot of mercury inventories of the two events in excess over the general pollution in 11 sediment cores versus distance from the Rhone mouth.

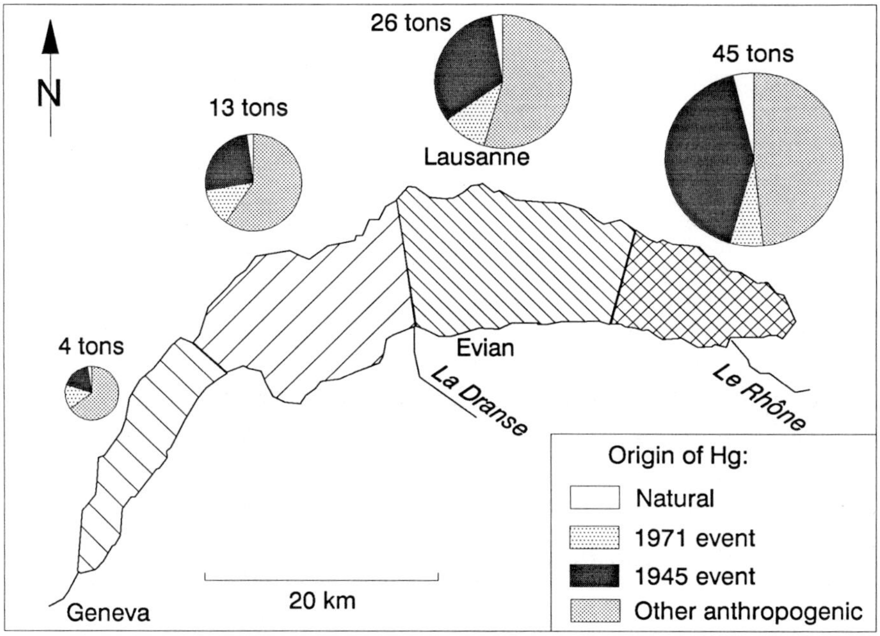

Figure 13: Mercury inventories and their origin in different parts of the lake.

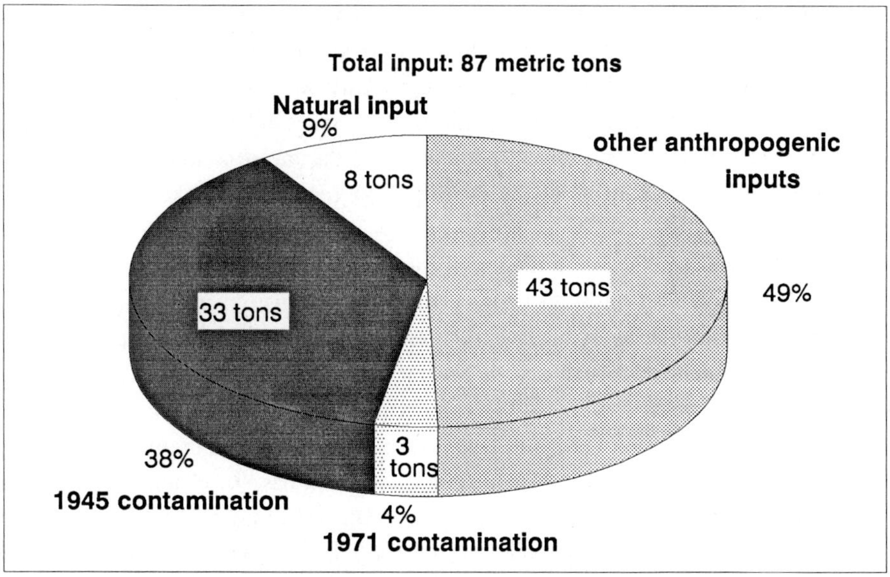

Figure 14: Estimated mercury inventory in Lake Geneva sediments since 1900 along with the contributions of different origin.

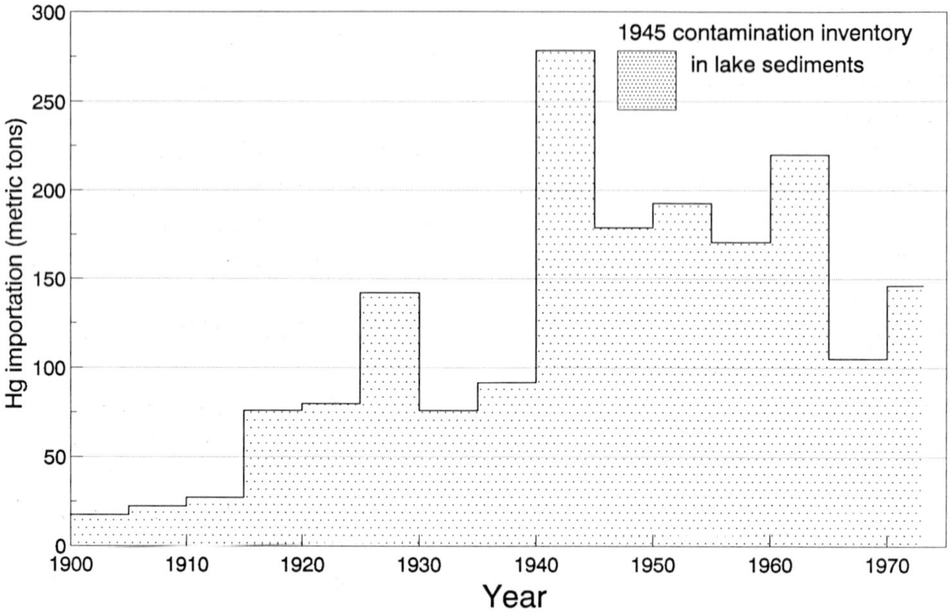

Figure 15: Mercury importation to Switzerland for the period 1900-1973 according to Scolari and Vernet (17). The estimated loading of Hg resulting in 1945 contamination is given for comparison.

In 1971, the mercury concentration of the Rhone sediments at the river mouth was about 2 $\mu g/g$, resulting in a flux of 0.8 ton/year to the lake. If we consider all other parameters being equal, the comparison of mercury flux with mercury concentration between 1971 and 1945 results in a concentration at the mouth in 1945 rising up to 8.4 $\mu g/g$, that is 100 times the natural content. This concentration is twice the mean content observed in the heavily polluted bottom sediments of the Detroit River (connecting Lake St-Clair to Lake Erie) in 1970 (18).

The amount of mercury delivered in the mid-forties to Lake Geneva is quite impressive but the impact of this pollution on biota is speculative, because we do not know the rates of methylation and demethylation occurring in the sediments at the time of the pollution. The relation linking the methylation rate with the mercury concentration in the substratum depends on a set of parameters such as pH, microbial activity, redox potential and other parameters (19), which are not known in Lake Geneva for the period in question. As a first approximation, assuming that the value of these parameters has not change between 1945 and today, and taking into account "that an increase with ten times of the concentration of inorganic mercury in the substratum will lead to a doubling of the methylmercury output" (16), we estimate a mean concentration of mercury in lake perch to be approximately 0.5-

0.6 $\mu g/g$ (wet weight) at the end of the forties. This value could be even higher if we assume according to D'Itri (20) that mercury level in fishes is generally higher in oligotrophic (Lake Geneva in 1945) than in eutrophic (Lake Geneva in 1971) lakes, at the similar Hg loading.

The survey of perch in Lake Geneva since 1973 (1) shows that their Hg content decreased four times between 1973 and 1980, whereas Hg concentration in surficial lake sediments decreased only two times during the same period. Using these proportions, the estimated mercury content in perch in 1945 would be as high as 3.4 $\mu g/g$ (wet weight), a value 7 times higher than the Swiss legal limit.

The mercury content in burbot are usually twice those in perch which can be readily explained by the higher exposure of bottom dwelling burbot to the mercury accumulated in sediments. The mean mercury content of 66 burbots of the same weight class (50 - 100 g) sampled in 1987 decreased to 0.12 $\mu g/g$ (21), but the distribution of results is distinctly asymmetric with a number of concentrations between 0.3 and 0.4 $\mu g/g$. These highly biased concentration may be in relation to the bio-erosion observed in parts of the lake with the pillow-hollow morphology of bottom sediments.

Both the direct observations and the results of radioisotopic measurements strongly suggest the resuspension and local erosion of sediments by bottom dwelling burbot.

These erosion events, reconstructed in figure 16, are not synchronous at different hollows. At one hollow site the last erosive event occurred shortly before sampling, removing 3.6 g/cm^2 (or 12 cm assuming the same porosity as in core SM4) of sediment, an equivalent of about 30 years of the normal sedimentation. Consequently, there is a hiatus in the mercury contamination record from 1987 to 1958, at this site. At another hollow site the most recent sediments deposited after 1974 are preserved. The last erosive event, which occurred about 13 years before sampling date, removed 2.7 g/cm^2 (or nearly 8 cm, under above-mentioned assumption) of sediments. This represents about 22 years of sedimentation between 1974 and 1952. The older sediments appear to be well preserved at all sites.

This example clearly shows that benthic fish may disturb the sedimentary record, remove the surface sediment and therefore expose the older layers, heavily contaminated with mercury, to direct contact with lake water and biota. Brendl (22) measured a higher oxidizing potential in surface sediments at hollow than at pillow sites and suggested that this effect is related to resuspension of sediments in hollows by benthic fish. This may also impl that the initially buried Hg-rich sediments are exposed to the aerobic conditions resulting in a higher rate of methylation of Hg (19). It is very difficult to estimate the regional importance of this phenomena but the bio-erosive structures in the form of trenches were also observed in Lake Superior and attributed to burbot activity (23).

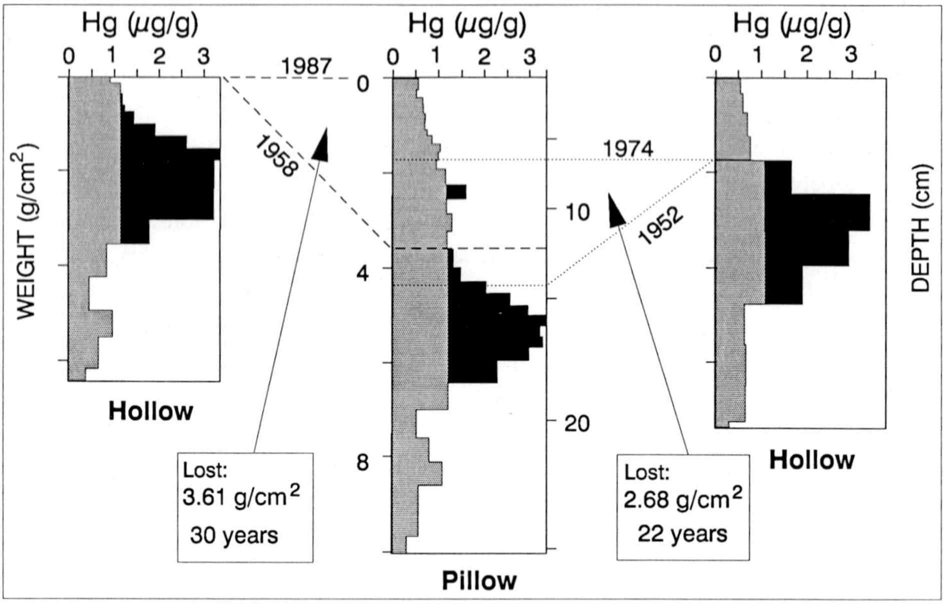

Figure 16: The interpretation of erosional events in hollow sites based on combined porosity, radioisotopic and mercury data and their impact on the Hg profiles in sediments.

Large scale perturbations of sedimentary records resulting from depositional and or erosive effects of turbidity currents are frequent and may lead to serious misinterpretation of pollution history. However, in favorable circumstances, the joint use of radioisotopic methods and careful examination of sediment lithology and porosity can result in a quite precise chronology as it has been shown in figure 11. It is interesting to note that the turbidity current, which is capable to transport sandy sediments, removed a thinner sediment layer than the bio-erosive action of burbot. Moreover, the deposition of rather compact turbidites has a beneficial effect, as they can rapidly and cut out the direct contact between bottom water and severely contaminated sediments, whereas the sediment resuspension by burbot has the opposite effect.

5. SUMMARY

Mercury contamination, being considered as the major environmental problem in this lake, has been monitored in the alpine Rhone River and Lake Geneva surface sediments since 1970. Dated sediment cores taken between 1976 to 1988 revealed the earlier history of mercury contamination which started at the end of 19th century. A major mercury pollution episode in Lake Geneva occurred in the mid-1940's. ^{210}Pb and ^{137}Cs dated cores provide sedimentation rates from which mercury fluxes could be computed. In certain regions the sedimentary record is disturbed by turbidites. Moreover, decimetre-scale morphological structures resembling pillows and hollows, covering a large area of Lake Geneva bottom have been observed with a submersible. Core pairs taken selectively from pillows and adjacent hollows show different records of Hg contamination. A careful examination of the vertical profiles and inventories of natural (^{210}Pb, ^{7}Be,^{234}Th) and artificial (^{137}Cs, ^{134}Cs) radionuclides in these cores reveals erosional episodes in hollows. These episodes in the different hollows are not synchronous and are attributed to the sediment resuspension by bottom dwelling burbot. This form of bio-erosion may bring the older, contaminated sediments into direct contact with lake water and biota.

From the sedimentary record, we broadly estimate that about 80 metric tons of anthropogenic mercury have been discharged into Lake Geneva since 1900. The major contamination in mid-forties alone is responsible for about 40 % of this input. The inventory of mercury in the sediments decreases with distance from the Rhone River mouth, suggesting that this river was the major source of the pollution.

ACKNOWLEDGEMENTS

The authors wish to thank Prof. J. Piccard and the crew of submersible "F.-A. Forel" for their help in coring. Figures 1, 2, 7-10 and 12 are reproduced with some minor changes from the extended abstracts (24, 25) of International Conference on Heavy Metals in the Environment, September 1989, Geneva (courtesy of CEP Consultants, Edinburgh). The manuscript benefited from the review of Dr. R. R. De Vitre. This study was partly supported by Fonds National Suisse de la Recherche Scientifique.

REFERENCES

1. P.-Y. Favarger and J.-P. Vernet, in: J.-P. Vernet (Ed.), Heavy Metals in the Environment, Abstract vol. 1., Geneva, 1989, pp. 126-129.
2. J.-P. Vernet and R.L. Thomas, Eclogae geol. Helv., 65 (1972) 293-306.
3. H. Howa, D. Arbouille, D. Span and J.-P. Vernet, in: J.-P. Vernet (Ed.), Heavy Metals in the Environment, Abstract vol 1, Geneva, 1989, pp. 239-242.
4. J.-P. Vernet and P.-Y. Favarger, Hydrobiologia, 92 (1982) 643-649 .
5. J.-P. Vernet, A. Chappuis, P.-Y. Favarger and E. Davaud, Bull. A.R.P.E.A. Hauterive, 74 (1976) 23-44
6. J.-P. Vernet, E. Davaud, P.-Y. Favarger and A. Chappuis, Bull. A.R.P.E.A. Hauterive, 86 (1977) 19-30.
7. J.-P. Vernet, Bull. Soc. Vaud. Sci. Nat., 69 (1966) 287-294.
8. M. Sturm, A. Zwyssig and J. Piccard, in: J.-P. Vernet (Ed.), 3rd Intern. Symp. Interactions between Sediment and Water, Geneva, abstract, 1984, p. 126.
9. S. Krisnaswamy, D. Lal, J.-M. Martin and M. Meybeck, Earth Planet. Sci. Lett., 11 (1971) 407-114.
10. J. Dominik, A. Mangini, and G. Müller, Sedimentology, 28 (1981) 653-677.
11. J. Dominik, D. Burrus, and J.-P. Vernet, Earth Planet. Sci. Lett., 84 (1987) 165-180.
12. A Murdoch, L. Sarazin and T. Lomas, J. Great Lakes Res., 14 (1988) 241-251.
13. J. Dominik, D. Span and J.-L. Loizeau, submitted to Climate Dynamics.
14. Ch. Schuler, J. Dominik, A. Leueck, A. Losher and P.H. Santchi, in: Symp. on Functional and Structural Properties of Large Lakes, Konstanz, September 1987, abstract, 1987, p. 113.
15. J. Dominik, and J.-L. Loizeau, P.-L. Portier and J.-P. Vernet, in: Conf. inter. limnologues d'expression française, Aussois, May 1988, abstract, 1988.
16. V. Gandais, Cahier de la Faculté de Sciences, Université de Genève, 19 (1989) 75-82.
17. G. Scolari and J.-P. Vernet, Rapport of B.R.G.M. 74SGN298 MAR, Orléans, 1974.

18. Y. Hamdy and L. Post, J. Great Lakes Res., 11 (1985) 353-365.
19. A. Jernelöv and K. Johansson, in: G. Müller (Ed), Heavy Metals in the Environment, Heidelberg, 1983, pp. 727-740.
20. F.M. D'Itri, C.S. Annet and A.W. Fast., MTS Journal, 5 (1971) 10-14.
21. C. Corvi, in: Rapp. Commission Internationale pour la Protection des Eaux du Léman, (CIPEL), Campagne 1987, Ed. CIPEL, Lausanne (1988) 134-140.
22. Brandl, H., Inaugural-Dissertation, Philosophischen Fakultät II der Universität Zürich (1987).
23. Boyer, L.F. and Whitlatch, R.B., in: 32nd Conference on Great Lakes Research, May 30 - June 2, 1989, Madison, abstract (1989) p. 31.
24. J.-L. Loizeau, J. Dominik and J.-P. Vernet, in: J.-P. Vernet (Ed.), Heavy Metals in the Environment, Abstract vol. 1., Geneva, 1989, pp. 341-344.
25. J. Dominik, J.-L. Loizeau, B. Gallerini, J. Piccard, R. Monod and J.-P. Vernet, in: J.-P. Vernet (Ed.), Heavy Metals in the Environment, Abstract vol. 1., Geneva, 1989, pp. 461-466.

Chapter 9
Wastewater

FORMATION OF SURFACE COMPLEXES BETWEEN HEAVY METALS AND SLUDGE PARTICLES

Ching-Tzone Tien and C.P. Huang

SUMMARY

The surface properties of sludge solids from an extended aeration plant were investigated. Surface acidity constants were determined by the alkalimetric titration method. The specific surface area of the sludge solids measured by the BET method are 84 m^2/g. The surface complexation model was used to describe the adsorption of seven heavy metals onto sludge solid surfaces. The speciation of the adsorbed metal species onto the sludge solids surface at various pHs was made. It was found that the adsorption energy ($-\Delta G_{ads}$) calculated from the stabilization constants follows the order: Hg(II) > Pb(II) > Cd(II) > Zn(II) > Co(II) ~ Ni(II) ~ Cu(II). The competitive adsorption of these heavy metals on the sludge particle surface was also investigated. Under the experimental conditions of this study, results indicate that metal adsorption onto the sludge solid surface is not affected by the presence of other metals.

1. INTRODUCTION

The presence of heavy metals in the municipal wastewater treatment plant effluent may have adverse effects on the quality of the receiving waters. In light of public health protection, studies concerning the sources of heavy metals, the mechanism of heavy metals removal in the waste water treatment plant and the fate of heavy metals discharged into surface streams and/or land are deemed necessary.

In the activated sludge system, adsorption by the bioflocs, intracellular uptake and the separation of the metal-rich biomass in the secondary sedimentation tank are the major heavy metal removal mechanisms. It has been reported that the activated sludge system is more efficient than primary sedimentation in the removal of metals, especially chromium, copper and lead. Brown et al. (1) found that the removal of total chromium is 27% and 58%, respectively, from the primary and the secondary plants. In studying the relative contributions of surface adsorption and intracellular uptake of soluble metals by sludge solids, Sterritt and Lester (2) found that the decrease in metal removal by formaldehyde-treated inactive cells versus active biomass was 5% Cd, 8% Co, 12% Cu, and 30% Ni. Thus, they concluded that the active intracellular uptake of heavy metals was insignificant and that the majority of metal adsorption was due to the surface adsorption process (2).

Activated sludge solids consist of a mixture of mainly bacteria, protozoa, fungi and algae. Many different species of bacteria isolated from activated sludge have been shown to be able to produce polymer (ECP), or capsules. This is typically thicker and less dense than the cell wall and is, as a rule, not easily visible with the electron microscope (3). In a mixed culture of bacteria and algae, Tenney and Verhoff (4) found that the relative chemical compositions of extracellular polymers after 96 hours of growth were 31.5% polysaccharide, 42% protein, 14.4% RNA 14.4%, and 12% DNA.

It has been reported that different metal adsorption sites exist on neutral polysaccharide and anionic polysaccharide. The hydroxy groups (R-OH) of the hexose and the carboxyl groups of nucleic acids, proteins, and carbohydrates with uranic acid residues were thought to be the metal binding sites of sludge solids (5). Broadbent and Bradford (6) reported that 55% of the cation exchange capacity of soil organic matter is from carboxyl groups. Carboxyl group with a pK_a of 5 exhibit weak acid properties in aqueous solution. The dissociation of H^+ from the carboxyl group will result in a potential binding site ($RCOO^-$) for heavy metals. The formation of protein involves the combination of a large number of amino acids by peptide bonds. However, only these acid/base groups which are present as side chains on amino acids are capable of contributing to the metal binding capacity of proteins. In addition to the carboxyl groups of amino acids, the basic amino groups also ionize in solution and contribute to the metal binding capacity. Understanding the interaction mechanism of metal ions with the surface of sludge solids is essential to the control of heavy metals in the aquatic system. Several adsorption models have been developed to interpret and predict the adsorption behavior of metal ions from dilute aqueous solutions. Some models emphasize the importance of the electrical double layer structure and the physical interaction in controlling the distribution of metal ions in the solution; others stress the specific chemical or coordinative interactions of metals with surface sites. These models include: [1] the Gouy-Chapman-Stern-Grahame model which accounts mostly for electrostatic adsorption and considers the columbic attraction between metal ions and the hydrous solid surface; [2] the ion exchange model wherein protons are replaced by metals upon the adsorption of metal ions on the solid surface; [3] the surface complex formation model which emphasizes the specific adsorption of metal ions on the deprotonated surface (7); [4] the combined surface ionization and complexation model proposed which considers both the electrostatic and the chemical interactions simultaneously (8); and [5] the James and Healy model which includes hydration energy in addition to columbic and chemical energy as the total free energy of adsorption (9).

In this research, the adsorption behavior of Cu(II), Co(II), Cd(II), Hg(II), Ni(II), Pb(II), and Zn(II) onto sludge solid surfaces was investigated. The surface complex formation model was selected due to its simplicity in application. It must be mentioned that the approach presented in this study by no means depicts the mechanism by which metals are attached to the heterogeneous sludge solid surface or the microscopic surface properties of the sludge particle. Rather, the method provides a simple means for quantifying the removal of metals by a specific type of sludge particle using solution pH as the master variable.

2. Background

2.1 Surface Acidity

A weak acid is one that has a weak proton donation tendency; it takes a higher free energy to transfer protons from the acid to the aqueous solution. The functional groups of a sludge solid surface are believed to possess the nature of a weak acid. The strength of an acid (HA) relative to H_2O as a proton acceptor is often expressed by an equilibrium acidity constant, K_a. Proteins and polysaccharide contain many weak-acid-type functional groups and each has its own "microscopic" acidity constant (K_i).

Tanford (10) reported that in most common proteins, one out of every three or four amino acid residues contains a titratable acidic or basic group. The number of titratable groups per protein molecule ranges from ca. 20 to 250 for the many common proteins with molecular weights below 100,000. The number is even larger for proteins with higher molecular weight. According to Tanford (10), the titratable groups of the protein molecules include, but are not limited to the following: the carboxyl groups of glutamic acid (pK_a = 4.37-4.6) and the aspartic acid (pK_a = 4.5) side chains; the amino group (pK_a = 10.4) of lysine side chains; the guanidyl groups (pK_a = 12.0) of arginine side chains; the imidazole group (pK_a = 6.15) of histidine side chains; the phenolic group (pK_a = 9.7) of tyrosine side chains; and finally, the terminal amino and carboxyl groups of nearly all polypeptide chains. The difficulty of determining the microscopic constants increases rapidly as the number of acid groups increases. To simplify the process for this research, an averaging intrinsic constant (K_{int}) was determined by the method of potentiometric titration developed by Huang and Stumm (7). The amphoteric properties of amino acids along with carboxyl and other functional groups on the surface of sludge solids are similar to that of hydrous α-Al_2O_3. For simplicity sake, let S-OH_2^+, S-OH and S-O^- represent the positive, neutral and negative states, respectively, of the functional groups. The deprotonation reactions are:

$$S\text{-}OH_2^+ = S\text{-}OH + H^+ \; ; K_{a1} \qquad [1]$$

$$S\text{-}OH = S\text{-}O^- + H^+ \; ; K_{a2} \qquad [2]$$

The procedures of potentiometric titration include the addition of an aliquot of acid or base to the sample and the recording of subsequent changes in pH. In titrating sludge samples with an acid or base, the surface charge density can be calculated from the following charge balance (or proton condition) equation:

$$(C_B - C_A + [H^+] - [OH^-])/S = \sigma_\circ/F = \Gamma_{OH^-} - \Gamma_{H^+} \qquad [3]$$

where C_B and C_A are concentrations of strong base and strong acid added during titration, (M); S is total specific surface area of sludge sample, (m^2/g); F is Faraday constant, (96,500 coul/mole); $\sigma°$ is surface charge, ($\mu col/cm^2$); Γ_H^+ and Γ_{OH}^- are the amounts of proton or hydroxide adsorbed on surface of sludge sample.

Experimentally, the surface charge ($\sigma_\circ$) can be calculated from the following formula:

$$\sigma_\circ = ((\Delta V)\cdot N\cdot 96{,}500\cdot 100)/(S\cdot X\cdot V) \qquad [4]$$

where $\sigma_\circ$ is surface charge in μ coul/cm^2; ΔV is volume (in ml) of acid or base consumed between a point at a given pH in the titration curve and the point at which the titration curves are intersected, called the zero point of charge, pH_{ZPC}; N is normality of titrant; X is concentration of sludge sample (in g/l); V is volume of sludge sample (in ml); and S is total specific surface area of sludge sample.

By applying the Gouy-Chapman version of the electric double layer theory (11), the surface charge ($\sigma_\circ$) and the surface potential ($\phi_\circ$) can be related as follows:

$$\sigma_\circ = 2(RT\epsilon I/2\pi)^{1/2} \sinh^{-1}(z F \phi_\circ / 2RT) \qquad [5]$$

where R is the gas constant (1.987 cal K^{-1} $mole^{-1}$); T is the absolute temperature (°K); I is the ionic strength (mole $liter^{-1}$); ϵ is the dielectric constant (8.9 x 10^{-11} coul $v^{-1}cm^{-1}$); z is the valence of inert electrolyte ions.

In the presence of an inert electrolyte, the surface potential ($\phi_\circ$) is close to the potential (ϕ_d) at the plane of the closest approach (the outer Helmholtz plane). After rearranging equation [5] and substituting the corresponding constants at 20°C, the equation becomes:

$$\phi_d = \phi_\circ = 0.05 \sinh^{-1}(\sigma_\circ/[11.74\ I^{1/2}]). \qquad [6]$$

According to the Botzmann equation, the concentration of H^+ in the immediate proximity of the charged surface, $\{H^+\}$, is related to that of the bulk solution $[H^+]$ by the formulas:

$$\{H^+\} = [H^+]\exp(-F\phi_d/RT) \qquad [7]$$

If the concentration of each component involved in the reaction is measured at the immediate proximity of the charged surface, the equilibrium constants, referred to as intrinsic acidity constants (K_{a1}^{int} and K_{a2}^{int}), can be represented in the equations:

$$K_{a1}^{int} = \{S\text{-}OH\}\{H^+\}/\{S\text{-}OH_2^+\} \qquad [8]$$

$$K_{a2}^{int} = \{S\text{-}O^-\}\{H^+\}/\{S\text{-}OH\} \qquad [8a]$$

The total exchangeable site, $\{x_T\}$, at the solid surface is:

$$\{X_T\} = \{S\text{-}OH_2^+\} + \{S\text{-}OH\} + \{S\text{-}O^-\} \qquad [9]$$

For pH < pH_{zpc}, the surface of the sludge solid sample is positively charged. Since $\{S\text{-}O^-\}$ is insignificant, equation [9] becomes:

$$\{X_T\} = \{S\text{-}OH_2^+\} + \{S\text{-}OH\} \qquad [9a]$$

The positive site is further designated by the following:

$$\{X_+\} = \{S\text{-}OH_2^+\} \qquad [9b]$$

For pH > pH_{zpc}, $\{S\text{-}OH_2^+\}$ is insignificant and equation [9] then becomes:

$$\{X_T\} = \{S\text{-}O^-\} + \{S\text{-}OH\} \qquad [9c]$$

The negative site is then represented in the following:

$$\{X_-\} = \{S\text{-}O^-\} \qquad [9d]$$

After substituting $\{X_T\}$, $\{X_+\}$ and $\{X_-\}$ into equations [8] and [8a], the following equations can be obtained:

$$1/\{H^+\} = [1/\{X_+\}]\ \{X_T\}/K_{a1}^{int} - 1/K_{a1}^{int} \qquad [10]$$

$$\{H^+\} = 1/\{X_-\}\ K_{a2}^{int}\ \{X_T\} - K_{a2}^{int} \qquad [10a]$$

By plotting $1/\{X_+\}$ versus $1/\{H^+\}$ and by utilizing the relationship of the variables in equation [10], K_{a1}^{int} and $\{X_T\}$ can be obtained from the intercepts of the $1/\{H^+\}$ and $1/\{X_+\}$ axes, respectively. Similarly, K_{a2}^{int} and $\{X_T\}$ can also be obtained from the intercepts of the $\{H^+\}$ and $1/\{X_-\}$ axes, respectively.

The positive and negative sites $\{X_+\}$ and $\{X_-\}$ can be calculated from the surface charge ($\sigma_\circ$) by the equations:

$$\{x_+\} = \sigma_\circ^+ / (96{,}500 \cdot 10^6) \quad [11]$$

$$\{x_-\} = \sigma_\circ / (96{,}500 \cdot 10^6) \quad [11a]$$

where $\sigma_\circ^+$ and $\sigma_\circ^-$ are in μ coul/cm^2 and $\{X_+\}$ and $\{X_-\}$ are in equivalent/cm^2.

2.2 Complex Formation

According to the concept of the complex formation model, a set of possible reactions between the metal species M^{2+}, MOH^+, $M(OH)_2$ and $M(OH)_3^-$ with the functional groups at the surface of the sludge solids can be expressed by the equations:

$$S\text{-}OH_2^+ + M^{2+} = MSO^+ + 2H^+;\ K_1 \quad [12]$$

$$2S\text{-}OH + M^{2+} = M(SO)_2 + 2H^+;\ K_1' \quad [12a]$$

$$2S\text{-}O^- + M^{2+} = M(SO)_2;\ K_1'' \quad [12b]$$

Similar reaction equations can be written for other metal hydroxo species including MOH^+, $M(OH)_2$ and $M(OH)_3^-$. In addition to investigating the adsorption behavior of the bare metal species, M^{2+}, as proposed by other investigators (8), we have incorporated all possible metal hydroxo species into the present complex formation model. The total adsorption density (Γ_T) was set to equal the sum of the adsorption density of each individual adsorbed species, i.e. $\Gamma_{M^{2+}}$, Γ_{MOH^+}, $\Gamma_{M(OH)_2}$ and $\Gamma_{M(OH)_3^-}$ in the following equation:

$$\Gamma_T = \Gamma_{M^{2+}} + \Gamma_{MOH^+} + \Gamma_{M(OH)_2} + \Gamma_{M(OH)_3^-} \quad [13]$$

The total adsorption density, Γ_T, can be obtained from an adsorption experiments. As mentioned later, a 2:1 stoichiometric relationship between hydrogen and copper ions was observed. The concentrations of the surface functional groups $S\text{-}OH_2^+$, S-OH and $S\text{-}O^-$ can be determined from the formulas:

$$\{S\text{-}OH_2^+\} = \{X_T - 2\Gamma_T\}\ \alpha_+ \quad [14]$$

$$\{S\text{-}OH\} = \{X_T - 2\Gamma_T\}\ \alpha_\circ \quad [14a]$$

$$\{S\text{-}O^-\} = \{X_T - 2\Gamma_T\}\ \alpha_- \quad [14b]$$

where X_T is the total number of surface sites in equivalent/liter and α_+, $\alpha_\circ$ and α_- are the fractions of the positive, neutral and negative groups, respectively. The alpha (α) terms are determined by:

$$\alpha_+ = [H^+]^2/D \quad [15]$$

$$\alpha_\circ = K_1[H^+]/D \quad [15a]$$

$$\alpha_- = K_1\ K_2/D \quad [15b]$$

where $D = [H^+]^2 + K_1[H^+] + K_1 K_2$, and K_1 and K_2 are the first and second microscopic acidity constants which can be calculated from the intrinsic acidity constants:

$K_1 = K_a^{i1nt} \exp(F\phi_{\circ}/RT)$ [15c]

where $F/RT = 39.63$ volt^{-1} at 25°C.

The stability constant of the reaction formulated in equation [12b] can be expressed as:

$K_1'' = [M(SO)_2] / ([SO^-])^2 [M^{2+}])$ [16]

Similar equations can be formulated for $K_2''_{(MOH+)}$, $K_3''_{(M(OH)2)}$, K_1 [12] and K_1' [12a]. Not all reactions between metal species and surface functional groups will occur at a given pH. To calculate the stability constants K_1'', K_2'' and K_3'' at a given pH, the concentration of each metal specie at a given pH must be determined first. For example, the concentration of a "bare" metal specie is determined by:

$[M^{2+}] = [M_T - \Gamma_T] \alpha_{M^{2+}}$ [16a]

where M_T represents the total metal initial concentration and α_{M2+}, which can be obtained from the metal speciation diagrams, represents the fraction of total metal in the form of M^{2+} at a given pH.

3. EXPERIMENTAL PROCEDURES

3.1 Surface Area Measurement

Moisture free and structurally undisturbed sludge solid samples were used for the specific surface area measurement. The following procedures suggested by Weber et al. (12) were used to prepare dried sludge solid samples for surface area measurement. All sludge samples used in this study were taken from a regional extended aeration wastewater treatment in north-eastern Maryland. The following procedures were followed in the preparation of sludge sample: [1] Filter sludge sample through a No. 40 sieve to remove grits and floating matter. [2] Immerse sample in a 2% para-formaldehyde and 2.5% glutaraldehyde solution with a 0.1 M sodium phosphate buffer at pH 7.3. The para-formaldehyde will quickly kill bacteria cells as well as reduce cell swelling or shrinking whereas the glutaraldehyde will cross link with proteins in the cell membrane to form rigid networks for preserving fine structures. [3] Transfer and immerse sample in 1% OsO_4 with a 0.1 M sodium phosphate buffer at pH 7.3 for 3 hours. OsO_4 is used to maintain the shape of the liquid portion of the cell membrane. [4] Sequentially rinse sample in graded ethanol (50%, 70%, 80%, 90% and 100%) to replace water in or on the cell. [5] Rinse sample in a 1:1 100% ethanol-amylacetate solution and let stand overnight to dehydrate and to replace ethanol with a nonpolar solvent for further preservation of the fine cell structures. [6] Place sample into a Dupont Sorvall CO_2 critical point drying system for further dehydration.

The moisture free sludge solid samples prepared above were placed in the Quantasorb Model LMFC-4 for outgassing. The surface area was measured by the same Quantasorb system and nitrogen gas was used as adsorbate molecules.

3.2 Titration of Sludge Solids

Instead of using the conventional acid and base titration method, the following procedures were followed for the titration of the sludge solids. [1] Filter sludge sample through a No. 40 sieve to remove grit and debris. [2] Transfer sample to a container (4 liters) filled with distilled water and aerate with a ball diffuser for 10 minutes. [3] Turn off the air supply to the diffuser and allow the suspended solids to settle at a room temperature of ca. 21° C. [4] Decant the supernatant and refill the container with distilled water to the 4 liter mark; caution should be taken to minimize the loss of solids during the decanting process. [5] Reaerate the sample for 10 minutes and measure the conductivity of the mixed liquor. [6] Repeat steps [2] through [5] until conductivity of the sample approaches to that of distilled water and then measure the concentration of the suspended solids. [7] Divide the 4 liter sample into two 2-liter containers and adjust the ionic strength in each container to 10^{-1} M and 10^{-2} M ($NaClO_4$), respectively. [8] Distribute sample in each 2 liter container to a set of 17 bottles, each bottle containing 100 cc of the sample (total 34 sample bottles). [9] To 8 of the 17 bottles in each set, add 0.1 ml aliquot of 0.1 M $HClO_4$. To another 8 samples, add 0.1 ml aliquot of 0.1 M NaOH. One sample in each set received no pH adjustment. [10] Place sample bottles in a shaker for 12 hours at room temperature of ca. 21° C. Shake samples moderately without diffused air. [11] Measure and record the pH of each sample after 12 hours of shaking.

3.3 Adsorption in Single Metal System

There were a total of eight series, each including one of the 7 heavy metals (Hg, Pb, Cd, Zn, Co, Ni and Cu) and one blank series with no metal addition. The procedure for the adsorption test was carried out as follows: [1] Prepare samples as stated in steps [1] through [6] of the titration section except increase sample volume to 8 liters and adjust the ionic strength of the 8 liter sample was adjusted to 5 X 10^{-2} M. The suspended solids concentration was adjusted to 1.1 g/l. [2] Distribute the sludge sample to 72 plastic bottles (9 bottles for each of 8 series), each bottle containing a 90 ml sludge sample. [3] Add various quantities of 1 M NaOH or 1 M $HClO_4$ to each individual sample bottle of the same series for pH adjustment. [4] Add 10 ml of a metal perchlorate solution (10^{-3} M) to each of 9 bottles in the same series. [5] Place sample bottles in a shaker for 12 hours at room temperature of ca. 21°C. [6] Measure and record the pH of each sample. [7] Filter 10 ml sample from each bottle with a Millipore filter (0.45μm) and measure the remaining metal concentration in the filtrate by an atomic adsorption spectrophotometer (Perkin Elmer Model Zeeman 5000).

3.4 Adsorption in Multi-metal System

The procedure for this experiment is similar to the previous section (Adsorption in Single Metal System) except for step No. 4 where solutions of mixed metals were utilized. These solutions contained 10 ml of a mixture of the 7 heavy metals that ranged in concentration from 20 μM to 100 μM after being added to each of the five bottle series. One series was again left without any metal addition for background purposes.

4. RESULTS AND DISCUSSION

4.1 Surface Area of Sludge Solids

The specific surface area of sludge solids was measured by the BET method developed by Brunauer, Emmett and Teller (13). An inert gas such as nitrogen or helium is often used as an adsorbate. The specific surface area of sludge solids used in this research was measured at 84 m^2/g. Lamanna and Mallette (14) reported the specific surface area of 5.6 m^2/g for Escherichia coli. Soil organic matter was determined to have a specific surface area of 558-803 m^2/g (15). Sludge solids from rotating biological reactors were found to have a specific surface area of 2 m^2/g (16) while the specific surface area of kaolinite ranged from 10 to 20 m^2/g (17). Huang et al. (18) have recently determined the specific surface area of fungi to be 35, 34 and 44 m^2/g, respectively for Aspergillus oryzae, Fusarium solani and Trichoderma sp.

4.2 Amphoteric Points

The results of the alkalimetric and acidimetric titration of the sludge solids is shown in Figure 1. The pH_{zpc}, as defined by other investigators (11, 19), is determined to be 5.8. However, electrophoresis measurements of zeta potential of the same treated sludge solids suggest an isoelectric point (IEP) of approximately 1.6 (20). The difference in determining surface charge of egg albumin over a wide range of pH by electrophoresis and titration methods has been reported by Overbeek (21), and it was found that the electrophoresis charge is only 60% of that determined from titration. This would tend to lower the value of IEP. Hiemenz (22) noted that the discrepancy between the two methods resulted from the fact that electrophoresis measures the net charge inside the surface of the shear plan whereas titration binds the H^+ or OH^- onto the surface of the molecule. It has also been speculated that impurities due to the specific adsorption of cations on the surface of sludge solids may also result in this discrepancy. Breeuwsma and Lyklema (23) observed that Ca^{2+} would make the hematite surface more positive. Keng and Uehara (24) further concluded that a more positive hematite surface will shift the pH_{zpc} to a more alkaline region. Pyman et al. [25] reported that the pH value at the point of intersection of two or more proton titration curves using different ionic strengths depends strongly on the concentration of any specifically adsorbing ions present in the swamping electrolyte solution. Parker et al. [26] have called this point of intersection the point of zero salt effect (PZSE). If significant permanent surface charge is present, an ion exchange will occur between protons added by titration and metal cations adsorbed on the permanent charge sites. This phenomenon will affect the measurement of pH_{zpc}. Benjamin and Leckie [27] attributed the difference between the IEP (2.0) and the pH_{zpc} (7.0) of α-FOOH to the inclusion of oxyanions of sulfur while the solids were being prepared by the oxidation of iron sulfide.

Measurements of background concentrations of heavy metals released into distilled water from the sludge solids used in this study after 12 hours of shaking time are: 0.124 mg/l Hg, 0.08 mg/l Ca, 0 mg/l Ni, 0.033 mg/l Cd, 0.166 mg/l Zn, 0.187 mg/l Cu and 1.75 mg/l Pb. The background concentrations were residual metal concentrations remained on the solid surface after sample preparation with distilled water, but were released after 12-hr shaking. The presence of trace metals on the solid surface may be partially responsible for the difference between pH_{zpc} and IEP.

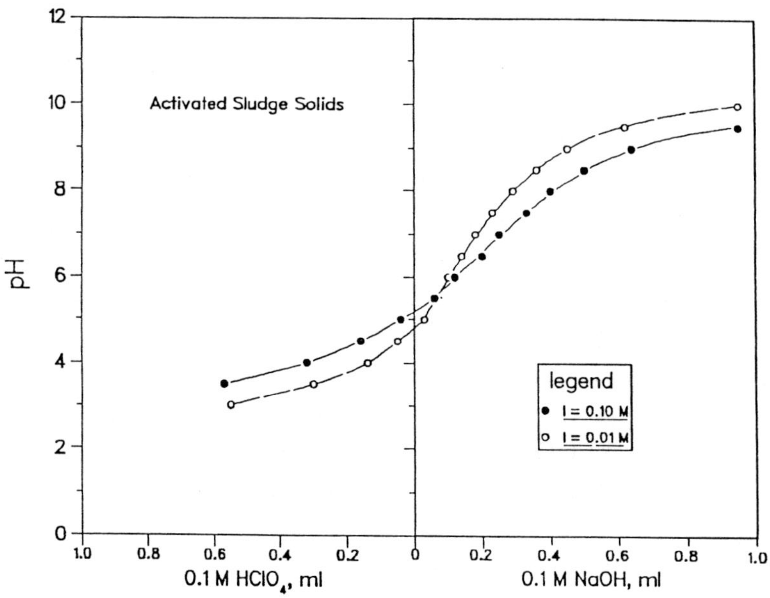

Fig. 1 Alkalimetric and Acidimetric Titration Curves of Sludge Solids
Experimental conditions: solid 1.35 g/L; $NaClO_4$ as ionic medium.

4.3 Surface Acidity

The intrinsic constants of sludge solids were determined by graphing the relationship developed in equations [10] and [10a]. The intrinsic acidity constants, pK_{a1}^{int} and pK_{a2}^{int}, as determined from the graphs shown in Figure 2 are 5.3 and 6.32, respectively for I = 0.1 M. The values change to 5.78 and 6.0 for I = 0.01 M.

The intrinsic constants and surface properties for various solids are included in Table 1 for comparison. The free energy of deprotonation includes the free energy of dissociation of H^+ from the surface sites, as measured by K_{a1}^{int} and K_{a2}^{int}, and the free energy of proton removal from the sites of dissociation into the bulk solution. The free energy (ΔG) of deprotonation is related to the intrinsic constant (K_{int}') by the following equations:

$$-\Delta G = RT \ln K' \qquad [17]$$

where $K' = K_{int}' \exp(F\phi_\circ/RT)$

or

$$-\Delta G = RT \ln(K_{int}' \exp(F\phi_\circ/RT)) \qquad [18]$$

$$= RT [\ln K_{int}' + F\phi_\circ/RT] \qquad [18a]$$

$$= F\phi_\circ + RT \ln K_{int}' \qquad [18b]$$

From the above equations, it can be concluded that the free energy of deprotonation ΔG is proportional to the pK_{int} value. With respect to all of the intrinsic constants shown in Table 1, the surface acidity of solids or the easiness of proton dissociation from the surface sites have the following sequence: Activated carbon (Nuchar SA) > TiO_2 ≈ Activated sludge solids > γ FeOOH.

TABLE 1
Comparison of Surface Properties of Sludge Particles with Various Hydrous Solids

	$X_T{}^a$	pK^1_{int}	pK^2_{int}	S^b	pH_{zpc}	I	Reference
TiO_2 (CLD 640)	12	3.0	8.6	15	5.8	KNO_3 (10^{-3} M)	(9)
Activated Carbon (Nuchar SA)	--	2.27	5.76	1351	4.0	$NaClO_4$ ($5x10^{-2}$ M)	(18)
γ-FeOOH	10	--	--	22	7.0	--	(27)
α-FeOOH	10.98	6.99	8.4	71	--	KCl (0.1 M)	(34)
γ-Al_2O_3	6.43	7.7	9.3	117	--	NaCl (0.1 M)	(7)
Activated Sludge	12.9	5.5	6.1	65-104	5.8	$NaClO_4$ (0.05 M)	This work

a in sites/nm^2; b in m^2/g

4.4 Stoichiometric Relationship

The stoichiometric relationship described in equations [12] through [12b] between the amount of metal adsorbed and the number of protons released into the solution was verified. The results displayed in Figure 3 indicate that for each mole of Cu adsorbed, an average of 1.8 mole of H^+ is released. Therefore, a 2:1 stoichiometric relationship between hydrogen and copper ions is used in equations [14], [14a] and [14b]. The stoichiometric relationship determined in this research is in agreement with the results found by other investigators.

Perona and Leckie (28) determined the proton stoichiometry for the adsorption of Cd^{2+}, Cu^{2+}, Pb^{2+} and Zn^{2+} by amorphous hydrous ferric oxide and reported a 2:1 (proton to metal) relationship for the pH range of 6 to 7. Corapcioglu and Huang (29) found that for each mole of metal ions adsorbed on activated carbon in the low pH region (pH<7), two moles of H^+ were released. From the utilization of the Kurbatov method (30), Park and Huang (31) determined a 1:1 ratio for the adsorption of heavy metals onto the hydrous CdS(s) surface.

4.5 Stability Constants

The stability constants of the reaction between sludge solids and the seven heavy metals used in this research were calculated according to the complex formation model. The speciation of the sludge

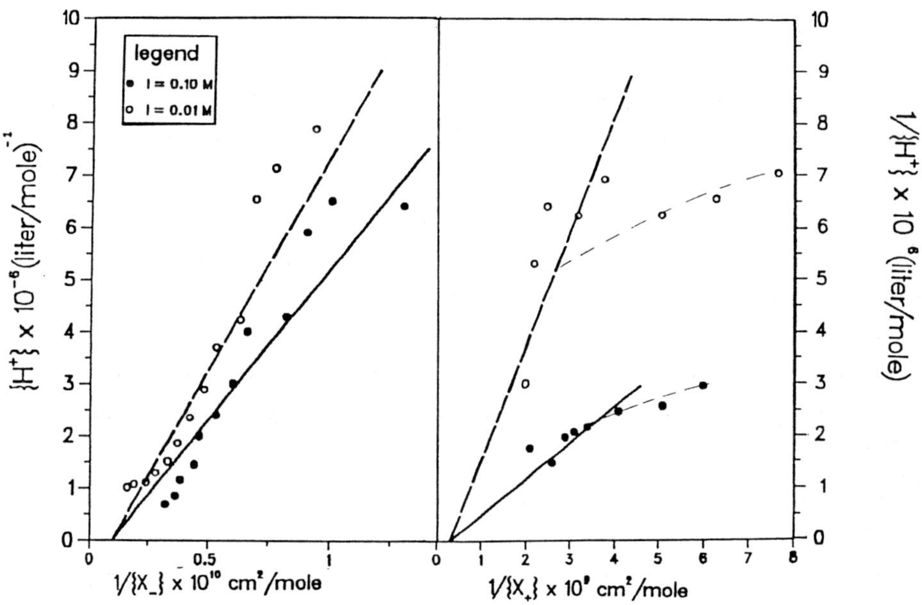

Fig. 2 Determination of Surface Acidity Constants.
Righthand figure is plot of Equation [10]; lefthand figure is plot of Equation [10a].

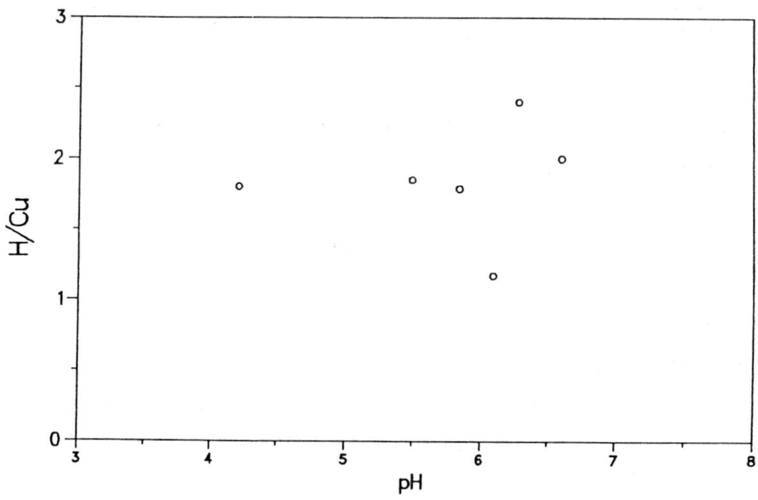

Fig. 3 Stoichiometric Relationship Between H^+ released and Cu^{2+} adsorbed as a function of pH. Experimental conditions: $Cu(ClO_4)_2$= 10^{-4} M; $NaClO_4$ = 5 x 10^{-2} M; sludge solid= 1 g/L.

solid surface is presented in Figure 4 whereas the speciation of the various surface complexes can be found in Figure 5. A listing of stability constants and adsorption energy between the sludge solids and the seven heavy metals are shown in Table 2.

The energy of specific chemical adsorption calculated from the stability constants ($-\Delta G_{ads}$ or ΔG_{chem} = -1.363 logK (Kcal/M)) is significantly higher (approximately two to four times in magnitude) when compared to the electrostatic energy (4.6 Kcal/M) of adsorption for a divalent metal ion onto a surface of counter charge against a potential drop of 100 mv. Therefore, the assumption that specific chemical adsorption outweighs electrostatic adsorption in the surface complex formation model is readily justified. The stability constants of Cu^{2+}, Zn^{2+} and the sludge solids determined from this study are one to two order of magnitude higher than those obtained by Sarzanini (16). One reason for this difference may be due to the fact that the Sarzanini used the Langmuir isotherm with a 1:1

TABLE 2
Stability Constants of Surface Metal Complexes

		Cu	Pb	Zn	Ni	Co	Cd	Hg
M^{2+}	[a]$LogK''_1$	7.28	8.27	6.96	6.36	6.44	7.28	14.50
	[b]$-\Delta G_{ads}$	9.92	11.27	9.48	8.67	8.77	9.92	19.76
MOH^+	$LogK''_2$	6.91	8.62	9.49	6.37	6.82	6.91	11.9
	$-\Delta G_{ads}$	9.42	11.75	12.93	8.68	9.30	9.42	16.2
$M(OH)_2$	$LogK''_3$	6.65	7.17	9.40	6.27	6.63	6.65	6.09
	$-\Delta G_{ads}$	9.06	9.77	12.81	8.55	9.04	9.06	8.3
	[c]$LogK_1$	5.26	---	5.15	---	---	---	---

[a] $M^{2+} + 2SO^- = M(SO)_2$; K''_1
$MOH^+ + 2SO^- = ((SO)_2 - MOH)$; K''_2
$M(OH)_2 + 2SO^- = ((SO)_2 - M(OH)_2)^{2-}$; K''_3

[b] $-\Delta G_{ads}$ (K_{cal}/mole) = 2.303 RT $\log K''_i$ = 1.363 $\log K''_i$ at 25°C
Ionic strength: 0.05 M $NaClO_4$; surface loading: 0.8 μ mole/m^2

[c] Reference (16).

stoichiometry to evaluate the metal adsorption, e.g. Cu(II), Zn(II) and Cu(II) onto sludge solid surfaces grown from a rotating biological reactor. Another reason contributing to the difference could be the lower specific surface area (0.5 to 2 m^2/g) measured in the Sarzanini study.

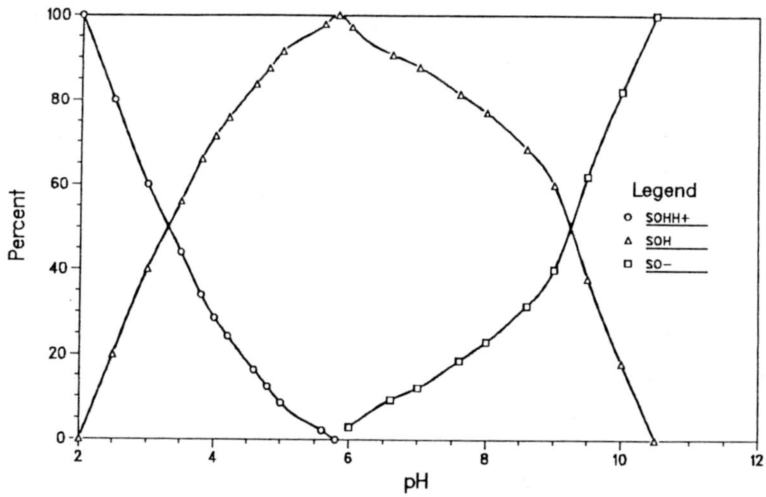

Fig. 4 Speciation of Sludge Solids Surface. For ionic strength of 5 x 10^{-2} M.

The data listed in Table 2 suggest that the magnitude of adsorption energy ($-\Delta G_{ads}$) of the seven heavy metals (M^{2+}) used in this study results in the following order: Hg(II) > Pb(II) > Cd(II) > Zn(II) > Co(II) ≈ Ni(II) ≈ Cu(II).

This sequence of adsorption energy of metals onto sludge solid surfaces predicted from the surface complex formation model is generally in agreement with the sequence of metal affinities for activated sludge found by other investigators. Brown and Lester (32) indicated that the relative affinities of four heavy metals for binding sites on extracellular polymer surfaces were in the order of Cd(II) > Co(II) > Ni(II) > Mn(II). The affinity of metals on activated sludge was reported by Nielsen and Hrudely (33) to be in the order of Pb(II) > Cu(II) > Cd(II) > Zn(II) > Ni(II).

4.6 Effect of Metal Type

The adsorption of seven heavy metals onto sludge solid surfaces for pH ranging from 3 to 10 are summarized in Figures 6 and 7. The data shown in Figure 6 were collected from a system that involved the adsorption of seven mixed metals. The data shown in Figure 7 represent the results of the adsorption of individual metal species. The reader may notice a difference in the adsorption behavior of Cu(II) in Figures 6 and 7, This difference may be attributed to the fact that soluble organics from the deterioration of the sludge solids has a higher affinity to the copper ions (19). The pH adsorption edge of Figure 7 generally results in the following order: Hg(II) > Pb(II) > Cu(II) > Co(II) > Cd(II) > Zn(II) > Ni(II). Slight changes in the pH adsorption edge were found in Figure

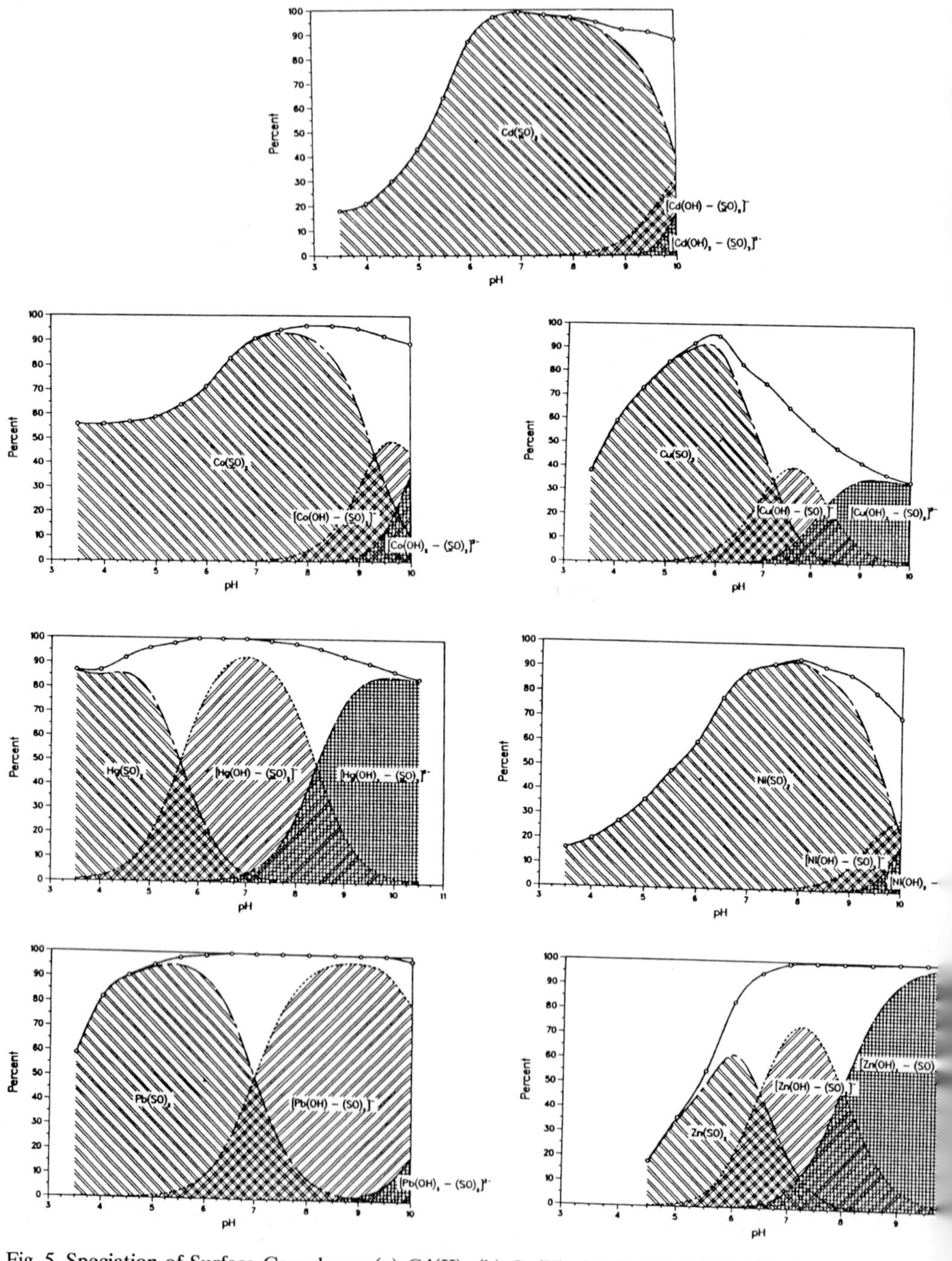

Fig. 5 Speciation of Surface Complexes: (a) Cd(II), (b) Co(II), (c) Cu(II), (d) Hg(II), (e) Ni(II), (f) Pb(II) and (g) Zn(II). Conditions: $Me(ClO_4)_2 = 10^{-4}$ M; $NaClO_4 = 5 \times 10^{-2}$ M; sludge solid= 1 g/L.

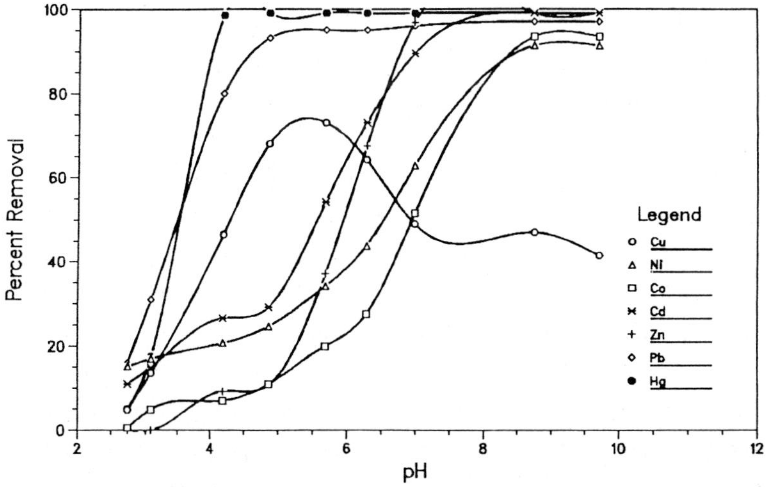

Fig. 6 Adsorption of Heavy Metals in a Mixture of Seven Metals. Experimental conditions: solid 1.19 g/L; metal (as ClO_4 salt) 10^{-4} M; ionic strength 5 x 10^{-2} M($NaClO_4$).

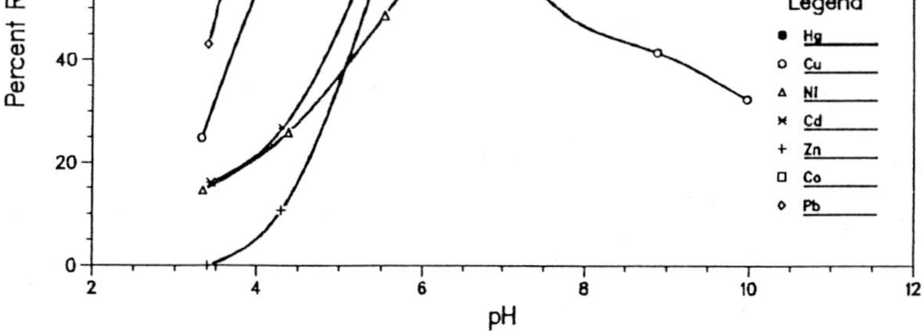

Fig. 7 Adsorption of Heavy Metals in a Single Metal System. Experimental conditions: solid 1.19 g/L; metal (as ClO_4 salt) 10^{-4} M; ionic strength 5 x 10^{-2} M ($NaClO_4$).

6; the trend is Hg(II) > Pb(II) > Cu(II) > Cd(II) > Zn(II) ~ Ni(II) > Co(II). Except for Co(II), the extent of metal adsorption onto sludge solid surfaces appears not to be affected by the presence of other metals in the solution.

5. CONCLUSIONS

The surface acidity constants of sludge solids can be determined by the alkalimetric titration method. The acidity constants determined in this study are $pK_{a1}^{int} = 5.5$, $pK_{a2}^{int} = 6.1$ and $pH_{zpc} = 5.8$. Also in this research, the specific surface area of sludge solids was measured by the BET method. The CO_2 critical point drying process was used for dehydrating the sludge sample, and the specific surface area was determined to be 84 m^2/g. The isoelectric point (IEP = 1.6) of sludge solids determined by electrophoresis measurements is lower than the pH_{zpc} (5.8) determined by the alkalimetric titration method. The discrepancy between IEP and pH_{zpc} could be attributed to the following: (1) electrophoresis measures the net charge inside the surface of shear plan whereas titration binds the H^+ or OH^- onto the surface; and/or (2) impurities due to specifically adsorbed cations occurs.

The surface complexation model which considers all metal hydrolysis species as adsorbates in an adsorption process provides a simple method to predict the distribution of the metal species adsorbed at various pH. This model was used to determine the stability constants of seven heavy metals adsorbed onto sludge solids. The adsorption energy ($-\Delta G_{ads}$) calculated from the stability constants suggest the following trend: Hg(II) > Pb(II) > Cd(II) > Zn(II) > Co(II) ~ Ni(II) ~ Cu(II). The sequence of the adsorption energy of metals onto sludge solid surfaces predicted from the surface complex formation model is generally in agreement with the sequence of metal affinities for activated sludge found by other investigators.

The competitive effects of seven heavy metals on the adsorption sites were also investigated. The results indicate that the extent of metal adsorption onto sludge solid surfaces appears not to be affected by the presence of other metals in the solution.

ACKNOWLEDGMENTS:

This work was supported in part by an EPA U-911732 Drinking Water Fellowship and National Science Foundation through Grants No. CEE8104728 and CEE83-13290. Support from the Department of the Environment of the State of Maryland is also greatly appreciated. Any opinions, findings, conclusions or recommendations expressed in this publication are those of the authors only and do not necessarily reflect the views of the funding agencies.

REFERENCES

1 Brown, H. G., et al. Environ. Lett., 5, (1973) 103-114.
2 Sterritt, R. M., and Lester, J. N. Proceedings, International Conference on Heavy Metals in the Environment, Amsyterdam, NL, 491, 1983.
3 Stanier, R. Y., et al. The Microbial World, Prentice Hall, Inc. New York, 1970.

4 Tenney, M. W., and Verhoff, F. H., Biotech. and Bio-engineering, (1973), 1045-1073.
5 Brown, M. J., and Lester, J. N., Water Research, 13, (1979) 817-837.
6 Broadbent, F. E., and Bradford, G. R., Soil Science, 74, (1952) 447-457.
7 Huang, C. P., and Stumm, W. J. Colloid & Interface Sci., 43(2), (1973) 409-420.
8 Davis, J. A., James, R. O., and Leckie, J. O., J. Colloid & Interface Sci., 63(3), (1978) 480-499.
9 James, R. O., and Healy, T. W., J. Colloid & Interface Sci., 40, (1972) 42-81.
10 Tanford, C. Advance in Protein Chemistry, 17, Academic Press, New York, 1972.
11 Stumm, W., and Morgan, J. J., Aquatic Chemistry, Wiley Interscience, New York, 1980.
12 Weber, W. J., et al., Environ. Science and Technology, 12(7), (1976) 817.
13 Brunauer, S., Emmett, P. H., and Teller, E., J. Am. Chem. Soc., 60, (1938) 309-319.
14 Lamanna, C., and Mallette, M. F., Basic Bacteriology, The Williams and Wilkins Company, Baltimore, 1959.
15 Mortland, M. M., and Kemper, W. D., Methods of Soil Analysis, Amer. Soc. Agron. Monog., 9, Part 1, (1969) 532-544.
16 Sarzanini, C., Separation Science and Technology, 18(1), (1983) 1-4.
17 Bohn, H. L., McNeal, B. L., and Oconnor, G. A., Soil Chemistry, John Wiley and Sons, Inc., New York, 1979.
18 Huang, C. P. et al., Particulate Science and Technology, 6, (1988), 405-419.
19 Parks, G. A., Advances in Chemistry Series, 67, (1967) 121-160.
20 Tien, C. T., and Huang, C. P., J. of Environ. Eng., ASCE, 113(2), (1987) 285-299.
21 Overbeek, J. T. G., Advances in Colloid Science, 3, (1950) 123-135.
22 Hiemenz, P. C. Principles of Colloid and surface Chemistry, Marcel Dekker Inc. New York, 1977.
23 Breeuwsma, A and Lyklema, J, Discuss, Faraday Society, 52, (1971) 324-333.
24 Keng, J., and Uehara, G., Soil Crop Sci. Soc. Fla. Proc. 33, (1974) 119-126.
25 Pyman, M. A. F., Bowden, J. W., and Posner, A. M., Aust, J. Soil Res. 17, (1979) 191-195.
26 Parker, J. C., Zelazny, L. W., Sampath, S. and Harris, W. G., Soil Sci. Soc. Am. J. 43, (1979) 668-674.
27 Benjamin, M. M., and Leckie, J. O., Environ. Sci. Technol., 16(3), (1982) 162-170.
28 Perona, M. J., and Leckie, J. O., J. Colloid & Interf. Sci., 106(1), (1985) 64-69.
29 Corapcioglu, M. O., and Huang, C. P., Water Research, 21(6), (1987) 1013-1024.
30 Kurbatov, M. H., Wood, G. B., and Kurbatov, J. D., J. Phys. Chem., 55, (1951) 258-269.
31 Park, S. W., and Huang, C. P., J. Colloid & Interf. Sci., 117(2), (1987) 431-443.
32 Brown, M. J., and Lester, J. N., Water Research., 16, (1982) 1539-1560.
33 Nielsen, J. S., and Hrudey, S. E. Environmental Engineering Technical Report 80-2, Department of Civil Engineering, University of Alberta, Alberta, Canada, 1981.
34 Atkinson, R. J., A. M. Posner and J. P. Quirk, J. Phys. Chem., 71, (1976) 550-562.

KINETICS OF HEAVY METAL ADSORPTION ON SLUDGE PARTICULATE

Ching-Tzone Tien and C.P. Huang

SUMMARY

The rates of metal adsorption on sludge solids were investigated. The adsorption reaction follows a second order expression. The adsorption and desorption rate constants were determined for seven heavy metals, namely, Hg(II), Pb(II), Cd(II), Cu(II), Zn(II), Ni(II) and Co(II). The equilibrium adsorption constants calculated from the rate constants were compared with the stability constants calculated from batch equilibrium adsorption studies. The equilibrium constants calculated by these two methods are in the same order of magnitude. The sequence of the equilibrium constants follows the order: Hg(II) > Pb(II) > Cd(II) > Cu(II) > Zn(II) > Ni(II) > Co(II). The maximum adsorption densities were determined and compared with those reported by other investigators. The film diffusion coefficients (D_f) and the pore diffusion coefficients (D_p) were determined. It appears that surface reaction rather than mass transport processes determines the rate of adsorption reaction.

1. INTRODUCTION

Heavy metals are of concern to man and his environment due to their high toxicity potential. The United States Public Health Service adopted a drinking water standard which imposed upper limits on the concentration of heavy metals as early as 1962(1). Although data concerning the efficiencies and factors affecting the removal of heavy metals in the wastewater treatment plant are readily available in recent literature, information is rare concerning the rate of adsorption and desorption of heavy metals on the sludge solid surface and the determination of adsorption and desorption constants. The objective of this study was to investigate the rates of adsorption and desorption of seven heavy metals associated with sludge solids, and to determine the adsorption and desorption constants of metal ions onto the surface of sludge particles.

2. RATE OF METAL UPTAKE BY MICROORGANISMS

In much of the research in this area, it has been reported that two distinct steps are involved during the removal of heavy metals by microorganisms. Failla et al. (2) indicated that the starved cells of **Candida utilis** (yeasts) accumulated Zn(II) by two different processes. The first was a rapid energy-independent and temperature-independent system that probably represented binding to the cell surface. The second process was involved with intracellular transport which was affected by the energy source, pH and temperature. Failla (2) further pointed out that Zn(II) was rapidly bound during the initial five minutes after the addition of Zn(II), Co(II) and Ni(II) to the solution containing the yeast cells. Furarmann and Rothstein (3) revealed that in well-starved cells without substrate, small but significant amounts of Co(II) and Zn(II) were taken up mostly within the first five minutes of exposure; however, no significant uptake of Ni(II) was measured. The initial uptake was solely due to the binding of cations on the surface anionic sites. The uptake rate increased five-fold to eighteen-fold for cells pre-treated with glucose and phosphate. Nelson et al. (4) studied the kinetics of metal-ion uptake by activated sludge in a batch reactor over a twenty hour period. They found that adsorption occurred swiftly and nearly to completion within the first hour of contact with a minimal subsequent uptake. Neufeld and Hermann (5) observed that the initial uptake of Hg(II), Cd(II) and Zn(II) tends to level off after about three hours of contact between metal and biomass. It was further concluded that the initial metal removal process is not a biological phenomenon, but is related to the physical, chemical and probably the surface properties of the biological mass. The metal uptake process was found not to be affected by the metal concentration and the organism viability. Cheng et al. (6) reported that under aerobic conditions, metal uptake by the biomass was essentially completed after 10 minutes of contact time. This was followed by a long-term, slow-phase uptake. Nelson (7) indicated that a two-phase uptake scheme was involved during the removal of heavy metals by microorganisms. The rapid first phase was related specifically to the immediate surface characteristics of the microorganisms. The first phase reached a state of semi-equilibrium with active surface sites initially available. The second phase of uptake was largely controlled by the metabolic condition of the culture as manifested in the growth rate of the microorganisms.

In this research, a kinetic study of the metal uptake by sludge solids was focussed on the rapid extracellular adsorption rather than on the slower intracellular transport. A continuous-flow reactor was used to observe the metal adsorption behavior of sludge solids. The reactor was designed to provide a short contact time between the sludge solids and the heavy metals. Samples were collected at different time intervals for the purpose of determining the metal concentration change during the adsorption and desorption processes. Data collected was used to calculate the metal adsorption and desorption rates along with various kinetic constants.

3. ADSORPTION ISOTHERMS AND STABILITY CONSTANTS

The Langmuir isotherm, Freudlich isotherm and their modified versions were used by several investigators to describe the adsorption behavior of metals with sludge solids and to determine the stability constants. Gould and Genetelli (8) indicated that the modified Freundlich isotherm, which incorporates the effects of hydrogen ion concentration, was superior to the Langmuir and Temkin isotherms in describing the adsorption behavior of Zn(II), Ni(II), Cu(II) and Cd(II) onto anaerobically digested sludge over the range of pH from 5 to 8. However, the Langmuir isotherm was used to calculate the stability constants. Nelson et al. (4) used the surface complexation model to determine the surface complexation (adsorption) constant and concluded that the adsorption isotherms used in their study adequately represented the distribution of metals between bacterial solids and solution phases. Neufeld et al. (9) pointed out that the kinetic and equilibrium relationship between the seven heavy metals used in their study and the activated sludge acclimated to the synthetic nutrient was adequately described by the Langmuir isotherm. In the study of the removal of Cu(II) and Cd(II) by metabolically controlled activated sludge, Casey and Wu (10) determined the kinetic constant (K and n) from the Freundlich isotherm and calculated the ultimate adsorption capacity (Γ_m) from the Langmuir isotherm. Cheng et al. (6) determined the complex formation constant for Cu(II), Pb(II), Ni(II) and the organic matter of the supernatant as well as the sludge solids based on the cation-exchange technique.

In this research, adsorption and desorption experiments were conducted with a completely-mixed continuous flow reactor. Data collected from these experiments were analyzed to determine the adsorption and desorption rate constants.

4. MATERIALS AND METHODS

4.1 Theoretical Background

Tien and Huang (11) demonstrated that for each mole of Cu(II) adsorbed, an average of 2 moles of H^+ is released; therefore, a 2:1 stoichiometric relationship was used to derive the second order reaction equation. The formula for the adsorption and desorption of metals by sludge solids can be expressed as follows:

$$M + 2S \underset{k_d}{\overset{k_a}{\rightleftharpoons}} MS_2 \qquad [1]$$

where M, S, MS_2, k_a and k_d are free metal ions, free surface adsorption site, metal-sludge association (surface complex), adsorption rate constant, and desorption rate constant, respectively.
Accordingly, the adsorption rate can be expressed as follows:

$$d\Gamma/dt = k_a M (\Gamma_m-\Gamma)^2 \qquad [2]$$

where Γ is the adsorption density and Γ_m is the maximum adsorption density.
By rearranging equation [2] it becomes:

$$d\Gamma/(\Gamma_m-\Gamma)^2 = k_a M dt \qquad [3]$$

From mass balance relationship, it is known that $M = M_o - a\,\Gamma$, where M_o and **a**, are total metal concentration and conversion constant (from solution concentration to solid surface concentration), respectively. In a continuous flow reactor such as the one used in this study, the concentration of metal at any time was close to the original metal concentration, i.e. $M = M_o$. Integrating equation [3] with the initial conditions $t = 0$ and $\Gamma = 0$, it yields:

$$1/\Gamma = 1/(\Gamma_m^2 k_a M t) + 1/\Gamma_m \qquad [4]$$

By plotting $1/\Gamma$ versus $1/t$, $1/\Gamma_m$ can be determined from the intercept of the $1/\Gamma$ axis. The constant k_a can be determined from the slope.

Likewise, the rate of desorption of the heavy metals from sludge solid surfaces is proportional to the adsorption density which is the amount of metal already adsorbed per unit weight of sludge solid. The relationship can be expressed as follows:

$$-d\Gamma/dt = k_d \Gamma \quad [5]$$

By integrating equation [5] with the initial conditions $t = 0$ and $\Gamma = \Gamma_m$, the result is:

$$1n(\Gamma/\Gamma_m) = -k_d t \quad [6]$$

By plotting $1n(\Gamma/\Gamma_m)$ vs. t, k_d can be determined from the slope. Note that during the desorption process, Γ will gradually decrease due to the desorption of metal from the sludge solid surface. The amount of desorption (Γ_d) can be determined from the desorption experiment; therefore, $\Gamma = \Gamma_m - \Gamma_d$. Substituting $\Gamma = \Gamma_m - \Gamma_d$ into equation [6] and rearranging, Γ_d can be expressed as follows:

$$\Gamma_d = \Gamma_m (1 - \exp\{-k_d t\}) \quad [7]$$

This implies that $\Gamma_d = \Gamma_m$ when time approaches infinity.

With the adsorption rate constant, k_a, and the desorption rate constant, k_d, so determined, the equilibrium constant, K, can be calculated by the following relationship:

$$K = k_a/k_d \quad [8]$$

5. MATERIALS AND METHODS

5.1 Preparation of Sludge Solid Sample

All samples used in this research were taken from the Highlands extended aeration plant, Cecil County, Maryland. The following procedures were followed in sludge sample preparation: [1] Filter sample through a #40 sieve to remove grits and floating matter. [2]Transfer sample to a 4 liter container and aerate with a ball diffuser for 10 minutes. [3] Turn off the air supply to the diffuser and allow the suspended solids to settle under room temperature, ca. 21°C. [4] Decant the supernatant and refill the container with distilled water to the 4 liter mark (caution should be taken to minimize the loss of solids during the decanting process). [5] Reaerate the sample for 10 minutes and measure the conductivity of the mixed liquor. [6] Repeat steps [2] through [5] until conductivity of the sample approaches that of distilled water, and then measure the concentration of the suspended solids. [7] Adjust the ionic strength in the sludge sample to 5×10^{-2} M ($NaClO_4$). [8] Adjust the pH of the sample to pH ca. 6.

The particle size distribution of the prepared sludge particles was also determined. A Hiac Particle Counter was used to determine the distribution of different particle sizes. The average size was found to be 20 μm.

5.2 Preparation of Metal Solutions

The following procedures were used to prepare metal solutions: [1] Dissolve the metal perchlorate salt in distilled water to make a concentration of 5×10^{-4} M. [2] Adjust the ionic strength of the metal solution to 5×10^{-2} M ($NaClO_4$). [3] Adjust the pH of the metal solution to pH ca. 6 except for Hg(II) and Pb(II) which pH was adjusted to ca. 4.5 to prevent chemical precipitation.

5.3 Kinetic Adsorption Experiments

A complete mix continuous flow (CMCF) reactor was assembled for the adsorption experiments. A schematic diagram of the reactor used in this study is shown in Figure 1. This reactor is similar to the miscible displacement reactor used by Carski and Sparks (12). It consists of a 5 mL mixing chamber, a magnetic mixer and a filter at the outlet of the reactor. A 5 mL sludge solid sample at concentration of 1 g/L was injected into the mixing chamber by a plastic syringe. A variable speed pump (LKB Model 2132) was used to deliver the metal solutions to the mixing chamber at a rate of 1 mL per minute. A fraction collector (LKB Ultrorac II 2070) was used to collect the effluent sample from the reactor at 2 minute intervals. The metal concentration of the effluent sample was measured by an atomic adsorption spectrophotometer (Perkin Elmer Model Zeeman 5000). The above experiment was repeated for each of the seven heavy metals used in this study.

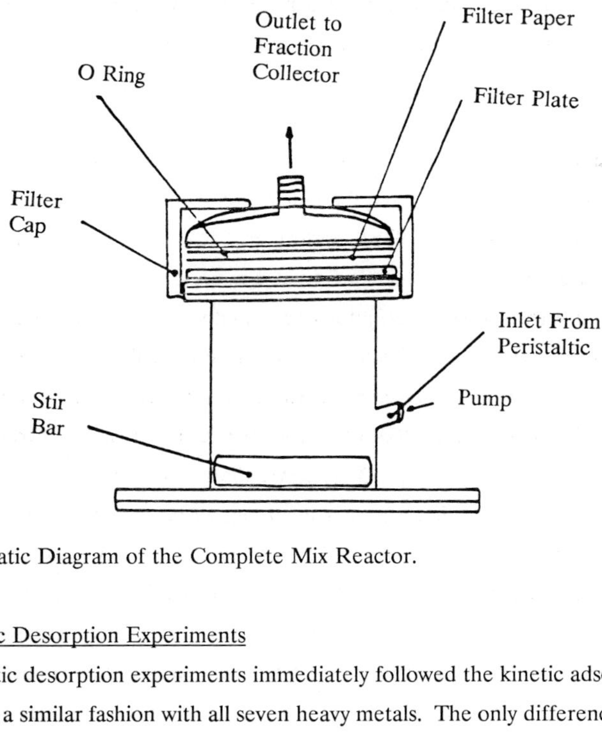

Fig. 1 Schematic Diagram of the Complete Mix Reactor.

5.4 Kinetic Desorption Experiments

The kinetic desorption experiments immediately followed the kinetic adsorption experiments and proceeded in a similar fashion with all seven heavy metals. The only difference involved replacing the metal solutions with constant ionic strength solution. The constant ionic strength solution was prepared as 5×10^{-2} M ($NaClO_4$) with pH adjusted with strong acid/base to that of the metal solutions used in the adsorption experiments.

5.5. Background Tests

In order to correct for the amount of metal adsorbed on the filter and on the reactor wall during the adsorption and desorption experiments and to eliminate the effect of dilution in the reactor, background runs were performed. The procedures for conducting the adsorption and desorption background tests for all seven heavy metals were similar to the experimental procedures described above except for the following difference. Distilled water of the same ionic strength and pH of the sludge solid sample was used to replace the sludge sample in the reactor.

6. RESULTS AND DISCUSSION

Figure 2 shows typical results of the adsorption experiments exemplified by Cu(II). The concentration of adsorbed copper at various time intervals was determined by the difference between the curve representing the copper concentration in the effluent of the continuous-flow reactor without sludge solids (background) and the lower curve representing the same conditions with the sludge solids. These concentration differences at various time intervals were then re-plotted (Figure 3). The area under the curve in Figure 3 represents the amount of copper adsorbed on the sludge solid surface at various time intervals. The amount of copper adsorbed at various time intervals divided by the weight of the sludge solids represents the adsorption density (Γ) as shown in Figure 4. As time increases, the rate of adsorption density decreases as represented by the decreasing slope of the curve in Figure 4. This suggests that the surface sites are becoming saturated with copper ions and that the adsorption density reached its maximum level at ca. 32 minutes. The data shown in Figure 5 was used to determine the adsorption rate constant (k_a) and the maximum adsorption density (Γ_m) according to the relationship established in equation [4]. The Γ_m determined above is 35.7 mg/g which is somewhat higher than 31 mg/g observed from Figure 4 at time greater than 32 minutes.

Figure 6 shows typical results of the desorption experiments also exemplified by Cu(II). The concentration of desorbed copper at various time intervals is similarly determined by the difference between the upper curve of Figure 6, representing the copper concentration in the effluent of the continuous-flow reactor containing the sludge solids, and the lower curve, representing the same conditions without sludge solids (background). These concentration differences were then re-plotted (Figure 7). The area under the curve represents the amount of copper desorbed from the sludge solid surface at various time intervals as shown in Figure 8. This curve indicates that the amount of desorption (Γ_d) approaches a steady value of 25 mg/g at ca. 17 minutes. However, the adsorption density (Γ) shown in Figure 4 approaches a steady value of 31 mg/g at ca. 32 minutes. This suggests that all of the adsorbed copper will not completely desorb from the surface sites. Those undesorbable copper ions may be permanently adsorbed or perhaps transported intracellularly. Sterritt and Lester (13) found that the amount of Cu(II) uptake intracellularly is 12% of its total removal. By examining the adsorption and desorption data shown in Figures 4 and 8, respectively, it may then be concluded that the adsorption of Cu(II) by sludge solids is an irreversible process. The desorption rate constant (k_d) is determined according to the relationship established in equation [6] and shown in Figure 9. The procedures for analyzing the adsorption/desorption data for the remaining six heavy metals were similarly accomplished according to the process described above. All of the results can be located in Table 1.

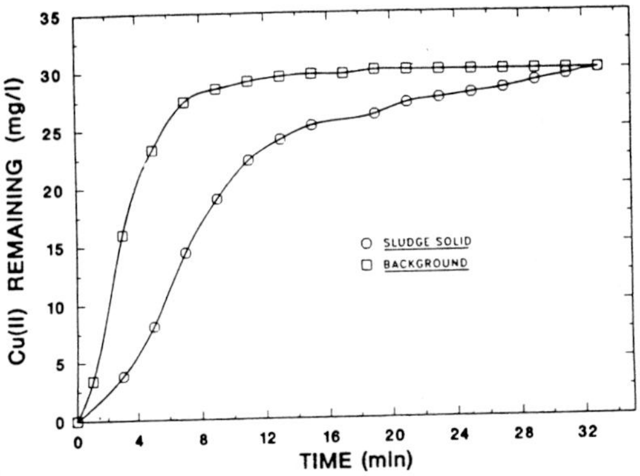

Fig. 2 Concentration of Cu(II) Remaining in the Reactor Effluent During an Adsorption Experiment. Experimental Conditions: Solids = 1.21 g/L, I = 5 x 10^{-2} M ($NaClO_4$), pH = 6.1, $Cu(ClO_4)_2$ = 5 x 10^{-4} M (31.7 mg/L).

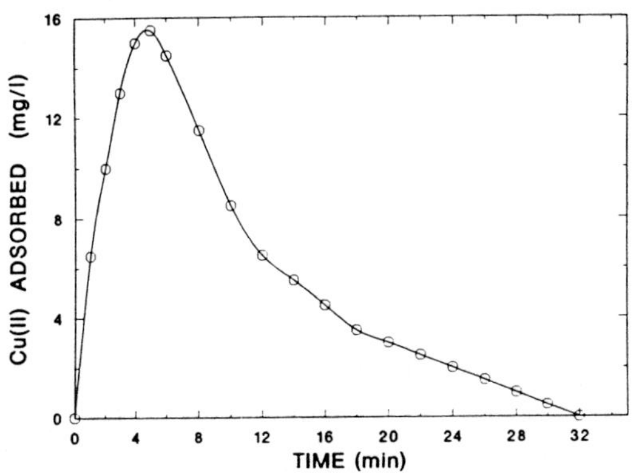

Fig. 3 Concentration of Cu(II) Adsorbed by Sludge Solids at Various Time Intervals.

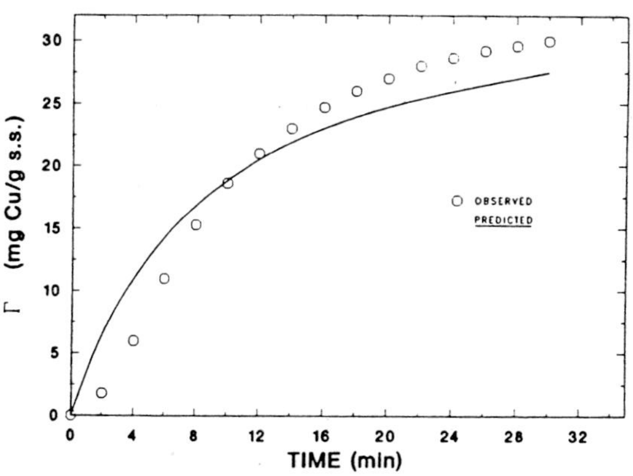

Fig. 4 Comparison Between Observed and Predicted Adsorption Densities (Γ).

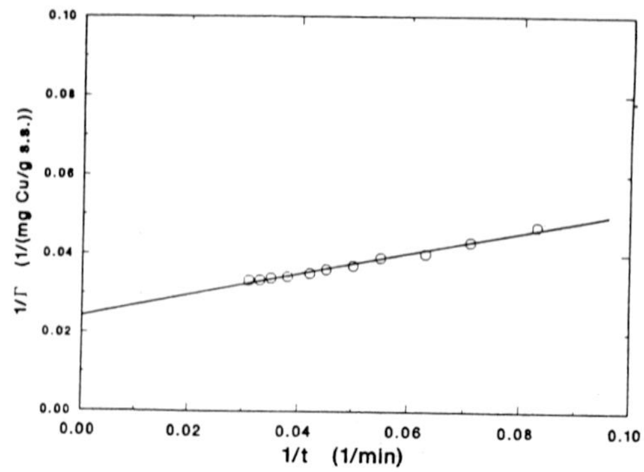

Fig. 5 $1/\Gamma$ Versus 1/t Curve for Determining k_a and Γ_m.

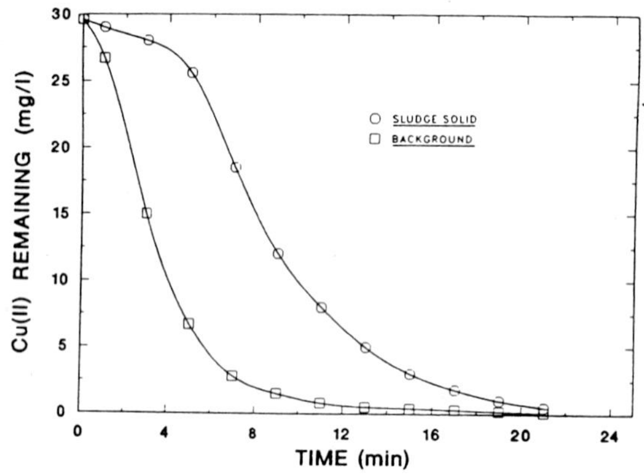

Fig. 6 Concentration of Cu(II) Remaining in the Reactor Effluent During a Desorption Experiment. Experimental Conditions: Solids = 1.21 g/L, I = 5 x 10^{-2} M ($NaClO_4$), pH = 6.1, $Cu(ClO_4)_2$ = 5 x 10^{-4} M (31.7 mg/L).

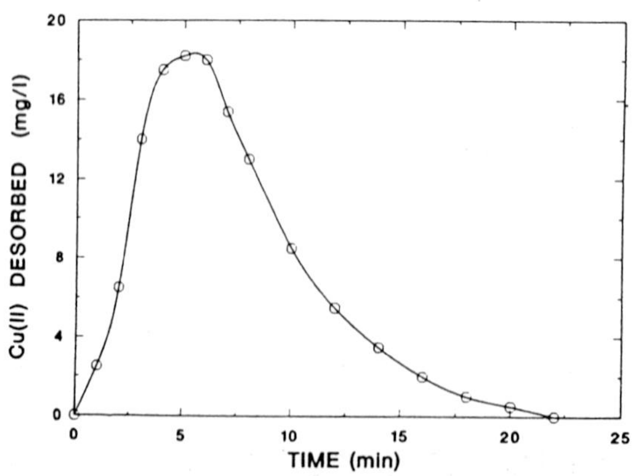

Fig. 7 Concentration of Cu(II) Desorbed by Sludge Solids at Various Time Intervals.

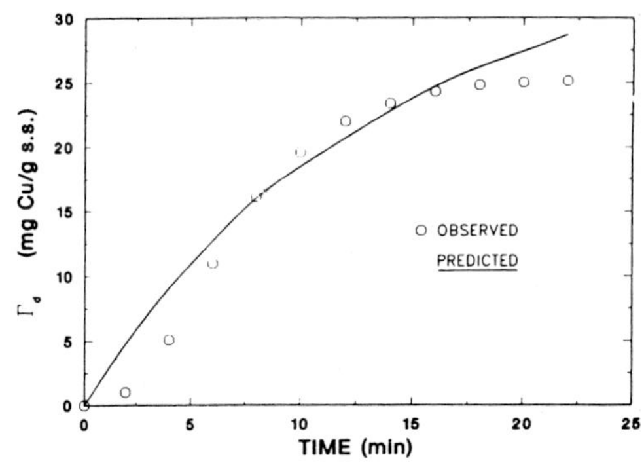

Fig. 8 Comparison Between Observed and Predicted Amounts of Desorption (Γ_d).

TABLE 1
Summary of Adsorption/Desorption Constants

Metal	k_a[a]	k_d[b]	Γ_m[c]	log K[d]	log K_1[e]	R(Å)[f]	D_f[g]	D_p[h]
Hg	$1.7x10^7$	$5.7x10^{-2}$	131.0	8.5	14.5	1.49	4.0	5.6
Pb	$9.4x10^5$	$1.0x10^{-2}$	90.9	8.0	8.3	1.47	4.1	5.9
Cd	$5.6x10^5$	$7.6x10^{-2}$	44.9	6.9	7.3	1.48	2.3	6.7
Cu	$4.8x10^5$	$7.4x10^{-2}$	35.7	6.8	6.3	1.17	1.5	5.6
Zn	$2.0x10^5$	$4.0x10^{-2}$	20.0	6.7	7.0	1.25	0.4	3.5
Ni	$1.7x10^5$	$5.0x10^{-2}$	40.8	6.5	6.4	1.15	1.0	4.4
Co	$1.85x10^4$	$5.5x10^{-3}$	9.1	6.5	6.4	1.16	0.1	1.2

[a] k_a in $(M)^{-2}(min)^{-1}$; [b] k_d in $(min)^{-1}$; [c] Γ_m in (mg/g); [d] $K = k_a/k_d$ (Determined from this study.)
[e] Determined from the surface complex formation model (11); [f] R is covalent radius (Å) (28);
[g] D_f in 10^{-11} (cm^2/sec.); [h] D_p in 10^{-11} (cm^2/sec).

6.1 Rate Limiting Step

Crank (14) indicated that when diffusion takes place in a semi-infinite medium, the amount adsorbed and desorbed is directly proportional to the square root of time ($t^{1/2}$) during the earlier stage of the adsorption and desorption process. McKay et al. (15) suggested that intraparticle diffusion is characterized by a dependence on the ($t^{1/2}$), and this was assessed from plots of adsorption density versus $t^{1/2}$.

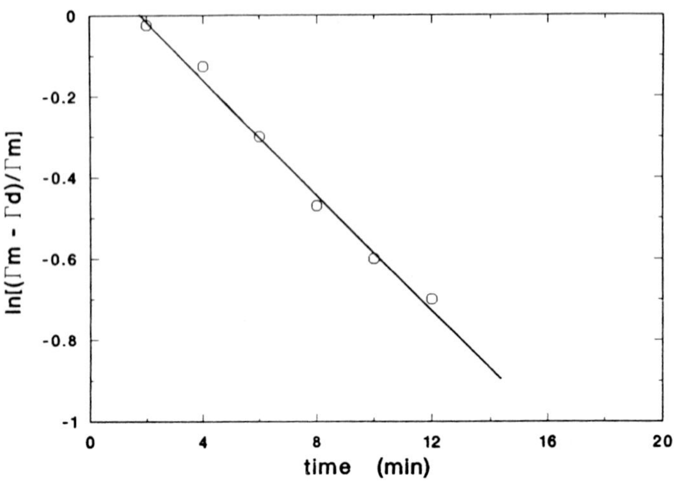

Fig. 9 Determination of Desorption Rate Constant (k_d) versus time (t).

They indicated that if intraparticle diffusion is a rate-controlling process, these plots are linear, and the slopes of the plots are defined as intraparticle diffusion rate parameters. Helfferich (16) used the following half time equations to determine the film diffusion coefficients (D_f) and the pore diffusion coefficients (D_p).

$t_{½} = 0.03\ r^2/D_p$ (pore diffusion) [9]

$t_{½} = 0.23\ (rl/D_f)(C_A/C_B)$ (film diffusion) [10]

where $t_{½}$, r, l, C_A/C_B are time needed to achieve half of the maximum adsorption density, radius of the adsorbent, thickness of the diffusion film, and loading of the adsorbent at equilibrium, respectively. Michelson et al. (17) reported that for film diffusion to be a rate limiting step, the value of D_f should be in the range of 10^{-6} - 10^{-8} cm^2/sec and for pore diffusion to be rate limiting, D_p should be in the range of 10^{-11} - 10^{-13} cm^2/sec. Bhattacharya and Venkobachar (18) in the study of Cd(II) adsorption onto coconut shell and bituminous coal with fixed beds found film diffusion to be the rate limiting step for both adsorbents.

Using an average size of 20 μm for sludge particulates and a film thickness of 10^{-3} cm (Michelsen et al. (17), the diffusion coefficients for the seven heavy metals studied were determined from the approach described above. The results are summarized in Table 1. According to the range of diffusion coefficients given by Michelson et al.(17), it appears that surface reaction rather than mass transport processes i.e. film and pore diffusion is the rate limiting step for the adsorption of the metals.

In the adsorption experiment of this research, the adsorption of heavy metal was undertaken in a complete mix reactor. Metal ions were distributed evenly throughout the reactor as soon as they entered the reactor. Therefore, the metal concentration gradient in the reactor was minimal, if any. This could be the reason for the lower D_f values shown in Table 1 when compared with those of Michelson et al. (17). Weber (19) indicated that when a reactor provides a high degree of mixing, film diffusion is often not rate limiting.

Specific surface area of sludge solids used in this research was measured to be 84 m^2/g using the BET method (11) which is much less than that of porous activated carbon (1,351 m^2/g) (20). The major portion of the sludge solids consists of bacteria which may have cell membrane and cell wall. Although there exist hypothetical pores in the cell membrane for the purpose of transporting molecules into the cell (21), metals transported intracellularly are minimal compared with those adsorbed on the cell surface (13). The small specific surface area of sludge solids suggests that there are less torturous pores to which metals can diffuse.

6.2 Prediction of Adsorption Density

As previously stated, Tien and Huang (11) used a 2:1, i.e. 2 sites occupied by 1 Cu(II) ion, stoichiometric relationship to evaluate the adsorption behavior of heavy metals onto sludge solids. This ratio was also verified by measuring the amount of Cu(II) adsorbed and the number of protons released into the solution.

In this research, the second order reaction equation was used to describe the kinetic behavior of metal adsorption onto sludge solids. Equation [4] was used to predict the adsorption density. The results were compared with the observed adsorption density and shown in Figure 4. The second order reaction equation provided a fairly good prediction for the adsorption density of Cu(II) and for the remaining heavy metals used in this research. This further validates the 2:1 stoichiometric relationship of the adsorption of heavy metals onto sludge solids. It must be mentioned that a 1:1 stoichiometric relationship was also tested, but the results of adsorption density calculated with this relationship did not fit the experimental data well.

6.3 Kinetic Constants

The kinetic constants determined from the experiments described in this research are helpful in understanding the adsorption and desorption behavior of heavy metals with sludge solids. In integrating equation [3], the temporal metal concentration (M) was approximately constant at M_o. This is readily justified by the fact that fresh metal solution was constantly pumped into the reactor. The adsorption rate constants (k_a) calculated from equation [4] for seven heavy metals are shown in Table 1 and follow the order: Hg(II) > Pb(II) > Cd(II) > Cu(II) > Zn(II) > Co(II) > Ni(II). The preference of sludge solids for the adsorption of one metal ion over another appears to be affected by the covalent radius. Bohn et al. (1979) indicated that ions of smaller dehydrated radius have a greater density of charge per unit volume, hence attracting waters (hydration) more strongly. As a result the radius of such a hydrated cation is larger than that of a cation with larger ionic (unhydrated) radius. An ion with a smaller hydration radius (or larger ionic radius) is more attracted to the sludge solids than is an ion of larger hydration radius. Bonding between heavy metals and functional groups on the surface of microorganisms are believed to be brought by covalent bonding. In studying the nature of chemical bonding between metallic ions and algae cell walls, Crist et al. (23) suggested that the covalent bonding of amino and carboxyl groups is responsible for the adsorption of Cu(II) ions by the **Vaucheria sp.** cell wall.

The desorption rate constants determined from equation [6] are also included in Table 1 and are in the following order: Cd(II) > Cu(II) > Hg(II) > Ni(II) > Zn(II) > Pb(II) > Co(II). Metals with higher adsorption rate constants, such as Hg(II) and Pb(II), do not necessarily have higher desorption rate constants. Higgins and Burns (24) indicated that both mercury and lead are highly toxic elements with no known beneficial biological function. Mercury has been shown to be amenable for biological methylation. McEntire and Neufeld (25) reported that mercury is methylated by microorganisms as a means of detoxification. The methylated mercury is believed to be in the form of monomethyl or dimethyl mercury which may not be as easily desorbed as the Hg(II) ions. This however is not the situation of the present study.

The equilibrium constants (K) determined by the ratio of k_a to k_d are also listed in Table 1. A larger equilibrium constant implies a higher adsorption rate constant which suggests a greater extent of metal adsorption by sludge solids. Results from this kinetic study indicates that the equilibrium constants are in the order: Hg(II) > Pb(II) > Cd(II) > Cu(II) > Zn(II) > Ni(II) > Co(II). This sequence is in general agreement with the results of other investigators. From a kinetic study using a continuous-flow reactor, different from the present reactor, Neufeld et al. (9) concluded that the overall affinity of biomass for heavy metals decreases in the order: Pb(II) > Cd(II) > Hg(II) > Cr(III) > Cr(VI) > Zn(II) > Ni(II). However, in another study by Neufeld and Herman (5), the initial uptake of metals by the biological floc two hours after metal dosing is in the following order: Hg(II) > Cd(II) > Zn(II). The difficulty in accurately measuring the concentration of Hg(II) could be a factor affecting the adsorption sequence of Hg(II) in both of Neufeld's studies. The Hg(II) could have evaporated due to its relatively high volatility. This could also be the reason for the difference in equilibrium adsorption constant, K, of Hg(II) determined in this present study and that determined in batch experiments by Tien and Huang (11).

The equilibrium constants, K, of all seven heavy metals, except for Hg(II), determined from this present study are in the same magnitude as the stability constants determined by the surface complex formation model(11). Both the equilibrium constants and the stability constants are referred to as reaction constants of heavy metals and sludge solids at equilibrium conditions. The reaction constants found in Table 1 can be accurately determined by either the kinetic approach as used in this study or by the surface complex formation model as reported by Tien and Huang (11).

6.4 Maximum Adsorption Density

The maximum adsorption densities, Γ_m, of seven heavy metals predicted from equation [4] are summarized in Table 1. Yamada et al. (26) determined that the equilibrium solid phase concentration of $HgCl_2$ was 96 mg/g for dry sludge solids which is lower than the results from this study (131 mg/g). Casey and Wu (10) reported that the maximum adsorption capacity based on the Langmuir isotherm ranged from 80-400 mg/g-VSS for copper and 72-1250 mg/g-VSS for cadmium. These values are dependent upon the COD/P ratio. The adsorption densities for copper and cadmium included in Table 1 of this study are 35.7 mg/g and 44.9 mg/g, respectively. These values are smaller than those reported by Casey and Wu. It must be mentioned that the adsorbent in their study was measured in terms of grams of suspended solids instead of grams of volatile suspended solids used by Casey and

Wu. Typically, the volatile solid in an activated sludge system ranges from 65-70% of the total suspended solids (27). Moreover, the removal of some metals as reported by Casey and Wu could be caused by chemical precipitation, as they used a high metal concentration in the feed. Nevertheless, Casey and Wu (10) attributed the high adsorption density of copper and cadmium in their study to the activated sludge solids grown under a phosphorus-limited condition. They further concluded that the capsulated materials produced under severe restrictions in phosphorus greatly improved the metal adsorption capacity of sludge organisms.

Cheng et al. (6) reported that the maximum adsorption density of nickel at initial concentration of 22.2 x 10^{-5} M by sludge solids acclimated in the synthetic substrate at concentration ca. 1.5 g/L and pH 6.5 is 2.2 mg/g. This is much lower than the Γ_m value for nickel shown in Table 1 of this research. The different surface characteristics of the sludge solids grown in the laboratory reactor may be the reason for the lower nickel adsorption in Cheng's study. Gould and Genetelli (8) studied the heavy metal complexation behavior in anaerobically digested sludge and found that the zinc adsorption density is ca. 18 mg/g under a zinc concentration of 5 x 10^{-4} M and pH 6.1. This is of the same order of magnitude as observed in this study (Table 1).

7. CONCLUSIONS

Among the seven heavy metals studied, mercury has the highest adsorption density (131 mg/g) whereas cobalt has the lowest adsorption density (9.1 mg/g). The equilibrium adsorption constants for each metal calculated from the kinetic adsorption and desorption experiments, except Hg(II), are in agreement with the results determined by batch adsorption experiments, i.e. from the surface complex formation model. The affinity of sludge solids for seven heavy metals based on the kinetic observations are in the following order: Hg(II) > Pb(II) > Cd(II) > Cu(II) > Zn(II) > Ni(II) > Co(II). This order also agrees with that of the previous batch adsorption experiments (29). The kinetic approach used in this study provides another alternative for verifying equilibrium adsorption constants determined from batch experiments.

The second order reaction equation was used to describe the kinetic behavior of metal adsorption onto sludge solids. The results clearly indicate that second order reaction can predict the adsorption density well. This provides further evidence on the premise that one adsorbed metal ion occupies two surface sites.

The adsorption of heavy metals by sludge solids in a complete mix reactor used in the kinetic study was found not to be controlled by mass transfer process, i.e. film or pore diffusion. Adsorption of heavy metals on the exterior surfaces of sludge solids appears to be the rate limiting step.

The desorption data collected from this research demonstrate that the majority of the metal adsorbed is desorbable. The undesorbable portion could be due to intracellular uptake or strong bonding to the surface sites.

ACKNOWLEDGMENTS:

This work was supported in part by an EPA U-911732 Drinking Water Fellowship and National Science Foundation through Grants No. CEE8104728 and CEE83-13290. Support from the Department of the Environment of the State of Maryland is also greatly appreciated. Any opinions, findings, conclusions or recommendations expressed do not reflect the views of the funding agencies.

REFERENCES

1 U.S. Public Health Service, Drinking Water Standards, Revised, U.S. Dept. of Health, Education and Welfare, Public Health Service, Washington, D.C., 1962
2 Failla, M. L., et al., J. Gen. Micro., 94, (1976) 23-36.
3 Fuhrmann, G. F., and Rothstein, A., Biochem. Biophys. Acta., 163, (1968) 325-330.
4 Nelson, P. O., et al., J. Water Pollut. Control Fed., 53, (1981) 1323-1333.
5 Neufeld, R. D., and Hermann, E. R., J. Water Pollut. Control Fed., 47, (1975) 310-329.
6 Cheng, M. H., et al., J. Water Pollu. Control Fed., vol. 47, No. 2, (1975) 362-376.
7 Nelson, P. O., The Fate and Distribution of Cadmium and Zinc in the Aerobic Aquatic Environment, Ph.D. thesis, Cornell Univ., Ithaca, N. Y., 1976.
8 Gould, Ms S., and Genetelli, E. J., Water Res., 12, (1978) 505-512.
9 Neufeld, R. D., et al., J. Water Pollut. Control Fed., 11, (1977) 489-497.
10 Casey, J. D., and Wu, Y. C., Proc. 32nd Ind. Waste Conf., Purdue Univ., 141-151, 1978.
11 Tien, C. T., and Huang, C. P., in: J. P. Vernet (Ed.), Heavy Metals in the Environment, Elsevier, Amsterdam, 1991.
12 Carski, T. H., and Sparks, D. L., Soil Sci. Sic. Am. J., 49, (1985) 1114-1116.
13 Sterritt, R. M., and Lester, J. N., Proceedings, International Conference on Heavy Metals in the Environment, Amsterdam, NL, 491, 1983.
14 Crank, J., The Mathematics of Diffusion, Clarendon Press, New York, 1976.
15 McKay, G., et al., Immobilization of Ions by Bio-sorption, John Wiley & Sons, 59-69, 1986.
16 Helfferich, F., Ion Exchange, McGraw Hill Book Company, Inc., N. Y., 1962.
17 Michelson, L. D., Gideon, P. G., Page, E. G., and Kutal, L. H., Office of Water Research & Technology, Bull., No. 74. 1975.
18 Bhattacharya, A. K., and Venkobachar, C., Jour. of Environ. Eng., ASCE, 110(1) (1984) 110-122. (1984),

19 Weber, W. J. Jr., Physicochemical Processes for Water Quality Control, Wiley-Interscience. 1972.
20 Huang, C. P., et al., Particulate Science and Technology, 6, (1988) 405-419.
21 Novikoff, A. B., and Holtzman, E., Cells and Organelle, Holt, Rinehart and Winston, Inc., 1976.
22 Bohn, H. L., et al., Soil Chemistry, John Wiley & Sons, Inc., 1979.
23 Crist, R. H., et al., Environmental Science & Technology, 15, (1981) 1212-1217.
24 Higgins, I. J., and Burns, R. G., The Chemistry and Microbiology of Pollution, Academic Press, 1975.
25 McEntire, F. E., and Neufeld, R. D., J. Water Pollut. Control, 74, (1975) 465-470.
26 Yamada, M., et al., J. Ferment. Technol., 47, (1969) 155-160.
27 ASCE (1977), Wastewater Treatment Plant Design, Manuals and Reports on Engineering Practice, No. 36, New York.
28 Sargent-Welch Scientific Company, Table of Periodic Properties of the Elements, Catalog No. S-18806, Skokie, Illinois 1979.
29 Tien C. T. and Huang, C. P., J. environmental Engineering, ASCE, 113(2), (1987) 285-299.

REMOVAL OF HEAVY METALS BY FUNGAL (*ASPERGILLUS ORYZAE*) ADSORPTION

Jih-Pin Huang, C.P. Huang and Allen L. Morehart

SUMMARY

The removal of some heavy metals by Aspergilus oryzae mycelia was studied. Metal removal was determined by several factors with pH being the most important parameter. Physical and chemical treatment modified the surface of the A. oryzae mycelia and changed the metal removal characteristics. Growth conditions also affected surface properties of mycelia. At low C/N ratio, the N-content of mycelial wall increased yielding more binding sites for metal ions. Fungal mycelia of different growth stages exhibited different metal removal capacities. The N-content of young cells was greater than the mature hyphae. As a result, young cells showed a metal removal capacity greater than the old fungal mycelia. Modified Langmuir adsorption isotherms, incorporating the pH effect, could be used to desceibe the metal removal characteristics of fungal mycelia.

1. INTRODUCTION

The treatment and disposal of heavy metal bearing wastes presents a unique challenge to the environmental engineer. Unlike many organic wastes, heavy metals cannot be degraded biologically into harmless products. Heavy metals removed from polluted water must be recycled for industrial use or disposed in a way that presents little or no threat to humans or the environment.

Chemical contaminants at low concentrations are difficult to remove from aqueous solutions. Chemical precipitation, reverse osmosis, and any other methods become inefficient when contaminants are present in trace concentrations. The process of adsorption is one of the few alternatives available for such situations. Numerous adsorbent materials have been tested for their capacity to remove heavy metals, including: activated carbon, activated alumina, ion exchangers, tree bark, straw, peanut husks, peat moss, crushed coal, and insoluble starch xanthate. Some of these materials, such as ion exchange resins are effective but expensive; other adsorbents such as coal and straw are inexpensive but ineffective. Activated alumina is effective in removing heavy metals. However, alumina is readily soluble under extreme pH conditions (1). Only activated carbon has been recognized as a highly effective adsorbent for the treatment of heavy metals in water and wastewater so far. It was found that among all brands of activated carbon, only the L-carbon shows the good removal capacity of heavy metal (2-4). The L-carbon is an acidic activated carbon and is generally a powder. This has limited the application of activated carbon process to a continuous mixed reactor and required a laborious solid separation. In light of these findings, the search for an inexpensive, and yet effective adsorbent is well justified.

In the United States, large volumes of fungal biomass waste are produced annually, including biological waste treatment sludge, pharmaceutical manufacturing wastes and fermentation industrial wastes. Some of this enormous volume of biological waste solids could be recycled as commercial adsorbents, thus serving the dual purpose of utilizing materials that would otherwise be wasted.

Biotechnological processes have been applied to the recovery of heavy metal from mines and industrial wastewaters (5-9). Research in recent years indicates that many microorganisms can accumulate high concentrations of metals (10-19). Attention is being focused on the development of new microbial processes with the potential to recover valuable metals or to treat waste streams. Interest also lies in selecting microbial species to provide more practical and economical engineering systems for the efficient use and reuse of the biosorbent.

Microorganisms, including algae, bacteria and fungi, are typical biological solids that exhibit surface specificity toward heavy metals. Bacteria are less resistant to toxic metals than algae and fungi. Fungi and algae can be grown under extreme environmental conditions; i.e., low pH and high temperature. Algae, however, are autotrophic microorganisms which require a constant supply of photoenergy, oxygen and a carbon source. Moreover, due to the small size, algae as well as bacteria are difficult to harvest and require laborious solid-liquid separation. Consequently, fungi may be better suited as an adsorbent for metal removal than any other microbial mass because of their great tolerance toward toxic metals and other adverse environmental conditions such as low pH. They also grow rapidly, have high wall binding capacity, and often exhibit high intracellular uptake values of living cells. Besides, fungi are microorganisms that can be economically cultivated in mass quantities under various environmental conditions.

Recently we have investigated the heavy metal removal characteristics of about a dozen fungal species and found that Aspergilus oryzae and Saccharomyces cervisiae were amomg the most effective (20-21). The objective of this study was to describe the phenomenon of metal ion adsorption to Aspergillus oryzae mycelia. Growth status and conditions and regeneration effect of this fungus on the metal-binding capacity also were considered.

2. MATERIALS AND METHODS

2.1 Test Fungus

Aspergillus oryzae is an asexual filmentous fungus of world-wide distribution. It is used commercially in the production of soy sauce and fermented soybean products. Cultures of A. oryzae were purchased from the Carolina Biological supply Company, Burlington, NC (15-5950) and maintained on Sabouraud's dextrose agar at 26 °C. Mycelia of A. oryzae were obtained by growing shake cultures in 250 ml flasks containing 10 ml of Czapek's liquid mineral medium (pH 4.5, 25 oC, 125 rpm, 3 days), inoculated with mycelia and conidia taken from 7-day-old slant cultures. These pellets served as starter inoculum for a bench scale 14 liter fermenter (New Brunswick Scientific

MICROFERM Fermentor). Ten liters of Czapek's liquid medium in the fermenter vessel was inoculated, aerated with filtered air at an impeller speed of 400 rpm. After six days, the mycelial pellets (1-3 mm) were harvested by vacuum filteration, and washed three times with sterile distilled water. Mycelia prepared in this manner were used for all heavy metal experiments.

2.2 Characterization of Fungal Mycelia

Characterization of the surface properties of fungal mycelia is an important part of understanding the process of heavy metal adsorption by fungi. Fungal mycelia surface properties such as surface area, surface acidity, zeta potential and the nature of the chemical composition of the mycelia were determined. The specific surface area of fungal mycelia was measured by three different techniques: N_2 gas adsorptions, dye adsorption and permeametry method. A Quantasorb Model QS-7 (Quantachrome Corp.) continuous flow surface area analyzer was used to determine the specific surface area of A. oryzae mycelia by the BET adsorption. Prior to BET measurements, critical point drying (Sorvall Critical Point Drying System, DuPont Instrument) method was used to prepare the fungal mycelia for BET measurements. A common commercially available dye, New Coccine Acid Red #18 (Aldrich Chemical Company, Inc.) was used as the adsorbate. The molecular weight and area are 604 and 196 A^2, respectively (22). Calculation of surface area by dye adsorption was based upon an assumption of monolayer coverage. A third method for determining the specific surface area of fungal mycelia was permeametry which is based upon the Carmen Kozeny filtration principle (23):

$$S^2 = A\, g\, \rho_l\, H\, E^2 / K\, \eta \rho^2_s\, Q\, (1 - e)^2 \qquad [1]$$

where S, A, g, ρ_l, H, e, K, η, ρ_s, and Q are the specific surface area of fungal mycelia (cm^2/g), the cross sectional area of the filter column (cm^2), gravity constant (980 cm/sec^2), density of fluid (g/cm^3), constant head of filtration (cm), porosity of filter bed, shape factor, viscosity of fluid, density of solid (g/cm^3), and flow rate (cm^3/sec), respectively.

Electrophoretic mobility measurements of fungal mycelia particles were made with a Zeta-Meter ZM-80 (Zeta-meter Inc.).

Fungal surface chemical composition analysis of chitin (N-acetylglucosamine) content of A. oryzae was performed. Mycelia samples were dried at 103 °C for at least one hour, cooled, and transferred to screw top test tubes and immersed in 0.5 ml distilled water and 0.1 M borate buffer pH 9. The mixture was heated in boiling water bath for 10 minutes then cooled to 5 - 10 °C. Two ml of dimehylaminobenzaldehyde (Fisher) and 4 ml formic acid (J.T. Baker Chemical Co.) were added slowly with a dropper. The mixture was shaken one minute by hand and incubated at 37 °C for twenty minutes. Actual extracted chitosan concentration per gram dry weight was determined by spectrophotometer measurement at 540 nm (24).

X-ray energy dispersion analyses (EDAX) were performed on an EDAX 9100 with an ECON windowless detector. Table 1 summarizes typical surface properties of the fungal mycelia studied.

2.3 Physical and Chemical Treatments of Mycelia

Mycelia were washed with 5×10^{-3} M $HClO_4$ for 10 minutes to remove surface impurities before each experiment except as otherwise mentioned. This washing reagent also served as a regenerant during the regeneration experiment. Mycelia were exposed to extremes of temperature by autoclaving

Table 1
Surface Properties of A. oryzae Mycelia

Specific Surface Area (m^2/g)	
BET	35
Dye Adsorption	32
Permeametry	08
pH_{zpc}	< 2
Chitin Content (w/w)	30%

120 °C, 15 psi, for 15 minutes, and freeze-drying 24 hours at -65 °C. Chemical treatment of mycelia was achieved by exposure, on a v/v basis for 20 minutes each, to the following chemicals: 70% ethanol, 10% formalin, and 5% KOH. After pretreatment, fungal mycelia were collected by centrifugation and resuspended in metal solution for metal uptake experiments.

The mass of the fungal solids utilized were measured on the basis of fresh weight for consistency of comparison. However the dry weight taken either in 103 °C oven for 1 hrs or in the air for 24 hrs was recorded.

2.4 Culture Conditions

The influence of nutritional status was examined relative to carbon/nitrogen (C/N) ratio. Nitrogen ($(NH_4)_2SO_4$ and KNO_3) and carbon (sucrose) sources were added to Czapek-Dox broth to make up a specific C/N ratio of the nutrient. The original C/N ration of the Czapek-Dox broth is 25.5 (w/w).

2.5 Batch Uptake Experiments

Batch uptake experiments of a number of heavy metal by untreated and treated fungal mycelia were conducted using both pH and initial metal concentration as the master variables. Metal containing solutions were prepared by adding given amounts of metal perchlorate salts to distilled water. Uptake experiments were conducted under either constant pH or metal concentration. For constant metal concentration experiments, the final test solutions had a pH range of 2 to 9, the total metal concentration was 5 x 10^{-5} M, the ionic strength was 5 x 10^{-2} M added as $NaClO_4$ and a 1 g/l of filamentous fungus (on a dry weight basis). For constant pH experiments, the initial metal concentration varied from 2 x 10^{-6} M to 1 x 10^{-4} M, all below the solubility limit.

Both uptake experiments were performed by suspending the mycelia in 100 ml of metal perchlorate concentration and mixing for 10 min to attain the pseudo-equilibrium (only surface binding occurs) or mixing on a reciprocating shaker at slow speed for three hours at ambient temperature to include intracellular uptake.

The cells were removed by filtration through a 0.45 μm Millipore membrane filter, and the filtrates were analyzed for residual metal concentration by atomic absorption spectrophotometer (Perkin-Elmer Model Zeeman 5000).

2.5 Regeneration of Mycelia

The capacity of an adsorbent to be regenerated for subsequent reuse is important in determining the ultimate economic engineering value of that adsorbent. Mycelia of A. oryzae were subjected to repeated adsorption regeneration cycles. Acid treatment (5×10^{-3} M $HClO_4$ or H_2SO_4 for 15 minutes) was used to desorb metals from the metal-laden mycelia.

3. RESULTS AND DISCUSSION

3.1 Effect of Physical and Chemical Treatment

Figures 1 to 5 show the effect of pH on the removal of Cu(II), Zn(II), Ni(II), Cd(II) and Pb(II) by fresh, regenerated, and heat-treated mycelia. In general, the amount of metal removal increased with increasing pH.

Figure 5 shows that Pb(II) removal by fresh and acid-washed mycelia are pH-independent and are almost completely removed by the acid-washed mycelia over the pH range 3 - 5.

It should be noted that Pb(II)-laden mycelia were not regenerative by acid-washing apperently due to irreversible adsorption. Generally, the regenerated (acid-washed) mycelia, showed a much higher metal removal capacity than either the fresh or the heat-treated mycelia over the acidic pH range. The relation among the trends of metal uptake by either fresh, regenerated (heat-treated A. oryzae) mycelia are similar to that by those untreated.

Results shown in Figure 2 indicate the significance of physical and chemical treatment in modifying the metal adsorption capacities of fungal mycelia. It is noted that acid treatment greatly increased the metal removal capacity of fungal mycelia whereas heat treatment inhibited the metal removal capacity of fungal mycelia. Treatment of engineered fungal mycelia may offer advantages in the development of engineered fungal mycelia adsorption systems. Mycelia that can be preserved and thus cultivated, processed, shipped and stored off-site, offers a distinct advantage over mycelia systems that use on-site cultivated and processed mycelia. Furthermore, if small scale cartridge type engineered mycelia systems can be developed, they could be utilized in small plating operations which represent the vast majority of metal plating operations in the United States.

Large volumes of fungal biomass waste are produced annually in the United States including biological waste treatment sludge, pharmaceutical manufacturing wastes and fermentation industrical wastes. Some of this enormous volume of biological waste solids may be recycled as commerical adsorbents, thus serving the dual purpose of utilizing materials that would otherwise be wasted.

Portions of fresh A. oryzae mycelia were treated by several chemical and physical techniques in order to evaluate the potential for alterating the mycelia surface to enhance adsorption and for mycelia preservation. Figure 2 shows the removal of Cd(II) as a function of pH for chemically and physically treated fungus. In each case, treated fungus is compared to untreated fresh A. oryzae mycelia. A 5 g/l portion fungal mycelia was used for each test.

3.1.1 Physical-treating Process

Freeze-drying of A. oryzae mycelia has shown to enhance uptake of Cd(II) from solution at low pH. At high pH, removal is indistinguishable from the untreated mycelia. If during the

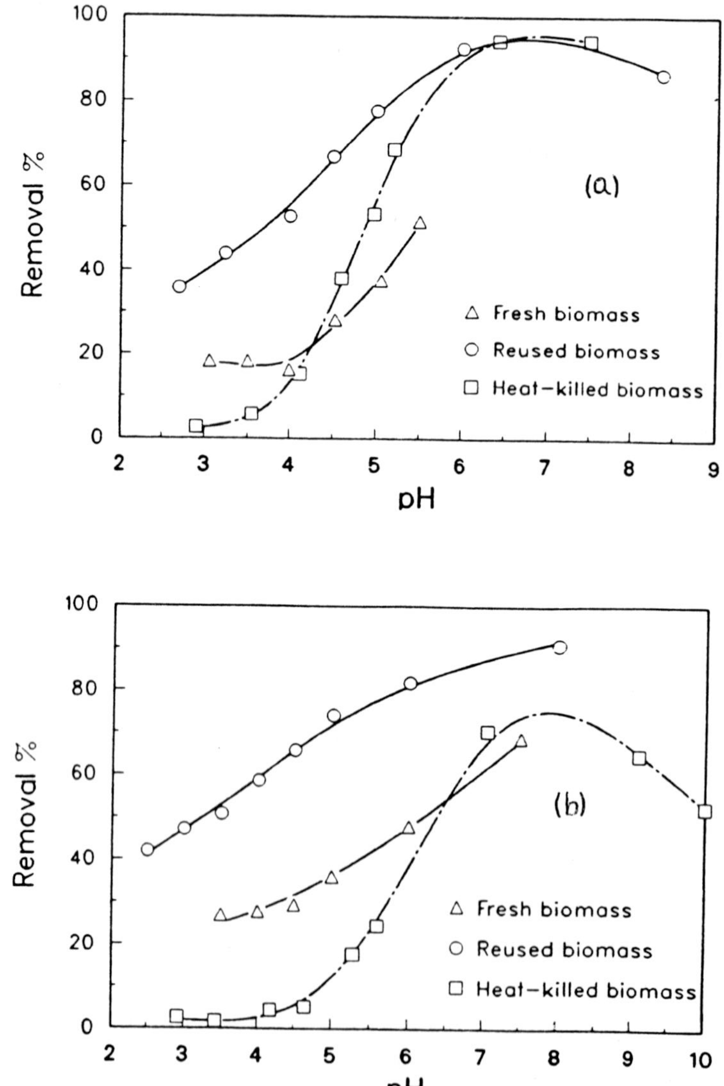

Figure 1 The pH Effect on Metal Removal by Fresh, Reused, Acid-washed or Heat-treated A. oryzae Mycelia. (a) Cu(II), (b) Zn(II)

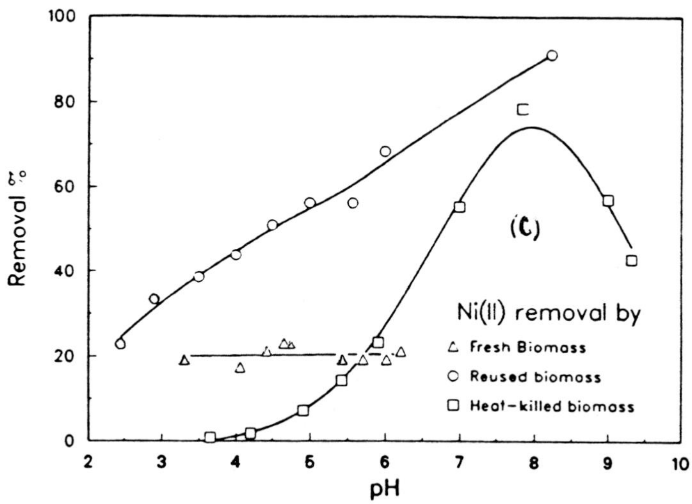

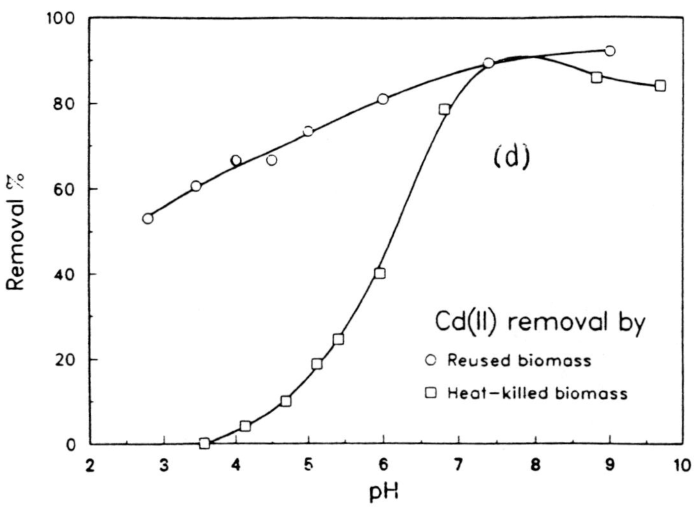

Figure 1 The pH Effect on Metal Removal by Fresh, Reused, Acid-washed or Heat-treated A. oryzae Mycelia. (c) Ni(II), (d) Cd(II).

freeze-drying process major cell surface functional groups are left essentially intact, then the enhanced Cd(II) uptake may be related to the difference in true weight between the frozen-dried mycelia solid and the fresh vaccum filtered A. oryzae solid. During both thermal and freeze-drying process, approximately 80-90 % solid mass was lost. Dehydrated fungal mycelia effectively adsorbed heavy

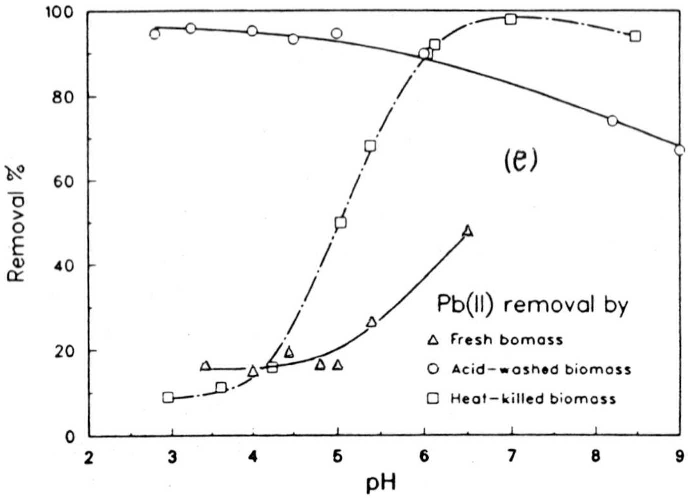

Figure 1 The pH Effect on Metal Removal by Fresh, Reused, Acid-washed or Heat-treated A. oryzae Mycelia. (e) Pb(II).

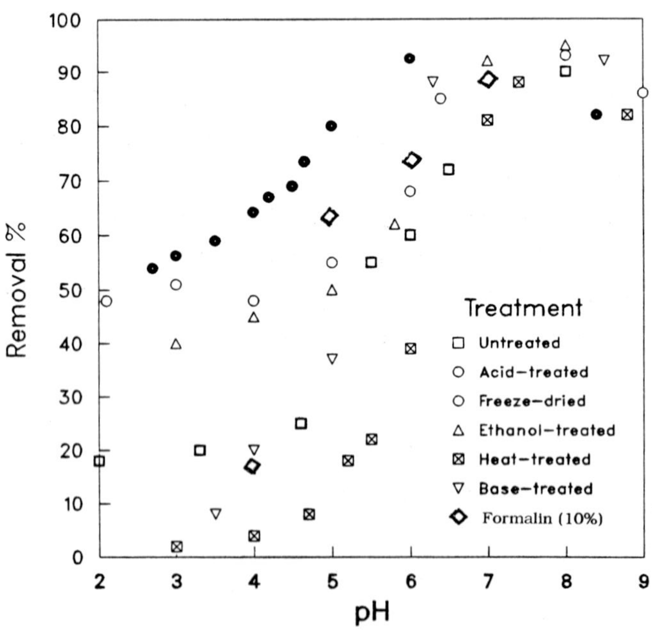

Figure 2. Removal of Cd(II) vs pH for Various Treated A. oryzae Mycelia Experimental Conditions: Cd(II) = 10^{-5}, I = 5×10^{-5} M $NaClO_4$, Mycelia mass = 5 g/l.

metals (25). However, since thermal processing is generally expensive, careful economic analysis may be required to assess its engineering feasibility efficiency. However, thermal processing does offer two distinct advantages: [i] Reduction in total mass and volume of biomass reguired for adsorption; [ii] Preservation of mycelia offering greater economy in storage and shipping of adsorbent.

Heat-treated mycelia exhibited much less adsorption of Cd(II) at low pH than untreated mycelia. At elevated pH, removal of Cd(II) was esseentially identical to untreated mycelia. Heat-treatment of fungal mycelia induces severe change to cell surface functional groups. Galun, et al. (25) have reported an enhanced uptake of U(VI) by boiled P. digitatum mycelia. Autoclaving differs substantically from boiling as a physical process due to the addition of high pressure.

3.1.2 Chemical-treating Processes

Figure 2 illustrates the adsorption of Cd(II) from solution by mycelia treated with 5% KOH, 10% formaldehyde and 70% ethanol, respectively. Chemical treatment of fungal mycelia offers the possibility of enhancing adsorption capacity through manipulating the surface characteristics of mycelia.

Gross surface characteristics such as dewaterability and structural integrity of the mycelia were greatly altered by treatment with KOH. Uptake of Cd(II) is not enhanced by treatment with KOH, particularly at low pH. As with other chemical treatment systems, uptake of Cd(II) is not affected significantly by formaldehyde, particular at low pH. Formaldehyde effectively kills A. oryzae cells and reportedly denatures, crosslinks, and masks cell surface binding sites (6).

Treatment with 70% ethanol is a widely accepted method of sterlizing or killing mycelia. Ethanol treatment of A. oryzae mycelia does not significantly affect uptake of Cd(II), or perceptively change the dewaterability or "feel" of the mycelia.

Treatment of A. oryzae mycelia by different physical and chemical methods did not significantly enhance uptake of Cd(II) with the exception of freeze-drying techniques. Various chemical and physical treatments kill fungal mycelia without causing a significant change in the adsorption capacity. This is an obvious advantage in the eventual development of full scale engineered bioadsorbent systems.

Treatment of A. oryzae mycelia by two standard sterilization techniques, autoclaving and 70% ethanol treatment do not enhance or impede Cd(II) uptake at pH > 5. Use of non-viable mycelia, particularly freeze-dried mycelia offers the advantages of potential long-term storage and handing.

3.2 C/N Ratio

Table 2 lists the results of metal uptake by untreated and acid-washed mycelia under growth conditions with various C/N ratios.

Table 2 indicates that for the untreated mycelia, when the C/N ratio decreases to 3.0, the amount of metal uptake increases markedly.

There was no significant difference in the capacity of metal removal by the mycelia with C/N ratios from 25.5 to 10. Compared to the untreated mycelia, the metal uptake by the acid-washed mycelia was increased markedly and unaffected by the C/N ratio of growth conditions. Both untreated and acid-washed A. oryzae mycelia with three different C/N ratios have been analyzed with EDAX to identify their surface composition and results are shown in Figures 3 and 4. With the higher C/N ratios, the surface of fungal mycelia contained a lower amount of nitrogen than that with the lower C/N ratios. This implies that an abundance of nitrogen compounds in the growth medium will generate mycelia with larger amounts of nitrogen on the cell surface.

Table 2
Uptake of Heavy Metal by A. oryzae Grown in Nutrients of Various C/N Ratios[a]

C/N ratio		25.5	15.0	10.0	3.0
Mycelias generated[b] (g-fresh wt./l)		15.0	14.0	15.6	10.5
		Uptake Capacity (μmole/g-dry wt)			
Cu(II)	untreated	218 ± 5	225 ± 1	228 ± 15	480 ± 15
	treated[c]	500 ± 20	525 ± 20	536 ± 15	
Cd(II)	untreated	198 ± 20	193 ± 12	212 ± 30	415 ± 10
	treated	383	---	514 ± 7	495 ± 5
Ni(II)	untreated	89	96	100 ± 5	255 ± 7
	treated	211	---	382 ± 9	360 ± 7
Pb(II)	untreated	92	---	80 ± 30	554.5
	treated	550	---	533 ± 1	560
Zn(II)	untreated	202	---	0	253 ± 5
	treated	420 ± 10	---	400 ± 14	420 ± 9

[a]: pH = 5.0 ± 0.02; initial metal concentration = 5×10^{-5} M, and A. oryzae = 5 g/l fresh-wt for uptake experiment
[b]: in the 400 m/l Czapek-Dox broth with adjusted C/N ratios for 6 days incubation
[c]: the mycelia were acid-washed twice for 15 min each.

Similar results also have been reported by Tien and Huang (26). Therefore, a larger amount of nitrogen compounds on the surface offered more binding sites for metal ions that lead to a marked increase in biosorption. The acid-washed mycelia show almost the same amount of nitrogen compounds on the surface regardless of culture conditions. This indicates that acid treatment may dissolve polysaccharide compounds in the outerlayer of the fungal wall, and hence expose additional binding sites (e.g. amino groups) that were previously masked. This gives a reasonable explanation about the C/N ratio-independence of metal uptake by the acid-washed mycelia.

A significant case is the Pb(II) binding which showed a low affinity with the untreated mycelia harvested from high C/N ratio; however, the Pb(II) binding shown a markedly strong affinity with the untreated mycelia harvested in C/N ratio of 3, as well as all kinds of the acid-washed mycelia harvested from the various growth conditions. This indicates that Pb(II) has an affinity for amino functional groups much stronger than other functional groups. Hence, Pb(II) can be effectively adsorbed only when the nitrogen-abundance layer of fungal wall is exposed to the solid/liquid interface. In addition, the total weight of mycelia harvested from the C/N = 10 nutrient is the largest; however, growth of mycelia was reduced in the presence of abundant nitrogen (C/N=3.0).

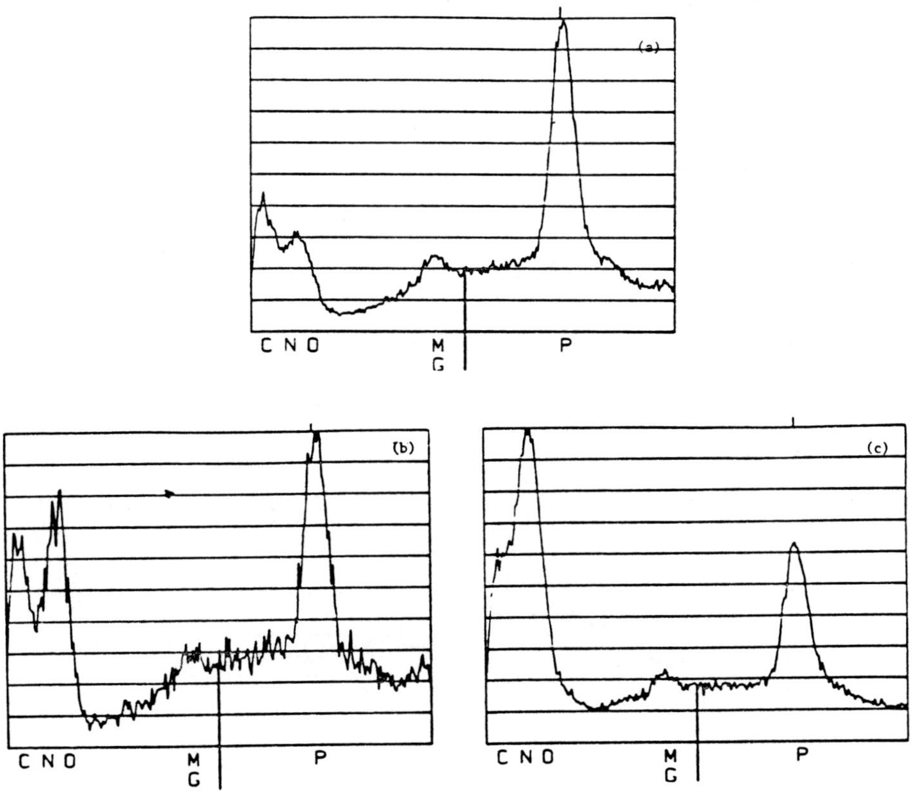

Figure 3 X-ray Energy Dispersion Analysis Spectra of Untreated A. oryzae Mycelia under Various C/N ratios of Culture Conditions, (a) C/N = 25.5; (b) C/N = 10; (c) C/N = 3.

3.3 Mycelia Growth Stage

Cu(II) uptake by A. oryzae mycelia harvested from the various growth stages are presented in Figure 5. Figure 6 presents typical X-ray EDA spectra indicating the varieties of C, N and O components corresponding to the growth stage. During the growth period of the fungal mycelia, the amino groups are the major components on the fungal surface. Then polysaccharide groups form as a major part of the outerlayer. When the growth ends, the mycelia autolyze and the polysaccharide compound in the cell wall.

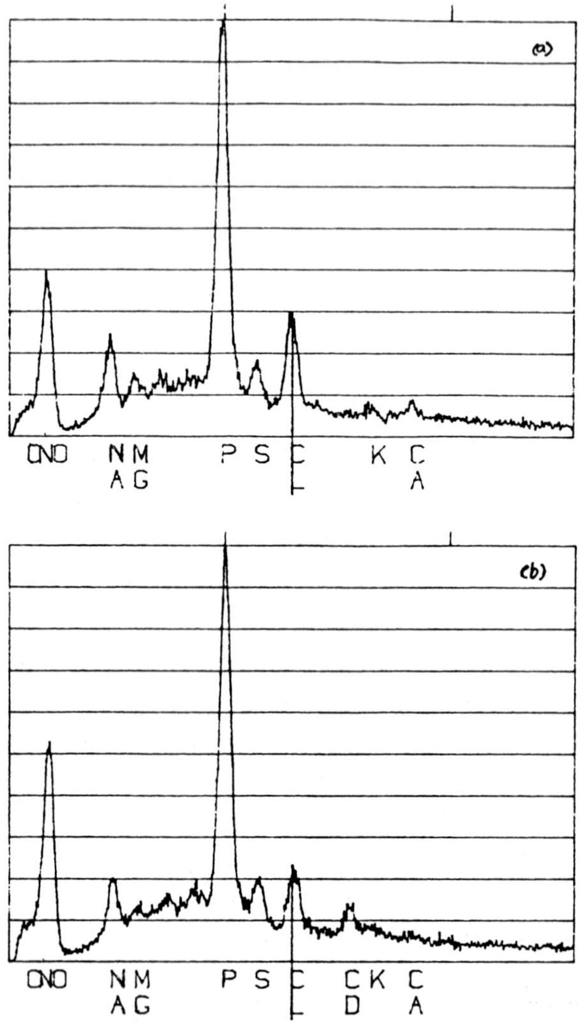

Figure 4 X-ray Energy Dispersion Analysis Spectra of Acid-washed A. oryzae Mycelia under Various C/N ratios of Culture Conditions, (a) C/N = 10; (b) C/N = 3.

It is evident from EDAX spectra shown in Figure 6 that nitrogen compounds are abundant on the fungal surface during the lag phase of the growth period, the carbon compounds become predominant during the linear phase of the growth period. Finally, in the decline phase, the carbon compounds disappear leaving the nitrogen compounds on the fungal surface. The young mycelia can adsorb more Cu(II) ions due to N-site abundance. However, regardless of N-site abundance, the older mycelia with less capacity for binding Cu(II) was probably due to the decay of some surface functional groups. As shown in Figure 6, the young mycelia with high nitrogen content exhibited a greater metal uptake capacity than the older mycelia.

3.4 Inorganic Electrolyte

Figure 7 shows the effects of types of ionic strengths on Cd(II) removal by the A. oryzae mycelias. The types of chemicals used to adjust the ionic strengths are different. However, the actual ionic strengths are all adjusted to 5 x 10^{-2} M. As can be seen in the plot, Cd(II) removal capacity in the NaClO4 as a background electrolyte is almost the same as that in $NaNO_3$ and Na_2SO_4. But Cd(II) removal capacity in NaCl is hindered significantly. This may be attributed in part to the formation of Cd(II)-chloro complexes which are not adsorbed.

The effect of electrolyte concentration has been examined by the sorption of Cu(II) using the untreated and the acid-washed mycelia. The results (Table 3) show the type of electrolyte ions can significantly interfere with the Cu(II) adsorption. There are two ways to explain this phenomenon. According to simply diffuse-layer model as applied by Stumm (27), it is assumed that only hydrogen ions as well as metal ions interact specifically with surface sites. The simple electrolyte ions are assumed to interact with the surface only by accumulation (e.g. electrostatic attraction) in the diffuse layer. Under this assumption, increasing electrolyte concentration in the solution not only compresses the electric double layer, but also encourage the accumulation of electrolyte ions on the outside of the Stern layer, which in turn blocks the interface, resulting in less metal attaching to the surface sites.

In triple layer model and the Stern model, a layer of ions, including metal ions and electrolyte ions, share the binding sites on the surface. The locus of the center of these ions is known as the inner Helmoltz plane (IHP). Yates et al.(28) have found it is necessary to consider simple background electrolyte ions to be at least weakly specially adsorbed, and an intrinsic ion-binding equilibrium constant. Under this assumption, not only hydrogen ions but also sodium ions compete with metal ions for the sites. However, hydrogen ions as well as sodium ions will be much less strongly bound to all sites than metal ions. Therefore, even Na(I) has a concentration 1000 times that of Cu(II), the effect of Cu(II) adsorption from Na(I) competition still is insignificant.

3.5 Other Competing Cations

Cation competition is of great importance in metal ion adsorption. Calcium and magnesium are of special interest because they are present in most waters as major contributors to hardness. Unlike Na(I) which is an indifferent cation in coordination chemistry, Ca(II) and Mg(II) can specifically adsorb on to the biosurface. The results shown in Figures 8 and 9 indicate that both Ca(II) and Mg(II) have little effect on Cd(II) removal at molar concentrations of 10^{-5} and 10^{-4} M. However, at 10^{-3} M both Ca(II) and Mg(II) exhibit significant effect on Cd(II) removal in the range between pH 2 and 7. In addition, it can be seen that Cd(II) removal capacity is the same in both cases. The affinity toward fungal surface exhibited by Ca(II) and Mg(II) are almost same. It must be noted that the affinity of Cd(II) and Mg(II) toward the mycelial surface is weak compared with that of heavy metals.

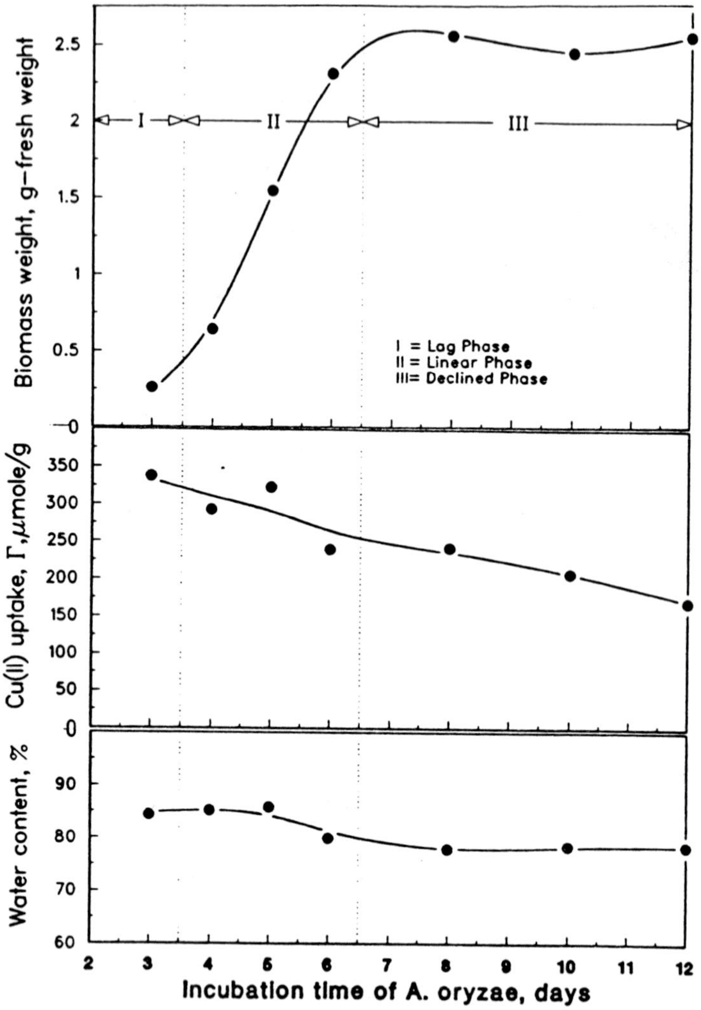

Figure 5 CuII) uptake by A. oryzae under Different Growth Stages of, Uptake Conditions: Initial Cu(II) concentration = 5 x 10^{-5} M; Mycelia = 5g/l fresh-weight; Final pH = 5.00 ± 0.02; I= 5 x 10^{-2} M $NaClO_4$

3.6 Adsorption Isotherms

Modified Langmuir adsorption isotherms were used to describe the pH-dependent of metal adsorption. Depending to the mode of proton interaction with metal adsorption process, three modified forms of Langmuir adsorption equations that incorperate the effect of proton adsorption are possible: competitive, uncompetitive and noncompetitive. The general form of the Langmuir equation is:

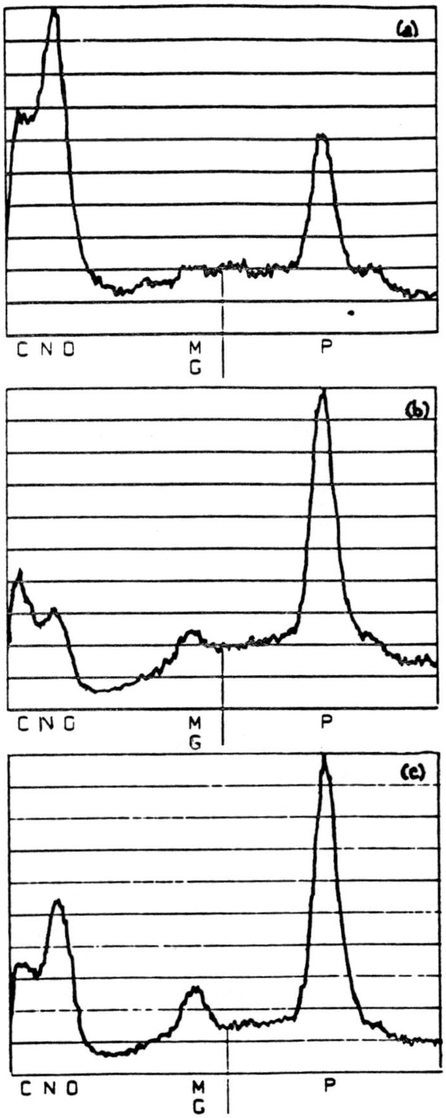

Figure 6 X-ray Energy Dispersion Analysis Spectra of A. oryzae under Different Growth Stages, (a) Incubation time = 3 days (lag phase); (b) Incubation time = 6 days (linear phase); (c) Incubation time = 12 days (decline phase)

$$\Gamma = \frac{\Gamma_m K_m [M]_e}{K_m + [M]_e} \qquad [2]$$

which can be linearized as

Table 3
Cu(II) Adsorption Capacity (Γ) of Untreated and Acid-washed Mycelia under Different Concentration of Electrolyte[a]

Type of The Mycelia	Ionic Strength(M)		
	5×10^{-1}	5×10^{-2}	1×10^{-2}
Untreated	131 ± 31	206 ± 10	283 ± 22
Acid-washed	260 ± 10	402 ± 5	448 ± 30

[a] : Γ in μmole/g, pH = 5.0; Initial Cu(II) Concentration = 5×10^{-5} M; Mycelia = 0.09 ± 0.005 g-dry weight/l

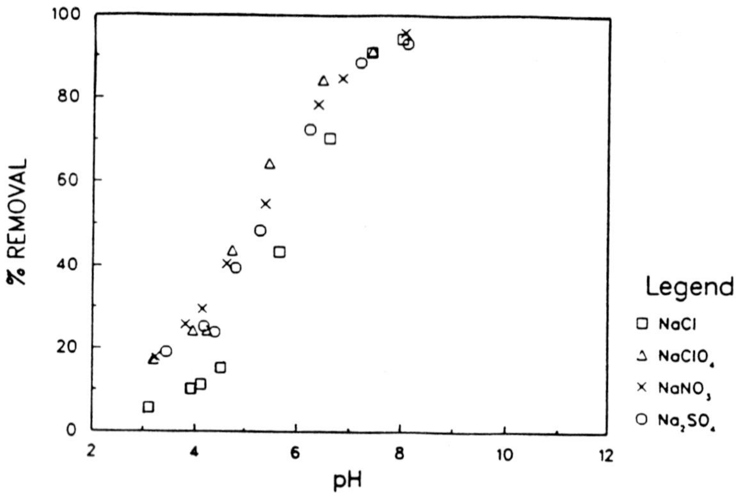

Figure 7 Removal of Cd(II) by A. oryzae in Different Forms of Ionic Strength. Experimental conditions: 10^{-5} M $Cd(ClO_4)_2$, 5×10^{-2} M Ionic Strength.

$$\frac{1}{\Gamma} = \frac{1}{\Gamma_m} + \frac{K_m}{\Gamma_m [M]_e} \quad [3]$$

where Γ, Γ_m, K_m, and $[M]_e$, are adsorption density, adsorption density at the monolayer coverage, equilibrium constant, and the equilibrium concentration of metal ion, individually.

The proton competitive Langumiur equation describes a situation where both metal ions and protons are competiting for the same adsorption sites. The slope of the general Langmuir adsorption is midified to include the effect of pH:

$$\frac{1}{\Gamma} = \frac{1}{\Gamma_m} + \left[\frac{K_m}{\Gamma_m [M]_e}\right]\left[1 + \frac{[H^+]}{K_H}\right] \qquad [4]$$

where $[H^+]$, K_H are proton concentration and proton interaction constant, respectively.

In proton uncompetitive adsorption, metal ions are adsorbed prior to the formation of metal-proton complexes. The linearized isotherm takes the form:

$$\frac{1}{\Gamma} = \frac{1}{\Gamma_m}\left[1 + \frac{[H^+]}{K_H}\right] + \frac{K_m}{\Gamma_m [M]_e} \qquad [5]$$

When proton is noncompetitive with metal ion adsorption, the interaction of metal ions with surface sites is independent of the proton interaction. The Langmuir adsorption isotherm is modified as:

$$\frac{1}{\Gamma} = \left[\frac{1}{\Gamma_m} + \frac{K_m}{\Gamma_m [M]_e}\right]\left[1 + \frac{[H^+]}{K_H}\right] \qquad [6]$$

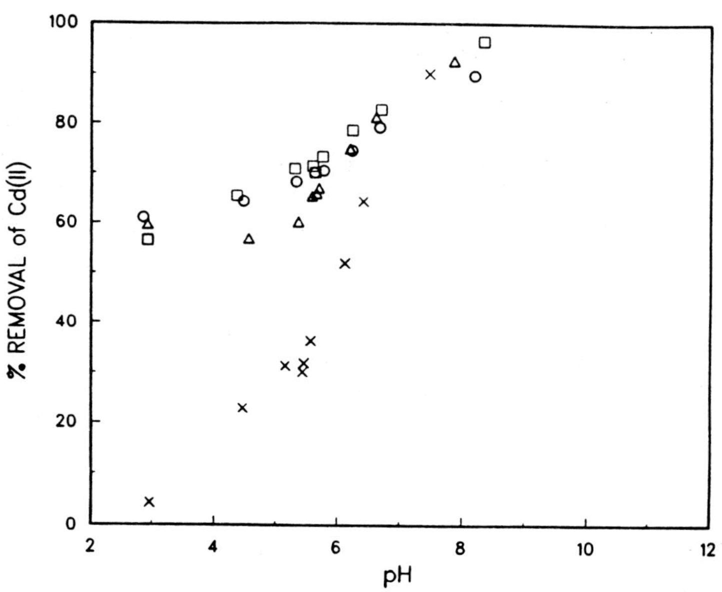

Figure 8 Cd(II) Removal Inhibition by Ca(II). Cd(II) = 10^{-5} M; I= 5 x 10^{-2} M $NaClO_4$; $CaCl_2$ Concentration: 10^{-5} M (□), 10^{-4} M (△), 10^{-3} M (X), Blank (O).

Figure 10 summarizes the linearlized plots of the modified Langmuir adsorption isotherms.

The three modified Langmuir equation appear in distinct forms when plotted as l/C_e versus l/Γ. These plots (Figure 11 indicates that a proton-competitive adsorption exists between most metals and heat-treated pelletized A. oryzae mycelia. Uncompetitive adsorption was found at the Cu(II) (Figure 11a) uptake onto yeast cells. Table 4 shows adsorption constants obtained from the modified Langmuir isotherms.

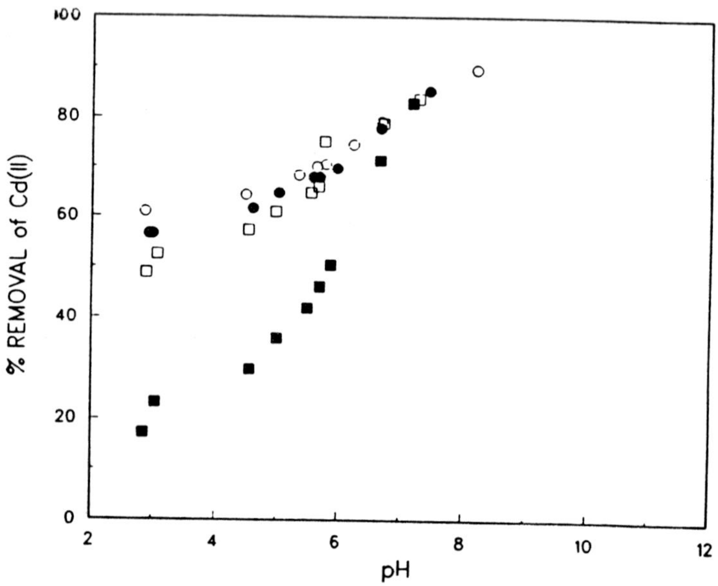

Figure 9 Cd(II) Removal Inhibition by Mg(II). Cd(II) = 10^{-5} M; I= 5 x 10^{-2} M $NaClO_4$; $MgSO_4$ Concentration:10^{-5} M (●), 10^{-4} M (□), 10^{-3} M (■), Blank (O).

Table 4
Modified Langmuir Isotherm Parameters for the Adsorption of Several Metals onto Treated A. oryzae Mycelia[a]

Metal	Isotherm	Γ(μM/g)	K_m (μM)	K_i (μM)
Cd(II)	Competitive	43.0	10.0	31.0
Cu(II)	Competitive	71.5	1.0	0.6
Pb(II)	Competitive	80.8	1.7	0.3
Ni(II)	Competitive	37.9	2.2	4.6
Zn(II)	Competitive	42.5	31.0	25.0
Co(II)	Competitive	21.2	41.3	7.4

[a]: Oven dried at 103 °C for one hour. Fungal mass was on dry weight basis.

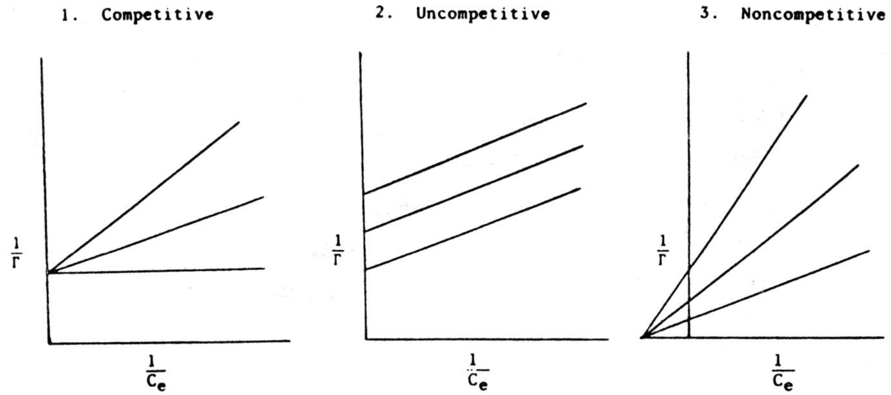

Figure 10 Summary of Linearized Modified Langmuir Adsorption Isotherms Incorperating the Effect of Proton Interactions

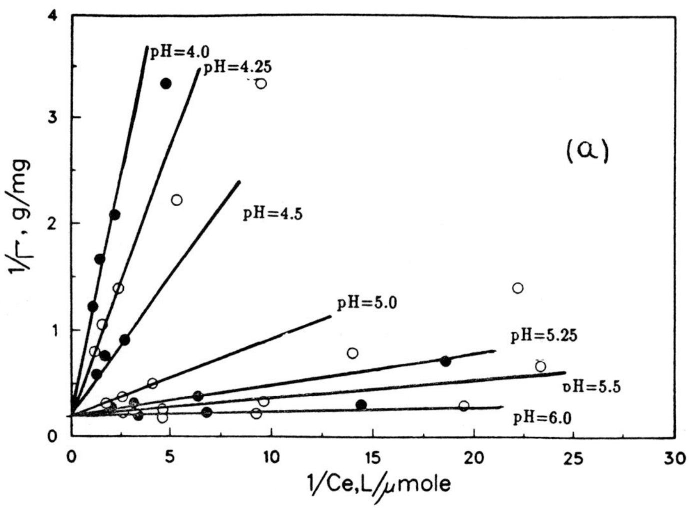

Figure 11 Reciprocal Adsorption Density vs. Reciprocal Equilibrium Metal Concentration for A. oryzae. (a) Cu(II).

The values obtained for Γ_m (maximum adsorption density mg/g) show that the six metal ions tested have relatively similar affinity toward the A. oryzae mycelial pellets; however, rating the metal ions based on Γ_m (mg/g) yields the following order: Pb(II) > Cu(II) > Cd(II) $\approx$ Zn(II) > Ni(II) >

Co(II). According to both of Γ_m and K_m, the affinity of metal for yeast cell clearly follows the same order as that of A. oryzae. Huang et al. (29) have reported the adsorption of heavy metal by activated carbon, Nuchar SA increases in the order: Pb(II) > Cu(II) > Zn(II) > Ni(II) > Co(II) > Cd(II) (2). Tein and Huang (26) have reported the adsorption of metal ions onto the activated sludge particles. follows the order: Hg(II) > Pb(II) > Cu(II) > Co(II) > Cd(II) > Zn(II) > Ni(II). The order of metal ions adsorption onto Al_2O_3 is: Hg(II) > Cu(II) > Pb(II) > Zn(II) > Ni(II) ∝ Co(II) ∝ Cd(II). While the order of metal ions affinity towards the fungal surface generally following the same order, except perhaps Cd(II), it is not possible to assert the determining factors. Though the order appears to be governed by the acidity of the metal ions and the acidity of the adsorbent.

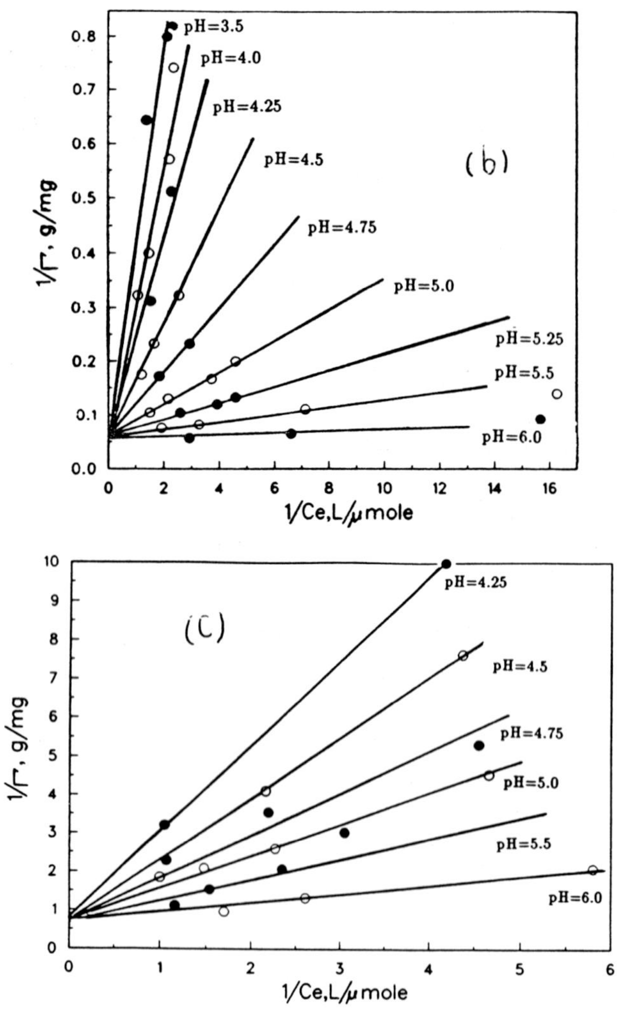

Figure 11 Reciprocal Adsorption Density vs. Reciprocal Equilibrium Metal Concentration for A. oryzae. (b) Pb(II), (c) Co(II).

ACKNOWLEDGMENT

This work was supported by a research grant No. 14-08-001-G1292 from the US Geological Survey. Content of this publication does not necessary reflect the views and policies of the United State Department of the Interior, nor mention of trade names or commercial products constitute their endorsement by the U. S. Government.

REFERENCES

1. Huang, C. P. and E. Rhoads, J. Colloid and Interface Science, 131, (1989) 289-306.
2. Corapcioglu M. O. and C. P. Huang, Water Research, 21, (1987) 1031-1044.
3. Huang, C. P. and Wirth, P, K., J. Environ. Eng., ASCE, 108 (1982) 1280-1299.
4. Huang, C. P. and Blankenship, D. W., *Water Research,* 18, (1984) 37-43.
5. Darnall, D. W., Environ. Sci. Technol. 20, (1986) 206-208.
6. Tsezos, M. and D. M Keller, Biotechnology and Bioengineering, 25, (1983) 201-214.
7. Tsezos, M. and B. Volesky, Biotechnology and Bioengineering, 24, (1982) 385-401.
8. Tsezos, M and B. Volesky, Biotechnology and bioengineering, 23, (1981) 583-604.
9. Friis, N. and P. Myers-Keith, Biotechnology and Bioengineering, 28, (1986) 21-28.
10. Mowell J. L. and G. M. Gadd, J. Gen. Microbiology, 130, (1983) 279-284.
11. Mowell J. L. and G. M. Gadd, J. Gen. Microbiology, 130, (1984) 279-284.
12. Norris, P. R. and D. P. Kelly, Developments in Industrial Microbiology, 20, (1979) 299-308.
13. Norris, P. R. and D. P. Kelly, J. of Gen. Microbiol, 99, (1977) 317-329.
14. Paton, W. H. and K. Budd, J. Gen. Microbiology, 72, (1972) 173-184.
15. Roomans, G. M., Theuvent, A. P. R., Van den Berg T. P. R. Borst-Pauwel, G. W. F. H., Biochimica et Biophysica Acta, 122, (1979) 118-125.
16. Norberg, A. B. and H. Persson, Biotechnology and Bioengineering, 26, (1984) 239-246.
17. Wood J. M., in H. Sigel (Ed), Metal Ions in Biological Systems, 1984, Vol. 18, Ch. 7, 223-246.
18. Wood, J. M. and Wang, H. K., W. Stumm (Ed), Chemical Processes in Lakes, 1986, John Wiley Sons, New York, Ch.4, 81-95.
19. Huang, C. P. Westman, D., Quirk, K., Huang, J. P. and Morehart, A., Wat. Sci. Tech, 20 (1988) 369-376.
20. Huang, C. P., Westman, D., Quirk, K., J. P. Huang, and Morehart, A. L., 6,(1988), 405-419.
21. Huang, J. P., Huang, C. P., and Morehart, A. L., *Water Research,* 24(4), (1990) 433-439.
22. Smith, P. G., and Coackley, P., Water Resources, 17, (1983) 595-604.
23. Clyde, Orr, Jr., and Dallavalle, J. M., Fine Particle Measurement, The MacMillan Co., New York, 1959.
24. Su, Hui-Jung, The Removal of Heavy Metals from Water by Crab Shell Biosorbent, BS Thesis, University of Delaware, 1989.
25. Galun, M., Keller, P., Malki, D., Feldstein, H., Galun, E., Siegel, S. M. , and Siegel, B. Z, Science, 219, (1983) 285-287.
26. Tien, C. T. and C. P. Huang, J. of Environmental Engineering, ASCE, 113, (1987) 285-299.
27. Stumm, W., and Morgan, J. J., Aquatic Chemistry, John Wiley Co., New York, 1981.
28. Yates, D. E., Levine, S. and Healy, T. W., J. Chem. Soc. Faraday, I(70), (1987), 1807-1819.
29. Huang, C. P., Hsieh, Y. S., Park, S. W., Corapcioglu, O. M., Bowers, A. R., and Elliott, H. A., in Patterson. J. W. and Passino, R., (Eds), Metal Speciation, Separation and Recovery, Ann Arbor Sci. Publ., Ann Arbor, MI, 1987, 437-465.

TRACE METAL ATTENUATION RESULTING FROM IN-SITU DEGRADATION OF SANITARY LANDFILL LEACHATE

Ronald J. Scrudato and James J. Pagano

1. ABSTRACT

Select trace metal concentrations in leachate collected from a large upstate New York sanitary landfill during a twelve month period were found to vary in relation to leachate 5-day biochemical oxygen demand (BOD), acetic acid concentrations, and pH. Similar relationships were found in leachate subjected to laboratory anaerobic digestion in a modified upflow anaerobic sludge blanket digestor (UASB) operated under mesophilic conditions. Results indicate trace metal concentrations in sanitary landfill leachate will vary depending on the stage of refuse decomposition. Leachate from landfills in which active anaerobic decomposition has occurred will contain lower trace metal concentrations as a consequence of the reduction in volatile fatty acids and corresponding increase in pH. These results also indicate that in-situ leachate degradation can reduce the trace metal concentrations of landfill leachates. By promoting in-situ degradation in the design and operation of sanitary landfills, the trace contaminant concentrations of leachate can be significantly reduced.

2. INTRODUCTION

Until the early to mid-1970s, there was little distinction made between municipal and industrial solid and hazardous waste. Wastes were routinely co-mingled and disposed of in common landfills. Wetlands, abandoned quarries, mined out sand and gravel pits were used for disposal of mixed industrial and municipal solid wastes. Waste materials were covered on a daily basis to reduce odors, exposure to animals and also to prevent blowing wastes from littering the countryside.

These early waste management practices were generally accepted as economical, environmentally sound and frequently constructive, particularly when unusable wetland areas were reclaimed for the siting of railroads, roads and industrial sites. As these mixed wastes decomposed, free liquid (leachate) was generated reflective of the composition of the co-mingled wastes. With the infiltration of surface waters through permeable caps and intrusion of groundwater into the waste material, leachate was formed which reflected the more soluble and mobile fractions of the waste materials. Offsite migration of leachate led to the contamination of area surface and groundwaters and the anaerobic decomposition of the degradable fraction resulted in the release of CH^4, H_2S, CO_2 and the more volatile fractions of the wastes including volatile organic compounds (VOCs).

The release and offsite migration of contaminants from poorly sited, designed and operated waste management facilities resulted in sweeping changes in federal and state policies. Many former sanitary landfills have been classified as state or federal superfund inactive hazardous waste sites and thousands of landfills were closed or required to implement closure procedures. In addition, significant changes in landfill design were promulgated and implemented which focused on minimization of leachate production. Double composite liners, tight impermeable caps and separation of the refuse from nearsurface groundwater are design features incorporated to minimize leachate production by restricting infiltration of surface and/or groundwaters. These new landfills are also required to include effective leachate collection systems and require gas collection facilities to manage the products of methanogenesis and other volatiles emanating from the decomposing refuse. The leachate that is produced and collected from the newly designed landfills is normally transported to nearby sewage treatment plants for disposal. These recently mandated landfill regulations have significantly reduced the quantity of leachate, but have also reduced the rate of waste biodegradation and therefore waste stabilization. Additionally, the disposal of collected leachate in sewage treatment facilities contributes to additional loading of contaminants to receiving waters and sewage sludge.

Municipal solid waste begins decomposing shortly after burial through a complex series of biological and physico-chemical processes. Moisture, nutrients, temperature, pH and the character of the waste directly affect decomposition (1,2). Of these parameters, moisture has been determined to be the most significant to refuse decomposition (3). Landfill leachate composition reflects the character of the refuse and the landfill site characteristics.

During the early stages of decomposition, aerobic processes are prevalent (4,5). During aerobic decomposition, the refuse and leachate temperatures will increase. As the available oxygen is depleted, anaerobic decomposition becomes dominant and the temperature of the refuse and leachate will decrease.. During the earlier stages of anaerobic decomposition, volatile fatty acids (VFA) are produced causing a reduction in the pH of the leachate. As methanogenic bacteria consume the available VFAs, the pH of the leachate becomes progressively more alkaline. In general, older leachate is more alkaline than leachate from younger sanitary landfills (6). As methanogensis progresses and the pH becomes more alkaline, the trace metals complex and are more readily adsorbed onto particulates and thereby concentrate in the sludge and refuse residue.

3. FIELD AND LABORATORY INVESTIGATIONS

As part of a field and laboratory investigation of leachate characteristics at an upstate New York sanitary landfill, we assessed the changes in select trace metal concentrations in relation to BOD and select VFA reductions and resultant changes in pH. We also measured trace metal concentrations in leachate and sludge samples from a laboratory-scale mesophilic UASB digestor during a ten week period. Changes in leachate chemistry from the field and laboratory investigations were similar and provided significant insight to landfill processes.

The Seneca Meadows Sanitary Landfill is located in upstate New York and manages about 600,000 tonnes of municipal and non-hazardous industrial wastes annually. The 45-ha facility consists of an older, unlined section bordered by nine discrete cells subdivided into two to three subcells, each with discrete leachate collection systems. The subcells are lined with 0.3-1.0 m of compacted clay and contain discrete leachate subdrains and collection wells. A herringbone network of gravel-filled trenches drain leachate at the base of the refuse fill to a peripheral sump where the leachate is recovered. Each of the peripheral cells was subdivided into two to three subcells with separate and discrete leachate collection sub-drainage systems (Figure 1). The older portion of the landfill was filled during the period of the late 1970s to 1983. This portion

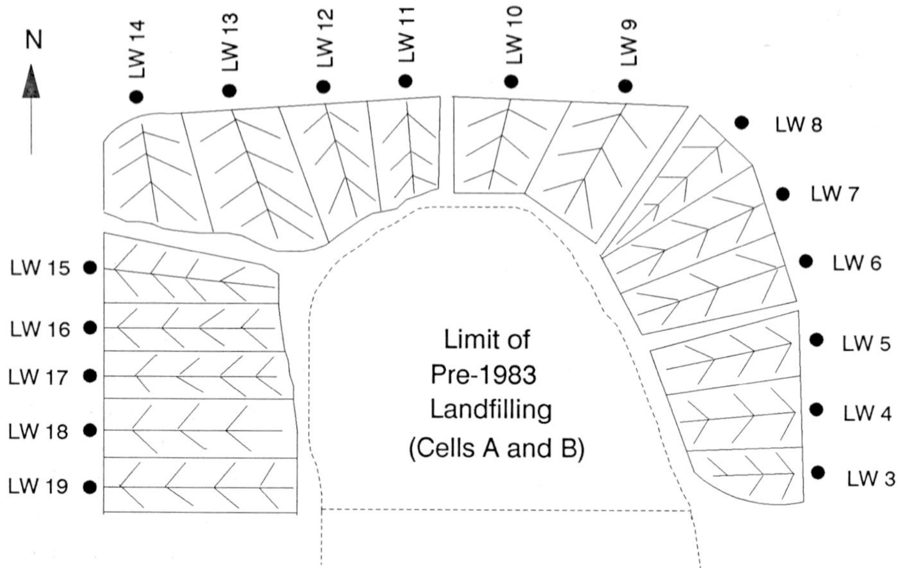

Figure 1. Older central portion (pre-1983) of the Seneca Meadows Landfill bordered by peripheral cells with herringbone leachate drainage systems.

of the facility did not include clay liners or leachate collection systems. The peripheral cells were constructed and filled during the period 1984-1990. The herringbone sub-drainage systems were constructed to facilitate the migration of free liquid to a collection sump located at the outer, downslope perimeter. Each of the subcells was filled with refuse prior to filling of adjacent subcells. Filling progressed counterclockwise from the southeast (leachate well 3). At the time of filling of the subcells, the cap from the pre-1983 portion of the landfill was removed to facilitate leachate migration into the subcell leachate collection systems.

In the field study leachate samples were collected bi-weekly for a one year period and less frequently for an additional seven months from the Seneca Meadows Sanitary Landfill leachate collection wells using an all teflon baler. Temperature and specific conductance were measured in the field using a YSI 3000 TLC meter. Five-day BOD (7) and VFA concentrations were measured using ion exclusion chromatography (8). Select trace metal analyses for elements with atomic numbers >20 were conducted using proton-induced x-ray fluorescence with sensitivity of about 0.5 mg/l (9).

Specific conductance, temperature, BOD, pH and VFAs were measured from 17 leachate collection wells during the period July 1987 to July 1988. In addition, trace metal analyses of three leachate well samples were conducted during the period October 1987 to February 1989.

4. LABORATORY INVESTIGATION

In the laboratory phase of the study, leachate from an active Seneca Meadows Landfill leachate well (well 19) was subjected to anaerobic digestion utilizing a 10 l modified UASB mesophilic digestor (Figure 2). Flow rates varied from 0.6-2.3 l/day with associated hydraulic retention times (HRT) of approximately 18-4 days, respectively. Select trace metal analyses were conducted of effluent digestor leachate and granular sludge collected from within the digestor, the sidewall biofilm, and bottom portion of the digestor. The biofilm and bottom sludge samples were collected at the termination of the experiment. VFAs and BOD were measured weekly using the previously described methods.

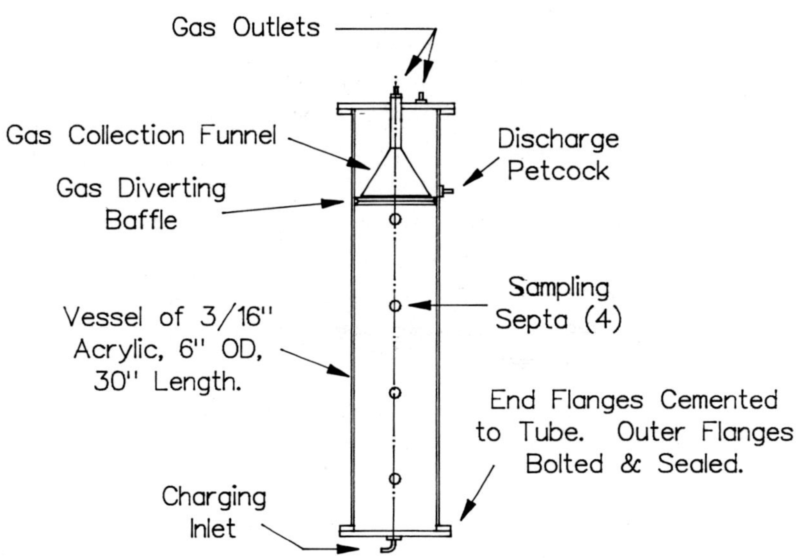

Figure 2. Schematic of the 10 l upflow anaerobic sludge blanket (UASB) digestor utilized for the laboratory phase of the study.

5. RESULTS

Figures 3, 4 and 5 illustrate the changes in BOD, acetic acid and Ph for leachate collected from wells 5, 7 and 16 during the study period. As can be seen from these illustrations, leachate from these three wells had elevated initial concentrations and experienced significant reductions in BOD, acetic acid and Ph changes of 1.5 units.

The high BOD and VFA leachate derived from the pre-1983 portion of the landfill. Subcells 5, 7, and 16 tapped into high strength organic leachate which drained into the subcell collection sumps. Several of the subcells contained high strength inorganic leachate. The specific conductance for all of the 17 well samples analyzed ranged from about 770 to >33,000 umhos/cm. Subcells located in the northern portion of the landfill exhibited BOD and specific conductance values typical of area groundwaters (<10 mg/l and 1500-3000 umhos/cm, respectively). Subcells with low BOD yet elevated specific conductance suggests that the leachate which derived from the older portion of the landfill was depleted in the more readily biodegraded organics or that the older leachate derived from inorganic soluble salts. The high BOD leachate of wells 5, 7 and 16, however, originated from the older section of the landfill and on entering the subcell environment experienced rapid degradation of the biodegradeable leachate fraction. As can be seen from Figure 3, subcell 16 experienced a change in BOD from in excess of 7000 mg/l to less than 200 mg/l in about a 2 1/2 - 3 month period. These rates of biodegregation occurred at temperatures which averaged between 10-12°C for most of the subcell leachate samples. Leachate temperatures were found to vary seasonally, but at no time exceeded 17.4°C during the study period.

Trace metal analyses of leachate samples collected during the period November 1987 to February 1989 demonstrated significant reductions in leachate metal concentrations as illustrated in Figures 6 and 7. These changes in trace metal concentrations paralleled an increase in the pH of the leachate during the same time period as illustrated in Figure 5.

Similar changes in BOD, VFA and pH concentrations were noted in the laboratory phase of the study (Figures 8,9). Leachate collected from well 19 was subjected to mesophilic conditions in a 10-l UASB

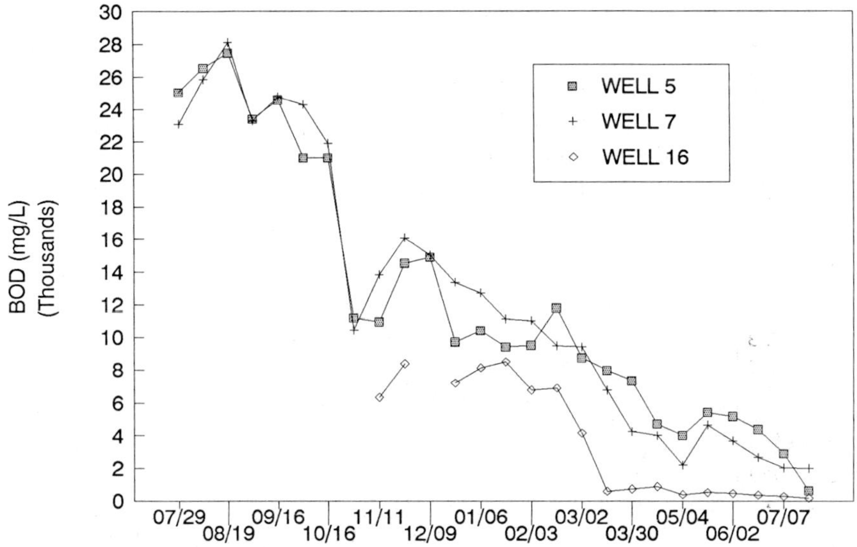

Figure 3. Biochemical Oxygen Demand (BOD) changes in leachate from wells 5, 7 and 16 at the Seneca Meadows Landfill during the period July 1987-July 1988.

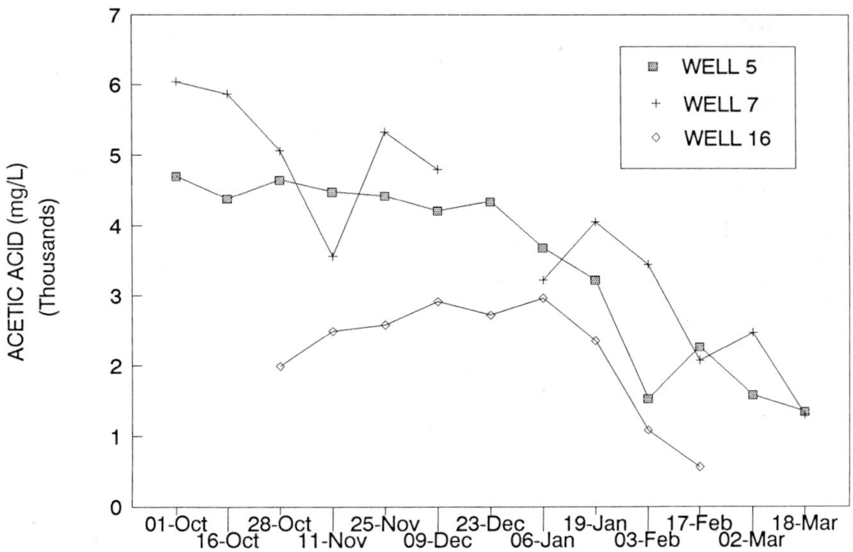

Figure 4. Acetic acid changes in leachate from wells 5, 7 and 16 at the Seneca Meadows Landfill during the period October 1987-March 1988.

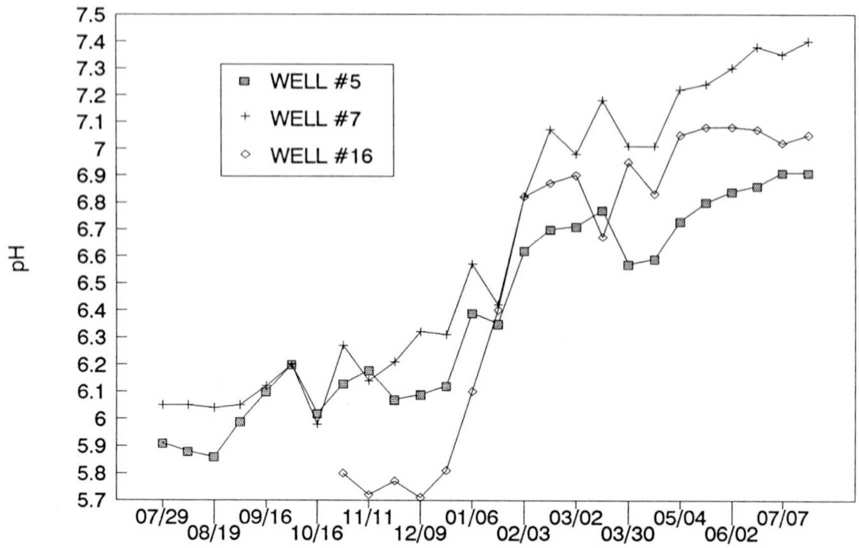

Figure 5. Changes in pH in leachate from wells 5, 7 and 16 at the Seneca Meadows Landfill during the period July 1987-July 1988.

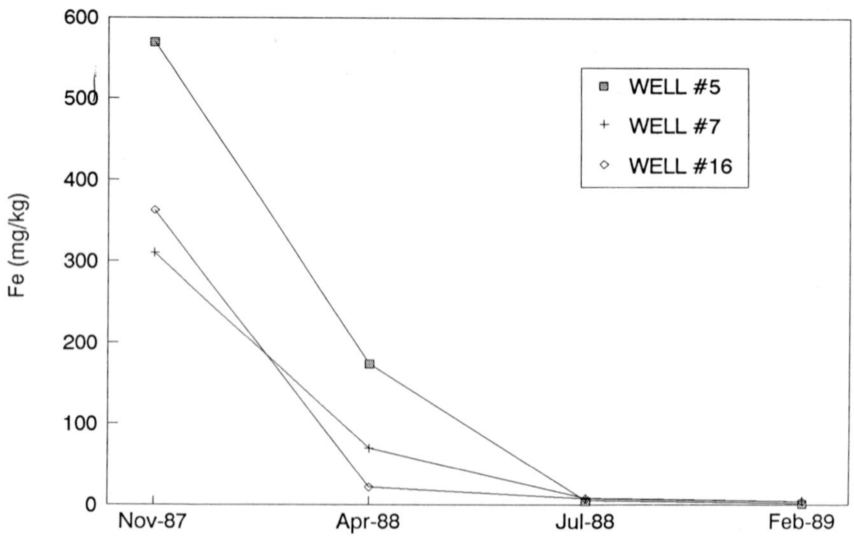

Figure 6. Iron concentrations in leachate from wells 5, 7 and 16 at the Seneca Meadows Landfill for the period November 1987-February 1989.

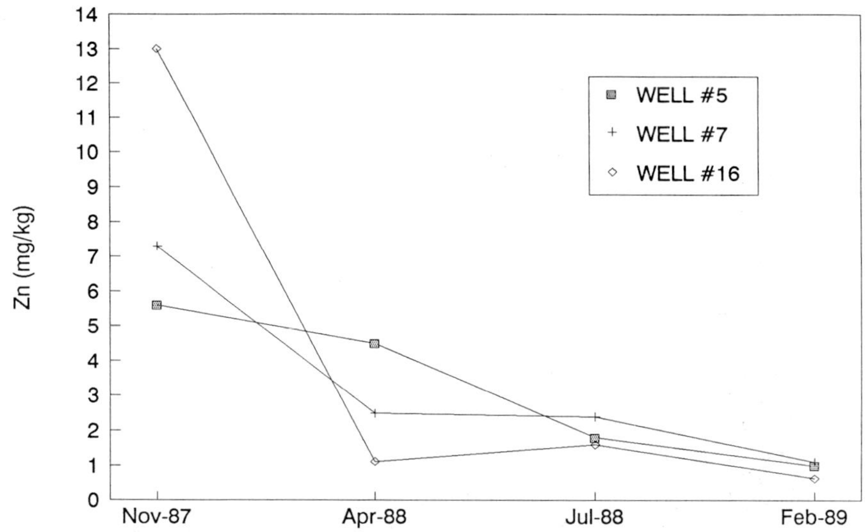

Figure 7. Zinc concentrations in leachate from wells 5, 7 and 16 at the Seneca Meadows Landfill for the period November 1987-February 1989.

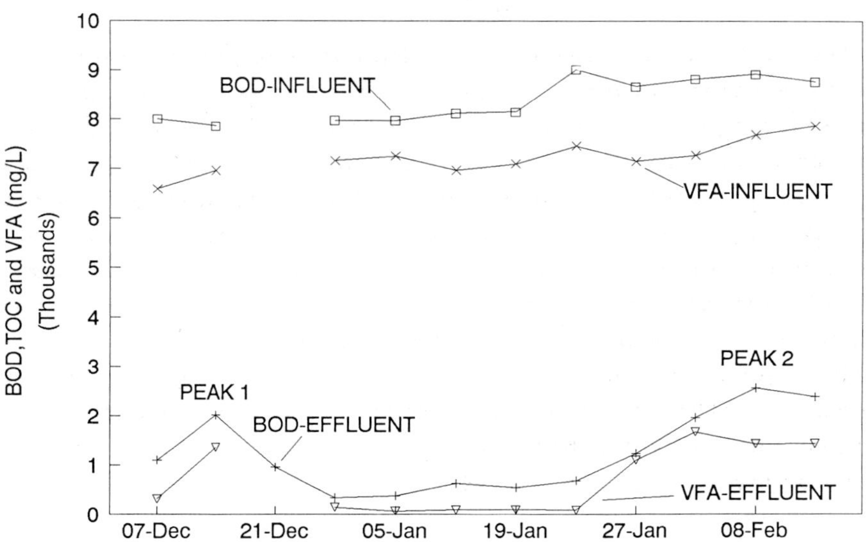

Figure 8. Comparison of volatile fatty acid (VFA) and biochemical oxygen demand (BOD) concentrations in influent and effluent leachate from well 19 after biodegradation in a UASB reactor.

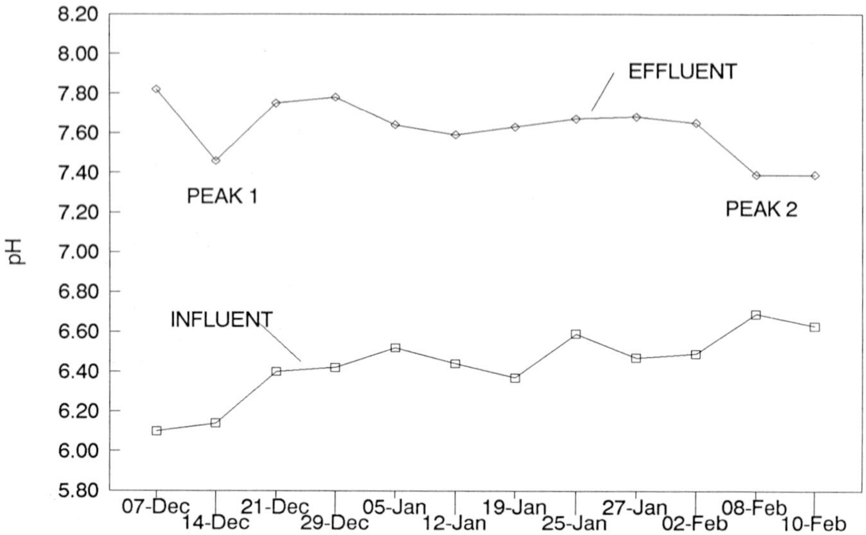

Figure 9. pH over time in influent and effluent leachate from Seneca Meadows Landfill (well 19) subjected to treatment in an UASB reactor.

reactor during a ten week period. During that period the hydraulic retention times ranged from 18-4 days and the reductions in BOD, VFAs and concomitant increase in pH amounted to 85%, 91% and 1.4 units respectively (Table 1)(10). Leachate and sludge samples were collected and analyzed for trace metal concentrations. As can be seen from Table 2 (11), the concentrations of select trace metals were significantly different in the influent and effluent samples. The effluent concentrations were significantly lower in iron and manganese. The concentration of trace metals increased appreciably in the anaerobic sludge particularly in the sidewall and bottom sludges. Although lead and mercury were not detected in the influent, effluent leachate or free sludge, both metals were found in the sidewall and bottom sludge.

Table 1. Chemical composition of well #19 leachate from the Seneca Meadows Landfill before and after treatment in a UASB reactor (10).

SAMPLE	Fe	Zn	Mn	Pb	Hg
Leachate (influent)	140	3.8	9.2	nd	nd
Leachate (effluent)	24.5	3.1	<0.5	nd	nd
% removal	82	19	>95	-	-
Sludge (free)	24	4.5	1.4	nd	nd
Sidewall sludge	176	8.0	49	1.2	0.6
Bottom sludge	420	15	80	2.5	2.8
% increase	300	395	842	<250	<280

Table 2. Influent and effluent well #19 leachate (mg/l) and sludge (mg/kg) concentrations of Fe, Zn, Mn, Pb and Hg from the Seneca Meadows Landfill treated in a UASB digestor (11).

PARAMETER	n	AVERAGE (mg/L)	RANGE (mg/L)	AVERAGE CHANGE	OPTIMAL CHANGE
VFA (influent)	11	7226	7871-6591	-	-
Formic	11	81	109-0	-	-
Acetic	11	1883	2066-1632	-	-
Propionic	11	1338	1655-1173	-	-
Butyric	11	3875	4217-3467	-	-
VFA (effluent)	11	659	1675-72	91%	99%
TOC (influent)	12	4856	5594-4456	-	-
TOC (effluent)	12	921	1546-520	81%	89%
BOD (influent)	11	8406	9025-7975	-	-
BOD (effluent)	11	1234	2565-343	85%	96%
pH (influent)	12	6.44	6.69-6.10	-	-
pH (effluent)	12	7.62	7.82-7.39	1.2 units	1.4 units

6. CONCLUSIONS

Based on the one year field study conducted at the Seneca Meadows Landfill, it is clear that leachate underwent rapid biodegradation once the leachate from the older portion of the landfill entered the subdainage collection system. These results indicate that in-situ treatment is a viable options for reducing the more biodegradable fractions of landfill leachates.

It is also well known that landfill leachate trace metal

concentrations will vary relative to its age. As evidenced by the results from the UASB reactor studies, select trace metals will be concentrated in the sludge as the leachate undergoes biodegradation and the pH of the liquid increases. Similar processes are also operative in landfills where trace metals will also become concentrated in the produced sludge and other solids. Although we did not attempt to characterize the presence or concentrations of trace organic contaminants in this study, it is likely that trace hydrophobic, organic contaminants will also become attenuated in the anaerobic sludge produced during methanogenesis.

These results suggest that landfill designs should consider enhancing in-situ biodegradation of leachate. This direction is compatible with the bioreactor concept advocated by Ham (12). In-situ treatment may also prove beneficial in attenuating trace metals thereby reducing contaminant loadings to sewage treatment plants which are currently being utilized for the disposal of sanitary landfill leachates. If specific treatment facilities are to be used to degrade sanitary landfill leachates as suggested by Ross (13), they should be designed, constructed and operated to complement enhanced in-situ biodegradation.

7. ACKNOWLEDGEMENTS

This project was partially supported by the New York State Energy Research and Development Authority. Dr. Roger Hinrichs conducted the metal analyses and professor Harry Hawkins assisted with the design and construction of the anaerobic digestor.

8. REFERENCES

1. J.F. Rees, J. Chem. Tech. Biotech., (1980) 458p.
2. J.F. Rees, Eff. Water Treat. J., 22 (1982) 457p.
3. M.A. Barlaz, M.W. Milke, and R.K. Ham, Waste Management Res. (1987) 27-39.
4. A.J.B. Zehnder, Water Poll. Micro., 2 (1978) 349-376.
5. R.K. Ham, M.A. Anderson and R.Stegman, EPA-600/1 (1979) 79-109.
6. J.M. Lema, E. Ibanez and J. Canals, Env. Tech. Letters, 2 (1987) 555-564.
7. APHA, AWWA, WPCF, Standard Methods, American Public Health Association, Washington, DC (1985).
8. W.H. McDowell, S.M. Stedman and J.J. Pagano, in: Advances in Ion Chromatography, Century International, Medfield, MA (1990) 387-399.
9. V. Yalkovich, Nuc. Instr. Methods, 114 (1974) 573-579.
10. J.J Pagano, W.H. McDowell, R.J. Scrudato, (submitted for publication).

11. R.J. Scrudato, W.H. McDowell, R. Hinrichs and J.J. Pagano, in: Proc. of Intern. Conf. on Trace Metals in the Env., 2 (1989) 048-051.
12. R.K. Ham, Proceedings of Institute of Waste Management of South Africa Biennial Congress, (1988).
13. W.R. Ross, Factors influencing the chemical characteristics of landfill leachates. Water SA, 16, 4 (1990) 275-280.

Chapter 10
Analytical and general methods

ANALYSIS OF THE BINDING HETEROGENEITY OF NATURAL LIGANDS USING ADSORPTION DATA

M.M. Nederlof, W.H. van Riemsdijk and L.K. Koopal

1. INTRODUCTION

The binding of metal ions and other species to ligands in natural environment is important with respect to the bioavailability and mobility of the species present. In the last two decades much attention has been paid to the problem of heterogeneity in relation to ion binding to polyfunctional ligands with contributions of various authors (1-16). In most cases a binding curve (or adsorption isotherm) is measured on the basis of which information is obtained with respect to the binding characteristics of the underlying binding sites. A useful binding curve can be obtained by titration of a fixed amount of ligands with an adsorbate (protons or metal ions) at otherwise constant solution composition.

Insight in the binding processes is hard to gain since natural systems are poorly defined and they possess different types of reactive groups. In addition, many natural colloids exhibit variable charge characteristics, leading to 'conditional' or 'apparent' parameters, when the charge effects are not taken into account explicitly. In principle it is possible to consider the variable charge effects separately from the chemical heterogeneity effects (16,17). In any case, heterogeneity analysis is a valuable tool to get insight in the binding behaviour and it may provide arguments for choosing a realistic model for the description of the binding.

In general the methods for heterogeneity analysis can be divided in two groups. The first starts with the assumption of a discrete heterogeneity, examples are the classical method of Scatchard (10,18) that determines a limited set of affinity constants and fractions, and the numerical method of Brassard et al. (19) that is able to determine a large set of affinities and their fractions. The second group of methods assumes a continuous distribution of affinities. Apart from the physical arguments in favour of a continuous distribution (1-4,11,12) a practical argument is that for the description of the discrete case a large number of parameters is required, i.e. for each site type an affinity constant and a fraction. However, wide and smooth distribution functions can often be characterized by only a few parameters. In addition, on the basis of the final result of the heterogeneity analysis it can still be decided whether the surface has the characteristics of a discrete distribution (several sharp peaks) or that of a smooth continuous one.

Examples of the second group of methods, well known in the field of ligand binding, are the Affinity Spectrum (AS) methods (9,14) and the Differential Equilibrium Function method (DEF) (1-4,7,8). Also in the field of physical chemistry both purely numerical and semi-analytical methods have been published that deal with heterogeneity analysis (20-23). So far none of the purely numerical methods have been applied to the problem of ion binding to natural ligands. Examples of semi-analytical methods are the Condensation Approximation (CA) (25,26), the Asymptotic Correct Condensation Approximation (ACCA) (27,28), the Rudzinski-Jagiello method (RJ) (29,30) and the recently developed LOGA method (23).

In this paper we will present and discuss the main semi-analytical methods for heterogeneity analysis and discuss the problems that arise when these methods are applied to experimental data. In section 2 some general binding equations are given that apply to heterogeneous ligand systems in general. In section 3 three groups of semi-analytical methods are discussed: the Local Isotherm Approximation (LIA) methods, the Affinity Spectrum (AS) methods and the Differential Equilibrium Function method (DEF). It will be shown that for each method one or more derivatives of the overall binding function are required. For experimental data reliable determination of such derivatives is impossible, unless some preprocessing of the data is done, in which the effects of experimental errors are minimized. For preprocessing we have chosen the smoothing spline technique (31-35) which has the advantages that no type of functionality is assumed a priori and that the amount of smoothing can be tuned to the magnitude of the error (section 4). The main problem is the determination of the smoothing parameter. The way to obtain this smoothing parameter is discussed for the case that the experimental error is known, using this error, and for the case it is not known, with Generalised Cross-Validation (36). In both cases physical constraints are needed to obtain a suitable value of the smoothing parameter. Section 5 deals with the application of the presented methodology to a data set of Copper binding to natural, mainly organic ligands as present in Lake Tjeukemeer (The Netherlands) (37). Finally some concluding remarks will be made in section 6.

2. BINDING EQUATIONS FOR HETEROGENEOUS LIGAND SYSTEMS

First the entity to which binding takes place has to be discussed. In the most simple case this entity is a small molecule with one reactive group or one binding site only. One step more complicated is a homogeneous colloid characterized by a number of identical binding sites. In these cases the reactive entities are chemically homogeneous. The reactive entity may also be heterogeneous, characterized by the presence of different site types. Schematically for a heterogeneous ligand system two extreme situations can occur. The equivalent sites can either be grouped together in large patches (patchwise heterogeneity) or be mixed randomly with the other groups (random heterogeneity). The sites can be present on polyfunctional ligands (random or patchwise) or on a series of in itself homogeneous ligands (patchwise). For polyfunctional ligands the locations of the binding sites are fixed and one speaks of localized binding. In this situation strong lateral interactions may occur and influence the binding behaviour. Moreover, the topography of the surface (random or patchwise) becomes important (21,38). Not taking into account these complications properly will lead to 'apparent' affinities that are a function of both the 'intrinsic' chemical heterogeneity and for instance electrostatic effects. In this paper we will neglect these complications and assume a random topography.

The binding of a metal species or protons to a heterogeneous ligand system may be considered as the sum of the contributions of the different site types present. For one such site type i the binding of species M can be written as:

$$S_i + M \rightleftarrows S_iM \qquad [1]$$

where S_i is the free and S_iM is the complexed site. By application of the law of mass action to this equilibrium under the assumption of ideality (both in solution and at the ligand system) the affinity constant K_i may be written as:

$$K_i = \frac{\{S_iM\}}{\{S_i\}[M]} \qquad [2]$$

where $\{S_iM\}$ and $\{S_i\}$ are the concentrations of the complexed and free sites respectively and [M] is the concentration of the species M. The fraction of site type i occupied by M (indicated by θ_i) is defined as:

$$\theta_i = \frac{\{S_iM\}}{\{S_i\} + \{S_iM\}} \qquad [3]$$

It follows from [2] that θ_i may be written as:

$$\theta_i = \frac{K_i[M]}{1 + K_i[M]} \qquad [4]$$

Equation [4] is called the *local binding function* or *local isotherm* and is equivalent with the well known Langmuir equation. It applies to both a solution with a large number of equal single site ligands and to one with homogeneous polyfunctional ligands each containing a number of equal sites. This is a consequence of the assumed ideality.

For a heterogeneous ligand system the overall binding function (or overall adsorption isotherm) is obtained by summation of the local binding functions weighted by the site fractions:

$$\theta_t = \sum_i f_i \theta_i \qquad [5]$$

where θ_t is the fraction of all sites occupied by M, and f_i is the fraction of site type i with respect to the total concentration of sites of the system. Equation [5] applies to a heterogeneous system with a discrete number of site types. When the difference between the K_i or rather the $\log K_i$ values become infinitesimal Eq.[5] becomes an integral equation:

$$\theta_t = \int_\Delta \theta f(\log K) d \log K \qquad [6]$$

where θ is the local binding function (the kernel of the integral equation), f(logK) is the distribution function and Δ denotes the range of logK values present. The product $f(\log K) d \log K$ is equivalent to the fraction f_i in the discrete case.

Equations [5] and [6] are equivalent expressions for the generally used ideal overall binding function for a heterogeneous ligand system and form a starting point for a further analysis. We assume here a large number of different site types and then [6] provides a basis to obtain the distribution function if the overall binding function θ_t and the type of local binding function θ are known. The local binding function θ follows from theoretical considerations and θ_t is experimentally accessible when the total number of sites $\{S_t\}$ is known, it can be expressed as:

$$\theta_t = \frac{\{LM\}}{\{LM\}+\{L\}} = \frac{\{LM\}}{\{S_t\}} \qquad [7]$$

where $\{L\}=\sum\{S_i\}$ is the total concentration of free sites of the ligand system and $\{LM\}=\sum\{S_iM\}$ is the concentration of metal bound to the ligand system. When it is difficult to determine the total site concentration also absolute binding values can be used. It follows from [7] that $\{S_t\}$ is essentially a normalization constant, without knowing this constant the non normalized distribution F(logK) can be obtained as $F(\log K)=\{S_t\}f(\log K)$.

Gamble and Buffle and co-workers use a different starting point for the heterogeneity analysis. Their approach is based on the application of the mass action law to the overall ligand system. The metal binding to the ligand system as a whole is in this case represented as:

$$L + M \underset{\leftarrow}{\overset{\rightarrow}{}} LM \qquad [8]$$

and for each value of [M] the law of mass action is applied:

$$\overline{K} = \frac{\{LM\}}{\{L\}[M]} \qquad [9]$$

$\overline{K}$ is called the weighted average equilibrium function (7,8). Gamble and later Buffle have elaborated this concept and analysed $\overline{K}$ or quantities derived from it in great detail (1-4,7,8). We will discuss these shortly in section 3.4

3. MATHEMATICAL TECHNIQUES FOR HETEROGENEITY ANALYSIS

3.1 Introduction

Equation [6] is the basic binding equation for ion binding to heterogeneous systems. Experimentally one may determine the concentration of free metal [M] and the total amount that is bound, leaving the distribution function f(logK) as the unknown. When the integral equation can be solved for f(logK) information can be obtained about the (chemical) heterogeneity on the basis of an adsorption experiment, without making a priori assumptions with respect to the distribution function. However Eq.[6] is a so called Fredholm integral equation of the first kind which is difficult to solve exactly (numerically and analytically) (39,40). Mathematically the problem is ill-posed, that is to say, small errors in θ_t show up in the distribution function as large spurious peaks. In literature several numerical methods (20,24,41) are presented that render good results for data that cover a large range of concentrations and surface coverages. In practice, however, often only a 'data window' can be measured and the experimental error is relatively large. Therefore approximations were sought in order to be able to solve Eq.[6] analytically, without making assumptions with respect to the distribution function.

Here we concentrate on the semi-analytical methods and we will treat successively the Local Isotherm Approximation (LIA) methods, the Affinity Spectrum (AS) methods and the Differential Equilibrium Function (DEF) method. The first two are directly based on the general binding equation, Eq.[6], the latter method starts with the average equilibrium function, Eq.[9], and uses the binding equations [4] and [5] for the derivation of the expression for f(logK).

The various methods will be applied to an ideal synthetic data set with a known underlying distribution in order to be able to compare the different methods with respect to their ability to recover distributions from a 'measured' binding curve.

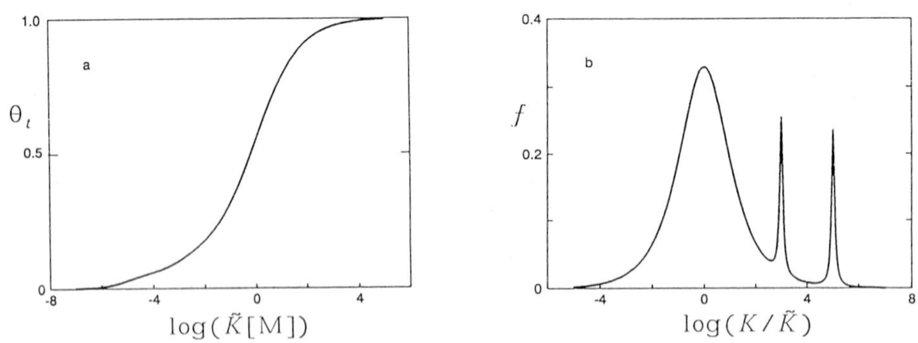

Figure 1: Overall binding function (a) and distribution function (b) corresponding with the synthetic data set, for details see text.

Figure 1a shows the binding function which is the result of a weighted addition of three so called Langmuir-Freundlich isotherms that are defined by:

$$\theta_t = \frac{(\tilde{K}[M])^m}{1 + (\tilde{K}[M])^m} \qquad [10]$$

where m is a measure for the heterogeneity and $\tilde{K}$ is the average affinity. The distribution function corresponding with Eq.[10] is given by Sips (42):

$$f(\log K) = \frac{\ln(10)\sin(\pi m)}{\pi\{2\cos(\pi m) + (K/\tilde{K})^{m} + (K/\tilde{K})^{-m}\}} \qquad [11]$$

When m=1 Eq.[10] is equal to the homogeneous (Langmuir) binding function [4], the smaller m (0<m<1) the larger is the heterogeneity. The distribution function given by Eq.[11] is a semi-gaussian symmetrical peak. The overall distribution function underlying the binding function shown in figure 1a is a combination of three such distributions with peak positions at $\log\tilde{K}$ = 0, 3 and 5 respectively, m values of 0.5, 0.95 and 0.95 and weight fractions 0.9, 0.05 and 0.05. So two small and narrows peaks are superimposed on a large broad peak. Figure 1b shows the result, where $\tilde{K}$ is the mean affinity of the large broad peak. The chosen example corresponds with a situation with a large majority of 'background sites' represented by the broad peak and a few nearly discrete 'dominant site' types superimposed on it. The example is similar to the synthetic data set used by Buffle (4).

3.2 Local Isotherm Approximations (LIA)

3.2.1. Basic principle

The basis of the Local Isotherm Approximation (LIA) methods is that the local binding function Eq.[4] is replaced by another function, 'the LIA isotherm', in order to solve Eq.[6] analytically for f(logK) (23). The more closely the LIA isotherm, indicated by θ_{LIA}, follows the true local isotherm, the more closely approximates the calculated distribution function, indicated by f_{LIA}(logK), the true one. In general Eq.[6] is effectively replaced by:

$$\theta_t = \int_{\Delta} \theta_{LIA} f_{LIA}\, d\log K_{LIA} \qquad [12]$$

The different LIA isotherms are all of the form:

$$\theta_{LIA} = \alpha_1 (K[M])^{\beta} \quad \text{for} \quad K[M] \le K^*[M]^* \qquad [13a]$$

$$\theta_{LIA} = 1 - \alpha_2 (K[M])^{\beta} \quad \text{for} \quad K[M] > K^*[M]^* \qquad [13b]$$

where α_1, α_2 and β are parameters that determine the shape of the LIA isotherm and $K^*[M]^*$ is the 'breakpoint' in the isotherm. The fact that the local isotherm is split in two parts has the consequence that the integral equation is also split in two parts which are no longer Fredholm integrals of the first kind. For each concentration $[M]^*$ one may write:

$$\theta_t([M]^*) = \int_0^{\log K^*} \theta_{LIA,1} f_{LIA} d\log K_{LIA} + \int_{\log K^*}^{\infty} \theta_{LIA,2} f_{LIA} d\log K_{LIA} \qquad [14]$$

where logK* is related to the breakpoint in the local isotherm, which is determined by a best fit criterion and is a function of the concentration [M]*. Depending on the values of α_i and β, expressions can be found for the distribution function on the basis of Eq.[14].

3.2.2. Condensation Approximation.

A classical and very simple LIA method is the Condensation Approximation (CA) (25,26). In this method the θ_{LIA} is a step function. This is equivalent with assuming that the sites are filled sequentially in the order of decreasing affinity, that is the strongest sites are occupied first and the weakest sites the last. This means also that at an infinitesimal increase in concentration only one site type is involved that goes from completely empty to completely covered. In the Condensation Approximation θ_{LIA} becomes:

$$\theta_{CA} = 0 \quad \text{for} \quad K[M] \le K^*[M]^* \qquad [15a]$$
$$\theta_{CA} = 1 \quad \text{for} \quad K[M] > K^*[M]^* \qquad [15b]$$

the breakpoint is uniquely found to be $K^*M^* = 1$ (26), for θ given by Eq.[4] (see Fig. 2a). The relation $K_{CA}=1/[M]$ also gives the transformation from the log[M] axis to the logK axis.

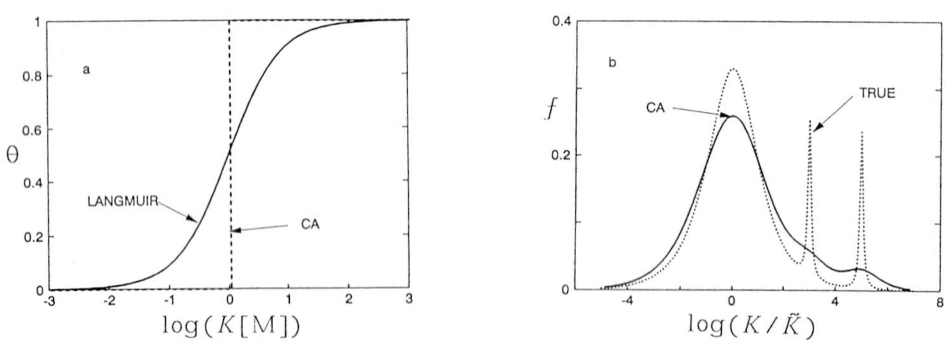

Figure 2: Condensation Approximation. a) The CA isotherm and the true local isotherm. b) The CA distribution for the synthetic data set compared with the true distribution.

By substituting θ_{CA} for θ_{LIA} in Eq.[14] one obtains:

$$\theta_t = \int_{\log K^*}^{\infty} f_{CA} d\log K_{CA} \qquad [16]$$

where $K^* = 1/[M]^*$ is any value of K_{CA}. Note that in reality the sites are filled in the opposite order as the integration of Eq.[16], namely the highest affinity sites first. Mathematically, however the resulting integration is the same. Inversion of the integral is simply done by taking the first derivative with respect to log[M]:

$$f_{CA}(\log K_{CA}) = \frac{d\theta_t}{d\log[M]} \qquad [17a]$$

with

$$\log K_{CA} = -\log[M] \qquad [17b]$$

Application of Eq.[17] to the binding function of figure 1a results in the f_{CA} shown in Figure 2b.

Compared to the true distribution the CA distribution is too wide and flat. The first high affinity peak is smoothed but recovered, the second high affinity peak has nearly disappeared and the broad peak is rather well recovered. Taking the simplicity of the method into account the result is not bad. The result is typical in so far that the CA method is not suited to reproduce small details in a distribution.

3.2.3. The logarithmic symmetrical approximation (LOGA).

A fairly good approximation of the true local isotherm as given by Eq.[4] is found with the LOGarithmic symmetrical Approximation (LOGA) for which $\theta_{L/A}$ is given by (23):

$$\theta_{LOGA} = 0.5(K[M])^{\beta} \quad \text{for} \quad K[M] \le K^*[M]^* \qquad [18a]$$

$$\theta_{LOGA} = 1 - 0.5(K[M])^{\beta} \quad \text{for} \quad K[M] > K^*[M]^* \qquad [18b]$$

were $\alpha_1 = \alpha_2 = 0.5$ is chosen to obtain a continuous local isotherm at the breakpoint ($K^*[M]^*=1$) and β can be adjusted to obtain a close fit between the true and the approximate local isotherm. By using a least sum of squares method the value $\beta = 0.7$ was found. The θ_{LOGA} using the above set of parameters is indicated by LOGA-1 in Figure 3a.

Inserting Eq.(18) in [14] results in the LOGA distribution function (23):

$$f_{LOGA}(\log K_{LOGA}) = \frac{d\theta_t}{d\log[M]} - \frac{0.189}{\beta^2}\frac{d^3\theta_t}{d\log[M]^3} \qquad [19a]$$

with

$$\log K_{LOGA} = -\log[M] \qquad [19b]$$

In literature similar expressions for the distribution function have been reported. For $\beta = 0.79$ Eq.[19] corresponds with the result obtained by HSU et al. (29,30) and for $\beta = 1$ a limiting case of the second order affinity spectrum is obtained (23). The latter method will be discussed in more detail in the next section.

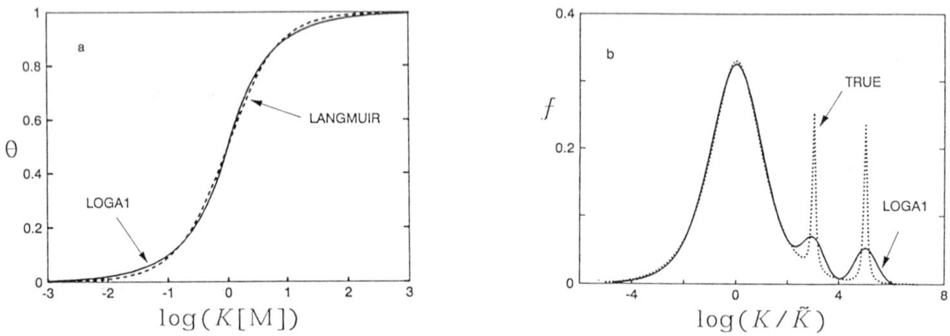

Figure 3: LOGA-1 Approximation. a) The LOGA-1 isotherm and the true local isotherm. b) The LOGA-1 distribution for the synthetic data set compared with the true distribution.

Application of the LOGA method to our synthetic data set leads to the result shown in Figure 3b. With the LOGA-1 method all three peaks are reasonably well recovered. Despite the fact that the LOGA isotherm follows the true local isotherm rather closely, still a non-negligible difference remains with respect to the true distribution. Compared with the CA distribution a considerably better result is obtained. In order to achieve this improvement, besides the first, also a third derivative of the overall binding function is needed. When the method is applied to experimental data, which are prone to errors, this may be a serious drawback as will be illustrated below.

3.3. Affinity Spectrum (AS)

3.3.1. Basic principle

A second way to solve Eq.[6] for f(logK) is presented in the Affinity spectrum (AS) methods. These methods have become well known as a tool to study ligand equilibria through the work of Hunston (9),Thakur et al. (14) and Shuman et al. (13). The work goes back to Nolle (43) and Schwarzl and Staverman (44,45), who studied the viscoelastic behaviour of materials based on an expression similar to the general binding equation given by Eq.[6].

The basic idea of the approach is that the integral equation is transformed in such a way that the kernel in Eq.[6] resembles a Dirac-Delta (indicated by δ) function as close as possible. In general notation this can be written as:

$$\tilde{\theta}_t = \int_\Delta \tilde{\theta} f(\log K) d\log K \qquad [20]$$

where $\tilde{\theta}_t$ and $\tilde{\theta}$ are transformations of θ_t and θ to achieve that $\tilde{\theta}$ resembles a δ function. In principle the distribution function is not changed by the transformation. However, an approximation of the true distribution is obtained when $\tilde{\theta}$ is approximated by a δ in order to solve Eq.[20] for f(logK). With this approximation f_{AS}(logK) at each value of [M] is obtained as:

$$f_{AS}(\log K) = \tilde{\theta}_t([M]) \qquad [21]$$

The transformation of log[M] to logK is depending on the function $\tilde{\theta}$.

3.3.2. First order Affinity Spectrum

A simple transformation that results in a bell shaped kernel with a weak resemblance to a δ function is taking the first derivative of Eq.[6] with respect to log[M]:

$$\frac{d\theta_t}{d\log[M]} = \int_\Delta \frac{d\theta}{d\log[M]} f(\log K) d\log K \qquad [22]$$

The kernel $d\theta/d\log[M]$ is shown in Figure 4a and indicated by AS1.

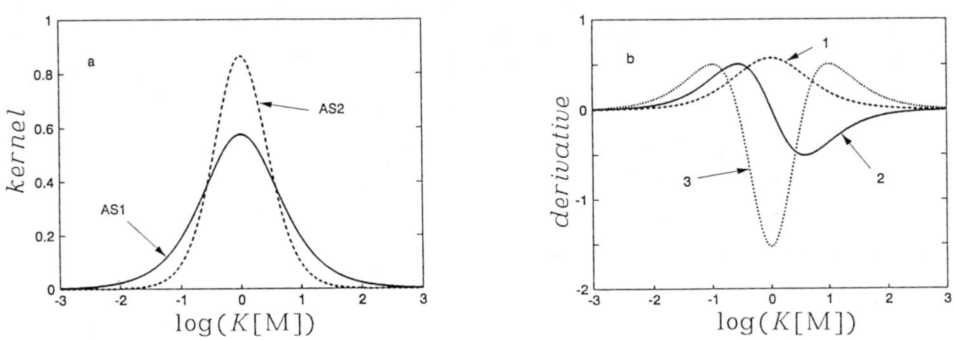

Figure 4: Affinity Spectrum method. a) The kernels of Eq.[20] for the first (AS_1) and second order (AS_2) Affinity Spectrum. b) The first three derivatives of the Langmuir isotherm with respect to log[M] indicated by 1-3.

Approximating this kernel by the δ function results in the CA distribution. The peak position of the kernel (K[M]=1) determines the position of the δ function and thereby the transformation of log[M] to logK. In essence this result was obtained by Nolle (43). Ninomiya and Ferry (46) contributed by following an analogous procedure avoiding derivatives of the overall binding function. They used linear combinations of shifted isotherms over a log[M] axis to transform the integral equation. For the first order Affinity Spectrum their result is:

$$f_{AS1} = \frac{\theta_t(\log[M] + \log\alpha) - \theta_t(\log[M] - \log\alpha)}{2\log\alpha} \qquad [23a]$$

with

$$\log K_{AS1} = -\log[M] \qquad [23b]$$

where log(a) is the distance between two equidistantially spaced data points. For $\log\alpha \to 0$ Eq.[23] reduces to [17], which is the CA result. When data are not equidistant some interpolation technique is needed to apply Eq.[23]. In practice log(a) acts as a primitive smoothing parameter and is advised to be in the range 0.2-0.4 to correct for small experimental errors (46). Since for our synthetic data set f_{AS1} is equivalent with the CA distribution the result is not shown in a separate figure.

3.3.3. Second order Affinity Spectrum.

As the AS_1 kernel is a poor approximation of the Dirac-Delta function, Schwarzl and Staverman (45) developed a procedure to obtain sharper kernels using the first derivative in combination with higher derivatives. Their third order method implies a linear combination of the first three derivatives of the overall binding function with respect to log[M]. The coefficients in front of the derivatives were determined by formulating a number of constraints on the resulting kernel of the transformed integral Eq.[20]. The constraints of Schwarzl and Staverman (45) reformulated for the present problem are:

1) The kernel should always be positive, since both the affinity distribution and the overall binding function are positive.

2) The kernel should be normalized to 1.

3) The kernel should have one maximum at $\log K = -\log[M]$ and become zero at $\log K = -\infty$ and at $\log K = +\infty$.

4) The kernel should stay symmetrical in order to obtain the sharpest possible kernel.

Applying the third order method to our problem with Eq.[4] as local isotherm results in:

$$f_{LOGA-AS}(\log K_{LOGA}) = \frac{d\theta_t}{d\log[M]} - 0.189\frac{d^3\theta_t}{d\log[M]^3} \qquad [24a]$$

with

$$\log K_{LOGA} = -\log[M] \qquad [24b]$$

This result is equivalent to a LOGA expression with $\beta = 1$ (see Fig. 5a).

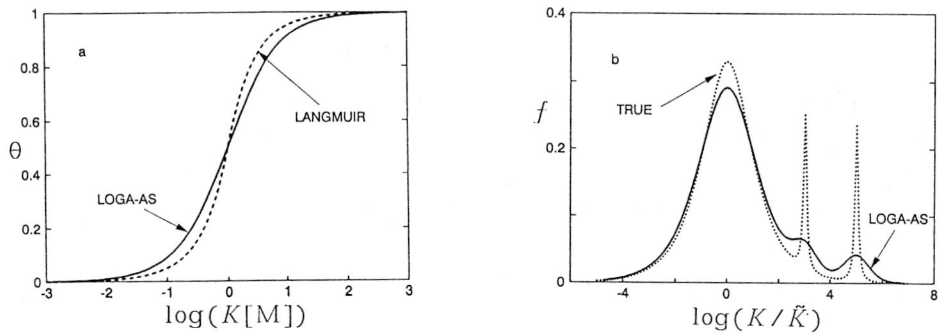

Figure 5: Second order Affinity Spectrum. a) LOGA-AS isotherm and the true local isotherm. b) The LOGA-AS distribution for the synthetic data set compared with the true distribution.

Striking in the analysis is that the second derivative disappears. This is understandable when the derivatives of the local isotherm are considered. In Figure 4b it is shown that the second derivative is asymmetrical and thus constraint 4 is not obeyed.

Ninomiya and Ferry (46) again avoided the use of derivatives and obtained what is known as the second order affinity spectrum:

$$f_{AS2}(\log K = -\log[M]) = \frac{\theta_t([M]\alpha) - \theta_t([M]/\alpha)}{2\log\alpha} -$$

$$\frac{\alpha}{(\alpha-1)^2}\frac{(\theta_t([M]\alpha^2) - \theta_t([M]/\alpha^2)) - 2\,(\theta_t([M]\alpha) - \theta_t([M]/\alpha))}{2\log\alpha} \qquad [25]$$

Similar as with the first order Affinity Spectrum Eq.[25] reduces for $\log\alpha \to 0$ to Eq.[24]. Note that also here a third derivative, or a numerical approximation of it is needed.

Application of the LOGA-AS method gives the distribution presented in Figure 5b. Since the coefficient in front of the third derivative is smaller than with LOGA-1 the details are somewhat less sharp. Further the result is nearly the same due to the similarity in both equations.

3.4. Differential Equilibrium Method (DEF)

3.4.1. Average Equilibrium Function

The DEF method is based on application of the Mass Action Law to the ligand system as a whole, resulting in Eq.[9], which equals in the normalized form:

$$\overline{K} = \frac{\theta_t}{[M](1-\theta_t)} \qquad [26]$$

For a homogeneous ligand system Eq.[26] equals [4] and the correct affinity constant is obtained. For a heterogeneous system with Eq.[26] at each value of [M] a $\overline{K}$ value is obtained that corresponds with an equivalent homogeneous binding function which, at the given value of [M], results in the same value of θ_t.

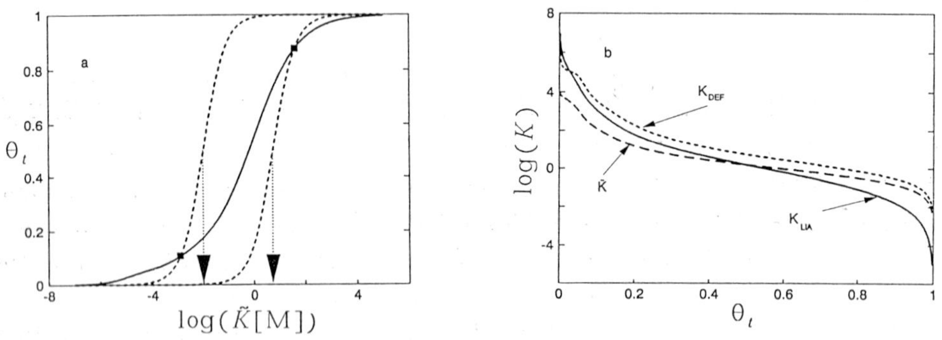

Figure 6: Differential Equilibrium Function. a) Graphical determination of $\overline{K}$ from a binding function for two marked points. b) 'Gamble' plots of $\overline{K}$, K_{LIA} and K_{DEF} as a function of θ_t.

This is illustrated in Figure 6a for two points of the synthetic overall binding function. The $\overline{K}$ value of the homogeneous isotherm that intersects the overall binding curve at the marked points can be read directly on the ordinate axis at the log[M] value corresponding with $\theta_t = 0.5$ as $\log\overline{K} = -\log[M]$, indicated by the dotted lines with the arrows. Note that because of the scaling of the ordinate axis in Figure 6a, the value of $\overline{K}$ is obtained as $\log\overline{K} = -\log\tilde{K}[M] + \log\tilde{K} = -\log[M]$. Calculation of $\log\overline{K}$ at each [M] of the synthetic data set result in decreasing values as a function of θ_t. Figure 6b shows the $\overline{K}$ in combination with K_{LIA} and K_{DEF} (section 3.4.2.). In this way a first impression of the heterogeneity is obtained.

3.4.2. Differential Equilibrium Function (DEF)

A disadvantage of the average equilibrium function is that the total number of sites of the ligand system should be known. Moreover there is no explicit relation between the individual K_i values and $\overline{K}$. To overcome these problems Gamble (7,8) introduced a differential parameter K_{DEF} (see Fig. 6b). The function K_{DEF} was derived by realizing that $\theta_t = \sum f_i \theta_i$ (Eq.[5]) and that

$$\theta_i = K_i[M](1-\theta_i) \qquad [27]$$

which follows from Eq.[4]. Then Eq. [26] may be written as:

$$\overline{K}(1-\theta_t) = \sum K_i f_i (1-\theta_i) \qquad [28]$$

which gives an implicit relation between $\overline{K}$ and the individual K_i values. Then Gamble (8) implicitly assumed that at a small change of $\overline{K}(1-\theta_t)$ only one site type is involved by stating:

$$\Delta(\overline{K}(1-\theta_t) = \Delta\{K_i f_i (1-\theta_i)\} = K_i f_i \Delta(1-\theta_i) \qquad [29]$$

where i indicates that only the coverage of site type i changes at a certain concentration. This is only true in the case that the individual logK values are separated over a large distance (about 4 logK units). For strongly heterogeneous cases, that are considered here, Δ should be taken infinitesimally small and then Eq.[29] implies that sites are filled and emptied sequentially, that is to say, at an infinitesimal decrease in [M] only one site type goes from completely occupied to completely empty and thus $d(1-\theta_i) = 1$. Since in reality the sites are not filled and emptied sequentially K_i and f_i are replaced by their approximated values $K_{i,DEF}$ and $f_{i,DEF}$. For a continuous distribution $f_{i,DEF}$ should be replaced by $f_{DEF}d(\log K_{DEF})$ and Eq.[29] is written as:

$$d\{\overline{K}(1-\theta_t)\} = K_{DEF} f_{DEF} d\log K_{DEF} \qquad [30]$$

The fraction of sites $f_{DEF}d(\log K_{DEF})$, which is occupied, is responsible for the change in $1-\theta_t$:

$$d(1-\theta_t) = f_{DEF} d\log K_{DEF} \qquad [31]$$

Combination of Eqs. [30] and [31] gives the differential equilibrium function of Gamble (7,8):

$$K_{DEF} = \frac{d\{\overline{K}(1-\theta_t)\}}{d(1-\theta_t)} \qquad [32]$$

Note that in order to arrive at this result, $1-\theta_t$ in the right hand side of Eq.[27] is in effect replaced by an approximation (sequential filling). The consequences of this aspect will be discussed in the next section. Application of Eq.[32] to our synthetic data set results in the K_{DEF} plot shown in Figure 7b.

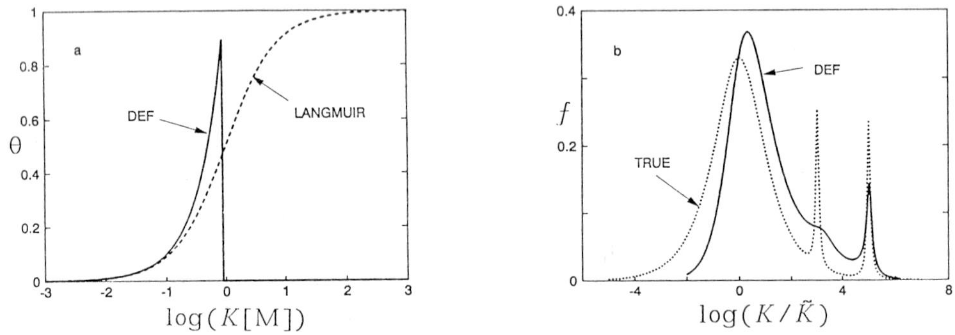

Figure 7: Differential Equilibrium Function. a) The LIA interpretation of the DEF method. b) The DEF distribution for the synthetic data set compared with the true distribution.

Buffle (1-4) has extended the method by defining the distribution function $f(\log K_{DEF})$, which follows directly from [31] and can be written as:

$$f_{DEF}(\log K_{DEF}) = -\frac{d\theta_t}{d\log K_{DEF}} \qquad [33]$$

In order to calculate f_{DEF} first $\log K_{DEF}$ is calculated with Eq.[32] for each value of θ_t, then the derivative of θ_t with respect to $\log K_{DEF}$ is taken. Figure (7b) shows the DEF distribution for the synthetic data set. The high affinity peak at logK=5 is recovered extremely well. The second peak at logK=3 is reproduced very weakly and the broad peak is shifted considerably to higher affinities. So the DEF method seems especially sensitive to find the peak of highest affinity sites.

3.4.3. Interpretation of DEF

Nederlof et al. (47) found that the DEF method can also be interpreted within the LIA concept. Introducing the sequential filling concept in the right hand part of Eq.[27] gives the θ_{DEF} function (see Fig. 7a):

$$\theta_{DEF} = K_{DEF}[M] \quad \text{for} \quad K[M] \le K^*[M]^* \qquad [34a]$$

$$\theta_{DEF} = 0 \quad \text{for} \quad K[M] > K^*[M]^* \qquad [34b]$$

Since $K^*_{DEF}[M]^*$ is not constant, the breakpoint varies with K_{DEF}. Substitution of Eq.[34] in [14] and dividing by [M] gives:

$$\frac{\theta_t}{[M]} = \int_0^{\log K^*_{DEF}} K_{DEF}\, f(\log K_{DEF})\, d\log K_{DEF} \qquad [35]$$

The distribution function can be obtained from Eq.[35] by taking the first derivative with respect to $\log K_{DEF}$:

$$f(\log K_{DEF}) = -\frac{d(\theta_t/[M])}{d\log K_{DEF}}\frac{1}{K_{DEF}} \qquad [36]$$

Combination of [36] with the definition of K_{DEF} (Eq.[32]) results in the distribution function as given by Buffle (Eq.[33]).

According to the analysis above, the DEF method can thus be interpreted as another approximation of the local isotherm. The main difference with the methods presented in the LIA section is however that the coverage of θ_{DEF} drops to zero at high concentration, which is physically not realistic.

4. HETEROGENEITY ANALYSIS ON DATA WITH A RANDOM ERROR.

4.1 Introduction

From section 3 it turns out that for all semi-analytical methods for heterogeneity analysis, derivatives are needed of the binding function. In addition, especially in the case of the affinity spectrum the data should be equidistantially spaced on the log[M] scale. As derivatives are rather sensitive towards random errors in the data, we will study the effect of a random error in the data on the distribution function.

Let us assume that the equilibrium concentration of free M is determined experimentally and that the total amount of metal present ($[M]_t$) is known. The amount of metal bound to ligands ($[M]_b$) can be obtained as $[M]_t$-[M]. The error in log[M] is assumed to be randomly distributed with a variance of $\sigma^2 = 4.10^{-6}$, which corresponds with an error in the concentration of .5 percent. The error in $[M]_t$ is assumed to be negligible. This means that the data are of extraordinary good quality. We simulated 71 ($[M]_t$,log[M]) data points, equidistantially placed on the log[M] scale over 7 decades, using the binding function of Fig. 1a, leading to a new synthetic data set. The error in log[M] was obtained with a random generator.

The new data set thus consists of log[M] values as a function of the total metal concentration $[M]_t$. Since the amount bound metal can be obtained as $[M]_t$-[M] the total metal concentration should be preferably (much) higher than the free concentration in order to obtain a reliable value for $[M]_b$. In practice therefore only a part of the concentration range as presented in Figure 1a is accessible in one experiment. The concentration range below $\log(\tilde{K}[M]=0)$ was used for this example as indicated in the figure. This concentration range covers the high affinity sites.

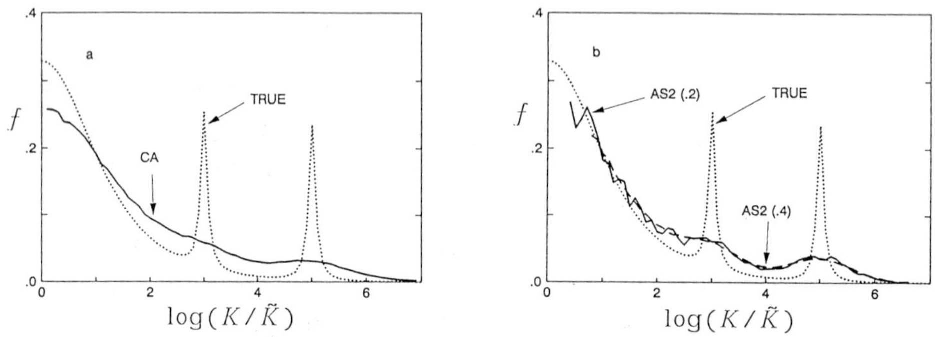

Figure 8: Heterogeneity analysis result, based on the synthetic data set with a small random error (with variance $\sigma^2 = 4.10^{-6}$ in log[M], 71 data points). a) The CA distribution and the true distribution. b) The second order Affinity Spectrum for log(a)=0.2 and 0.4 together with the true distribution. For log(a)=0.2 the most fluctuating result was obtained.

Since it is to be expected that most problems will occur at the higher derivatives, first the CA method is applied. Numerical calculation of the first derivative gives the result shown in Figure 8a. Since the error is small, the distribution function has no unrealistic peaks. However, when the LOGA is applied large spurious peaks result due to the third derivative. In general it can be said that the higher the derivative used in the calculation, the more spurious peaks occur. Applying the AS2 method with log(a)=0.2 and log(a)=0.4 results in Figure 8b. Note that 'real' data are rarely equidistant on the log[M] scale, at the most on the $\log[M]_t$ scale, so that in such cases an interpolation technique is required to create an equidistant data set. Since the error is small and the difference between $[M]_t$ and [M] is sufficiently large, the smoothing of the Affinity Spectrum method with log(a)=0.4 is enough. However, for larger errors a more sophisticated smoothing procedure is required. In addition it would be useful to adjust the amount of smoothing to the magnitude of the error. Fish et al. (6) and Turner et al. (15) who had problems with the application of the affinity spectrum to experimental data reached a similar conclusion.

4.2 Smoothing techniques.

4.2.1. Introduction

The type of smoothing should be independent of the type of binding function, and only dependent on the magnitude of the error. For very accurate data little smoothing is needed, for large errors much smoothing will be necessary. In addition it is very useful to have a procedure available for which the magnitude of the error is not needed beforehand.

A well known method for describing data is fitting with a polynomial. A drawback of this technique for the present problem is that a certain type of function is imposed, and thus a certain type of distribution function. Therefore the smoothing spline technique is to be preferred, which makes no a priori assumptions with respect to the type of functionality.

4.2.2. Smoothing splines.

An 'ordinary' spline is a sequence of polynomials that connect so called 'knot positions', here the experimental data will be taken as knots. The polynomial been two successive knots is fitted in such a way that at the knots the spline is continuous up to the vth derivative, where v is the order of the spline. An vth order spline corresponds with a (2v-1) order polynomial function between each pair of knots (33). When the data are not exact, it is not realistic to develop a spline through the measured points, since there is a range of uncertainty around them. For the case of fitting a smooth curve through non exact data, the smoothing spline was introduced (31-35). The amount of smoothing is controlled by the smoothing parameter p. The value of p regulates the trade off between the goodness of fit and the smoothness of the curve. The goodness of fit is calculated as the weighted sum of squares of the difference between the smoothed spline values (indicated by $s_p(x_i)$) and the measured values $y_i(x_i)$. The smoothness of a vth order spline is obtained as the integral of the vth derivative $s_p^{(v)}(x)$ over the whole x range. For a given smoothing parameter p the spline coefficients are obtained by minimizing the function C_p:

$$C_p = \sum_{i=1}^{n} w_i\{y_i - s_p(x_i)\}^2 + p\int_{-\infty}^{+\infty} | s_p^{(v)}(x) |^2 dx \qquad [37]$$

where w_i is a weighting factor for the case the variances of the individual points are not the same and n is the number of data points. The higher the value of p the more weight is given to the smoothness of the curve compared to the goodness of fit. For low orders of the spline, higher derivatives become discontinuous or zero. The derivatives of the spline needed for a heterogeneity analysis method should be continuous in order to obtain continuous and smooth distribution functions. We suggest a quintic spline (v=3), all methods presented can be used in combination with this type of spline, since a quintic spline is continuous up to the fourth derivative. By this choice, the order of the spline is fixed and independent of the heterogeneity analysis method to be used. The smoothing parameter, however, should be coupled to the magnitude of the experimental error.

4.3. Determination of the smoothing parameter.

4.3.1 Introduction

As mentioned above, the smoothing parameter determines the trade off between the smoothness of the curve and the goodness of fit. Ideally the smoothing parameter is large enough to assure that fluctuations as result of experimental error are smoothed out but small enough that inflections due to functional groups are conserved. However, when the latter inflections are only as large as the deviations resulting from experimental errors, no method can discriminate between the two and the inflections will thus be smoothed away.

In addition, the spline should fulfil some 'physical' constraints which help to find the appropriate smoothing parameter. Firstly, the binding function should be monotonically ascending since at increasing concentration the amount complexed sites will be higher or at least stay equal. This means that the first derivative of θ_t with respect to the log[M] should be zero or positive. The remaining constraints depend on the type of local isotherm. By accepting the Langmuir isotherm as local isotherm Merz (48) has shown that the binding curve plotted as $\theta_t([M])$ should be concave. This constraint is called the 'concavity criterion'. Mathematically this constraint requires that the second derivative of θ_t with respect to [M] should be negative or zero. The third constraint follows from the affinity spectrum derivation, but is generally valid, it states that:

$$\frac{d\log\theta_t}{d\log[M]} - 0.189\frac{d^3\log\theta_t}{d\log[M]^3} \geq 0 \qquad [38]$$

The way in which the smoothing parameter is found depends on whether or not the magnitude of the error is known beforehand. However, the physical constraints should be fulfilled in both cases.

4.3.2 Determination of smoothing parameter for known error.

The known error can be used to obtain the correct smoothing parameter. For each p an estimate of the 'true mean squared error' is obtained that corresponds with the spline obtained with that value of p. The absolute value of the difference between the calculated variance based upon the spline procedure and the known variance as a function of the smoothing parameter can be minimized to obtain the optimum value of the smoothing parameter. In Figure 9a the difference function (ϵ) normalized by the known variance for our synthetic data set (variance $\sigma^2 = 4.10^{-6}$) is shown.

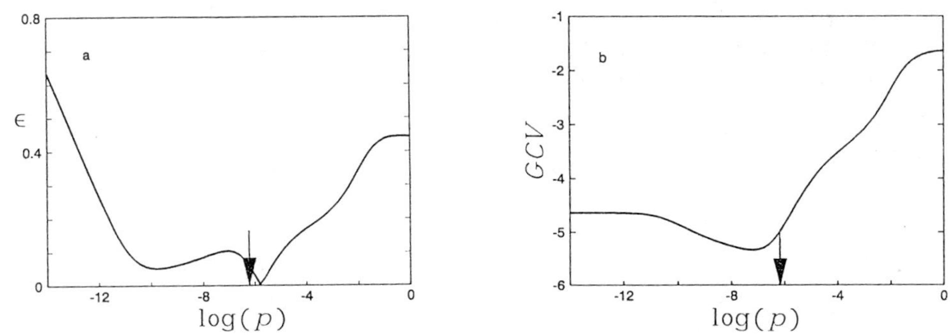

Figure 9: Determination of the smoothing parameter. a) For a known error (variance $\sigma^2 = 4.10^{-6}$) log(p) is determined by the difference function $\epsilon = (\log\hat{\sigma}^2 - \log\sigma^2)/\log\sigma^2$ which is the relative difference between the spline estimate of the variance and the true variance. The minimum log(p) value corresponding to the spline that fulfils the three physical constraints is indicated with the arrow.

This procedure works well for large data sets where the minimum in ϵ also satisfies the other constraints. For small data sets the procedure might be less reliable. In general the physical constraints should be used to check whether or not the obtained smoothing parameter is large enough. For the chosen example, with 71 data points, the smoothing parameter (log(p)=-5.8) obtained from the difference function was appropriate, the thus obtained smoothing parameter was large enough to fulfil the physical constraints (see Fig. 9a). Figure 10 shows the CA, LOGA, AS2 and DEF distributions based on the resulting spline.

Comparing this result with the ones obtained for exact data shows first of all that only that part of the distribution is obtained that corresponds with the concentration range of the experiment, furthermore it becomes clear that, due to the presence of a small error, some detail is lost.

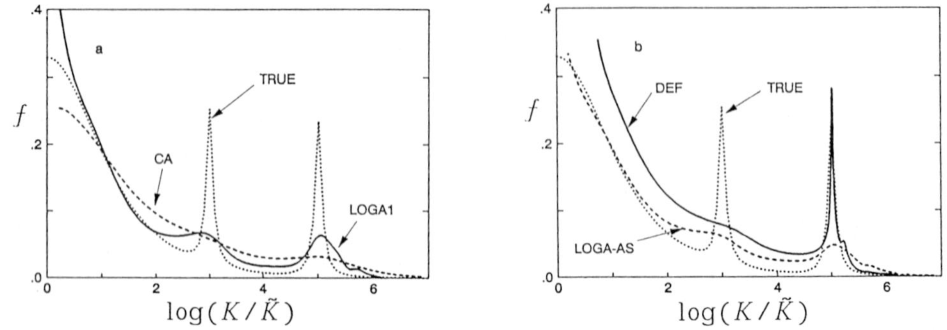

Figure 10: Heterogeneity analysis based on the constrained smoothing spline, for the synthetic data set containing a small random error (variance $\sigma^2 = 4.10^{-6}$ in log[M], 71 data points). a) The CA and $LOGA_1$ distributions, b) the AS_2 and DEF distributions compared with the true distribution.

The above result was obtained with an error of only 0.5 percent in log[M], which is hard to realize. More realistic is a relative error of 5 percent. A data set composed of the same 71 points, but now with an error of 5 percent, leads to the distributions shown in Figure 11. These distributions show considerably less detail than those shown in Figure 10.

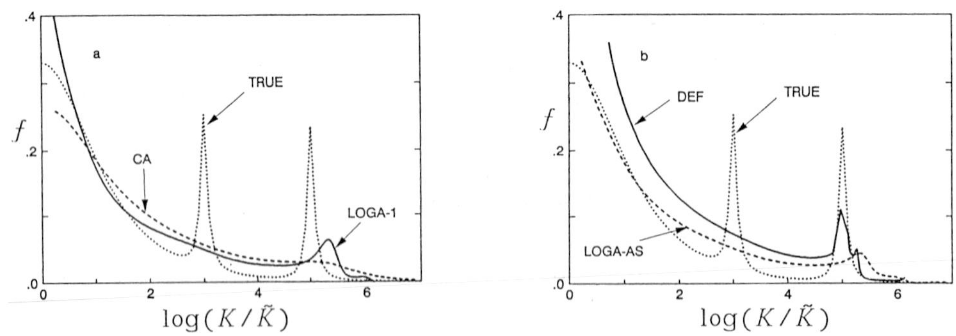

Figure 11: The same as figure 10, but now for data containing a large random error (variance $\sigma^2 = 4.10^{-4}$ in log [M]).

The high affinity peak at logK=5 is still reasonably well recovered, but the peak at logK=3 has disappeared. Thus, in order to recover a detailed distribution function, accurate data are required.

The examples presented in the figures above, were made with 10 data points per decade, 71 in total. In literature often data sets are found that contain only 20 data points or even less. This is also the case with the example we will treat in detail in section 5. When the spline procedure is applied to small data sets rather smooth distributions result, the lower the number of points the more detail may be lost. It is important to notice that with small data sets, small and narrow peaks can never be obtained reliably.

Finally, the distribution obtained covers only logK values that correspond with the experimental concentration range. For instance, in the examples chosen very low affinity sites could not be detected. In reality also very low concentrations may not be measurable. In that case also the high affinity part of the distribution function remains undetected.

4.3.3. Determination of the smoothing parameter for unknown error.

Often the magnitude of the experimental error is not known or only roughly known. In such cases methods are needed that determine the amount of smoothing independently. Craven and Wahba (36) showed that a suitable criterion to obtain the smoothing parameter is to minimize the so called Generalized Cross Validation (GCV) function. Without giving details the main principle is as follows. First a spline function is obtained with a given value of the smoothing parameter, but one data point is left out. This spline can be denoted as $s_p^i(x)$. Since a spline is able to interpolate between the original data points, an indication about the amount of smoothing needed, is obtained from the ability of the spline to reproduce data point i that was left out in the calculation. This procedure has to be repeated for each data point i. The Cross Validation (CV) choice of the smoothing parameter is the value of p which minimizes the function:

$$CV(p) = n^{-1} \sum_i w_i \{y_i - s_p^i(x_i)\}^2 \qquad [39]$$

For computational reasons Craven and Wahba (36) developed a related criterion which they called the Generalized Cross Validation (GCV), which will not be explained in detail here. Instead of the CV, the GCV function is minimized to obtain the optimal smoothing parameter p. Figure 9b shows the GCV function for the nearly exact data set (0.5 percent error, 71 points).

With the GCV method often a too low value for p is obtained. Comparison of the $\log(p_{GCV})=-7.2$ (GCV smoothing parameter) with the $\log(p)=-5.8$ found in section 4.3.2. shows that the difference is considerable. A similar result was found by Merz (48). A way out of this problem is again taking into account the physical constraints derived in section 4.1.. The initial smoothing parameter (p_{GCV}) is raised until all the constraints are fulfilled, then a value of $\log(p)=-6.2$ results, which is close to $\log(p)=-5.8$. For the systems studied above, based on a relatively large data set, it appears that the final smoothing parameter is approximately the same as that found for the case that the error is known, since minimizing the difference function resulted in only a slightly larger value of log(p) than the minimum log(p) obtained from the constraints. However, this result is not general as will be seen in section 5.

In this way the GCV criterion in combination with the physical constraints gives a tool to obtain an optimally smoothed spline when the error is not known. It would be most convenient when constraints are rigorously integrated in the spline software. To our knowledge such programs are not yet available, although constrained splines have got some attention in the mathematical literature (49). As long as these specific spline algorithms are not available, the present sequential approach has to be followed.

Finally it should be noted that the average error in the data, estimated by the spline program, can be compared with a rough error estimate on the basis of the experiment. The result should not be trusted when the spline estimate of the error is much smaller than what seems reasonable.

5. A PRACTICAL EXAMPLE OF HETEROGENEITY ANALYSIS.

5.1. Introduction

In this section we will apply the methodology presented in sections 4 and 5 to a set of real experimental data and discuss the whole procedure to be followed. Figure 12a shows the results of a typical metal binding experiment. The data were taken from Verweij et al. (37) who studied the copper binding characteristics of humic ligands present in Lake Tjeukemeer (The Netherlands). A 0.2 μm filtrate of the lake water was titrated with a copper solution, the free metal concentration was measured with a ion selective electrode (ORION). The pH was kept at 6.0 to prevent copper precipitation, the ionic strength was constant (0.01 M (KNO_3)). The calibration curve of the copper electrode was found to be linear in the range $3.1*10^{-7}$ to $1.5*10^{-4}$ M Cu^{2+}, so the measured potential (mV) was linearly translated into $\log[Cu^{2+}]$ units. For further details see the original paper (37). Important for the moment is that the basic data consist of $\log[Cu^{2+}]$ measured as a function of the total amount copper added to the solution expressed or the total copper concentration $[Cu]_t$. The experimental error in the free copper concentration is likely to be much higher than that in the total copper concentration.

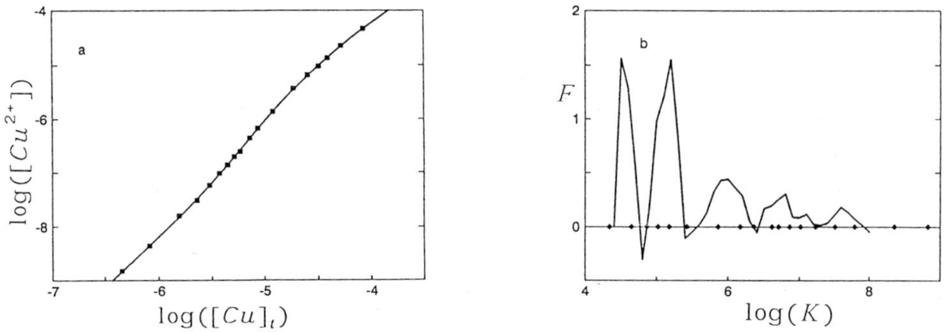

Figure 12: Copper binding to humic ligand material of Lake Tjeukemeer. a) Basic data as obtained from the experiment and the obtained spline function. b) Second order Affinity Spectrum as obtained by Verweij et al.(37).

The binding curve (see Fig. 13b) was obtained by plotting the amount bound copper ($[Cu]_t-[Cu^{2+}]$) as a function of $\log[Cu^{2+}]$ (free copper concentration). To obtain the distribution function Verweij et al. interpolated the points of the resulting binding function linearly to obtain equally spaced data and applied the second order Affinity Spectrum method. This leads to the result shown in Figure 12b. On the ordinate axis the positions of the data points (according to $\log K=-\log[Cu^{2+}]$) are marked. Several sharp peaks result, including some negative ones. Mostly the peaks cover only two or three data points, and it may be questioned whether they are realistic. In order to judge their results more carefully, a rigorous analysis based on the constrained smoothing spline approach will be done.

5.2. Analysis based on a constrained smoothing spline.

Since the basic experimental data are $\log[Cu^{2+}]$ as a function of $[Cu]_t$ we propose to apply the spline procedure to these basic data. When the electrode has a constant error over the whole concentration range there will be a constant standard deviation in the measured potential and thus in $\log[Cu^{2+}]$. This means that the weighting factors w_i in the spline procedure may be set to 1. Note that the electrode was calibrated in the concentration ($3.05*10^{-7}$ to $1.5*10^{-4}$) so that the low concentrations are not covered by the calibration. An estimate of the error for this experiment is hard to give. It seems reasonable however to assume a minimal error in the potential of 1 mV, which corresponds with an error of .03 in $\log[Cu^{2+}]$ ($\sigma^2=9.10^{-4}$). Since this is only a rough estimate of the error, the constrained GCV procedure was applied. The resulting GCV plot is shown in Figure 13a. Two minima occur, namely at log(p)=-6.4 and at log(p)=-3.6. The latter minimum is however deeper and leads to an appropriate value of p. The effect of the physical constraints on log(p) is indicated by the arrow.

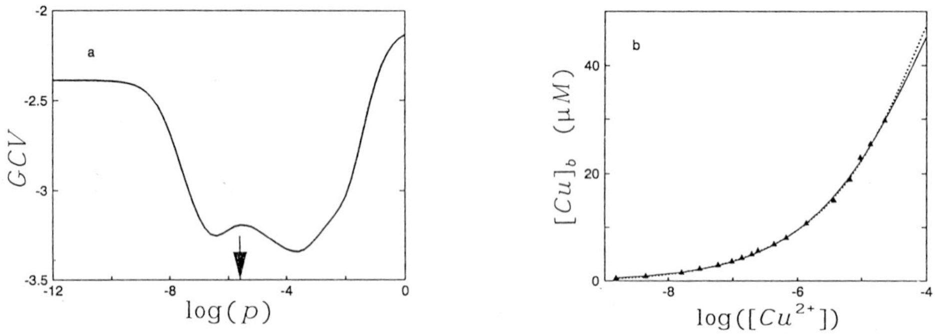

Figure 13: Application of the smoothing spline technique to the Tjeukemeer data set. a) GCV function; the minimum log(p) value corresponding to a spline that obeys the physical constraints is indicated by the arrow. b) The resulting binding function based on the smoothing spline with log(p)=-3.6 and a fit of the Langmuir Freundlich equation (Eq.[10]) (dotted line). The symbols indicate the experimental points.

The constraints require a smoothing parameter of log(p)=-5.6. However the lowest GCV value is found at log(p)=-3.6, and this value fulfils also the physical constraints. Taking the value of log(p)=-5.6 results in a spline that is not smooth enough. The LOGA distribution function based on this spline function is similar with that obtained with the second order Affinity Spectrum method. The error estimate corresponding to log(p)=-5.6 is somewhat low ($\sigma^2=1.10^{-4}$) and corresponds with an error in the potential of 0.3 mV. For log(p)=-3.6, corresponding with the deepest minimum in the GCV function, an error of ($\sigma^2=2.10^{-4}$) is found, corresponding with 0.5 mV, which is still lower than the minimum value estimated above.

For further analysis we prefer the spline obtained with log(p)=-3.6 (see Fig. 12a). Since the spline can easily be interpolated, a large data set (150 points) was generated to obtain the binding curve, bound copper as a function of the free copper concentration (Fig. 13b). Since we have a large number of 'spline data points' the derivatives needed for the heterogeneity can be obtained numerically. The resulting distribution functions are shown in Figure 14.

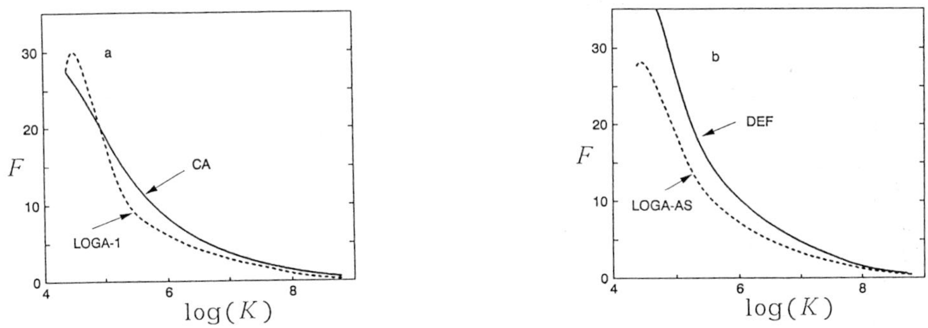

Figure 14: Non normalized distribution functions of the ligands present in Lake Tjeukemeer calculated on the basis of the spline (log(p)=-3.6). a) CA and $LOGA_1$ distributions, b) DEF and LOGA-AS distributions.

With all methods similar, rather smooth distribution functions result that give no indication of individual sharp peaks. This is in agreement with the results obtained with the synthetic data set in section 4, which indicated that details are lost very quickly as the error becomes too large. In view of the above we have to conclude that the distribution function obtained by Verweij et al. should be disregarded.

The smooth and continuous distribution functions as shown in Figure 14 resemble the right part of a gaussian peak, suggesting that a 'single smooth peak' model can be fitted to the binding data of Figure 13b. We tried the Langmuir-Freundlich equation [10] and found a good fit (by minimizing the sum of squares) for m=0.44, logK=3.7 and $\{S\}_t=1.1*10^{-4}$ as shown by the dashed line in Figure 13b. The site concentration is slightly higher than the value estimated by Verweij et al., who found a range of $4.8\text{-}8.5*10^{-5}$. However it should be noted that this site concentration is very hard to determine. So the heterogeneity analysis shows that a simple model is sufficient to adequately describe the metal binding to the ligand system in Lake Tjeukemeer for the given conditions (pH, ionic strength, experimental error). For a more detailed description better data are required.

6. CONCLUDING REMARKS

6.1. Resolution power for highly accurate data.

The ability of the semi-analytical heterogeneity analysis methods to reproduce (known) affinity distributions was tested in the absence of experimental errors . With the LIA methods the result strongly depends on how close the local isotherm is approximated. The CA method is simple and gives reasonable results for wide and smooth distributions. Narrow peaks in a distribution are strongly flattened, but the peak positions are at the correct logK values. The LOGA method reproduces wide and smooth distributions almost perfectly, narrow peaks are somewhat flattened but much better recovered than with the CA method.

For the Affinity Spectrum methods it applies that the sharper the kernel of the transformed general binding integral equation (the kernel is the transformed local binding function) the better is the approximation of the true distribution function. In the absence of experimental errors the AS_1 result is identical to the CA distribution. The AS_2 distribution (for exact data indicated by LOGA-AS) shows somewhat less detail in comparison with the LOGA distribution since the factor in front of the third derivative is lower.

Since the K_{DEF} is a weighted average of the affinities the characteristics of the approximation of a true distribution are less predictable. In general the lower affinity parts of a distribution are shifted to higher affinities, but the method seems sensitive to recover the first high affinity peak. Sharp peaks with affinities lower than the first one are poorly recovered.

6.2 Resolution power in the presence of experimental errors

All the methods presented need the first and or higher derivatives of the overall binding function. Reliable determination of derivatives from experimental data is hardly possible, unless some preprocessing of the data is carried out. Therefore the LIA methods and the DEF method can not be used without preprocessing of the data with a smoothing technique. The Affinity Spectrum method is the only method that has some smoothing capabilities. For very small experimental errors the smoothing of the Affinity Spectrum is enough, for larger errors unrealistic spurious peaks result that can easily be misinterpreted in terms of functional groups. The AS method in fact uses numerical approximations of the derivatives, in principle the same problems arise as with the LIA methods. The sensitivity of K_{DEF} to experimental error is on itself not large since 'only' a first derivative is needed. The DEF distribution however is more sensitive to experimental error when the K_{DEF} remains approximately constant with increasing θ_t . The latter phenomenon occurs around the position of sharp peaks in the distribution, here the derivative $d(\theta_t)/d(\log K_{DEF})$ tends to unrealistic values even when K_{DEF} has only a small error.

6.3 Error treatment with the constrained smoothing spline technique.

The constrained smoothing spline technique is most appropriate to prevent the occurrence of spurious peaks in f(logK) that are the result of experimental errors. Advantages of this method, with respect to for instance curve fitting, are that no type of binding function is assumed a priori and that the amount of smoothing can be coupled to the magnitude of the error.

For known errors the smoothing parameter can be obtained by comparing the estimated error from the spline program for a given smoothing parameter with the known error, using a so called 'difference function' . For unknown errors the GCV technique provides a basis to estimate the smoothing parameter. For large data sets the smoothing parameter obtained with the GCV criterion is probably always smaller than that which would have obtained when the error had been known: the smoothing parameter is underestimated with the GCV criterion. Therefore, in addition to the GCV criterion also a number of physical constraints should be fulfilled. In practice, the procedure used is as follows. First an initial value of p is determined with either the GCV criterion (for unknown error) or the difference function (for known error), then the smoothing parameter is raised till also the physical constraints are fulfilled. It appeared for the test functions that the two approaches (for known and unknown error) result finally in approximately the same amount of smoothing, provided the physical constraints were applied.

The larger the experimental error, the more smoothing has to be applied and the more details in the binding curve are lost. That is to say that details, due to certain functional groups, that are within the error limits will be smoothed away. For small experimental errors the resulting distributions are only slightly less good as compared to the ones obtained with exact data. For larger errors much detail is lost and smooth distributions result. So for large experimental errors never a detailed distribution function can be obtained.

6.4. Application of the constrained smoothing spline to experimental data.

The spline procedure should be applied to the basic data, that is the data that were experimentally obtained without further processing. In this way the variance in the data points is approximately constant. When the experimental data are first processed to the amount of metal bound, large differences in the variance of the error in the data may be the result, which seriously complicates the smoothing procedure. In the example chosen we used log[M] as a function of $[M]_t$ as basic data. After application of the spline procedure the smoothed data should be transformed into a binding curve. On the basis of this binding curve a heterogeneity analysis can be done. One should be aware of the fact that small errors in the spline might blow up to large errors in the binding curve. This occurs for instance at high concentrations when [M] is almost equal to $[M]_{t,}$ and the subtraction $[M]_t$-[M] can not be done reliably. In order to improve the situation the concentration of sites $\{S_t\}$ should be increased to increase $[M]_b$. In practice this means that only a window of concentrations (and site coverages) is accessible. One should be aware of these complications in the design of an experiment. Apart from keeping the experimental errors as low as possible, a sufficiently large number of data points should be measured in order to be able to recover details in the distribution if present. For instance, a duplicate experiment may help to decide whether affinity peaks are due to heterogeneity or are artefacts of the procedure.

By way of example an analysis was made of data of copper binding to humic ligands from Lake Tjeukemeer. Rather smooth distributions resulted, much smoother than the Affinity Spectrum obtained by Verweij et al., that consisted of sharp peaks. The analysis shows that for this data set there is no reliable indication for the presence of sharp peaks in the distribution. On the basis of the calculated distribution function the Langmuir Freundlich equation was selected to describe the binding function. Indeed a good fit was found. This shows how the heterogeneity analysis may help to choose an appropriate binding model for the ligand system under consideration.

Acknowledgment

This work was financed by the Netherlands Integrated Soil Research programme under contract number PCBB 8948.

W. Verweij and H. de Haan (Limnological Institute, the Netherlands) are acknowledged for their willingness to provide us with the original data set of the Copper binding experiment and H.J. Woltring (Biomechanics Consultant) is acknowledged for the GCV spline package and his advice.

REFERENCES

1 J. Buffle, Complexation Reactions in Aquatic Systems: An Analytical Approach, Ellis Horwood, Chichester, 1988.
2 R.S. Altmann and J. Buffle, Geochim. Cosmochim. Acta, 52 (1988) 1505-1519.
3 J. Buffle, R.S. Altmann and M. Filella, Anal. Chim. Acta 232 (1990) 225-337.
4 J. Buffle, R.S. Altmann, M. Filella and A. Tessier, Geochim. Cosmochim Acta, 54 (1990) 1535-1553.
5 S.E. Cabaniss, M.S. Shuman and B.J. Collins, in : C.J.M. Kramer and J.C. Duinker (Eds.), Complexation of Trace Metals in Natural Waters, Martinus Nijhoff, The Hague, 1984, pp. 165-179.
6 W. Fish, D.A. Dzombak and F.M.M. Morel, Environ. Sci. Technol., 20 (1986) 676-683.
7 D.S. Gamble, A.W. Underdown and C.H. Langford, Anal. Chem., 52 (1980) 1901-1908.
8 D.S. Gamble and C.H. Langford, Environ. Sci. Technol., 22 (1988) 1325-1336.
9 D.L. Hunston, Anal Biochem., 63 (1975) 99-109.
10 I.M. Klotz and D.L. Hunston, Biochemistry, 10 (1971) 3065-3069.
11 E.M. Perdue and C.R. Lytle, in : R.F. Christman and E.T. Gjessing (Eds.), Aquatic and Terrestrial Humic Materials, Ann Arbor, 1983, pp. 295-313.
12 E.M. Perdue, in: G.R. Aiken, D.M. Mcknight, R.L. Wershaw and P. Maccarthy (Eds.), Humic Substances in Soil, Sediment and Water: Chemistry, Isolation and Characterization, Wiley Interscience Publ., New York, 1985, pp. 493-526.
13 M.S. Shuman, B.J. Collins, P.J. Fitzgerald and D.L. Olson, in: R.F. Christman and E.T. Gjessing (Eds.), Aquatic and Terrestrial Humic Materials, Ann Arbor, 1983, pp. 349-370.
14 A.K. Thakur, P.J. Munson, D.L. Hunston and D. Rodbard, Anal. Biochem., 103 (1980) 240-254.
15 D.R. Turner, M.S. Varney, M. Whitfield, R.F.C. Mantoura and J.D. Riley, Geochim. Cosmochim. Acta, 50 (1986) 289-297.
16 J.C.M. De Wit, W.H. Van Riemsdijk, M.M. Nederlof, D.G. Kinniburgh and L.K. Koopal, Anal. Chim. Acta, 232 (1990) 189-207.
17 J.C.M. De Wit, M.M. Nederlof, W.H. Van Riemsdijk and L.K. Koopal, Water Air and Soil Pollution (1991), in press.
18 G. Scatchard, I.K. Scheinberg and S.H. Armstrong (Jr), J. Am. Chem. Soc., 27 (1950) 535-540.
19 P. Brassard, J.R. Kramer and P.M. Collins, Environ. Sci. Technol., 24 (1990) 195-201.
20 W.A. House, in: D.H. Everett (Ed.), Colloid Science Specialist Periodical Report 4, Royal Soc. Chem., London, 1978, pp. 1-58.
21 M. Jaroniec and P. Bräuer, Surf. Sci. Rep., 6 (1986) 65-117.
22 L.K. Koopal, M.M. Nederlof and W.H. Van Riemsdijk, Progress Colloid Polymer Sci., 82 (1990) 19-27.
23 M.M. Nederlof, W.H. Van Riemsdijk and L.K. Koopal, J. Colloid Interface Sci., 135 (1990), 410-426.
24 C.H.W. Vos and L.K. Koopal, J. Colloid Interface Sci., 105 (1985) 183-196.
25 S.S. Roginsky, Adsorption Catalysis on Heterogeneous Surfaces, Academy of Sciences U.S.S.R., Moscow, 1948.
26 L.B. Harris, Surf. Sci., 10 (1968) 129-145.
27 G.F. Cerofoloni, Surf. Sci., 24 (1971) 391-403.
28 G.F. Cerofolini, Thin Solid Films, 23 (1974) 129-152.
29 C.C. Hsu, B.W. Wojciechowski, W. Rudzinski and J. Narkiewicz, J. Colloid Interface Sci., 67 (1978) 292-303.
30 W. Rudzinski, J. Jagiello and Y. Grillet, J. Colloid Interface Sci., 87 (1982) 478-491.
31 I.J. Schoenberg, Proc. Nat. Acad. Sci. U.S.A., 52 (1964) 947-950.
32 C. Reinsch, Numer. Math., 10 (1967) 177-183.
33 C. De Boor, A Practical Guide to Splines, Springer-Verlag, New York, 1978.
34 B.W. Silverman, J.R. Statist. Soc. B., 47 (1985) 1-52.
35 H.J. Woltring, Adv. Eng. Software, 8 (1986) 104-107.
36 P. Craven and G. Wahba, Numer. Math., 31 (1979) 377-403.
37 W. Verweij, H. De Haan, T. De Boer and J. Voerman, Freshwater Biology, 21 (1989) 427-436.
38 L.K. Koopal and W.H. Van Riemsdijk, J. Colloid Interface Sci., 128 (1989) 188-200.

39 N. Baker, The Numerical Treatment of Differential Equations, Oxford Univ. Press, Oxford, 1977.

40 B. Noble, in: D. Jacobs (Ed.), The State of the Art in Numerical Analysis, Academic Press, San Diego, 1979, p. 915.

41 W.A. House, J. Colloid Interface Sci., 67 (1978) 166-180.

42 R. Sips, J. Chem. Phys., 18 (1950) 1024-1026.

43 A.W. Nolle, J. Polym. Sci., 5 (1950) 1-54.

44 F. Schwarzl and A.J. Staverman, Physica, 18 (1952) 791-798.

45 F. Schwarzl and A.J. Staverman, Appl. Sci. Res., 4 (1953) 127-141.

46 K. Ninomiya and J.D. Ferry, J. Colloid Interface Sci., 14 (1959) 36-48.

47 M.M. Nederlof, W.H. Van Riemsdijk and L.K. Koopal, submitted for publication.

48 P.H. Merz, J. Comp. Phys., 38 (1980) 64-85.

49 T. Elfving and L. Andersson, Numer. Math., 52 (1988) 583-595.

COMPARISON BETWEEN THREE METHODS OF FLUVIAL SEDIMENT SAMPLING

J.-P. Vernet, P.-Y. Favarger, D. Span and C. Martin

1. INTRODUCTION

Every two months, or six times in 1990, samples were collected from six rivers to compare fluvial sediment sampling methods.

On each of the six rivers, a representative site was selected to collect total suspended sediments (TTS), bottom sediments (SED) and trap sediments (TRAP). The aim of this survey was to determine the validity and the representativity of the geochemical information obtained by each sampling method.

Our study is based on about 17 samples from each representative site of the six rivers which add up to approximately 100 samples. Chemical and sedimentological analyses were carried out on each sample for fractions less than 63 microns only.

Hydrographic stations are situated in the vicinity of every sampling site. We therefore know the daily water discharge near the sampling site for each of the six rivers.

2. SAMPLING METHODS

Suspended sediments were collected by approximately four hours of pumping in the river and continuous flow centrifugation to get a sample of at least 1 gram. Two centrifuges, a Westphalia and an Alfa-Laval, were operated in parallel. Water pumping was executed with particular attention. By a secondary pumping in an overflow splitter, water was maintained in a continuous and intense agitation. The performance of this system is the subject of a preliminary publication (Ref. 1) which deals with the problem of contamination. The first author, S. Santiago, considers that although each centrifuge has a recovery efficiency of 90 to 95 %, the whole system with all necessary manipulations shows a recovery efficiency of at least 75 %.

The method of sediment traps is less complicated. Traps are ordinary PVC cylinders (diameter: 6 cm, hight: 25 cm), ballasted with river pebbles, deposited on the bottom of the river and protected by surrounding pebbles. The fluctuations of the water level and the curiosity or carelessness of people passing by, are disadvantages of this method.

The third method is the easiest one. In a zone of finer grained sediments in the river bed, the first millimeters of deposited sediment are collected. A more sophisticated technique (Ref. 2) may be applied.

3. RELATION BETWEEN GRANULOMETRY AND DISCHARGE

The simple relation between the concentration of suspended loads and the water discharge (Ref. 3) shows a constant increase of concentrations corresponding to the increase of the water discharge. Figure 1 reflects this general relation which seems occasionally to be reversed, as noted for rivers in spate. In that case, the concentration of suspended sediments increases more rapidly than the water discharge at the beginning of a period of high flow (Ref. 4). This phenomenon is then inverted because the available stock of sediments in the basin is drained faster than the water inflow. The general model of relations is relevant for the two examples of Figure 1 shown together with the theoretical curve. The broken line curve stands for a river on the plain or for an area at medium altitude with a hydrological regime of periodic floods corresponding to the precipitations. The dotted line stands for a river originating in the Alps with a high summer flow due to the meltwater of upstream glaciers. Here the theoretical relation seems to be well respected, except for the decrease of suspended load concentrations in the second half of the flood period.

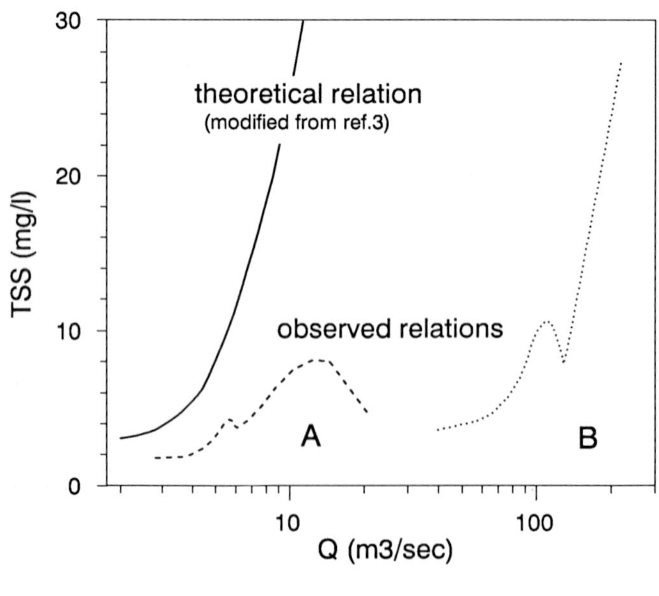

Fig. 1 Relation between discharge and total suspended solids.
A: storm-dependent river system
B: discharge controlled by the upstream glaciers

4. COMPARISON OF THE RESULTS

The only subdivisions introduced are relative to the three sampling methods and the particle size distribution which we suspect to be a determining factor (Fig. 2). For that purpose we took the results of the chemical analysis of two elements: copper and lead. Based on the results, the following comments can be made:

- we note very variable results and in most cases a higher content in the samples of suspended sediments.
- the results of trap and bottom sediments are similar and so is the period of integration of soluble elements in the environment.
- the trap samples show the best correlation coefficient followed closely by the bottom samples.
- granulometry: the fine silts, the predominant fraction, shows the best correlation coefficient.
- on the base of the mean diameter, one always gets an anti-correlation.

If this type of analysis is extended to six elements, taking into account the standard deviation, the general application and the validity of some of the previous findings becomes evident (Figure 3)

- for the relation between the results of trap and bottom sediments,
- for the results of suspended sediments which have generally a higher content and a higher standard deviation.

The chemical analyses were carried out only on the sediment finer than 63 micron which contains all adsorbed complexes. It is therefore necessary to comment on the grain size distribution outlined in Figure 4.

- the similarity of grain size distribution between trap and bottom sediment is striking.
- suspended solids have a smaller mean diameter and smaller grain size values with a higher percentage of clay (0-4 microns) and fine silts (4-32 microns).

These characteristics of suspended sediments explain the higher contents of heavy metals. The dilution is less due to the small percentage of coarse silts (32 - 63 microns) which usually hardly shows any adsorbed complexes. However, this observation doesn't explain the occasional high contents which are rather related to sporadic discharges of contaminants from industry or any other source connected with human activities.

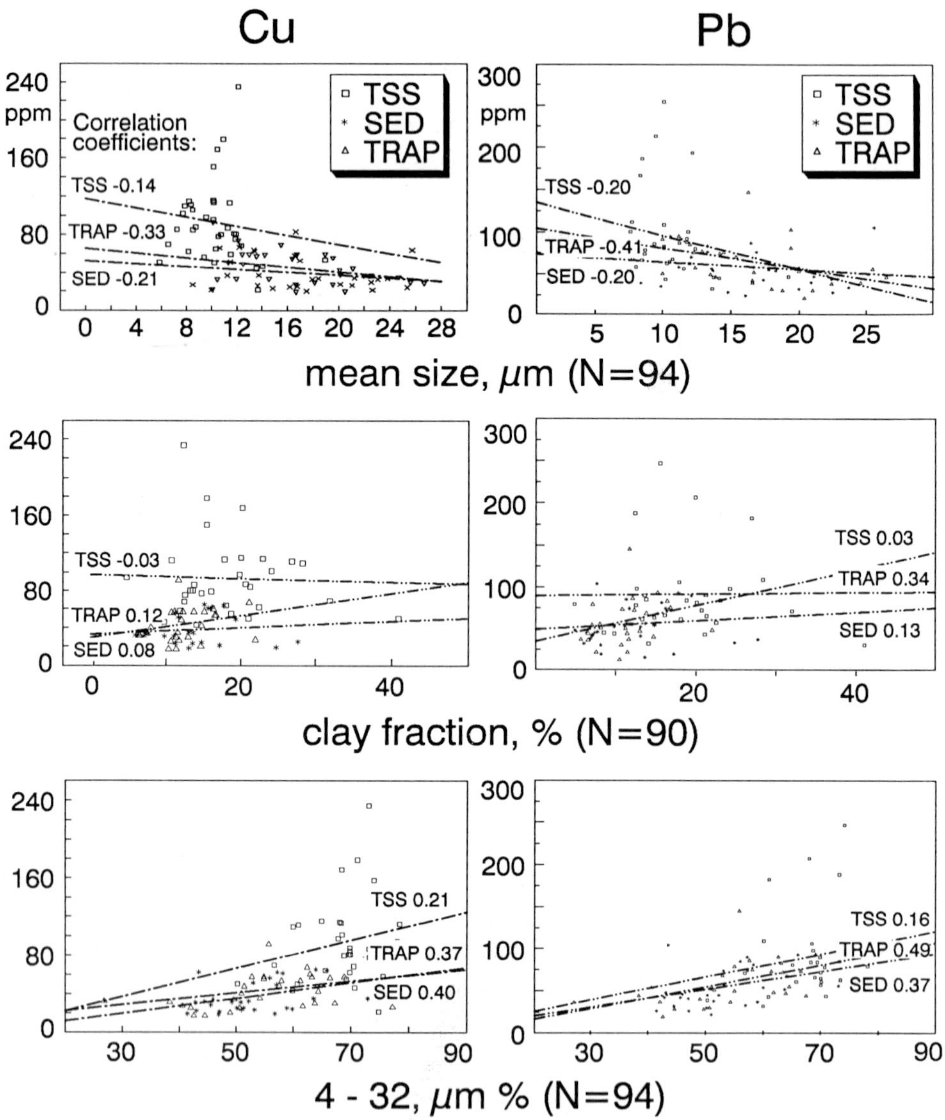

Fig. 2 Analysis of Cu and Pb concentrations on all samples according to the particle size distribution.

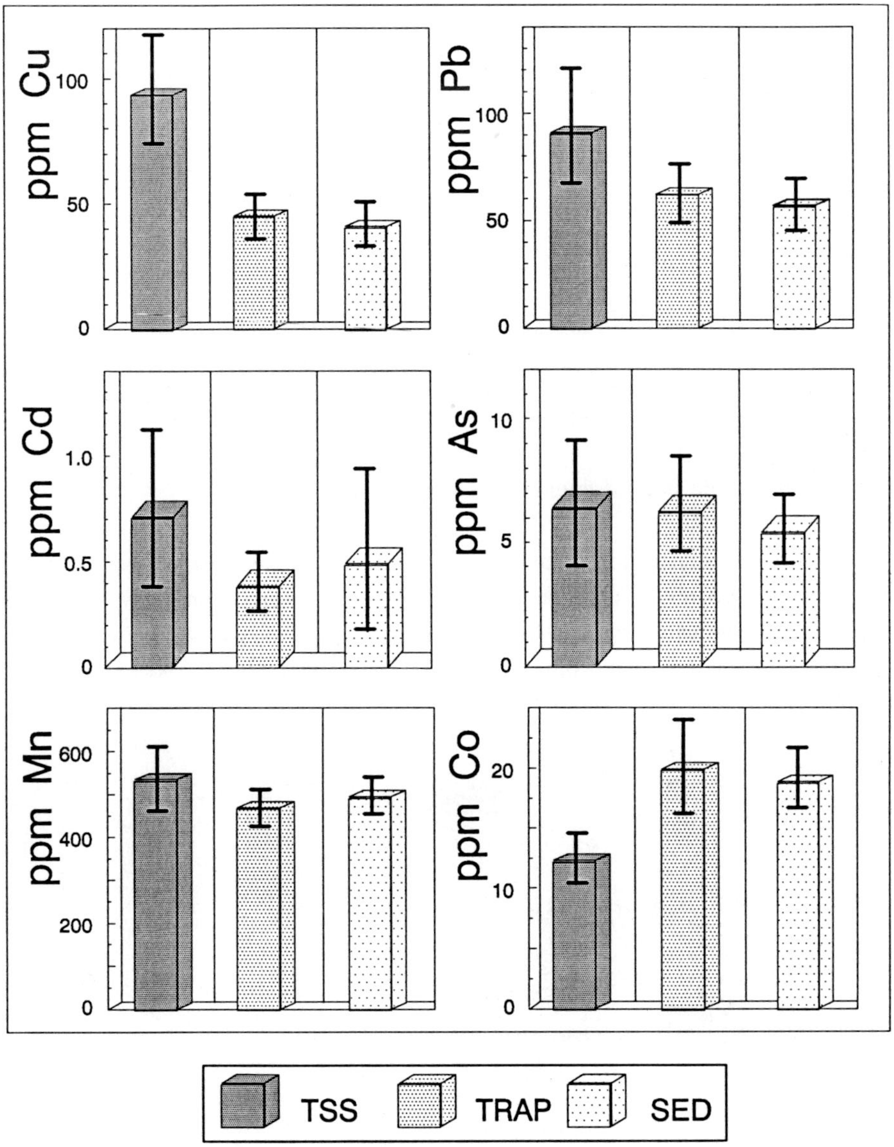

Fig. 3 Comparison of the mean concentrations of 6 metals on all samples according to the sampling method.

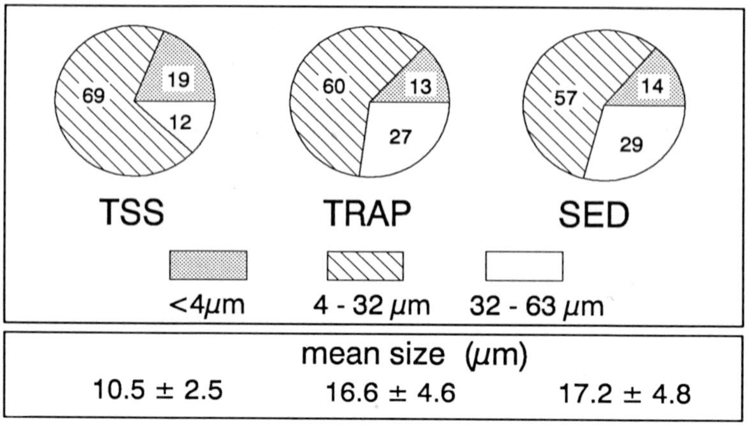

Fig. 4 Mean grain size distribution of ‹63 micron sediment.

Figure 5 is a schematic presentation of this situation with the different integration periods for the sediments of the three sampling methods, ranging from a few hours for the suspended sediments to the interval between two survey campaigns for the trap sediments and the unknown period for bottom sediments which can be longer or shorter than the one of the trap sediments.

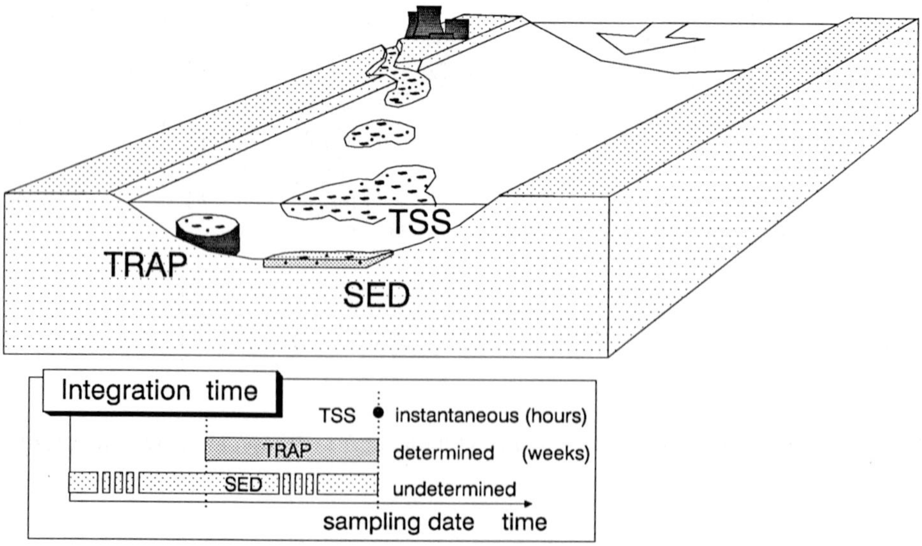

Fig. 5 Spacial and temporal comparison of 3 sampling methods

In an effort to make a synthesis of the geochemical results according to the collecting method and the particle size distribution, Figure 6 classifies the elements based on their correlation coefficient.

It clearly appears that the trap sediments, followed closely by the bottom sediments, show the best correlation coefficients for most heavy metals, and are therefore likely to be the most representative method while the suspended sediments remain more selective.

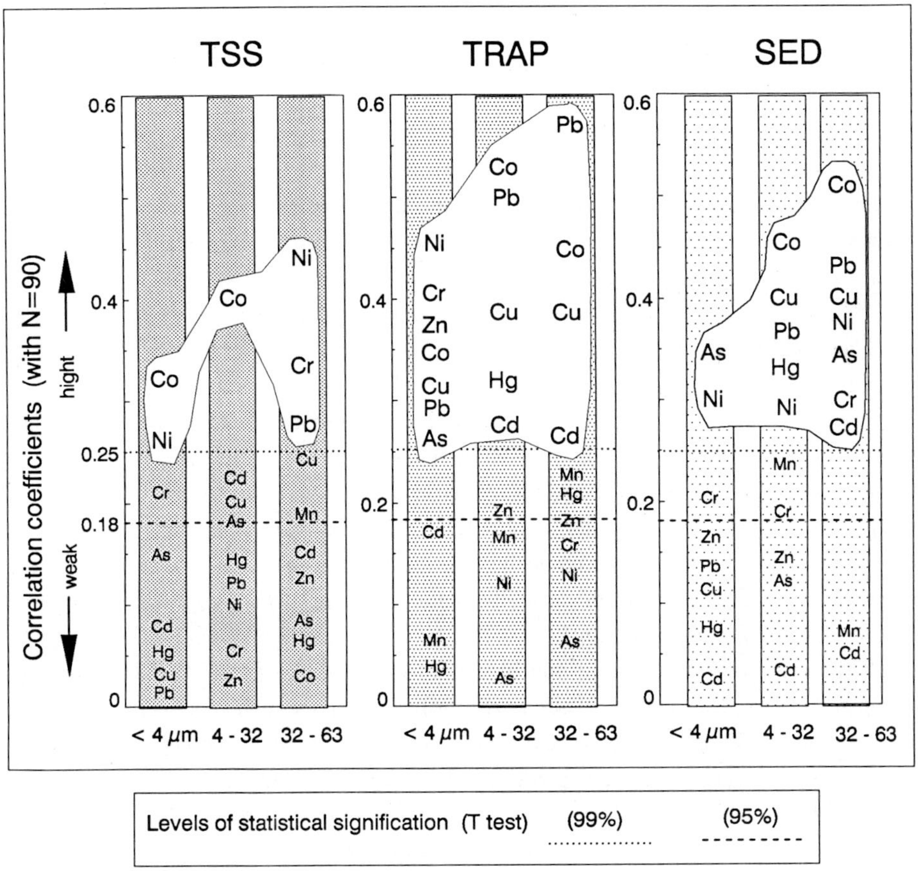

Fig. 6 Relationship of metal to sampling method according to the particle size distribution of the samples.

5. CONCLUSIONS

Despite the limited results and the imperfections of this study, some main conclusions become apparent:

1. The difficult and expensive collection of suspended sediments on a survey with a limited number of interventions gives scattered results with a low correlation coefficient. The values reflect the situation at a specific time and can be seriously modified if an illtimed discharge of contaminants coincides with the sampling period. Only repeated regular sampling over a definite period will ensure the validity of the results.

2. The results of analyses of trap and bottom sediments recovered in sites that favour such deposits are similar and have a nearly identical particle size distribution. Their correlation coefficients are higher than that of the suspended sediments, taking into account all studies of the metal content of all grain sizes. A number of elements follow the same model.

3. The evident conclusions are:

a) The flow of contamination loadings must be determined with suspended sediment collected at close intervals, taking into account the water discharge, and, if possible, the sampling should be function of the water discharge.
b) More sophisticated sediment traps than the one we used for this study are recommended and should be placed at the best locations in the river. We consider that the same vertical position of the traps during the entire sampling, which varies according to the water level, to be the best solution.
c) If the requirements under a) and b) cannot be fulfilled, one has to be satisfied with bottom sediments. The quality of information obtained from such sediments is very good and is usually sufficient. Regular sampling over a year with sampling periods chosen according to the water discharges gives valuable, even excellent results. These results are certainly more valuable than those due to a faulty execution of the method of suspended sediments. The choice of the method depends on the purpose of the analyses as well as on the budget and the local environment.

6. REFERENCES

1 S. Santiago, R.L. Thomas, J.L. Loizeau, P.Y. Favarger and J.P. Vernet, Hydrological Proc., 4 (1990).
2 C.P. Lambert and D.E. Walling, CATENA, 15 (1988).
3 A. Lerman, Geochemical Processes, J. Wiley, 1979.
4 L. Zhang, D.J. Gregor and J.P. Vernet, Revue des Sciences de l'eau, 2, 709-720 (1989).

AUTHOR INDEX

A CAPITAL'S CAPITAL

The Princeton Economic History of the Western World
Joel Mokyr, Series Editor

A list of titles in this series appears in the back of the book.

A Capital's Capital

Two Hundred Years of Wealth and Inequality in Paris

Gilles Postel-Vinay

Jean-Laurent Rosenthal

PRINCETON UNIVERSITY PRESS

PRINCETON AND OXFORD

Published by Princeton University Press
41 William Street, Princeton, New Jersey 08540
99 Banbury Road, Oxford OX2 6JX

press.princeton.edu

GPSR Authorized Representative: Easy Access System Europe - Mustamäe tee 50, 10621 Tallinn, Estonia, gpsr.requests@easproject.com

ISBN 9780691276113
ISBN (e-book) 9780691276137

Library of Congress Control Number: 2025947020

British Library Cataloging-in-Publication Data is available

Editorial: Joe Jackson and Rebecca Binnie
Production Editorial: Natalie Baan
Jacket Design: Karl Spurzem
Production: Danielle Amatucci
Publicity: William Pagdatoon
Copyeditor: Leah Caldwell
Jacket images: © SA Automobiles Citroën and Smith Archive / Alamy

This book has been composed in Adobe Text and Gotham

Printed in the United States of America

10 9 8 7 6 5 4 3 2 1

Il faut du temps pour être heureux. Beaucoup de temps.

—ALBERT CAMUS, *LA MORT HEUREUSE*

A . . . , pour la remercier de m'être si chère.

—Jules Supervielle

CONTENTS

ILLUSTRATIONS

Figures

Tables

Boxes

ACKNOWLEDGMENTS

This project began in the fall of 2001, and for almost a quarter century, we have benefited from the help of the Paris archives' staff. Vincent Tuchais, who oversees the reading room, repeatedly jumped in to simplify our access to needed records.

We were closely engaged with the data collection. Nevertheless, the scale of our dataset was only possible with a crew of dedicated research assistants. Laura Betancur, Maria Chichenkova, Maria Emanovskaya, Aliona Gloukhova, Melike Kara, Gulkan Kolay, Alexandr Mitrushkin, Namgyal Henry, Nicolas Pastore, Kadrye Razamanoglu, Esteban Reyes, Tania Shukan, Marie Carmen Smyrnelis, Stela Suhan, Asli Sumer, and Nazli Temir, all jumped into the task with tremendous care. Aleņa Lapatniovna was part of the team from the beginning to the end. She recruited most of its members and supervised the effort at the archives.

Pierre-Cyrille Hautcoeur, Angelo Riva, and David Smajda helped ease our use of their financial history dataset (dfih.fr). Olivie Accominoti and Kristian Rydqvist kindly shared some data for foreign assets. Arnaud Manas sent us the lists of Banque de France shareholders that we digitized. Jacques Friggit sent us his gold coin price series. Clément Dherbécourt collected a large part of the 1972 and 1977 data and shared it with us. Laura Taylor prepared all the maps.

The John Guggenheim Foundation, the National Sciences Foundation (US), the Agence Nationale pour la Recherche, the European Research Council, and the California Institute of Technology provided important funding support.

When writing up the findings, we benefited from comments and encouragement from our close colleagues: Jerome Bourdieu, Lionel Kesztenbaum, André Masson, Thomas Piketty, and Akiko Suwa-Eisenmann at PSE; Philip Hoffman and Tracy Dennison at Caltech; and many other scholars at seminars and conferences. Dennison organized a book workshop in May 2024 where Daniel Bogart, Mallory Hope, Pawel Janas, Peter Lindert, Jared Rubin,

Carol Shamas, Andrew Sinclair, and Bin Wong offered many comments that shaped the final version and challenged us to make the book readable. Oscar Gelderbloom, Pierre-Cyrille Hautcoeur, and Francesca Trivelatto sent us comments of great value. The three readers from Princeton University Press were most helpful in focusing the arguments.

Paula Scott edited the manuscript, providing a major advance in coherence and readability.

ABBREVIATIONS

ADP	Archives départementales de Paris
AGAP	Administration générale de l'assistance publique à Paris, 1851–1864
ASF	*Annuaire statistique de la France*
ASVP	*Annuaire statistique de la ville de Paris*
BSLC	*Bulletin de statistique et de législation comparée*
BSM	*Bulletin de statistique municipale*
CNRV	Caisse Nationale des Retraites pour la Vieillesse
IGR	*Impôt général sur le revenu*
IRVM	*Impôt sur le revenu des valeurs mobiliéres*
INED	Institut national d'études démographiques
JO	*Journal officiel*
MDH	*Mémoire des Hommes*
MDLC	*Morts dans les Camps*
MDLS	Mémorial de la Shoah
RMD	*Registres de Mutations par Décès*
SEF	*Statistiques et études financières*
TSA	*Tables de Successions et Absences*

Introduction

WEALTH AND INEQUALITY IN THE LONG RUN

How do economies marshal their wealth into productive capital? How are wealth and capital distributed among individuals and between the public and private sectors?

These are the broad questions we pose; finding answers requires a long-term view because successful economies sustain capital accumulation across generations, not for a decade or two. Capital accumulation leads to large increases in private wealth, which is not just a peripheral aspect but the very heart of our study. At the individual level, family wealth or the lack of it constrains opportunities for education and retirement. Scholars have also noted the connections among wealth inequality, political polarization, and public policy. Very high levels of inequality are often associated with limits on democracy and strict limits on progressive taxation, thus privileging private at the expense of public capital, and fee-based rather than tax-based public services (e.g., education and infrastructure). Therefore, any study of private wealth must be a study of the political economy of capital. Our analysis of two centuries of capital accumulation is made possible by the extraordinary records that survive in Paris and in France.

These records are useful well beyond thinking about important economic questions; they also inform the history of Paris and its relationship to France. The capital has always been at the center of most key political events even before the French Revolution and continuing to this day. As we shall see, Parisian wealth was consistently much larger and more unequally distributed than in the rest of the country. Paris was also a forerunner in economic transformations that appeared as new assets in the estates and new occupations of wealthy decedents in each period. These economic transformations include the rise of financial capitalism, aging, and the diffusion of pensions, as well as the impact of increases in public services. Political and economic trends and shocks are reflected, sometimes in dramatic ways, in the portfolios of Parisians. In many ways, change started in Paris, and other parts of the country followed a generation or more later. Finally, Paris has a unique history. The particular concentration of real estate ownership meant that small and medium fortunes were composed of other assets, unlike the countryside and most other cities in France and elsewhere.

Economists append "rich" to a person or place to denote either high income or high wealth. High-income societies have high wages. High wages, in turn, depend on abundant capital of many kinds, including human capital, physical capital, and financial assets. That capital has to be owned by someone. Rich societies have high levels of private wealth; they also have high levels of public property, a fact that underscores the role of state action in wealth accumulation. Many economists have thus devoted entire careers to questions of capital and the institutions and organizations that facilitate or hinder its accumulation. This is often a macroeconomic question because state action is critical for understanding why some societies grow wealthy and others do not. Nevertheless, the same issues arise in urban and regional economics.

In many economic growth and performance models, capital, not wealth, matters. To clarify that distinction, capital is either an input into production (as in a machine) or it directly produces valuable services (like a taxi). Wealth is a broader concept that includes capital and assets that produce only utility (like art). To illustrate this distinction, consider an individual whose wealth includes a home, business equipment, and a kilo of gold in a safe. The home is capital because it is producing housing services. The equipment is capital because it is an input into a productive activity. The gold may produce utility by reducing her fear of being without resources if the economy collapses, but it is not capital. It is just sitting in the safe. Yet the same wealth deposited in a bank becomes productive capital when lent out. We will use a definition of wealth that includes all assets, even those that do not produce income. We also define a narrower concept of productive capital: it leaves aside the kilo of gold and is narrower than capital because it does not count things like home furnishings. In practice, we will analyze the evolution of wealth and productive capital separately.

The income distribution matters most obviously because of its implications for consumption and welfare. We also care about wealth because it can be used to sustain consumption when income is low, yet wealth has many other implications. Being wealthy beyond one's human capital implies having control of some assets. To be rich translates directly into power and independence. These connections have long been understood, even before the French Revolution, and made stark by Marx. The yeoman farmer, the artisan, the *rentier*, the industrialist, and the banker are all wealthy. They also have income. Income, particularly labor income, is bounded by the ages at which one enters and exits the labor force. It is also bounded by the finite time we have to work each month or year. Wealth does not have such bounds. A young child endowed with assets by his parents or grandparents is wealthy and thus has property income even though he does not work. A worker who retires can enjoy a pension and a house she bought. Moreover, the amount of wealth any individual can control does not have an obvious upper bound. One has to have income to live, but one does not have to have wealth, and the range of upper wealth levels is always far wider relative to its average than labor income. In short, whether one is interested in growth or distribution, accurate measurements of wealth and capital are necessary.

Although in some places, particularly in Southern Europe, registries known as cadasters and wealth tax rolls with detailed information on individuals' assets have survived from the late medieval to the early modern period, most societies have failed to record individual wealth except at death (Herlihy and Klapisch-Zuber 1985; Alfani 2023). Most such records are local, limited to a village, a town, and exceptionally to a small region. They offer an accurate representation of local wealth, particularly for real estate. The smaller the area surveyed, the bigger the problem for top wealth holders who often have estates in many locations. Moreover, in France, at least, some forms of aristocratic property were not recorded because they were not taxed. This makes an accurate measurement of top fortunes extremely challenging. As an alternative, scholars have relied heavily on probate records, private documents drawn up at the time of death that evaluate wealth.[1] Probates are complicated because different societies have different rules for deciding when estates are probated and what wealth is detailed in the probates. Moreover, small estates with a single adult heir are unlikely to get probated. Both wealth registers and probates have the additional problem that they do not take the whole population into account, in particular those without wealth.

Luckily, France has a more complete source for information on wealth over the long run than probate records. During the French Revolution, the legislature enacted an inheritance tax law that applied to the first franc of every estate. The tax remains in place today, and we have exploited its records. For this book, we collected thirty-two complete cross sections of estates in Paris. The reasons to focus on Paris are simple: the records are all in one location, and many top wealth holders resided in the capital in the past. To the individuals with estates, we add the population of individuals who died without an estate using published totals of deaths by age and sex for a full representation of wealth at death for each cross-section year. The dataset has a cross section every five years from 1807 to 1972, only skipping 1917 and 1967.[2] We also have a complete enumeration of all estates in the top three deciles of 1977, the most recent set of records held by Paris archives as of 2024. The resulting dataset contains observations about the wealth at death of 776,162 Parisians, 248,892 of whom died with a positive estate. For a stratified sample of 46,380 individuals with a positive estate, we have collected full details on their personal and (if relevant) community assets. The asset details files contain 1,153,505 entries needed to establish the value of the sampled individual's estates (it also contains another 352,738 entries accumulated for individuals not in the sample). These datasets, along with a wide variety of other datasets, are available for further research (for details, see appendix 1).

The Parisian data show that between 1807 and 1912, private wealth grew rapidly, and so did the 1% wealthiest individuals' share of wealth. In contrast, during this period of wealth growth, the share of individuals dying with no estate was almost constant at 70%. Then, from 1912 to 1947, wealth fell precipitously, returning to a real level close to what we found in the 1820s. During these fateful decades, the wealth share of the top 1% fell from 65% to 55%, but the share of individuals leaving an estate rose to nearly half of all adults. After 1947, wealth rebounded, with the top 1% of wealth share stabilizing at about 50%, and the share of individuals leaving an estate continued its rise, reaching about 60%. At all times, Paris was much richer than the rest of France and also much more unequal (see chapter 10).

These broad results are consistent with the figures reported by Thomas Piketty (2001, 2013) and his arguments for France. Piketty participated in the initial stages of the present research (Piketty et al. 2006, 2014). Piketty (2001) used the tabulated data by inheritance bracket published by the tax administration at various points throughout the twentieth century to measure

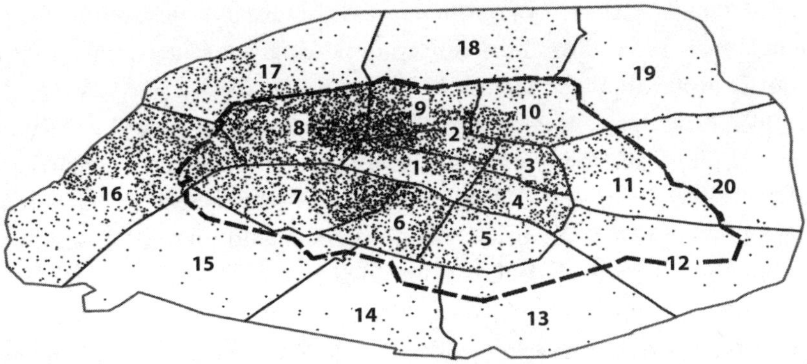

FIGURE I.1. The quartier of residence of the top 1% richest decedents in Paris, 1807–1972. *Note*: Each dot represents a member of the 1% who died in one of our cross sections. Each decedent is randomly distributed within the quartier where he or she legally resided. Since 1860, Paris has had 20 arrondissements, each divided into four quartiers (neighborhoods). The dashed line delineates the boundary of the city before 1860. *Source:* Estates dataset (see appendix 1).

wealth inequality in France. Our complete individual-level Parisian records data show that published tabulations tend to underestimate top-end wealth concentration (see chapter 3). As a consequence, our estimates of top-end wealth concentration are even higher than those previously published, and the mid-twentieth century decline is more moderate, especially in Paris. More importantly, because we have individual data, we can carry out a variety of analyses to better understand how wealth and its distribution evolved. In particular, we can disentangle men's and women's wealth. It is also necessary to take into account who controls wealth because, in nearly all cases, married women surrendered the management of their assets to their husbands while maintaining separate ownership of their personal property (Bessière and Gollac 2022). Our data are rich enough to tally community property and personal assets and thus to measure the inherited wealth share for individuals who die married, extending work published earlier (Piketty et al. 2014). As we show, the inherited wealth share is large, and wealth is persistent across the generations. We also show that despite a high inherited wealth share, Paris had a lot of social and geographic mobility. Using the addresses of the richest centile among decedents, we can show that inequality was also severe within Paris. Figure I.1 distributes these very rich decedents within the quartiers as defined after the city doubled its size in 1860. (The old boundary is the dashed line in the figure.) The map shows most of the top centile of wealth holders concentrated in just a few of the 80 quartiers of the city over a century, even though those in the arrondissements numbered above 12 accounted for the majority of the population by 1900.

Overall, we draw one key conclusion about wealth and inequality. In many ways, private wealth is fragile; it is easy to destroy (real wealth per rich decedent fell by 59% between 1912 and 1922). Yet inequality is robust and hard to control (the top 5% share stayed the same between 1912 and 1922). Although the timing of the wealth decline from 1912 to 1947 coincides with the drop in wealth inequality highlighted by Piketty and his coauthors, in our view, two distinct phenomena are at work in World War I and during the interwar period. The first matters at the top; it involves difficult economic times, inflation, and taxation that hit the very rich harder than everyone else. These factors are the key drivers of the wealth decline. The effect of adverse economic shocks on inequality was muted, however, because the rich enjoyed systematically higher capital returns, and their portfolios had fewer bonds than the rest of the population. The second phenomenon involves an increasing fraction of individuals leaving bequests, driven by secular increases in age at death, income, and the rise of pension programs.

We also draw some important conclusions about the role of taxation in limiting inequality. For Paris, at least, the French estate tax regime had a limited impact on wealth distribution. However, more generally, the fiscal regime played an important role. After WWI, income taxes reduced the consumption value of capital income. Because capital was so concentrated in Paris, progressivity increased the fiscal burden of Paris more rapidly than that of the rest of the country. Although taxes began to rise around 1910 to pay for military expenses, they also changed the funding of public services from fee-for-service to open access. This second change had massive and direct benefits for the education and health of Parisians, particularly the poorer ones. This reduction in inequality is obviously not captured directly in any wealth-at-death statistic. Thus, we emphasize the importance of measuring the many different facets of inequality that matter to welfare.

This Project in Context

Debates about the evolution of wealth and its distribution have a long history. Recent publications by Alfani (2023), Atkinson (2015), Piketty (2001, 2013), and Waldenström (2024) provide extensive discussions of those debates and how they have influenced efforts to measure income and wealth distributions. More closely related to our efforts, Soltow and Van Zanden (1998) and Alfani (2023) have sought to understand the extent to which inequality varied from the Middle Ages to the Industrial Revolution. Kuznets's (1955) inequality hypothesis is one of the main motivations for their work. This hypothesis focuses on sectoral inequality. It posits that inequality rises in periods of economic change because new sectors offer higher returns or incomes. Later in the process of change, inequality decreases once most of the capital or labor is in the new sectors. Another motivation of Alfani, and Soltow and Van Zanden, comes from the general decline in the skill premium observed in wages after the Black Death (1346). One last motivation involves the rise of new wealth forms, which could be held either more or less widely than the original asset (land), depending on local circumstances. Peter Lindert's work on English wealth (1986) is particularly important for us because land in England, like in Paris, was owned by very few people and was an important driver of total wealth inequality.

Among the key issues relevant to our project is the connection between economic growth and wealth accumulation by the poorer half of the population. In one view, prosperity lifts all boats. Others have suggested that structural or behavioral impediments limit wealth accumulation so that large parts of the population remain poor. A second issue involves public policy. Again, some scholars have argued that tax regimes should be used to reduce inequality, while others believe the cost of foregone prosperity is too high. A third issue involves the political ruptures necessary to move from a low-tax regime to a progressive one. Indeed, in societies like Paris that are very unequal, progressive taxation is required to improve the welfare and human capital of most of the population. Some view this process as driven by demographic change, while others argue that international relations and large economic shocks largely drove fiscal innovation. As noted earlier, private wealth's trajectory unfolds over generations, making the empirical evaluation of these broad questions difficult if one focuses only on a few decades. We start this introductory section, therefore, by considering the practical and empirical challenge of finding good sources to measure wealth and its distribution. We then discuss the specific attributes of the source we have chosen: fiscal records of wealth at death for Paris.

The dearth of longtime series has complicated the analysis of the intertwined issues of wealth and its distribution. States have largely eschewed the responsibility of tracking wealth and its distribution precisely. Government officials have taken the implicit position that the value of such data to public policy is less than the value of preserving individual privacy. Some governments have, to be sure, produced detailed surveys of wealth, like the United States' Survey of Consumer Finance. Unfortunately, such samples are too small to track the wealthy, partly because they are extremely unlikely to participate. Atkinson and Piketty championed using fiscal data, whose publication started at about the same time states instituted progressive income and estate taxes to measure wealth and income distributions. Together with a large group of scholars, they reconstructed the evolution of income and wealth distribution for many countries in the twentieth century (Atkinson and Piketty 2007; Piketty and Atkinson 2010). Thomas Piketty used the French estate tax data tabulations from 1902 to 1964 in his work on high-income French people (2001). The efforts of the scholars of this group, known as the World Inequality Lab, revitalized the study of long-run inequality (Alvaredo et al. 2018), and our book participates in that effort. At the same time, we have taken a radically different approach to these questions. First, we focus on wealth at death in Paris rather than the whole of France. Second, we collect individual-level data rather than tabulations. Therefore, as noted before, each measured estate is attached to the social characteristics of the decedent (including their age, sex, address, occupation, marital status, and type of heirs). We also collected the composition of their estates for large samples of decedents. Not only do we have detailed individual data, but we also follow the wealth of the capital's inhabitants over a long period—from the First Empire (1804–1815) to the Fifth Republic (1958–present). Our dataset thus covers the century before the French treasury began publishing tabulations in 1902 and another fifteen years after it stopped in 1964. The two centuries over which we have collected individual data allow us to provide a better sense of the dynamics of wealth and its distribution in France than the terse published tabulations previously used.

That said, Piketty was not the first to use French estate data, nor were we the first to use Parisian archival data. Adeline Daumard (1963, 1973) compiled two cross sections of all Parisian estates filed in a year (1821, 1847). Her pioneering work was part of a project to study social hierarchies launched by Ernest Labrousse in the 1950s. Later, Daumard added 1911 to parse wealth holdings' evolution across social categories. Carried out by hand, her effort was remarkable. Unfortunately, her results, while uncontested, did not spark further inquiry. Also, she grouped individuals by social categories in her published work, and her individual observations were lost.

Starting in the 1980s, a completely different project focused on 3,000 families chosen in the 1820s whose last names begin with the letters TRA and whose descendants' civil records (births, marriages, and deaths) were traced to the 1960s. Bourdieu et al. (2014) collected the wealth at death of a large fraction of individuals in the TRA project and any other individuals whose last names started with the letters TRA they found. Their research, which continues, focuses heavily on the evolution of wealth as France became less rural. Their sample, however, seems to underrepresent large estates, particularly Parisian fortunes. Like Daumard's, our project focuses on Paris; like the TRA project, it spans the nineteenth and twentieth centuries. It links with Piketty's work because what happens in Paris—France's capital—matters for France.

Benjamin Franklin reportedly quipped that death and taxes are the only things that are certain. Both may be inevitable, but they are far from fixed. Moreover, where wealth is concerned, how the wealth at death relates to the wealth of the living requires some elaboration. Going back to Mallet (1908), Séailles (1910), Strutt (1910), Mallet and Strutt (1915), and Lampman (1962), economists have relied on multiplier techniques to compute the wealth of the living in a year from that of the dead. The basic intuition is that the wealth of the living in a year is the wealth of the deceased in the same year multiplied by the inverse of mortality risk. The only issue is how narrowly to define mortality risk (by age, or by age and sex, or by age, sex, and occupation). Our long-run study faces two challenges. First, the relationship between those who die and those who survive is influenced by immediate factors that are sometimes dramatic aggregate shocks. For Paris, these include, as we shall see, the cholera epidemic of 1832 and the deportation of Jews in 1942. Second, the relationship between wealth and death has evolved over two centuries. In 1807, as in 1847, most of the dead were of working age, and they died in relatively large numbers in every decade of age. After 1945, most of the people who died in Paris were retired. Although there were many individuals in Paris aged 20–29, few died so young. The same can be said about the "certain" taxes. While inevitable between 1807 and 1972, they were far from constant. Avoiding estate taxes may always have been desirable, but the incentive to escape the taxes was, on average, ten times higher after 1920 than before 1900. Again, we consider these changes in the tax regime to understand what wealth-at-death data tell us about wealth. These issues highlight the need for a quantitative and historical approach—one which we strive to provide in this volume—to the collection of evidence and the evolution of wealth, capital, and inequality.

The challenges of time and sample years are not specific to our study; they are inherent in any long-term study that seeks to understand both economic structure and change. They are also what makes such work important. Whether one comes at these issues from the left (seeking to find ways to improve the welfare of the general population) or the right (seeking to find ways to reduce state influence and allow markets to improve the productivity of wealth or, more bluntly, to preserve social order and the existing hierarchy), change over time is an essential element of what is at stake. Over two centuries, the level and distribution of wealth, its taxation, and the wealth at death we observe were all altered by financial market development, capital accumulation, urbanization, technological change, the politics of taxation, and public spending on education, health, and infrastructure. In each period, the level and distribution of wealth, as we shall see, also contributed to financial market development, capital accumulation, urbanization, technological change, and the politics of public policy. Wealth is a fully endogenous variable. Individuals and organizations can alter the level and characteristics of their wealth quickly. Organizations can do so by borrowing more or less, issuing additional equity claims, distributing or retaining net income, etc. National or local agencies can claim certain services as their monopoly or leave them to the market and, even then, choose to regulate them or not. Individuals have choices over the size of their bequests and who receives them. Individuals also decide where to live, whether to remain single or marry, and how many children to have. We thus implicitly face problems of causality. For instance, increasing longevity matters to changing the share of individuals who leave a bequest. At the same time, having wealth is beneficial for longevity. We chose not to attempt to purge the "endogenous" phenomena as is now popular in economics. Indeed, doing so would have forced us to focus on the short-run evaluation of the impact of events when we are interested in long-run change. Instead, we make endogenous evolution one of our objects of study. This way, we hope to raise questions that matter well beyond Paris. As it turns out, the long span of time we examine allows us to decode many important questions because the timing of demand-side (individual) shifts often lags or precedes supply-side shifts (institutional change) by a decade or more. In Paris at least, two patterns recur time and again. First, private supply-side innovation did little to change the wealth distribution, though it mattered tremendously to wealth accumulation. Second, public intervention mitigated the welfare consequences of wealth inequality but its deployment depended on a political economy that involved France rather than the city. Time, it seems, is our great resolver.

A Capital's Capital

Our effort to reconstruct private wealth and its distribution for Paris is connected to our long-term interest in what generally drives prosperity. Our previous research has considered the evolution of credit markets over several centuries (Hoffman et al. 2000, 2019) and how Paris played a preeminent role both before the French Revolution and after 1860. Credit markets were connected to wealth because mortgages were the primary form of outstanding debt. Those without real estate could not easily borrow, and the option to take on debt increased with real estate wealth. This was one of the reasons we became interested in wealth inequality. Rosenthal's work on the formation of businesses in Paris (Lamoreaux and Rosenthal 2005) also raised questions about inequality: Was entrepreneurship an important path to wealth? Were joint stock companies important to capital (and wealth) formation? Postel-Vinay's work on large-scale agriculture enterprises highlighted large landowners' financial connections to urban (especially Paris) credit markets (Moriceau and Postel-Vinay 1992). For all these reasons, developing additional evidence on the wealth of France and Paris seemed imperative if we seek to understand prosperity.

In imagining such a project, one might intuitively want to sample across France rather than focus on the capital city. A national sample would certainly have some desirable properties. However, we had to leave that option aside because of practical constraints. Outside Paris, France has about 90 *départements*, each with between 10 and 20 offices where estate taxes are paid. We would have run into a quandary: How could we ensure a representative sample at each wealth level? The first problem is that the rich are not randomly distributed across space. Moreover, as we will see, the range of wealth is very wide in the top centile and wider in the top thousandth of the distribution; thus, an accurate measure of their wealth shares requires that we capture nearly all the individuals in these sets. Rather than sample a few locations, we chose to focus on one place, the capital, which we knew was home to many very wealthy people. We also chose not to sample among the decedents but to consider all decedents of a given year. More precisely, unlike Daumard, who focused on all the estate declarations filed in a given year, independent of the year the individuals died, we collected the filings of all those who died in a given year. Until 1912, the two methods provide similar results. Once inflation set in during WWI, however, our approach provides more accurate estimates of wealth and inequality (see chapter 3 for more details). Paris has the advantage that its 57,000 volumes of estate records are all housed in its archives, and those volumes contain all the estate records from 1807 to 1960 and have only limited gaps from 1960 to 1977.

Studying a city over the long term also helps us understand the long-term dynamics of wealth today since the globe's population is now heavily urbanized. Those dynamics include both the importance of financiers' wealth and the capital demand of the cities themselves. Let us start with Paris's capital market. By the French Revolution, Paris had been France's capital and its largest and richest city for the preceding millennium. Paris has been the country's financial center for at least two hundred years since the decline of Lyon as a banking center. The French Revolution further concentrated legislative, judicial, and financial activity in Paris. Therefore, Paris is the core of various webs of French capital flows. The central bank, the Banque de France, was established there, and although it opened branches throughout the country, the discount policy was decided in Paris. All the key investment banks had their offices in the city. When universal branch banking diffused after 1870, the most important of these firms were also headquartered in the capital. Although there were stock exchanges in a half dozen cities in France by the mid-nineteenth century, the Paris Bourse dwarfed all the others, and it would progressively take over most of the trading in securities. That, in turn, gave Parisians access to a greater variety of investment options than most other people in France. In 1818, the first savings bank opened in Paris. Its success spurred the opening of similar institutions in hundreds of locations across the country. When postal savings were initiated, Paris again led the way.

Paris also featured prominently in France's participation in the massive international capital flow of the first golden age of capitalism (1870–1913), most famously related to the funding of the Suez and Panama Canal companies. The capital dominated both the registration of companies doing business abroad and their share and bond issues. The same can be said of colonial investment. Banks like the Crédit Foncier Algérien, the Banque d'Indochine, and many others were headquartered in Paris and issued their securities there. This is the reason why these capital flows would be visible in Parisians' portfolios. We also want to know how far down the wealth distribution foreign assets, such as the infamous Russian bonds repudiated in 1918, reached.

Financial modernization is a significant theme in economic history and development economics. Tracking its progress for France requires taking Paris into account. As we shall show, the new financial institutions just mentioned were successful because Parisians had a great demand for financial services and a great appetite for new investment opportunities. At the same time, Paris differed from the rest of France in savings opportunities.

Starting in the 1820s, Parisians had access to new forms of life-cycle savings, including insurance company policies (life policies and tontines). Then, in the second half of the nineteenth century, retirement programs spread, covering first the public sector, then railroads, and large industrial firms. The organizations that made retirement contributions for their workers early on were overrepresented in Paris. Thus, some inhabitants of the capital experienced new forms of life-cycle savings relatively early.

At the same time, some Parisians' desire for modern investment came from being denied access to real estate investment—the initial form of wealth for most households in most economies. Paris housed a large population within a small area in multifamily buildings. Most households rented. The city extended its territory in 1860, doubling in size. Yet the pattern of multifamily buildings let out to households continued with little change, and Paris remains extremely dense. Since the French Revolution and until the 1960s, when condominium associations diffused, less than 10% of its households could imagine owning real estate in the capital. Since then, individual apartment ownership has become more prevalent.

Paris also had a unique relationship with other forms of capital. The royal road system, and later railroads, were built as star systems with Paris at their center. As it grew, the city's demand for infrastructure rose as its leaders became ever more starkly aware of Paris's unhealthy environment. Improving water supply, building sewers, and paving streets were all expensive investments. However, French cities could only borrow or raise their tax rates if they secured the approval of the minister of finance. Because this approval was rarely granted, cities generally lacked capital and, instead of borrowing, resorted to concession systems to produce local services (which were then paid for by user fees). Paris was different because of its close relationship to political power and its significant revenues from excise taxes. The capital was managed by prefects named directly by the French executive, and the finance minister regularly approved their requests to borrow funds for infrastructure. Thus, Paris could sell massive bond issues to the public. The need for more real estate finance blossomed in the 1850s with the rebuilding of Paris under Baron Haussmann (prefect 1853–1870). A special bank was created: the Crédit Foncier de Paris. Promptly renamed the Crédit Foncier de France, it concentrated a huge share of its activities in Paris (Hoffman et al. 2000, chap. 11). Up to WWI, half of the value of all mortgages it granted were given out in Paris and backed by Paris properties. The mortgage-backed securities it sold to fund the mortgages were also issued in the city.[3]

Paris also has a unique place in France's economic history because large population flows were needed to sustain a city with high mortality and low fertility. As we shall see, most individuals who died in Paris were not born there. These migrants were mostly poor, came before marrying, and took on a variety of unskilled and semiskilled occupations. Other newcomers, like writers (Balzac or Zola), musicians (Berlioz or Chopin), engineers (Eiffel), and painters (Max Ernst), came to Paris with their talent. Some of the migrants made fortunes, while others died with little wealth. However, all contributed to the cultural life of the capital. Another set of individuals arrived as political entrepreneurs who used their political or military service as a springboard to wealth (examples include Napoleon's marshals or Adolphe Thiers). The last set of migrants comprised those who came with their wealth already established at various stages of life and were attracted to the city because of its economic opportunities or because of its social and political importance (for instance, the Roederer champagne family or the bankers Goudchaux, and Casimir Perier the prime minister).

Wealth requires us to consider reverse flows, like the Schneiders, a Parisian banking family that bought the Creusot iron works in the Saône-et-Loire. They funneled large amounts of capital they raised in Paris into their steel works. Their capital, in a sense, migrated from Paris to the provinces. The Wendels started their iron works in Lorraine and later moved to Paris. Securing capital and contracts for their business was one motivation for having a foot in the capital. Because Paris was so rich and unequal, the migrants mostly came from poorer and more equal places. Another dimension of Parisian exceptionalism is that a relatively large fraction of the individuals who died as residents in Paris and left a bequest did not have direct descendants.

In all these ways, Paris, over the past two hundred years, is a laboratory for thinking about today's global cities that include capitals like Beijing or Mexico City, but also other leading metropolises like New York City, Kolkata, São Paulo, or Milan. As noted above, today's global cities share many of the characteristics of Paris: they are rich and unequal. Global cities drive cultural and financial change, and the individuals who reside there greatly influence public policy. These cities have financial sectors of considerable size and scope. If capital matters, then global cities matter because they organize the flow of capital. In places like New York or London, the financial industry is the principal engine of economic activity and attracts large numbers of migrants from within the country and abroad. In many global cities, the rise of social services has eliminated the urban penalty in life expectancy, which was long standard for all large cities. Moreover, they also often have access to a broader range of public capital and services than the rest of the country. More important, although much of the focus has been on national aggregates (for fiscal reasons), these hide tremendous geographic heterogeneity. That heterogeneity includes much higher levels of private wealth and public services in the top cities than elsewhere. It also reflects the fact that these places are highly unequal.

In some cases, one can quantify some of these differences among global cities over the past few decades. One could measure the importance of the financial sector or count the number of top corporations headquartered there. One could also measure their cultural amenities. However, a good measure of income and wealth distribution, including fine detail at the top, is almost always missing.[4] The advantage of our hand-collected data is that we can compute wealth estimates for any desired group.

French Inheritance Taxation

The French Revolution created an inheritance tax through two laws. The first, enacted on December 5, 1790, recreated the system of taxation of private property transfers (including inheritances) that had prevailed under the Old Régime. A second law, enacted on December 12, 1798, organized the inheritance tax proper. As Daumard details, it took some time for the system to reach its steady state structure (1973, 4–6). By the end of the First Napoleonic Empire (1815), however, the key pieces were in place.

The French tax set in 1790 was an inheritance tax rather than an estate tax (as prevails in Great Britain and the United States). An estate tax is assessed on estates of a relevant size independent of who receives the estate. For 2024, US estate taxes will be paid on any part of an estate over $13.6 million, and the lowest bracket is 18%. Inheritance taxes, in contrast, are assessed on the heirs. They thus vary depending on who receives the estate, and because the progressive schedule applies to each heir, the tax burden is reduced when there are more heirs. In France in 2024, heirs who are children, mothers, fathers, and grandchildren of the deceased each are exempted from the tax on the first 100,000 euros they receive, and the lowest rate is 5%. The top rate (estate value above 1.8 million euros) pays 45%. The exemption for wealth transmitted between siblings is about 15,000 euros, and the lowest bracket is 35%. Nieces and nephews have an exemption of less than 8,000 euros, and the tax is 55%. Everyone else pays 60% with no exemption. To levy the tax, now and in 1807, estate declarations first established the total value of the estate and then detailed what went to different types of heirs.

When the inheritance tax was enacted in 1790, the only assets exempted were the bonds of the public debt, and the rates were low. Heirs in the direct line of descent (children and grandchildren) paid 1% on real estate and 0.25% on movables. Individuals not related to the deceased paid 10%. Over time, the tax evolved. In 1850, the exemption for public debt bonds was eliminated, and the 1% rate was applied to all assets in the direct line of descent. Progressive rates were imposed starting in 1902 (law of February 25, 1901). For more details on the period after 1901, see chapter 3.

Collecting the tax proved a challenge. France was divided into several thousand bureaus (about one per dozen municipalities by 1900). To simplify things, taxes on real estate were paid in the bureau that contained the asset. Taxes on movable wealth were due at the bureau of the individual's legal residence. Because progressivity required the consolidation of all assets, the 1901 law mandated that all assets be declared at the bureau where the deceased had his or her legal residence.

The administration faced several problems. To start, it had to find the individuals who had left estates. It then had to collect information about the values of bequests. Finally, it had to record these values and the taxes to be paid. It solved these by creating two record sets. First, the Tables de Successions et Absences (TSA) provided monthly lists by the first letter of the last name of every person who resided in the bureau or had real assets there. These tables were maintained until 1969. Each individual's details were then recorded on a personal sheet (*fiche*) and placed into a binder. For individuals with an estate, the TSA volumes and the fiches provide the date of their estate filings. This record set for Paris comprises 5,240 volumes, and we opened most of them to collect basic information about the deceased and where to find the estate.

A second set of volumes, the Registres de Mutations par Décès (RMD), contains the estate declarations. Before 1902, the clerks transcribed the estate declaration onto the registers once they had accepted the submission from the heirs' representative and computed the tax. After 1902, the heirs presented estate documents on preprinted forms, which the clerks marked up. They also calculated the tax owed on the last page of the declarations and then filed them in binders chronologically. The filings themselves come in at 53,181 volumes for Paris alone; we opened roughly half of them (for more details, see chapter 2).

France has a marriage law with a default regime distinguishing between personal and community assets. Before 1965, real estate was a personal asset. All other assets the individual brought to a marriage went into the community. After 1965, the definition of personal assets changed, and all assets owned by a spouse before marriage or inherited during the marriage were recognized as that spouse's personal property. None of the individuals in our data married after 1965, yet they could depart from the default regime by writing a contract. Indeed, the law allows extensive deviation from that regime. Spouses have widows' portions and usufruct rights, either set by law or chosen as part of a marriage contract. France also has an inheritance law that limits testamentary discretion when there are children or other heirs in direct line of descent. Then, the estate must be divided into a number of shares equal to the number of children plus one, the reserve share. Before death, an individual can write a will to decide what to do with the reserve share. Each child or heir line has the right to one share of the estate.

Thus, an estate filing for someone who dies married begins with a gloss of the marriage regime and testamentary dispositions of the deceased. The filing then lists the heirs and their relation to the deceased, all of which would be unnecessary in an estate tax regime. These details provide us with the date and place of marriage, the existence and type of a marriage contract, and the spouse's place and date of birth (to establish the value of the life interest). We also have the list of children and individuals receiving bequests from the reserve share. The estate filing then proceeds to describe the deceased's personal property and some of the spouse's to establish their claims on the community property. The estate document continues with a list of community assets and their values. Having established the community's value and netted out various claims, the community is then divided, and the deceased receives his or her share, most often half. The estate is then computed by adding his or her share of the community to his or her personal assets (for more details, see chapters 2 and 7). Someone who dies single or unmarried has a simpler filing. It begins by establishing that the individual is not currently married. It continues with the list of heirs and finishes with the list of assets. That is all that is needed to establish a single individual's wealth.

What makes these records unique relative to many other types of wealth evaluation is that they are not just a static image of wealth at death. For those individuals who died married (52% of the sample), we also have access to an evaluation of their inherited assets (the personal property described above). These data are precious because they allow us to investigate issues of social mobility without dealing with matters of selective matching (as would attempts to match parents and children). A second feature of the filings is just as valuable. The tax authorities insisted that unpaid pension income be included in estates. That allows us to accurately measure who had pension or retirement income and how that changed over time. Our data is too coarse to capitalize the pensions, but it allows us to track the rise of life-cycle savings.[5]

Despite some lacunae, one of the most valuable uses of the data is a critical reevaluation of the thesis that a key aspect of modern finance and wealth management is the replacement of bequests with life-cycle savings. Elaborated by Franco Modigliani (Ando and Modigliani 1963), the hypothesis is that, in traditional societies, individuals relied on their children for old age security and "paid" for their services with bequests. When new financial products designed to meet the needs of retirees emerged, children were left free to move on to more rewarding occupations, and parents had more choices regarding how to spend their old age. Beyond the wealth and personal choice issues, Modigliani (1986), Ransom and Sutch (1986), and Sutch (1991), who tried to apply the theory to the United States, highlighted a virtuous cycle. Life-cycle savings suggest saving a constant rate of one's labor income to pay off debts from education and accumulate the capital needed for retirement. Those savings go into productive assets. Because life-cycle savings increase with labor income, they provide what is needed to keep the economy's capital growing over time. By contrast, in the view of Modigliani, bequests had a target aspect largely independent of the economy's demand for capital. In our view and, in the Parisian data, however, life-cycle savings are not a substitute for bequests but a complement. The reason is simple: given sufficient income and lifespan, many individuals want to insure for old age (engage in life-cycle savings) and transmit wealth to their offspring (save for bequest).

Paris is an ideal venue to examine this and other hypotheses about modernization. The revolution essentially wiped out financial savings, so the nineteenth century started with a relatively blank slate. Moreover, most Parisians were salaried or piece-rate workers forced to be "modern" since they had no business or farm to bequeath. As a result, policymakers and philanthropists alike sought to increase the savings rate of Parisian wage earners as early as the 1820s. They wanted workers to save more for reasons that combined welfare, fiscal, and political motivations. First, of course, individuals with savings would have better lives. In addition, they would also be less likely to be on the dole and less likely to riot or, worse yet, become revolutionaries. Of course, by the 1930s, retirement programs had diffused significantly in Paris, creating a social process for life-cycle savings. However, it remains an important question to understand how social policy interacted with income and life expectancy over the long term to transform estates.

Death and Wealth Distribution over the Long Term

The economic history of wealth and inequality drawn from estate tax records in Paris can be summarized in the four figures below. These show that a simple, unilinear modernization story explains very little about the path of wealth and its distribution. We learn a lot from starting around 1800 rather than 1900 when public tabulations became available. They also raise important questions that the chapters that follow will investigate.

Figure I.2 displays the total Parisian population and the number of deaths each year in large age categories up to 1977. Paris experienced massive growth in its population, including a jump around 1860 when it doubled its territory. Paris's population then stabilized near 3 million from 1910 to the early 1950s before entering a steep decline as the suburbs grew quickly. The city had lost almost a third of its inhabitants before its population stabilized at around 2.1 million from the 1980s to the present. Up to 1910, the city was a growing share of the French population. Since then, it has been a falling share, which matters when we consider Paris's share of French wealth and its role in French inequality. Deaths also tell a two-part story but with a different chronology. From 1817 to 1880, deaths were increasing in number and were highly volatile due to massive mortality shocks like the cholera epidemics of 1832 and 1849. Deaths among those younger than 20 accounted for 40% of all deaths, with infant deaths accounting for almost as many deaths as those aged 1–19. The elderly (aged 60 or more) accounted for a quarter of all deaths. The second regime began in the 1880s and featured declining total deaths, largely driven by a fall in deaths for those younger than 20, who accounted for about 2% of all deaths in the 1970s (and about 1% around 2010). Conversely, the share of individuals who died aged 60 or older grew over time, and by the 1970s, it accounted for 80% of all deaths (and those 80 or older nearly 40%). This broad demographic sketch has implications for wealth inequality because it takes time to accumulate wealth, and at the same time, living to be very old consumes resources. Moreover, as we shall see, there were large age differentials between rich and poor decedents.

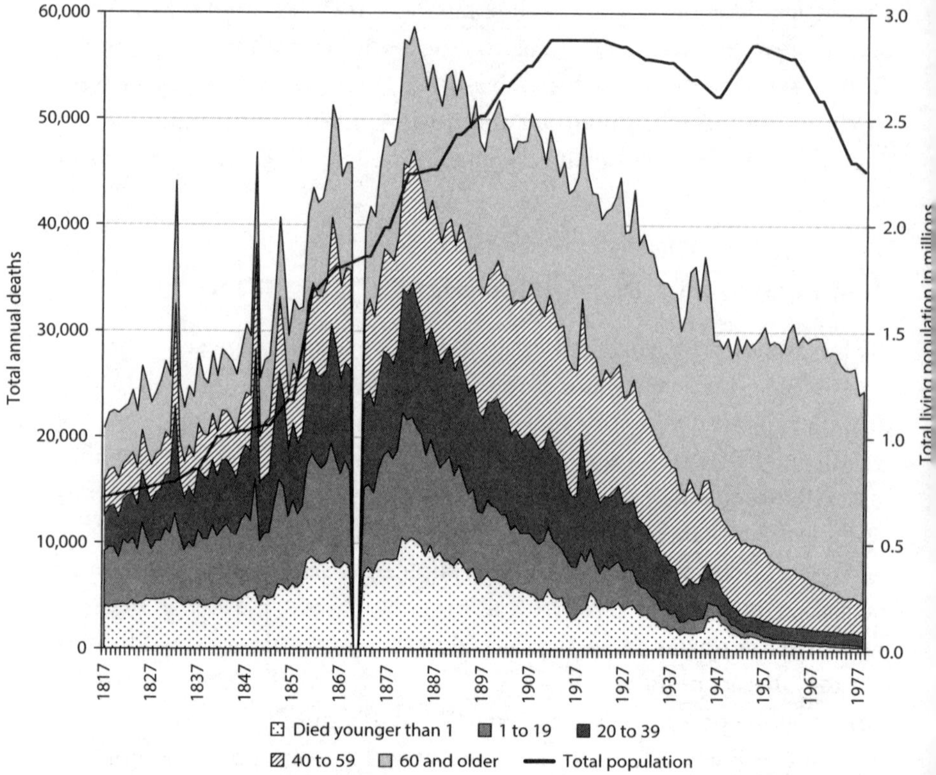

FIGURE I.2. The population of Paris and deaths by major age group, 1817–1977. *Note*: There are partial death totals for 1870 and 1871; however, the death registration system did not operate for several months starting in November 1870, and we do not reproduce the partial tallies. For WWI, the figures omit Parisians who died in the trenches, while for WWII, the figures omit both war casualties and those who died in deportation. *Source:* ASVP and INED.

Figure I.3 shows that Paris experienced massive growth in wealth between 1807 and the end of the nineteenth century. This was followed by a precipitous decline from 1912 to 1947, with real wealth among those with an estate lower than what it had been at its early nadir in 1812. The average wealth of all decedents was only 20% higher in 1947 than in 1817. The post–World War II period saw a massive rebound in wealth, but by 1972 real wealth had only returned to 63% of the pre-WWI peak. Wealth is indeed fragile and takes time to rebuild. The period from 1807 to 1912 is broadly consistent with various arguments from Marx to Kuznets to Piketty. They see inequality increasing early in the development process. They also argue that wealth accrues principally to a few individuals well positioned

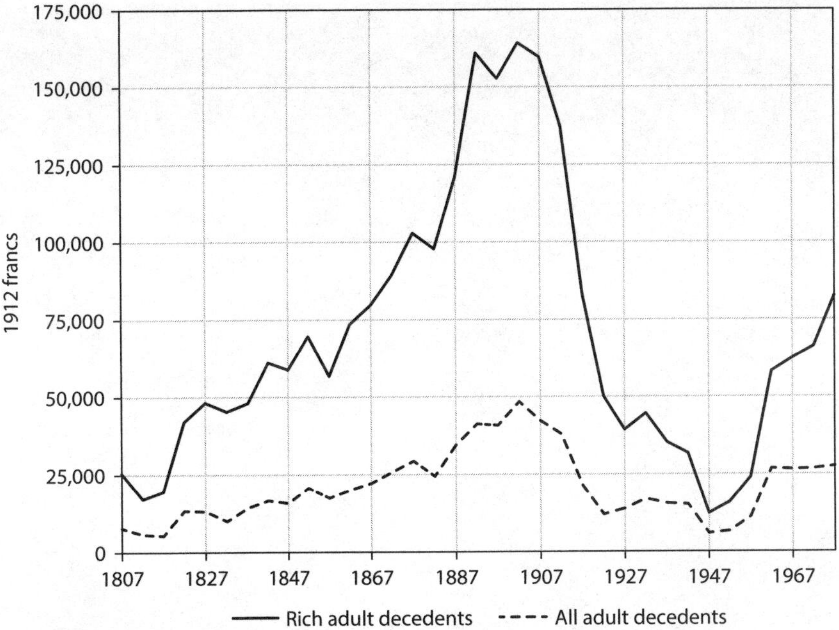

FIGURE I.3. Average wealth at death of Parisians in 1912 francs, 1807–1972.
Source: Estates dataset (see appendix 1).

in the process of change. The twentieth century tells an entirely different story, one that requires different economic arguments and a more nuanced historical approach.

Figure I.4 provides a summary of the distribution of wealth. At the bottom, we have the wealth held by every adult below the top decile. That share was less than 5% before WWI, but it increased slowly and steadily starting in the 1920s, reaching 13.5% in 1972. That allowed the top decile to hold about 90% of all the wealth over nearly the whole period. Within the top decile group, the bottom 9% (P90–98) had about 45% of all the wealth for most of the nineteenth century, and their share dropped a little between the last three decades before 1914 and post-WWII. The top centile (henceforth the very rich) controlled at least 50% of all wealth annually except in 1957. Its wealth share increased from about 50% in 1807 to more than 60% in 1914. From 1922 to 1972, it fell by about 9%. The top thousandth (henceforth the super rich) saw no change between the period of high inequality (1882–1912) and the post-WWI period. In both cases, they controlled 23% of the wealth. Thus, nearly all the decline in wealth concentration comes from a fall in the rest of the top centile.

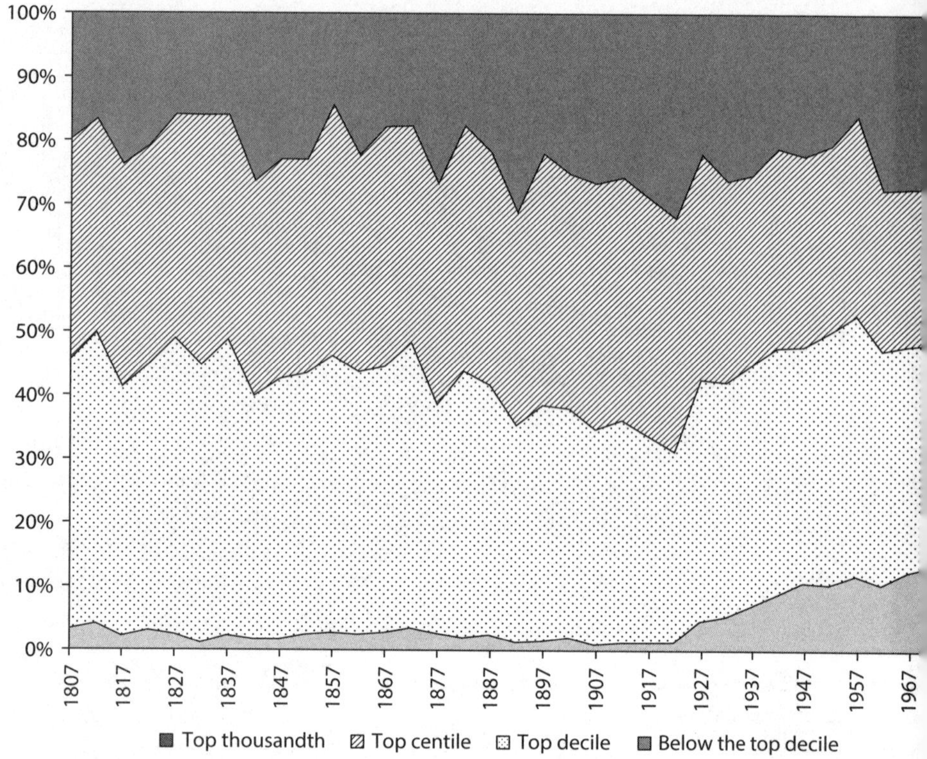

FIGURE I.4. Wealth concentration in Paris, 1807–1972. *Note*: Figure I.3 is cumulative for the top three categories. The top centile is the sum of the top two areas: top thousandth and top centile. The top decile is the sum of the top three areas. *Source:* Estates dataset (see appendix 1).

To these very broad findings, we add two other dimensions of inequality. Figure I.5 displays the share of decedents with an estate, and the ratio of men's to women's wealth. Unlike wealth shares, which vary over the nineteenth century, the share of adults who die with an estate is essentially constant from 1807 to 1922. After 1922, the share of individuals leaving an estate began to increase and reached almost 60% by the 1970s. The growth of bequests coincides with the reduction of inequality and the broad diffusion of life-cycle savings programs (retirement, war pensions, etc.). This major innovation coincides neither with the mortality transitions of the 1880s nor with the wealth maximum of the 1900s; it has its own story. The figure also displays the ratio of the average estate of men versus women. Generally, men have larger estates than women, particularly from the 1820s to 1870s, when men's advantage is 60%. From 1882 to 1912, men were only 15% richer

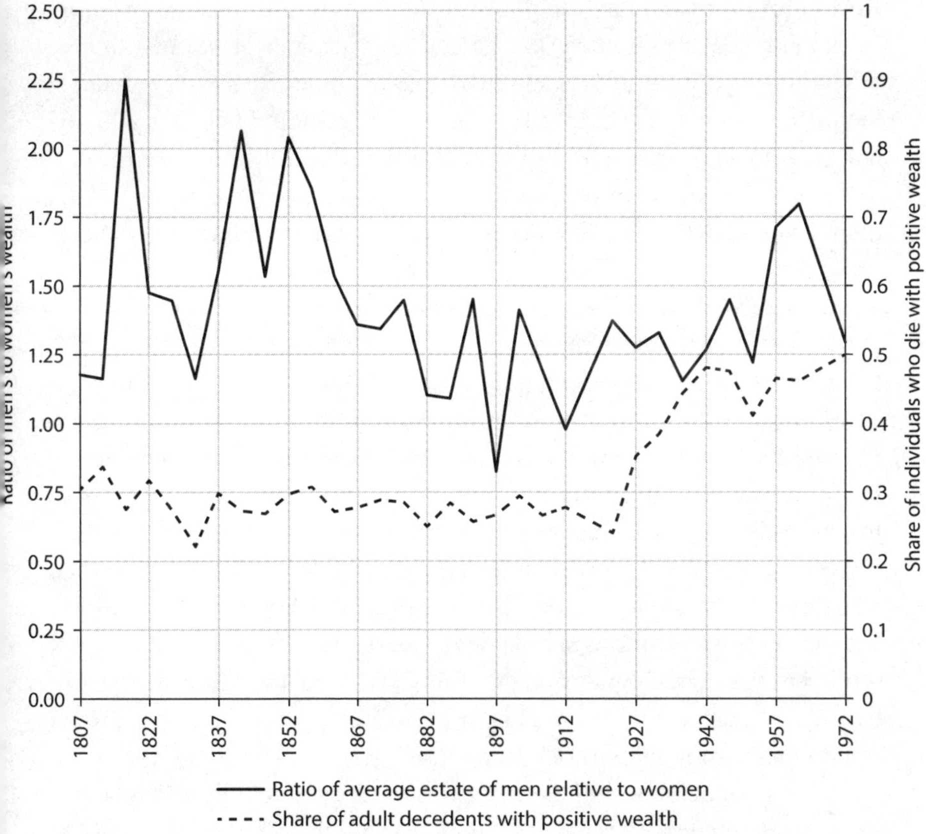

FIGURE I.5. Other dimensions of wealth inequality in Paris, 1807–1972.
Source: Estates dataset (see appendix 1).

than women. During the twentieth century, nearly all the advantages to men came from the higher share of men in the top 0.1% of the wealth distribution. Leaving the super rich out of the computation leaves men with an advantage of 3 to 5% over women.

The four figures above also show tremendous change over time and, at the same time, the fundamental structure of wealth inequality in Paris. Understanding change and structure led us to divide the book into two parts.

Chapters 1 to 6 are chronological. Their goal is to tell the story of wealth and inequality in Paris from 1807 to 1977 (the book's conclusion examines what we know about wealth and inequality since 1977). The first two chapters take on the period before WWI. Chapter 1 deals with the recovery of wealth at death and the growth of inequality following the end of the French

Revolution. It focuses on the political nature of wealth inequality before 1870. Chapter 2 deals with the first golden age of capitalism and asks, among other questions, whether financial market development increases or reduces inequality. We devote chapter 3 to the fiscal transition that began with the law of 1901 and the rise of progressive taxation. Chapter 4 analyzes the dramatic wealth collapse of 1912–1947, while chapter 5 takes on the evolution of inequality during the same period. Chapter 6 deals with the post-WWII experience.

In the second, more thematic part—chapters 7 to 11—each chapter deals with an essential theme of the book over the whole period from 1807 to 1977. Chapter 7 deals with inherited wealth and mobility. Chapter 8 examines the super rich (the top 0.1%) to understand why their wealth share is so resilient. Chapter 9 takes on the problems of why so few people left an estate before 1927 and why small bequests increased rapidly thereafter. Chapter 10 seeks to put Paris in its French context, asking, among other questions, how a society as unequal as France maintained a low tax equilibrium for more than a century. This stability is astonishing after 1870, when universal suffrage and competitive elections should have favored redistribution. Chapter 11 returns to the spatial dimension of wealth. We argue that the spatial distribution of wealth within Paris is highly correlated with other dimensions of welfare; affluent neighborhoods have more good attributes and fewer bad ones, and poor neighborhoods are the reverse. However, the extent of those differences depends on public spending because the poor cannot afford to procure these attributes at market prices. In the twentieth century, a major component of the decline of inequality in Paris is, therefore, public spending on public services.

1

Dukes and Indigents

INEQUALITY IN THE SHADOW OF THE OLD RÉGIME, 1807–1867

Our first cross section of individual wealth comes from individuals who died in 1807. It has four key features. Seventy percent of all individuals had no wealth at death, the top 1% wealth share was 54%, the average wealth for women was 80% of that of men, and there is little evidence of any age-wealth relationship past age 40. As we shall see, there was little movement in the share of individuals who died without an estate or in the top wealth share. The relative wealth of women was volatile and remained low before 1867. The one indicator that changed was the age-wealth profile. By the 1860s, those who died at age 70 or older were about twice as wealthy as those who died in their fifties, who themselves were much wealthier than those who died in their forties. This chapter explores the forces behind these stark facts. Of particular interest is how much these patterns reflect the fundamental wealth structure of a capital city and how much they reflect the upheaval of the French Revolution.

This raft of results comes from our ability to compile complete cross sections of wealth at death for Parisians. Death occurs in an instant, but even for the poor, it sets in motion a complex machinery. First, a medical officer declares a cause of death.[1] Second, municipal officials create a civil death record.[2] In a third step, the deceased has to be buried either at the estate's or the public's expense. These steps are quasi-universal. What makes France and Paris different is one final step that arose because the state wanted to collect inheritance taxes. To do so, it had to track deaths and bequests, which it did in registers called the Tables de Successions et Absences (henceforth the TSA). For each estate tax bureau, the TSA volumes compile the names of all individuals who died there as well as all individuals who died elsewhere and had taxable assets in the bureau. The TSA volumes also contain the dates at which estate taxes were paid for those individuals who left a positive estate. Until 1870, they also provided a summary record of wealth at death broken down by movable and real estate wealth. After 1870, the TSA registers lacked the summary information on wealth and only recorded the dates of the declarations. Thus, 1867 forms a good boundary for this chapter.[3]

The primary tasks of this chapter are to analyze the structure of inequality and the recovery of wealth in postrevolutionary Paris. However, we start with a brief history of the French estate tax. How the estate tax was administered proves useful for understanding how the data was generated more than two centuries ago. The administrative structure frames what lessons can be drawn from them. We then use the data extracted from the TSA from 1807 to 1867 to measure wealth and wealth inequality. We find that although Parisians were relatively poor at the end of Napoleon's empire, wealth grew rapidly under the Restoration and the regimes that followed. As noted at the outset, wealth was very unequally distributed. The richest 5% of Parisians owned at least 85% of all wealth, and the bottom 60% owned nothing. To explain these basic findings, we return to history and analyze the context within which the individuals who died between 1807 and 1867 accumulated their wealth (which we take as the thirty years before they died). The last section focuses attention on wealth by sex and age.

While the data on inequality before 1800 are sparse and coarse, the consensus of economic historians is that inequality likely increased from the Black Death (1347) to the eve of the Industrial Revolution (1750) due to a secular trend in the concentration of real assets, and the growth of urbanization (Soltow and Van Zanden 1998; Alfani 2023). Most importantly, under the Old Régime, wealth largely came from two sources: birth and politics. One could either be born a duke or be made a duke by royal grace; either way, one would be wealthy and privileged. Short of birth or success in government or royal service, the path to wealth was exceedingly narrow. In contrast, the kind of sectoral change that began with the Industrial Revolution was supposed to have broadened the path to wealth. Industrial and financial fortunes went to new people. However, before the Industrial Revolution began to have significant effects on France, the

French Revolution occurred. It was a critical event for wealth and the institutions that allow us to measure it.

From 1789 to 1825, a massive churning process unfolded that affirmed the primacy of politics over economics in different ways. The French Revolution abolished the Old Régime privileges that gave the elite income and other benefits. These included feudal obligations that members of the third estate were required to perform for the church or the aristocracy (e.g., entry fines or corvée labor) and an exemption from many royal taxes, including the principal one, the *taille*. As Saint Jacob (1960) showed, in Burgundy, privileges had become more rather than less valuable over time. At the same time, the revolutionaries were committed to private property. Therefore, the revolution tried to implement a compensated abolition of feudal privileges and an uncompensated abolition of fiscal privileges in August of 1789. However, as the political situation worsened, compensation became infeasible, and it was abandoned in 1793. The revolutionaries were forced to innovate financially because contestation led to a massive drop in tax revenue by the fall of 1789. Over the next several years, three additional great transfers of property occurred: one from holders of public debt to the state (the state reduced its obligations by two-thirds in 1797), another from private debtors to creditors (via the assignat inflation that completely eroded the value of the unit of account between 1792 and 1797), and the last one from the church and aristocrats to other property holders through the confiscation of real assets that were then sold at auction. While the extent to which wealth inequality changed between 1789 and 1797 has yet to be measured carefully, these four transfers created massive churning at the top of the wealth distribution. It is likely that both the uncompensated abolition of privileges and the cancellation of debts through inflation helped the young and less rich. Then, as Napoleon consolidated power, he began distributing titles and wealth to those whose loyalty he valued—creating new members of the elite. Titles came with property, most often confiscated in territories Napoleon had annexed. In addition, many of Napoleon's generals joined the ranks of the wealthy, having found ways to profit from the wars they helped pursue up to 1815. The Restoration government did not contest the properties of the Napoleonic elite (save for a few who had too obviously reneged on their oath of loyalty in 1814). Instead, the early 1820s were consumed with compensation issues for aristocrats whose property had been confiscated when they went into exile during the revolution. In 1825, the French legislature voted to increase the national debt by a billion francs (more than it had paid the allies who deposed Napoleon) and handed the bonds to émigrés who had been expropriated during the revolution. France also insisted on compensation for owners of property in Haiti, the lucrative colony France had lost after its enslaved people revolted and gained their independence in 1804. Haiti was forced to accept liability for 150 million francs. With all these tumultuous events, these were times of extraordinary volatility in wealth.

Going back to 1789, it should be noted that revolutionary leaders were committed to abolishing old feudal privileges and designing a fiscal system that would operate without regard for social status. That starkly contrasted to the prerevolutionary system, where the elite paid far less taxes. One of the key commitments in this effort was a complete revamping of direct taxation. The effort produced four direct taxes: one on real property, a second on doors and windows, a third based on residence rental value and occupation (*contribution mobilière*), and the fourth on businesses (the *patente*). These taxes were all assessed and collected annually (Woloch 1994). There were also transfer taxes. These included taxes on real property sales, donations, and inheritance. It is this last innovation that concerns us. Scholars seeking details on the inheritance tax can consult Bourdieu et al. (2014) and Daumard (1973, chap. 1). For our purposes, a summary of its structure suffices. The tax was a proportional inheritance tax whose rates depended on the relationship between the recipient and the deceased. Children faced the lowest rates (in the first third of the nineteenth century, .25% on movables and 1% on real estate), and individuals or entities with no relationship to the deceased faced the highest rate (7%). The tax was also universal; all estates were liable, and only holdings of the national debt were exempt until 1850. Third, the tax on movable wealth was paid at the bureau that comprised the legal residence of the deceased, but the tax on real estate was paid at the bureau where it was located. These rules and rates essentially remained in place until a major reform in 1901 that introduced progressivity required a single filing for the whole estate. Fourth, absent any other documentation, an estate tax filing was proof of property (I own this house because I paid the estate tax on it when I inherited it from my uncle). The proportional nature of the tax and its low rate meant that this was a revenue source, not a policy instrument aimed at inducing investment in one asset or another.

Whatever the legislators who devised the inheritance tax may have thought, administering it posed a daunting challenge for the bureaucracy. Around 1800, about three-quarters of a million French people died each year; some 400,000 of these were adults. How, then, could a quill and paper administration keep track of who had paid, who had not, and who should be investigated? Setting up an estate tax administration took more than a decade, and it was complicated through the 1820s because the estate of an individual who died married was governed by marriage agreements. Early on, marriage rules depended on regional customs from the Old Régime and marriage contracts. Over time, however, a recording system was devised and implemented. It remained in place from 1812 until the late 1960s. This system divided France into bureaus (roughly at the canton level, the first jurisdiction above the municipality) where heirs would pay the inheritance tax. Before 1901, inheritance taxes on each piece of real estate were paid in the bureau where the asset was located. This was efficient because local officers had better information about the value of real estate in their jurisdiction than those in the bureau where the individual had a legal residence. The scheme also saved on expensive correspondence, and since the tax was strictly proportional, it was feasible to apply. Meanwhile, all movable wealth had to be reported to the bureau officials where the deceased had his or her legal residence.

In the capital, the scale of this bureaucratic enterprise was significant, as each year more than 20,000 Parisians died, roughly half of whom were adults (by the early 1900s, more than 30,000 adults died each year in Paris). Recordkeeping was further complicated because, as we will see in chapter 4, most Parisian decedents were migrants, and although, in the nineteenth century, the Parisian estate tax administration was only responsible for collecting taxes on wealth located in Paris, additional communication with a decedent's place of origin was often required. In addition, over time, a growing number of individuals who were legally resident in Paris died outside the city (in care facilities in the suburbs, in second homes, and so forth). Maintaining the TSA lists required a massive amount of correspondence. Moreover, some nontrivial fraction of Parisian real estate belonged to individuals whose legal domicile was not Paris. Finally, in the system's early years, administrators were hampered by the fact that under the Old Régime, Paris was exempt from transfer taxes (Hoffman et al. 2000, 28). Thus, the collectors had little information to cross-check the heirs' declarations. Despite these challenges, it seems that the system was put in place relatively quickly after the revolutionary inflation ended. The city was divided into six bureaus (each overseeing estate declarations for two of the twelve pre-1860 arrondissements), and by 1807 Paris's inheritance tax bureaucracy was humming along.[4]

1.1. Constructing a Dataset

As noted above, our initial effort focused on producing statistics of wealth at death for Paris in the nineteenth century because the state published no such data, but we were not the first. Starting in the late 1950s, Adeline Daumard (1963, 1973) collected three cross sections of wealth at death data for the city (1821, 1847, 1911). For all her cross sections, Daumard used both the TSA and RMD.[5] She also consulted notarized probates for the nineteenth century. We decided to replicate her effort for 1847, but rather than follow her painstaking procedure, we relied on the TSA's summary reports. To save time (and thus research funds), we collected only age, gender, marital status, and movable and real estate wealth for all adult decedents (aged 20 or older).

To link real estate assets not located in the bureau of the deceased's legal residence, we also collected social information for all individuals with movable wealth above a threshold and with any real estate wealth. The procedure detailed in appendix 2 allowed us to allocate nearly all real estate to a Parisian with movable wealth or to someone who was not a legal resident of the city. Using these rules, we reconstructed Paris's wealth distribution for 1847 in a few months. We then extended this procedure to every year starting with a 2 or a 7, beginning in 1807 and stopping in 1867.[6] We collected the information provided by the TSA only for those who die with some wealth (henceforth often referred to as "the rich," "the wealthy," "who die wealthy," or "who leave an estate"). By symmetry, we refer to the other population as "the poor" or "who die leaving no estate."

What about the poor? We considered linking the estate tax record to death certificates. Unfortunately, Paris's civil registration records from 1792 to 1860 were lost in the fire that destroyed the Hôtel de Ville in 1871. We are fortunate that starting with 1817, the civil registration authorities produced and published annual tables of reported age at death by sex for the entire population (ASVP 1880). We also have total deaths that go back to 1807, and we assume that the gender and age breakdowns for 1807 and 1812 are proportionally similar to those in 1817.[7] We estimate the number of poor decedents by simply subtracting the number of rich we found from the total deaths. We then add the appropriate number of adults by age group and sex to the dataset (see appendix 2). While this procedure is adequate to measure wealth by sex and age, the consequence is that we have no evidence about the other social characteristics of those who died poor (e.g., occupation, place of birth, marital status). These procedures produce our first data of 13 annual cross sections. These data contain 232,206 individuals, of whom

65,642 died with some wealth. The sample represents 20% of adult Parisians who died between 1805 and 1869.

1.2. Basic Results

These data produce several significant findings. First, the average estate (deflated to 1912 francs) grew by 183% in real terms from 1807 to 1867 or 1.8% per year, about twice as fast as the rate of growth of income per adult or wealth per adult in France between 1810 and 1870 (Piketty and Zucman 2014, table FR.2). We must deflate our series because, even though the gold and silver values of the franc were fixed, price changes were sufficient that a franc of wealth in 1807 had a different purchasing power than a franc in 1867, for example. We chose 1912 as the benchmark year, first because it is about the midpoint of the data we will examine, and it is also the last cross section when France was still on the metallic standard for the currency that had prevailed for a long time.[8] The growth in wealth is impressive but not so surprising since the nineteenth century's economic growth favored cities. Moreover, this wealth was starkly distributed. Over the 13 cross sections from 1807 to 1867, 72% of the individuals who died left no estate at all.

Even among the minority who left an estate, inequality was very high. As figure 1.1 shows, in every cross section, the richest 1% (henceforth, the very rich) controlled half or more of the wealth, and individuals below the top 5% had less than 15%. The very top of the distribution comprises the richest 0.1% (the super rich). Its share of wealth was extremely stable over the century, even if it is more unstable year to year. The 232 individuals that comprised that group from 1807 to 1867 held about 18% of all the wealth at death. Even leaving aside those who died without bequests, the average wealth of the top 1% wealth holders was at least one hundred times that of the average member of the ninth decile (P80–89). Whether this wealth produced income for consumption or capital for production, the means of the few was astounding, with capital income for the top 1% amounting to hundreds of times the average labor income. Moreover, this inequality showed no tendency to decline despite the rapid rise in wealth over the period covered by figure 1.1.

Economists have become fond of using a synthetic measure of inequality, the Gini index, that evaluates societies against a benchmark of perfect equality (Gini coefficient of zero) (e.g., Alfani 2019; World Bank Economic Indicators). The Gini rises with inequality, reaching 1 in a society where a single individual owns everything. As figure 1.2 shows, the Gini is very high at above .95 for every cross section. It also has no particular interpretable

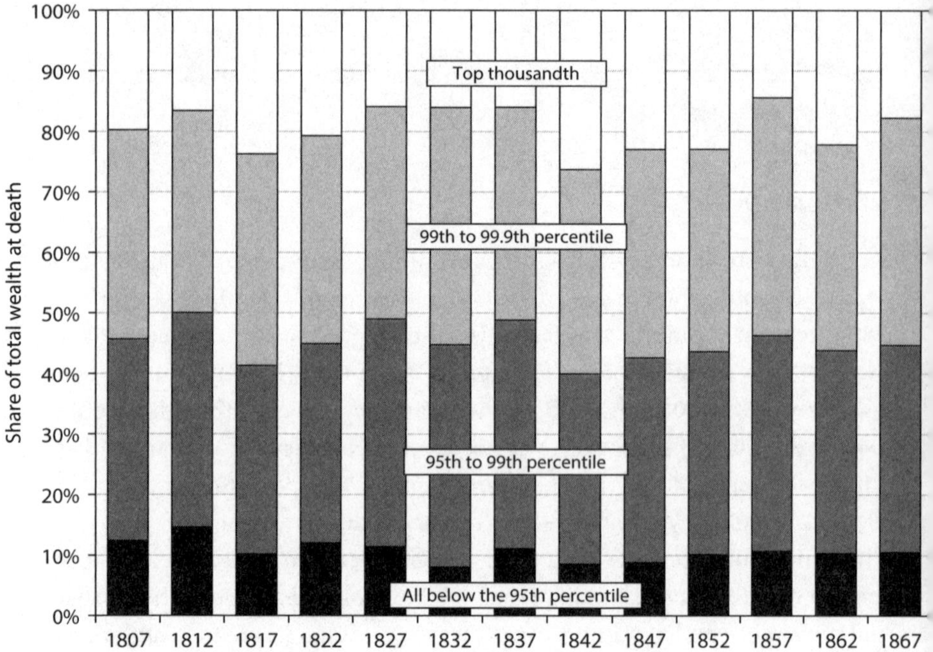

FIGURE 1.1. The distribution of wealth at death in Paris, 1807–1867.
Source: TSA dataset (see appendix 1).

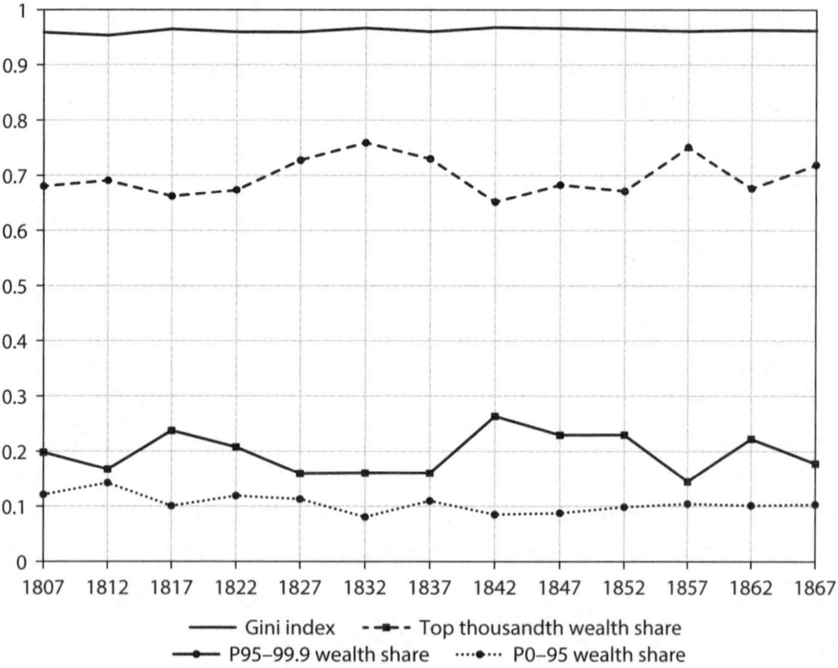

FIGURE 1.2. The Gini index versus the fractiles of the distribution of wealth at death in Paris, 1807–1867. *Source:* TSA dataset (see appendix 1).

connection to the evolution of the wealth share displayed in figure 1.1. This is why we favor looking directly at the fractiles of the distribution.

The uninformative behavior of the Gini index requires a digression in the form of a caveat. The Gini is heavily influenced by the wealth share of the top 0.1%, accounting for a fifth of all the wealth on average. The super rich are erratic in their appearance in the data. Although we collected the entire population for each year, the datasets are small enough that the top of the distribution is unstable. This instability comes from two sources: to start, the sample is small (the top 0.1% numbers only between 13 and 27 individuals), furthermore, the variance of its wealth is huge. Each member of the top 0.1% contributes more than a half percent of total wealth, but the range within that group is wild. The poorest member of that elite was Jean Jacques Bérard, a wholesale merchant. He was 66 when he died in 1817 and left 576,000 1912 francs. The richest was André Auguste Dassier, a banker who was 72 when he died in 1862, leaving 25 million 1912 francs. The wealth ratio between the richest and poorest in the top 0.1% runs from a low of 2.8 (in 1812) to a high of 17 (in 1822). All of this is just the noise of the uncertainty of death among the super rich. However, we cannot ignore that group because its share in wealth is so large. By contrast, over the 13 cross sections, the ratio between the richest and poorest members of the P99–99.9 group varies between 3.6 (in 1842) and 5.3 (in 1812 and 1817). Because there are so many more members of that group (at least 110 a year), the wealth of the richest member matters much less. We will leave the 0.1% out of the discussion for much of our social analysis. We devote chapter 8 to a closer analysis of top wealth holders over the whole period.

At the other end of the distribution, we must consider the mass of individuals with no estate. Information about them is scant. In December 1842, five individuals were entered into the TSA for the eighth bureau, dealing with deceased individuals whose names started with DA.[9] Pierre Danet, age 70, was a resident of a different neighborhood who was recorded with property worth 300,800 francs. Antoine Damiron, a rentier, died at 75 in the institution Sainte Perrine (a private and fee-charging retirement home) with an estate of 182.5 francs. The other three left no estate. Marie Pauline Dalbaret died at age 34 and was buried at the city's expense; Emilie Josephine Daujon, a laundress, died single, aged 25, in a maternity ward. The last one, Pierre Victor Dauty, a tailor, died single, aged 31. In other places in the registers, some of the individuals without estates are noted as indigent, as certified by the municipality, though this is, in fact, rare. So, what is the meaning of having no estate?

Daumard (1973, 122–124), while acknowledging the problem, adopts the nineteenth-century term "indigent." Indigency, in the nineteenth century, referred to individuals who had died in complete poverty and were receiving poor relief. She then returns to the relative safety of the wealthy. Dalbaret, Daujon, and Dauty suggest that the poor died young and worked (two of the three had an occupation listed). To better understand the mass of individuals with zero wealth, we collected all the death records for January 1862 from the civil registration of deaths (État Civil). This dataset contains 4,859 individuals (2,095 of whom were younger than 20, 2,759 adults over the age of 20, and five adults whose age was not reported). For those 20 or older and whose last names started with the letters A, B, C, D, and T we collected names, marital status, occupation, place of death, and residence (for more detail, see appendix 1). We then matched these data with the estate tax data to produce two populations: the rich (died and left an estate) and the poor (died without an estate). Simple comparisons are telling. The celibacy rate is high, and it is higher for the poor than the rich (20% versus 18%), because they died far younger than the rich and, to a smaller extent, because they were less likely to marry at any age.

The poor were workers. We coded without an occupation all those listed as *rentiers* (individuals living from their asset incomes, rents, interests, and dividends); *propriétaires* (property owners); *sans profession* (without occupation); or no reported occupation (and the few who were retired). The others had an occupation listed, and we presume they were working near the time of their death. More than 90% of men who died younger than 60 have an occupation, while twice as many women who died poor (71%) have an occupation than women who died rich (34%). The poor, it seems, worked until they died, a topic we will revisit in chapter 7. This conclusion is strengthened when we look at those who died at age 60 or older; half of rich men still report an occupation, while two-thirds of poor men have an occupation. For women, the gap is even more significant: 40% of poor women have an occupation, but only 10% of rich ones do. The rich died essentially at home, and only 5% are reported to have passed away in a hospital. At the time, hospitals did little to prolong life. Instead, they were hospices and places where the city sought to isolate contagious individuals. Almost 40% of the poor died in a hospital (unsurprisingly, the proportion is higher for those who were unmarried than those whose spouse was alive).

Interestingly, the share of adults who died with an estate was not larger in the wealthier arrondissements (7th, 8th, 16th) than in the rest of Paris.

The reason is the large number of servants. Wherever the top wealth holders congregated, the demand for personal services inside and outside the home was high. The people who served the rich either lived with their employers or nearby. These occupations attracted many migrants to the city, but they were not a path to rapid wealth accumulation. The sample is too small for us to accurately measure the 1% wealth share of the top arrondissements, but it is quite likely that inequality in these districts was even more stark than in Paris as a whole.

That so many of the poor had an occupation at their death is partly related to changes in the welfare system in the early nineteenth century. Before the French Revolution, the city and the Catholic Church had run a complex system of welfare relief. The revolutionaries expropriated the church, and later regimes disemboweled the city's ability to aid the needy by moving welfare under the direct control of the state. Meanwhile, the immediate expropriation of the church without alternative provision of education, poor relief, and hospice care was driven by financial expediency. These reforms, however, were particularly severe for the needy in Paris. They had long been accustomed to a vast network of rich churches and convents for aid prior to the revolution (Lyon-Caen 2018; Gribaudi 2014, 152–163).

Paris suffered another, much more long-lasting, blow to its ability to determine expenditures on social services. Napoleon Bonaparte finalized the reforms that divided France into different administrative levels, with municipalities as the lowest level. Most municipalities were administered by a locally elected municipal council that grew more independent over time. Paris was and continues to be different in many ways. Most important, the state directly administered the capital with a prefect rather than a mayor. It was the prefect who made decisions as to how to allocate Paris funds. His primary concern was public order, and a mass of poor, indigent individuals was a threat to social order (a threat that had been fully realized between 1789 and 1793). After 1815, prefects decided that rather than placate the indigent with welfare, they would invest in repression and encourage unemployed migrants to return to the provinces. As a result, poor relief was not a priority, and the number of individuals classified as indigents and receiving poor relief plummeted from about 20% to about 5%. Poor relief did not begin to rise until after the advent of the Third Republic in 1870 (AGAP 1862). The poor of Paris for most of our period died working (like Daujon and Dauty) or ill (in hospitals).

Nevertheless, it is essential to keep in mind that with no social safety net, no real estate holdings in the city, and minimal access to insurance or precautionary savings, the vast majority of the city's dwellers lived day to day. When they died, they could not afford a grave, and many had to be buried at the city's expense. By the 1870s, individual graves accounted for 35% of burials. The other dead were buried in common plots reserved for paupers. In this latter period, the municipality also tallied who was buried at the city's expense (including expenses on a religious ceremony). Through the end of the 1880s, more than half of Parisian decedents were deemed poor enough for this benefit.[10] The city totals include children; thus, our fraction of destitute adults may appear large, but dying poor does not imply a city-paid burial because, when possible, the authorities would turn to children and relatives to bear these costs. We will return to the issue of the persistence of a substantial fraction of individuals dying with no wealth in chapter 9.

1.3. Context

From 1807 to 1867, the variations in wealth shares over time were small relative to the variation in average wealth. Given the small sample instability of the top part of the wealth distribution, it's possible that inequality did not change over these six decades. The growth in wealth at death, however, implies that Parisians who died under the First Empire were poor relative to those who died under the Second Empire.

Parisians of the first quarter of the nineteenth century were also likely to have been poor relative to their predecessors of the Old Régime. While we have limited data, the fortunes of the Old Régime's elite seem to have been much larger than what we observe in the first four cross sections after the revolution. For instance, the Duc de Penthièvre and the Duc d'Orléans each had incomes over 3.5 million francs, and their wealth was evaluated at above 100 million in 1789 (Duma 1995, 52, 175).[11] In contrast, in our samples, we do not see any estates worth 100 million francs until the 1890s. As discussed in chapter 8, outside our sample years, we found one estate above 100 million francs under the Second Empire, James Rothschild.[12]

Four possible explanations for the relative poverty of the first half of the nineteenth century are asset destruction, asset values, changes in the rich's locational preference, and redistribution among either private wealth holders or between the state and private wealth holders. As we argue below, all are part of the explanation for the low levels of wealth we see after the French Revolution.

Asset destruction occurred in finance because of inflation. Indeed, to keep itself afloat, the revolutionary government issued a paper currency, the assignat, that was legal tender. It lost more than 99% of its value in five years. The revolution's inflation would seem to be a good candidate for destroying wealth as it brought French credit markets to a standstill and allowed long-term debtors to pay off their loans at a tiny fraction of their initial value (Hoffman et al. 2000, chap. 8). From a macroeconomic perspective, the wealth losses of creditors were directly balanced by wealth gains by debtors. At first glance, therefore, total wealth should have remained the same (because bonds are always in zero net supply). However, a closer look at the estate tax administration's records suggests otherwise. The estate tax was assessed on gross wealth. On the one hand, the heirs of creditors were required to report loans that the deceased had made to other individuals (and our evidence for 1822 is that they did so with great frequency). On the other hand, debtors just reported their assets. So, any debt erased by inflation implies an equivalent drop in gross wealth. In prior work, we estimated that the debt stock in Paris fell from 1 billion *livres*/francs in 1780 to 500 million in 1807 (see Hoffman et al. 2000). Because the rich and the old are generally lenders to the less rich and the young, and because cities were net lenders to the countryside, Parisian decedents would have been "losers" during the revolution's inflation. Total gross wealth reported to the estate tax would likely have fallen by tens of millions of francs per year even though net assets in the population remained unchanged.

The other avenue for asset destruction involves the loss of tangible assets during the revolutionary conflicts (these were intense in places like the Vendée and Lyon, among others). In Paris, at least, the impact of the revolution on real assets was multifaceted. Few private buildings were destroyed, though substantial private and all church property (the *biens nationaux*) was confiscated. Although some palaces were put to public use (the National Assembly building and the National Archives, among others, are former princely palaces), most of the real estate seized from aristocrats was put back into private hands by auction. Again, absent changes in real estate values, these political operations should have had little impact on average private wealth. Some individuals lost wealth, but others gained. Because of the inflation of the assignat, the buyers of *biens nationaux* were able to pay off their liabilities to the state at a fraction of the cost. The transfer of aristocratic buildings might have led to small declines in wealth (due to poor maintenance and other transaction costs), but not enough to explain the low levels of wealth of Parisians between 1807 and 1827.

Of course, another redistribution of property was of an entirely different magnitude: the sale of church property. The intent was to sell these assets.

With political instability, however, people preferred to use gold and silver to store value rather than as a medium of exchange, so there was little coin to facilitate transactions. At the same time, the state faced a decline in revenue. The solution was to print money. That was the origin of the assignat, which was backed by the nationalized property. The state printed assignats and used them to pay its bills. The assignats were to be retired when buyers of auctioned church property presented them to pay for their purchases. In theory, at least, the net effect on prices should have been nil. In practice, the state printed more assignats than the auction process could absorb, and inflation raged. In this context, the sale of church property created massive opportunities for speculators. In the long term, this effort should have led to an increase in private wealth at death since most of the church's tangible assets were sold on credit to private individuals, and they could acquit their liabilities in depreciated assignats. In Paris, there was a glut of real assets because the Old Régime city was ringed by a belt of monasteries, convents, and other religious properties, representing about 13.7% of the buildings in 1749. These were suddenly available for occupation by private individuals (Lyon-Caen 2018). As Americans experienced in 2009, real estate markets are inelastic; small changes in the supply of real estate for sale can drastically affect prices. Paris took several decades to absorb the expanded supply of real estate due to the revolution. One important reason for the slow adjustment was that the capital's population declined significantly between 1789 and 1815 (from about 640,000 to less than 560,000 in 1801), rebounding to 700,000 in the early days of the Restoration. Real estate values were thus likely to be low. Although the process of selling *biens nationaux* should have led to an increase in wealth, the price shock to real assets, in the short run at least, is likely to have been negative.

Turning to how the estate tax administration evaluated real estate generates additional insights into the gyrations of wealth. In part because of the inflation, but more generally because the political uncertainties of the French Revolution endured, investors may have required higher yields and lower capitalization factors (i.e., lower value-to-income ratios). Throughout the nineteenth century and beyond, estate tax officials sometimes relied on the building's purchase price to decide its value when the deceased had acquired it recently. However, officials more often calculated value as the annual rental income multiplied by 20 (Eichholtz et al. 2021). They assumed that real estate investment had an effective gross yield of 5%. As a result, the low real estate values that we saw between 1807 and 1827 come from low rental income, not low capitalization rates. These low rental values are not surprising. Most dwellings were rented with oral agreements, the market was competitive, and rents had to be low in the early days of the First Empire before starting to rise as the city's population began to grow again. For most other assets, the impact of the revolution on valuations through capitalization rates is not clear as we do not observe prices for most assets early on. For others, like peer-to-peer mortgages, after 1807, the yields were constant at 5% per year. The national debt was the one regularly traded asset, and from 1795 to 1825, it experienced significant capital gains. Its market price more than doubled between 1810 and 1820. Unfortunately, the national debt was exempt from inheritance taxes until 1851, and changes in its value cannot directly account for changes in average wealth at death over the first half of the nineteenth century.

Estates could also have been low from 1807 to the 1830s because the rich eschewed Paris for the provinces because they could not abide by either Napoleon (for some) or the Restoration kings (for others). That had certainly been the case between 1789 and 1800. We have little evidence on this matter, except that the share of the city's real estate held by Parisians did not change much from 1807 to 1867. The flight to the provinces hypothesis might hold for the First Empire years, but by 1827 the regime had cobbled together an uneasy alliance between Napoleon's military elite and the old aristocracy. Each had a profound vested interest in maintaining their accumulated properties.

The last candidate for the low wealth values we observed in the aftermath of the revolution was losses due to direct government intervention in asset markets. The first of these is the essentially uncompensated end to venality—the sale of government office. Typically, each government official had to buy his office from his predecessor, often for tens if not hundreds of thousands of *livres*. The value of all offices was about a billion *livres* in 1789. The offices were only partially reimbursed, with payments in devalued assignats, when venality ended. The office market was heavily intermediated because offices could be mortgaged just like real estate, and the most important and valuable offices were in Paris. The preservation of venality would have raised average wealth values and contributed to a larger long-term debt market, therefore raising average wealth values.

A second, larger cut involves the national debt. In 1797, after the assignat inflation had fully eroded the public debt, the state restructured its obligations, assuming only one-third of the face value of its debts. The bankruptcy of two-thirds was a particularly severe wealth shock in Paris because Parisians held a big fraction of the debt and would continue to do so in the nineteenth century (Hoffman et al. 2000). Napoleon managed to fund his military campaigns with little recourse to debt. Hence, the national debt in 1815 was quite small. Then the Bourbons made two significant additions: 700 million francs in 1817 to buy off the coalition that put them back on the throne of France and 1 billion francs in 1825 to compensate the émigrés for their lost property. Even then, the debt in 1830 was much smaller than in 1789. Returning to our estates, it is unclear how the default of two-thirds affected other privately held assets. It should have increased the demand for assets not subject to inflation (real assets and other equities). At the same time, the state default should have reduced estate values, and, combined with the inflation, it might have also lowered asset values by reducing discount rates. Another way that private assets may have been impacted is through declines in financial incomes: declines in financial transactions (either debt issues or secondary transactions) over a generation would have reduced the wealth of financiers.

Overall, these four different shocks (inflation, asset destruction, nationalization, and default) produced a fall in the wealth of Parisians between 1789 and 1807. Moreover, the preceding discussion illuminates how the revolution promoted wealth redistribution on a grand scale. Wealth flowed from the private sector to the state and back, from individuals who fell afoul of the revolution to those who gained from it, from lenders to borrowers. As such, one might have expected inequality to fall, and it likely did at the very top. In Paris, where at least two-thirds of the population died with nothing to their names and where 10% of the population controlled 95% of the wealth, it is unlikely, however, that the structure of inequality in the early ninetenth century was very different from what had prevailed in the decades preceding 1789. Instead, the revolution ruined some individuals, made millionaires of others, and redistributed positions within the top decile while significant wealth was lost along the way. Among the winners were those speculators like the Comte Roy or the Mallet banking family who made fortunes by winning bids in the state's land auctions and then paid for their bids in highly depreciated assignats. Others were made rich by providing services to the state, from banking to army provisions.

Among the most dramatic winners of the revolution were the top military and administrative officials of Napoleon who survived past 1815. The emperor had endowed these men with aristocratic titles and property confiscated throughout the empire, pensions, and other emoluments. After 1815, Louis XVIII executed or exiled a few of the most visible supporters of Napoleon's attempt to return to power. Louis XVIII realized, however, that he could not run the country without the help of Napoleon's officials. For example, Talleyrand was a preeminent diplomat who served Napoleon, Louis XVIII, and Louis Philippe. Most striking are Napoleon's 26 marshals (18 of whom were promoted in 1804), who were all made dukes or princes, most with large attendant wealth holdings. In 1815, the Bourbons hesitated on how far to purge the leadership of the army, but political immediacy forced them to compromise, and the marshals' booty was secure.

However, when examined from the perspective of wealth at death, the impact of the 1804 promotions is dispersed since they did not die all at once. Seven of Napoléon's marshals died before the end of 1816, mostly on the battlefield. The others enjoyed long lives: nine died before 1830, eight during the July Monarchy, and the last two passed away in 1850 and 1851. One reason is that half of the marshals had been under 25 in 1789, but another is that a full twelve died older than 75. As this extreme group suggests, massive events like the French Revolution and the First Empire reverberated into the wealth-at-death statistics into the 1850s. Although few people lived into their eighties in the early nineteenth century, the rich, as we shall see, were long-lived.

Napoleon's marshals provide a good jumping-off point for us to dive deeper into our data analysis. Their longevity raises important issues that we illustrate by considering the savings and consumption experiences of the youngest members of the 1807 and 1812 cross sections. Seventeen percent of the adults who died then did so between the ages of 20 and 29. Born between 1778 and 1782, they became adults twenty years later, between 1798 and 1802. On the one hand, they arrived at adulthood after the worst dislocations of the revolution. On the other hand, they had a very short time to accumulate or inherit wealth before their demise in 1807 or 1812. It is not surprising that 91% of them died leaving no estate. Other cohort members would die over the next fifty years and have more time to accumulate and consume. The evidence in table 1.1 shows a strong age-wealth accumulation pattern. Wealth increases with lifespan to such an extent that the average estate of the 80-year-olds is 40 times larger than that of those who died in their twenties in the waning days of the First Empire. The cross-sectional age pattern is more muted, but wealth still increases with age, and the 80-year-olds who died under Napoleon were 10 times wealthier than the 20-year-olds.

Napoleon's marshals raise some critical questions beyond the obvious questions of age and wealth. They were promoted for success on the battlefield, and few came from rich or powerful families. Napoleon made most of them wealthy. Those who lived long continued to use their political might to

TABLE 1.1. Cross-sectional and cohort age-wealth at death patterns for the 20- to 29-year-olds who died in 1807 or 1812

	Share of deaths in year of death	Share of deaths in the cohort	Share who die with wealth	Real wealth	Year of death
BIRTH COHORT 1780–1789					
20–29	14.5%	9.2%	10.5%	1,254	1807–1812
30–39	12.7%	10.2%	27.1%	5,889	1817–1822
40–49	14.3%	18.6%	25.8%	8,870	1827–1832
50–59	12.9%	18.1%	28.4%	14,119	1837–1842
60–69	14.9%	17.1%	34.0%	23,436	1847–1852
70–79	12.1%	17.5%	35.3%	32,252	1857–1862
80+	5.2%	9.3%	37.1%	52,190	1867–1872
1807–1812 CROSS SECTIONS					
20–29	17.0%		8.9%	990	1807–1812
30–39	12.0%		25.4%	3,999	1807–1812
40–49	12.7%		38.8%	7,394	1807–1812
50–59	15.0%		42.4%	8,000	1807–1812
60–69	19.5%		34.4%	6,318	1807–1812
70–79	16.9%		40.0%	8,297	1807–1812
80+	6.9%		42.4%	10,989	1807–1812

Source: Estates dataset.

Note: The real wealth value is computed as the 1912 CPI deflated wealth. Column 1 of the top panel does not sum to 100% because the denominator changes with every row.

secure additional riches. Moreover, one reason that individuals have estates is because they want to pass on wealth; this wealth has dynamics that extend well beyond the age at which one's life ends. We continue to encounter descendants of Napoleon's marshals into the 1950s, and though they are no longer at the apex of the wealth distribution, they have not sunk into the middle class either. The marshals were obviously all men. Even in societies where women's property rights are weak (e.g., England in the same period), a focus on male wealth misses a lot. In the case of Paris, ignoring gender is simply impossible. Our dataset is nearly evenly split between men and women (49.4% of all adult deaths were women, and the same is true for individuals with a positive estate). Given the strictures society places on women, it is critical to understand the dynamics of wealth by sex.

1.4. Inequality Under the Microscope

Around 1800, most of the Anglo-Saxon world featured unrestricted bequests and weak property rights for women. Absent prenuptial agreements, a woman's property would pass to her husband at marriage, and the estate of a married man would be taken to be the household's wealth. France before 1789 was a hodgepodge of different jurisdictions where the extent of married women's property rights and freedom to testate varied (Goy 1988). The Code Civil, enacted in 1804, aligned the country with Paris's custom. It featured strong property rights for women and equal treatment of children regardless of gender. Wives' dowries were senior debts in a couple's community property, and wives and husbands had equal claims to community property. The revolution also restricted the freedom of bequests. Individuals were obligated to divide two-thirds or more of their estate equally among their legitimate children without regard to gender. One can ask, did these formally equitable rules matter? Over the period from 1807 to 1867, the proportion of women with positive estate was the same as that of men. However, at the same time, it should be noted that the value of women's positive estates was only two-thirds that of men's. As table 1.2 shows, women were relatively poorer than men because they were significantly underrepresented in the top 1%. They were only 30% of the 0.1%, and their share of decedents by fractile reached parity only below the 95th percentile. If we eliminate the top 1%, women's wealth reaches 75% of men's.

Over time, the ratio of women's to men's estate values developed into a rough U-shaped pattern. Women's estates were near parity with men's in 1807–1812, falling irregularly to about 55% around 1847, and then women's estates grew faster than men's and reached 113% in 1867.[13] Attributing the decline and its reversal to the persistent differential effects of the revolution is tempting. To the extent that the sexist nature of society tilted women's portfolios toward bonds and within equity toward real estate rather than businesses, the revolution's redistribution would have hit them particularly hard. However, the early parity of men's and women's fortunes is inconsistent with this hypothesis. The period of high mobility that spans 1789 to 1825 clearly could not be gender-neutral since the opportunities for advancement were overwhelmingly reserved for men. In the early cross sections, the bulk of the wealth came from individuals in their forties or fifties at the time of the revolution and thus mostly suffered wealth losses between 1789 and 1797. By the 1820s, the cohorts that dominated were those in their twenties during

TABLE 1.2. Wealth by sex (1807–1867)

	Fractiles of the wealth distribution						
	All	P70–79	P80–89	P90–95	P95–99	P99–99.9	P99.9–100
	Average wealth						
Men	18,207	217	3,420	23,319	131,853	595,807	3,074,313
Women	11,687	209	3,257	22,593	124,458	589,202	2,359,977
	Women's share of decedents in fractile						
	50.5%	52.1%	50.8%	49.1%	45.8%	40.4%	31.1%
	Women's share of wealth in fractile						
	39.6%	54.3%	51.2%	49.5%	48.5%	44.3%	40.1%

Source: Estates dataset.

Note: The fractiles are computed each year, so the P99.9–100 column reports the values for the population of the top 0.1% of individuals as defined by the year they died.

the revolution and First Empire; they inherited less, but the men could seize the opportunities presented by regime change. By the 1850s, the dominant cohorts were the children of the same period who then reached the rough wealth parity implied by the egalitarian marriage and inheritance regimes. We will repeatedly return to the interaction between gender and wealth in subsequent chapters. Let us now turn to age. As we discussed briefly in the previous section, the relationships between age and wealth are strong, though complicated.

The factors that lead individuals to accumulate wealth are varied but fall broadly into a few categories: investment, insurance, life cycle, and bequests. The investment motivations are on the production side of the economy. When capital markets are imperfect and information asymmetries are high, securing outside financing is difficult, so the owner's equity will dominate most businesses' capital. If firms are profitable, the owners will have to accumulate wealth. Capital market imperfection will also make it difficult for owners to sell equity as they age.

On the household side, wealth accumulation may also have to do with capital market imperfection. If these are severe, real assets may be the cheapest way to deal with labor market or enterprise risk. More importantly, households accumulate wealth as part of the life cycle. The perfect life-cycle saver is born with nothing, goes into debt for her education, then repays her debts and saves for retirement during the active phase of her work life. Finally, she dissaves during retirement, ending up with no wealth when she dies. Realizing such a program would require either extraordinary foresight or extraordinary luck. The uncertainty of age at death means that individuals must save more than they need to avoid running out of money if they live long. As a result, those who die retired but young die wealthy. Lifespan risk implies that precautionary savings will likely peak right about retirement age, creating an inverted U-shape to wealth at death by age.

A different motivation for saving is passing on wealth to one's progeny or other worthy recipients. At the extreme, consider the dynast who inherits a property from his parents and maintains it over his lifetime but consumes the capital income or services the property provides and does not actively save. Absent capital gains, he would die with roughly the same wealth at any age. Taking into account capital gains, the older the dynast dies, the more wealth he leaves behind. In cross-section, there may be little increasing wealth except for, the older that one dies, the more likely one is to have inherited wealth. One can also imagine someone whose dynastic motivation is to save a constant fraction of income. That strategy could only be deployed by individuals with sufficiently high income and thus more likely to be available to those who become wealthy early on in life. More importantly, such dynasts have more wealth as they get older because they have more time to save, not just because of capital gains.

Let us examine the wealth by age data in cross-section as displayed in figure 1.3. The figure displays average wealth by age group (deflated to 1912 francs) and follows its distribution over time by distinguishing four periods (1807–1822, 1827–1837, 1842–1852, 1857–1867). We excluded the top 0.1% to make the figure more intelligible and to avoid building steep age profiles simply because the wealthiest Parisians died old.

In each of our periods, the average wealth of those who died in their twenties is low—less than 4,000 francs (though to be fair, it grows by 70% over sixty years). As we move across the age groups in the first period, wealth jumps quickly to about 6,000 1912 francs and stays there until a bump brings the 80+ group up to 10,000 francs. While this is not the expected pattern for life-cycle savers, it could result from dynastic accumulation. The next three periods show the same pattern, except that the age-wealth curve steepens as we move further away from the revolution. Individuals who died in their fifties were nearly three times as wealthy as those who died in their twenties in the 1807–1822 period. In 1827–1837, those who died in their fifties were more than four times as wealthy as those who died in their twenties, and

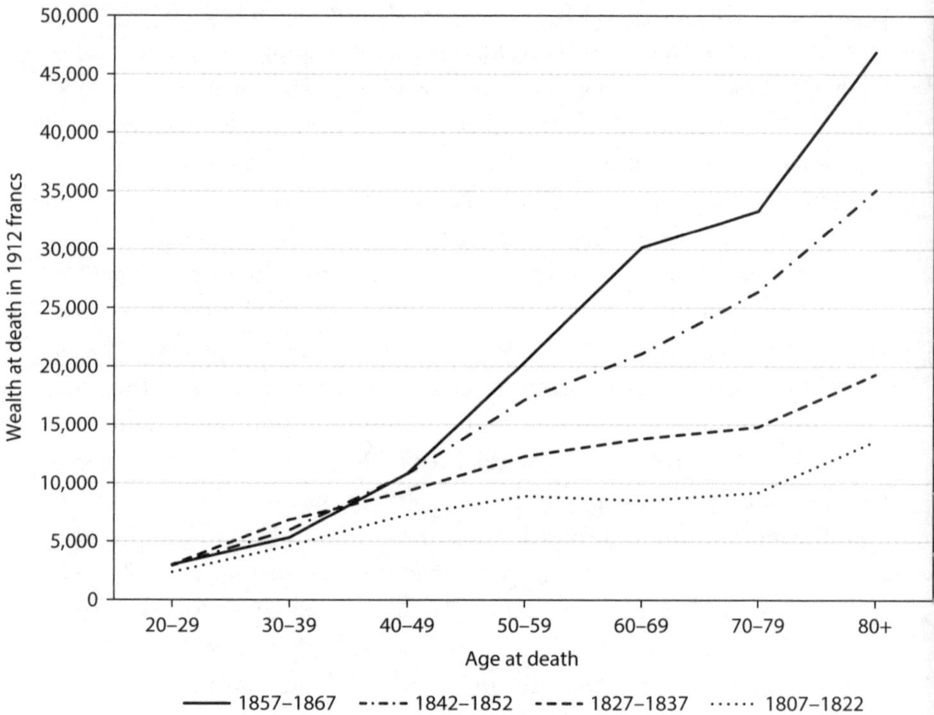

FIGURE 1.3. Age-wealth profiles for Parisians in 1912 francs, 1807–1867 (P99.9–100 excluded). *Source:* TSA dataset (see appendix 1).

they were almost six times as wealthy in the last two periods. Those who died between 60–79 were somewhat wealthier than the 50–59 reference group before 1842, but the age-wealth profile then steepened again. By the last period, those who died at age 80 or older were 12 times wealthier than those who died under 30. These results were derived after we excluded the top 0.1%. Moreover, the group that dies 80 or older contains more than 2,000 individuals in each period, so these steep age profiles are not a statistical fluke. The steep age profiles of the middle part of the nineteenth century are halved if we exclude the top 5% (who own 85% of the wealth), but they remain. These data do not support either the life-cycle saver model or that of the simple dynast.

Instead, the flat age profiles of the early period combined with the low levels of wealth we uncovered suggest that part of the rapid growth of wealth between 1807 and 1867 was a recovery from the revolution's turmoil. If we take the steep age profiles of 1857–1867 as the standard, we can apply them to the earlier cross sections. We use a given age group as the benchmark and then adjust the average wealth by age to get the same age-wealth profiles as for 1857–1867. It turns out that whether we benchmark at 20 or 50 years old does not matter. The first two decades of the nineteenth century were missing about 30% of their wealth because the old were not rich enough.

Now, let us consider the same data by birth cohort. To do so, we deflate all the reported wealth to 1912 francs again and consider age by quinquennium. This means that the individuals aged 40–44 of 1822 are in the same cohort as the 35–39 of 1817 and the 20–24 of 1807 and corresponds to individuals born from 1778 to 1782. The oldest members of the cohort born between 1778 and 1782 died in 1867, aged 80–84. Each cross section contributes, at most, one observation per birth cohort. Although we have 13 cross sections, the cohort born from 1783 to 1787, who died aged 20–24 in 1807, is the only one we observe through the ages of 80–84. Life expectancy at 20 was closer to 40 than 60, so one might wonder about the accuracy of measuring wealth at such a high age. In fact, the 80+ group comprises about 7% of our data every year except 1832 (a year of a cholera epidemic that was particularly harsh on younger adults).

Because cohorts are incomplete, we display the wealth ratio by twenty-year intervals (20 to 40, 40 to 60, and 60 to 80) and again exclude the top 0.1%. Table 1.3 shows that the gains associated with long life were dramatic. For the cohort born between 1783 and 1787, those who died at 80 or older were 22 times wealthier than those who died young (5,65*1,47*2,65 = 22). The other cohorts are incomplete; nevertheless, the connection between age and wealth is striking. Within cohorts, living longer was associated with significant differences in wealth. Even for the young, and eliminating the cohorts who died young under Napoleon, those who made it to 40 were, on average, 4.5 times wealthier than those who died before 25. Wealth more than doubled within cohort between those who died in their forties and those who died in their sixties and more than doubled again between 60 and 80. Excluding the very rich, reaching old age in a cohort was associated with wealth about 20 times greater than dying among its youngest members.[14] These effects are more than twice as big as those we observe in cross-section. They are also inconsistent with a simple dynast who passively holds on to the assets he or she inherits, or with a simple life-cycle savings model.

The changes in wealth we observe seem to occur because of the decisions of a subgroup of the population: the successful savers. These are part of a larger group that seeks to save to produce bequests, the bequest savers. Some of these have enough adverse shocks that they die with zero wealth. The rest of the population living in Paris consumes but does not save (this includes,

TABLE 1.3. Wealth ratios by age within cohorts

Year cohort reached 20	Age at death 20–40	40–60	60–80	Year cohort reached 80
1747			1.71	1827
1752			2.89	1832
1757			4.21	1837
1762			2.21	1842
1767		1.62	1.83	1847
1772		2.09	3.90	1852
1777		3.28	2.37	1857
1782		1.70	2.26	1862
1787	5.65	1.47	2.65	1867
1792	4.30	3.59		1872
1797	7.00	2.44		1877
1802	2.28	2.17		1882
1807	2.58	2.89		1887
1812	3.63			1892
1817	2.98			1897
1822	3.54			1902

Source: Estates dataset.

Note: Each cell in the table reports the ratio of wealth between two groups in the same birth cohort that died 20 years apart; thus, cell 1.62 in the line "1767" reports the wealth ratio of those who were born between 1767 and 1771 and who died between 40 and 44 years old to those born in the same quinquennium who died 20 years earlier aged between 20 and 24. The empty cells refer to data we do not have because individuals in the denominator or the numerator (or both) would have died before 1807 or after 1867.

of course, many individuals whose income is too low to permit savings). The data are consistent with the idea that there are both savers and nonsavers in the population. Figure 1.4 shows the fraction of individuals with a positive estate by age across four broad periods. These shares increase with age, likely because wealth helps us live longer, something we will return to later. More importantly, the fraction that dies with wealth varies little over time (the early and late periods are hard to distinguish, and the drop in the second period results from the high number of poor who died during the cholera epidemic of 1832). The share of 50-year-olds who die wealthy is always about one-third, even though the supply of savings vehicles grew over time. This

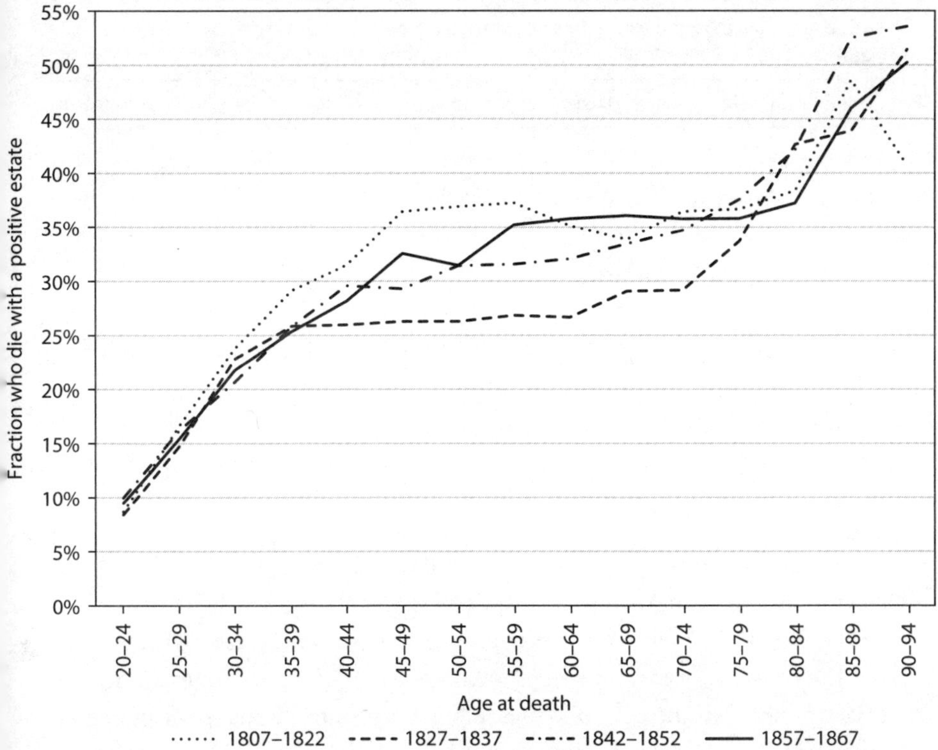

FIGURE 1.4. Fraction who die with a positive estate by age group. *Source:* TSA dataset (see appendix 1).

suggests that the high share of zero-wealth individuals is likely due to many Parisians not being interested in constituting bequests. We will return to these questions in chapter 9. In death, Paris was an unequal world dominated by some very wealthy and, largely, elderly men.

Even with our partial reassembly, real estate was more concentrated than wealth in general. The richest 5% of decedents owned at least 91% of the real estate in Paris but held at most 77% of movables. Because the real estate share of total wealth was about a third, its distribution greatly contributed to inequality. The concentration of real estate is owed to Paris's historical experience as a fortified city: over time, land inside the city walls became progressively scarcer, and individual parcels were expensive. Moreover, capital and legal constraints made joint or corporate real estate ownership difficult. All this comes through when we examine the fraction of estate owners by fractile. For anyone below the top decile, the likelihood of owning real estate was negligible (less than two people in one thousand did); it rose

TABLE 1.4. Share of real estate in real estate owners' portfolios by fractile of total estate

Decade of death	P70–80	P90–99	P99–99.9	P99.9–100
1800	92.6%	81.4%	59.3%	41.2%
1810	85.7%	77.0%	63.3%	32.0%
1820	81.4%	76.9%	58.6%	43.5%
1830	94.4%	84.9%	66.7%	55.9%
1840	85.9%	81.8%	63.2%	44.2%
1850	90.2%	78.3%	57.6%	39.3%
1860	85.9%	76.7%	62.8%	48.3%

Source: Estates dataset.

Note: Since we do not observe rural real estate, both the sample restriction (real estate owners) and the ratio are based on Parisian real estate only.

monotonically with wealth. Half the individuals in the top 5% owned real estate. That fraction rose to 70% for the top percentile of the distribution. At the very top (0.1%), the share of real estate owners is also 70%.

The propensity to own real estate increases with wealth overall. In contrast, the share of the portfolio invested in real estate among those who own some strictly decreases with wealth. In most decades, individuals below the 90th percentile who own Parisian real estate have 90% of their wealth in that asset. For individuals in the top decile but not the top percentile, that share falls to 75%. The top centile, excluding the top thousandth, is at 60%, and the top 0.1% holds less than 40%. The variation over time does not appear significant (see table 1.4). The declining share of real estate with wealth, combined with the fact that the share of individuals who own real estate in Paris increases with wealth, suggests that individuals relied on different strategies to build their wealth. Moreover, for the top 1% at least, real assets were generally part of a portfolio strategy.

Our closer examination of wealth closes with a consideration of a particular form of social status: nobility. The nobility under the Old Régime was already a heterogeneous elite. It included a landed elite with close connections to the military whose titles were often ancient. It also included the top of the administrative elite who acquired their noble titles by virtue of their office, in some cases recently. It also included a hodgepodge of individuals whose social status reflected some personal or ancestral close connection to the crown. During the First Empire, Napoleon created a couple thousand new titles of nobility, and the regimes that followed would also create some

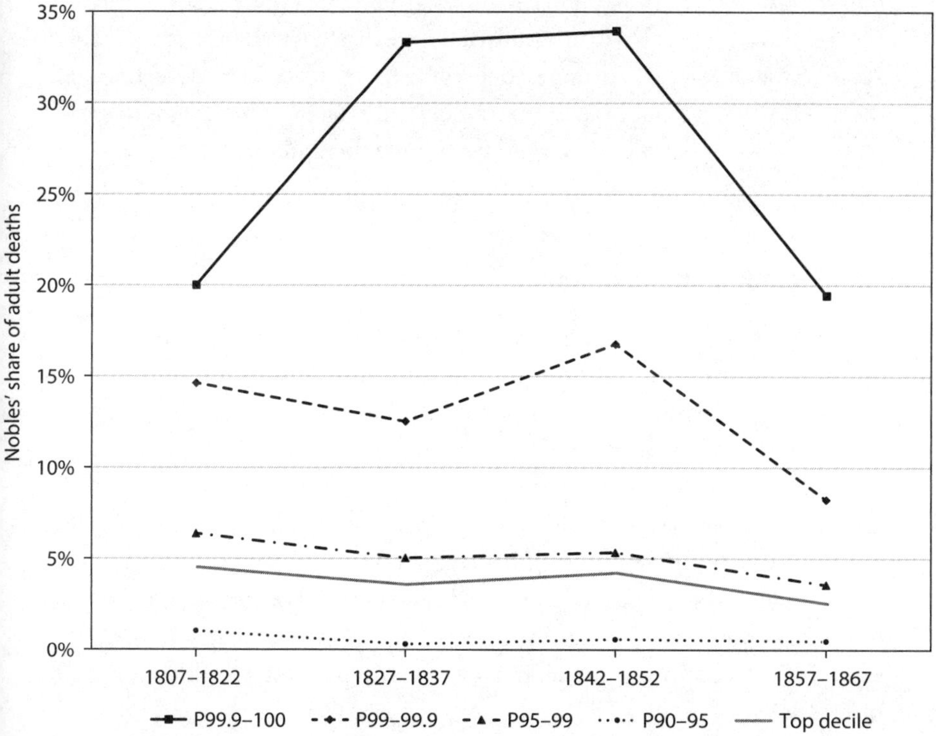

FIGURE 1.5. Nobles' share of adult deaths by fractile. *Source:* TSA dataset (see appendix 1).

new ones. We developed two lists of nobles. The first includes all individuals with a title or whose name made them known as nobles. The second includes all individuals with a particle (names with de or du followed by a name, such as de Bourrée de Corberon). Neither list is perfect. The first one is too limited to capture all the nobles, as titles are not systematically included in the TSA. The second is surely too large, as many individuals with a particle were not noble. In addition, because we do not have the names of every person who died, we can only identify a fraction of the wealthy nobles. That said, our data capture noble wealth well because we have name information for the richest 10% who own at least 90% of the wealth. If we use our narrow definition of noble, the wealth share of that group in the top decile ranges between 2.6 and 4.5% with a tendency to decline over time (see figure 1.5). In every period, the share of nobles is rising with wealth. On the one hand, nobles made up less than 2% of the French population and hence of decedents (Higgs 2019, chap. 1). On the other, they still accounted for a disproportionately high 15% of Parisian wealth in the 1860s. The overall picture we observe is that nobles amounted to nearly 20% of the wealth before 1850, then a decline set in so that nobles had dropped below the 15% threshold in the last period. All in all, however, nobles were extremely rich: the broader list implies they were 5.5 times wealthier than average in the 1860s and those with identified titles 8.2 times wealthier.

1.5. Wealth of the Living

Going back over one hundred years, economists have considered using wealth at death information to infer the level and distribution of wealth among the living through the estate multiplier method (used by Atkinson and Harrison 1978 for the UK, and Kopczuk and Saez 2004 for the US). The first approach simply assumes that death is random relative to wealth, and then the wealth of the dead is a proper but unbalanced sample of the living. It would be unbalanced because mortality is increasing with age. The sample can be rebalanced by reweighting the observation with the inverse of death probabilities by age and sex.[15] This approach, which follows the older literature, produces the first set of estimates (estate multiplier lines in figure 1.6).

Unfortunately, the links between age and wealth do not run simply from age to wealth (as in life-cycle savings or dynastic models). Wealth also serves to preserve life. Thus, we also need to consider the fact that at all ages, the wealthy die at lower rates than those who have no wealth-differential mortality. We take a simple approach and estimate a model that assumes that the poor (those with zero wealth) die at twice the rate as the rich (those who die with positive wealth).[16] This level of differential mortality is consistent with the range of mortality estimates we observed in Paris in the 1880s before the diffusion of sewers (Kesztenbaum and Rosenthal 2011, 2017). We then apply this method to produce a second set of estimates of the wealth of the living (differential mortality at 100% in figure 1.6).

Figure 1.6 displays wealth shares for the top 1% and for those between P75 and P95. It shows that these two procedures lead to similar estimates of inequality among the living. The basic estate multiplier does produce more inequality among the living than among the dead (simply because the living population includes many young people who are unlikely to die and are overwhelmingly poor). The differential mortality approach produces virtually the same distribution as the one we obtain from the estates themselves. The estate multiplier method tends to give a smaller wealth share to

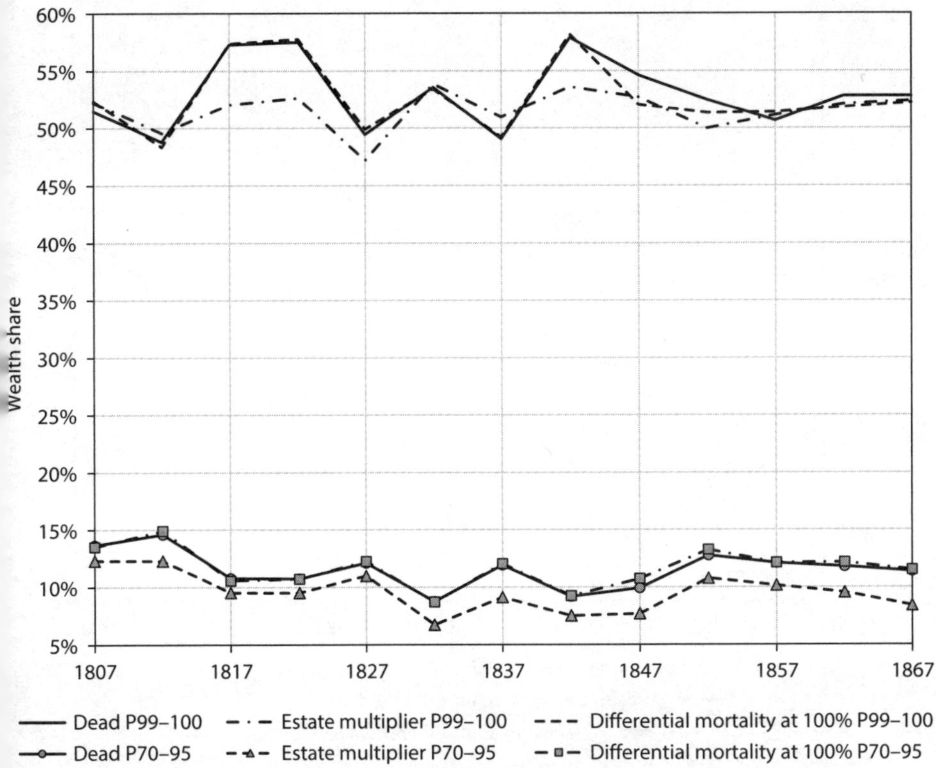

FIGURE 1.6. Inequality among the dead and the living. *Note*: The figure displays three estimates of the wealth shares for two groups: the top 1% and those below the top decile. The "Dead" estimate is the wealth share we compute directly from the estate. So, "Dead P99–100" is the wealth share among the dead of the top 1%. The estate multiplier P99–100 reports the top 1% wealth share estimate for the living, assuming mortality is random relative to wealth. The differential mortality P99–100 reports the estimate for the living, assuming a differential mortality of 100% between those with a positive estate and those who die with no wealth. *Source:* Estate multiplier dataset.

the top 1% in favor of the next 4% of the distribution. The reasons for these differences are that the simple estate multiplier approach tends to underweight the old (because there are few of them and many of them die) and overweight the young (who are numerous and have low mortality rates). Since the 1% are old, they suffer from the first effect, while the middle range of the distribution, which is younger, gets a relatively higher weight. When we add differential mortality, we reverse this phenomenon because, as we saw earlier, the older age groups have few individuals with no wealth. Overall, we prefer the differential mortality approach because, as noted above,

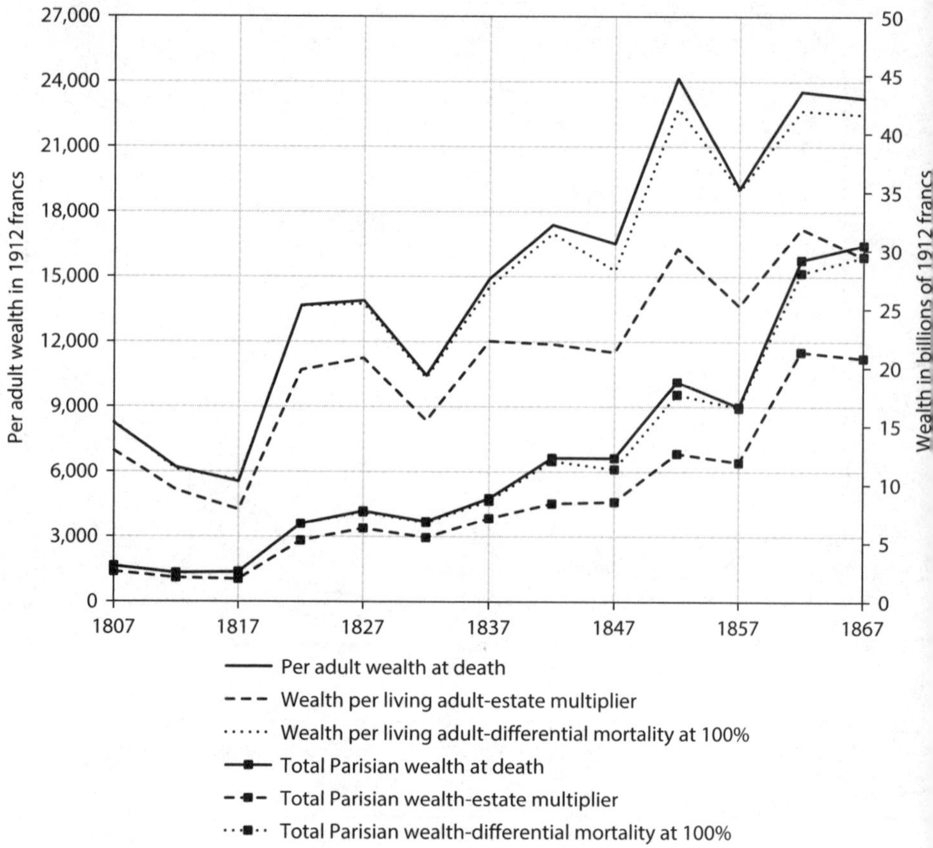

FIGURE 1.7. Wealth in Paris, 1807–1867. *Source:* Estate multiplier dataset and TSA dataset (see appendix 1).

the evidence that wealth prolonged life cannot be ignored. In either case, the conclusions that Paris was very unequal and that the distribution did not change much over the first six decades of the nineteenth century hold for both the dead and the living.

The initial goal of the estate multiplier method was not to estimate inequality but wealth. Figure 1.7 shows that levels of wealth depend on the method used. The figure shows both per adult wealth and total Parisian wealth estimated via three methods: "wealth at death" assumes that the dead are a representative sample of the living, the estate multiplier adjusts for age-specific mortality, and differential mortality corrects for the rich dying at lower rates. As in figure 1.6, the differential mortality approach produces estimates of average wealth similar to those derived directly from estates.

The estate multiplier approach is more pessimistic. It suggests that per adult wealth in 1912 francs grew by 128% between 1807 and 1867, and total wealth grew by 713% (because the population grew by 256%). If we take the more realistic estimates produced by the differential mortality approach, per adult wealth grew by 171%, and total wealth was multiplied by almost ten, to some 30 billion francs. Some Parisians, it seems, took Prime Minister Guizot's famous phrase to heart: "Enrichissez-vous."

1.6. Conclusion

By 1867, the individuals who had lived through the French Revolution or Napoleon's empire as adults had all passed away. The individuals leaving bequests in that year had lived and invested in a world that witnessed significant regime changes (the fall of Napoleon's empire in 1815, the Revolution of 1830, the Second Republic, 1848–1851). Until 1848, the connections between wealth, status, and politics were impossible to escape. Titles came with political rights and wealth requirements, and voting rights also depended on tax payments. If one were male, meeting the threshold for voting would be easiest through land ownership. Women, of course, did not have the right to vote until 1944.

After 1848, all men could vote, but the connection between politics and wealth remained in various ways. One that is particularly salient for capital is that listing firms on the Paris Bourse depended on government approval. Moreover, many listed companies started as concessions (e.g., railroads, canals, and corporate banks). Some scholars have followed Alexander Gerschenkron in dating French industrialization to the 1850s, with the government's authorization of large corporate banks. Yet the rise of the equity market started earlier, and the transformation of investment opportunities for Parisians was quite gradual. As we shall see in chapter 2, the share of all wealth invested in financial assets grew from 1822 to 1872, and the process began well before the creation of large banks under the Second Empire.

However, the transformation of the investment landscape mattered only to a fraction of the population. The opposition between the haves and the have-nots central to many political change models (e.g., Acemoglu and Robinson 2006) does not fit France well as a whole because land ownership across the country was widely distributed. These models, however, have a powerful empirical echo in the Parisian experience. Acemoglu and Robinson examine politics and public goods when a wealthy minority chooses the regime (democracy or autocracy) and a poor majority chooses to acquiesce or revolt. The poor, thus, are a danger to wealth holders. Every group that came to power between 1789 and the 1880s had to be concerned with the Parisian political powder keg (Chevalier 1958). All sought the security of a large contingent of troops stationed in the capital, but those measures did not prevent a series of uprisings. Many of these uprisings failed, but during the revolutions of 1830 and 1848, the huge mass of individuals with no wealth who lived in the eastern and northern parts of Paris drove regime change. Examining this period brings us to two conclusions. The first concerns wealth, and Paris shines as a success story: over the first three decades of the nineteenth century, real per capita wealth was multiplied by 2.8 among the dead (and 2.7 among the living). If we add to that population growth in Paris, the aggregate inheritance flow was multiplied by nearly ten (as was estimated wealth). This evidence partly explains why the Haussmann rebuilding of Paris took place in the second half of the nineteenth century: by then, Paris was rich enough to reinvent itself.

Our second conclusion concerns inequality. There was a long-run equilibrium where the top 0.1% owned a bit less than a quarter of the wealth, the whole of the top 1% about half, and the bottom nine deciles next to nothing. Moreover, it was always the case that more than two-thirds of the Parisian population died with no wealth. Estimates of the distribution of wealth among the living are similar. Even with the whole population, variations around that equilibrium seem circumstantial. For instance, the cholera epidemic of 1832 increased the fraction of Parisians who died with nothing. To give another example, in 1822, the death of two unusually rich individuals, Baron Dopff, the Paris agent of the bankers who negotiated the first Bourbon loan in 1817, and Louise Therèse d'Orléans, pushed inequality up. In the low-tax environment of the nineteenth century, it seems that wealth inequality was persistently high even in the country's financial capital. Taking these two conclusions together, inequality and wealth accumulation seem almost entirely disconnected. However, the high levels of inequality were not without costs. Consider age at death by fractile (figure 1.8). The lowest line is the average age at death for adults in the six lowest deciles (groups that always die with nothing). Their death age starts at about 50, then falls to about 46. It does not return to 50 until the 1880s. One reason for the decline is, of course, the worsening of living conditions in Paris before the expansion of the city and the rise of infrastructure investment of the 1860s. Another involves the dramatic decline in welfare spending in the capital that we discussed earlier. The second group, P70–89, includes mostly wealthy people (except in 1832, at least

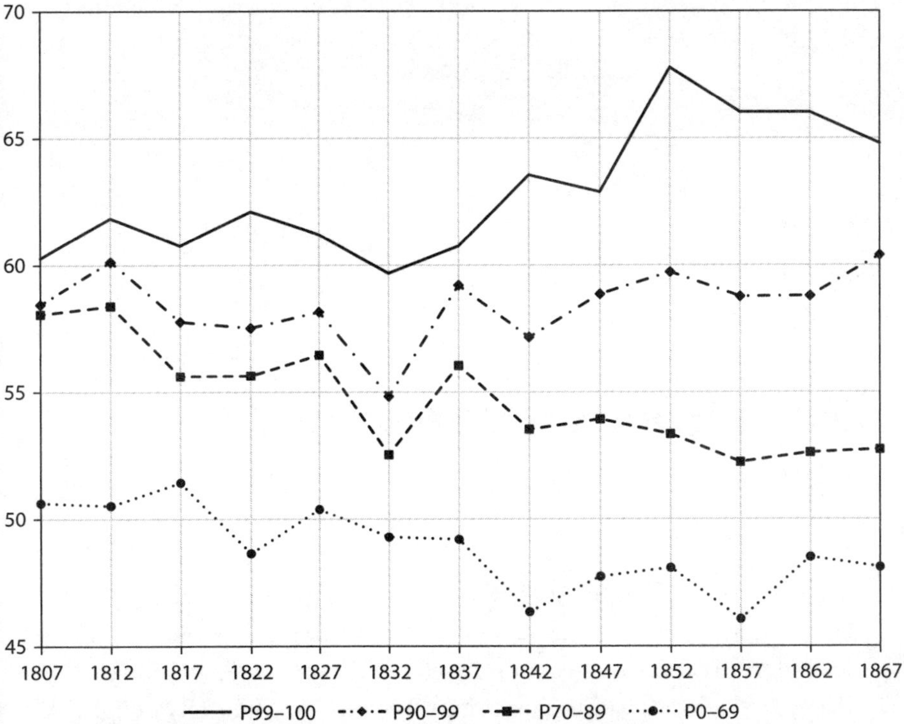

FIGURE 1.8. Age at death for adults in Paris, 1807–1867. *Source:* TSA dataset (see appendix 1).

80% of this decile dies with some assets). This second group is four to seven years older than the truly poor. It also experiences a decline in average age at death of four years. The top decile (which owned at least 88% of the wealth) begins with an average age at death similar to the P80–89 group. Its average age at death grows by two years over six decades and ends up with an average at death of 60. The top 1% starts with an average age at death above 60 and ends with an average age of 66 and has the highest increase in age over the period.

The TSA registers have taken us quite far; they allowed us to reconstruct wealth and its distribution for the first six decades of the nineteenth century when detailed data is scarce anywhere in the world. The results of the analysis raise questions that we explore in subsequent chapters. These include the extent of social mobility and the interaction between shocks and investment strategies. However, to do so, we needed new data, which is the first task of chapter 2.

2

Belle Époque Capitalism, 1852–1912

For France as a whole, the Belle Époque runs roughly from 1890 to 1914 and refers to the burst of economic activity that preceded World War I. For Paris, as we shall see, this chronology does not hold: the city of lights saw a remarkable process of growth that started before the Second Empire (1852–1870) and may have reached its apogee a decade earlier than the rest of the country. This development is revealed in the rapid growth of the wealth of Parisians. During the first five decades of the century, while the population of France as a whole grew by less than a third, the capital's population more than doubled. During this period, its polymorphous industrial and commercial activities broadened. The thousands of small- and medium-sized specialized companies that had traditionally animated the capital's economy were joined by new, and often large, firms. These new firms (including railroads, metal works, banks, and theaters) reinvented industry, services, and culture (Hugo 1867; Benjamin 1982). From 1807 to 1870, the revenue of the business tax (the *patente*) rose almost four times (from 4 million to 18 million francs) (Passy 1874, 391).[1] Moreover, many of these firms' activities were carried out beyond the city's walls, in the suburbs, in the rest of the country, and for a few in the colonies and beyond. Paris was changing France. The city's growth in all these ways called for additional infrastructure of all kinds. That, in turn, created yet more demand for capital in the capital.

Often called the first golden age of financial capitalism, the second half of the nineteenth century saw the blooming of large-scale financial institutions (exchanges and banks) with a global reach. Banks were newly large, and the scale of some industrial and transport enterprises reached unprecedented levels in terms of capital and employment (Berlanstein 1991). These phenomena gave rise to two intertwined notions of capitalism. The first highlights the scale of industrial enterprises, the attendant financial resources they require, and the disassociation between the ownership of the enterprises and the individuals who provide the labor (and one should add management). The other focuses more strictly on the financial sector's growing economic and political importance. Both have long lineages in social science, with the first going back at least to Marx. In a controversial and celebrated article (1951), Alexander Gerschenkron proposed a model that linked these aspects of capitalism to explain the accelerated economic growth of certain countries in the later nineteenth century. France was one of the cases that Gerschenkron put forward. In his view, the acceleration of economic activity depended on an acceleration of capital investment that raised labor productivity. In economies that were not too poor (e.g., France or Germany), a reorganized financial system with new large banks provided the resources for the industrial transition. Since then, scholars have debated whether France does or does not fit this model and the extent to which specific financial organizations (banks versus stock exchanges) are more important in fostering economic growth. These debates carry forward to current policy and academic debates about what financial systems are most apt to produce economic growth (e.g., Levine 1997; Laporta et al. 1997, 1998).

Going back to Alexander Gerschenkron's famous article, the literature has often cast France's financial model as bank-dominated (Neal 2000). In Gerschenkron's case, this characterization was made largely out of convenience to fit France between England's financial market model and Germany's bank model. In an important but often ignored article, Roehl (1976) cast doubt on where France had been placed. For Roehl, all the evidence pointed to France looking much more like Britain than Germany. Later, economists seem to have extrapolated the financial structures that followed WWI (heavily bank-based) to the earlier periods. As we shall see, Parisians had important positions in banks, but most of their wealth was held directly in stocks, bonds, and real estate.

The sectoral transition models of economic change espoused by development economists and made prominent by Simon Kuznets (1955) draw in part from the same assumptions about capital deepening and labor productivity. In this view, traditional assets (land and real estate, family enterprises, and public debt) are replaced by modern ones that embody new technologies (the equity and debt of railroads and factories). In Kuznets's view, the new assets had high returns, while the older assets were depreciated by economic change. In these sectoral models, individual investors have portfolios either in the old or the new economy. Those investors and entrepreneurs in the new economy (the capitalist part) see their wealth rise rapidly. The wealth of those in the old economy falls or stagnates. Our samples of Parisian portfolios dramatically demonstrate the contrast between an Old Régime mode of investment that we still find in 1822 and the more modern portfolios of the two decades that preceded WWI.

Another strand of the literature has focused on international capital. In the second half of the nineteenth century, Western Europe emerged as the dominant source of external capital for infrastructure and export-oriented activities globally. Because this period also saw a massive expansion of colonialism, the interconnections between politics, territorial expansion, and financial capitalism have been hotly debated. Davis and Huttenback (1986) explored these issues extensively in the case of Britain. France amassed the second-largest colonial empire in the nineteenth century, covering vast parts of Africa and smaller domains in Asia and the Americas. After 1870, it has been argued that political motivations explain large-scale investment in the Ottoman empire and Egypt. Most prominently, the desire to secure an ally to the east of Germany led the French to provide much financial support to Russia. For a long time, the consensus was that the close regulation of financial markets in France enabled the government to achieve these goals with private funds (Lévy-Leboyer 1977). Recent work has cast doubt on these views. There was far less actual regulation than scholars had thought. Moreover, capital exports were more broadly distributed than politics suggested (Parent and Rault 2004).

In all these cases, the key to the puzzle is, who owned what? The first issue is, who? It is a question that extends well beyond Gerschenkron and Kuznets. Indeed, every scholar of capitalism has been interested in who owned the enterprises of economic transformation, and the components of individual wealth holdings are fodder for research in fields as varied as political economy and business and social history. This matters in the context of industrial and financial capitalism. One can ask questions like, was shareholding broadly distributed? Or was equity held by a small elite group? One can also ask about ownership of financial firms. To some extent, the high concentration of wealth in Paris already tells us that neither industrialization nor financial development led to the democratization of wealth. However, as we shall see, the details of wealth holdings rule out many theories of change. In Gerschenkron's thesis, the state at least partly funded the new banks. In the French case, there is little evidence of such support: private ownership of financial assets (whether invested in government-supervised entities like the Caisses d'Épargne or Crédit Foncier or in private entities) was the rule.

The second issue is, what? Evaluating the Kuznets hypothesis (and, in particular, its impact on inequality) requires assessing who owned the capital of the modern sector. The most important of these sectoral issues is the decline of traditional assets (e.g., land) in favor of new assets (e.g., financial or industrial capital). However, there are more subtle issues as well. For instance, in the bank versus market debate, one would like to know whether French individuals placed their funds with banks who then directed funds or whether they owned stocks and bonds directly. Similarly, figuring out who benefited from capital exports by investing in foreign assets can help resolve different arguments about the political economy. This chapter begins to answer these questions for Paris. Of course, Paris is not France. However, as the political and economic center of the country, it is an excellent place to start asking questions about the interlinkages between markets and politics.

Because of these questions, collecting data directly from the estate filings was less of a chore and more of a voyage of discovery. Indeed, rather than try to match what the TSA reported (movable wealth and real estate), we decided to collect a stratified sample of estates. As noted in the introduction, the stratified sample aims to provide complete asset details for all the richest 2% of Parisians who died in sample years and a large proportion of other estates. Using these stratified samples, we document that financial innovation had a complex relationship to wealth. On the one hand, the share of wealth invested in financial assets rose more for the bottom part of the distribution than for the top; on the other, the very wealthy continued to have a heavier share of their investments in higher-risk, higher-return assets than the middle class. Moreover, despite all this financial innovation, there was little movement in the share of Parisians who left an estate or were in the wealth share of the top 1%.

This chapter proceeds in five steps. We start with a description of the source that we will rely on: the inheritance tax filings that allow our analysis to run all the way to 1972. We describe the critical data collection decisions we made. The evolution of the estate tax system requires us to consider in detail the extent to which the changes in the source can explain the rapid growth of wealth from 1807 to 1912. At worst, the declarations' increasing coverage of assets would translate into higher taxable wealth with little real change in wealth. As we will argue, the growth we observe in wealth is real, and our inequality measures are robust to different approaches. While we view this source-critique section as essential, readers who are impatient for results may decide to skip ahead to section 2.2, where we detail the evolution of wealth at death, its impressive growth over the last half of the nineteenth century, and its ever-strengthening relationship with the age of the deceased. The third section considers the distribution of that wealth. The fourth takes advantage of French marriage law to estimate the share of wealth an individual bequeathed that could have been created from the wealth he or she had inherited. Section 5 considers the evolution of the portfolios of wealthy Parisians.

2.1. From the TSA to the RMD

We rely on two sets of French sources to assemble our dataset. The first is the Tables de Successions et Absences (TSA) that we discussed in chapter 1. The TSA is a large record set for Paris alone (more than 5,500 volumes).[2] The inheritance filings recorded in the Registres de Mutations par Décès (RMD) are an archive an order of magnitude larger than the TSA (some 52,000 volumes).[3] The information in these registers is also massively richer (though age is often missing). Because the French estate tax is assessed on heirs and differs by heir type, each filing begins with the social description of the decedent and the beneficiaries of his or her estate. They also contain lists of community and personal assets. The lists detail each financial asset, including bank accounts, stocks, and bonds, as well as other forms of private investment; every piece of real estate comes with an address and an evaluation of its rental income; furnishings are not described individually, although extensive collections of art or jewelry may be evaluated separately. In the 1820s, filings longer than two pages were rare, but by the 1910s, filings could run to twenty pages or so—and occasionally much more. Collecting information from the RMD presents several challenges: some are mundane, and others force us to make crucial decisions—these are detailed in appendix 2. Our efforts, however, produced an evaluation of movable and real estate wealth for every Parisian who died in a year that ended in 2 or 7 from 1872 to 1972, except for 1917 and 1967. Second, they also produce a stratified sample that details the wealth holdings of 46,380 Parisians (about 8% of decedents each year).

Following wealth at death from 1807 to 1912 requires one more adjustment. Before the 1901 law that instituted progressivity, inheritance taxes were assessed on gross wealth. Thus, before 1901, the estate declarations of individuals who were not married at death were just a list of assets. For those who were married, only family liabilities were accepted (we discuss this in detail in section 2.4). After 1901, the tax was assessed on net wealth, limiting liabilities to consolidated debts. Over time, liabilities came to include health and burial expenses and tax liabilities for the year of death. Most declarations only provide a total for liabilities, and we net those out to recover a consistent definition of gross wealth from 1807 to 1972.

More details on how the RMD dataset is constructed can be found in appendix 2. It is important to note that the sampling approach to estates is neutral to most issues. Given the dramatic concentration of wealth in Paris, our stratified sample includes details on 70% of all the wealth transmitted. To the extent that our decision to only open the volumes for thirty months creates a bias, it is a downward one for both wealth and inequality because delay in filing is correlated with wealth (see chapter 3). However, we pursue the analysis as if the sample was unbiased. Since we want to show that Paris was rich and unequal, it is the appropriate assumption to make.

Overall, the resulting dataset has many virtues. It allows us to construct a consistent measure of wealth (gross assets net of non-Parisian real estate) from 1807 to 1972. Second, we cover the entire population for each sample year and fully sample the top 1%. Thus, we have a very accurate measure of the share of wealth of the richest Parisians. Third, the dataset provides a representative picture of wealth portfolios throughout the distribution. We start the analysis with the evolution of average wealth.

2.2. Wealth

The 1912 cross section, the last one we consider in this chapter, is more than a century distant from the 1807 cross section that opens our study. Over that time, the units we wanted to measure changed. This involves the definition of wealth, how it is valued, and the definition of taxable wealth. This section, thus, begins by explaining how we construct a consistent series of wealth as defined by gross assets net of real estate outside Paris and then analyzes the evolution of wealth from 1807 to 1912. Doing so involves dealing with three complications. The most obvious is the change in source from real and movable values reported in the TSA registers to the detailed asset lists of the RMD volumes. There are also two important innovations: first, in 1850, the state decided to include the public debt in taxable wealth; second, in 1901, the estate tax was made progressive, and the estate tax filings at the bureau of an individual legal domicile had to include the entirety of his taxable estate. Thus, from 1902 onward, we observe the real estate Parisians held outside the city.

We start with the raw data (line "raw" in figure 2.1), which is the average wealth per Parisian decedent in 1912 francs. Then, to tackle the consistency of the TSA and RMD series, we can compare the TSA cross sections of 1822, 1842, 1852, and 1862 with those of the RMD because we collected both. First, the 1822 RMD series has 160 more rich individuals than we recovered from the TSA, while the 1852 RMD has 226 fewer decedents and the 1862 RMD 373. The 1822 difference may come from individuals whom the RMD reported as dying in one year, but the TSA inscribed them in a different year.[4] The 1852 and 1862 shortfall likely comes from a poor transcription of the RMD numbers. The differences are tiny, given that each cross section has between 4,000 to 6,000 estates per year.[5] Second, when we come to average wealth, the TSA produce an average wealth that is about 7% higher. The difference arises because, for many estates of individuals who died married, the TSA reports the total value of the community property rather than the individual's share. However, when we open the RMD, we always recover the estate. The problem is limited because, by definition, it cannot arise for single individuals (never married or widowed, who account for about 48% of wealthy decedents and half the wealth left behind). Moreover, it affects estates where there is no personal property. These are smaller than those of individuals with a mix of community and personal assets. While it is clear that the TSA source is sometimes inexact, it is crucial to keep in mind that these registers were finding aids. Once the heirs of the individual at hand had paid the tax, the tax authority's interest declined, and whether the clerk noted an estate of 320 francs as 320 or 640 was of little significance. Tax officials may well have wanted to compare the filing of a spouse to that of her predeceased partner. However, then again, what the clerk noted would not matter much because the official doing the comparison had access to the filing itself. For real estate, one might even make the case that the value of a building is more informative than the value of the heir's share. In any case, our procedure deflates average wealth in the TSA to 91.8% of their reported values (the average reduction necessary to bring the TSA of 1822, 1852, and 1862 back in line with the RMD). This correction produces the line "reduce community assets in TSA" in figure 2.1.

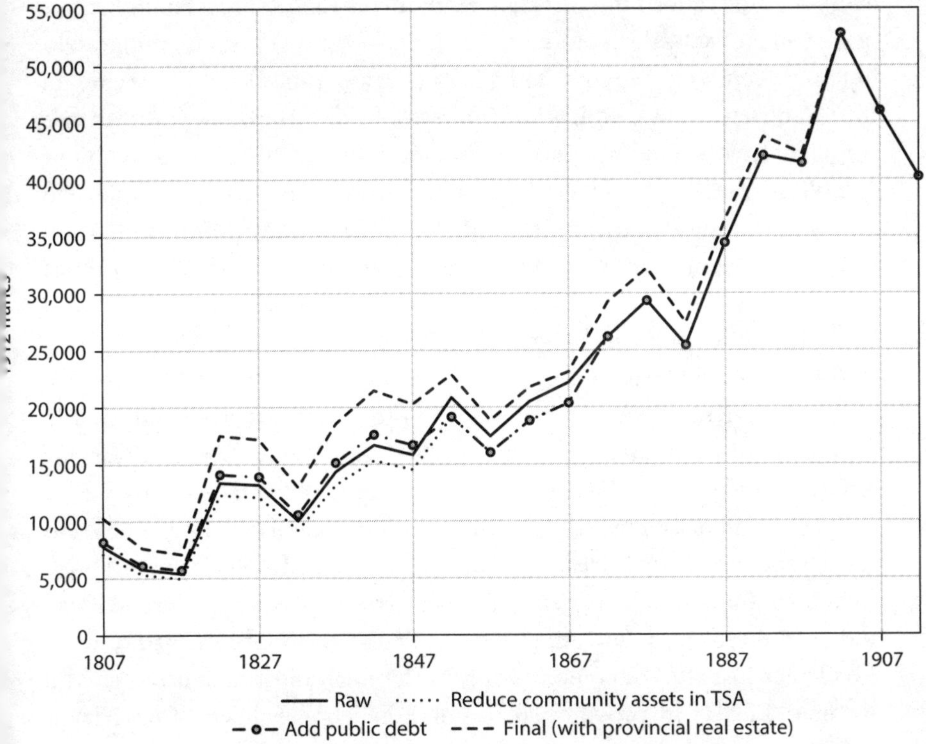

FIGURE 2.1. Average wealth per decedent in Paris, 1807–1912, in 1912 francs. *Source:* TSA and estates dataset (see appendix 1).

Public debt is the second issue that makes the series not immediately comparable. Until 1850, the capital invested in public debt annuities was exempt from the estate tax. However, as Daumard observed, the accrued interest on these annuities was not. Individuals rarely managed to die when they received the last payment on their annuities, and their heirs had to report the interest the state owed the deceased. That report nearly always includes enough information to evaluate the capital value of the annuities. For instance, consider the estate of Marie Anne Sophie Aujollet (d. March 11, 1822, and her heirs filed a declaration on August 7, 1822). The heirs reported that in her estate of 17,639 francs, 838.35 francs were arrears of interest payments on a government annuity (*rente*) of 1,784 francs.[6] The nominal capital value of a 1784 *rente* at 5% is 35,680 francs. At market prices in 1822, the bonds were worth 31,398 francs or more than the value of her taxable estate. Aujollet is an extreme case. We can apply the same procedure to everyone who reports interest from public debt. Thus, there is enough information in the estate declarations to estimate the size of Parisians' public debt holdings accurately. We collected this data for a stratified sample of estates in 1822. A quarter of individuals with an estate held public debt, and Madame Aujollet notwithstanding, public debt amounted to 20% of total wealth or 23 million francs. We collected the same information for 1842 and got two additional observations from Daumard for 1820 and 1847. Between 1820 and 1852, public debt declined from between 17 to 20% of all wealth, to about 15%.[7] Putting the public debt back into aggregate Parisian estates produces the "add public debt" line in figure 2.1. That is the core series of gross assets net of real estate outside Paris.

Finally, we must estimate the value of provincial real estate outside the capital held by Parisians from 1807 to 1897. Indeed, such wealth was not reported in Paris until the centralized filings of the twentieth century. Fortunately, Daumard painstakingly matched probates and estate tax records to get an estimate of non-Parisian real estate for 1820 and 1847 (Daumard 1973, 230–234). In 1820, non-Parisian real estate was 19.9% of total gross assets and 17.3% in 1847. By 1907, that share had fallen to about 10%, half of what it had been just after Napoleon's demise. This final adjustment produces the estimate "Final (with provincial real estate)" in figure 2.1. In fact, from trough (1817) to peak (1892), the series that includes provincial real estate grows at about the same rate as the one that does not include it (2.6% per annum versus 2.8, respectively). So, adding provincial real estate in 1901 to Parisian filings does not explain the prolonged growth of wealth over the nineteenth century. The raw, constant 1912 francs, series of average wealth per adult decedent has a low of 5,400 francs in 1817, a peak at 47,500 in 1902, and then a steep drop to 38,000 francs in 1912. The final series also has its nadir in 1817 at 7,217 francs, a peak in 1892 at 50,200 francs, and ends at 38,317 francs in 1912.

In retrospect, our measurements of wealth are remarkably robust. The range across the different measurements of average wealth per deceased for a cross section is at most 17%. Over the whole period, leaving aside the exceptionally high prices of 1817, the price level varies from a low of 61 (1822) to a high of 100 (1912), or by 75%. In contrast, our constant 1912 francs series shows average wealth growing more than seven times from 1817 to 1912. Most of the change over time comes from growth in wealth itself.

The nominal per capita wealth growth rate was about 2.4% a year from 1807 to 1892, followed by a 1% annual decline in the last two decades. In real terms, however, the growth of wealth occurred in two key phases. First, there was a recovery from 1817 to 1852 when real wealth per capita grew at a 3.7% rate per year, and age-wealth profiles began to steepen. A second expansion phase ran from 1852 to 1892, growing at 1.7% per year. For Parisians, at least, wealth growth rates are inconsistent with a "take-off" of the economy after 1852 (cf. Gerschenkron and proponents of finance-led growth; chap. 1). One might believe that estate tax data smooths the data to the extent that it would be impossible to detect a sharp break in wealth accumulation because of variation in the length of savings interval across individuals. However, even converting the data into wealth of the living through different estate multiplier methods fails to produce any take-off for wealth. As we shall see later, even if the aggregate data is not consistent with take-off theories, the portfolios of Parisians do reveal profound changes in the structure of wealth over the century. Those changes brought about a remarkable growth in the annual inheritance flow. It jumped from under 66 million real francs in 1817 to a 1907 peak of 1.3 billion because a tripling of the Paris population compounded the growth in individual wealth. Wealth in Paris was simply much greater in the second half of the nineteenth century than it had been under Napoleon.

Wealth has numerous purposes. Most basically, an investor buys a stock or bond to transfer resources into the future when they will be consumed. Using wealth as deflated by the consumer price index (CPI) suggests that, at its peak in 1892, average wealth in Paris produced seven times more consumption opportunities than in 1817. Let us consider, however, other ways of evaluating the value of average wealth. The ideal point of comparison would be labor earnings in Paris. Such data do not exist, and the wage history of Paris remains to be written. Thus, we rely on data for France as a whole. The first way to evaluate average wealth is to compare it with the average French adult labor income computed by Piketty and Zucman (2014), which is the labor share of GDP divided by the adult population (henceforth adult labor income). Average wealth in Paris was about 16 times adult labor income before 1825. It then jumped quickly to 24 times adult labor income by the 1830s and then stayed at that level through the Second Empire. Another jump after 1870 brought the wealth-to-adult-labor-income ratio to about 34 for the rest of the period. We can also compare average wealth to an annual wage earner's income that we recomputed from Bayet to obtain the cost of a 55-hour, 50-week worker (something like the cost of a full-time domestic servant).[8] We view this wage as relevant both to the cost of consumption for the very rich, who spend much of their income on personal services, and to the incomes of most of the population. The pattern is very similar: average wealth grows from 14 times the Bayet wage before 1827 to more than 25 by

1867 and ends at nearly 32 after 1882. Our reader might object to comparing wealth to annual labor incomes. This is a fair point since wealth is a capital measure. A better wage analog is lifetime labor income. Consider that the average age at death of the poor in Paris remained below 55 before 1870. Supposing that such individuals had started working at 15 and thus had a 40-year working life earning the Bayet wage, average wealth equals a third of their *undiscounted* lifetime earnings before 1820 and three-quarters in the last four decades before WWI. The conclusion is inescapable: wealth was growing faster than labor income.

Since the purpose of wealth is to transfer consumption into the future, one can ask what incomes would be associated with this ability. Over the whole nineteenth century, the only asset returns we know of are those for government bonds. If we take them as a proxy for returns to financial assets, average wealth at death produces an interest flow equivalent to adult labor income with little change over time. Having average wealth at time of retirement would thus allow an individual to replace their average wage. There is little change over time because the massive growth in wealth was offset by a decline in yields from about 6% before 1820 to about 3% around 1900. Thus, the capital income from a given wealth fell. Moreover, wages rose as well. In any case, from a consumption perspective, the impact of wealth is stark: an individual with an average job and average wealth could consume twice as much as a counterpart with the same job and no wealth. It is also clear that starting from no wealth and an average job would require a very high savings rate (30% or more) to reach the average wealth at age 60 in one's cohort for a person investing solely in government bonds.

2.3. Distribution

While informative, the discussion of average wealth, nominal or real, has serious limits given that it was so unevenly distributed. Indeed, in Paris during this period, the median wealth at death was zero. Thus, we can only understand the importance of wealth if we consider its distribution. We measure the distribution of wealth directly because we do not sample the wealthy within a year but try to collect data on everyone. Figure 2.2 shows the steady growth of the wealth share of the top 1% (the sum of the top two groups in the figure). In the first half of the nineteenth century, the top 1% wealth share was about 54%, which varied between 49 and 58%, but there was not much of a trend. From 1852 to 1882, the wealth share of the top 1% was above 51% (it averaged 56%). The real jump occurred in the last two

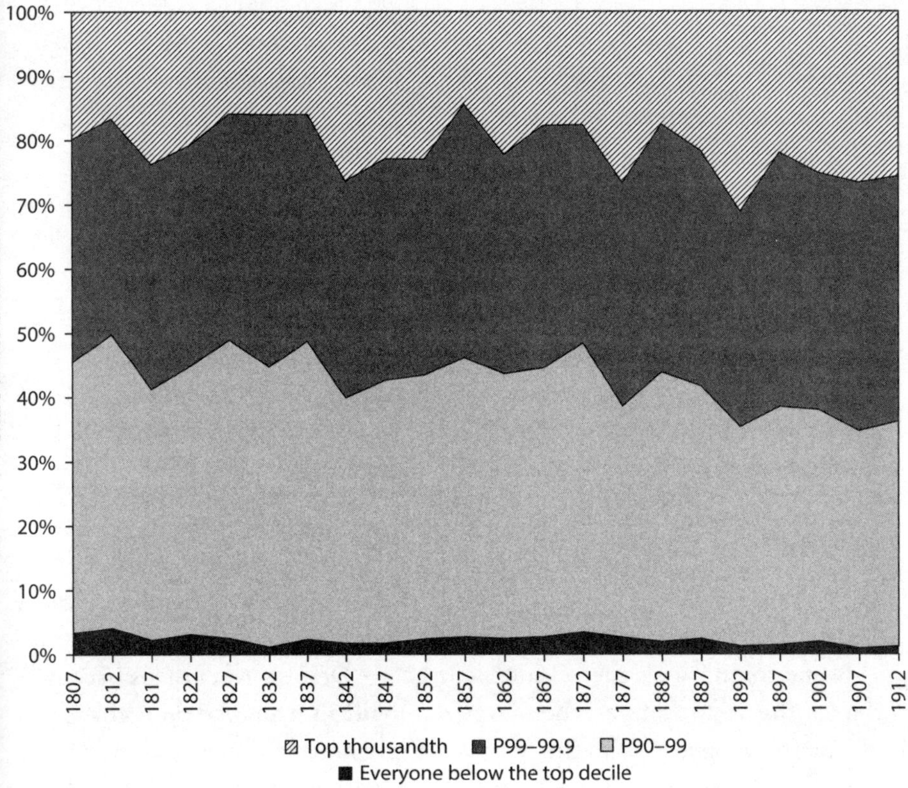

FIGURE 2.2. Wealth shares in Paris, 1807–1912. *Source:* Estates dataset (see appendix 1).

decades before WWI when the top 1% wealth share had a low of 61% and averaged 62.3%. Much of the growth in the wealth of the top 1% accrues to the very top (0.1%), whose wealth share before 1850 was about 19%, but after 1892 it averaged more than 25%. The very top of the distribution garnered 77% of the gains from increased inequality.

Since these cohorts accumulated their assets during the first golden age of financial capitalism, it is hard to argue that financial development necessarily reduces inequality. By financial capitalism, most scholars refer to the period of rapid expansion of international capital flows (see Davis and Huttenback 1987 for Great Britain; and Lévy-Leboyer 1977 for France) and the equally rapid expansion of the number of financial assets traded on centralized exchanges such as the Paris Bourse. We also include the development of local financial intermediaries such as banks (e.g., Crédit Lyonnais, Société Générale, savings banks [i.e., *caisses d'épargne*]), mortgage banks

TABLE 2.1. Age-wealth profiles for Parisians (top thousandth excluded) in 1912 francs

	Age at death						
Birth cohort	20–29	30–39	40–49	50–59	60–69	70–79	80+
1700–1719							5,505
1720–1739					3,904	6,878	15,693
1740–1759			12,560	5,733	10,389	13,488	20,775
1760–1779	616	3,266	8,711	11,431	12,954	20,977	37,848
1780–1799	3,455	6,746	8,504	13,940	25,026	32,139	49,910
1800–1819	2,751	5,209	9,775	18,109	33,092	48,531	77,846
1820–1839	2,604	4,685	11,963	24,159	42,696	58,727	62,449
1840–1859	3,636	6,311	11,936	26,071	31,598	23,986	22,583
1860–1879	4,752	6,617	9,528	10,787	14,840	13,274	17,626

Source: Estates dataset.

(e.g., Crédit Foncier de France), and insurance companies that offered both life and tontine insurance (Hautcoeur 2007). Despite, or perhaps because of, all these innovations, the share of individuals who died poor changed little. It averaged 72% in the first half of the nineteenth century and 73% between 1892 and 1912. As we saw, the wealth share of the middle-class P90–99 declined from 45% to 37%.

Nineteenth-century Paris was thus very unequal and composed of three groups. The first was the top 1% who controlled more than half the wealth. A sizable group of about a quarter of the population shared the rest of the wealth. Finally, the largest group, about two-thirds of the population, died with nothing. The differences in average wealth are obvious. The first group had, on average, 25 times the wealth of the second, and by the end of the nineteenth century, more than 30 times.

The pattern of steeply increasing wealth at death by age that we described in chapter 1 continues after 1867 and, if anything, strengthens (see table 2.1). Removing the top 0.1% from the dataset does reduce the steepness of the age-wealth profile but does not eliminate it. For instance, if we take individuals born between 1820 and 1839 and compare those who died aged 80–89 to those who died between 50–59, the full dataset tells us the octogenarians were three and a half times wealthier than the quinquagenarians (99,500 1912 francs relative to 27,500 francs). If we remove the top 1%, the ratio falls to 2.6 (62,500 to 24,150 francs), but the pattern remains.

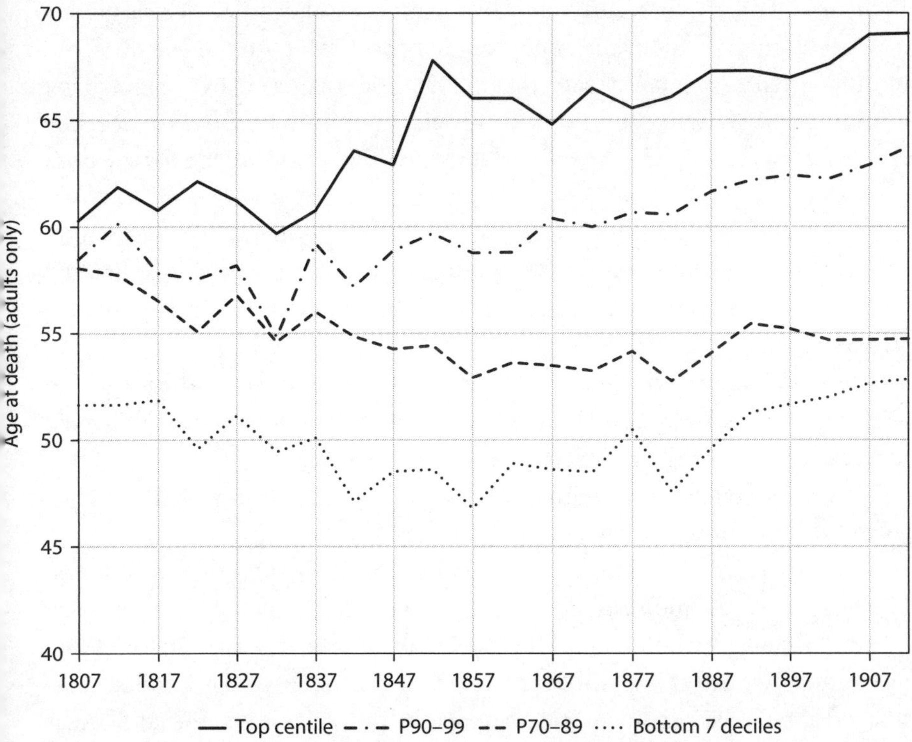

FIGURE 2.3. Average age at death per intermediate fractile in Paris, 1807–1912.
Source: Estates dataset (see appendix 1).

As we saw in chapter 1, age-wealth profiles are steep. There is no evidence that economic change redistributed wealth from the old (with traditional assets) to the young (with modern assets).

If we directly consider the age at death of the Parisian population, the second half of the nineteenth century reveals a marked change. As figure 2.3 shows, the age at death among the adults who died poor (P0–70) reaches a nadir around 1857 at 47. It then rises fitfully and eventually converges on the age at death of those who died with little wealth (P70–89). Those two groups still had an average age at death of less than 55 in 1912. Over the nineteenth century, the poor gained nearly four years in adult life expectancy. It is important to note that, as we will discuss later, this is not the result of a general aging of the city due to a decline in the rate of immigration. As Kesztenbaum and Rosenthal (2017) have shown, the increase in life expectancy was closely connected with the diffusion of better water infrastructure, in particular general sewerage. For the top 1%, the nineteenth century is a history of nearly continuous improvement in age at death, reaching almost 70. Although the poor did show some improvement, on the eve of WWI, differential mortality was such that the rich could expect to live much longer than the poor. Given that to enter the data, one had to have already reached age 20, an average age at death of 53 involves an adult lifespan for the poor of 33 years.

By contrast, the top 1% enjoyed nearly another half adult lifespan. In 1807, they died at about age 60, a full decade older than the bottom 70% of the population. In the decade before WWI, the richest Parisians died at almost 70, or 17 years later than the bottom 90%. Lifespan inequality rose and then fell over the century. Indeed, the top 1% lifespan advantage had been less than 40% before 1840, but it jumped to 60% or more over the next decades, before falling back to about half after 1892.

The late nineteenth century shows a second major break in the trends we observe in the wealth of women relative to men. Consider first the average wealth of women to men among rich decedents (the evolution of average wealth among all decedents is very similar). We observe two distinct periods, with 1860 marking the break between the two. Before 1860, women's wealth was significantly less than men's. A woman's estate averaged about 65% of a man's wealth at death. The series is volatile and ranges from highs of about 85% in 1827 and 1832, to lows near 50% in 1817, 1842, and 1852. Then, after 1860, women's estates began to grow faster than men's and, except for 1892, never dropped below 83% of men's estates. Women's wealth was higher than men's in 1887, 1897, and 1912. If we remove the top 0.1% of estates, we observe the same pattern with less volatility: women's wealth at death grew faster than men's from the mid-nineteenth century and reached rough parity by the 1880s. This evolution raises several questions. In particular, to what extent did France's egalitarian inheritance and the community property marriage regime contribute to this gender equality? A full treatment of these questions would require a dataset that is more detailed than ours. The information about how a deceased's estate was divided among his or her children is available in the estate tax records precisely because the tax was an inheritance tax. Similarly, because spouses' names are part of the estate tax documents, it is possible to link the filings of husbands and wives. We did not record this data to save resources because our initial focus was on measuring inequality. In a doctoral thesis that utilized a part of our dataset, Clément Dherbécourt investigated the intergenerational transmission of wealth for the top 1.5% of decedents in the last third of the nineteenth century. He proceeded in two steps: first, he investigated what each heir got, and second, he sought out the heirs' estate tax filings. For Paris, at least, the evidence is strong that estates were split equally among all the children (Dherbécourt 2013, chap. 3).[9]

Even if parents' estates were relatively equally distributed among children regardless of gender, there are other dimensions where such equality did not hold. The most obvious was labor markets and education, where women were discriminated against across the board. This likely had important consequences on differences in wealth mobility between men and women. However, the peculiarities of French marriage law, with its heavy focus on community property, meant that married women had claims on the wealth accumulated from both their own and their husband's earnings. Again, our emphasis on cross-sectional inequality means that our data, at best, hint at the tempering effect of community property on wealth differences between genders. Another issue we will return to is how the portfolio composition of women's assets differed from men's. Such differences could arise because men tend to overestimate their ability to manage risk and thus have more aggressive portfolios (Waldman 1994). It could also arise from the political economy of the household where the woman's wealth plays the role of a buffer (e.g., real estate or public debt). In contrast, the man's wealth is invested directly in higher-return assets.

Beyond family law's likely tempering effects on the very real discrimination against women in labor markets (e.g., entire professions were closed to them), we also need to consider wealth management. While there were many women entrepreneurs in France, the ranks of managers of the largest public and private firms were exclusively male. So, the only place women could hope to even the playing field was with investments and, in particular, the portfolios of inherited wealth. To address these issues, we will first consider inherited wealth and then analyze portfolios in the last substantive section of the chapter.

2.4. Inherited Wealth: A First Look

In most societies, measuring social mobility requires matching two individuals at different times (parent and child at the same age) or the same individual in two different documents (first when the individual inherits, second when he or she dies). We could have taken this approach with the Parisian data. Indeed, the names, addresses, and relations to the deceased of all the recipients of an estate are included in the declarations precisely because French estate taxation was, from its inception, an inheritance tax.

Clément Dherbécourt (2013) attempted to complete the matches for the very wealthy. His success was mixed because, over the 25 years, which is the standard intergenerational interval, heirs move around quite a bit.

Instead, we rely on the details of French marriage law that lead individuals to distinguish between inherited and accumulated assets. Throughout this section, we will use the specific example of Antoine Adrien Cousin to illustrate the details of French family law. Cousin died in 1852, and we have recovered his 1812 marriage contract with Suzanne Letissier. We have also collected his and his wife's estate filing and his and his wife's parents and their children's estates.

Even before the French Revolution, Paris was quite different from the Anglo-Saxon regime of bequest and coverture laws where, by default, married women owned no property. Enacted in 1804, the Code Civil laid out a default marital regime, giving couples and their families wide latitude to contract. In this default regime, a couple had a community account on which both husband and wife had equal claims as well as separate personal property accounts (*propres* or *reprises*). The community account was funded initially by the movable assets the couple brought to the household (e.g., Suzanne Letissier, for instance, brought a dowry of 200,000 francs; see box 2.1). Following marriage, all income flows went into the community account, including labor income, rents, interests, and dividends from community and personal assets. The personal property of each spouse fell into two accounts. The first tallied the assets that the spouse owned at marriage or inherited subsequently that remained unsold (*reprises en nature*). The second tallied the nominal value of assets that had been sold or arrived in cash (*reprises en deniers* henceforth: personal property in cash). This account was a claim on the community that was repaid upon dissolution of the marriage and paid no interest. Finally, should the community engage in expenses for the maintenance of a spouse's assets, that expense was booked as an interest-free loan from the community to the personal account of that spouse (*récompenses* henceforth: personal liabilities). Thus, a household had up to seven accounts (community, wife and husband's personal property accounts, wife and husband's interest-free loans to the community, and wife and husband's liabilities to the community). At dissolution, husband and wife each received half the community net of the interest-free loans. Couples, or more likely their parents, could modify this arrangement before marriage in various ways by going to a notary and signing a marriage contract.

Box 2.1 details the clauses of Antoine Cousin and Suzanne Letissier's marriage contract. The key clause for us is that each kept all their inherited

BOX 2.1 Marriage contract for Antoine Cousin and Susanne Letissier

Cousin and Letissier wed on September 26, 1812 (he was 27, and she was 17). Before the civil and religious ceremonies, they and their parents signed a marriage contract before notary Massé on September 16 (the seven-page contract involved thirteen articles that departed from the standard regime.

Dowry clauses

Antoine Cousin brought his notary office and its archive. He owed 240,000 francs on the office. He also brought 24,000 francs, the value of his bond for the office, and 6,000 francs of furniture.

Suzanne Letissier brought a dowry of 200,000 francs and put 10,000 francs in the community. 108,000 francs was paid immediately, 92,000 francs would be paid in the next three years and her parents would pay 5% interest. All cash payments were to go to paying off the notary office acquisition.

Additional clauses

- Each party was solely liable for his or her debts contracted before the marriage.
- The survivor would have a life interest in the other's estate and a 15,000-franc widow's portion.
- Dowries would be returned to the parents if a spouse died without children without interest.
- If Cousin died first, Suzanne would be entitled to a life annuity of 4,000 francs.
- If Letissier died first, Cousin would keep the notary office and furniture.

Source: AN/MC/ET/LXIV 588.

assets (movable and real) as personal property and was solely liable for their debts incurred before marriage. This common contractual modification led to the so-called community limited to acquisitions regime.[10] Another common clause of marriage contracts involved a widow's portion (*préciput*). This clause we find in the contract between Cousin and Letissier. There was also a life interest in some or all of the community assets (*douaire* or *usufruit*). One clause reflected the specifics of our example couple. Cousin had bought his notary office in 1810 for about 300,000 francs and still owed 240,000 francs. Letissier's dowry would be invested in the notary business. The contract further specified that, if she died, Cousin would not have to sell the business to settle her estate.

Modifications were common. Over the cross sections of 1822, 1852, 1862, 1872, 1882, 1892, 1897, 1907, and 1912, we collected the type of marriage of 28,373 individuals. Fifty-three percent used the default regime, another 39% modified by contract, only 7% opted for separation of all assets, and finally, just 14 couples chose the universal community regime. The estate of a person in a separation regime would only report personal assets, which may have been inherited. However, we will leave these aside. Similarly, the universal community regime would report only community assets. However, a contract was standard for the top 1% of wealth holders, among whom only 4.8% were married in the default regime. Most couples and their families kept all inherited assets (not just real estate) as personal assets. In this sense, Cousin and Letissier's choices were relatively standard.

Leaving aside valuation problems, the first definition of inherited wealth is the deceased's personal assets plus his or her net cash claims on the community. Our data give us directly an account of inherited wealth for nearly all individuals who die married and wealthy. To the extent our sources have a bias, they underestimate inherited wealth for individuals in the default regime because heirs, or their agents, may have lumped into the community any asset they could not determine was personal. To illustrate these distinctions, we turn to Cousin's estate declaration. Box 2.2 details his personal assets in three parts. The first includes all the cash contributions to the community account. For most of these, a date is provided for the transfer. The second part includes the personal assets that remained in kind at the time of Cousin's death. These are real estate and financial assets (including a bond owed by the purchaser of one of Cousin's inherited houses). The third part involves Cousin's personal liabilities. These are sums he owed the community, including legal fees for settling his parents' estate and two clauses of the marriage contract. The first was that he would not be reimbursed the 10,000 francs that he had invested at marriage into the community; the second clause stated that the first to die would pay the children's dowries.

Box 2.3 details the community assets of Cousin and Letissier. We have broken these into three categories as well: (1) furnishings and cash, (2) rents, interest, and dividends, and (3) real estate. Interestingly, the community has no financial assets of its own.

BOX 2.2 The personal assets of Antoine Cousin

Part 1: Transfer to community (cash and assets realized)	
Dowry (1812)	134,000
Received from inheritance Mad. Laredet	33,724
Houses (sold 1847)	24,000
House rue du Caire 2 (sold 1849)	30,000
Interest	1,823
Cash	12,180
Cash on account	2,505
17 certificates debt of Kingdom of Two Sicilies (sold 1850)	22,538
3 City of Paris bonds (paid off)	3,000
13 shares of the Caisse Hypothécaire (sold 1850)	1,820
Peer-to-peer loan to Simon (from mother, paid off)	30,000
Peer-to-peer loan to Petit (from mother, paid off)	10,000
Peer-to-peer loan to Lobry (from mother, paid off)	20,500
5 certificates of Egyptian debt (sold 1850)	5,980
Total	332,072
Part 2: Personal assets in kind	
Small debts	9,222
10 shares Trois ponts sur la Seine	9,005
4 shares Manufacture Saint Gobain	80,000
54 shares of the Banque de France at 2,753 francs	148,652
Peer-to-peer loan	120,000
Debt from the buyer of house rue du Caire 2	160,000
French *rente* (coupon 17,340)	409,108
Houses rue Saint Honoré 207, rue Deux écus 5	302,000
House rue Saint Honoré 274	484,340
House rue Neuve Saint Marc 11	180,000
Farms near Boissy Saint Léger	335,700
Total	2,238,027
Part 3: Personal liabilities	
Investment in community	–10,000
Payments to notaries Thiac and Agasse	–4,350
Children's dowries	–240,000
Total	–254,350

Source: ADP DQ7 3918 #784 (October 19, 1852).

BOX 2.3 The community assets of Cousin and Letissier

Part 1: Furnishings and cash		
Home furnishings in Paris		54,371
Home furnishings in l'Artoire castle		18,131
Cash		41,405
Part 2: Rents, interest, dividends		
Rue du Caire 13, 15, 25, rue de la Feuillade 5	W	1,999
Land in Les Essarts Le Roi (l'Artoire)	W	3,082
Forest in Rambouillet	W	721
Farms in Orne and Mayenne	W	6,316
Farms in Orne	C	5,863
Rue Saint Honoré 207	H	971
Rue Saint Honoré 274	H	1,648
Rent arrears	H	4,911
Rue Neuve Saint Marc 11	H	975
Interest on debt 160,000 francs	H	628
Rue des Deux Écus 5	H	200
Account with Fremyn farm manager	C	2,659
Farm at Noisy	H	522
Land in Villeneuve Saint Georges	H	1,301
Interest on public debt (17,340)	H	6,503
Interest on public debt (12,395)	W	1,549
Dividend shares St. Gobain	H	434
Interest on debt Becker	H	7,501
Rue Saint Antoine 201, 203, 205	H	3,208
Remainder of country house St. Hubert	C	7,000
Work on house	C	600
Debt Allain heirs	C	200
Part 3: Real Estate		
Houses rue Saint Antoine 201, 203, 205	C	522,000
Woods near Boissy Saint Léger	C	2,400
Land and castle of l'Artoire	C	170,860
Land in the Orne	C	3,000
Dowries		240,000
Total		1,110,962

Source: See box 2.2.

Note: H = income from husband's (Cousin) asset; W = income from wife's (Letissier) asset; and C = community asset. See box 2.2 for sources.

BOX 2.4 Computing Antoine Cousin's estate

STEP 1: THE NET COMMUNITY COUSIN-LETISSIER	
Community property	1,110,962
Personal assets in cash, Cousin	−332,072
Personal assets in cash, Letissier	−611,613
Personal liabilities, Cousin	254,350
Personal liabilities, Letissier	10,000
Widow's portion	−15,000
Net community	416,627
Half goes to each spouse	208,313
STEP 2: COUSIN'S ESTATE	
Share of community	208,313
Personal assets in cash	332,072
Personal assets in kind	2,238,027
Personal liability	−254,350
Total estate	2,524,063
Inherited assets: transfers to community + personal assets in kind + personal liabilities	2,570,099
Implicit inherited wealth share	101.8%

Source: See box 2.2.

Armed with these two lists, the heir's agent started computing the net community property by deducting the transfers from Cousin and Letissier, the widow's portion, and adding them back to their liabilities. Box 2.4 reports that computation. It starts with the value of community property from box 2.3, 1,110,962 francs. It deducts 958,685 francs owed to Cousin and Letissier and then adds 264,350 francs they owed the community. The community, on net, is worth 416,627 francs, of which Cousin's estate receives half or 208,313 francs. The next step is to compute Cousin's estate. To Cousin's share of the community, we add his transfers to the community, net out his liabilities, and finally add his personal assets. The result is an estate valuation of 2.52 million francs.

We can make this computation for every married person we sampled. This approach to inherited wealth has one important advantage, it requires only data available in the estate tax record. However, we must acknowledge two serious drawbacks. First, it neglects all capital income that accrues to the inherited assets. For instance, suppose the dowry had been invested in government bonds in 1812, and they would have returned better than 5% over the next four decades before Cousin died. This first definition of inherited assets ignores the capital income flows. The second issue involves the valuation of the claims on the community; given that these are historical values, they are net of the capital gains (or losses) that ensue from the time the resources were put into the community account to the death of the individual. Though convenient, this first approach significantly understates inherited wealth relative to the proper, capitalized definition we deploy in chapter 7. Here, we focus on a few important lessons from the uncapitalized direct approach.

Figure 2.4 shows the results of this computation. Practically, we compute the inherited share as the ratio of inherited assets as valued in the declaration to the estate. We recode to 100% all individuals who, like Cousin, have more personal assets than estate and produce a weighted average for each wealth group. For everyone below the top decile, the reported inherited wealth share is less than a third, with little change over time. For both the top decile and the top 1%, the inherited wealth share starts low, at about one-third, like everyone else. Then it grows quickly, and for the top 1%, it is above half, and at times, it is near 60%.

The average inherited wealth share for the whole population is about 25% and follows a pattern close to the ninth decile. If we weigh by wealth, the reported inherited wealth share is above 30% after 1842. We clearly see that inherited wealth mattered. This is not because of some computations made by economists in the twenty-first century; it is a measure directly contained in the estate filings. Other indications show that the individuals we studied were aware of its importance.

As we noted earlier, until 1914, a bit more than half of the individuals who died married had set up their marital property according to the default regime; the other half chose to draw up a customized contract. If inherited wealth were unimportant, then the estate of the first person to die in the couple would be close to half the community property. In 1822, 84% of individuals married in the default regime had an estate value within 3% of half the value of the community property. Thus, they inherited little. Meanwhile, the average estate for individuals married in the default regime was relatively modest, at 3,500 francs. By contrast, the average estate for those who, like Cousin and Letissier, had chosen a modified community property regime was eight times larger at 29,000 francs. This ratio would become greater than ten times in the late nineteenth century, illustrating the divergent estate strategies employed by individuals from different backgrounds.

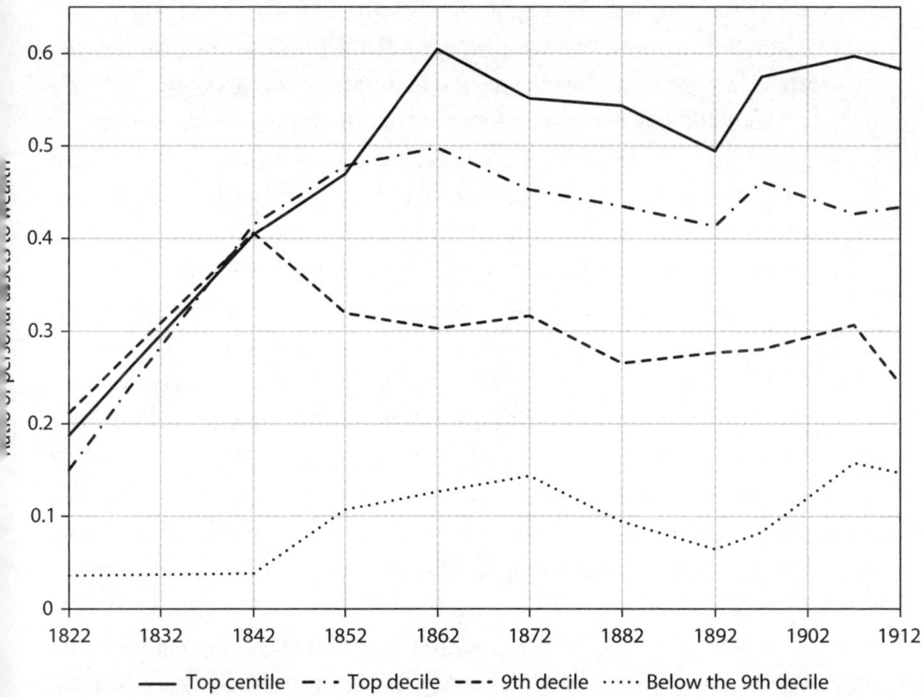

FIGURE 2.4. Reported inherited wealth in Paris, 1822–1912. *Note*: To compute inherited wealth shares, we use a two-step procedure. For each individual, we measure inherited wealth in bequests as the minimum of reported personal assets and wealth (because the share of inherited wealth in one's bequests cannot be greater than 1). Second, we sum inherited wealth and divide it by total wealth to arrive at an aggregate inherited wealth share. *Source:* Estates dataset (see appendix 1).

To be sure, couples who inherited only cash assets and in similar amounts would also see their estate be close to half the value of the community property. However, such couples were rare even in an assortative marriage society like Paris. By contrast, in 1822, only 13% of individuals married by contract had an estate valued within 3% of half the community property. We can repeat this procedure for all the years in our sample. The share of individuals whose estate is near half the community property is always very low for those who contract (16% on average) and high for those who did not (65% on average). The share of individuals whose estate is close to half the community property falls in 1907 and 1912, likely due to the inclusion of non-Parisian real estate in what we can observe. Of those who married with a contract, 63% have estates larger than 53% of the community.

These coarse measures highlight the existence of two distinct populations within the wealthy; some individuals started their working and saving trajectories with little wealth, and they expected to inherit little. Marrying under the default rules made sense since there was nothing to protect. To the extent that members of this group ended up wealthy, it was due to effort or luck. Those who married with a contract often received dowries and anticipated benefiting from additional intergenerational transfers. Given the uncertainties as to whether married couples would have children who survived to adulthood, it is not surprising that parents wanted to ensure that wealth would be returned to the family if no heir was produced. Moreover, families also wanted to protect their daughters' dowries from the possible foolhardiness of their husbands, who had sole authority over all household assets. To the extent that such couples ended up wealthy, their favored start must have helped produce their estates. The Cousin family is a good example of starting their married life with wealth. As noted above, Cousin and Letissier's combined dowries were 334,000 francs in 1812, when the threshold to the top 1% was 120,000 francs. They proved less generous with their children when both married in 1844. The son received a dowry of 20,000 francs and an advance of 200,000 francs so he could take over his father's notary office, and the daughter received only the 20,000 francs dowry.

While both inheritors and noninheritors are represented in every fractile, contract type and deviation from half the community property strongly correlate with wealth at death. In the nineteenth century, 82% of individuals who died wealthy but not in the top two deciles did not draw up a contract; in the ninth decile the proportion is 61%; for the top decile the proportion is less than a quarter, and for those in the top 1% it is only 7%. The group of individuals whose estate values are close to half the community property is most prevalent for fractiles below the top two deciles. Nearly three-quarters of the individuals below the ninth decile had an estate within 3% of half the community; in the next decile, the proportion is always less than 60%; it continues to fall as one goes up in wealth, dropping to less than 20% in the top 1%.

Given these two populations and their relative representation in the fractiles, it is important to ask what is the ratio of inherited assets to gross wealth. For the weighted sample of Parisians who died married in the second half of the nineteenth century, inherited assets represented, on average, 46% of their gross estate. The share of inherited assets is 19% for those who married without a contract and just shy of 50% for those with a contract. Given the large difference in the size of estates for those married with and without a contract, it will come as no surprise that 93% of inherited assets were left behind by individuals who had a marriage contract. Moreover, the 1% held about 10% more of the personal assets of Parisians than it did of total assets. Inherited wealth was an important component of wealth, and it was unequally distributed.

We close this section by considering some aspects of social mobility. It would be wrong to consider societies as unequal as Paris as frozen with two or three groups that are self-reproducing classes. This is wrong because Paris, through the nineteenth century, could not sustain its growth without large-scale infusions of newcomers. In fact, in the later decades of the nineteenth century, about 80% of the adults who died in Paris were not born there (ASVP 1881, 282; 1885, 238). This was a very broad phenomenon, as two-thirds of those who died rich were also not born in Paris, and even about half of the top 1% were born elsewhere.

As we saw earlier, more than half of all Parisians who died married between 1822 and 1912 had been married under the default regime, and for nearly all of those, their only bequest was their half of the community property. It was possible, however, to go from having very little to almost the very top of the wealth distribution. Aristide and Marguerite Boucicaut are good and famous examples of such trajectories. Aristide was the son of a small shopkeeper in the Orne, while Marguerite, the daughter of an unwed mother, came from even poorer origins. Yet Aristide Boucicaut created the first modern department store, the Bon Marché, and was the tenth-wealthiest person to die in 1877. A decade later, Marguerite was the second-richest person to die in 1887, in part because she was Aristide's heir. Both were among the top 200 wealthiest individuals who died in Paris before WWI. Examples like this can be found in most of our cross sections (recall Napoleon's marshals in chapter 1). In the nineteenth century, however, 80% of the top 1% of wealth holders, and 87% of the top 0.1%, had inherited assets. Unsurprisingly, we can document that 95% of them had marriage contracts. The Boucicauts are rare examples of individuals who vaulted to the top without help from family or a marriage contract.

2.5. Portfolios

Recall from chapter 1 that in all our cross sections, the aristocracy was overrepresented in the top 1% of wealth holders in Paris. While this overrepresentation persisted into the twentieth century, it declined over time. That the aristocracy was wealthy is hardly surprising, but its progressive eclipse raises several issues. Given the Old Régime aristocracy's putative bias for owning real assets and the important role of public debt and real assets in the endowments of the Napoleonic aristocracy, one might wonder whether these groups missed the boat on enrichment possibilities created by financial and industrial capitalism. One way to answer this question is simply to separate assets into modern and traditional and compare the portfolios of nobles to those of commoners.

The relative decline of noble wealth is just one example of several issues that can be illuminated by understanding the composition of Parisian wealth. Let us start with some definitions of assets. We follow the same classification that we used in Piketty et al. 2006 with some additional refinements. We begin by distinguishing between real and financial assets. We then separated financial assets into bonds, equity, and deposits. We also distinguished between assets where the income produced was domestic (e.g., the Paris-Lyon-Mediterranée railroad) versus international (e.g., the Suez Canal) without concern for whether the issuing entity was legally French or whether the assets were publicly traded or not. We also tracked if they were listed on the Paris Bourse.

For this section, we turn to the asset files drawn from the estates that contain the specific information on 245,000 assets held by Parisians along with another 27,000 items that describe the deceased's claims on their communities (i.e., personal assets in cash, personal liabilities, dowries, widows' portions and the like). From the asset lists, we develop portfolios for the sampled individuals that include personal assets and half the value of every community asset.

Our samples of portfolios cover almost 19,150 individuals from 1842 to 1912. They include a growing number of estates over time, from 1,407 individuals in 1842 to 2,775 in 1912. These samples are very large relative to most historical studies of wealth based on probates or other summaries of wealth at death. At the same time, before WWI, on average, the portfolios of Parisians only had a dozen assets around 1882 (7 in 1842 and 17 in 1912). Below the top decile, individuals only have one or two assets (see figure 2.5). Creating very fine categories would mean that nearly all categories would be empty for most people. Because most portfolios only include a few assets, aggregating by fractiles produces "representative" highly diversified portfolios that no one holds. Instead, we opted for a sequential approach where the categories are defined by the question we are asking. Generally, we consider the investment portfolios of three groups: the top 1% (who own at least half the wealth), the next 9% wealthiest, and the rest (who never

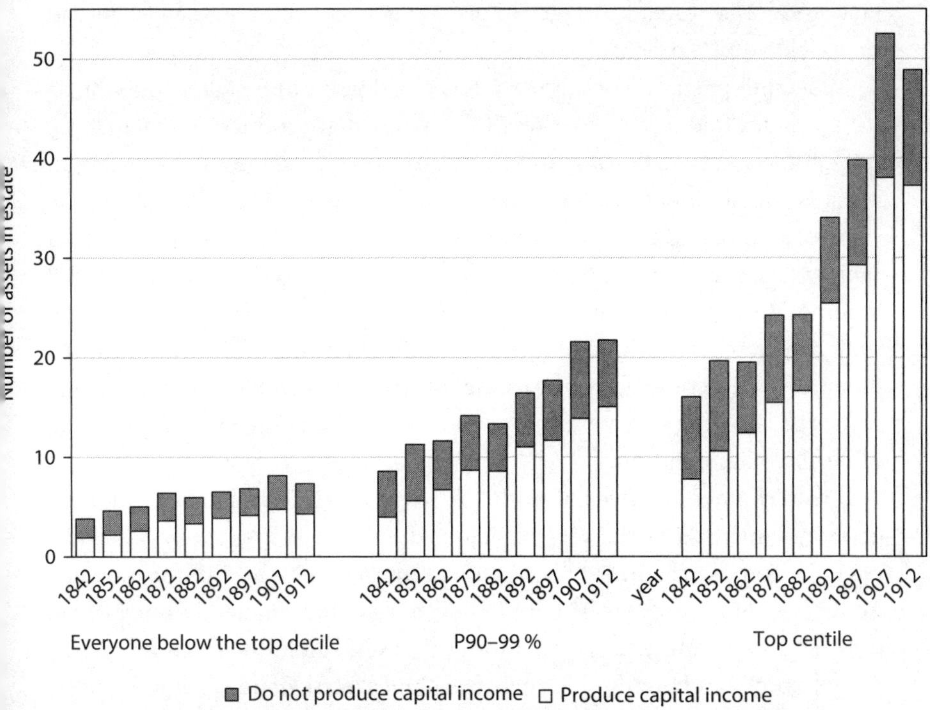

FIGURE 2.5. Number of assets in Parisian portfolios, 1842–1912. *Source:* Assets dataset (see appendix A2.3).

own more than 10%). In almost all cases, the patterns we describe are monotone with wealth (for instance, the share of assets that produce income or are international is increasing with wealth).

The low diversification of most Parisians' portfolios is not surprising given that there were no mutual funds and that the price of most bonds and stocks was in the hundreds of francs. These indivisibilities have important consequences. First, for everyone in the top decile, wealth was large enough that individuals could be diversified enough to avoid a trade-off between individual asset risk and return. That fact was even more true for individuals in the top 1% who held dozens of assets. For everyone below the top decile, investments were so few that diversification was limited. Most faced the choice of a low-risk and low-return asset (i.e., public debt) or high-risk and high-return assets (e.g., a single stock). Over time, the top decile became even more diversified while the rest of the population experienced much less growth in its assets. Recall that the fraction of individuals who die wealthy is essentially constant during this time. These two facts show that the advantages of greater access to financial markets did not trickle down much.

A second pattern from figure 2.5 is worthy of note: richer individuals have a larger fraction of their wealth in assets that produce capital income rather than assets that do not, including cash and furnishings, some claims on other family members, or bank accounts. The effect appears in terms of the share of assets and value. Individuals below the top decile have about 40% of their assets, representing 60% of their wealth, in assets that produce capital income. At the other extreme, individuals in the top 1% have two-thirds of their assets, representing 73% of their wealth, in assets that produce capital income. In this context, even if capital returns and savings rates were equal across groups, the rich would have more capital income and a higher rate of wealth accumulation.

From the perspectives of both Kuznets and Gerschenkron, and most economic historians, the second half of the nineteenth century was a key period of rapid urbanization and industrialization on the European continent. These two processes, coupled with legal and financial innovations, created a whole new range of "modern" assets, including foreign investments, stocks and bonds of French companies, bank deposits, and so forth. These can be contrasted with traditional assets such as real estate, public debt, private peer-to-peer loans, cash and bullion, and home furnishings that had been the principal components of wealth before the French Revolution and before industrialization.

The distinction between traditional and modern assets allows us to establish that there was a radical transformation in the wealth held by Parisians. As figure 2.6 shows, in 1822, 88% of the wealth of Parisians was composed of traditional assets. That share drops steadily to about 55% by 1900. In the aggregate, Kuznets's model of sectoral transition is important for understanding the nineteenth century. By 1912, French equities and foreign assets constituted 34% of all Parisian wealth. Kuznets's model also had implications for the distribution of wealth. To stay close to his formulation, one would want to classify individuals by occupation into "traditional" or "modern." However, we recognize that individuals may hold their wealth in a variety of assets and thus that a large part of their portfolio may be in activities quite different from their occupations, so we classify the assets directly. If we look at the prevalence of modern or traditional assets across the three key fractiles of the wealth distribution, in 1822, all groups looked pretty much the same, with about 90% of investments being traditional assets. Then, all groups move an ever-larger share of their wealth into modern

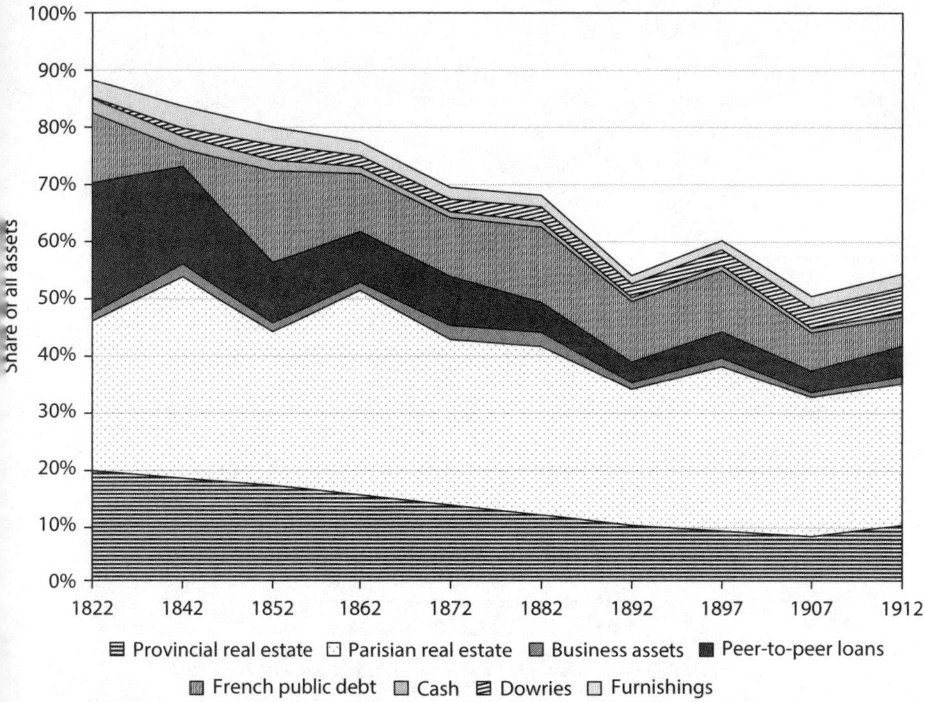

FIGURE 2.6. The decline of traditional assets as a share of all assets. *Source:* Assets dataset.

assets. The move starts earliest for the top 1% and is more pronounced for that group. The other two groups show a substantial decline in traditional assets, but they continue to represent more than 65% of all wealth toward the end of the period. Again, these patterns reveal an advantage for the rich: to the extent that the new economy has higher returns than the traditional economy, not only will the rich have higher wealth accumulation because more of their wealth is in productive assets, but their wealth is also more invested in high-return assets.

One can also compare men's and women's investments. It is often assumed that women are either directly more conservative in their investments, or that because of the legal structure of French marriage, women's wealth was parked in low-return assets. As figure 2.7 shows, although women had slightly more traditional assets than men in 1842, the share of modern assets rose at the same rate for both sexes.

Another way to parse the data is to isolate individuals from noble families from commoners. Being an aristocrat was no hindrance to being

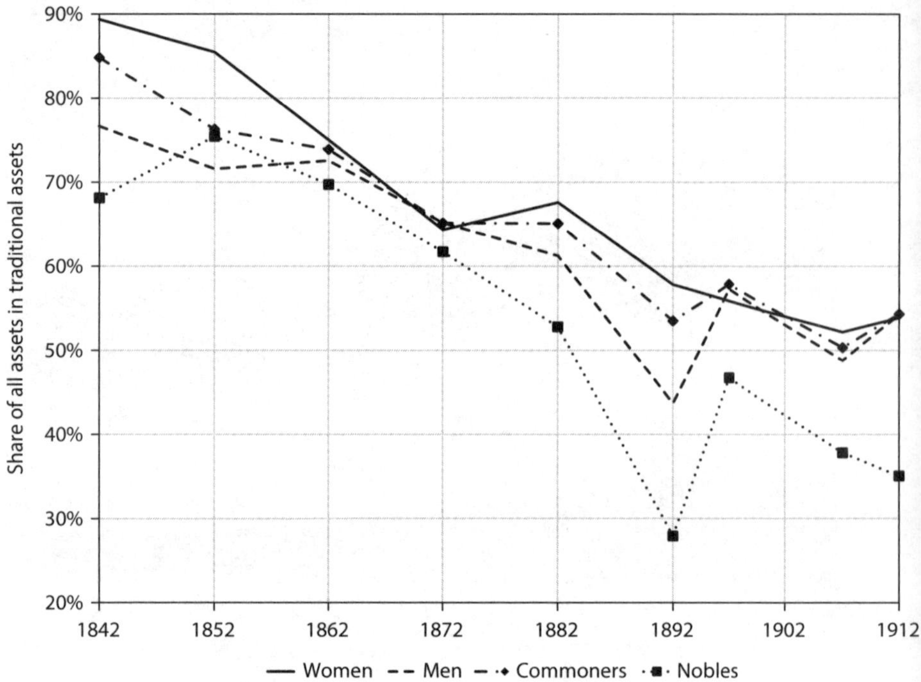

FIGURE 2.7. Share of traditional assets in portfolios by social status.
Source: Assets dataset.

forward-looking in one's investment. As figure 2.7 shows, nobles always have a smaller share of their Parisian wealth invested in traditional assets. One might worry that this is simply the mechanical effect of nobles being wealthier than commoners, but this difference is roughly the same within fractiles. Perhaps nobles were more likely to hold provincial real estate and thus seem more forward-looking because they were richer than we measure. Moreover, their rural real estate provided a good buffer against any additional risk from investing in modern assets.[11] Still, their wealth at death does not suggest that, as a group, they made suboptimal investments.

Theories of economic transformation put a heavy emphasis on investment in specific sectors like industry or railroads. Another cut through the data looks at activity with a special focus on finance, industry, infrastructure, and railroads. Some caveats, however, apply. To start, we try to classify firms' equity and debt into ten activity sectors. It is easy to identify railroad holdings because even if the name of the security in the estate does not mention the word "railroad," as in "2 actions Danube Save Adriatique," the geographic definitions almost always tip us off, e.g. "Lérouville à Sedan." It is also easy to identify sectors for firms listed in stock market directories. Other enterprises can be identified because their sector is included in the title as "Mines de" or "Banque de." In other cases, the name itself is sufficient to allow identification of "Alais Froges" (chemicals) or "Cail" (metallurgy). In many cases, however, firms are only identified by the names of their owners, and these firms drop into the unknown category. These firms are most often smaller noncorporate entities and their share of assets falls over time.

For the 1842 sample, we identify the sectors for less than half the equities and bonds in portfolios. This is largely because, at that time, there were essentially no railroads (though the Paris to Versailles line had been open long enough that French explorer Dumont d'Urville and his family could die in an accident and appear in our data), and the huge investments in urban infrastructure that would characterize the second half of the nineteenth century had barely begun. In 1842, a lot of equity was still in sole proprietorships and partnerships, and these enterprises were difficult to characterize by type of business. For 1852, we had more success and can attribute a sector to more than 70% (by value) of stocks and bonds in the estates of Parisians. After 1882, our success hovers around 90%.

The first observation from this exercise is that there were two key sectors over the whole period: railroads and public debt (see figure 2.8). Public debt was never less than 22% of the equity and bonds of Parisians and, most of the time, about a third. Railroads started small (less than 2% in 1842), but by midcentury they accounted for about a quarter of Parisians' wealth and half the equity and bonds. Finance runs a distant third at about 10%. Mines and utilities each account for 8% of wealth. Returning to Gerschenkron and theories of bank-led development, there is little evidence that the Parisians particularly channeled their investments into deposits at banks or savings institutions. Such deposits never account for more than 5% of total wealth. Instead, Parisians' wealth was held directly in stocks and bonds. Although the certificates might be deposited with a banker or a broker, usually they seem to have been in the possession of the deceased. The various forms of bonds and debts were an important, though declining, share of all financial assets. Bonds, including peer-to-peer loans, were more than 80% through 1852 and then drop to 60% on the eve of WWI. Those bonds were held directly, not on the balance sheet of banks. Thus, firms with large-scale capital needs used financial intermediaries to place bonds with the public, not with banks. In nineteenth-century Paris, financial transformation (from deposits to stocks or bonds) was limited.

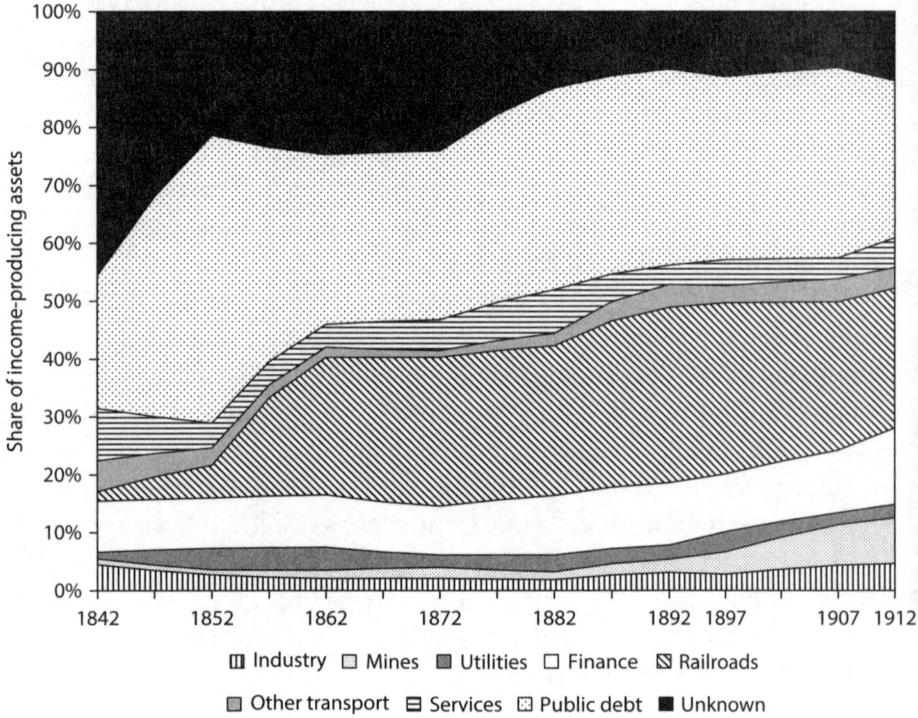

FIGURE 2.8. Investment by sector. *Source:* Assets dataset.

A second observation is that the growth of French wealth, as seen through the portfolios of Parisians, was, except for railroads, a relatively smooth process. There is no moment of take-off for utilities, industry, mines, or finance. The wealth of Parisians became more diversified over time. For railroads, the pattern is quite different: they accounted for about 4% of the bonds and equities of Parisians in 1852; by 1862, that number had jumped to 16%. From then to WWI, railroads accounted for a quarter of all the equities and bonds held by Parisians. As we shall see below, the difference between railroads and every other sector becomes even more apparent when we break out foreign assets. We speculate that for France as a whole, the process of wealth transformation was even smoother and slower because the initial share of land and peer-to-peer loans was so much larger outside Paris (we will consider the Paris versus France issue in more detail in chapter 10).

The rise of financial assets (stocks and bonds) in Parisian wealth antedates the nineteenth century (Hoffman et al. 2019). As the financial literature has shown, those older developments were massively amplified after the 1820s by the rise of investment banks whose job was to find individual holders for all these financial claims and of stock exchanges that allowed individuals to alter their portfolios much more easily than had been the case with traditional assets (Hautcoeur 2007). We cannot say much about investment banks because the securities we observe are already in individuals' portfolios. We have, however, integrated information on what assets were publicly traded (listed on the Paris Stock Exchange, henceforth the Bourse). After timid beginnings in the first half of the nineteenth century, listings on the Bourse expanded rapidly until WWI. As many have noted, listing on the Bourse was more difficult than listing on the London Stock Exchange, and many securities were traded in an over-the-counter market known as the Coulisse (Davis et al. 2003; Lagneau-Ymonet and Riva 2012).

Given the growing importance of publicly traded assets, we look at how listing on the exchange impacted our decedents' estates and measure what proportion of wealth was in securities listed on the Bourse over time. We find that most of Parisians' wealth was not traded on the market (these assets include real estate, peer-to-peer loans, bank accounts, insurance policies, and cash and furnishings). Among financial assets (equity and bonds), however, the share of wealth that was listed rose from less than 60% to about 80% by 1872 and stayed near that level until WWI. Breaking the data down by fractiles shows that the poorer groups had about the same share of traded assets as the richer ones. This similarity masks two opposing forces. The portfolios of the eighth and ninth deciles have fewer income-producing assets, but among those, a large share was in publicly traded government bonds. Conversely, the top 1% has a smaller share of publicly traded financial assets because many have large positions in privately held businesses (e.g., the Rothschilds or the Mallets own their eponymous banks). If we make the denominator gross assets rather than financial assets, the reverse pattern holds because of the important role of cash and furnishings in the wealth of the eighth and ninth deciles. Because of the extraordinary concentration of wealth in the hands of the top 1%, most of the benefits from the rise of the Bourse accrued to the very wealthy. This conclusion reverses Neymarck's (1903, 1908) arguments that financial innovation offered investment opportunities to an ever broader set of the population. While this is a tempting argument, it has little empirical content, as the share of individuals who died wealthy did not change from 1852 to 1912 and the wealth share of the 1% tended to grow rather than fall.

The last important dimension of asset holdings involves the geographic location of the capital in Parisians' estates. Recall that, following Davis and

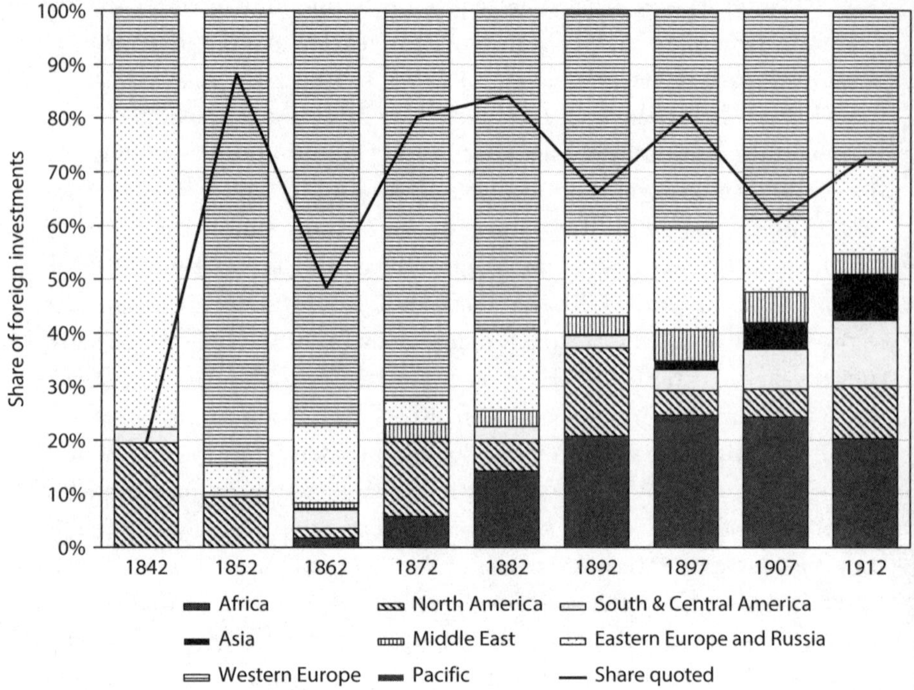

FIGURE 2.9. Geographic distribution of foreign investments. *Source:* Assets dataset.

Huttenback (1986), we attribute a location to the productive activity that provides the return on capital. Case in point, the Suez Canal was legally a French corporation (also registered in Egypt) with its headquarters in rue de Cléry in Paris. The canal, whose shares and bonds were traded in Paris and Cairo, derived its income from providing transport services, so we coded it as Egypt and Africa. In the nineteenth century, businesses tended to operate in a small number of locations, usually one country, so assigning a location was relatively straightforward, and we gave a place to more than 95% of all financial assets. We find that the share of Parisian financial assets that were French was more than 90% through 1862, and though it fell over time, it was never less than 70%. If we zoom in on the international part of the Parisian portfolio (see figure 2.9), Western Europe (including Germany, Switzerland, and the Austro-Hungarian empire) was always the most favored investment region. After 1892, Western Europe fell to about 40% of all foreign assets as Eastern Europe and Russia grew to about 20%, and Africa also reached a similar share.[12] The early twentieth century saw a clear rise in investments in the Americas. A line in the graph reports the share of foreign assets that were traded on the Bourse, and it is clear that the internationalization of Parisian portfolios was closely connected with financial innovation. Going all the way back to 1852, foreign assets were always more likely to be traded than French financial assets.

Here again, we find that the wealthiest individuals were much more geographically diversified than either the rest of the top decile or the rest of the individuals who died with wealth. After 1892, it is striking that almost half of all the assets of the top 1% wealth holders were international, and the foreign assets of the eighth and ninth deciles never breached 20%.

In appendix 2, we break international issues down by sector. We find that in every cross section, nearly 75% of these assets were either public debt or railroads. Railroads were negligible until 1862, when they jumped to 40% of all international holdings, before entering a slow decline to a smaller but still substantial 20% of international holdings by 1912. Mines grew rapidly after 1897, as Parisians showed an affinity for South African gold and diamond endeavors.

Overall, the portfolios of Parisians show two important patterns. First, they are increasingly financially sophisticated and thus very responsive to legal and financial change. The wealth holdings of Parisians make clear that the second half of the nineteenth century was indeed the first golden age of financial capitalism. Signs of the transformation of wealth can be found in the number of assets, the sectors where wealth was held, the rise of secondary markets, and the destination of capital. At the same time, for every one of these indicators, the richest Parisians (the 1%) who left at least 50% of the wealth behind were the most rapid adopters of new technologies and the greatest beneficiaries. The extent of these differences is such that they could not but contribute to the growing differential in wealth across the fractiles.

Finally, it is important to note that the massive and dramatic financial development we document did not make it any easier for individuals below the eighth decile to accumulate bequests. Financial development had no direct consequence on the share of the population that died poor. That said, to the extent that financial innovation facilitated the development of life-improving infrastructure (e.g., clean water and sewers), there were broad indirect benefits that we saw in the data on age at death. In the case of Paris, these kinds of investments were handled by a mix of private and public organizations. Haussmann's rebuilding effort required municipal investment, for example, to buy out landowners and rebuild and pave new streets. However, the buildings that lined his new boulevards were privately owned (either by individuals or by corporations like insurance companies). For water, the city financed the sewers and the water infrastructure, and private companies dealt with retail customers. But poorer neighborhoods were generally much slower to see improvement (Kesztenbaum and Rosenthal 2017). For gas, tramways, the subway, and electricity, private companies with concessions ruled the day. All these organizations issued financial assets. Our data show that rich Parisians owned these and profited from all these new investment opportunities while their poorer neighbors were left out. Of course, those who benefited most were those who, like Louis Alfred Daval, rebuilt the old city and led its expansion into the areas annexed in 1860. When Daval died in 1902, he owned some eighty properties, nearly all in the Clignancourt and Goutte d'Or neighborhoods of the 18th arrondissement. This massive real estate portfolio of 20 million francs was in part funded by 7 million francs of mortgage debt taken out with four different banks. Daval was a self-made man, the son of a café waiter. When he married his first wife, Mathilde Bauche, in 1872, he did so without a contract. When she died in 1892, neither he nor she had personal property, and the community property included 4.4 million of Parisian real estate (ADP 27689 #393; for other examples, see Yates 2015).

2.6. Conclusion

Paris in the second half of the nineteenth century was, in many ways, a precursor to the wealth dynamics that most of the Western world's population experienced since WWII and, more so, after the 1980s. Paris was wealthy, inherited wealth was important, and a large fraction of that wealth was held in financial assets. The main difference between our world and the later nineteenth century is that taxation rates were very low. Nothing we see then reflects tax avoidance or tax optimization. During this period, as the last section showed, there was massive innovation in finance and the economy more broadly. In 1807, the speed of travel over land for people and news was mostly constrained by the speed of a horse. By 1912, travel from Paris to most of France could be counted in hours instead of days, and news traveled at the speed of light. Much of the technical change was embodied in physical capital that required financing when deployed. Nowhere was this more evident than in the immense importance of railroads in fixed capital formation, as documented by economic historians who work on the macroeconomic aspects of growth and the evolution of financial markets (Lévy-Leboyer and Bourguignon 1990). Industrialization and the transport revolution required new investments all over France and in Paris. They also accelerated a trend underway long before in Europe: urbanization. Urbanization required investments in infrastructure within cities (e.g., housing, water, lighting, gas). The growth of cities like Paris (from 650,000 in 1817 to 2.8 million in 1911) was also dependent on investments in infrastructure to reduce the cost of providing food, fuel, and other inputs. Whether publicly (water, lighting) or privately (housing, gas) provided, these investments produced financial activity in the form of municipal bonds and the equity and debt of private companies.

All these developments are consistent with increases in wealth and have much to do with the sevenfold increase in the real value of estates from 1807 to 1907. When we consider inequality, it is tempting to return to a two-sector model, a traditional one where labor and capital incomes are low and a modern one where labor and capital incomes are high. The persistent influx of population into Paris (from 1807 to 1912, about two-thirds of all the Parisians who died in Paris were not born there) is consistent with the two-sector model for labor. The rapid rise of Paris (and other large cities) likely contributed to increased French wage inequality in the nineteenth century, consistent with the Kuznets framework. Despite considerable work on migration and the sectoral transformation of the workforce (Magnac and Postel-Vinay 1997; Marchand and Thélot 1991; Postel-Vinay 2002; Sicsic 1992), the evolution of wage distribution in France has been understudied for lack of systematic sources.

We know that capital income plays a very large role in income inequality. That was one of the reasons we were drawn to work on wealth inequality (Piketty et al. 2006). However, when it comes to wealth, we must recognize its fluid properties that break down the analogy with income inequality. Individuals are not required to hold their wealth in the firm they work in or near where they live. Consider a model economy that is initially egalitarian without capital markets (the yeoman economy). Then, someone who wants to start a modern enterprise must sell all their traditional wealth and invest it in the new sector. The Kuznets framework works well in this setting because capital incomes will be higher in the new sector than in the old one. For nineteenth-century Paris, however, there were two problems: there were very active capital markets, and it was not an egalitarian economy.

When there are capital markets, someone working in the traditional sector can invest in the modern one, while someone who toils in a modern factory may invest in rural real estate. Even in an egalitarian yeoman economy, capital markets make the impact of new sectors on wealth inequality much more difficult to parse than on wage income. Indeed, individuals are likely to want to diversify investments across sectors, even absent return differentials, if only to mitigate risk. If individuals were sufficiently diversified, new sectors would not impact wealth distribution.

Of course, even in 1807, Paris was not an egalitarian yeoman economy: the top 1% of wealth holders owned half the wealth, and many of them had inherited substantial wealth. Further, the financial markets of the nineteenth century made it difficult for individuals to be well diversified. Financial innovation, thus, likely produced two effects. For the bulk of the wealthy population (between the 70th and the 95th percentile of the distribution), innovation increased inequality between those who remained in the traditional sector (say, public debt) and those who invested in the new sector. However, for most of these individuals, diversification was limited. The succession of agricultural crises in the last third of the nineteenth century hurt individuals who had made peer-to-peer loans to farmers, while those who invested in modern endeavors like the Suez Canal or department stores did well. However, being in the modern sector was not without risk, as the crashes of the Panama Canal and the Union Générale remind us. Nevertheless, financial markets permitted many to invest in sectors uncorrelated with where they worked. As the example of nobles shows, it also allowed individuals without an occupation (rentiers) to invest in various activities.

At the top, most individuals were well diversified and increased their investment in high-return assets over time. Within the top 1%, the Kuznets sectoral effect has little bearing because individuals held assets that spanned both sectors. The distribution of wealth was unequal in 1807, and it is not surprising that it was also unequal in 1912. More importantly, Kuznets's sectoral model made predictions about two phases of economic change. The first, which we have just analyzed, confirms that inequality increases because of economic innovation. Kuznets also hoped that inequality would fall once the traditional sectors became small enough. Paris in the nineteenth century shows no sign of such dampening effects. On the contrary, the succession of innovations seems without end. One could start with insurance and canal companies early, follow those with French railroads and coal mines, continue with international sovereign bonds and railroads, move to gas works and tramways, and then to electricity, gold, and diamond mines. One does not get an impression of a radical burst of change (the first and second industrial revolutions) from Parisian wealth portfolios. Instead, successive waves of new ventures just kept arriving as technology changed and the geographical scope of investment grew ever larger. Before WWI, the second part of the Kuznets dynamic never operated in Paris. Instead, the innovation and the growth of large-scale firms meant wealth inequality in Paris remained very high.

Wealth inequality, far from declining, was increasing. This increase is easiest seen when we isolate the top 1% (representing some 250 adults on average among the dead or 13,000 among the living per year) from the rest of the population. On average, these individuals owned about 53% of the wealth in our cross sections. In the first two decades of the nineteenth century, their share was about 47%, while in the two decades before WWI, that share was 61%. The top 1% had increased their wealth share by 30% over a century. As we have seen, one element that allowed the top 1% to increase their wealth faster than the rest of the wealthy involves differential returns. The primary driver of these differentials was the share of wealth comprising assets that produce capital income. In the first two decades of the nineteenth century, about 70% of adults who died had an estate so small that it failed to attract the attention of the tax authorities. Despite massive financial innovation, that number had risen to 72% in the two decades before WWI. Waves of individuals moved to Paris in great numbers, likely attracted by higher wages, and these individuals consumed their earnings rather than saved. They came to Paris and became renters, not homeowners. They were workers, not entrepreneurs, who died without real estate or business wealth. Those who died in 1902 had failed to save just as those who died in 1822, even though the range of savings destinations was larger in 1902 than in 1822 (we return to this issue in chapter 9).

Given this constancy in the share who died with little or no wealth, the path of inequality is entirely driven by the distribution of wealth among the top three deciles. Within this smaller group, the seventh and eighth deciles generally had about one-third of their wealth in furnishing, bank accounts, and cash compared to 15% in the top decile, creating a substantial difference in return to wealth. Added to this difference among deciles is a break between the top 1% and the rest, whereby the richest group invested much more abroad than anyone else. Given differences in yields, the rich could enjoy high consumption rates and maintain their wealth with modest savings rates.

Taking the findings from chapters 1 and 2 together leads to one last lesson. The low tax, low redistribution environment of nineteenth-century Paris means that inequality tends to persist or increase in part because it is easy to transfer wealth through the generations and to accumulate wealth by saving from capital income if one inherits wealth. After all, the French

Revolution, political transitions, wars, epidemics, and financial crises did little to undermine the durability of family wealth. What they accomplished was a partial replacement of who was at the top, much like what the more positive innovations in finance and technology of the second half of the nineteenth century. With its trio of calamities and complex tax history, the twentieth century tells an entirely different story. To start, we will turn to taxes, which turn out to be more uncertain than death.

3

The Laboratory Evolves

PROGRESSIVE TAXATION, 1901–1972

In the most dramatic reform of the inheritance tax since the French Revolution, the law of February 25, 1901, instituted a progressive inheritance tax, whose broad outlines remain in place today. This and a plethora of legislative decisions over the following decades led to a more than tenfold rise in both average and marginal rates of taxation in direct line of descent. This chapter breaks with the chronological structure to describe this momentous change and how it affects how we measure wealth inequality. We devote a second section to the question of whether taxation reduces wealth accumulation and close with a consideration of the impact of progressive taxation in Paris versus France. The reader who is not concerned with such matters can proceed directly to chapter 4.

3.1. The Rise of Progressivity and the Centralization of Estate Taxation

Many scholars (e.g., Lindert 2004; Piketty 2001, 2014; Scheve and Stasavage 2012) have discussed the massive increase in fiscal pressure in most Western countries that marked the beginning of the twentieth century. Here, we focus on estate taxes, leaving the discussion of capital income and general income taxation for chapters 4 and 5. The reason for breaking down fiscal change this way is simple. The move to progressive estate taxation changed how individuals interacted with the estate tax administration; other reforms did not.

The progressive regime instituted in 1901 required establishing the total value of wealth devolving to each heir. The filings of the estates had to be centralized at the bureau where the deceased had his or her legal residence. As discussed in chapter 2, starting with the 1902 cross section, we thus observe all real estate holdings of decedents, not just those located in the capital. Another element of the reform was that the tax was assessed on net estates, so consolidated liabilities were also reported. Moreover, the reform did not change the fact that a large fraction of all state revenues were collected from estates not inherited by direct heirs.

More importantly, the law raised tax rates. In direct line of descent, the tax rate doubled (from 1.1% to 2.5% for the parts of estates greater than 100,000 francs). This rate increase potentially mattered to the top 5% of estates in Paris in 1902, but we should keep in mind that the tax was an inheritance tax. For example, two children splitting an estate of 200,000 remained under the top marginal rate. For estates in the top 0.1% (4 million and above), with most having between two and three children in direct line of descent, the average inheritance tax was close to the maximum marginal rate. For bequests to individuals beyond six degrees of kinship, the tax increase was abrupt: marginal rates continued to rise past 100,000 francs and topped out at 18.5% for bequests above 1 million francs. For Paris, at least, only about 60% of all decedents had children or grandchildren. Most of the other estates went to kin, but about 5% went to individuals categorized as "not related" to the deceased. So, progressivity and the differential treatment of individuals based on different degrees of kinship mattered.

More estate tax laws progressively raised the marginal tax rates over the next four decades.[1] On the eve of WWII, the maximum marginal rate in direct line of descent was 47%, and 80% for nonkin. The Vichy government raised these top rates to 66% and 88% in 1940. By the mid-1950s, the trend reversed, and maximum marginal rates began to fall. By 1959, they were down to 15% in the direct line of descent and remained so until our study's end. Nonkin rates remained high throughout, and by the end of the 1950s, progressivity disappeared in favor of flat rates between 40% and 60%.

Some economists will see rates above 50% as confiscatory. However, the law of 1934 introduced maximal average rates of 25% for children and 40% for nonkin (something ignored by Scheve and Stasavage 2012). Figure 3.1 provides both top rates and maximal average rates. Curiously, many changes to the top rates that followed 1934 were of no practical significance since the

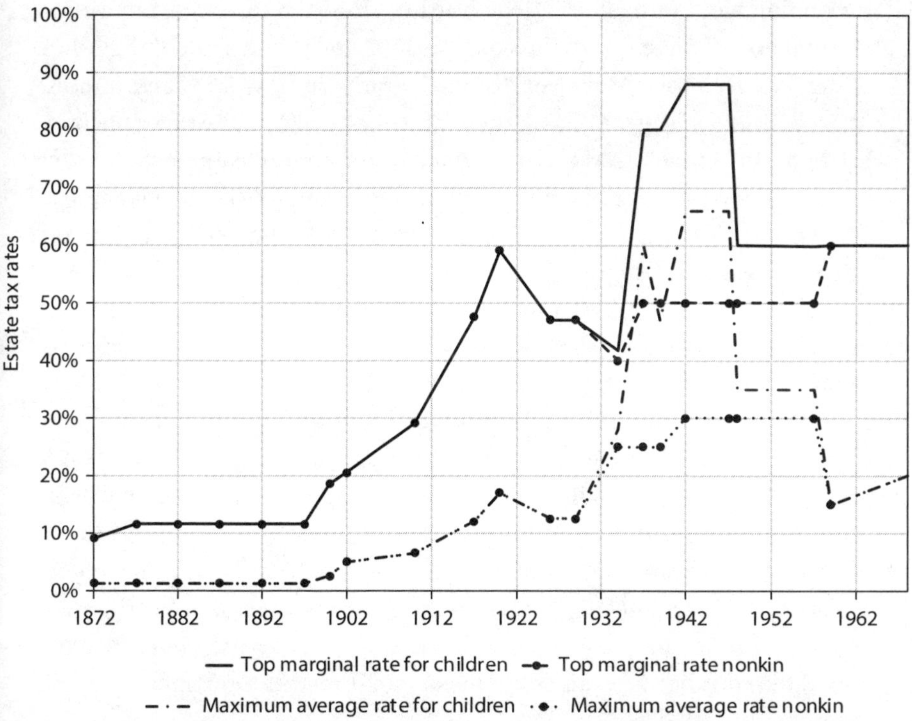

FIGURE 3.1. Inheritance tax rates, 1872–1967 (%). *Source:* JO For details, see appendix 3.

maximal average rate intervened much earlier. As an example, the law of 1937 has a top rate of 60% in direct line of descent (for the part of estates above 150 million francs). If we take as an example an estate of 5 million francs that devolved to a single daughter and apply the schedule, she would have faced a maximum marginal rate of 35% and paid a total tax of 28.5% (just under 1.43 million francs). Instead, she paid 1.25 million francs because the maximal average rate was 25%. Starting in 1942, the tax authorities further refined these maximal rates to differentiate across the number of direct-heir lines (so that estates divided across three children or more had a maximal rate of 20%, and those going to only one child had a maximal rate of 30%).

Figure 3.1 shows that the maximum average tax rate was much lower than the marginal rate. In fact, as the 1940 estate tax schedule reported in table 3.1 shows, the maximal average rates demonstrate that all the top marginal rates were, in fact, moot. When the Vichy government raised the top rate on nonkin from 80% to 88%, it was without any fiscal effect since once an estate hit the 53% marginal rate, the maximal average rate for nonkin went into effect. Vichy was not alone in publishing rates of no consequence. Starting with the schedule 1937, the top two rates for any line of descent are moot. However, some governments saw it in their interest to advertise lowering or raising these illusory brackets. The schedule published for 1934 that introduced maximum average rates featured no illusory brackets. However, every schedule published through 1952 included brackets of no significance for some sets of heirs because the maximum average rate superseded those rates.

As we have emphasized, the French estate tax system was an inheritance tax in which the relationship between the heir and the deceased mattered. Figure 3.1 also charts the top marginal rate and the maximal average rate for estates going to nonkin; these were always substantially higher than those going to children, with a maximal average rate of 50% following 1937. The schedule was generally very steep for nonkin, and the maximal average rate was reached quickly. Finally, the reform of 1959 put an end to the maximal average rates and lowered the top marginal rate for children to 15%. It also instituted a linear tax on estates going to nonkin of 60%. This 45% gap in tax burden based on relationship to the deceased was larger than the 42% difference between an estate passing in direct descent and one going to nonkin in a 1920 law.

The system did not stop at distinguishing between children and everyone else. Going back to the nineteenth century, there were always at least four categories: children, parents and spouses, close cousins, and everyone else. Table 3.1 provides the different categories for the 1940s. The 1940 law maintained the long-standing distinctions in the inheritance tax structure but added new categories as progressivity magnified legislators' inventiveness. The 1940 law featured 18 progressive schedules, and table 3.1 reproduces the 9 most relevant. The least onerous applied to children who are inheriting as part of a group of three or more heir lines, and the most severe to nonkin. After WWI, legislators included as an heir line any child who had reached age 15, not just those who were alive or who had had children, so the rate faced by inheritors was reduced if their parents had lost a child in the trenches. In the absence of direct heirs, tax rates were much higher—including on small estates—and always rose with kin distance. In the 1902 schedule, the authorities recognized as a kin heir anyone separated by less than 12 degrees of kinship; by 1944, kin stopped at second cousins. Again, it is striking to consider how these details play out in the extent of progressivity. If it is possible, albeit unlikely, that a concierge could trace their ancestry back 12 generations (to Louis XIV's days), aristocrats surely could

BLE 3.1. The 1940 estate tax schedule

	Direct heir lines			No direct heirs					
	3	2	1	Spouse	Parents	Siblings	Kin 2 degrees	Kin 3 degrees	Beyond 3 degrees
	Marginal tax rate (%)								
	2.2	1.4	1.1	6.1	6.6	19.8	24.5	31	35.5
000	3.3	2	1.7	11	12.1	22	27.5	33	40
,000	7.2	4	2.8	17.6	19.8	27.5	33	40	44
,000	11.6	6.6	5	22	24.5	35.5	38.5	44	48.5
0,000	15.4	8.8	6.6	27.5	30	42	44	49.5	53
0,000	19.8	12.1	8.8	32	35.5	46.5	49.5	53	55
0,000	28.5	17.6	14.3	37.5	41	51	53	55	59.5
000,000	34	23.5	17.6	42	46.5	55	57	59.5	66
000,000	38	26.5	19.8	47.5	52	59.5	62	64	70.5
000,000	44	30	24.5	52	57.5	64	66	68.5	75
,000,000	48.5	36.5	29	55	64	68.5	71.5	75	79.5
,000,000	57.5	46.5	37.5	64	71.5	73	77	79.5	82.5
0,000,000	66	52	44	74	77	79.5	82.5	86	88
aximum average rate	20	25	30	35	35	40	48	50	50

urce: JO, December 10, 1940 (law of December 9, 1940), p. 6041.

te: Marginal rates that do not apply because the maximum average rate binds are shaded in gray.

do so—or genealogists could do it for them. Moreover, a substantial fraction of the Parisian population died childless, so the schedules for individuals without direct heirs were important. The schedule assessed different rates for 13 different levels of wealth (from 1–5,000 francs all the way to estates larger than 150 million francs). This creates 117 (9 multiplied by 13) marginal rates. A full 57 of these were illusory because the maximal average tax rates superseded them.

The reader will surely not have missed the pronatalist aspect of these policies. Families of three rather than two children who transmitted estates paid lower rates because each inheritance was smaller, but they also paid a lower rate. Take the case, for example, of the estate of Édouard Huot de Saint Albin, who died at age 53 on April 5, 1947 (ADP DQ7 36422 #932). Huot left an estate of 60 million francs to be divided among his heirs (his widow and three adult children). The Huots had married under separate property, but she had a right to a life interest in a quarter of his estate. That right was evaluated at 5.99 million francs. Each child received 18 million francs. The widow's tax bill of 1.1 million was assessed at the maximum average of 20%, net of 100,000 francs credit for three children. Her effective tax rate was 18.3%. Each child paid 3.5 million francs in tax for an effective rate of just under 19.5% (they were assessed at the maximum average tax rate of 20% after a 500,000 francs reduction [*abattement*] for being a group of three siblings). The state thus collected 11.6 million or 19.3% from the Huot estate. Had the full progressive schedule been applied to the widow and the three children, the total tax bill would have been 15.5 million or nearly 26% of the estate. Now, suppose Huot and his wife had only had two children. The reductions would have disappeared, and the children would have faced a 25% effective tax rate. The state would then have collected 14.4 million francs. If the Huots had had only one child, the total tax bill would have come to 17.4 million, or just shy of 30%.

We can also compute these counterfactuals for an estate of 6 million francs (representing someone in the 98th percentile of the distribution in 1947) to be divided between a widow and up to three children. Applying the maximum average rate to the widow and the progressive schedule to the three children gives a tax burden of 8%, while the two-child family pays 19%, and the one-child family pays 27%. The same calculations for an estate of 600,000 francs (representing someone in the top decile of Parisians wealth holders) are less interesting because the per-child deductions meant that no tax is owed in the three children case. The two-child family spends 5%, and the one-child family pays 11%. While progressivity is preserved, the maximum average rate massively dampens it. The reductions were generous and increased progressivity, but only for the relatively rare families with three or more children.

France long had a famously low rate of demographic growth, and its low natality was an important matter of political debate, including in the interwar years. However, one wonders about the value of estate tax policies in spurring natality. By 1930, the average age of decedents was well past childbearing and rearing years. One suspects these complex and variable rules reflect a profound attachment to patrimony and a partly hidden gift to the wealthy. It was politically desirable to advertise both the very high marginal rates that would hypothetically take big bites out of large estates, and the reductions for large families. It was equally politically desirable to ignore that the very rich were more likely to have children and downplay the reality that maximal average rates preserved wealth for the richest families. That said, there is no question that the 1901 reform, with the higher rate of inheritance, was, in fact, a break from the low nineteenth-century rates.

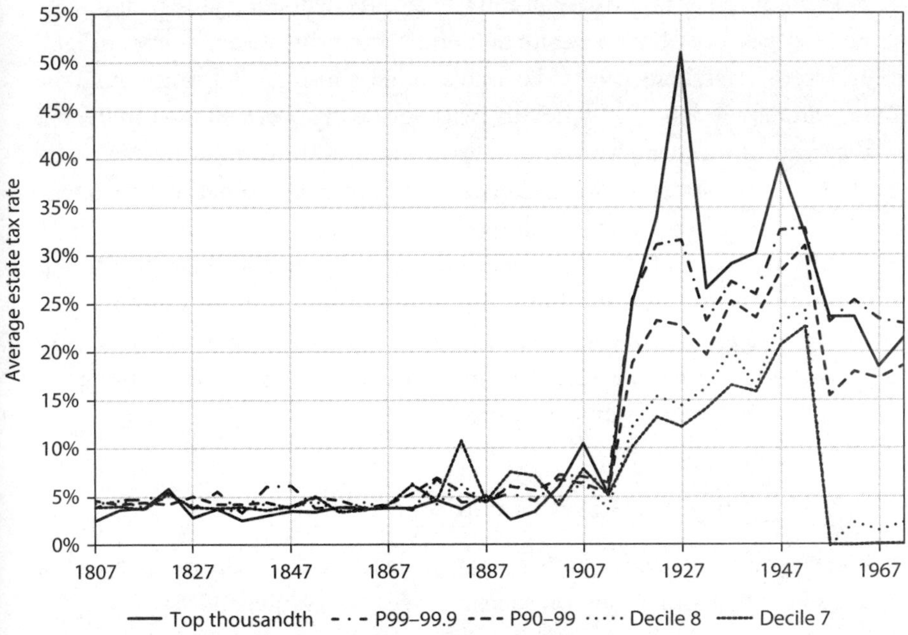

FIGURE 3.2. Average share of each estate paid in inheritance taxes by fractile. *Note*: We apply the tax schedules in each year to each estate then we average the tax burden by fractile. *Source:* Estates dataset (see appendix 1).

The rise of progressive taxation was particularly important for Paris because the city was rich and unequal, and its denizens were even less inclined to produce direct heirs than the rest of the country (Brée 2016). We can apply the tax rate schedules to our estates, using information on the number of heirs to create a simulated set of average tax rates by fractile. These rates are not exact. Indeed, we make several simplifying assumptions. To compute the number of direct descendants, we take the number of children reported and one-third of the number of grandchildren (because this is the number of grandchildren per heir line when our data collection is sufficiently precise). When our notes mention children plural (*enfants*), we assume that that involves two children. We also assume that when there are no children, there is only one heir of the type reported (niece, nephew, grand-uncle . . .), even though these estates are often divided among many claimants. Thus, our measured tax rates are somewhat biased upward. Figure 3.2 displays average tax rates by fractile. Before 1914, the variation across fractiles was trivial and related to the extent to which decedents of a particular fractile died without any heirs. Those estates had to pay the highest possible rates, as in the case of the Boucicauts, who died in 1877 and 1887. Each was among the top wealth holders. Marguerite died after her husband and son and left her wealth to be used to build a hospital. Starting in 1912, the tax rates are ascending by fractile, with jockeying between the top 1% and the top 0.1% due to high taxes on indirect descent. The collapse of the tax rate after 1959 on estates below the top decile comes from a radical restructuring of the inheritance law the same year.

The reform of 1901, by its centralizing of estate tax filings, made our research easier. The more important impact was a large increase in most estates' taxation rates. Higher rates prevailed for everyone from 1915 to 1959 and have continued since for the top decile of estates and for all inheritances not received by children or spouses.

3.2. The Impact of Estate Taxation on Estates

Many economists will assume that estate taxation shifts the relative cost of transferring wealth through legacies relative to consuming it (Masson 2023). Suppose individuals are optimizers and set their bequests so that the marginal benefit of resources given to the next generation equals their marginal value of personal consumption. Then, when bequests become more costly because of higher rates of estate taxation, they should decline as a fraction of individual wealth and lifetime income. The general path of wealth at death from 1912 to 1947 described in the introduction is broadly consistent with the view that estate taxation matters; the rapid rebound from 1947 to 1962 is not.

We can refine the question of the deleterious effect of estate taxes on bequests in two ways. We start by considering types of heirs. As discussed earlier in the chapter, the pronatalist bent of French fiscal policy meant that tax rates increased as the heirs became more distant from the decedent. Even after the late 1950s tax cuts, cousins and relatives were taxed at between 40% and 50%, and nonkin paid 60%. If the fiscal cost of legacies mattered, estates in direct line of descent should fall less than those in indirect line of descent or nonkin—something we can check. We limit the sample to women decedents who died wealthy, over the age of 50 and whose wealth did not place them in the top thousandth—this group is large (more than 3,000 in each cross section from 1902). We focus on women above 50 because there is little chance they had more children over the previous decade in response to changes in the tax law, but they could adjust their legacies. We must leave out women who died without wealth because we do not observe whether they had children or not, but that omission is

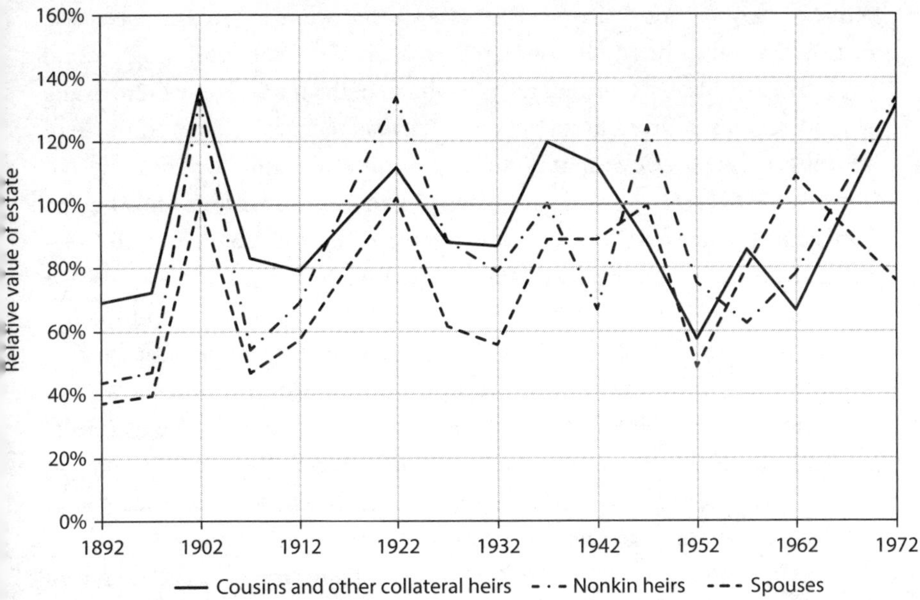

FIGURE 3.3. Value of estates by heir type relative to estates that pass to a single child (women aged 50+ not in the top 0.1%). *Source:* Estates dataset (see appendix 1).

of limited consequence. Indeed, at the same time as tax rates rose, the share of individuals who left estates was increasing, so we are not facing a problem of selection where the individuals most sensitive to tax rates decided to dispose of all their wealth before death.

We look at the value of estates whose principal heir is not a child. As we noted in chapter 1, French law requires that no less than half of any person's estate be transmitted in direct line of descent, and when there is more than one child, each must receive a minimum portion. So, if there are children or descendants of children, they will appear in the documents. We scale those estates by the average value of the estates to a single child each year to consider the variation in estate value. Estates going to an only child are the largest single transmission category in Paris, rising slowly from 25% to 30% over the first six decades of the twentieth century.

We break out estates whose principal heir is not a child into four categories of principal inheritors: parents, spouses, collateral (cousins, nieces and nephews, aunts and uncles), and nonkin. We leave aside parents representing less than 5% of inheritors and focus on the other three categories. Figure 3.3 displays the ratio of the average value of estates by their type. We find no evidence that high tax rates on inheritors other than children reduced bequest motivations. The estates that went to cousins, spouses, or nonkin were not smaller relative to those that went to a single child in the 1940s, '50s, or '60s than they had been in the 1890s. After 1920, estate taxation made indirect heirs poorer (because the state captured more of the inheritance flow). Still, it did not substantially reduce accumulation or bequests. While this is not conclusive proof that estate taxes had few behavioral consequences for Parisians, these data lead us to doubt that estate taxes had a large impact on the willingness to bequeath.

We also consider estates in the direct line of descent. The combination of progressivity and a tax on inheritances (rather than estates) meant that the tax's share of the estate was smaller for the same estate when it went to more heirs. Other forces may have been operating as well. For example, having more children might increase an individual's urge to accumulate in order to provide bequests, and wealthy individuals could afford to have more children. However, if we accept that these forces were roughly constant over time, we would expect estates to be proportionally larger for large families than for small ones in the high tax period. Figure 3.4 (on the next page) reports the ratio of estates for individuals with two, three, and four to six children relative to estates going to a single child (we leave out estates going to seven or more children as they represent less than 1% of estates). The figure shows that individuals with three or more direct heir lines had estates that became proportionally larger in the 1930s than those with one or two direct heir lines. For individuals with two direct heir lines, it seems safe to conclude that their estates were 50% larger than those with only one direct heir line, but this ratio did not change much over time. Overall, it seems that estate taxation had little effect on the accumulation motivations of a large part of the population of decedents.

3.3. Conclusion

The rise of progressive taxation created profound changes in how citizens understood inequality in their countries. Indeed, whether it be on income or wealth, when the government started to collect progressive taxes, they also published tabulations that provided the first real glimpses of inequality in many societies (for details, see Piketty 2001; appendix 3). Someone whose parent left her a fortune of 5,000 francs in 1902 could have learned from those published tables that 41% of all estates in the Seine were larger than her inheritance (average value 205,000 francs). She could have also looked at France as a whole and seen that about one-seventh of all estates were larger

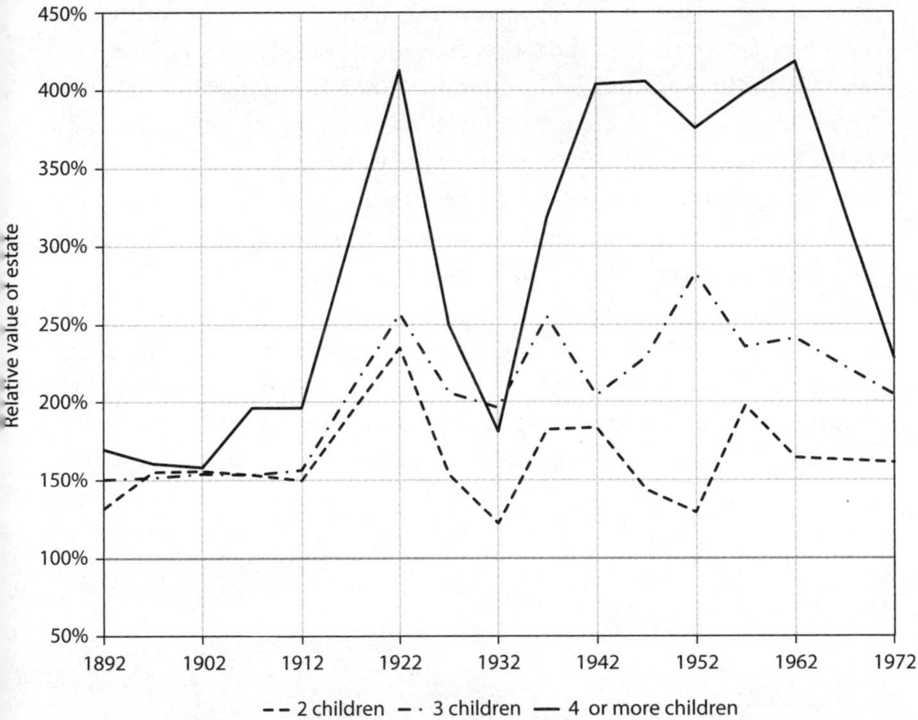

FIGURE 3.4. Value of estates by number of children relative to estates that pass to a single child (women aged 50+ not in top 0.1%). *Note*: "Children" here refers to direct heir lines (children and the descendants of children who have died). *Source:* Estates dataset (see appendix 1).

than hers, with an average value of 75,000 francs. Even that quick inspection would have convinced her that the département of the capital was unequal and wealthy. Had this inquisitive person decided to apply Pareto interpolation, she would have noticed that the top 0.1% wealth at death share in the Seine in 1902 was slightly lower than that of France (22.1% vs. 24.2%) but that the 1% share was larger in the Seine than in France (58.7% vs. 54.4%). In the Seine, the top 10% concentrated essentially all wealth (96%), while in France, it was only 88.5%. Had she been able to plumb the Parisian data, she would have found that the capital's wealthiest thousandth held a quarter of all the wealth, and the top 1% had 62%. The crumbs left to the lower 90% of the distribution were 2% of the wealth. One might think that the difference of 8% (62% vs. 54.4%) in the wealth share of the top 1% is small, but that is wrong. Reducing the top centile wealth share of Paris to that of France would have been enough to give every Parisian above the median the same wealth as that observed for the ninth decile. France was unequal, and Paris was both the richest and the most unequal of all places in France. As we shall discuss in more detail in chapter 10, the combination of higher wealth, higher inequality, and a larger proportion of indirect transmission in Paris relative to the rest of France meant that the estate tax increases that started in 1901 were more intense in the capital than in the provinces.

The next three chapters describe the evolution of wealth and inequality in the twentieth century. We emphasize the importance of the macroeconomic environment over tax rates for the evolution of wealth. As expected, periods like WWI or the Great Depression were not conducive to creating large bequests. The evolution of inequality, as we shall see, will require us to pay close attention to taxation, but again, it plays a limited role, and other factors, including changes in pensions and human capital, are more important.

4

The Collapse of Wealth, 1912–1947

For historians and economic historians, what is known as the long nineteenth century ended with the conflagration that began in August 1914. With the start of World War I, a new and challenging era had begun. For the issues relevant to this book, WWI marked the start of an entirely new period marked by fundamental changes in private wealth accumulation and distribution. As figure 4.1 shows, although private wealth peaked in the 1890s, in the three decades that followed 1914, it fell precipitously. By 1947, average wealth in Paris was only 18% of what it had been in 1912, and only 14% of what it had been in 1902 in real terms. In Paris, private wealth at death per capita fell so dramatically that, by 1947, in real terms, it was at levels close to what it had been in the first two decades of the nineteenth century. This chapter chronicles this apocalypse. The next chapter will examine the evolution of inequality during the same three and a half decades.

To understand why Parisians in 1947 were only about as wealthy as they had been in 1822, we must sift through half a dozen plausible explanations. The great shocks of the twentieth century—the two world wars and the Great Depression—play a critical role in the literature's explanation of the moderation of inequality by the 1950s. They were also the immediate and ultimate causes of the collapse of wealth. We, however, want to break down the wealth loss to understand it better. After all, Parisian fortunes neither vanished nor evaporated. Moreover, given what we observed in the nineteenth century and absent policy innovations, we would have expected these shocks to reduce wealth but then to be followed by a return to the previous

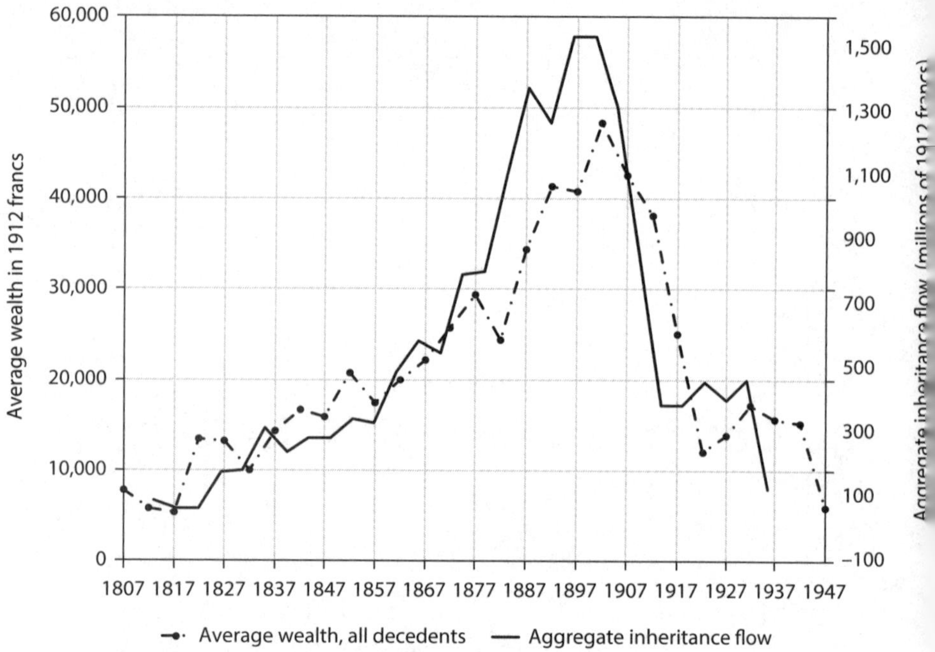

FIGURE 4.1. Inheritance in Paris, 1807–1947, in 1912 francs. *Source:* Estates dataset (see appendix 1).

equilibrium. Thus, we are interested in the policy and economic innovations that led these shocks to have different long-term effects from those of the nineteenth century.

Taken together, the three catastrophes involve dramatic changes in the population of Paris and, in different ways, of France. These demographic developments might have changed the distribution of income between capital and labor and, thus, the return to wealth. They may also have affected the intensity of bequest motives. The great shocks also produced the first significant inflation since the French Revolution (the price level rose by a factor of 63 between 1912 and 1947). The world wars destroyed assets, interrupted international capital flows, and led governments to repudiate foreign claims. Meanwhile, the Great Depression forced the liquidation of far more businesses than usual and made the return to the gold standard impossible. Finally, the French state intervened in the economy throughout these periods by substituting public capital for private capital.

History does not provide natural experiments to economic historians interested in long-term growth because the responses to random shocks are all endogenous. Individuals had many options when faced with the adverse events that started in 1914 and the public policy changes they induced. For instance, they could have tried to maintain the real value of their bequests by changing their portfolios or saving more. They could, equally possibly, have decided that the tax regime favored current consumption over bequests and thus left less to their heirs than they would have without the tax hikes that started in 1910 and would continue through the 1920s. Moreover, when we consider the valuation of assets, prices partially reflect the monetary, fiscal, and policy circumstances of the moment and investors' expectations of future economic growth and public policy. Given that asset prices reflect events and expectations, our analysis of these different factors will provide, at best, orders of magnitude for the impact of one or the other on the decline in wealth. In the short run (e.g., 1912 to 1922), it may be possible to estimate the impact of inflation on portfolios by estimating how much inflation reduced the consumption value of a 1912 portfolio held to 1922. In the long run, however, history intervenes, and new events muddle any evaluation. For example, we cannot compute the welfare gains associated with individuals rebalancing their portfolios toward equities in the roaring twenties to escape inflation. Those that did so experienced large gains in the first decade after WWI. However, they were much more exposed to the equity price falls of the Great Depression than they would have been had they maintained portfolio structures closer to those that prevailed before 1914. More interesting to our minds is the political economy analysis that forces us to chart the interconnection between the different shocks of the period. The establishment of various rent control measures starting in the 1910s is less of a surprise when one considers the importance of nominal pensions (for retirement, to veterans, or to their widows). Government-induced inflation was shrinking the government's pension largesse, and rent control was a way to dampen this connection. Should we attribute the loss in real estate values driven by rent control to inflation or rent control? Suppose we were to take representative portfolios for each fractile. In that case, one might seek to decompose the wealth fall into subcomponents (say X% to inflation, Y% to asset destruction, Z% to taxation). However, as we shall see, that is simply impossible at the individual portfolio level because some were entirely exposed to inflation (those where domestic debt dominated). In contrast, others suffered massively from wars (e.g., deported Jews).

Supply-side economists will be quick to draw inferences about the negative impact of taxation growth on aggregate private wealth. After all, the sevenfold increase in real wealth from 1807 to 1907 occurred in a period of linear and negligible estate and income taxation. Over the next five decades, this was followed by an equally astounding decline that brought real average wealth at death in 1947 to what it had been just after Napoleon's abdication. As discussed in chapter 3, estate taxation explains only a small part of the puzzle. Indeed, an array of fiscal measures moderated what may have seemed to be punishing marginal tax rates, and very few estates paid the maximum average rate of 25%. Moreover, as we shall see in chapter 6, estates recovered quickly after WWII in an equally severe tax system. However, there is no doubt that estate and income taxation depressed capital income.

The chapter quickly reviews the path of wealth before moving to a history of changes in the Paris population between 1912 and 1947. Important changes in migration followed the massive casualties of WWI. The arrival of large numbers of migrants from Central Europe and the former Ottoman empire fed into long-standing antisemitism that, in turn, led to the deportation of tens of thousands of Parisian Jews during WWII, many of them children. While these population losses are not unique to the history of Paris, they raise important questions about the evolution of wealth. We then consider the evolution of private wealth in Paris between 1900 and 1947. We proceed with a brief review of the standard four sources of wealth decline (taxation, inflation, destruction, and nationalization). The last section, as in chapter 3, examines the evolution of the consumption value of wealth over these three and a half decades.

4.1. Wealth, 1900–1947

As noted above, wealth collapsed between 1912 and 1947. Returning to figure 4.1, the decline in average wealth is mirrored in our estimates of the total inheritance flow. Moreover, our estimates of the wealth of the living also show the same decline. Average wealth at death could have fallen either because total wealth dropped (which is what the aggregate inheritance flow shows) or because of a surge in mortality among the poorer parts of the population. In Paris, at least, the second explanation can safely be put aside, as adult deaths peaked in the decade before WWII, and mortality was falling over time. A more subtle demographic hypothesis would be that given the magnitude of the mortality shocks of the world wars, bequest motives were dampened or eliminated for many older individuals due to the lack of (male) heirs. This hypothesis must also be put to the side since the share of decedents leaving a bequest rose throughout the interwar period (an issue we return to in chapter 5). We thus focus on wealth itself.

The most obvious candidates, however, involve the private returns to private wealth. Parisians' exposure to adverse shocks depends on how much

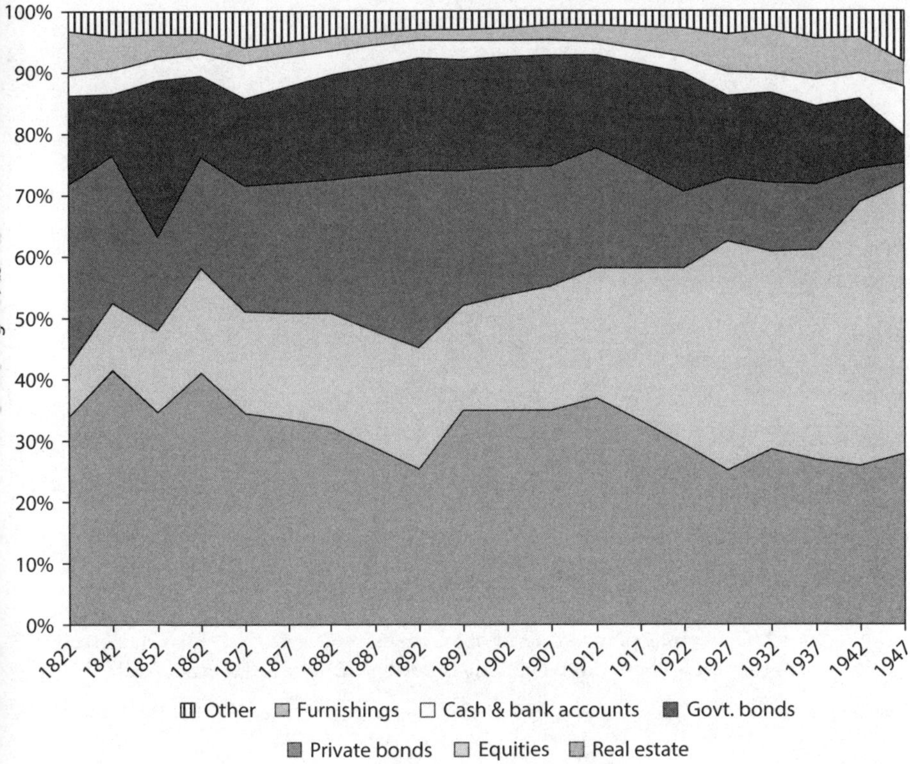

FIGURE 4.2. Aggregate wealth portfolio for Parisians' estates, 1822–1947. *Source:* Assets dataset.

their portfolios are diversified. Figure 4.2 shows that before WWI, real estate accounted for about a third of total wealth, equities accounted for about 20%, and bonds were often more than a third of total wealth. By 1947, bonds had shrunk to only 7% of Parisian portfolios, real estate was down to 29%, and equities had jumped to 45% of all wealth.

Parisians' portfolios were also quite widely distributed, with 40% of wealth invested in assets in the city itself and a similar amount for the rest of France. Figure 4.3 shows that 20% of all wealth in Parisian estates was invested in foreign assets. Over the twenty years following the Great Depression, there was much instability, but the foreign share remained at roughly 20%. The big change involved a decline in investment located in the city itself. Foreign investments evolved in favor of colonial investments, leading to a steady rise in the share of investments in Africa to about 35% in the 1930s and 1940s (see figure 4.4). The focus on colonial investments offset a decline in European and Russian investment. Eastern Europe and Russia's

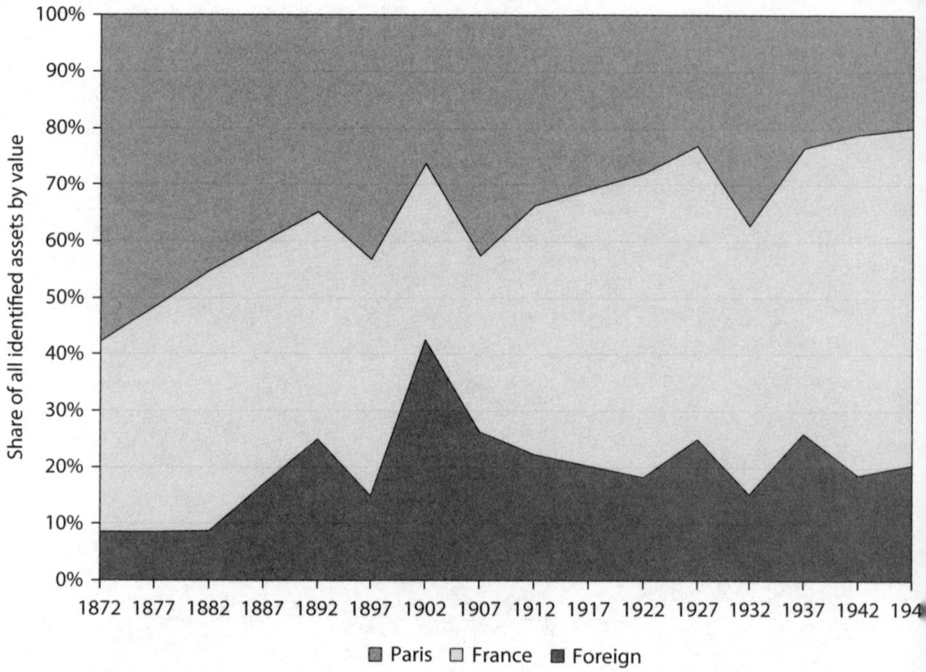

FIGURE 4.3. Geography of investments of Parisians' estates, 1872–1947. *Note:* Assets located in the place where capital income is produced. *Source:* Assets dataset.

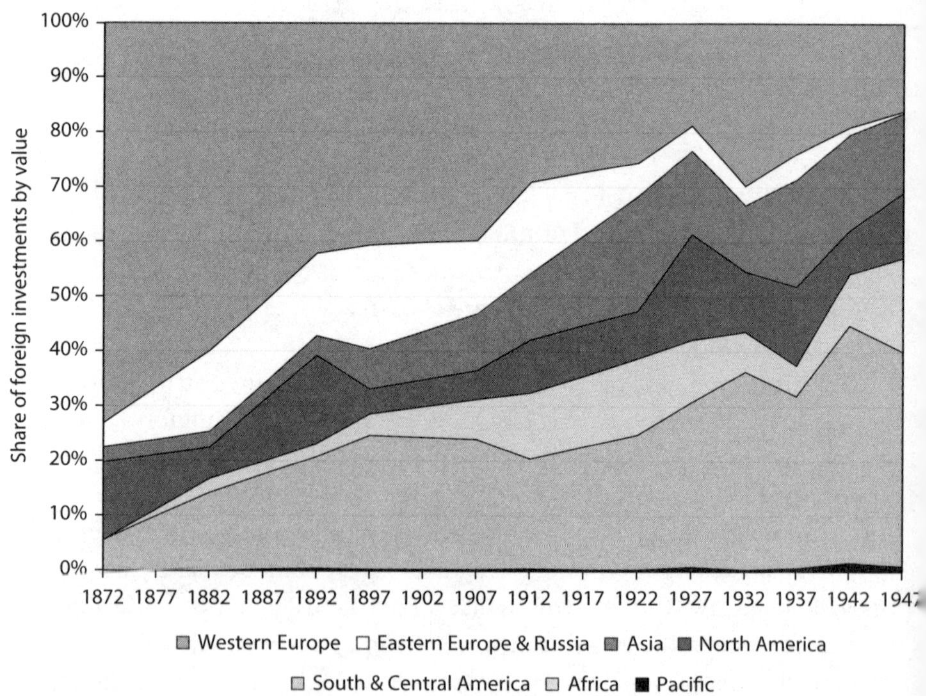

FIGURE 4.4. The distribution of foreign investments in Parisians' estates, 1872–1947. *Source:* Assets dataset.

shares had been about 17% of the foreign portfolio before WWI, dropping to 7% in 1922 and to about 3% after 1932. Portfolios evolved because Parisians continuously reinvested their wealth and invested new savings, including in securities doomed by inflation. Although, as we shall see, wealth and income taxation increased markedly over the first half of the twentieth century, tax optimization was not the cause of the changes we see. Indeed, estate and income taxes were nearly independent of the type of investment.

The first half of the twentieth century did, however, feature a plethora of wealth shocks that mattered to one type or another of Parisians' investments in specific locations. To take two examples, the rent control measures that started during WWI and continued in the 1920s mattered for French urban real estate income but less so for rural real estate and not at all for other forms of investment. Similarly, the Russian repudiation of all foreign financial claims that followed the 1917 revolution mattered only to Russian assets. Thus, most of the shocks we discuss had well-defined sectoral or geographical limits and their impact was dampened by the highly diversified Parisian wealth portfolio. In some cases, the shocks were more of a slow-moving disaster, as was the case for investments in French public debt, which totaled 31 billion francs in 1912, of which 21 billion were in perpetual bonds (*rentes*) (ASF 1913, 265). The value of these *rentes* was essentially eliminated by inflation by 1947 (their real value was about 1% of what it had been in 1912). If Parisians had been purely passive, their investments in the national debt in 1947 would have been about 1% of their wealth, yet it was almost 7%. Parisians kept buying public debt and other bonds and experienced new losses throughout the period. So, just measuring the impact of inflation on the 1912 portfolios undercounts the losses they experienced. Even in 1912, French government bonds accounted for less than 7% of Parisians' portfolios, so the inflation shock to pre-WWI French government bonds alone cannot explain the 82% collapse of average wealth. Instead, we have to follow the historical sequence of shocks. That task will occupy most of the rest of this chapter.

4.2. Population

The period from August 1914 to August 1944 featured two large shocks to the population of Paris and of France in general. About 100,000 Parisian men died during WWI, while in WWII tens of thousands of Parisians, mostly foreign-born Jews, were deported and killed in concentration camps. Smaller numbers died fighting the Germans or as civilians. Large mortality shocks were not unknown to the city prior to these events. The cholera epidemic of 1832, which lasted more than a month, doubled the number of deaths in Paris, relative to the prior decade's annual average. In 1870–1871, the siege of Paris and the repression of the Commune led to tens of thousands of fatalities. In each case, the city recovered its population quickly. For example, the population of Paris in 1836 was 9.8% smaller than what the census of 1831 had reported. This loss was more than made up for in the next five years, however, and the city's population in 1841 was 8% larger than in 1831. Similarly, the 1872 population of Paris (after the Commune) was 2% larger than that of 1866. Paris suffered no long-term consequences from demographic shocks because it was an open economy. As early as 1866, the census reported that two-thirds of all Paris's residents were not born there, and that share varied little before WWI. Moreover, between 5% and 7% of all residents were born abroad. This was nothing new—as Piette and Ratcliffe (1993) have argued, high levels of migration dated back to the early nineteenth century and Sussman (2006) shows it was true in the Middle Ages. When demographic shocks hit Paris, the city just pulled in more migrants.

The losses from the world wars were also made up promptly. However, their particular characteristics make them worthy of discussion, as they surely had consequences for the evolution of wealth in the ensuing decades. France lost some 1.4 million citizens fighting WWI. They received the mention "died for France" and are recorded in a dataset assembled by the ministry of defense and posted online. This dataset is precious to us because it includes nearly everyone's places of birth and residence. Almost all were men who had reached the age of 17 by 1918 and were younger than 45 in 1914; thus, they were eligible for military service. The data for 90% of the sample reported that they were indeed soldiers or officers. Key for us is that 85,400 of the war dead were recorded as residents of Paris. We use information about the place of birth and recruitment to infer the likely place of residence for those for whom it is not reported. The total likely comes to just under 100,000 men (+/–2,000 depending on how one accounts for individuals without a known residence).[1]

Paris's 100,000 war dead corresponds to 23,000 fatalities per year, to which we must add 4,200 more men aged 20–44 who died in Paris. This brings the total number of men aged 20–44 who died each year to about 27,200. In contrast, each year during the decade before World War I, about 6,700 men between the ages of 20–44 died annually in Paris. Combining these data leads to an excess annual mortality figure of 20,500, which over the whole of the war runs to 89,000 men or 13% of men in the relevant age group in the 1911 census. The 1921 census has a total Parisian population nearly identical to that of the 1911 census, but that masks a 5% drop in the male population offset by a 5% increase in the female population. Taking the men alone, we divide them into three groups. Let us start with boys (those younger than 20 in 1921), who were unlikely to have served during the war; their numbers declined by 10% between 1911 and 1921. A similar decline is seen for girls. The low number of children in Paris reflects the long-term drop in French fertility and the low birth rates of the war years. For those aged 20–44 in 1921 who were very likely to have served in the war, their numbers fell by 10% between 1911 and 1921. That steep decline is far less than the excess mortality noted above. Men aged 44 or older were 11% more numerous in 1921 than in 1911. The same pattern holds for 1926, though the number of men likely to have served in the war was off only 7% (and the number of men old enough not to have served rose even more, by 13%). By 1931, the number of men likely to have served in the war (now aged 30–54) was only 4% less than what we would have expected absent the war.

Why were there so few missing men in Paris? The compensating factor was, as we saw before, migration. Migration was a long-standing phenomenon in Paris, both to fuel the population growth over the long nineteenth century and to offset the net negative population dynamics of the native population. Within this general pattern of migration, an important factor during the interwar period was a rise in the proportion of foreign-born newcomers.

The migration of the interwar period was a key factor in the particular impact of WWII on Paris's population because international migrants were targets of Nazi and Vichy arrests. A ministry of defense website, Mémoire des Hommes (MdH), provides a dataset for WWII deaths of active duty military and a separate one for civilian casualties of war in France. Unfortunately, unlike the WWI database, there is no information on the victims' residences. Moreover, as we shall see, the Mémoire des Hommes datasets are incomplete. For instance, some sources put the number of civilian casualties at 350,000 when the database reports only 104,000. As with WWI, nearly all these fatalities occurred outside Paris but over a much wider territory than that spanned by the trenches in 1914–1918. Individuals born in Paris's share of military casualties was about the same as the share of Paris in the French population. Paris's share of civilian casualties was lower than its share of the population because the capital (unlike its environs) was spared from severe bombing, and it was not a location of major fighting. Indeed, the number of all French military fatalities in 1942 was small (10,279), and only a few hundred of those were likely to be Paris residents. The same was true for civilian casualties. Nevertheless, military and civilian casualties are not likely to appear in the total number of Parisian decedents published in the *Annuaire statistique de la ville de Paris* (ASVP) so our population at risk is somewhat smaller than it should be.

The major problem involves the thousands of Jews deported from Paris and killed in 1942, who are nowhere reported in a way that is easy to integrate into our research. We started with two datasets: the ministry of defense's record of individuals deported to concentration camps and a separate source that lists the death records of all individuals who died after being deported from France. This data source exists because, in the 1980s, an effort was made to append such records with the mention "died for France." As appendix A4 describes in more detail, we merged the two deportee datasets, which allowed us to remove duplicates and improve the place of death and birth relative to either source. Individuals born in Paris accounted for 11% of deportation fatalities, more than twice of Paris's share of the French population. A full 43% of all deportees born in the city were younger than 19, three and half times the proportion of all deportees and almost five times the proportion for those born outside the capital. The reason was that most of them were children of foreign-born parents. Scholars familiar with Vichy's bargaining with the Nazis will not be surprised. Foreign-born Jews and their children (born abroad or in France) were the primary targets of large-scale roundups and subsequent deportation and death at the hands of the Nazis (Joly 2022).

To get a precise sense of how many Parisian adult Jews were deported and killed in 1942, we needed finer data than the merged database described above and in appendix A4. Indeed, we needed to know residence at the time of deportation, which is not reported in either dataset (though it would be in the original death records). We relied on the passenger lists for the 80 deportation trains that left France for Auschwitz and other camps, carrying more than 75,000 Jewish victims. These lists were compiled by the authorities in French internment camps in collaboration with the Nazis and are available from the Mémorial de la Shoah website. Merging all three sources for the individuals whose last or maiden name starts with either A or T, led to a dataset of 10,401 individuals, of which 4,698 could be matched to a convoy and 5,703 (mostly individuals born in France and without a Jewish name) that could not be matched.

We focus on 2,237 individuals who were at least 20 in 1942 and could be matched to a convoy that left France before December 31, 1942. Within the sample, 18% had no address, 62% resided in Paris, and 19% lived elsewhere in France (for more details, see appendix A4). The high proportion of Parisians is not surprising since the Parisian authorities arrested huge numbers of foreign-born Jews and their children in 1942, in particular with the infamous July 16–17, 1942, Vél d'Hiv roundup (because the victims were initially detained at a cycling track, the Vélodrome d'Hiver). By the narrowest definition, the share of Parisians among adult deported Jews is 50.8%. It restricts the numerator to the 1,137 adult Jews for whom we found a Parisian address, while the denominator is all adult Jews in convoys 1–45. A more satisfactory estimate uses the share of Jews for whom a death date is known to add another 259 individuals of the 269 Parisians Jews for whom no death date is known. This implies that 62.4% of all adult Jews deported in 1942 were Parisians and died in 1942. Finally, we could assume that not having an address implies that one is not Parisian (this biases the share of Parisians down).

Combining the 62.4% with information on the proportion of deported adults in the first 45 convoys leads to 21,493 adult Parisian Jews deported and killed in 1942. These individuals were not reported in the death totals for Paris published in the ASVP because they did not die there. They must be added to the 33,208 adults who had died in Paris that year. Given that we have the population of individuals for whom estate taxes were filed, we generate two measures for the fraction who die without wealth. Without the deported Jews, the share of individuals who die with nothing is 55%. If we include the deported Jews who legally died poor because they had been despoiled, then that share jumps to 63%. This second measure is equivalent to the share of individuals who died poor in the nineteenth century. Though Vichy and the Nazis had despoiled them, were these Jews really all poor?

The assumption that only deported Jews for whom an estate tax filing was recovered were wealthy is untenable because tax filing requires heirs, and heirs who know how to find the deceased's wealth. The Vichy government confiscated most Jews' assets, the Holocaust destroyed entire families, and financial institutions' secrecy after the war all made it difficult for distant heirs to learn about relatives' wealth. There are data, however, that suggest that many deported Jews had wealth, and some had quite large amounts before Vichy and the Holocaust made them poor. The best data involve lists of seized bank accounts. Indeed, in May 1941, the German occupying authorities required that all Jewish bank accounts be impounded.

As part of the Commission Mattéoli's effort to quantify the financial losses suffered by Jews in France during WWII, Claire Andrieu and her team collected a sample of deported Jews (those whose last names started with BA) (Andrieu 2000, 177–180). They estimated that 16% of the individuals who had their bank accounts seized were deported if we take both certain and likely matches. While informative, the share of deportees among seized bank account holders is not what we seek. We need the share of Parisian deported Jews who had bank accounts because it can be used to estimate the share of Jews who were wealthy.

We merged our dataset of 1,396 Parisian deported Jews with the list of seized bank account holders whose last names started with T or A established by the Commission Mattéoli. The merge produced 162 individuals with the same first and last name and address in the two sources and another 42 for whom last names and addresses match (for extra confidence, we leave these 42 aside). The 162 fully matched individuals represent 14.2% of the 1,137 adult deportees who lived in Paris and died in 1942. In the 1942 Parisian estate dataset, 52% of individuals with wealth have financial accounts.[2] Assuming the same proportion for deportees leads to 27.4% of deportees, or 5,889, who likely had an estate.

The next question is how many deported Jews came to the attention of the estate tax authorities. In principle, the TSA listed all legal residents of Paris when they died and were thus potentially liable for the estate tax. The registers should, therefore, include all Parisian deported Jews who died in 1942, whether or not there was wealth to transmit. We sought out who among the deportees in our sample was inscribed in the TSA by examining all the volumes for individuals whose last names started in A and T and whether or not they had a tax filing. We collected all the names that appeared at the end of each month's list from March to December 1942. Some entries mention that the person died in Auschwitz or Drancy, but most do not. In fact, the TSA includes only a tiny fraction of those who died in a concentration camp (in France or abroad).

The procedure yielded 308 individuals. This list included 43 children, one person who was deported by convoy 68 and thus could not have died in 1942, and one person who appears in the MdlS as not deported. That leaves 263 adult Parisian Jews in the TSA, and 246 were matched to our deported dataset. We left the 17 that we could not find aside as they could have died in France rather than being deported. The 246 adults with a TSA represent 17.5% of the 1,406 deported Jews with a Paris address, leading us to estimate that out of the total 21,493 deported Jews, some 3,757 could have been inscribed in the TSA. Of the 246 we found in the TSA, only 30 (2.1%) ever had a filing of any kind, a stark contrast with the nearly 45% filing rate for the nondeported population. It also implies that about 458 deported Jews (21,493*2.1%) should appear in the full dataset we collected for Paris in 1942. In any case, the share of deported Jews for whom an estate was filed seems to be on the order of a tenth (2.8%) of what the bank account sample would lead us to expect (27.4%). Thus, and unsurprisingly, the tabulations of estates, like the death totals, are inaccurate because both omit nearly all Jews killed as part of the Holocaust.

In sum, the deportees and their lost wealth make it impossible to compute the share of the 1% of wealth at death in Paris for 1942. We can do so for the 33,125 individuals who were not caught up in the Nazi horror, and we took great care to do so. We report those numbers, but we urge the reader to keep in mind that 21,493, or about 40% of the total adult deaths of Parisian residents, are not taken into account. Further, even if we had data to include the deportees, their connection to the wealth of the living would be even more complicated because the standard approaches of selective mortality that we deployed in chapter 1 cannot fix a holocaust. So, despite our best efforts, some important questions cannot be answered.

This history also raises questions about how one considers long-term history based on building samples of cross sections. The long-term evolution of wealth and wealth inequality in France might seem best charted by ignoring year-specific shocks in favor of smoothed secular-term trends. In that light, it might have been better to select 1941 rather than 1942 for our sample. However, doing so would do violence to history because of the elision of WWII's human and wealth tragedy. It would also do violence to economics because the wealth distribution at any point in time is the result of both long-run dynamics in accumulation and disaccumulation of assets as well as annual shocks that interfere with our measurement of the long-run processes.

4.3. Four Horsemen for a Wealth Apocalypse

Private wealth can be destroyed by taxation and inflation, physical catastrophes, depreciation without destruction, and the loss of control of assets. In our period, the state was one horseman for the wealth apocalypse because wealth taxation was significant, and inflation at times was severe. Both redistribute private wealth or income to the state. A second horseman involves natural disasters and human conflicts that destroy wealth. Houses swept up in a flood or blown up by bombardments lose their economic value. Physical destruction is but one of the ways private wealth vanishes. Economic change, in general, and technical change, in particular, are a third threat to the value of many assets. In such instances, the physical assets remain but are dramatically depreciated because they have little economic value, as in abandoned farmsteads or obsolete industrial plants. The final horseman involves loss of control over assets. This can happen either internationally, like Russian railroad shares after 1917, which were confiscated by the Bolsheviks, or nationally when the Vichy regime confiscated the assets of Jews. In these cases, the assets continue to exist, but investors lose their claims because of state action.

Private wealth levels fell because of processes that reduced wealth (e.g., bombings, inflation, or defaults) and because other processes reduced the consumption value of that wealth (e.g., taxation of capital income, inflation in the case of nominal long-term bonds). In both cases, one would expect asset values to fall. Moreover, we must worry about the direct effect that reduces asset values and the indirect effect that potentially, at least, reduces asset accumulation. To take a building as an example, its private value may fall because of a fire, rent control, or government taxation of rental income. At the same time, the owner's desire to invest in his or her building may fall if he or she fears expropriation from invaders, further taxation, or the imposition of rent control. We use our data to evaluate the direct effects; these are, after all, quite mechanical. When the Soviets nationalized private enterprise and repudiated their debts, the value of Russian assets fell to zero. That is the direct or mechanical effect. In this case, the indirect impact is straightforward, as Parisians would not accumulate new Russian assets for a while. In other cases, where the incentive effects are not so stark, it is more difficult to evaluate the indirect effect without other sources of variation.

Finally, it is important to note that by the 1920s, the French state intervened to dampen the impact of the destructions on private wealth with programs that subsidized rebuilding and reinvestment. As a result, in many instances, there is a stark difference between national wealth—which is immune in the short run to processes that redistribute wealth from bondholders to debtors or from taxpayers to holders of destroyed assets—and private wealth or even individual wealth. Moreover, during the period of calamities from 1914 to 1947, the portfolios of Parisians, with their bias toward financial assets and foreign investments, were more at risk than the average French portfolio. Finally, the portfolios of Parisians were far more exposed to the destructions of 1914–1947 than the physical city.

4.4. Taxation

As is well known, the world wars were responsible for increased fiscal pressure and the rise of progressive taxation as the state faced daunting budgetary problems during and after the wars. In chapter 3, we detailed the experience of estate taxes. To recap briefly, the long nineteenth century featured low inheritance taxes, on average about 3%. The differences between the top 1% (who own more than half the wealth) and everyone else were negligible. In the interwar period, inheritance tax rates rose quickly, and in Paris, these taxes shunted about a quarter of wealth transmitted through the generations to the state. Absent other changes, and given intergenerational intervals of thirty years or so, an increase in individual rates of saving could have mitigated the private wealth decline from taxation, and therefore consumption from inherited wealth would have fallen.

Of course, whether individuals adjust their savings rate depends on whether they are sensitive to the disincentive to save that comes from taxation or whether they have target bequests they seek to achieve. Since everyone only dies once, we cannot hope to observe the same person in both low and high tax regimes. However, we can put some bounds on the role of taxes in reducing wealth because inheritance taxes were much lighter for wealth that went to children than for indirect transmissions (see chapter 3). Estates passing to children (direct estates) did not pay more than 12% before the 1940s, while those passing to more distant relatives (indirect estates) paid between twice and five times the direct line of descent rate. If taxes led to reduced taxable bequests by consumption or through tax avoidance strategies, we should expect indirect estates to collapse far more than direct estates. The effect can be measured, and it is small. Estate size did not depend on the presence of heirs before WWI. After the war, indirect estates were 10% smaller on average than those where at least one child would inherit. Behavioral responses, thus, are also of limited importance to determining the size of bequests (we will return to the issue in chapter 9).

Taxes on wealth were not the only ones to change over time; there was considerable innovation in income taxation. The history of income taxation is detailed in Piketty (2001, chap. 4), and our work does not add to his analysis, so this section is a simple summary that provides context. As noted in our introduction, direct income and wealth taxation in the nineteenth century was mostly low and proportional. The first seeds of change came with enacting a tax on capital income in 1872. The *impôt sur le revenu des valeurs mobilières* (IRVM) was assessed on dividends and interest at an initial rate of 3%. It was paid by the issuers of the securities, not their owners. French government bonds were, of course, exempt. The rate was raised to 4% in 1890. The twentieth-century innovations got underway in 1914 when capital income from foreign entities had to pay 5%. Then, the rates rose progressively to 25% on foreign capital income by 1926 and 18.5% for domestic capital income in 1928. Finally, the 1952 fiscal reform folded the IRVM into the general income tax. The IRVM was surely an irritant to individuals who depended on capital income for their lifestyle. Its chronology matches the chronology of the estate tax with a significant break toward higher rates around WWI, followed by a decline in the 1950s. Because the very rich have large capital incomes, the abolition of the IRVM in 1952 was, in effect, a tax cut and a reduction in overall progressivity.

The other important tax was the *impôt général sur le revenu* (IGR), which took effect in 1915. Initially, it targeted the top part of the income distribution. Households with incomes less than 2,000 francs were exempt (Piketty and Zucman's [2014] estimate of per capita income was 1,600 francs at the time). The top marginal rate of 2% on incomes above 25,000 francs implied that the tax was not onerous. From there, rates rose. The next year, the top rate was 10% (on income above 150,000 francs). By 1926, the top marginal rate was just shy of 60% (on income above 550,000 francs) and reached 70% in the 1940s. As with the IRVM, the end of WWII saw a massive reduction in the complexity of the IGR, though the top rate remained high (60%) to the end of our period.

To evaluate the impact of these taxes on wealth at death, we deploy a simple strategy. We evaluate the income flow (rents, interest, and dividends) of estates before and after each tax is paid; we also compute the income value of the estate when all relevant taxes are paid (the estate tax, the capital income tax (IRVM), and finally the income tax (IGR). We normalize these four series by the pretax income of the estate. We use capital income flows because they are a good measure of consumption possibilities for individuals with bequest motives. Moreover, some taxes were assessed on capital income, and shocks like consumer price or wage inflation are best understood from an income rather than wealth perspective. We neglect capital gains and losses because we do not know when they were taken, and we suspect that individuals often did not factor these gains and losses into their consumption. For instance, consuming the capital gains on real estate that one transmits to the next generation is difficult. Focusing on flow income alone is also consistent with the literal interpretation of a rentier—someone who lives off the rent of his or her wealth. The evidence will show a dramatic decline in the consumption value of wealth.

For each estate, we consider the estate's value to the decedent and to a hypothetical heir who had paid the inheritance taxes (thus, both the value of the estate and the value of the estate net of estate taxes). For estates that pass in direct line of descent, we evaluate the cost of inheritance taxes based on the actual number of direct heir lines, but we still treat the estate as if it had only one heir. Essentially, our procedure focuses on the modal group: households where both parents have roughly the same wealth and two children.[3] For estates with other principal heirs (spouse, parents, relatives, nonkin), we apply the appropriate schedule as if there were a single heir (which is often not the case). This simplification allows us to consider the estate's value to the person who controls it (the decedent) and to the next generation (the heir). In most cases, these procedures lead us to bias the tax burden on estates upward.

We then compute the estate's flow income (rents, dividends, interest). Because we want to abstract from other changes, we impute these capital incomes for each estate by attributing to it the average flow return for portfolios in the appropriate fractile for the five years centered on the year of death (e.g., for the year 1912, we use 1910–1914). This means we neglect income differences over time that come from differences in rates of return. We use flow returns to real estate computed by Eichholtz et al. (2021) for real estate, the dividend yield computed by Lebris and Hautcoeur (2010) for equities, and the yield on government bonds for bonds (Homer and Sylla 1996). We rely on relatively liquid and low-risk assets in each case to determine yields that minimize risk premia. We also have a final category that includes cash, bank accounts, and movables, to which we attribute a zero yield. As an aside, it is clear that capital returns are increasing in wealth mostly because the zero-yield part of the portfolio declines with wealth. Neglecting risk and liquidity premia likely understates the extent of the wealthy's return advantages.

For each fractile, we take the ratio of the income after estate taxes have been paid (essentially the year after death) to pretax income. The fractiles are plotted in figure 4.5, which is consistent with our discussion of estate taxes in chapter 3. The "death tax" reduces capital income by less than 5% before 1900, and the variations across fractiles and over time come from the randomness of the distribution of heirs not in direct line of descent. The change in the first decade of progressivity is noticeable in that the fractiles are ordered, with the richest paying the most. After WWI, the state's take

FIGURE 4.5. After-estate-tax income as a share of pretax income. *Source:* Estate tax dataset.

increased, reaching nearly 30% for the top 1% and 20% for the eighth and ninth deciles. Then, the reforms of the late 1950s had two effects: the eighth and ninth deciles essentially paid no tax, and the taxes on the top decile were cut substantially. Had this been the only innovation, we would be writing a history of a temporary fiscal adjustment due to the great shocks of the first half of the twentieth century that lasted less than a generation after the end of WWII.

Let us start with the tax on dividends and interest, the IRVM. To evaluate its impact, we return to our samples of asset portfolios and compute the relevant share of assets liable for the IRVM. By definition, cash and various financial accounts with zero capital income are not liable. Real estate, peer-to-peer loans, and French government bonds were also exempt from the tax. Only 40% of all assets were liable for the tax on average. The data from our detailed samples show that the differential rates between domestic and foreign bonds, combined with the greater international exposure of richer portfolios, made these linear taxes weigh more on the very rich than on the rich. By the 1930s, the IRVM took a substantial bite out of capital income (10% or so below the top 1% and about 15% for the richest hundredth). As

FIGURE 4.6. After IRVM and IGR income as a share of pretax income. *Source:* Estate tax dataset.

with the estate tax in direct line of descent, its increased intensity reflects the growth of revenue needs for the state that started in WWI.

The impact of the income tax, the IGR, is quite different. Because we do not know the labor income of individuals, we compute the income tax payments from the flow of rents, interest, and dividends. Thus, we assume that individuals decide their labor force participation as if labor income faced the top marginal rate. While we adopted this solution because we generally know nothing about labor income, it is also consistent with many economic models of labor-leisure trade-offs. There, the individual considers the value of after-tax employment income relative to the value of leisure. We deploy the tax schedules in force from 1912 to 1972 (Piketty 2001). Figure 4.6 shows that the IGR took a very progressive form immediately after its inception. The capital income of Parisian decedents below the top decile of wealth made them essentially exempt from the IGR. On the other hand, for the top 1% and even more for the top 0.1%, the IGR transferred a very large fraction of their capital income to the state.

Figure 4.7 brings all these taxes together. We compute the after-tax income of the heir relative to the income the decedent would likely have

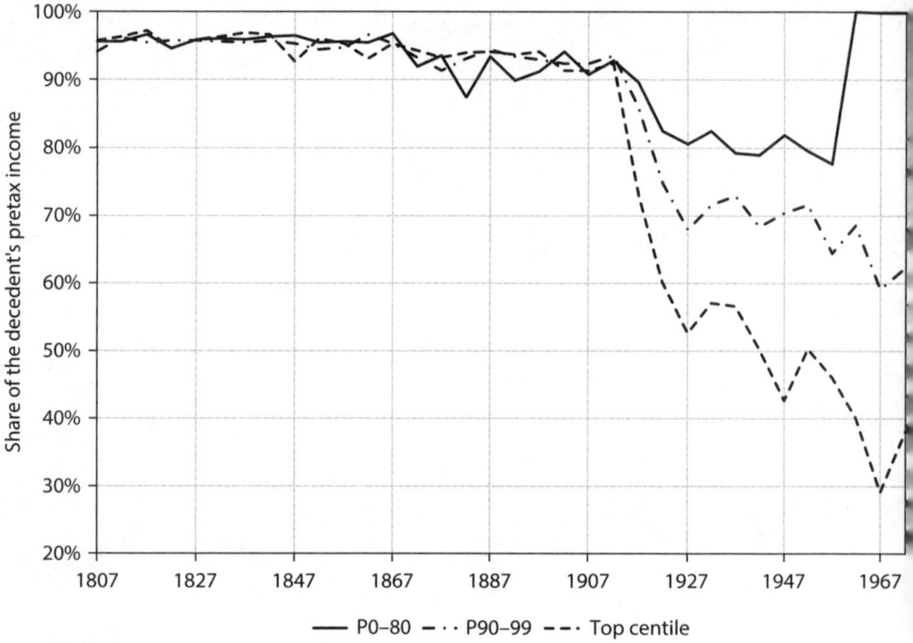

FIGURE 4.7. After-tax income of inheritor as a share of pretax income of the decedent. *Source:* Estate tax dataset.

had in the year before his or her death. Thus, we consider the income of the heir who controls the estate after paying the inheritance taxes, net of the IRVM, and paying the income tax on the residual. The figure paints a stark picture of the division of capital income between the state and individuals. Until 1912, figure 4.7 essentially reproduces figure 4.5 because there was no income tax, and the IRVM was light (at most 4% on a fraction of all assets). Then, the curves take abrupt dives as the state's bite into capital income grows. The state's share in private capital income rises all the way until 1947.

To the extent that the valuation of assets reflects after-tax rather than pretax income, the rise of the IRVM and IGR should have reduced the valuation of assets subject to these income taxes. The IRVM's 25% tax on capital income should have led to a 25% decline in the valuation of the assets it covered to maintain a constant private capital income. At the same time, the IGR might have had the same impact on all capital income. In other words, increased income taxes (IRVM and IGR) may have led to a fall in private wealth on the order of 30% or 40%, a far cry from the 90% decline we observe.

4.5. Inflation

In the second half of the nineteenth century, inflation was negligible. From a trough in 1852 to 1912, prices rose by only 0.6% a year. France was on a metallic standard then; ten paper francs could be exchanged for 45 grams of silver or 2.96 grams of gold. Then, with WWI, France abandoned the gold standard and abrogated all the contractual clauses that gave lenders the option to get paid in bullion (gold clauses) for domestic assets (as we shall see, those did not include the Suez Canal). Between 1912 and 1947, prices rose by an average 12% a year. Inflation was not constant. WWI increased prices by a factor of three and a half, then from 1922 to 1937, prices did not quite double, and in the last decade, 1937–1947, which includes WWII, prices rose by a factor of ten.

How taxation impacts the real value of wealth is relatively easy to compute. Indeed, before the 1950s, taxes were mostly independent of portfolio choices. The other shocks of the twentieth century are more complicated because their impact on an individual's fortune depends on their portfolio choices. Take inflation and consider some hypothetical and unrepresentative individuals who all inherited a million francs in 1912, all died in 1947, and consumed all their capital income each year. A million francs in 1912 meant they were among the top 1% wealth holders. Our first individual, who illustrates the raw impact of inflation, had placed the million francs in French *rentes*. Inflation eroded the real value of those 1 million francs by a devastating 97.8%. The real capital value in 1947 was 13,800 1912 francs, less than the interest payment our investor would have received in 1912 (33,500 francs). French bonds suffered the same fate as French *rentes*, so we leave them aside.

There were better options but few good ones. A second illustrative individual might have contemplated living off the dividends of the CAC40 (the Bourse's stock index extended back in time by Hautcoeur and Lebris). A million francs invested in CAC40 shares in 1912 could have been sold for 2.37 million francs in 1947 or 37,000 1912 francs, slightly more than the prewar dividend yield of 34,000 francs. While this broad portfolio produced a wealth two and half times larger than investing in bonds, it was still a loss of 96.3% of the wealth that person started with.

Very few individuals held a portfolio of stocks wide enough to approximate the CAC40, partly because there were no mutual funds in the first half of the twentieth century. Instead, they chose a few stocks, like the Suez Canal. A million-franc placement there would have left the holder with 193,000 1912 francs in 1947 (12.1 million nominal). One million put into foreign investments, like Norwegian nitrates or Swedish ball bearings, would have left our investor with about 58,000 1912 francs in 1947 (Data for Financial History database [DFiH]; Rydqvist and Guo 2021).

The range of outcomes for foreign bonds is extremely wide. On the one hand, if sold, the same million held in 4% 1893 Russian bonds would have fetched about 3,000 1912 francs in 1947, with no interest payments since 1917. It is hard to think of a worse investment strategy. On the other hand, the winning investments over this period were probably Suez Canal 3% bonds. Our investor could have bought a face value of 1.184 million francs in 1912 and earned 35,500 francs in interest in 1912. These bonds could have been sold in 1947 for 71 million francs or 1.1 million 1912 francs. The excellent inflation hedge of the Suez Canal bonds comes from its gold clause not being abrogated when the French went off the gold standard. Even though the Suez Canal was a French company registered in Paris and traded on the Bourse, it was treated as a foreign company and was required to abide by the terms of the bonds. This was, of course, good for bondholders and bad for shareholders. It should be remembered that the primary shareholder of the canal was the British crown, whose interest the French state felt little need to protect.

Another individual might have put her million francs in real estate. Real estate proved to be better protection against inflation (with an 88% decline in value rather than 97%), but it still left our hypothetical investor quite poor by 1912 standards. The last asset to consider is gold. Going back to the nineteenth century, the French have had a long affection for holding gold (Flandreau 2004), and we find varying amounts in many top estates. Using a series put together by Jacques Friggit (2007) for 20-franc gold coins, we can construct a price of gold for both before and after the end of the gold standard in 1914. Gold was a good inflation hedge until 1937 (its value in francs is close to the price level). It became a great investment during WWII, when its real price increased. Someone who had buried their treasure in 1912 to unearth it in 1947 would have earned a nearly 3% real return on the condition of not drawing on it for income. Overall, simple, passive management of any portfolio except Suez bonds without savings from capital income was insufficient to produce even the very low wealth levels observed in 1947. This was in stark contrast to the nineteenth century, when individuals could consume much of their capital income and still expect real wealth increases.

Beyond the mechanical part of inflation, we need to consider two other issues. First, inflation raised the cost of borrowing simply by increasing uncertainty. High borrowing costs may have reduced the return to equities, not just bonds. Second, even if debt holders were the primary victims of inflation in the decade following WWI, we must distinguish between public and private debts. Unanticipated inflation primarily redistributes wealth from the private sector to the public sector by depreciating public liabilities (bonds and cash), thereby reducing private wealth. Unanticipated inflation also redistributes wealth from creditor to debtor, but total private wealth remains unchanged in the short run.

These issues played out in bleak terms during the interwar period because Parisians' portfolios before WWI were bond-heavy. About half of all bonds held by Parisians involved public debt. For bonds, inflation that is not factored into the rate of return certainly reduces private wealth. Inflation also leads to changes in distribution if these bonds are held differentially by one group or another. In 1912, every fractile held about 15% of its assets in government bonds, suggesting inflation's primary effect was to reduce wealth at every level. However, this intuition is not quite correct. Below the top centile, individuals were poorly diversified (see chapter 2). Some held bonds and suffered, while others favored equity, and their wealth was far less at risk. Inflation led to a rise in relative standing for those who held few or no bonds and a decline in relative standing for those whose portfolios were primarily in bonds. This churning inside the wealth distribution can only be understood if we take into account the fact that no one holds his or her fractile of the distribution average portfolio.

The same logic applies to the debt obligations of individuals and companies but with an additional twist. Private bonds are in zero net supply in the economy (someone owns every bond, and someone else owes the bond). Consider for a moment private peer-to-peer mortgages. When the value of a lender's claim falls because of inflation, their borrower's wealth should go up by the same amount. The same logic applies to securities issued by a firm. When the real value of the bonds issued by the firm falls due to inflation, the value of its equity should rise. Someone who holds the right proportion of bonds and stocks of that firm should see no change in his wealth. Again, this neutrality holds in the aggregate, but inflation favors debtors and holders of equities over bondholders at the individual level. In the longer term, however, inflation may have nefarious effects by increasing risk and, thus, the willingness of potential creditors to place their wealth in bonds.

The reverse phenomenon applies to all individuals who had accumulated long-term fixed interest rate liabilities on the eve of WWI (*rentes* or bonds with maturities of more than a decade). Consider the case of the national mortgage bank, the Crédit Foncier (CFF). In the decade before WWI, more than 80% of the CFF's loans and 92% of the capital were in self-amortizing mortgages with a fixed coupon and final maturities of at least 60 years. As a result, nearly all the loans issued between 1910 and 1914 were outstanding at the beginning of the conflict. Over those five years, the CFF had made 5,600 loans worth 449 million francs; it issued bonds to fund the mortgages. We assume a 4% interest rate and 60-year repayment schedule to keep things simple and compute the amount outstanding each year. In early 1915, at least 408 million were still outstanding. By 1947, the nominal value of the outstanding loans was 264 million, while its real (1912) value was only 3.6 million francs. CFF borrowers passively realized a gain of at least 260 million, while CFF bondholders had lost the same amount. When we think of aggregate wealth, this transfer does not matter, as what is gained by one side is lost by the other. The CFF itself was largely insured from the losses, though the revenue from its slice of the interest payments also shrank to next to nothing.

We evaluate the impact of inflation from 1914 to 1947 on portfolios in two steps. If inflation had been the only problem between 1914 and 1947 and individuals ceased buying bonds after 1914, then the bond positions would have melted away. The 33.6% position in bonds in 1912 with no reinvestment would have led to a 1.11% share of bonds in wealth by 1947. Surprisingly, as late as 1937, bonds were 23% of Parisian wealth, and in 1947, 7.6%. These levels suggest that individuals kept buying bonds throughout the period even though inflation was higher than the interest rate both during WWI and in the 1940s.

The second effect involves the impact of inflation on private wealth. There, we need to take into account that about half of those were government bonds. In terms of gross wealth, whether the bonds are private or public does not matter, and gross wealth should have fallen by about 31.9% as bonds were wiped out by inflation. In terms of net wealth, the effect would have been less because inflation redistributed wealth from lenders to borrowers (seen as a decline in liabilities). So, only the losses on public debt should be counted. These were about 14.25% of the 1912 portfolio.

In terms of gross wealth the losses are very large indeed, and even in terms of net wealth, they approximate the impact of large stock market crashes. The distinction between gross and net wealth helps to remind us that these catastrophic periods engender complex dynamics within the distribution of wealth. The younger half of the wealthy population is more likely to be indebted and thus to benefit from inflation, while the older half is more likely to hold bonds and to suffer. We return to these distributional issues in chapter 5.

4.6. Destruction

Parisian real estate's share in Parisian wealth was 25% in 1912, and it dropped to 11% in 1947. Our real estate index adjusted for inflation shows a decline of 87% from 1912 to 1947. This decline is a bit less than the aggregate fall of wealth, so we would have expected the share of real estate to remain roughly constant. Its decline reflects a variety of forces. Let us first turn to the losses that occurred from the physical destruction of property during the two world wars (the Depression was, of course, not a factor here). Scholars have documented the toll of the wars on real estate. Relative to areas where actual fighting occurred, or which were targets of strategic bombing, the capital got off relatively easily as far as destruction goes. One way to document the losses is to look at records of compensation requests for damaged property for both wars now held at the Paris Archives.

During WWI, the trench lines at the front were close enough to Paris that the German army could shell Paris, yet they were far enough that the number of shells that could be fired was small. After the war, the French government allowed individuals to make claims to receive compensation for damages incurred from bombardments. As a consequence, some 7,900 individuals or companies filed for damages at 4,790 street addresses.[4] The city itself also filed 439 requests for compensation for damage to its buildings and streets at 457 locations. Other sources make it clear that few buildings were actually destroyed by shells from the Big Berthas or bombs dropped from airplanes.

For WWII, an official tabulation completed on July 1, 1944, put the number of buildings seriously damaged or destroyed at 248, less than 10% of the destruction in the département of the Seine as a whole.[5] As was true for WWI, after 1945, the government created a compensation fund for war damages. As a result, 8,323 compensation claims were filed by individuals and firms who had suffered property or material losses from the war—1,430 claims involved real estate damage to 743 addresses.[6]

Before the nineteenth-century fortification lines were dismantled in the 1920s, Paris had counted about 90,000 parcels (each could include more than one building), so at most 5% of the buildings were hit by bombs in WWI and a trivial number in WWII. The indemnity requests tell the same story: the 10,000 requests for compensation from war damage concern about 1%

of the data. Building destruction and theft by occupying forces in Paris were not a major source of wealth decline during catastrophes.

To the extent that Parisians' real estate wealth declined, it is more likely due to institutional changes that reduced incentives for property owners to maintain the quality of the housing stock.

One possibility involves changes in ownership structures that created commons' problems for maintenance. Until the 1920s, nearly all Parisian real estate assets were held as complete buildings by specific individuals and their spouses. Then, the interwar period saw a rise in the number of privately held real estate companies (*sociétés civiles immobilières*, or SCI) that owned multiple buildings. These companies' functions were tax avoidance, as individuals could declare their equity at nominal value. Published lists of real estate owners in Paris (Mémento-Guide du Propriétaire 1951–1952) show that this transformation was slow in coming. Even in 1951, SCI accounted for less than 5% of all buildings held directly by individuals (i.e., excluding institutions and organizations). The larger transformation involved the transfer of property to the public sector to build out public housing in the interwar and postwar periods. Because the city and the state issued bonds to fund the purchases, these market transactions reduced the portfolio share of real estate. Finally, in the aftermath of both wars, rent control measures that remained in place for decades directly reduced the returns to investment in real estate quality. Unlike the war destruction that had a huge impact on a small number of buildings, rent control measures mattered to many properties, and their intensity depended largely on the realized inflation during the period they were in place.

That the share of real estate in wealth fell throughout the first half of the twentieth century probably had little to do with physical destruction. Rather, it reflected various state policies that imposed rent control for subsets of apartments and a transfer of property in Paris from the private to the public sector as warehouses and factories were turned into public housing and administrative buildings. Finally, in periods of high inflation like the 1940s, the inheritance tax authorities' method of using rental value to infer values for individual buildings led to systematic biases. Indeed, leases had been agreed to sometime earlier, and their value was less than if they had been negotiated very close to the time the owner died.

Outside Paris, both wars led to extensive physical damage. During WWI, the départements where the trenches ran and those in the east and north that Germany occupied suffered extensive damage. Piketty and Zucman have estimated the destruction caused by the war (with negative rates of investments). Their real national wealth series falls by 42% during WWI and 53% during WWII (Piketty and Zucman 2014, table FR.5a col. 8). This includes dramatic losses due to war destruction in some years and negative investment rates of −25%, partly compensated during WWI by positive investment to help the war effort. These huge declines left wealth per capita in 1947 at about one-third of what it had been in 1912. Another source to estimate the war's destructive toll on property is the wealth at death totals compiled by the estate tax administration for France as a whole. By 1947, it falls in real terms to 20% of what it had been in 1912. On a per adult decedent, wealth falls to 22% of its prewar totals.

The costs of this destruction were borne in complicated ways because of a wide variety of programs set up by the French government during and after both wars. For example, by the time the trench lines stabilized at the end of 1914, nine départements were under German control, including some of the most industrialized areas of the country. As a consequence, France launched nationwide investment programs to rebuild the industrial infrastructure necessary for the war effort. These were funded with low-interest loans and were sufficiently successful that, during WWI, there was an attempt to tax wartime profits. After the war, the government also set up a series of subsidized reconstruction programs to rebuild the "devastated départements" and Alsace-Lorraine. While some of that money went to churches and other monuments, most went to industrial and mining firms. Rich Parisians who had had significant investments in the northern départements benefited from the reconstruction efforts. However, as holders of vast amounts of national debt, they also paid in part for the effort as these bonds lost value due to inflation.

The extent of this redistribution is apparent in figure 4.8, which displays the value of bequests divided by the number of adult deaths for the Seine, the nine war-torn départements, the rest of France without the Seine, and France as a whole in the years before and after WWI. All inheritances are scaled to the average value of wealth in France, not including the Seine. Before WWI, the soon-to-be war-torn départements were richer than the others by about 20%. After WWI, the average wealth per adult decedent in war-torn départements converged to the French average and remained that way until WWII. Given the extent of destruction along the trenches and the clear lack of diversification of most individuals' portfolios, one would have expected a far steeper wealth loss before 1925 than we observe. The wealth equalization across France shows how much the state pushed resources to rebuild the war-torn départements. The more striking phenomenon

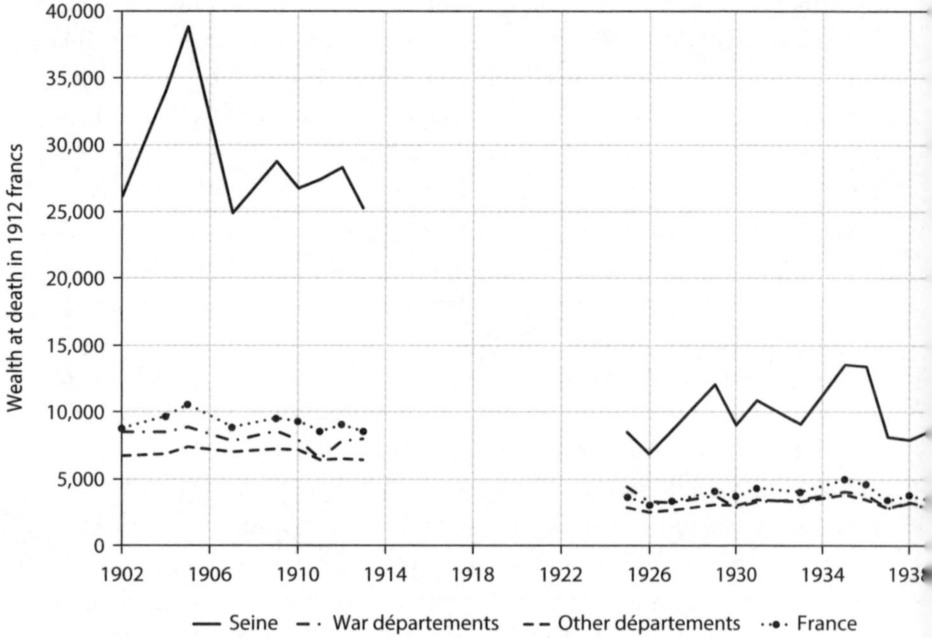

FIGURE 4.8. Average wealth per adult decedent in different parts of France in 1912 francs. *Source:* BSLC estate tax by département reports (Piketty 2001, 749).

in figure 4.8 is the extent of the drop in the Parisian wealth advantage, which happened before the drop in inequality that followed the Great Depression. For most Parisians, in the short run at least, reconstruction programs were a net drain on wealth. Indeed, they owned very few real assets in the war-torn départements, though they had invested in many important companies like the Chemin de Fer du Nord, coal mines, and textile plants in the northeast.

So far, we have shown that bonds, real estate, and most publicly traded assets failed to keep up with inflation. One key element to wealth formation has long been business formation, particularly privately held enterprises. Of the three shocks of the period, the Great Depression comes to mind most as a serious adverse shock on small- and medium-sized Parisian enterprises. The heavy toll of WWI on working-age men, however, cannot have helped business formation in the 1920s. It surely also raised many complications for family firms that lost one or more heirs in the wars. The despoiling and murder of Jews in WWII also cannot have helped recovery after 1945.

There is no direct measure of the extent of business activity in Paris, but many series offer glimpses of what happened to enterprises between 1914 and 1947. The first involves multi-owner enterprise formation. In France, these data cover corporate and noncorporate (partnerships and the like) businesses. After a sharp contraction during WWI, the number of new firms in Paris rose sharply in the decade following 1918. In this series, as in other markers of economic activities, the 1920s seem to be a period of recovery, much like what Paris experienced after the Commune. Unfortunately, the recovery from WWI was halted by the Great Depression. Indeed, in the decade after 1929, the number of new firms fell by 20% and did not recover until Paris was liberated in 1944. The next two years saw a dramatic surge in new firm formation. A broader series comes from the number of entities paying the business tax (*patente*) in Paris; over the long run, that number grew steadily from 175,000 in 1900 to 244,000 in 1928. Yet despite the overall positive trend in these decades, the number of businesses liable for the *patente* was extremely sensitive to war. There was a 25% decline in the number of establishments paying the *patente* during WWI. Whatever recovery had occurred up to 1928 was undone during the Great Depression, and by 1937, the number of businesses paying the tax was the same as it had been in 1912 (199,000). WWII then led to another 25% decline in the number of establishments paying the *patente*.

The death of enterprises can be tracked through three different series. The first is the number of multi-owner firms dissolved by agreement among equity holders. This series does not show much change over the decade that followed the crash of 1929, though it has a sharp spike from 1933 to 1935. We also have the number of judicial liquidations and bankruptcies in Paris from 1880 to 1940 (these series do not distinguish between personal and business insolvency). Both of these series show an absence of activity during WWI and a resumption of bankruptcy proceedings in 1919 as businesses that had limped along during the war were shuttered. Then, there was a substantial and long-lasting increase in the number of insolvencies in 1927 and again in 1930, both anticipating and following the stock market crash of 1929. One last series is available for Paris: the number of private businesses (*fonds de commerce*) sold in Paris. The number of transactions was high only before WWI and during the roaring twenties. During both wars and the Great Depression, the number of transactions was only about half what it was during the better times. The great shocks interfered with the formation and transmission of businesses in ways that are quite likely to have had long-lasting effects.

Given the extraordinary range of capital and employment in different businesses, these counts leave something to be desired as small enterprises dominate them, and it is known that in bad times some businesses are formed by former employees of firms that have either shuttered or shed labor. These businesses are likely to be small. Still, when we look at the value of businesses sold in 1912, it was about 12,000 francs, meaning the average owner had to be in the 85th percentile of the wealth distribution to afford such a purchase. Prices fell by 25% during WWI and recovered rapidly after the war, peaking in real terms in 1930 (about the ninetieth percentile threshold of wealth). Then, the value of private businesses began to fall. It continued to do so in real terms until a few years after the end of WWII. There is no doubt that business losses were steep and contributed to the decline in wealth. In real terms, businesses lost two-thirds of their value over the terrible decades from 1912 to 1947. However, this dismal performance still makes them one of the better inflation hedges relative to bonds, real estate, and many publicly held stocks, which fared far worse.

No such detailed data exist for businesses in France as a whole. The tax system changed in 1917, and the *patente* became a local tax, replaced nationally by a tax on business profits. Only Paris introduced a tax on business transfers like that for the *fonds de commerce*. We have access to bankruptcy filings for France and new multi-owner firm formation and dissolutions. Even though national series are interrupted during wartime, they tell a similar story. As for Paris, we see large drops during the crises followed by recovery.

Beyond business losses in Paris or France, the interwar period featured a rate of unprecedented financial destruction because of regime change in some countries or the extraordinary difficulty of dealing with the debt overhang that followed 1918. A large share of financial investments was destroyed, even leaving aside the bonds that melted away due to inflation. The most notable were foreign investments in places like Russia, Central Europe, and the Ottoman empire, which lost nearly all their value after WWI.

One set of investments suffered a complete loss—claims on private and public organizations in Russia following 1917. To avoid complex border issues, we look at the share of total wealth invested in companies whose principal business was in Eastern and Central Europe or Russia in the estate tax data. The share of Parisian wealth in Russian assets had a local maximum in the first decade of the twentieth century when it reached 3.6% of the value of assets; it then experienced an abrupt fall to 1% in 1922 and then slowly declined to a tenth of that value before disappearing from the records after WWII. If we look at foreign assets only, Eastern Europe represented 15% of the Parisian portfolio in 1907 and 1912, dropping to 5% in 1922 and less than 1% by 1932. These are huge and rapid declines, but they concern a tiny part of the total wealth of Parisians. A similar story can be told for the Middle East, where the decline started after 1922. The Russian revolutionary government's decision to repudiate its debts in February 1918 looms large in financial terms because it was radical. It was also politically expedient for French officials to blame foreigners for losses in the near term, even though French tax and monetary policy were engineering a much larger transfer of resources from private capital into the public sector.

4.7. Nationalization

A final piece of the loss of wealth puzzle relates to the impact of transfers of private equity to the public sector via nationalizations in France. When considering a period of wealth decline and in the shadow of both the 1917 Russian revolution and the whole dispossession of Jews in WWII, the reader may jump to the conclusion that nationalizations reduced private wealth. This section outlines why such hasty conclusions are unwarranted. Indeed, one motivation for nationalization can be to bail out bondholders in failing businesses. In this case, the nationalization of a business at market prices should leave equity holders indifferent and avoid losses to the bondholders. The losses to investors would come not from the price at which the state acquired the firms but from later inflation that devalued the bonds issued to pay for them.

In France, bailout nationalizations began in the nineteenth century, and they would reach their apogee after WWII. In the early twentieth century, the state took over the Chemins de Fer de l'Ouest, which went bankrupt and was bought out by the state in 1909. Then, in 1937, following massive losses at the remaining private companies, the state formed the National Railway Company (SNCF) (Jones 1984, chap. 3). In return for assuming the companies' debts, the SNCF received a majority stake in the five major railway networks (Est, Nord, Midi, Orléans, and PLM). A full nationalization occurred after WWII. Railroads were a very large share of all financial investments we observe in the inheritance data (more or less 20% of all financial investments, and French railroads were about half that total), so changes in those assets' value mattered. Because the bonds of the SNCF were traded, we can estimate the value of both equity and bond investments. By 1947, shareholders had lost at least 98.2% of the 1912 value of their investments, while bondholders had lost 98.4%.

In this light, there was little to choose from for long-term wealth preservation between the expropriation of Russian railroads in 1917 and the nationalization of French ones in 1937. However, while both ended poorly for private investors in 1947, the two events are, in fact, quite distinct. First, between 1917 and 1947, bondholders and shareholders received interest payments, while those who held Russian securities lost everything. Second, after nationalization, investors in French railroads had the option to rebalance their portfolios, which was impossible in the Russian case.

French railroad shares look so much like bonds because, after WWI, that is effectively what they were. The reason lies with the specifics of the concession contracts drawn up by the railroads and the French government in the mid-nineteenth century. To accelerate the build-out of the French network, the government guaranteed interest payments on the bonds and minimum dividend payments on the equity of the railroad companies through the end of the 99-year concession. The state reserved the right to claw back guarantee payments when the railroads were profitable. If railroads can, on average, make more than the minimum dividend payments, then the equity is priced like classic shares. If, however, there is a sufficiently long period of losses that are transferred to the state through the guarantees, then shares behave like bonds, and the government is the equity owner in practice if not in fact. That was the situation in which France's Front Populaire government found itself in 1937. Rather than take over the companies, it created an equity-for-debt swap and effectively turned the railroad shares into bonds. By 1944, when equity quotations stopped for most companies, investors had lost 94% of their investment relative to 1912 prices. Inflation between 1944 and 1947 further ate into the value of railroad assets.

An alternative interpretation of equity would say that the railroads were insolvent by 1937 and the equity should have just been wiped out. To be sure, assuming only the debt would have been better for the government. But even the leftist Front Populaire respected the contractual terms that bound it to the railroads, and as long as the guarantee clauses were observed, the state had to carry out the conversion. In sum, the nationalization itself is not the culprit of the wealth loss but rather the realization of a transformation necessitated by capital losses incurred between 1914 and 1936.

A second wave of nationalizations occurred at the end of WWII for many different reasons (Andrieu et al. 1987). In the case of Renault, the state assumed the ownership of a company that had visibly supported the Nazi war effort, while the rest of the automobile sector remained in private hands. The financial sector and major insurance companies were nationalized for a combination of reasons, including problems of solvency and a desire by the government to control investment flows during reconstruction (Andrieu 1993). The coal industry, electricity, and natural gas were nationalized because the sectors needed massive new investment that only the state could provide, and the state wanted to rationalize these sectors (we return to these issues in chapter 6).

The last transfer from private to public wealth arose in real estate with an effort to create affordable housing in the capital. Although these efforts sowed the seeds of much more massive public housing programs after 1945, in our period, Paris planned to build 62,000 dwelling units as low-rent housing (we can only be sure that 15,000 were built by 1931). These are tiny numbers compared to the capital's million dwelling units. The more important measures affecting housing in the city from 1914 to 1948 were severe rent control measures, which reduced the need for the state to reevaluate a variety of pensions. Overall, between inflation and taxation, the interwar period produced a massive redirection of capital income from private hands into the public purse.

4.8. Conclusion

What, then, to make of all these negative shocks? None of them individually can explain the aggregate decline in wealth. The dramatic decline in wealth at death resulted from not one catastrophe or one fiscal policy but the relentless erosion of capital returns over three and a half decades. The 1920s, which followed a huge wealth decline during WWI, saw a 42% increase in average real wealth. Continuing recovery at this rate for another two decades would have led average wealth to return to its 1912 levels around 1947. Of course, we know that the Great Depression and WWII intervened and that a further and more massive decline ensued. Similarly, if we look at specific causes, like inflation or taxation, none alone explains more than a quarter of the decline in private wealth.

The list of adverse events we have covered is partial and could easily be extended. Most notably, it is tempting to attribute part of the decline in reported wealth at death to the rise of foreign tax havens after WWI. Cummins (2022) estimates foreign wealth holdings as a residual from a counterfactual accumulation model. He finds substantial missing wealth for the very rich in England even before WWII, which he argues was held in offshore accounts. In appendix 3, we show that offshore investments were not quantitatively important before WWII but were more so after 1970. Still, we are dependent on scarce evidence on offshore assets, so our estimates are lower bounds. Cummins's approach suffers from the reverse problem, as it attributes all the differences between observed and counterfactual wealth to tax evasion. Nonetheless, we use a similar procedure to produce an upper bound to tax evasion. Let us start by ignoring taxation and take an average 1912 portfolio with one-third real estate, one-third equity, one-third bonds, and cash equivalents (close to the actual distribution of assets). If we attribute the nominal price changes for each asset stream, the real value of the estate in 1947 is 6% of its 1912 value. So, bequest-motivated savers who received an average inheritance in 1912 could not produce an average bequest in 1947 if they consumed all their capital income because real wealth in 1947 was 18% of what it had been in 1912. To generate an observed wealth in 1947 that is 18% of its 1912 level, individuals need to add 3.1% of their wealth each year from savings capital or labor income. As we shall see in the next chapter, average flow returns to wealth are between 3% and 4%. So, Parisians needed to reinvest most of the capital income from the inheritance they received to reach the observed level of reported wealth. The computation suggests that there was very little capital income available to send to Switzerland or other tax havens. In other words, there was little missing wealth during this period.

One might object that the flow returns are too low, that income from foreign assets (e.g., US or Scandinavian) was much higher, and that privately held businesses held their value much better than publicly traded equities. This may all be true, but the reader must remember that the computation above ignored taxes, which reduced income flows that could be directed to build wealth. That average wealth in Paris was 18% in 1947 of what it had been in 1912 is testimony to massive efforts at savings toward bequests that would be legally declared in France.

This chapter has examined the evolution of aggregate wealth and, thus, the average wealth of a representative Parisian. However, we know that individual portfolios looked nothing like the average, that individuals at different levels of wealth had different investment approaches and different returns and faced different marginal and average tax rates. It is thus not surprising that the first half of the twentieth century saw major changes in wealth inequality, the topic of the next chapter.

5

Inequality When Wealth Collapses, 1912–1947

As we saw in chapter 4, our data show that wealth at death per capita and the wealth of the living in Paris dropped dramatically between 1912 and 1947. That drop erased nearly all the gains accrued since the First Empire (1804–1815). The decades marked by the two world wars and the Great Depression were replete with negative shocks. Not surprisingly, scholars, including Thomas Piketty, have identified this time as the great period of wealth equalization. Others like Walter Scheidel (2018) have also argued that negative shocks have long been the primary source of wealth equalization. Moreover, many have made the case that difficult times are moments of social renewal in frames that are not very different from Schumpeter's arguments about creative destruction (Schumpeter 1950). In this view, mobility is heightened because one group's decline makes way for another's upward rise. Our data confirm that inequality dropped in Paris from 1922 to 1947, though if it came at the cost of the huge wealth decline, it seems to have been a very expensive and limited leveling. In fact, as we saw in chapter 4, wealth did fall dramatically in this period. In real terms, wealth per adult decedent fell by 76% from 1907 to 1947. As we shall see here, the top 1%'s share of wealth fell from 65% to 50% over those decades. In this chapter we will explore the relationship between wealth decline and the fall of inequality. While it is common to link the two, the evidence that the wealth drop itself was directly responsible for reducing inequality is, at best, circumstantial. Instead, we privilege fiscal innovations as the primary forces reducing wealth disparities. As we shall see, the largest absolute drop in wealth occurred with World War I, yet the 1922 distribution of wealth was as unequal as any that came before. The individual-level dataset for Paris allows us to measure how much the dramatic drop in wealth reduced inequality and the extent to which mobility increased.

The chapter begins by quantifying how much wealth inequality changed between 1922 and 1947. In section 5.1, we examine coarse statistics like the Gini coefficient and the fraction of individuals for whom heirs produced a tax filing and then detail the evolution of the wealth share of top fractiles. For the first time since 1807, the wealth share of the top 1% started to drop steadily, beginning in 1927 and continuing to 1947. We then turn to two standard explanations for changes in wealth distribution. In section 5.2, we discuss the two variables that push inequality up or down. Differential returns, on their own, in the long run, create inequality, while progressive taxation, on its own, tends to reduce disparities. In section 5.3, we examine how shocks shaped wealth mobility because of poor diversification. We conclude that while negative shocks played a critical role in the overall fall of wealth, as discussed in chapter 4, fiscal innovations were the primary driver in reducing the wealth share of the top 1%. The race between taxation and return produced a decline in relative wealth for the top wealth holders, as documented by Piketty (2001). Section 5.4 is devoted to a contrast between the top 0.1%, whose wealth share was essentially unchanged between 1912 and 1947, and the rest of the 1% who experienced an important decline in wealth share. Section 5.5 is devoted to the second key change of the interwar period: the rise in the share of individuals who were wealthy at death. Before WWI, less than 30% of decedents were wealthy enough to have a taxable estate. By 1947, almost half of all Parisians who died had a taxable estate.

Overall, we argue that the evolution of the top 1% (the very wealthy) and the rise of estates below the top quartile of the distribution were largely driven by independent phenomena. For the very wealthy, the competition between returns and taxes mattered most. A different dynamic was at play for the smaller fortunes that were never heavily taxed: increasing life expectancy and mandatory retirement programs increased savings to the extent that an ever-larger share of the population had a taxable estate.

5.1. Shocks and Inequality in Wealth at Death

Figure 5.1 presents the Gini coefficient computed for each cross section in our datasets of deceased Parisians; it also shows average wealth at death and the top 1% wealth share from 1807 to 1947. Relative to the other two series,

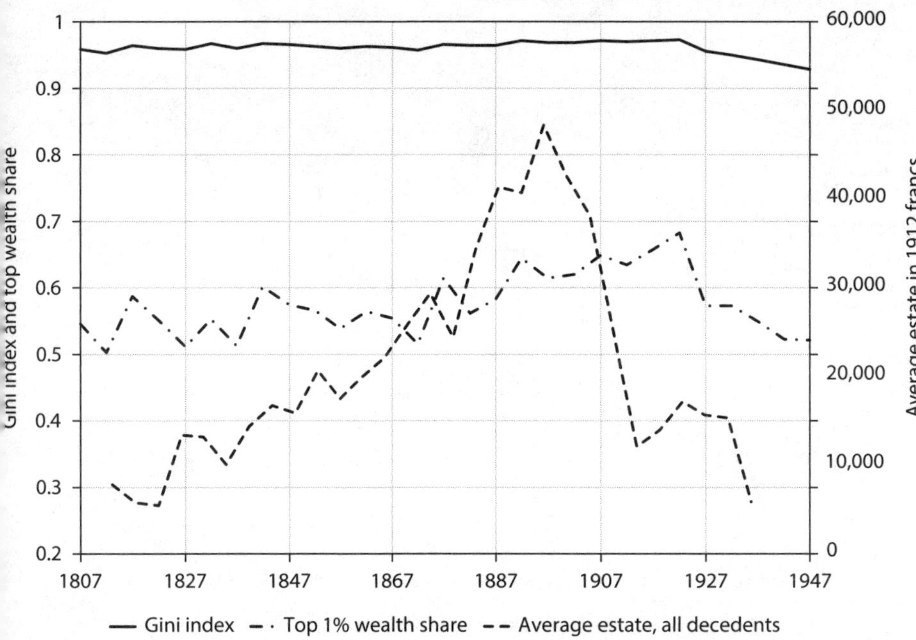

FIGURE 5.1. The Gini index and wealth at death. *Source:* Estates dataset (see appendix 1).

the Gini coefficient seems remarkably stable even though it does fall from a high of 97.3 in 1922 to 93.2 in 1947. Three observations about the Gini's value are in order. First, a century of low taxes and wealth accumulation leaves the Gini almost unchanged (it rises from .96 to .97 between the 1810s and the 1910s). In Paris, at least, average wealth per decedent understates the growth of aggregate wealth as the population of Paris had more than doubled, and so had the annual number of decedents. Thus, it is clear that in this dimension, aggregate wealth growth does not lift all boats. Second, after WWI, a trend sets in that leads to a decline in the Gini, just as wealth goes into a tailspin. Those who believe in the leveling effects of large shocks will at least find some comfort for their argument here. Finally, inequality is always very high (a Gini of .93 can be obtained, for instance, when only 6% of the population has any wealth and it is equally distributed among the rich, but also with a quite different distribution like the one that prevailed in Paris). Thus, variation in the Gini coefficient is, at best, only a start in the measure of inequality.

Figure 5.2 displays the fraction of the population that left a positive estate. Like the Gini coefficient, this measure of inequality is stable throughout the

FIGURE 5.2. Wealth at death and the fraction who die rich. *Source:* Estates dataset (see appendix 1).

nineteenth century. It remains stubbornly below a third and does not begin to rise until after WWI. When change starts, it is swift. The fraction of the population that dies wealthy rises from its near all-time low of 24% in 1922 to nearly 50% in the 1940s. Thus, this second extremely coarse statistic confirms the failure of the first golden age of capitalism (1860–1914) to boost the share of individuals who left bequests. It also confirms the notion that there was an important change in the distribution of wealth that followed WWI. At the same time, it raises questions about why the number of bequests rises in a period so adverse to wealth formation (negative returns and increasing wealth taxation). One might suspect that the "new" rich had little wealth. French inheritance taxes bind at the first franc, and the heirs had to file an estate tax document to receive any financial or real asset. As a result, individuals with life-cycle savings, including those who received pensions from WWI, got caught in the state's tax net. However, as we shall show, the increase in the fraction of individuals who died wealthy was not due simply to the rise of a new class of retirees whose sole wealth was a retirement account.

Precisely because we have the entire population of decedents each year, we can accurately reconstruct the evolution of the wealth-at-death

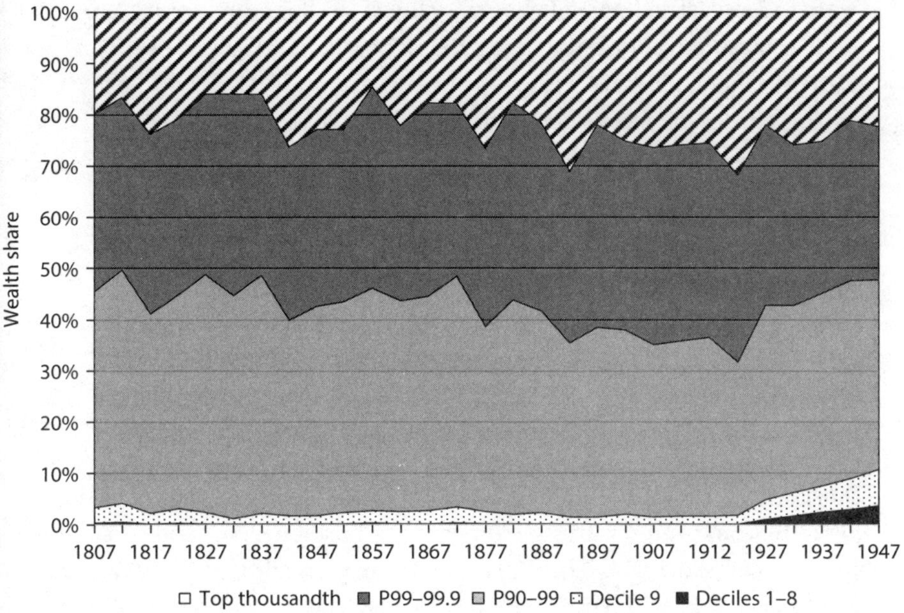

FIGURE 5.3. The distribution of wealth at death in Paris, 1807–1947. *Source:* Estates dataset (see appendix 1).

distribution. The only caveat is that since everyone only dies once, and since our wealth data are repeated cross sections, it is, in fact, extremely unlikely that we will observe the same estate in two adjoining cross sections. Figure 5.3 displays the distribution of wealth by top fractile. Before 1927, the history of wealth at death is the history of the top decile wealth holders (who are shown as the top three areas in the graph). That top group never accounts for less than 96% of the wealth left to heirs in any cross section. Within the top decile, the split between the 1% (the top two areas) and the rest of the top 10% is progressively more favorable to the top 1% since the rest of the top decile drops from 45% of the wealth to about 30%. In the first two decades of the twentieth century, the 1% wealthiest decedents accounted for 60% to 65% of all wealth bequeathed. In the stratosphere of the top 0.1%, wealth was remarkably high as the thirty-odd members of this group commanded 25% of the wealth left behind. As for the Gini and the share of individuals who die with nothing, 1922 marks an apex for wealth concentration. Then, the path of concentration reverses, and the top 1% share drops back to about 50%, a share not seen since the 1870s. Conversely, the rest of the top decile jumps from 30% to 38% of the wealth by the end of WWII. The sixth, seventh, eighth, and ninth deciles, which had all been irrelevant before WWI, begin to increase in importance. These lower deciles jump from 2% to 10% of the wealth. Without anticipating chapter 6, this marks the most dramatic change in inequality that we have seen in Paris.

This chronology leads us to return to the three great shocks of the twentieth century (WWI, WWII, and the Great Depression). In real terms, WWI led to the largest absolute fall in average wealth, from about 40,000 to 15,000 1912 francs. Yet despite the massive drop in wealth, WWI had little impact on inequality. The fraction of individuals who died wealthy was essentially unchanged; the average wealth share of the top 1% was 63% in 1922, the same as in 1912. The top decile share was 98% in both years. To some extent, this is a feature of the way we measure wealth at death—before inheritances are assessed. Most estates filed for decedents in 1922 and 1927 were last transmitted in the late nineteenth or early twentieth century when taxes were low. Thus, the wealth we see in the 1922 and 1927 cross sections had yet to suffer any reduction from progressive inheritance taxation. A second reason for the persistence of high inequality is that the main wealth loss of WWI came from bonds hurt by inflation, and bonds were the most widely held assets. It is not surprising that inequality was relatively insensitive to the Great War. The next two shocks (the Great Depression and WWII) combined to bring average wealth down to 7,000 1912 francs, and top wealth shares fell substantially. Overall, the period between 1914 and 1947 is the one that sees the most dramatic changes in wealth at death over the whole period of this study. On one level, the decline in inequality is small compared to the wealth drop of nearly 88% for the inheritance flow and 82% for the average decedent. In contrast, the wealth share of the top 1% falls by 16% (from 64% to 54%). From another perspective, a reallocation of 10% of the wealth to individuals below the top 1% is substantial, as the wealth share of all individuals below the top 5% was rarely more than 10% before 1947. There is also a dramatic change in the fraction of individuals who left bequests. Starting in the late 1920s and continuing to the end of our cross sections, we find wealth beyond the top two deciles for the first time. What is most surprising about these new estates is that they came about when the primary innovations affected life-cycle savings through the rise of mandatory pensions and not intergenerational transfers.

5.2. Competition Between Fractiles: Taxes and Returns

In chapter 2, we argued that, in a low-tax environment, it was not surprising that wealth was very concentrated and that bequests were concentrated among individuals who themselves had received sizable inheritances. The lack of fiscal redistribution meant that dynasties could transmit their wealth through time at a low cost. Further, the observed portfolio returns were increasing in wealth, which meant that the saving rate an individual needed to keep up with wealth growth was less on average for the very wealthy than for the middle rich. The first golden age of capitalism's return and tax regime allowed the very rich to have their cake and eat more of it than anyone else. After 1914, the advantages of the very rich dimmed. We saw in chapter 3 how the estate tax, in the long run, drove a sizable wedge between the deceased's wealth and what their heirs received. The interwar period was, in many ways, a transition: most of the estates had been received during the low-tax regime that prevailed before WWI, and only their heirs faced the higher taxes. In chapter 4, we saw that taxes reduced the after-tax income of the very wealthy. This section returns to this issue to understand how the fiscal and economic shocks of this period altered the distribution of wealth. Perforce, there is a bit of repetition in the exposition.

The higher returns of the top 1% wealth holders in the nineteenth century largely came from the much smaller share of assets that produced no capital income (furnishings, cash, bank accounts) in their portfolio compared with the average less wealthy decedent. This difference would be more significant if we could introduce risk premia for each portfolio because it is clear that more affluent individuals held riskier assets. We leave these issues aside because the difference in returns is quite substantial without detailed risk premia.[1]

For instance, the wealthiest 1% of Parisians were much more likely to hold emerging market assets with higher risk and higher returns. So, our return differentials are always downward biased. All these patterns persist well beyond WWI. They are a fundamental component of how individuals of different wealth make investment decisions. To measure differences in rates of return, we continue the strategy we used in chapter 2, dividing each portfolio into four categories: real estate, equity, bonds, and a composite category of non-income-producing assets. This is enough to account for the key portfolio pattern: the share of high-yield assets (shares and real estate) grows as wealth increases. This phenomenon is not particular to the period under study and is the persistent force behind the correlation between returns and wealth. This strategy allows us to produce a flow return to any individual's portfolio upon which we can assess income taxes and thus compute pretax and posttax income.

We compute flow returns as the trailing five-year average for each cross section, which an investor could have computed. Aficionados of rational expectation can use the next observations as the five-year forward-looking average.[2] Both before and after WWI, bonds had the lowest yields, followed

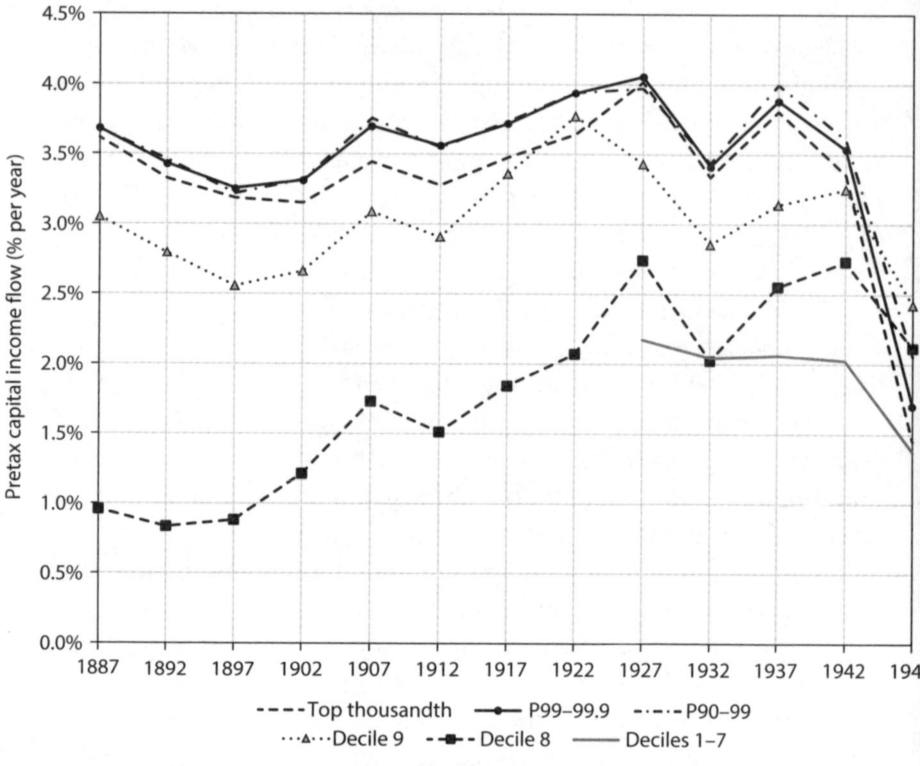

FIGURE 5.4. Flow returns by wealth fractile. *Source:* Estate tax dataset.

by equity (French shares had relatively high mandatory payouts to shareholders), and real estate was locked in at a yield close to 5%, given the data used to compute it by Eichholtz et al. (2021). As noted below and displayed in figure 5.4, the top 1% generally enjoy a significant (more than 100 basis points) pretax return advantage over everyone else. The very top has a lower rate of flow income because it is heavily invested in equities (rather than real estate) relative to the rest of the top decile. The lower deciles are generally ordered so that the poorer deciles have lower returns. The only deviation from this pattern arose during WWII (which appears in the 1947 returns) when dividend yields plummeted, and the upper fractile did less well in flow income than the relatively more bond-heavy middle class. From 1922–1947, inflation ran at almost 10% a year. The use of the printing press to cover government deficits hit everyone, but those portfolios with more equities were better protected than those with more bonds or cash. Thus, inflation was less costly to the top 1% than the rest of the population. Of course, with inflation running at 10% on average, all real returns were negative, which was an important source of the decline in wealth. Less negative returns, nevertheless, are better than more negative ones. Based on flow returns alone, we would have expected little change in the distribution of wealth over this period, even if wealth declined severely. Thus, changes in the distribution of wealth were primarily driven by the rise of effective progressive taxation. Adjusting the returns for the proportional taxes in no way changes these conclusions.

As discussed in chapters 3 and 4, the wars necessitated major fiscal adjustments. The estate tax and the income tax became quite progressive. As noted in chapter 3, the top marginal rates in the period were extremely high, and we see the returns of the top 1% drop dramatically. Because taxes were imposed both on wealth (through inheritance taxation) and on income (IRVM and IGR), we develop measures of capital income that consider both a particular decedent's income the year before he or she died and an imaginary heir who received the same wealth, paid the inheritance tax, and earned the resulting capital income. For both the decedent and the heir, we apply the linear rates of the IRVM on the appropriate assets, and then the IGR is applied to all the net capital income.

The impact of progressive taxation is partly cross-sectional and partly dynamic. Indeed, individuals whose income and wealth have been more heavily taxed are ceteris paribus going to die relatively poorer than those who have been less taxed. We observe this differential in the cross sections, and it drives the findings reported in section 5.1. There is also a dynamic aspect: it takes time for all the fiscal innovation to take hold. For instance, individuals who died with inherited wealth in 1922 and 1927 mostly received these bequests in the low-tax regime. We focus first on the cross-sectional aspects and reason in nominal terms.

Let us start with the super rich (top 0.1%). In 1937, the year before they died, a member in this group would have had an estate of 28 million francs or so. She would have received 1.1 million francs in gross income; the IRVM would have reduced that to 936,000 and the IGR to 604,000, for a total tax of 46%. Now consider the decedent's hypothetical sole heir. He receives the whole bequest and pays the average estate tax of the group (for more details, see chapter 3). The heir's wealth would have been just about 20 million, and his gross income would have been 801,000 francs or 72% of his forebear's gross income. After paying the IRVM and IGR on that income, the heir would have been left with 433,000 francs to spend or 39% of her forebear's income.

By contrast, take a representative member of the eighth decile from 1937; their estate was just about 20,000 francs. The gross income of the decedent would have been 1,896 francs. The IRVM would have reduced that to 1,747 francs, but they were exempt from the income tax at that level. The heir would have paid about 16% in estate taxes, leaving them with 1,588 francs of gross income and 1,462 francs of after-tax income or 77% of the forebear's gross income (total tax 23%). The state was taking a bite out of capital income at all levels of wealth, but its take from the eighth decile was just about half of what the super rich had to pay. Although the average rates at the top were very high, it is important to remember that in 1937, the average pretax income from a top 0.1% estate was almost 600 times that of an estate in the eighth decile. Progressivity cut that ratio by half. In the same year, the income of the average top 0.1% estate after all taxes was still 296 times that of the eighth decile. Interestingly, progressivity also matters to the ratio of the rest of the 1% to the top 0.1%. By 1937, the income of the average top 0.1% estate was 7.5 times that of the rest of the 1% and only five times after all taxes had been paid.

The extraordinary ratio between the average of the eighth decile and the top 0.1% leads us to another observation about Parisian society between the Commune (1870–1871) and WWI—the world of Zola and Proust. They describe, one more critically than the other, a tiny population of very wealthy individuals with extravagant lifestyles. A half-century later, individuals in the same wealth groups could afford fewer domestics. The data confirm this sharp contrast when we measure the consumption value of wealth using the Bayet wage series adjusted to a constant 2,750 hours of work a year. As discussed in chapter 2, we use this alternative measure of consumption because it reflects wealthy individuals' very high expenditure share on personal services before WWI. It also reflects the rising cost of labor services due to declines in time worked, both in hours per week and weeks worked. Before WWI, the gross capital income of the very rich (P99–99.9) decedents was about 45 times our gross labor income. Their heirs' after-tax capital income was consistently above 40 times labor income. Between 1912 and 1922, the decedent's gross income fell by two-thirds, and the heir's after-tax income fell by 75%. The decline continued, and by 1947, the gross income of decedents was only two times the average labor income, and the heir's after-tax income was about one times the labor income. The source of this radical reduction in the consumption value of wealth was, of course, a radical realignment of capital and labor income. Between 1912 and 1947, wealth at the top fell relative to the CPI while wages essentially doubled.

By our measurement, France's fiscal shocks from 1914 to 1947 were preeminent in Paris. This is because the radical inequality in Paris meant that there was a lot of wealth in brackets where the IGR was severe. Outside Paris, people were, on average, poorer, and wealth was more evenly distributed. For both these reasons, the fiscal shock was massively less severe in the provinces. Anticipating chapter 10, in 1937, 531 French estates worth more than 2 million francs were filed in France (out of 361,000 estates). These top estates represented 20% of the intergenerational flow. Now, let us split France between Paris and the rest of the country. In Paris, 268 of 13,565 tabulated declarations were greater than 2 million. These large estates accounted for 53% of the intergenerational flow. Outside Paris, there were 263 estates above 2 million, and they accounted for 9% of the intergenerational flow. These estates comprise the top 2% of estates in Paris, and they are the ones for which progressivity in direct line of descent was most severe. The conclusion is inescapable: for Parisians, the inheritance tax rate was particularly onerous. The heavy burden came in part because average wealth was higher in the capital than elsewhere but mostly because the wealth was much more unequally distributed.

Throughout the interwar period, the race between taxation and returns was won by taxation. Return differentials could not offset progressive taxation that took away more than 10% of the wealth transmitted at death and more than 25% of the income from that wealth. Even with the coarse lens we deployed in considering taxation, it is clear that effective average rates matter to both wealth levels and net incomes. It is worth noting that although our IGR and IRVM taxes are estimated (we do not see the income tax returns of either the decedent or heir), we can observe the tax payments on estates. As discussed in chapter 4, a small part of the decline in wealth we observe is likely due to tax evasion. The preceding suggests that the highest incentives to evade the tax authorities were at the top. If tax evasion had caused much of the observed decline in inequality or the loss of wealth, we would have expected the wealth drop to be proportionally largest at the top. However, the top 0.1%'s share does not drop much in the data. Instead, the rest of the 1% loses out to the lower part of the distribution.

5.3. Portfolio and Individual Mobility

The literature on inequality often assumes that very unequal societies have low mobility. Scholars have noted two key connections between extreme inequality and low mobility. First, being born wealthy is a clear advantage for someone who intends to die rich. Second, the wealthy will structure their society to protect their wealth and position at the top of the social order. In the same line of argument, prolonged periods of instability provide rare opportunities for social renewal. This was already evident during the revolutionary period when many of Napoleon's key military leaders came from modest backgrounds and ended up with exalted titles and great wealth. At the same time, many aristocratic families of the Old Régime lost wealth and power. Just how much the difficult years between 1912 and 1947 saw an increase in mobility can be measured in two ways using the sample of married decedents (recall that for this group, inherited wealth was detailed separately from community property). The first measure involves the share of married decedents who had separate assets. As we noted in chapter 2, married individuals' dowries and inherited assets were tabulated separately from community property in a separate asset account. We use the value of this personal property account to measure inherited wealth. The share of individuals with inherited assets falls only slightly—a 10% decline, from 53% before WWI to 49% between 1937 and 1947. Thus, about half the population of individuals who died wealthy had inherited something themselves. When estates were small, heirs other than the spouse, particularly children, may have taken little care to distinguish between personal assets and common property, allowing the surviving spouse to receive half the household wealth outright. From our point of view, the rough stability in the fraction of individuals with inherited assets, when the share of individuals who left a bequest was increasing rapidly, reflects a different phenomenon. At all times, some individuals who inherited something left no bequest, as they consumed their entire inheritance. When computing the inherited wealth share at death, they are unimportant because they transmit no wealth. The evidence suggests that the additional numbers of individuals with estates were roughly evenly split between self-made individuals and inheritors. In any case, the 10% decline in the share of inheritors among the wealthy is small compared to the 63% increase in the share of individuals leaving a bequest. If we look at the top 1%, the share of inheritors is extremely high throughout, although it does drop from 96% to 94.5%. We view this tiny drop as being within the error bands of the data. The rest of the top decile shows little trend, with more than three-quarters receiving a bequest. There are no substantive differences by sex: for both men and women, being rich is highly correlated with the likelihood of having inherited something.

A second approach uses regression analysis to measure the elasticity of estate size to the value of inheritances. The question is, if one increases a decedent's inherited wealth by 1%, how much does that increase their bequest? In a stable period, and if all decedents were dynastic altruists, the elasticity should be close to 1. In unstable periods, one can imagine that even a population of dynastic altruists will have great difficulty equating the wealth of its progeny with its own, and the coefficient should be much less than 1. In a period of significant wealth decline, one can imagine that the elasticity, though positive, would drop simply because our decedents are trying to protect their consumption. We thus estimated the elasticity of bequests to inheritance in a regression that also considers whether the decedent's age at death, sex, and type of marriage mattered to the size of his or her estate.[3] Age at death is always statistically significant and positive, though the coefficient drops by 25% (from 0.033 to 0.026) from before WWI to the 1940s. Neither marriage contract nor sex produces stable or statistically significant coefficients. The elasticity coefficient of inherited wealth on estate value is always highly statistically significant, but it drops by more than half from before WWI to the 1940s (0.32 to 0.14). Consistent with the earlier findings, inherited wealth increases bequeathed wealth but at a smaller rate than before. Finally, the predictive power of the regression also drops from .40 to about .27 over the same interval. All the changes occur slowly, which is unsurprising since each cross section contains individuals with different savings/dis-savings intervals. Overall, there is some evidence of increased mobility. Taken from a different perspective, seeing that more than 94% of all members of the 1% were inheritors themselves points to the fact that the circle of the very wealthy was as closed to newcomers in the 1940s as it had been in the 1890s. In retrospect, this is not surprising: even if the value of inherited wealth falls, it still makes it easier to constitute a bequest for individuals who want to do so.

When considering wealth aggregates in chapter 4, we emphasized that the high degree of diversification of aggregate Parisian portfolios meant that several fiscal and economic shocks were required to bring wealth down. When considering inequality and mobility, we need to reexamine the extent to which aggregate diversification is also found in individual portfolios. In the early twenty-first century, mutual funds have allowed individuals of modest means to replicate the market's average performance, but this is a fairly recent phenomenon. In Paris, as late as the 1970s, individuals were the direct owners of their equity and bond placements. Since share and bond values were often in the hundreds of francs, diversification was difficult for individuals of modest wealth, at least before WWI. After the war's inflation had eroded the real value of assets, large-denomination stocks and bonds were less of an issue.

However, the average portfolio of a fractile hides the poor diversification of everyone but the very rich. Because portfolios are concentrated, the individual rates of returns vary quite a lot. Return variance directly implies downward mobility for individuals whose portfolios overweighted Eastern European and Russian assets, or bonds in general, and upward mobility for individuals who favored equities or the debts of countries that remained on the gold standard. Because portfolio diversification increases with wealth, individual shocks are the largest for individuals below the top 1%. To make this case, we need to go beyond the data from estates we collected and look at asset prices to understand how asset values evolved. Fortunately, we can rely on the Data for Financial History database (DFiH) for information on all assets traded at the Paris Bourse.[4] We focus exclusively on asset prices and changes in asset values that matter for wealth inequality. We leave aside asset returns that we discussed in chapter 4. We will return to the consumption value of wealth in chapter 7.

In the difficult years from 1914 to 1947, the connection between inherited wealth and bequest might become weaker because individuals protected their own consumption at the expense of the inheritance they would transmit to the next generation. There are other underlying reasons to think that the elasticity of bequest to inherited wealth might decline. One of these is the poor diversification of most individual wealth portfolios. We detail this fact by looking at the average portfolio among the rich. Individuals in the eighth and ninth deciles have fewer assets and more concentrated portfolios than the average wealthy decedent. Individuals in the top 0.1% have more assets and are more diversified than the rest of the top 1%.

In the late nineteenth century, the average portfolio was composed of four or five income-producing assets and one or two assets that did not produce income (most often cash and home furnishings). The poorest among the rich had barely two assets, and even the very rich (top 1%) had less than 20. Continuing a trend that had begun in the 1860s (see chapter 2), over time, the number of items in portfolios grew for all wealth levels. The P80–89 (ninth decile) jumps from about three assets on the eve of WWI to almost ten by 1942. The top 1%'s average portfolio jumps from 30 assets to nearly 100. The total number of items in an estate, including assets that do not produce income, follows the same trends. These trends reflect two distinct processes. The first involves many fossil or ghost assets: shares and bonds that had lost almost all their value but continued to be listed in individual portfolios. Among many others, ghost assets include, for example, Panama Canal shares and bonds, Ottoman and Russian empire bonds, many South

African mines shares that had become worthless, the pre-WWI city of Paris or Crédit Foncier bonds, whose value in the late 1930s had been completely eroded. Alongside these older placements, there were new investments in equities. Both contribute to a growing number of assets.

Despite the increasing number of assets, portfolios continued to be highly concentrated. On average, portfolios before WWI had four assets, and their Herfindahl indexes are close to 60% (or a portfolio of two equally valued assets).[5] By the 1940s, portfolios had ten or more assets; the Herfindahl indexes only fell below 40% once in 1942. For the top 1%, the 1940s correspond to portfolios with 80 or more assets, but the Herfindahl indexes are slightly above 15% in all years except 1942. This is akin to portfolios of six assets equally distributed.

Another mark of the high concentration of individual portfolios is the share of total assets accounted for by the largest item. This was about 70% for the period before WWI, then declined to just less than half by 1942. For the top 1%, the largest asset represented, on average, 40% of all income-producing assets before WWI. It fell but still accounted for more than 25% every year except 1942.

The increasing diversification of portfolios with wealth has one important implication: the rich could afford riskier assets and, at the same time, experience less volatile returns. All things being equal, members of the richest centile were most likely to experience the average shock. For those in the ninth decile whose primary asset comprised 62% of their productive wealth, the flow returns and pricing of that asset were critical. For those who guessed right or were lucky (e.g., whose primary asset was Suez Canal gold bonds), their wealth fell more slowly than the average during the three decades before 1947. For those with large positions in Russian assets like the Krivoi Rog mines or the Société Russe de Construction de Locomotives et Mécaniques, their investments had vanished by 1930. Indeed, flow returns for both companies stopped with a last dividend in 1916, and stock prices plummeted to zero by 1930. The other assets in their portfolios would have had to perform remarkably well to maintain wealth like their luckier counterparts. This, we think, is a critical reason why the elasticity of bequest to inherited wealth was much smaller in the 1930s than in the 1890s.

To provide more specificity, we examine the holdings of Russian assets and Suez Canal bonds in Parisian portfolios from 1907 to 1947. Russian assets were the ruinous investment par excellence since the Bolsheviks repudiated all debts and nationalized all foreign equities. No compensation was paid before the fall of the Soviet Union. Nevertheless, Russian government bonds continued to be quoted on the Bourse through the 1940s (they lost 80% of their value by 1922 and almost 90% by 1932); other Russian securities fared even worse, dropping to 1% of par by 1932 and soon after they were removed from the price lists. In any case, Parisian estate documents continue to detail these investments through the interwar period. The reason, we think, is that heirs were hoping that the USSR would someday reach an agreement to resume payment on these financial assets. In this case, even if there was a substantial haircut, the only way to exercise a claim for the heir was to prove through estate documents that they had inherited the assets. Suez Canal bonds had the reverse experience, as we saw in chapter 4. Neither Russian assets nor Suez Canal bonds were a large share of income-producing wealth. They were an even smaller fraction of total wealth. Even before WWI, Russian assets never amounted to more than 4.35% of income-producing assets (4.5% of all assets by number), and that share plummeted after 1917 to 0.12% (2.71% in numbers). Suez Canal bonds and shares were sparser, amounting to 2% of Parisian portfolios before WWI, and although that share increased a bit after WWI to 2.6%, it was never large.

Among those who held such assets, they were much more important parts of their portfolios, amounting to about 5% of income-producing assets before WWI. Following the 1917 revolution, Russian assets went into a predictable tailspin while the Suez Canal remained important through the 1930s. After WWI, Suez Canal shares and bonds are 6% of the overall portfolio for those who own them. Because of the strong connection between wealth and diversification, the weight of Russian assets and Suez Canal bonds is declining in wealth (though the rich own most of those assets). For instance, before WWI, for those in the ninth decile who owned them, Russian assets represented almost 20% of their income-producing wealth. At the top, it was less than 1%. The same applies to Suez Canal bonds, and the proportions are similar. This implies that the very rich were protected from the vagaries of individual asset price shocks, even though the rest of the top decile was not. To make the point even more concrete, let us analyze three examples drawn from 1932. Léon Olry-Roederer, heir and owner of Roederer champagnes—the purveyors of bubbly to czars—unsurprisingly owned Russian assets when he died as the second-richest person in Paris that year. His fortune totaled 94 million francs, of which 13,998 francs were Russian assets.[6] They did not even amount to two thousandths of his wealth, hardly surprising since they traded at 4% of face value. Even if we reflate his Russian position to 1912 values, the resulting 350,000 francs would still only amount to 0.4% of his wealth. Olry-Roederer was diversified; he held 27 million francs or a quarter of his fortune in Roederer champagne and ancillary real estate, 6.3 million in shares of the Suez Canal, and 3.8 million francs of French government bonds. On this last position, if he bought most of them before the war, Roederer would have taken a capital loss of 20% or double the wipeout of his Russian investments.

Others were not so fortunate. Alice Bossy also died in 1932. She was the not-famous widow of Henri François Grignon, a retired notary who died in 1919. She left her brother an estate of 1.3 million francs.[7] In that estate, we find 76,500 francs of Russian assets (Imperial bonds and Krivoi Rog bonds and shares). Reflating the Russian assets as if the default had not occurred would lead to a value of 1.8 million francs, which would produce a Russian share in her estate of 59%. The jump in total wealth is large enough that, absent the default, she would have been a member of the top 1%. This is precisely the kind of large-scale personal impact that gets lost when considering the fractiles as aggregates.

Similarly, consider Christine Chevallier, who died on April 4, 1932. She had survived her husband, Emile Petit de Coupray, an artillery colonel, by 34 years.[8] When he died in 1898, they resided on the boulevard Malesherbes, and she subsequently moved to rue Cernuschi, both addresses for people of considerable means. Chevallier left an estate valued at 4.6 million francs, enough to put her comfortably in the top 0.5% of all decedents that year. Consider what might have happened had the government favored shareholders rather than creditors in the fight over the Suez Canal bond gold clauses. Her estate included 337,744 francs in Suez Canal bonds. In early April 1932, bonds of 500 francs were trading at 2,000, so she owned roughly 170 bonds with a face value of 84,500 francs. Had the gold clause been abolished, a reasonable approximation is that these bonds would have traded at the same discount as French government bonds—that is, at 72.45% of par for a market value of 61,220 francs—leading to a loss of 82% of her Suez investment. Her heirs would still have inherited a tidy sum, but the choice of Suez Canal bonds proved astute. Like Olry-Roederer, she had a large position in French government and railroad bonds (almost 2.5 million), and her bond losses were three times her gains in Suez Canal bonds.

We now perform these kinds of counterfactuals for all investors in Russia and Suez Canal bonds. We first ask what if these assets had traded at their prewar prices and compare them to prewar portfolios. In the 1920s, these reflated Russian portfolios would have been half those observed before 1917 and become smaller over time. Further, the higher the fractile, the further the reflated portfolios were from what was observed before 1917. It should be noted that each portfolio is observed only once, and no finance theory would ever predict that the portfolio share of a particular fractile should reflect the changes in asset prices. To the extent that Russian assets were deemed very risky, one would presume that they would have ended up in the portfolios of those most able to bear risk, but that is not the case since the top 1% shed Russian assets faster than anyone else.

We perform the reverse counterfactual for Suez Canal bonds and examine what would have happened to that wealth position if the gold clause had not been enforced and Suez bonds had performed like French *rentes* or railroad bonds. The counterfactual Suez bonds have a value of 20% of what we observe, and the portfolio share of the canal company collapses. At lower levels of wealth (below the top 5%), the abrogation of the gold clause would lead to losses ranging between 1% and 10% for investors holding Suez assets. For the top 1%, who own the majority of Suez bonds, these counterfactual losses are never more than 2% of wealth, another marker of the importance of diversification in stabilizing the portfolios of the rich.

Taking both together, it is clear that individual asset shocks create a form of mobility at lower levels of the wealth distribution that has to do with the large asset position of each individual combined with the radical idiosyncrasy of asset values. There were cases of individuals of great wealth who lost almost everything in the vortex of the first half of the twentieth century. None is more emblematic than André Citroën. The child of diamond traders who had migrated from Amsterdam, he made a massive fortune building a munition factory during WWI. He then transformed that factory into France's largest producer of automobiles by the mid-1920s. Between 1920 and 1930, he plowed over 65 million francs of his wealth into the firm. A relentless innovator, his retooling of the factory for the Traction Avant car model, combined with the collapse in demand from the Great Depression, led the company into insolvency (Broustail 2020). The Michelin tire company was its largest creditor and took it over. Citroën died in 1936 with 708,670 francs in assets and 2,003,951 francs in liabilities.[9] He is an extreme case of a rise followed by a fall, and others were ruined by one or another of the great shocks of this period. However, the evidence is clear that for most members of the 1%, diversification was such that absolute ruin was unlikely. To be sure, portfolios with higher realized asset values at the time of death led individuals to move up in the distribution, and those with overweighted assets with more relative losses moved down. In most cases, individual asset shocks did not threaten an individual's wealth. In the next section, we examine the portfolios of the very rich to understand what made them the winners of the wealth derby.

5.4. Wealth Within the 1%

One of the most striking phenomena from 1912 to 1947 is the decline of the wealth share of the top 1% from 63% on average in the three cross sections before WWI to 53% in the three cross sections from 1937 to 1947. By contrast, the top thousandth—the super rich—experienced little or no change. Its wealth share fell from 25% to 23% over the same years, and that drop was temporary. The decline in the wealth share of the rest of the 1% (from 38% to 30%) is quantitatively massive and consistent with the arguments we laid out above about taxation and asset destruction. The resilience of the super wealthy is a surprise.

We will explore the experience and composition of the top thousandth group in detail in chapter 8. Here, though, we take a first look at this surprising finding because the stability of its wealth share during the three great shocks raises several questions. It is generally thought that negative shocks and tax avoidance lead to declines in the taxable wealth of the best-off. The negative shocks reduce wealth at the top because the most wealthy hold more equities and these are supposed to be more sensitive to negative events than bonds. In progressive taxation regimes, the value and practice of tax avoidance strategies increase with wealth. Thus, we would have expected the decline in wealth of the super wealthy to be relatively larger than what we observe for the rest of the 1%.

Of course, dying among the richest thirty or so Parisians is an outcome that reflects a variety of processes. Membership in the super wealthy starts with earning high returns. WWI was terrible for most fortunes, but more than a few experienced large gains from the combination of massive demand for war materials and low-interest loans to build industrial capacity (Citroën is but one example of such a fortune). These gains were large enough that the French legislature passed a tax on war profits in 1916 (Hautcoeur and Grotard 2001). The Great Depression and WWII offered fewer opportunities for financial enrichment. Nevertheless, some members of the lower fractiles that were relatively spared by the catastrophes could have jumped into the 0.1%, while individuals like Citroën disappeared from the list.

It could also be that the true super rich had already stashed most of their wealth abroad and thus did not appear where they belonged in the distributions we compiled. If so, the difference between taxable and total wealth could be considerable. However, as noted in chapter 3 and appendix A3.8, while tax evasion began before WWII, the flow of funds abroad was relatively small into the 1950s. A second opposite force would be that individuals whose wealth was tied up in private enterprises in the nontradable sector would have had great difficulty sending their wealth abroad. Conversely, financial fortunes, numerous before WWI, would have been the easiest to move offshore. This hypothesis is testable.

We start with an example: the Hottinguer banking family. Jean Conrad Hottinguer was born in Zurich. He founded an investment bank in Paris that first appeared in the Paris city directory in 1800 and continues to operate to this day. He was a founding member (*régent*) of the executive committee of the Banque de France. The family's influence continued as family members succeeded one another as régent continuously until de Banque de France's nationalization in 1936. Each of these régents died a member of the super wealthy.[10] Jean Conrad died in 1841 with an estate composed of 2.6 million francs of movable assets, 546,000 francs of Parisian real estate, and 488,000 francs in his country estate in Boissy-Saint-Léger. His fortune of 3.6 million francs put him well above the 2.7 million francs average estate of the super rich. His son Jean Henri died in 1866, leaving an estate almost three times the size of his father's (even though he had siblings). The first Hottinguer to die after WWI was Rodolphe (Jean Henri's son), whose 1920 estate was valued at 30.7 million, nominally three times that of his father. In real terms, his estate was worth 9.2 million 1912 francs, only about 90% of what Jean Henri had left behind.

Jean Henri Maurice (Rodolphe's son), the last régent, died in 1943. His estate was complicated. The filing listed 214 assets totaling 73 million francs. He had also made three donations to his children, totaling 27 million francs when they were made. Our best estimate is that the donations were worth about 39 million in 1943 prices.[11] His estate filing included another 102 assets frozen by the war and not evaluated. We were able to find prices for 50 foreign securities, and using an interpolated exchange rate for 1943, these investments come to 20.5 million francs. Because we cannot evaluate 52 assets, our total estate of 132 million francs understates his wealth. The extent of this bias is likely small because we could find prices for the most important positions. A different and likely larger bias comes from using official exchange rates (50 francs to the dollar and 200 to the pound) that prevailed until the exchange was suspended in mid-January 1941. It is enough, however, to bring his wealth in 1912 francs to 75% of his father's.[12] This is roughly in line with the experience of the 1% and the top 0.1%

The trajectory of the Hottinguer family illustrates the rapid rise in the value of the super wealthy's estates up to WWI. The family resists the WWI shock, suffering a decline of 10% in real wealth when the average estate suffers a 28% drop. It also resisted the Great Depression and the first half of WWII, as between 1912 and 1942 the average top thousandth estate fell by 23.4% and the Hottinguers' by 24.8%. Thus, the family had an average fall in wealth—exactly what one would predict if they had a well-diversified portfolio. To the extent that other members of the top thousandth had a similar experience, we can start to understand the resilience of the super wealthy.

The Hottinguer family's strategy was closely connected to its history. As a Franco-Swiss banking family with close connections to the United States, they were ideally positioned to move their wealth offshore. That they remain steadfastly in the top 0.1% suggests that they did not do so wholesale. Moreover, the extent of evasion by not reporting assets abroad seems limited when we examine the estates of Jean Henri Maurice Hottinguer and his wife, Marian Hall Munroe. She was the daughter of John Munroe, an American banker principally active in Paris after 1860. When Marian Munroe died in 1927, she left an estate of 28 million francs, in real terms, just about the same as her husband's.[13] Her estate was mainly composed of two sets of items. First, her personal property came to 4.9 million. It was almost entirely composed of American securities. The second part of her estate was half the community property. As is often the case for the very rich, only 9% of the community property was in non-income-producing assets. Forty-five percent was in American securities (which in 1927 had done much better than French ones). Another 32% was in foreign securities. These were more of a mixed bag, with some investments in Russia, Central Europe, and South America that were all hit hard by WWI, and some like De Beers and Royal Dutch Shell that prospered. The French portfolio came to 13.6% of the community property. Given the preferences for holding foreign assets, had tax evasion been on Munroe's or her husband's minds, it would have been easy to implement. Indeed, the family reported bank accounts in Frankfurt, Vienna, Lausanne, London, and New York at different times.

The Hottinguers are outliers among the very wealthy because the amount of foreign wealth reported is vast compared to the average portfolio. Investments in the United States are important for all the Hottinguers who died in the interwar period. Their global diversification insulated part of their portfolio from French inflation and the disastrous performance of bonds and equity in Central and Eastern Europe. They are also unique in that their estates were significant, but not because they had large stakes in a massively valuable family business. The male members of the family systemically retained nearly all the equity in the eponymous bank, but it was not a very valuable business (unlike Sirop Ramy or Roederer Champagne).

The bank's nominal capital stock remained at 2 million francs before 1912 (profits were distributed to interest-bearing accounts rather than capitalized into equity) and may have risen to 6 million by 1943. Each of the successive régents could have afforded to own the entire equity of the family bank. Instead, for more than a century, they plowed returns to finance into a diversified portfolio. Finally, being wealthy allowed them to choose when to transfer assets across generations and thus reduce their tax exposure; the donations being valued at historical prices meant they were valued at only two-thirds of 1943 prices.

Estate taxes did weigh on the super rich because collateral successions were assessed at very high rates. The contrast between Marian Munroe and her brother-in-law Jean Conrad Hottinguer (great-grandson of the founder), who also died in 1927 but without children, is striking.[14] Jean Conrad left an estate of 44.5 million francs to be divided between his two brothers after paying a tax of almost 40%. In contrast, the régents and their wives all had at least one heir in direct line of descent, and they were effectively shielded from the highest burden of the estate tax by their demography. Before 1900, the children of the founder Jean Conrad and Jean Henri paid less than 1%. In the 1920s and 1930s, Rodolphe's and his wife's children paid 7% on his massive estate and 4% on her much smaller one. In 1927, the children of Marian Munroe paid a tax of 15.6%. Then, in 1943, the same children paid 20% to settle Jean Henri Maurice's estate. Their low tax rate was the result of a combination of the maximal average rate for someone with three children and that several heirs had many children themselves. Given the return differentials, a 20% tax rate was not punishing, although it would have reduced the consumption possibilities of the very rich.

The super rich are rarely as persistent as the Hottinguer family or as ephemeral as André Citroën. Still, there are two fundamental reasons for their resistance to the drop in wealth from 1914 to 1947. The first is that 1914–1947 was marked by severe shocks and also a period of considerable volatility. That volatility produced relatively outsized gains for individuals with the right asset positions. This outperformance was more likely for individuals who, like the Hottinguers, had large equity positions. The second is that the super rich who died in the interwar period had received inheritances that had not yet been severely taxed. The fiscal innovations of the 1920s, when top rates for estates going to siblings or outsiders first exceeded 40%, took a generation to diffuse. The estates we see in the 1920s and 1930s had been received mainly under much more benign tax rates, with top rates below 20%. What did hit quickly was the income tax. As we shall see in chapter 7, the income tax played a central role in redirecting capital income flows from the rich to the state.

5.5. The Diffusion of Small Fortunes

We have already noted that one of the startling phenomena of the interwar period is a permanent shift in the fraction of Parisians who leave an estate. As we shall see, this remains a conundrum, though we can rule out several hypotheses. The first is a shock to human capital. In the late nineteenth century, there was some growth in the fraction of students who completed the *certificat d'étude* (the primary school examination that was required for access to secondary education) and some increase in vocational schooling. Overall, however, the education of the cohorts who died in the twenties and thirties was only moderately better than those who died before WWI. Chapter 11 details the big break that came after WWII.

The rise in the number of small estates was also not due to some demographic shock that would have induced more individuals to leave bequests. Neither the share of individuals with direct heirs nor the number of children changed much. The increase in the share of individuals who left bequests occurred before the 1956 introduction of a threshold below which inheritances passed without being taxed. So, fiscal policy is not an issue. The most probable sources of change are retirement and disturbed dynastic planning. In both cases, they undermine the practical distinction economists make between bequests and other forms of savings. Bequest savings are complements to life-cycle savings in Paris. The shocks from 1914 to 1945 likely led families to reduce inter vivos transfers to preserve flexibility in old age.

The late nineteenth century saw a significant diffusion of mandatory retirement savings programs. These began with government employees and then spread to large enterprises and then to more and more employment sectors. The agricultural sector as a whole and entrepreneurs remained outside the system for most of our period. Nevertheless, increases in the share of individuals covered by retirement programs led to new filings because a retiree's estate was the beneficiary of any unpaid balance from retirement annuities. Other assets that had escaped taxation as too trivial, like furnishings, might also come to be listed in these estates. In practice, however, our retirees most often had financial or real assets beyond their retirement payment. As we shall discuss in more detail in chapter 9, the rise of mandatory retirement programs as a mechanism that led more individuals to have bequest savings is not straightforward. Indeed, at first glance, it would seem to lead to tiny estates, for why would an individual who does not intend to save for old age not only do so when forced but add to that effort voluntary savings for bequests? The share of retirees is measurable in two ways. First, we can tell because their heirs noted this status in estate documents. Second, retirees are identified in the documents by the existence of retirement assets in their estates. Below we examine both measures.

The other source of change does not involve interaction between life-cycle and bequest savings but between inter vivos and bequest motivations. Families, particularly those involved in small- and medium-sized enterprises, might have, before WWI, privileged inter vivos transfers to maximize the value of their business. Parents would have transferred those business assets to their children as they aged. The war massively disturbed these processes because military casualties overwhelmingly concerned the prime candidates to benefit from inter vivos transfers: men between the ages of 20 and 40. Parents facing the loss of a son may well have decided to hang on to the business much longer and thus die with their business assets. Again, these are outcomes we can observe in the estate (through the presence of business equity) and in the reported occupations. One caveat is important to note. There are reports of the distribution of occupations among the dead in Paris, and these can be contrasted with the distribution of occupation for those who died rich. This can produce a difference between the "rich" and the "poor," but we cannot determine the occupations of the sixth decile (P50–59) were in the decade before 1914. At the time, less than 30% of adults left a bequest; thus, the sixth decile cannot be distinguished from the lower ones. This obviously limits what we can learn about the expansion of wealth at death. Moreover, occupations for women are scarce. Given that many women were widows or single, we cannot use the occupations of the husbands they did not have. Instead, we observe groups when they enter our observatory: when they leave a bequest.

We start with a very simple split: on one side, we total the deceased who are reported as retired, without occupation, proprietors, or rentiers; on the other side, we have all the deceased for whom an occupation was reported. Consider first the aggregate population and a group eponymous of the social order of the first financial golden age: the rentiers—here we mean individuals wealthy enough to live from their capital income. About 9% of all wealthy decedents are reported by their heirs as rentiers before the war. That, however, is a last hurrah, and, as an identified social category, they disappear after 1914, dropping to 2.5% of the wealthy in the 1920s and an irrelevant 1% in the 1930s. For men, the decline in rentiers is partly offset by a rise in the number of men for whom no occupation is reported (sans profession). Taken together, the "idle" rich (rentier and without occupation) amount to about 20% of the wealthy after 1872. The share of retirees, however, roughly doubled from before to after WWI.

The other groups are those for whom an occupation is reported and who were implicitly still active when they died. Before WWI, the wealthiest 5% in our cross sections were mainly rentiers, individuals without an occupation or retired, in short, "inactive." The next 5% were much more likely to have an occupation, and the members of the eighth and ninth deciles of the wealth distribution have activity rates of 75% or more at death. Removing those individuals older than 65 from the active groups raises activity rates but does not change the rankings. This is no surprise since most men who died wealthy below the top decile were younger than 66 and had an activity. From then on, two conflicting forces seem to be at work. Increases in the average age at death reduced activity rates. Conversely, drops in the consumption value of wealth were increasing activity rates.

When the sixth and seventh deciles become visible, they look remarkably like the eighth and ninth deciles. Men with occupations dominate these deciles. Even then, their activity rates tend to fall over time due to increasing age at death. This broad finding suggests that retirement pensions are not directly driving the rise in the share of individuals with bequests. Moreover, the estates of men and women look similar.

As a second step, we focus on retirees. These are found because their occupations include "retraité" or "ancien" or "honoraire." For the fractiles we always observe (P70–100), the likelihood of an individual being marked as retired increases with wealth through 1932. Then, a reversal of sorts sets in, and the fraction of retirees drops at the top (P95–100) and stabilizes for P90–95 and P80–89. The bottom groups, which started with low numbers, see their retiree shares rise over time. However, even on the eve of WWII, they do not breach 15%. Retirement alone, thus, cannot explain the increased number of small fortunes in Paris. The same finding recurs when we look at the portfolios of individuals we sampled. Less than 20% of the individuals sampled have retirement assets (either retirement or life annuities).[15]

A few examples drawn from the bottom of the distribution can help illustrate these ideas. Among the smallest estates, we find Marie Dujat, who died in 1937 at 76. She is reported without occupation in the death records, but her estate tells us she sold fish. When she died, she had no business assets and left a gross estate of 4,019 francs (half an average worker's annual income that year and 655 1912 francs). The bulk of her wealth was a deposit account at the Caisse d'Epargne. She also had a life annuity of 937 francs on which she was owed 172.5 francs. Georges Gustave Leclère died at age 79 without an occupation and an estate of just under 1,000 current francs (a few months of a worker's income or 155 1912 francs). It was composed mostly of an account at the Caisse d'Epargne and a small pension, a *retraite ouvrière*, on which he was owed 105 current francs. Dujat and Leclère had their furnishings assessed at 5% of the gross estate. To find a pure retiree, we have to go even further down the estate distribution to Claudine Ménard, who died at 76 with a retirement from the Crédit Foncier, where she had worked and on which she was owed 203 francs (33 1912 francs). This was all her estate; she lived with her brother and sister, who claimed she had no movable assets.[16]

If we move into the eighth and ninth deciles, we find people like Arthur Jamet, who was a retired railroad worker when he died at age 66. His estate came in just short of 6,000 francs, including substantial movables, an account at the Caisse d'Epargne, and a pension from the Chemin de Fer du Nord. Pierre Gamon died at 84 with an estate of 84,000 francs. He was a retired jeweler with three different pensions. The first was a—very small—*retraite ouvrière*, the second was owed by the government because a child had died in WWI, and the last was privately subscribed at a tontine company called la France Prévoyante. His primary asset was a building on rue Charlot in the jewelry district of the 3rd arrondissement (likely where he lived and had worked).

Even Gamon had an estate of less than 14,000 1912 Francs. These small fortunes raise another possibility, namely that with the uncertainties of the war, a structural change in preferences over inter vivos and bequest transfers had occurred, privileging bequests. Over the longer term, WWI and the Great Depression might well have interrupted the normal transmission of assets. In the case of the war, the loss of life of the heir who was expected to take over the business might have led the deceased to hold on to it much longer (waiting for grandchildren) or, in a more radical fashion, to die with assets that they had planned to transmit. However, business assets are rare across the distribution of wealth below the top decile. Albert Aubry and his wife, Jeanne Léonie Réaux, were classic examples of such ownership. He died in 1937, at age 58, and she died the following year, at age 53.[17] They owned a charcuterie business on the boulevard Voltaire and lived at the same address. Their estates were filed one after the other. His estate came in at 453,000 francs, and hers at 484,000 francs. The difference in their wealth is due to their 1905 marriage contract, giving her all the furnishings and his estate owing the community one daughter's dowry. At the time of marriage, he folded the charcuterie business he had acquired in the same neighborhood, 60 rue de la Réunion, into the community.[18] He had bought it for 31,375 francs the previous year and still owed 23,000 francs (these values are very close to the 1912 francs values). Around 1922, he and his wife sold that business to buy another charcuterie at 130 boulevard Voltaire that they still owned when they died. Neither daughter was interested in picking up their entrepreneurial mantle. In 1938, the charcuterie was sold for 423,761 francs (69,129 1912 francs or more than double what the initial business had been worth). They had also bought a building in the 15th arrondissement and had leveraged themselves with 410,000 francs of debt at 5%. Over the 32 years of their marriage, their wealth had jumped from the 25,000 francs they had had when they wed to net assets of 1.1 million francs (180,000 1912 francs).

The Aubry-Réaux couple is an excellent example of how entrepreneurship can increase the size of bequests. In contrast to what we might have expected if business ownership were the dominant force for creating bequests at the bottom of the distribution, the frequency with which we encounter entrepreneurs increases with wealth. The interwar increase in the number of individuals with bequests was not the simple result of a shopkeeper's paradise. Shopkeepers were scarce below the top two deciles because becoming a shop owner required wealth, and succeeding as one meant accumulating substantial wealth. Parisian businesses were not cheap, and the individuals in the bottom part of the distribution of wealth were too poor to go into business. Moreover, as the Aubry-Réaux example details, successful shopkeepers in Paris had shops worth 100,000 francs or more, which put them far above the threshold wealth of the ninth decile of 1937.

What else might help explain the rise in the share of individuals leaving bequests? Having considered life-cycle and inter vivos transfers, we are left to consider increases in precautionary savings. We know of three forces that would have encouraged more precautionary savings: a growing likelihood of living to retirement age starting in the 1880s, savings drives during WWI, and growing macroeconomic uncertainty after 1914. Perhaps rather than executing inter vivos transfers, older households suspended them in order to preserve their independence. This argument is difficult to test with a single city in a relatively short period. However, it helps explain the rather diversified portfolio of the middle fractiles, which are inconsistent with the dominance of any alternative hypotheses. This precautionary motivation is also consistent with the fact that the gender distribution is relatively even in the lower fractiles. That would have been unlikely with retirement (a set of assets that went overwhelmingly to men early on) or with business assets.

The experience of the sixth and seventh deciles brings us back to the perspective of inequality being reduced by crises. Given the rapid rise in the share of individuals who left bequests from the end of the 1920s to the end of the 1930s, it might seem that the decline in the wealth share of the top 1% was making space for others. A closer examination of the data shows this was not the case. To begin, it is unlikely that the wealth share of the top 1% would have declined much, absent the real and fiscal shocks of the twentieth century, because the group of new wealth holders only possessed a tiny fraction of total assets. Moreover, one should remember that the rise in these lower fractiles resulted from savings behavior that began before WWI, when the 1% wealth share was at its peak. Indeed, those who died in the late 1920s in their early sixties had married in their late twenties between 1885 and 1895. These new wealthy were the least likely to have inherited assets, as they were married under the standard community property regime. Their accumulation is entirely to their credit. The decline in top fortunes between 1914 and 1943 is coincidental with the near doubling of the share of individuals who died wealthy, not causal.

5.6. Conclusion

One striking phenomenon not discussed in this chapter is the poor connection between the decline in wealth inequality and other measures of social disparity. In some cases, like life-expectancy inequality, disparities began to shrink much earlier and their decline continued long after wealth inequality moderated in the 1930s. For other types of disparities, such as access to secondary education, significant change did not come until the 1950s. As a result, examining different dimensions of inequality requires a particularly long view, and we devote chapter 11 to the topic.

This chapter comes to somewhat different conclusions from those of chapter 4. There, we emphasized the negative returns and fiscal shocks experienced by Parisians to explain massive declines in private wealth per capita. Here, we returned to the fiscal innovations central to understanding why wealth at the top fell faster than everyone else's. We then showed that other markers of inequality in Paris evolved over time almost independently of wealth inequality. These include age at death and the fraction who die with an estate. Age at death seems to be more closely related to public spending on infrastructure than private wealth. The individuals who composed the sixth and seventh deciles had to decide to save before the top wealth holders started to fade, and they maintained their small savings in the face of negative shocks. Moreover, neither the financial innovations of the nineteenth century nor the fiscal innovations of the twentieth century are likely to have been important in these behavioral changes.

This chapter and its examples raised a number of issues that we will reconsider over the nearly two centuries spanned by our dataset. The first involves intergenerational wealth transmission. The Aubry-Réaux couple were successful and bequeathed far more assets to their daughters than they had inherited. The dowries they had received helped them acquire their shop. We return to the problem of inherited wealth in chapter 7. The Hottinguer family represents an extreme case of persistence since the men are all members of the top 0.1% from the 1840s to the 1940s. We will consider the super wealthy in detail in chapter 8. Chapter 9 returns to individuals like Dujat and Gamon to consider the interaction between different forms of savings over the long term. Before that, in chapter 6, we turn to the wealth recovery that followed WWII.

6

Recovery, 1947–1977

The liberation of Paris in August 1944 and the end of World War II in May 1945 followed five long years of destruction. Thus, it is unsurprising that Parisians who died in 1947 were poor. The fall of Berlin did not end either economic uncertainty or the high taxes that weighed on wealth accumulation. Faced with an infrastructure in shambles and a financial system atrophied by inflation, the French governments of the late 1940s and the 1950s intervened relentlessly to put France back on a path to growth. As a result, the list of forces that could have impeded the recovery of private wealth is long. For certain forms of capital, the extent of government intervention was drastic, with results we can see in our data on Parisian decedents in the post-WWII cross sections.

For political economists, it is commonplace to argue that property rights insecurity is inimical to private wealth. Postwar France, in this respect, is an interesting case because the threat to private property was real. The Communist Party, committed to wholesale nationalizations, received the largest share of votes in the 1945 elections, and it remained an important political force after that. Additionally, the 1946 constitution of the Fourth Republic produced a parliamentary regime where no party ever had a majority in the lower house. Coupled with the challenges of decolonization, this political system yielded more than two governments a year until it was replaced by the presidential regime of the Fifth Republic in 1958 with De Gaulle at its head. Even the return of De Gaulle to power did not lead to a quick solution to the Algerian War of Independence.

Beyond politics, there were many economic problems. Inflation remained very high (nearly 30% annually from 1945 to 1952), then moderated until the 1970s, when inflation rates rose to double digits. Even in the moderate regime, inflation was 4% to 5% a year, or three to four times higher than it had been before WWI. This meant that in the 1950s, individuals who had accumulated bond portfolios had to reinvest a significant chunk of their coupon payments to maintain the real value of their portfolios.

As Éric Monnet (2018) details, the laws of 1945 nationalized much of the financial sector, including all the major banks and insurance companies. Initially, the nationalizations were intended to save a financial system brought to its knees by the war and tainted by collaboration with Germany. Most banks in 1945 were insolvent, and in a context where many financial actors' behavior under Vichy had come under a cloud, nationalization provided the opportunity to start anew. Nationalization gave the ministry of finance extensive control over the direction of investment. The state used its control of investment flows to focus resources heavily on infrastructure. The rapid growth of France after 1945 suggests that government officials successfully directed investment to high-return activities. Another consequence of nationalization matters to us. Financiers no longer collected the returns from financial intermediation. These now either accrued to the state or to the firms whose credit was subsidized. As a result, it was hard to make a large fortune in finance between 1945 and 1985.

At almost the same time it took over banks, the state nationalized a large chunk of the infrastructure sector, including electricity, gas, and local public transport. By definition, nationalized sectors do not issue equity to savers, and taking into account the creation of the French national railway company (SNCF) in 1937, this means that more than a third of the equity investment of Parisians in the 1930s had been transformed into bonds through nationalization. From the point of view of economic activity, the nationalization of infrastructure was quite a success. Indeed, the state poured enormous amounts of capital into these sectors, and they were thoroughly modernized. Yet, for many economists, this state intervention appears suspect mainly because the state cannot reproduce market signals when setting investment flows. Monnet (2018), however, argues that the state directed credit efficiently for several decades.

Other postwar interventions by the French government directed more of the returns from economic activity away from asset owners. In terms of distribution between capital and labor, new laws included the creation of a broad minimum wage (SMIG) in 1950 and its rapid increase after 1968. The state mandated the extension of paid vacation to three weeks in 1956 and four weeks in 1965 with little concern about the impact on firm balance sheets. The same can be said of the extension of retirement, unemployment, and health insurance that proceeded over the same period.

Property owners saw their rental income diminished with the 1948 enactment of strict rent control that applied indefinitely to the tenants in place at the time. Given that Paris was overwhelmingly a city of renters, the measure was particularly severe for the capital's real estate owners.

If you replaced Paris with a city in a low-income country, many economists would predict that the events of these decades would be inimical to private wealth accumulation. Paris, it would seem, was destined to be the capital of a grape republic. But that is not what happened. Instead, the economy grew rapidly, and, as we document, private wealth recovered, in particular below the top 1%. We split this chapter into three parts. In discussing the recovery phase from 1947 to 1962, we consider wealth and inequality together. We then proceed with an analysis of the lower part of the wealth distribution from 1962 to 1977. We argue that the same forces that promoted wealth accumulation after WWII continued to prevail in the 1960s and 1970s. We close with a reconsideration of top estates and the quality of the estate tax data. We show that it is very likely that, to a small extent in 1972 and a much larger one in 1977, we are missing wealth at the top. We show that the missing wealth comes equally from tax avoidance and tax evasion.

6.1. Evidence

Our estate tax data are entirely inconsistent with the hypothesis that nationalizations were inimical to the growth of private wealth. Instead, our data strongly indicate that in the thirty years following 1945, the wealth of Parisians grew rapidly. Nevertheless, as late as 1972, almost half of Parisian decedents had no wealth at death, and we cannot say anything about their life course. Among the wealthy, the rapid growth of wealth hides two distinct patterns: one for those below the 95th percentile and another for those above that threshold.

For those fractiles below 95% of the distribution, the shocks of 1912–1947 were severe. The average estate between the median and the 95th percentile in 1947 was only 34% of what it had been in 1912, but wealth bounced back after WWII, and that group's real wealth in 1962 exceeded its pre-WWI maximum by 57%. This rapid wealth accumulation continued from 1962 to 1977, and the group's real wealth was almost three times larger in 1977 than its pre-1914 peak. Within the P50–95 group, the growth rate of estate values was higher for small ones than large ones, likely because steady savings from labor income had a large impact on small estates. The three postwar decades, which came to be known as the "Trente Glorieuses," were excellent years for small and medium savers in Paris.

Between 1912 and 1947, the richest 5% of decedents had experienced a far more severe wealth loss than the rest of the distribution. Their 1947 wealth in 1912 francs had fallen by 88% from its pre-WWI peak. By 1962, the average real wealth of the top 5% was 487,000 francs.[1] Half a million francs is well above the same fractile's interwar maximum of 332,000 francs. It is, however, only a bit more than half its pre-WWI maximum of 885,000 francs. By 1977, the wealth we measure in the estate filings reached 527,000 francs, or 60% of the best figure before 1914. As discussed in appendix A3.8, there are good reasons to think that the super rich had some offshore wealth that escaped the tax authorities. Even taking into account the revision for tax avoidance, the top 5% only managed to reach 68% of the pre-WWI maximum. What is striking is that the great shocks of WWI, the Great Depression, and WWII led to similar losses (87% to 89%) of wealth for all groups in the top 5% of the distribution from 1912 to 1947. The recovery to 1962 is also comparable across the top fractiles. After 1962, the growth rate of wealth is inversely correlated with the fractile. Overall, however, the top wealth holders had a steeper hill to climb because the great shocks were worse for them (had their wealth grown at the same rate as it had for those below the 95th percentile, they would have just exceeded the pre-WWI maximum in 1977). The divergence arises because the reported estates of the top 0.1% eke out only an 11% real growth, or less than 0.7% annually, between 1962 and 1977,

The large difference in wealth growth rates below and above the 95th percentile continues and accelerates the changes in wealth distribution we observed starting in 1922 (see figure 6.1). Even so, in 1977, Paris remained dramatically unequal in wealth at death. Except for 1977, the bottom 95% of the population's bequests never amounted to more than a quarter of the total, and the top 1% almost always transmitted half of all the wealth in any given year. Nevertheless, a major shift in distribution occurred over the twentieth century, with an increasing share of individuals leaving estates. By the 1960s, the population in the sixth to eighth deciles was leaving nontrivial amounts of wealth. The data, thus, bring several surprises. We continue to argue for the relative independence of the factors that lead to wealth accumulation and the rise of bequests for individuals of modest means on

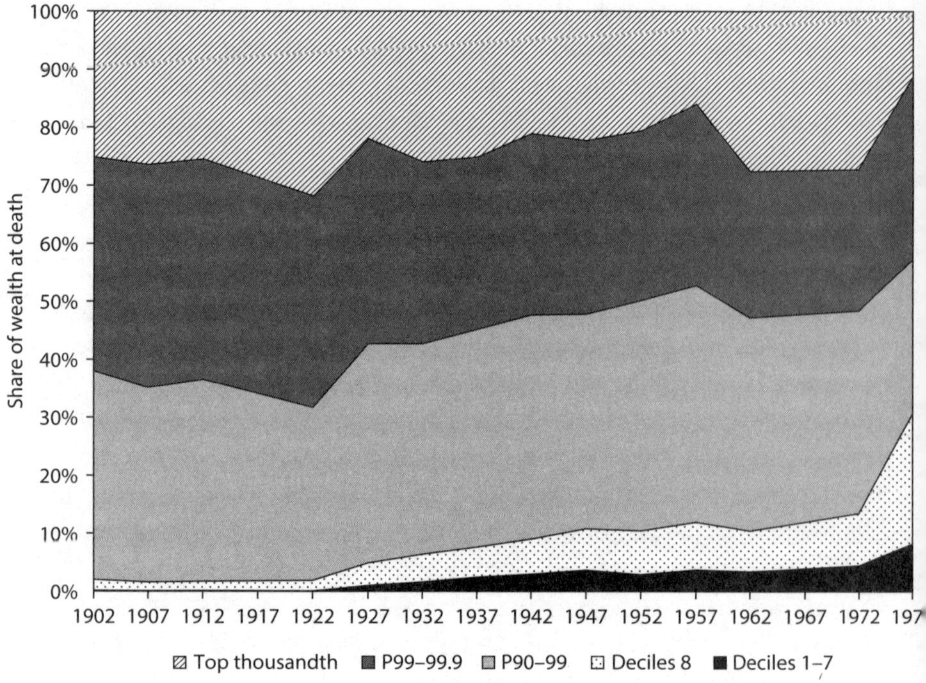

FIGURE 6.1. The distribution of reported wealth at death in Paris, 1902–1977.
Note: The figure is based on the data we extracted from the estate tax filings. The estates include both taxable wealth and investments in tax-exempt assets in France.
Source: Estates dataset (see appendix 1).

the one hand, and those factors that drove the level of bequests at the very top on the other.

6.2. A First Recovery: 1947–1962

In the 15 years between 1947 and 1962, real private wealth at death increased by 10% per year. As a result, part of the losses from the terrible decades between 1912 and 1947 were made up. Real average wealth in 1962 was 63% of its pre-1912 peak (and about 160% of the 1927 peak). As we saw above, the recovery was extensive. All the fractiles of the distribution had growth rates between 9.5% and 11.4% per annum. For all groups below the top 5%, recovery to pre-WWI levels had been achieved by 1962. Because the decline from 1914 to 1947 had been much more severe for top wealth holders than for the rest of the distribution, they only reached a bit past halfway to the pre-WWI peak. This is despite the fact that from 1947 to 1962, the very top

0.1% had the largest rate of wealth gain (11.4%). Because growth rates are similar, the distribution of 1962 is quite close to that of 1947 (see figure 6.1). The small changes we observe in 1952 and 1957 primarily reflect the random arrival of very large estates. The P50–95 share was big in 1957 because, by chance, the top estates were small, and the 0.1% share was consequently smaller than usual.

Explaining rapid wealth recovery with a stable wealth distribution is a challenge. In a Kuznetsian framework where investment and reallocation across sectors drive growth, periods of rapid growth are likely to be associated with increases in inequality as only a fraction of the society is likely to have its wealth in sectors that reap the rewards from change. A more historical view posits that, at the time, resources in France were primarily channeled to infrastructure by state-directed investment plans, which would lead us to expect a decline in inequality. Indeed, scarce capital should have constrained the returns to private entrepreneurship. An alternative to these two arguments sees public investment as complementary to private activity. For instance, the private sector was a major beneficiary of the rapid rebuilding of the electrical system, a task taken over by the public sector after 1945. We opt for this more nuanced view of the interaction between economic change and inequality.

Equally important is to take into account the nature of the data. The individuals who appeared in the 1962 cross section had accumulated their wealth in different decades from those of previous years and inherited different amounts of wealth. If we look back at 1947, the people we observe inherited mostly just after WWI. They then experienced the full brunt of the Great Depression and WWII before passing on. They had among the worst set of economic circumstances that we ever observe, and they were very poor. Those who died in 1962 were mostly born over the two decades straddling 1900. They lived through all the great shocks of the twentieth century and survived them. They then experienced a period of rapid economic expansion before leaving their estates. Their trajectory does not simply extend that of individuals who died in the 1950s. They were luckier, ending their accumulation period with 15 years of high growth rates of wealth. It is thus necessary to explore the range of forces that explain why those who died in 1962 were much wealthier than those who died 15 years earlier.

A straightforward explanation for the wealth recovery would be that asset prices far outstripped inflation. However, this was only partly the case. For bonds, we continue to benchmark their returns with French government bonds (5% nominal). As table 6.1 shows, these were a dismal

TABLE 6.1. Real annual growth rates of asset values and wages

	Bonds	Equity	Real estate		Fonds de commerce	Wages
			Paris	France		
1947–52	−19.8%	−11.4%	−2.1%	−2.9%	−9.5%	5.7%
1952–57	−1.4%	20.6%	34.0%	18.7%	31.4%	5.6%
1957–62	−2.7%	6.2%	11.2%	12.9%	13.3%	4.0%
1962–67					−0.6%	
1962–72	−4.3%	−4.7%	7.0%	7.1%		4.5%
1972–77	−7.5%	−14.3%	2.0%	3.0%		3.5%

Source: Bonds DFiH. For equity and real estate, see Piketty and Zucman 2014. For fonds de commerce, see ASVP. Wage is cost of 2,750 hours of labor per year (see chapter 2 and Bayet 1997).
Note: The 1972 data are annualized growth rates from 1962 to 1972.

investment throughout the period with negative real returns. We borrow Piketty and Zucman's 2014 asset price series for equity and real estate. Equity capital gains are volatile and do not show a long-run "better than inflation" performance. The Hautcoeur and Lebris CAC40 price index is even more pessimistic. There are two exceptions to this gloomy picture: real estate and private businesses. French indices suggest that returns to real estate in France were high, and real estate in Paris was an even better investment than elsewhere. In the nineteenth century, the real estate premium would have worked against any leveling tendencies because only the very rich could afford entire buildings. Starting in the 1930s, various innovations finally led to the codification of condominium associations (*co-propriétés*) in the law from July 10, 1965. These changes created opportunities for individuals to purchase apartments and broke the stranglehold of top wealth holders on Parisian real estate. Private business sales (*fonds de commerce*) show rates of capital appreciation similar to that of Parisian real estate.

Because real estate and private businesses were a minor part of all the wealth in Paris, the rapid wealth growth from 1947 to 1962 cannot be attributed to the impact of asset price recovery alone on the value of inherited wealth. This is particularly true for individuals below the top 5%. Indeed, most of these individuals had no inherited assets; they constituted their bequests by saving out of labor or entrepreneurial incomes. To make this case, we proceed in two steps. We first consider the share of individuals married with a marriage contract after WWI—it is broadly available and highly predictive of having inherited assets. Continuing a pattern we documented

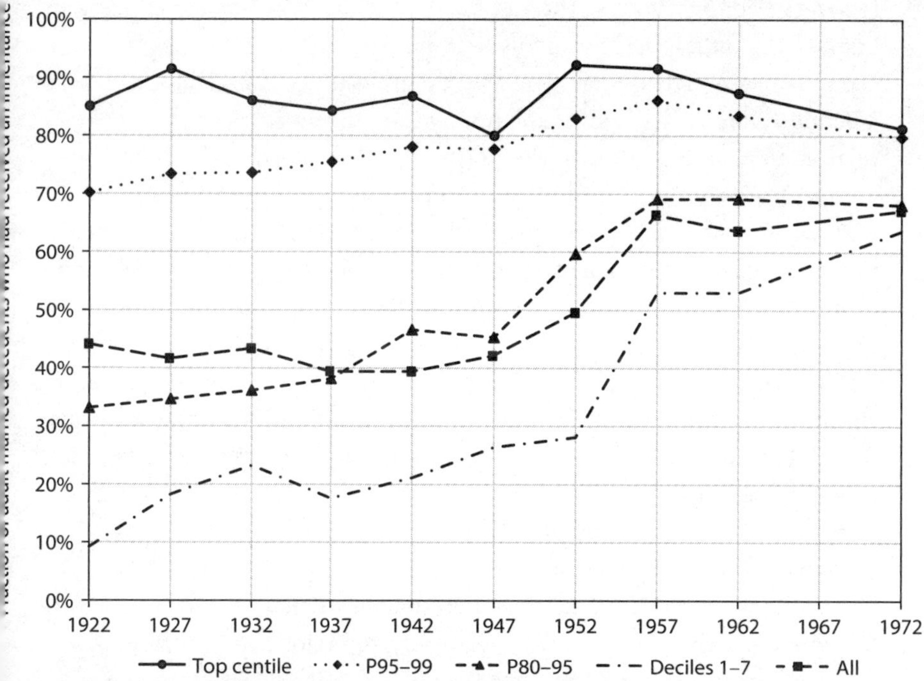

FIGURE 6.2. Fraction of individuals by fractile who had received an inheritance.
Source: Estates dataset (see appendix 1).

for the nineteenth century, despite some instability in the data, it is clear that the richer one died, the more likely one was to have a marriage contract. Moreover, the share of individuals with a marriage contract is low for anyone outside the top two deciles. Even for the P80–95 group, that share exceeded a third only in 1962.

Figure 6.2 reports the fraction of individuals who had inherited assets among those who died married. That share is typically smaller than that of individuals with a marriage contract and shows no trend before 1962. It does show, instead, a dramatic difference across the fractiles. In most years, individuals in the top 5% were at least three times more likely to have inherited assets through the 1930s than those in the sixth or seventh decile. Moreover, the wealthier an individual died, the more likely he or she was to have inherited assets. Among the population below the top 5%, three-quarters had married without a contract. Those with only a marriage certificate did not expect to receive a bequest, and only 10% had inherited assets in their estates.

If the bulk of the Parisian population could only constitute a patrimony by saving out of labor or entrepreneurial income, one of the drivers of wealth recovery had to be wages. The fact that half of the population died without bequests suggests that, for many, incomes were too low to allow them to accumulate enough savings to sustain themselves in old age, let alone leave a bequest. For the top half of the population, the more rapid rise of wages relative to capital income in the 1930s and 1940s allowed decedents who died in the 1950s and early 1960s to accumulate more. That is consistent with what we see among decedents after WWII. The long-term increase in the labor share of national income that followed 1918 could have provided more resources for individuals to save. As table 6.1 shows, wages measured as annual income at constant hours rose rapidly in the postwar era, making wealth accumulation possible. This positive performance hides stagnation at the bottom of the wage distribution. The real value of the minimum wage (SMIG) remained almost constant from its inception in 1950 until 1968 when it began a rapid rise. It is no surprise, then, that there remained a large group of nonaccumulators even in the 1970s.

What were the sources of income for individuals who died wealthy but did not make it to the top? The estate declarations provide two imperfect sources to help answer this question: professions and portfolios. The reported professions are imperfect because occupations for women were rarely noted. They are also imperfect for older men because they often omit occupation before retirement and provide a marker of social status (proprietor, retired, without occupation) instead. The latter problem was prevalent before WWI when many individuals, including those younger than 70, had no reported occupation. Over time, occupational detail improves, but it remains an issue (because many decedents are now just retired). To mitigate these problems, we focus on men who died younger than 70. For this group, the estate declarations mention an occupation for more than 80% of decedents in all years after 1900, except for the very top of the distribution.

Figure 6.3 shows the share of individuals by occupation in four panels that correspond to quite different wealth levels. The leftmost panel details those who are in the fractiles P50–80. For this group, the share of employees and workers is always substantial; they are also likely to be active but not business managers or owners. The next panel gives the occupational distribution for those in the P80–95 group; their wealth is several times the average annual labor income. They are much less likely to be employees or workers than those who died less wealthy. Instead, members of the P80–95 are more likely to be artisans and shopkeepers, what Marx had called *petits bourgeois* and others would call small-scale entrepreneurs. Few individuals were inactive (retired or without occupation), and the share of business

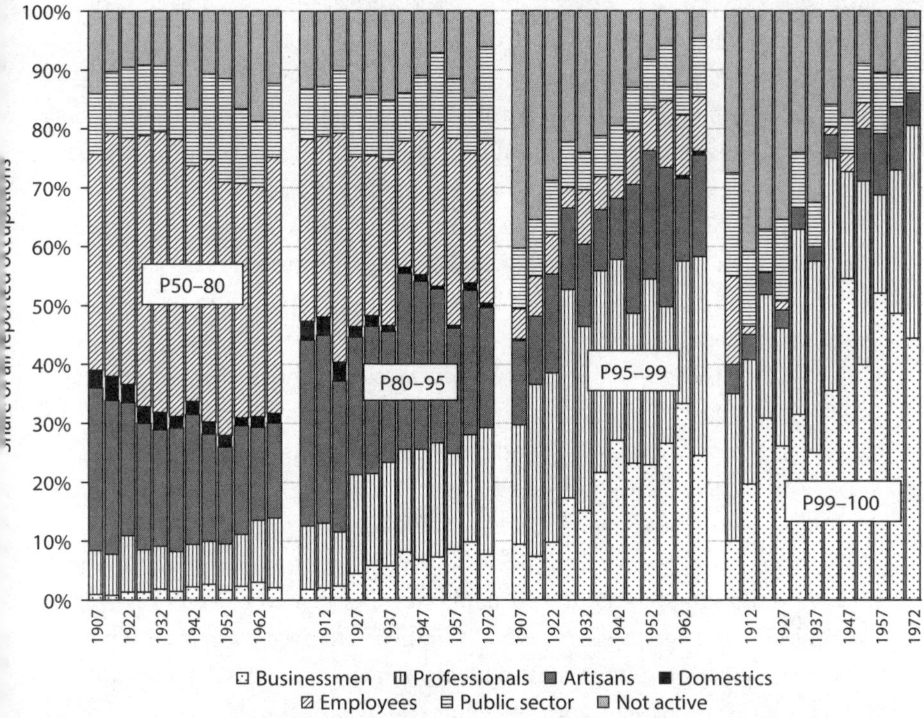

FIGURE 6.3. Occupational categories for men who died younger than 70 by wealth fractile in Paris, 1907–1972. *Note*: Individuals included are those in the detailed sample. *Source:* Estates dataset (see appendix 1).

managers or professionals, while still small, is systematically higher than for those in the P50–80 group. Finally, the occupational distribution of those two groups saw little change between 1907 and 1972.

The third panel of figure 6.3 shows the P95–99 occupations; these decedents are quite rich. The very rich (top 1%) appear in the rightmost panel. These two groups offer a stark contrast to those below the 95th percentile; they have different occupations and experience massive change over time. Both panels show that early in the twentieth century there were substantial numbers of men who died younger than 70 and had no occupation. Their share declined steadily and was less than a fifth after WWII, even at the very top. Like the rentiers, the idle rich went by the wayside in the interwar years. Instead, for the P95–99, close to 70% were either artisans, professionals, or business directors. For the P99–100, professionals and business directors alone comprise 70% of the data after WWII. These are distinct occupations from those heavily represented in the left two panels. All this supports the view that different processes were at play: one for most of the population, another at the very top.

The occupations reported in estate tax filings between 1947 and 1962 reflect decisions made in the 1940s at the latest, most likely in the 1920s and 1930s, well before the onset of the Trente Glorieuses period. The decedents with wealth in the sixth and seventh deciles may have changed jobs after WWII, but they were very unlikely to have attended high school. They acquired their skills and education in a context that featured universal primary education and quite limited access to high schools (for more details, see chapter 11). Secondary education opened the door to the professions, but until the 1950s it was tuition-based. We posit that, most likely, individuals without inheritance prospects came from families where credit constraints most often limited advanced education. The fact that professionals are scarce below the top 5% supports this view. A few additional findings are worthy of note. First, the public sector is a pretty constant share of the decedents' occupations because it includes both low-skill, low-paid jobs, as well as high-skilled, well-paid ones. For instance, a tribunal needs low-paid ushers and clerks as well as highly paid prosecutors and judges. All will get pensions if they stay in place long enough. Second, the P80–95 includes more small-scale entrepreneurs (artisans, shopkeepers) and business directors than the fractile below. It is important to remember that these occupations were not covered by retirement schemes during this period. They were obviously the beneficiaries of the rapid rise in equity values from 1947 to 1962.

The second source we have as to what made individuals wealthy is their portfolios at death. In chapter 5, we used these data to seek information about retirement pensions. Here, we are interested in what portfolios can tell us about either investment strategies or occupations. We again split the observation into four groups in figure 6.4. The leftmost panel shows portfolios of the P50–80 group. Its primary asset is real estate (through the 1950s, these were mostly outside Paris). Until the 1960s, the other noteworthy assets are life contingent. We split life-contingent assets into pensions and life insurance (a benefit that many enterprises had put in place for their employees). Equity and business assets have a limited role in this group's portfolio. The importance of these assets drops by about half after WWII. One reason for the fall in this category is that the P50–80 grows over time (from about half the eighth decile to include all of the seventh and part of the sixth). The primary influx involves employees and workers (who do not own business assets) but who own real estate.

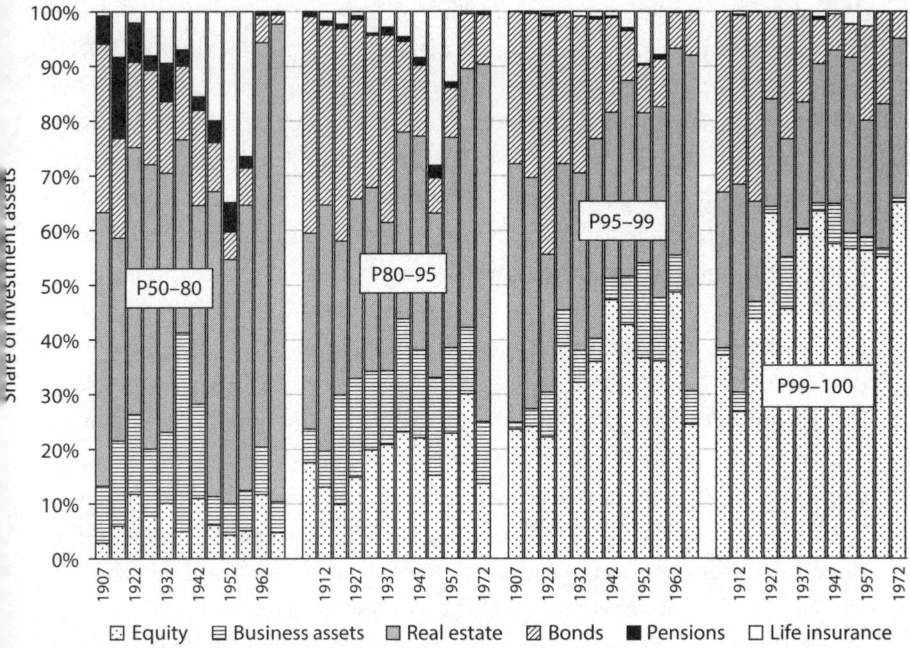

FIGURE 6.4. Portfolios for all men who died younger than 70 by wealth fractile in Paris, 1907–1972. *Note*: Individuals included are those in the detailed sample. Business assets are reported as either a whole business or a collection of items, including goodwill, inventory, and equipment. *Source:* Assets dataset.

The next group, P80–95, has a more standard portfolio. Except for 1952 and 1957, life-contingent assets play a small role in their wealth, and their aggregate portfolio is broadly distributed between equities, real estate, bonds, and other assets. The noteworthy element of this middle group's portfolio is the steady decline of bonds in total assets from nearly 40% up until 1922 to only about 10% in the last few cross sections. They are also a group that generally has twice as much of its wealth invested in business assets as the group below it. Consistent with occupational data, many members of this group constituted their bequests as small-scale entrepreneurs.

As with occupations, the right side of figure 6.4 is entirely different from the left. Life-contingent assets are nowhere to be seen. Equities (business assets and shares in companies) were very important in the 1920s and 1930s, and they grew to become the majority of the portfolio for all years that followed WWII. For individuals below the 95th percentile, the distinction between business assets and shares in companies is clear: they are too poor to own large stakes in large corporations even if they have management roles.

The business assets are assets in firms they are involved in and often bear their names. Above the 95th percentile, the distinction is less meaningful. Indeed, some family-controlled firms were corporations, and a subset of these had publicly traded assets (e.g., Michelin, Air Liquide, Le Creusot). Conversely, many highly valuable firms remained noncorporate, particularly in finance and luxury goods. High rates of wealth growth cannot come from savings out of wages because, for the very rich, wages are small relative to wealth; instead, they must come from reinvested capital income and capital gains.

Individuals who died very wealthy owned large equity positions in publicly traded or private corporations or family firms. In a sense, those who died in the 0.1% had won a capital gains race: the firms in which they held large positions outperformed the market. If we could examine the portfolios of those who died in 1962 a decade or two earlier, we would likely find that they both held a wide range of assets and, at the same time, they directed most of their wealth into a few companies. While we cannot be sure this is the case, we find the two patterns of diversification with large weights on a few assets in every cross section. This combination creates two opposing forces. The diversification means that they were more likely to receive the average return. That return, however, would need to be much higher to explain the rapid wealth recovery of the very rich from 1947 to 1962. Because none of them owned anything like a portfolio of the top 40 listed firms (the CAC40), they were each betting on a small number of sectors and a smaller number of firms in those sectors. They were thus exposed to the variance in returns of their narrow investments. For some, this was bad; for others, it was excellent. Those who guessed right or were lucky, all other things equal, ended up at the top. Given that we cannot roll back the clock ten or twenty years, we cannot find the individuals who twenty years before were just as rich as those who died in the 0.1%, but had had low returns and ended up elsewhere in the distribution of wealth.

A second force pushed up wealth at the top. It can be seen if we turn to the value of privately held businesses (*fonds de commerce*). The price of private enterprises is the one series (with real estate) with a growth rate high enough to explain the high growth rate of wealth at the top. One way to think about why fonds de commerce did so well is that their value combined standard returns for passive investors (what the Bourse delivers) and the control premium that tends to go to the individuals in charge of large corporations. In publicly traded firms, control premia were paid to directors, and these formally appear in our data as "tantièmes" and "jetons." These are not counted in equity return indexes, and they are concentrated in the portfolios of men in the top 1%. These payments are a tiny share of wealth. They are the visible elements of the benefits of control—most of which we do not see. The advantage of sole proprietorships and partnerships is that these benefits of control went to the owners and were thus capitalized in the value of these businesses.

Léon Bel, the founder of the Société Bel (the producer of La Vache qui rit and other processed cheeses), provides a good example of the private benefits of control that founders could exercise and the high degree of diversification of the very rich. Bel was the son of a cheese maker who, after WWI, decided to transform the family dairy to produce processed cheese for urban consumption. He and his brother started the business together, but after 1910, he was the business's sole owner. After WWII, he listed the company on the stock market as it continued to expand. He died in Paris in 1957, the eighth-richest person in the city. His estate included 8 personal assets, 6 of which were Société Bel securities and a claim of 7 million francs on the community property of his marriage due to the sale of a property in Lons-le-Saunier between 1948 and 1950.[2] The community's gross total was 872 million francs distributed across 266 items with a wide range of French and international assets. Of these, a small fraction, 19, were directly linked to the Bel enterprise. They amounted to 222 million or a quarter of the community property. Another 10 were in the cheese and dairy sector. They accounted for 36% of the community property. If we consider community and personal assets together, investments in the cheese sector were 65% of Bel's wealth. If the sector did well, so would the family. How well did Bel do? Establishing the real value of his inheritance is easy because the estate filing tells us exactly when cash contributions were made to the community property. 160,000 francs were put in at marriage in 1908. In 1922, he received his brother's estate, which was 81,000 francs. And then he netted 7.6 million francs from his building sale in Lons-le-Saunier. We assume the first two items were cash and should be reflated to eliminate inflation, and the building has to be reflated using the French real estate index. These assumptions lead us to compute a 1957 valuation of Bel's personal assets of 110 million (68 million of cash claims on the community and 42.6 million from in-kind assets). Still, his share of the community property (359 million francs) is more than three times his personal assets. Evidence of private benefits of control can be found in the Défossés stock manuals for 1947 and 1953: board members received 10% of the profits above the 5% mandatory dividends (and Bel was one of four members). Another 20% of the remainder went to holders of *parts de fondateurs*. Bel himself owned 1,600 of these shares.

We close this section with a reminder that wealth recovery did not imply a recovery in the consumption value of wealth. Capital income flows did not change much (between 4% and 5% of value), so the income value of wealth was growing at the rate of wealth. Below the 95th percentile, where neither estate tax nor income taxes bound, the rapid growth of wealth leads to a recovery in the value of capital income relative to wages. Above the 95th percentile, where the decline in wealth to 1947 was more severe, pretax capital income did not recover relative to wages, which grew much faster from 1922 to 1962. Both estate and income taxes mattered for the rich, and the consumption value of wealth after taxes shrank in years of labor income. In 1962, it was between a quarter (for decedents in the P95–99 range) and one-tenth (for heirs of the P99–100 group) of what the same fractile had enjoyed in the nineteenth century. The world had indeed changed.

6.3. The Continued Growth of Lower-Sized Estates

The reader will recall that 1962 is the first year our data are incomplete because from 1960 to 1977, some Parisian bureaus' estate tax filings were not deposited in the Paris archives. The lacunae were worse in 1967, so we did not collect data for that year. In 1970, the administration undertook the first major procedural reform since 1901, when taxes had become progressive. For over a century, estate tax administration had been located on the Place Saint-Sulpice, and it was then decentralized into up to four bureaus per arrondissement (one per neighborhood). More importantly, the administration changed its recordkeeping procedures, abandoning the TSA volumes and replacing them with binders that contained one or sometimes two preprinted forms (*fiches*) for each decedent. For those estates liable to be taxed, the fiches provide summary information about the taxable estate. The actual estate filings were deposited in the archives through 1977 but still need to be completed (for instance, we recovered only 18 of the 28 filings that the fiches suggest were in the top 0.1% for 1972). The filings contain the same detailed information about taxable and nontaxable assets as in previous years and thus are necessary to complete the fiches' aggregate taxable wealth value. We collected all the data from the fiches in 1972 and, to the extent possible, the stratified sample of estates for 1972.

At the bottom, estates below 10,000 francs were not liable for the estate tax. In these cases, the administrators often noted, "estate not taxable."[3] For individuals whose estate passed to children, there was an additional exemption for each child, which varied over time. These exemptions further reduced the number of individuals with a taxable estate. Thus, at the bottom of the wealth distribution, we understate the extent of inequality reduction and wealth growth. More precisely, we recovered 11,265 positive estates. For 410 of them, the fiches reported that the estates were below the threshold. Because the fiches are exhaustive and there were 27,825 adult deaths, we compute the fraction of adults who died wealthy to be 40.5%. We then add 5,024 retirees without wealth because the last pension payment was considered a form of wealth going back to the nineteenth century, bringing us to a share of 58.5% of decedents with an estate.

Interestingly, 49% of the 5,024 retirees with no estates evaluated in the fiches were women, almost the same share as for Parisian decedents (51%). Retirees with no taxed wealth are much more female than the rich (40%) and massively more female than the retirees with wealth (30%). The fact that older women in the 1970s were more often poor is not unique to Paris (McGarry and Schoeni 2000). The high pure retiree fraction for women may well have been exacerbated by the fact that at the modal age, they would have married in the decade after WWI when men were scarce. Retirees with no other wealth have an average age of 75 for men and 77.5 for women, which is a good deal older than the average wealthy decedent (70 for men and 74 for women). These pure retirees may well have consumed their wealth as they aged.

For 1977, the data include an enumeration of all the estates above 50,000 francs initially collected by Clément Dherbécourt. We completed this by adding a 20% random sample of the binders from which we collected information on all individuals with either no wealth or small amounts of wealth (below 50,000 francs). Once merged, these two datasets provide a reliable estimate of the wealth distribution in 1977. The share of individuals with retirement income and/or wealth is close to what we found for 1972. Relative to 1972, 1977 seems to have had fewer retirees with no wealth and more individuals with small amounts of wealth (below 50,000 francs); with notations in the margin "below the threshold." This is consistent with the impact of inflation following the first oil shock. In 1977, 44% of deceased adults had positive estates enumerated, and if we add retirees with no estates enumerated, we obtain 63.6% of deceased adults with positive estates.

That nearly 20% of individuals did not have a valued estate even though they had wealth is not surprising. To the extent that there is a surprise, as we saw in chapter 3, it is that after the creation of a lower threshold for the inheritance tax in 1956, estates below that threshold were overwhelmingly reported in 1957 and 1962 in Paris. It took more than a decade or so after the law passed before the number of such filings declined. For treasury officials dealing with such filings, the revenue from taxing inheritances worth less than 10,000 new francs was essentially trivial and likely not worth the effort. This was particularly true for Paris, where three-quarters of the wealth lay in the top 5% of the estates. Thus, the tax authorities developed the habit of noting “no taxable estate” rather than the exact value of small estates, which they should have established first. We cannot ignore small estates because they are essential to establishing the share of the population that died with some wealth to transmit.

Restoring small estates to our measures of inequality, as we did for 1972 and 1977, has three consequences. First, there is a large effect on the fraction of individuals who die wealthy, as seen in the preceding paragraphs. The second consequence is an understatement of the wealth of the people below the top four deciles because some wealth at death is not reported. That effect is small because, as figure 6.1 shows, 93% or more of all the wealth is in the top two deciles. Even if we attribute the sixth decile wealth to every person we estimate to be wealthy below that, it increases the wealth share of the group below the top 5% by less than 0.5%. The third consequence is a drop in the average wealth of those who die wealthy because those small wealth holders are below-average wealth. This last effect matters because the number of individuals who die with taxed wealth increased by about half in the twentieth century (from 25–30% to about 40%).

Despite this understatement, for everyone below the top 5%, the dynamics of wealth growth continued to play out in the next decade and a half. In these later years, the shift in the labor/capital distribution of national income that followed 1968 (Piketty 2001), the first oil shock of 1973, and subsequent inflation may have dampened the real growth rate of small and medium estates, but it did not stop it. Over the fifteen years between 1962 and 1977, the estates from the median to the 95th percentile had a real average wealth that grew at 4% yearly, and the real wealth of this group nearly doubled. The average wealth of the P50–95 group was small, however. It was only two and half years of labor income on average in 1977.

The distribution of wealth in the P50–95 group remains unequal: the seventh decile’s average wealth is about eight times that of the sixth decile, and the bottom half of the top decile has an estate 118 times larger than the average estate of the sixth decile. Except for the tenth decile, average estates amount to less than five years’ average labor income. As we shall argue in chapter 9, these modest assets were complements to retirement pensions rather than substitutes. It is easy to see how individuals might constitute such estates out of savings from wages.

To close this section, we return to one of the striking patterns in the portfolios of the lower part of the wealth distribution (P50–95). As time passed, this wealth acquired a singular focus on real estate, often on real estate outside Paris. By the 1970s, the P50–95 group was dying well past retirement age on average. Adding detail to figure 6.4, for the estates between the median and the 79th percentile, real estate increased from 60% to 80% of average wealth. Parisian real estate specifically jumped from a third of real assets to half. For those between the 80th and 95th percentile, real estate went from 47% to 66% of average wealth. Parisian real estate jumped from 20% of real assets to 41%.

There are two consequences to this predilection for housing assets. First, portfolios were poorly diversified because of fundamental indivisibilities in real estate investments. Most individuals below the top decile who owned real estate owned only one income-producing asset: an apartment in Paris or something outside the capital. The second is that, given how well real estate asset value performed relative to stocks or bonds between 1962 and 1977, this focus was enriching.

This importance of real estate in the portfolios below the P50–95 is unlikely to have resulted from some deep-seated understanding of future returns. Instead, it resulted from complementarities between investment (including for bequest motives) and consumption. By the 1970s, most individuals who died wealthy but below the top 5% had some kind of pension. While they died younger than the top 5%, their average age of death was 73, and almost 80% died older than 65, the standard age at retirement. When considering savings throughout their lifetime, they considered both bequests and the life cycle. The latter was primarily taken care of by the pension system; where to live during retirement was not. Investing in a home in which to retire was a sensible way of managing both a desire to transmit assets and reducing uncertainty about how to live while retired.

Looking back 100 years from 1977 highlights just how much had changed in age at death. In 1877 (and nearby years), the adult Parisians died on average at about 52 years, and less than a quarter died aged 65 or more. Even among the wealthy, those below the top 5% died young, at 54 on average, and only 30% died older than 65. In the mid-nineteenth century, retirement was an unlikely event. In the 1910 debates about the retirement law that provided benefits at 65, Jules Guesde, one of the key socialist leaders, denounced the measure. He called it *la retraite des morts* (pensions for the dead). He argued that almost no workers lived long enough to enjoy a pension—and he was not far wrong. By the mid-twentieth century, retirement was likely because all fractiles' age at death exceeded 60. The high real estate share of the wealth below the top 5% tells us that the bequests we observe seem to complement the life-cycle savings that are there but poorly observed.

6.4. Problems Within the 5% Richest Parisians

As noted at the outset of this chapter, between 1962 and 1977, the top part of the wealth distribution did not enjoy the growth of wealth that the lower fractiles enjoyed. The average reported wealth of this group multiplied by 2.6 in nominal terms between 1962 and 1977, but inflation transformed this growth into an 11% decline in real wealth over the same interval. Moreover, the performance of fractiles within the top 5% was directly inverse to its wealth. The P95–99 group saw its real wealth grow by a modest 0.7% a year between 1962 and 1977, for a 12% gain in real wealth. The P99–99.9 fractile only managed 0.1% a year for a gain in real wealth of 2%. Shockingly, the top group, P99.9–100, had an average wealth that fell by 4.2% a year, resulting in a loss of 47% of its real wealth. These super rich are the 25 wealthiest decedents on average and bequeath about 23% of all the wealth that passes from one generation to another.

Before 1977, Paris showed little of the moderation of inequality suggested by Piketty for France (2001, 2014). Inequality in France, an average between Paris and the rest of the country, could decline even if inequality did not change in Paris. This could occur if Paris's share of French wealth declined or inequality in the rest of the country dropped. Because 1977 is the end of our series, it threatens to play an outsized role in our conclusions. What could explain the decline in wealth at the top?

To check whether 1977 was an anomaly in terms of wealth at the top, we collected the 1% top wealth holders from the fiches available for the 7th, 8th, and 16th arrondissements for 1970, 1974, and 1976 and added them to the top 1% wealth holders for 1972 and 1977 to build an extended sample of the super rich. We chose these arrondissements because, after WWII, 101 of 120 (84.2%) of the members of the top 0.1% resided there when they died. The next most popular residence was the 17th, and it accounted for only seven top wealth holders, while no other arrondissement had more than three top wealth holders over all these years.[4] As this extended sample comes from the fiches, we have taxable wealth, not total wealth. When we compare the sample with the declarations for 1962 and 1972, to compare the comparable, we have thus to net out the exempt assets in those years.

Let us start with the very rich (top 1% excluding the top 0.1%). The sample confirms what we found for the very rich in 1972 and 1977. The very rich were unexpectedly poor in 1972, and their wealth grew only a little over the next five years. Between 1970 and 1977, the average wealth in this group fluctuated between 1.45 million 1912 francs and 2 million, given that, in 1962, the group's average wealth was 1.7 million; this represents, at best, a 1.1% growth rate per year. Within the very rich, we also find a definite pattern of declining growth rates with wealth after 1970. The wealth required to join the top 1% grew in real terms by 33% from 1970 to 1977. The median member of this group had 30% more wealth in 1977 than in 1970. The top quartile threshold only grew by 10%, and the richest member of the group (the individual not quite rich enough to make it into the top 0.1%) had essentially no wealth growth between 1962 and 1977. More important, the different years do not appear to offer significant variability. Had we ended our investigation in 1976 instead of 1977, or drawn 1970 instead of 1972, our conclusions for the P99–99.9 group would be identical.

We now focus on the super rich (the top 0.1%). They also fail to get richer, but their failure is much more dramatic than that of the very rich. Indeed, their share of reported wealth plummeted to 11% by 1977, far less than their long-term average wealth share of about 23%. This huge and late drop requires explanation. The first and obvious possibility was that the super rich decedents of 1977 were relatively poor by chance (their nominal wealth was, on average, 24.8 million, a drop from the 30.4 million of 1972). The range between the poorest and richest super rich is always extensive and quite variable from cross section to cross section. It ranges from 2.8 in 1817 to 33 in 1892. From 1947 to 1972, the ratio runs about 10 (with a low in 1957 of 4 and a high of 16 in 1962). The ratio from minimum to maximum wealth with the top 0.1% predicts the top group's wealth share very well (the correlation is .81). The variability in the min-max ratio itself is partly driven by the draw of the top wealth holder within this small group composed of about 25 super rich decedents, themselves drawn from a population of 500 or so living super rich. The top estate size in any given year was quite variable and dependent on exactly who died.

For example, 1977 had top estates much smaller than either 1972 or 1976 but in line with 1970 or 1974. In contrast, 1972 is unusual because the top two estates were very large (the largest estate is 22 times the smallest one in the top thousandth). Overall, had the min-max ratio in 1977 been 16 instead of 6, the top 0.1% wealth share would have been about 17% instead of 11%, halfway back to the long-term average of 23%. More importantly, the rest of the top wealth holders are also poorer relative to the next group (the rest of the 1%) than usual. Indeed, in 1977, the wealthiest 0.1% were only a bit more than four times as rich as the next group. This low number had only occurred once (in 1957) in more than a century. So, we cannot blame the low growth rates of the super rich solely on the absence of a great estate.

We have three paths to explain why the super rich had less wealth than expected in the 1970s. First, the poor performance of stocks and bonds kept asset prices low and likely held back capital appreciation. Second, the long shadow of the Great Depression and WWII, which many of the decedents in 1977 had experienced as adults. These shocks and the nationalizations that followed likely interrupted the accumulation trajectory of many would-be entrepreneurs, who then did not amass the large equity positions that would have been necessary to join the super rich. Finally, there may have been a sharp rise in offshore investments to evade taxes, which would have reduced taxable wealth growth for the largest estates (Zucman 2014; Cummins 2022).

As we discussed earlier, equity prices were in the doldrums in the 15 years between 1962 and 1972, with a negative real return of almost 8%. The 1970s were another lost decade for capital gains. Even forcing the top 1% to reinvest all dividends produces a negative real return of 5%. The portfolios of the very rich and super rich, with their massive (50% or so) weight on equity and a relatively small weight on real estate (30%), would have been particularly exposed to the poor performance of financial assets. We can test the importance of these channels by comparing the wealth performance of the top 0.1%, which is dismal, to the next group (the rest of the 1%). Given all the advantages the top group had in making profitable investments, a real decline in wealth for this group would make sense only if it was accompanied by a similar (and more likely larger) movement for other fractiles. That, as we have seen, was not the case. One might want to blame progressivity. If this involves estate taxation, it is hard to make the case. The parents of the individuals we observe in the 1970s died in the 1940s and 1950s when the maximum average tax rate was equally severe for the top 1% and the top 0.1%. In fact, the late 1950s saw a beginning in the moderation of the progressivity of the estate tax. The maximal marginal rate was lowered to 15% in 1959 (JO December 29, 1959, 12467–8). For indirect successions, taxes were now flat and high (from 45% for close relatives to 60% for people without family connections). Income taxes remained very progressive, thus, overall

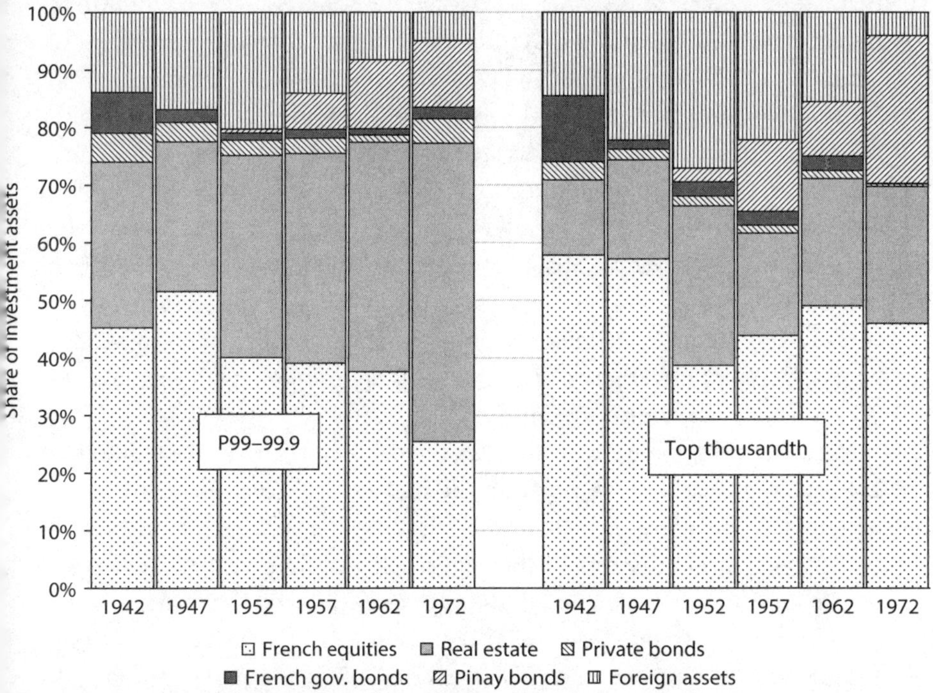

FIGURE 6.5. Portfolios for all men who died younger than 70 and in the top centile in Paris, 1942–1972. *Note*: Individuals included are those in the detailed sample. *Source:* Assets dataset.

tax rates were higher on the capital income of the top 0.1% than the rest of the 1%. However, because these individuals usually exercised control of an enterprise, they could choose when to distribute profits and thus mitigate the impact of taxation. Low returns and taxation may have played a role, but it is hard to see why they would affect the deceased of 1977 so much more than those of 1962 or 1972.

Figure 6.4 shows that the top 5% was heavily loaded with equities. Figure 6.5 breaks down the top 1% into the P99–99.9 and the top 0.1% from 1942 to 1972. The two groups have different investments, with the former having a more French-biased portfolio and a large bond position at all times. Similarly, the real estate share of the portfolio is smaller for the super rich than for the next group down. Finally, estate-tax-exempt assets (new real estate and government bonds knowns as Pinay bonds) increased over time for both groups and fastest for the top 0.1% (by 1972, these shielded assets amounted to 18% of their wealth). More important is the very large place of French equities in the portfolios of the super rich. Recall from table 6.1

that equities performed dismally between 1962 and 1977, even worse than bonds, due to interventions that tilted the share of firm returns that went to labor or reduced firm returns outright. Focusing on the top thousandth, in 1962 that group's average wealth was 20 million francs, and half its wealth was in equities. Consider what happens when we attribute the equity capital gains in table 6.1 to the top 0.1%'s French equity stake in 1962 to predict that group's wealth in 1977. We allow the rest of their portfolio to have the same wealth growth as we observe for P99–99.9 and reach a 1977 wealth of 34.6 million, much more than the 20.4 million francs reported in the fiches and the 25 million we estimate taxable and tax-exempt wealth might have been. The poor performance of equities is simply not enough to explain the very low levels of wealth we find for 1977.

The second explanation attributes the drop in wealth at the top to the cohorts of individuals who died in the mid-1970s. Selection is always an issue when considering the top 0.1%: they are the winners of the wealth race. About three-quarters of the super rich who died in 1977 were born before 1900 and thus old enough to have seen service in WWI and experienced the Great Depression and WWII. One version of selection is aging combined with retirement. As discussed earlier, the top part of the wealth distribution died old, and it is possible that over time, the super rich were spending more time retired, and were not enjoying the wealth accumulation benefits of being in control of an enterprise. If this kind of dynamic had taken hold in the 1970s, the individuals who died in 1977 or thereabouts would have long been retired and thus "enjoyed" only the passive capital gains or losses from their investments. One of the top wealth holders in 1977 provides just such an example. Mr. X followed his father into the profession of notary in the early 1920s and retired in the early 1950s, thus spending a quarter century just managing his investments. Although Parisian notaries pepper the ranks of the top 1%, it is very unusual for them to die as one of the super rich, confirming a lack of large entrepreneurial fortunes.

More generally, the cohort may well have lost, during WWI, many of its financially most promising members (young, educated men filled the ranks of junior officers who had the highest fatalities). Those who survived then went into less entrepreneurial careers. Because for most of the years in the 1970s, we only collected aggregate taxable wealth, we cannot use assets to infer individuals' careers. As a remedy, we decided to thoroughly examine the death and marriage records of the 157 individuals who comprised the super rich in the 1970–1977 extended sample. Because women are most often recorded as "without occupation," we gathered information on both the decedent and their spouse at the time of death, so we are considering 305 individuals.[5] The first observation is that spouses died over a wide span of time, from two who died in WWI to two who died in 2021. Top wealth holders and their spouses mostly died old. Women's interquartile age at death range is 72–89 (six centenarians), while men's range is 65–83 (one centenarian). Forty-six percent of the super rich were born in Paris, and 70% were married there. Of those for whom we could consult the marriage records, 64 out of 68 (94%) couples had marriage contracts, and 80% had drawn up their contracts in Paris (including some who did not marry there). From a demographic point of view, there is little evidence that the top part of the wealth distribution was different than it had been in the 1950s and 1960s. Like their predecessors, they died old, were more likely to have been born in Paris than the average, and even likelier to have married there, had high expectations of inheriting something, and thus drew up marriage contracts.

What of their economic life? The estate tax fiches are mostly uninformative: 90 of 173 fiches report either without profession or retired. This is particularly true for women, where 68 fiches of 75 have no helpful information. The death records are barely more telling: 99 of 212 have no helpful information. Here, however, the bias is entirely with women. Of the 125 men for which a death record could be accessed, only 20 do not provide occupation relative to 76 of 84 women's records.[6] Fifty-six men (45%) are business directors of some form or another, which is relatively high and consistent with what we observed in the prior period. If anything, these numbers are larger than expected. Marriage records also provide little information about the occupation of women (1 student, 1 secretary, and 68 without occupation). Of men, 13 of 70 are engineers, more than those who report themselves as without occupation (12); 8 had had military careers; and others are in business in various capacities but are not yet firm directors. Last, we can plumb the Legion of Honor files. These almost always provide some occupational information, but again they are heavily biased toward men. Taking all these sources together, we can piece together a fair amount of information about decedents' sectors of activity.

For women, we cannot attribute any occupation to 110, and we have 1 editor, 1 actress, 2 writer/painter, 1 artistic director of a significant jewelry firm, 3 firm directors, and 3 heiresses of significant French firms. This is not entirely surprising given how elite families treated women's participation in the labor force. Of the 139 men for whom we have at least 1 observation of occupation, the largest group is industry (58) followed by services (13), while commerce (11), unidentified business directors (11), and no occupation

(15) round out the large categories. In addition, there were 8 artists and art dealers. Civil servants, education, and agriculture each amount to no more than 6 individuals.

Finally, one striking phenomenon of the 1970s is the near absence of finance among the super rich's occupations. There were only 6 financiers among the 139 men whose occupation we found. What is missing in these lists are the family names of the elite investment banks of Paris: Goudchaux, Hottinguer, Rothschild, Stern, and others. Starting in 1862, that sector was very well represented among the super rich. If we take the distribution of wealth in 1977 as reflecting relatively proximate events, the reversal in wealth shares occurs in a period without a financial or political catastrophe. If, instead, we take the long view, the absence of bankers among the super rich is not surprising since they entered into activity after WWI and would have come to leadership after WWII, when most of the financial sector was nationalized. Nationalized banks could pay high salaries, but they did not distribute equity to their principals. In conclusion, it is unlikely that historical accidents adversely selected the very wealthy of the 1970s. However, the low fortunes of the 1970s might partly be explained by the absence of financiers at the top.

The third path to explain the lack of growth in the declared fortunes of the super rich is tax evasion. In particular, the data for 1972 and 1977 require us to return to issues we tackled in chapter 3's appendix A3.8. There, we argued that the major problem in the period that followed the first issue of Pinay bonds in 1952 was not so much illegal tax evasion as perfectly legal tax avoidance. Individuals could avoid the estate tax by selling their portfolios and buying Pinay bonds in anticipation of death. In 1972, assets shielded from inheritance taxes amounted to 22% of the top 0.1% wealth holders' fortunes. The problem with the 1977 data we collected is that they are drawn from the fiches that document taxable wealth, and they omit entirely exempt assets.

Like Cummins (2022), and as in chapter 4, we first compute the missing wealth of the super rich as a residual. Cummins's procedure measures missing wealth as the difference between observed wealth and a computed wealth level derived from estates in the past and allowed to grow over decades based on asset returns. We attribute the very rich's wealth growth rate (13%) from 1972 to 1977 to the super rich's wealth. Starting with 30.4 million in 1972 (24.9 taxable and 5.5 not taxed), their wealth should have reached 55.9 million in 1977 (this includes the wealth shielded through Pinay bonds and new real estate). We use the ratio of shielded to unshielded wealth in 1972 (22.6%) and apply it to the measured average wealth of the super rich in 1977 (20.4 million) to estimate 4.6 million francs of exempt wealth. That, in turn, suggests 30.9 million francs of missing wealth per super-rich decedent (55.3% of their total counterfactual wealth). More important, the super-rich Parisians had exempted 63.5% of their wealth from estate taxes. This estimate is very large because the difference between the counterfactual and what we see is attributed to evasion. It is a bit suspect because the counterfactual implies that there was no missing wealth before 1972. From our perspective, this is too radical a change in the tendency of the super rich to move assets offshore over five years.

An alternative is to compute missing wealth as what it had to have been if the top 0.1% wealth share had remained at its long-term average value of 23%. To get to such a share of wealth for the top 0.1% in 1977, we would need an average total estate of 59.5 million francs (a number even higher than the first counterfactual). That would comprise 20.4 million of taxable wealth as observed in the fiches, and 4.6 million shielded. Bringing average wealth to 59.5 million requires another 34.8 million francs to have been fully evaded offshore. If this were true, it would imply that the super rich had successfully shielded 7.7% of all their assets through legal tax avoidance and 58% through illegal evasion.

Because we worry that, by chance, the top wealth holder among the decedents of 1977 was not very rich, we created a joint distribution of the top 19 estates from our data on taxable wealth for 1974, 1976, and 1977 and inflated the 1974 and 1976 data to 1977 francs. This produces three estates larger than any estate observed in 1977 (two for 1976 and one for 1974). We randomly drew 19 estates from this dataset 200 times to create a distribution of average top-taxed wealth. The average simulated top wealth has a mean of 27.4 million francs, significantly higher than the 20.4 million we observed in 1977. Since we assume that the rate at which assets are shielded in France is 22.6% of taxable wealth, that implies 6.2 million francs were shielded in France and a residual 25.8 million in offshore investments. In this case, the top 1% would have shielded 53% of their assets through a combination of legal and illegal means. At the 75th percentile of the simulated distribution, taxed wealth is 30.9 million, 7.2 million is shielded onshore, and 21.4 million offshore. Even then, the top 0.1% of estates legally shielded 12% of their wealth and evaded the fisc by another 36%. In all these missing wealth computations, the larger the measured wealth in 1977, the larger the legal shielding (because it's a fixed proportion) and the smaller the offshore share of wealth.

We can compare these estimates to the ones produced by Alstadsæter et al. (2018) for France. They reported that 27% of the French top 0.01%'s wealth was held offshore in 1977. In the 1950s and 1962, the 0.1% wealth share for Paris was close to the 0.01% share for France. For this counterfactual, we assume that super-rich Parisians behaved like France's top 0.01%. The top group's measured average wealth, including legally shielded assets, is 25 million francs (20.4 million taxed and 4.6 million exempt). If 27% of total wealth is shielded abroad, that comes to 9.3 million francs. The total average wealth of the top 0.1% wealth then comes to 34.4 million francs, far lower than our two previous computations. The top thousandth's wealth share in this case would be 14.7%, not quite halfway between the reported 11% and the other counterfactuals at 23%. If Alstadsæter et al.'s estimates translate from France to Paris then 1977's top group suffers from severe adverse selection. For an offshore evasion of 27% of wealth to produce a wealth share of 23% for the super rich, their average taxed wealth would had to have been 33 million rather than 22 million. It is also possible that evasion in Paris was higher than average conditional on wealth simply because it had the largest number of intermediaries able to move wealth abroad stealthily.

While the range of evasion estimates is large (at least 36% and up to 55%), what is striking is that even if we take the most moderate one, the erosion of the tax base was very severe and change was very rapid. Nearly half the wealth (48%) of the super rich escaped the inheritance tax, up from 22% in 1972. The counterfactual based on Alstadsæter et al. has the minimum total evasion. It is still massive—the aggregate wealth that escaped estate taxes comes to 113 million through tax-avoiding investments like the Pinay bonds and 222 million evaded, 6% of all wealth in Paris. In other estimates, the super rich's offshore wealth is even larger. If we take the median estimate of 29.1 million francs of taxed wealth, then the super rich shielded 51% of their wealth. This corresponds to 11.2% of all the wealth transmitted in Paris in 1977 and 80% of the wealth that escaped taxes.

Let us note two consequences. The first is that despite a broad range of estimates, the extent of tax evasion (on the order of 40% to 50% of total wealth) meant that the tax rate for the super rich was likely about half what it was for individuals wealthy enough to face the maximum rate but who did not shield their wealth. In other words the progressivity of the inheritance tax that had prevailed after 1901 had been lost by the 1970s. Second, tax avoidance and evasion were much larger drains on inheritance tax revenues from the very wealthy than the impact of exemptions at the bottom. Even if we attribute to every individual who was noted as having insufficient wealth to be taxed or retired, the average wealth of the seventh decile, these exempt estates do not amount to 1% of the wealth. The tax regime mattered for the rich, the very rich, and the super rich.

Finally, we considered tax avoidance and evasion and showed that tax avoidance by the super rich was likely substantial in the 1970s. It involved, on average, a fifth of the taxed value of estates, and that proportion was at least twice as large as that for any other fractile of the distribution. We also computed plausible scenarios for tax evasion offshore wealth. They imply that about a third of the wealth of the top group was held offshore and not reported to the estate tax administration. Taken together, avoidance and evasion reduced taxes on the estates of the super rich by more than half. It should be noted that even if we add offshore wealth as we compute it, Parisians in 1977 were far from having reconstituted the private wealth levels that had prevailed in the late nineteenth century. The top thousandth's wealth at death was still only about 60% of what it had been in 1902.

6.5 Conclusion

For various reasons, 1977 marks the temporal end of our investigation. One reason is practical: in 2022, when we closed the data collection effort, the archives' run of records that began in the 1790s ended in that year. Another involves the changing geography of the capital: for the century following the last annexation of territory in 1860, Paris's suburbs could be ignored because government tabulations showed that very few top wealth holders lived just outside Paris. This fact allowed us to limit Paris to its legal boundary. The rise of western suburbs attractive to the very wealthy invalidates the key assumption that rich people who work in Paris largely live and die there. For these reasons, we did not continue this study into the second golden age of financial markets that, for France, started in the early 1990s during François Mitterrand's second presidential term.

The first six chapters of this book each cover changes in wealth and inequality over a well-defined period. However, assembling a dataset over seventeen decades opens opportunities for understanding the long-run evolution of wealth. The following five chapters take up this opportunity.

7

Capital and the Family

INHERITED WEALTH, 1822–1972

The issue of inherited wealth is critical for how we think about a society's opportunities. At one extreme, a static class-structured society can be dominated by inherited wealth because only a fraction of the population has any opportunity to accumulate. If the dominant asset is land passed through generations of wealthy individuals, inherited assets can easily dominate bequests. In this case, intergenerational mobility is low. At another extreme, capital markets are so efficient that talent gets fully rewarded in each generation, and innovation and individuals' personal preferences mostly drive wealth accumulation and bequests. In this case, inherited wealth is reduced by the dynamics of innovation, and intergenerational mobility is high. To parse where a society falls between these extremes, economists and other social scientists often correlate the wealth of parents and children.

Inherited wealth speaks to important questions about the interaction between taxation and justice. On one side, some argue that inherited wealth should be kept low by high taxes on bequests so that wealth accumulation can be the just reward for the efforts and talents of each individual. Others argue that the desire to bequeath is a fundamental preference that should not be tampered with, and a society that justly protects private property by keeping estate taxes low will be prosperous and have substantial inherited wealth.

For economic historians, there is an additional reason to track inherited wealth in the long run—changing motivations for savings. For Franco Modigliani and others in the mid-twentieth century, one of the hallmarks of modern societies is a large share of life-cycle savings. As more and more people add accumulating resources for retirement to their goals, on top of their precautionary savings, and fulfilling whatever bequests they want to make, the inherited wealth share should fall. So, early in the nineteenth century, when Paris was still a traditional economy, bequest motivations should have dominated, and the inherited wealth share should have been large.

Measuring inherited wealth and intergenerational mobility is complicated. The most common approach is to link parents to children and calculate the correlation between the wealth of two generations (e.g., Chetty et al. 2014a, 2014b). This approach has the advantage of measuring two comparable quantities and can be applied to every matched parent-child pair. It has drawbacks, particularly for historical data. The method is entirely dependent on the success rate of matching individuals across generations, with obvious concerns that the rich are always more likely to be found than the poor. There are additional complications: parents are not the only source of inherited wealth. As we have noted, this problem cannot be neglected for Paris, as more than 35% of the wealthy died without direct descendants. Moreover, individuals die at widely different ages, so the difference between the value of an inheritance at the time it was received and the value of the estate could be driven by a wide variety of phenomena. Our inheritance data help us resolve many of these problems because, as we saw in chapter 2, the French legal system distinguishes between community property and inherited assets. We return to this issue here and evaluate inherited wealth over the period for which we have detailed samples (1822–1972). To keep things clear, inheritance refers to what an individual receives from parents or relatives in cash or in kind, while the bequest or estate refers to what the same individual leaves to his or her heirs. The inherited wealth share is the fraction of an individual's estate that is made up of inherited wealth.

Measuring the inherited wealth share allows us to answer a series of questions: What share of bequests came from inherited wealth? How did the great shocks and the fiscal revolution of the twentieth century affect inherited wealth dynamics? How many of the wealthy were self-made? How many of the newly wealthy were so because of financial or entrepreneurial genius versus political success? How important was intergenerational transmission of wealth in the persistence of high inequality and limiting social mobility? Finally, from a more financial perspective, how much aggregate wealth variation was driven by inherited wealth? These questions are about long-term change because intergenerational intervals are a quarter century.

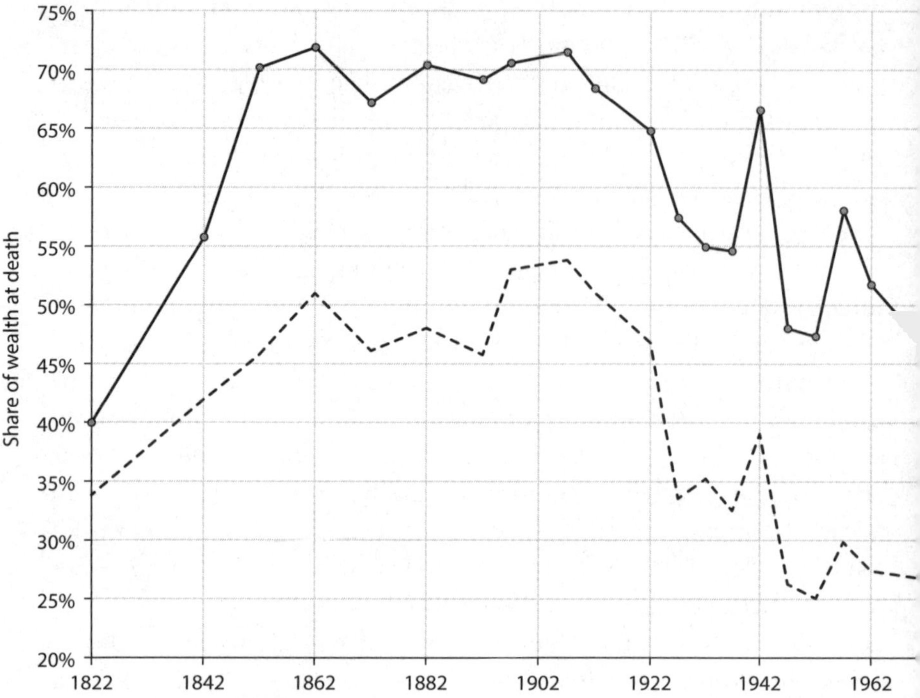

FIGURE 7.1. Inherited wealth shares in Paris, 1822–1972. *Note*: The reported inherited wealth share is the sum of personal assets and net claims on the community. The higher line adds capital gains and capital income, and they are computed using individual rates of return. The sample is restricted to individuals who died married with community assets. *Source:* Estates dataset (see appendix 1).

We start with the approach we used in chapter 2 by totaling the inherited assets as valued in the estate (see the lower line in figure 7.1). This series is too low for two reasons: (1) for all wealth received as cash (including dowries), the capital gains for the investments made during marriage accrue to the community, and (2) the community also receives all dividends, interests, and rents. Nevertheless, the ratio of reported inherited assets to total estate is never less than 40% before WWI and almost 30% after WWII. When we correct for missing capital gains and capital income (figure 7.1, upper line), the inherited wealth share jumps to 70% for every year from 1842 to 1912. So, inherited wealth is important. The decline after WWI is largely due to the fact that inherited assets were particularly hard hit by the negative shocks we discussed in chapters 4 and 5. There are very good reasons to believe that since 1972, the inherited wealth share has been on the rise.

This chapter begins with reported inherited wealth in an individual's estate and how it compares to what he or she received at the time their parents died. In section 7.2 we make the approach to inherited wealth precise, using Antoine Adrien Cousin as a guiding example. In section 7.3, we explain how we simplify the problem of dealing with capital gains and flow income using the example of the Mortier de Trévise family. Section 7.4 focuses on reported inherited wealth over the long run to demonstrate that inherited wealth is significant even if we neglect capital gains, interest, rents, and dividends. Section 7.5 takes on the problem of measuring capitalized inherited wealth for Parisians. In section 7.6, we move from wealth to its consumption value. In section 7.7, we return to the social implications of inherited wealth through the rise and fall of the rentiers.

7.1. Inherited Wealth in Individual Estates

In societies with coverture laws, universal community, or strict separate property marital regimes, as are common in Anglo-Saxon countries, there is never any reason to compute inherited wealth. Thus, this chapter's use of data may be unfamiliar to the reader. We use two families as examples to guide the reader through our construction of inherited wealth: Antoine Adrien Cousin and his relatives, whom we saw in chapter 2, and the family of Adolphe Mortier de Trévise. He was one of Napoleon's generals who, in return for a string of military victories, was made a marshal in 1804 and Duke of Treviso. With these titles came financial favors that he managed to keep after 1815, when Napoleon lost power. After the revolution of 1830, Louis Philippe, the king, came to rely on Mortier as army minister and later prime minister. He had just stepped down from chief minister when he died of wounds received when Joseph Fieschi tried to assassinate the king in 1835. Thanks to the riches he had received from Napoleon, he died very wealthy; his estate of almost 2 million francs would have placed him third among the Parisians of 1832 and fourth in 1837. Four more Mortier men would hold the title before the fifth duke died in 1946 without posterity, and the title lapsed.

The Mortiers illustrate both the power of intergenerational linkages and their pitfalls. Adolphe Mortier de Trévise was a self-made man, though his wealth came from military skill rather than business or financial acumen. All the dukes who followed inherited great fortunes. They also married wealthy women and left large estates. We know about the wealth of the Mortier de

Trévise dukes because they were all residents of Paris, and so were nearly all their wives, and their estates were filed in the capital. The Mortiers also had extensive real estate holdings outside Paris that were difficult to track but would undoubtedly add to the importance of inherited wealth in their estates. By another chance, all the dukes died before their wives, so their estate filings separately report community and inherited assets. This allows us to compare one duke's fortune to another and what a particular duke inherited to what he left his heirs. Indeed, we observe what a duke received from his parents as listed in the parents' estate filings and what the duke's heir claimed were his inherited assets in his estate filing.

Napoléon Mortier, duke no. 2, received about 836,000 francs when his father died in 1835 (his share of his father's movable estate plus his share of his parents' real estate) and 89,340 francs in 1855 when his mother died. This second duke left an estate valued at 2.6 million francs in 1869. So, his inherited wealth share could be computed as 36% if we use the value of the assets he inherited when he received them. We can also measure his inherited wealth as reported in his estate documents. His tax filing reports 1.2 million francs of inherited assets in Paris for a second inherited wealth share of 46%. Part of the difference between these two measures comes from when the inherited assets in kind were evaluated. In particular, the parental inheritance was mostly valued in 1835. By 1869, the same assets were worth 2.4 times more. If rural properties rose at half that rate, it would lead to a total set of passive capital gains of 678,000 francs to add to his inheritance, making the inherited wealth share 59%.[1] Moreover, nearly all the properties were rented (or could be). Even at 3%, the cumulative value of the flow income was large enough that Napoléon Mortier could easily constitute his entire bequest from inherited wealth. Napoléon Mortier reminds us that how we value inherited wealth is critically important.

Before going to the general patterns, we return to the Cousin example of chapter 2 because it offers other valuable lessons. The Cousins were not as grand as the Mortiers and much less inclined to hold wealth outside Paris. So, we collected the details from the estates of Cousin, his parents and sisters, his wife and her parents, and his three children. As we saw in chapter 2, when he died in 1852, Antoine Adrien Cousin left an estate of 2,524,000 francs, slightly less than his inherited assets valued at 2,570,000 francs (chapter 2, box 2.4). As box 7.1 details, Marie Suzanne Letissier, Antoine Adrien Cousin's wife, inherited 563,000 francs (152,257 + 410,975) from her parents and left an estate of 1.8 million francs to her children, more than three times her combined inheritance. Cousin's parents were wealthier and died later, leaving

BOX 7.1 Intergenerational wealth transmission for the Cousin family

		Death details				
		Year	Age	Wealth	Heir lines	Year of marriage
Marie Geneviève Colin	M-W	1824	62	152,257	1	1799
Pierre Letissier	F-W	1833	80	410,975	1	1799
Marie Adrienne Jandon	M-H	1847	88	1,227,256	3	1782
Jean Claude Cousin	F-H	1849	89	1,191,922	3	1782
Marie Suzanne Letissier	Wife	1860	65	1,814,881	2	1812
Antoine Adrien Cousin		1852	67	2,524,063	2	1812
Marie Rosalie Cousin	Sister	1849	62	908,385	2	1808
Marie Jeanne Victoire Cousin	Sister	1842	59	403,732	3	1804
Jean Adrien Cousin	Son	1866	53	2,512,259	3	1844
Suzanne Henriette Cousin	Daughter	1887	67	2,925,670	4	1844

Sources: ADP DQ8, 557 (Colin); 607, 702 (Letissier P); 802 (Jandon); 789 (Cousin JC) 1901, 1069, 1272, 906 (Cousin AA); 1687, 1100, 1211 (Letissier M) and ADP DQ7 3542 #1352 (Colin); 3557 #180 (Letissier P); 3755 #228 (Jandon); 3783 #838 (Cousin JC); 3918 # 614 (Cousin AA); 10731 #656 (Letissier M); 10396 #101 (Cousin JA).

Note: M = mother, F = father, W = wife, H = husband. The estates listed in box 7.1 are larger than what passes to the next generation because they count dowries, life interests, and legacies. Marie Adrienne Jandon's net estate was 1,221,890 francs. Net of her husband's life interest, each heir line received 319,929 francs. Net of bequests to domestics, Jean Claude Cousin's estate was 1,008,468 francs, and each heir line inherited 336,156 francs.

a combined 2.4 million francs. This sum was divided into thirds for each heir line (to Antoine Adrien, Marie Rosalie, and Marie Jeanne Victoire's children), so Antoine Adrien Cousin's inherited wealth as measured from his parents' estate was 800,000 francs, a considerable sum, similar to what his wife received.

Antoine Adrien and his two sisters left estates totaling 3.82 million francs, substantially more than what they had received, except for the oldest sister, who died before their parents (her children received her share of their estates). Jean Adrien and Suzanne Henriette Cousin, Antoine Adrien's children, left combined estates worth 5.43 million when their parents' estates totaled 4.3 million francs.

It seems tempting to combine these observations to conclude that this is a family of wealth accumulators. Cousin and Letissier received 1.36 million francs, left estates worth 4.3 million, and their children passed the 5 million francs mark. To do so would assume that all inherited wealth passes at the time of death and from parent to child. However, Marie Jeanne Victoire had an estate even though she died before her parents. Are we to conclude that she had no inherited wealth except for the dowry of 20,000 francs mentioned in the estate of her mother Marie Adrienne Jandon? The resolution to the problem comes from another item in Jandon's estate. Her children owed her and her husband two life annuities for a total annual income of 78,000 francs from donations in 1832 and 1847. We found the notarial contract for the 1832 donation, and in that contract, Jandon and her husband Jean Claude Cousin gave their three children real estate worth a bit more than a million francs. Then, in 1847, a second inter vivos gift was given for somewhat more than 1.5 million francs. With these two gifts scrupulously divided into thirds, Jandon and Cousin had transmitted 15 buildings, the entirety of their Parisian real property, to their children.

Box 7.2 details the importance of transfers over time. When his parents died, Antoine Adrien Cousin received only about one-third of his inherited wealth. Donations also explain why his sister Marie Jeanne Victoire had so much more wealth than expected, even though she died before her parents (she received the first of the two donations) and an estate so much smaller than her siblings (the rest of her share of her parents' wealth passed directly to her children). To take the ratio of wealth transmitted by the parents (or even the parents' estates) to Cousin's estate completely understates his inherited wealth.

Cousin's parents' donations were exceptionally large; donations themselves were common. As discussed in chapter 3, in 1898, there were 1,895 donations in the département of the Seine, most of them in Paris. For comparison, recall that there were 8,300 positive estates in Paris in 1897 with an average value of 129,000 francs, while the average value of donations in the département of the Seine averaged 73,500 francs. For France as a whole, there were 43,700 donations with an average value of 9,000 francs (BSLC 1899). Donations alone create some serious measurement errors for understanding inherited wealth using parent and child estate pairs.

Other forms of intergenerational transmission complicate the analysis. Take, for example, Aristide Boucicaut and Marguerite Guérin, the founders of the Bon Marché department store. His parents left estates worth about 20,000 francs when they died in 1839 and 1863; she was the daughter of a single mother who was famously poor and whose date of death is lost. Boucicaut died in 1877, leaving an estate of 9 million francs (nearly all in

BOX 7.2 Two views of inheritance flows to Antoine Cousin

Part 1: As observed outside Cousin's estate (at time of receipt)

Item	Year received	Value when received
(1) Dowry	1812	56,000
(2) Inter vivos gift from parents	1832	346,000
(3) Inter vivos gift from parents	1847	521,000
(4) One-third of mother's estate	1847	350,813
(5) One-third of father's estate	1849	336,156
Total		1,609,969
Total inherited from parents at death (4)+(5)		686,969

Part 2: Inherited assets in Cousin's estate filing

Item	Value at death
(1) Dowry	134,000
(2) Movable assets also in parents' estates	468,004
(3) Movable assets not in parents' estates	632,330
(4) Real estate also in parents' estates	335,700
(5) Real estate not in parents' estates	1,151,040
Total inherited wealth	2,421,074

Sources: Dowry AN/MC/ET/LXIV/558, Donation AN/MC/ET/XCVII/798. Estates ADP DQ7 3755, 3783, 3918.

Bon Marché equity and the buildings associated with it), a classic self-made man. Marguerite died in 1887 without heirs and donated the second-largest estate in Paris, 45 million francs, to build the Boucicaut Hospital. In 1887, she was the principal owner of the Bon Marché (the firm) and separately owned the store's buildings, each valued at more than 17 million francs. She inherited nothing from her mother (no personal property is mentioned in Aristide's estate). However, she inherited both from her husband (2 million francs) and from her son Antony (2 million francs), who died before she did. She exercised a clause in the partnership contract at Antony's death and retained all the family's equity in the Bon Marché. She, like Aristide, was a self-made entrepreneur (she ran the Bon Marché for a decade of rapid growth), but her inherited wealth from her mother (nothing) is not her total inherited wealth. Donations to charity on such a scale were extremely rare. Collateral inheritance, however, was common: 30% of all positive estates had no direct heirs (children or grandchildren) in Paris. For France, from 1826 to the 1870s, 30% of the wealth did not go to children or their descendants. After 1870, it's a third or more (ASF 1885; BSLC). Close family members (spouses, brothers and sisters, nephews and nieces, cousins, and so forth) were the usual recipients of the legacies of those without direct descendants.

Returning to the Mortier, their inheritances generally came from their parents. Yet there was a sizable exception. The third duke died without children. His title and much of his wealth passed to his brother. This, in part, explains the fourth duke's relatively large wealth as he inherited more than 1 million francs from his parents and another equivalent sum from his brother. For Cousin, Guérin, and Mortier, measuring inherited wealth from parents' estates to children's estates is problematic because, in each case, it misses assets that arrived through other channels. Moreover, the mismeasurement leads to a systematic understatement of inherited assets. Our approach selects the data on a different dimension—dying married—and relies on the fact that for those individuals, an accurate record of inherited assets was established to divide the couple's total resources into the community and the inherited assets of each spouse. This is the method we discussed in chapter 2.

7.2. Computing Cousin's Inherited Wealth

The ratio of personal assets to wealth at death is the starting point for measuring inherited assets. Yet the raw data cannot be the final answer for two reasons. To begin, a substantial share of all estates was composed of inherited assets that appear in estate filings just as cash or liabilities from the community account. These included dowries that were received in cash and inherited assets that were sold during the marriage. In both cases, this inherited asset component could have been invested and earned either capital gains or income, or both. We can make detailed computations for Cousin because the amount of information given in the marriage and the donation contracts is far more than usual. His total inherited assets in cash come to 332,072 francs: 134,000 were received in 1812 or 40 years before he died (see box 2.2), 30,000 came when he sold one of the properties he had received while his parents were alive, and the rest came to him later, most likely when his father died in 1849. All this cash could have been invested to produce income (rents, dividends, or interest) and experience capital value changes. For instance, had Cousin invested his dowry in 5% government bonds in 1812, he could have bought bonds with a face value of 162,000 francs that would have paid an annual interest of 8,121 francs and a total of 324,000 francs over the next 40 years. The power of compounding implies that had he reinvested these payments starting on his first wedding anniversary, his heirs would have inherited bonds with a face value of 1,322,000 francs. French 5% bonds had dropped massively during the 1848 revolution, but they were at par when he died, so that is a reasonable market value of the 134,000 francs his father and mother gave him to start in life. Cousin was not idle since he was a notary, a position expensive to buy but quite lucrative, so he had other income as well.

The second issue involves income flows from inherited assets. Beyond cash, the other important items Cousin received were houses, which earned rent when he received them. More generally, inherited assets that remain unsold produce capital income that should also be counted as part of inherited wealth. Here, and unlike the cash account, we have the terminal values, so we do not need to deal with capital gains. For Cousin, the houses he received in 1832 were earning rents worth 13,480 francs each year, and he had to pay an annuity to his parent of 10,000 francs a year. If we assume that 20% of the total rent went to upkeep, in the first years, Cousin was not making much money on these houses. By the time he died, the two houses were paying 31,767 francs in rent yearly, leaving him with an income of 25,000 francs (by then, his parents had died, leaving him free of the annuity). In the meantime, the houses had appreciated from 270,000 francs to 635,000. Had he invested the rent profits in government bonds, the first inter vivos gift donation would have ballooned to more than 1 million francs from 270,000 francs when he received it in 1832: capital gains and capital income mattered. The second inter vivos gift arrived much later, in 1847, and involved three houses earning rents worth 25,050 francs (net upkeep of 20,840), on which Cousin owed a 16,667 francs annuity. Given the difficult economic times of 1848–1851, we can assume that capital gains were relatively limited and that Cousin just received the net rents of 4,173 francs in 1848 and 1849, and 20,840 from 1850 to 1853 after his father died and the annuity stopped. We capitalize the inheritances received in 1847 and 1850 at 3%. Box 7.3 summarizes the results.

Although his estate was three and a half times larger than the bequests he received from his parents at their death, it was only slightly larger than the total uncapitalized inherited wealth recorded in his estate (134,000 + 468,004 = 600,004, cf. box 7.2). The capitalized value of his inherited wealth is, however, greater than his estate (3,661,548/2,524,063 = 1.45). This means that 100% of his estate is inherited wealth, implying that he and his wife

BOX 7.3 Computing Antoine Cousin's capitalized inherited wealth

Type of asset and year received	Uncapitalized value	Years held	Capitalized value
Dowry (1812)	134,000	39	1,322,000
From mother (1847)	319,929	4	360,083
From father (1849)	336,156	2	356,628
Donation (1832)	635,340	19	1,022,217
Donation (1847)	521,000	2	600,620
Other liabilities	−14,350	1	−14,780.5
Inherited wealth			
Total inherited wealth	1,932,076		3,661,548
Total estate	2,524,063		2,524,063
Inherited wealth share	76.5%		145.1%

Source: See box 7.2.

consumed a substantial fraction of the capital income from the assets he received from his parents. Remarkably, his estate came to only 69% of the wealth he would have transmitted had he lived only on labor income. Cousin, to be sure, is exceptional only by the magnitude of the inter vivos gifts he received. His family is outstanding because of the extreme longevity of his parents—Marie Adrienne Jandon died at 80, and Jean Claude Cousin died at 89, when the average age at death among the top wealth holders was 68. When they made their first inter vivos gifts in 1832, Marie Adrienne Jandon and Jean Claude Cousin were in their early seventies. Antoine Adrien Cousin would have had even more inherited wealth if his parents had had shorter lifespans. Cousin is not unusual in the magnitude of his inherited wealth. As we shall see, every year there was a sizable group of individuals whose estates were nearly entirely composed of inherited assets as reported in their estates.

7.3. Computing the Mortier Dukes' Inherited Wealth

While computing inherited wealth with the level of detail that we did for Antoine Adrien Cousin is possible, it is incredibly time consuming because, in reality, unlike in overlapping generational models, individuals do not die in lockstep (only Cousin, his mother, and daughter die in a year that ends in

2 or 7, everyone else had to be recovered separately) or in the same place. Collecting the estates of parents and children would require collecting data for the whole population of decedents. Recovering marriage and donation contracts would be even more complicated. So, we must simplify the computations. We use the Mortier family to explain how we go about this.

The Mortier family has a complete genealogy. Thus, it is easy to determine when individuals married and when their parents died. Unlike for Cousin, we do not observe inter vivos transfers except dowries. We assume that all inter vivos gifts other than dowries were received as inheritances, as we will do for the whole population.[2] This shortcut reduces the inherited wealth share because inter vivos gifts are, by definition, received before the parents die. Thus, they produce a longer stream of capital income than we measure by assuming all assets except dowries are received at the parents' death. The downward bias embedded in this simplification strengthens our demonstration that the inherited wealth share was quite high.

As we did for Cousin, we start by comparing parental inheritances and personal assets. Parental inheritances for a given duke are simply the parents' estates divided by the number of heir lines reported in parentheses after each spouse. Personal assets are what the heirs to the given duke reported as assets that lay outside the community property in his estate, be they inherited assets in kind or in cash (see table 7.1). Parental inheritances are evaluated when the parents died, while personal assets are partly assessed when the duke died, typically more than 20 years later. Since all the dukes only married once (between the ages of 24 and 32), we do not need to worry about their personal accumulation. We were not able to recover the wealth of the first duke's parents, though both of them had left an estate. As one can see in table 7.1, the second and third duke's inheritances and personal assets were quite similar in size. The two estimates do not match for the fourth duke, who inherited from his parents and partly from his brother, the third duke. The fifth duke also had a large deviation due to the massive inflation from 1912 to 1946.

We can now carry out the steps of adjustment described above for each of the four dukes that followed the founder. We start with personal assets in kind and personal assets in cash, as reported in the estate from table 7.1. The first step is to return the assets in kind to their value at the time they were inherited, which we did in Cousin's case by looking them up in his parents' estates and the inter vivos gifts. For the Mortier dukes, we use the capital asset values from three series to impute capital gains and capital losses: the CAC40 dividend return estimated by Lebris and Hautcoeur (2010), the

TABLE 7.1. Wealth and inherited wealth for the Dukes of Trévise

			Inheritance seen in		
				Own estate	
	Lifespan	Total estate	Parents' estates	Personal assets in kind	Personal assets in cas
Duke 1	1768–1835	1,999,477			
Spouse (4)	1779–1855	1,704,439			
Duke 2	1804–1869	2,630,665	925,979	1,415,454	−203,711
Spouse (4)	1808–1870	2,165,614			
Duke 3	1835–1892	1,184,203	1,199,070	732,183	452,018
Spouse (0)	1835–1916	2,584,131			
Duke 4	1845–1912	6,881,061	1,199,070	4,858,165	1,658,001
Spouse (3)	1855–1921	5,201,933			
Duke 5 (divorced)	1883–1946	13,011,572	4,027,665	13,600,000	622,767

Source: ADP DQ7, 3839 #59 (duke 1); 10516 #772 (spouse of duke 1); 10264 #570 (duke 2); 10624 #571 (spouse of duke 2) 10043 #900 and 12242 #1292 (duke 3); 31910 #1433 and 31916 #225 (spouse of duke 3) 31822 #1949 (duke 4); B7 31996 #466 and 32009 #365 (spouse of duke 4); 29090 #1792, 29097 #49, 2910 #2228 and 29148 #1343 (duke 5).

Note: For duke 5, who died divorced, inherited assets are the value of real estate found both in his estate and his parents' estate filings, and the cash value is his dowry as seen in his father's estate. We cannot count the inherited assets that were sold during his lifetime. The number next to the spouses is the number of children they had.

return to real estate computed by Eichholtz et al. (2020) and the yield on public *rentes* for bonds in Homer and Sylla (1996). To arrive at capitalized wealth, we assume that the inherited assets, including cash, experienced the capital gains and flow returns of the in-kind portfolio from when the duke inherited from his father to when he died. The computation produces the last column in table 7.2, capitalized inherited wealth.

We compare the different definitions of inherited assets to each duke's estate in table 7.3. Column 2 compares the nominal value of inherited assets when they were inherited to the estate the duke left behind. This column answers the question: What did the duke inherit relative to what he left behind? Even these numbers, which ignore real and nominal capital gains, are large except for the last duke. His tiny share is purely a product of the inflation between 1914 and 1946. Column 3 uses the value of reported assets in kind, leaving aside all the complications of the cash accounts.

LE 7.2. Adjustments to the Dukes of Trévise's estates

		Inherited assets				
		Cash (1)	In-kind, as reported (2)	In-kind, when inherited (3)	All, when inherited (4)	Capitalized inherited wealth
e 2	1804–1869	−203,711	1,415,454	1,250,440	1,046,729	3,547,253
e 3	1835–1892	452,018	732,183	674,885	1,126,903	3,208,920
e 4	1845–1912	1,658,001	4,858,165	3,667,808	5,325,809	16,703,621
e 5	1883–1946	622,767	13,600,000	1,566,100	2,188,867	56,147,707

ce: See table 7.1

: Column 3 shows the value of inherited assets when a duke inherited them, where we deflate the in- assets by the average nominal capital gains. Column 4 adds the values from columns 1 and 3.

TABLE 7.3. Inherited wealth shares for the Dukes of Trévise

			Inherited wealth share			
		Estate (francs) (1)	Valued at the time of inheritance (2)	In-kind as reported in estate (3)	Total as reported in estate (4)	Capitalized inherited assets (5)
			%			
Duke 2	1804–1869	2,630,665	39.8	53.8	46.1	134.8
Duke 3	1835–1892	1,184,203	95.2	61.8	100.0	271.0
Duke 4	1845–1912	6,881,061	77.4	70.6	94.7	242.7
Duke 5	1883–1946	13,011,572	16.8	104.5	109.3	431.5

Source: See table 7.1

The second duke has the smallest share of his estate, composed of assets in kind (53.8%). The last duke has more inherited assets in kind than the value of his estate. When we include the value of inherited cash (column 4), the second duke drops to 46.1% because his real estate dealings led him to have debts to the community; all the others have uncapitalized inherited wealth close to or larger than their estates. When we add the compounded capital income to their inherited wealth (column 5), it is clear that all the dukes constituted their estates from what they inherited with relative ease—they could consume significant amounts of the rents, interest, and dividends they received.

As we shall see for the population as a whole, the exact choice of return for the dukes' inherited assets does not matter very much. Given an average intergenerational interval of 25 years, a total return of 3% (capital gains and flow income) is enough to produce an inherited wealth share of 100% for anyone whose inherited assets at inheritance are 50% of their estate. Three percent is roughly the lower bound of capital returns for most of the period. Any return larger than 3% does not affect such an individual's inherited wealth share because it is 100% and cannot go above that. As we shall see, most people fall into one of two groups in Paris. Most, like the Boucicauts, have tiny amounts of inherited wealth, and only astronomical rates of return would bring their inherited wealth to 100%. A smaller but sizable group is like the Mortiers. They have inherited wealth larger than 50% of their estate, and once capitalized, it will hit the 100% threshold unless they die very young. We now turn to compute inherited wealth for the 23,957 individuals who were sampled and died married.

7.4. Reported Inherited Wealth in Paris, 1807–1972

The sample of individuals who died married is rather different from the whole population of Parisian decedents, who are evenly split between men and women. Our sample of married decedents is two-thirds men, like those who die married and wealthy. The very large difference in sex ratios has two sources: wealth and age. Those who died wealthy were a bit more likely to be men (53.4%), and the sampled population (unweighted) is also a bit more male (54%) because we massively oversample the richest decedents. Finally men tended to marry younger women, and men had shorter lifespans, and thus were more likely to die married. The uncapitalized inherited wealth shares are very similar for men (54%) and women (53%), and they track over time and across the fractiles. This is hardly surprising given the partible inheritance regime and the generally equal treatment of men and women by their parents. Cousin was an inheritor, and his capitalized inherited wealth was larger than his estate. Marie Suzanne Letissier, his wife, inherited less, but she also left an estate less than the capitalized value of her estate. Aristide Boucicaut was a self-made man, and though his wife inherited both from her husband and her son, her own stake in the Bon Marché was large enough that she was a self-made woman. For now, we leave aside differences between men and women.

Figure 7.2 displays the share of personal assets in the average estate broken down into three categories: real estate, other inherited assets in kind,

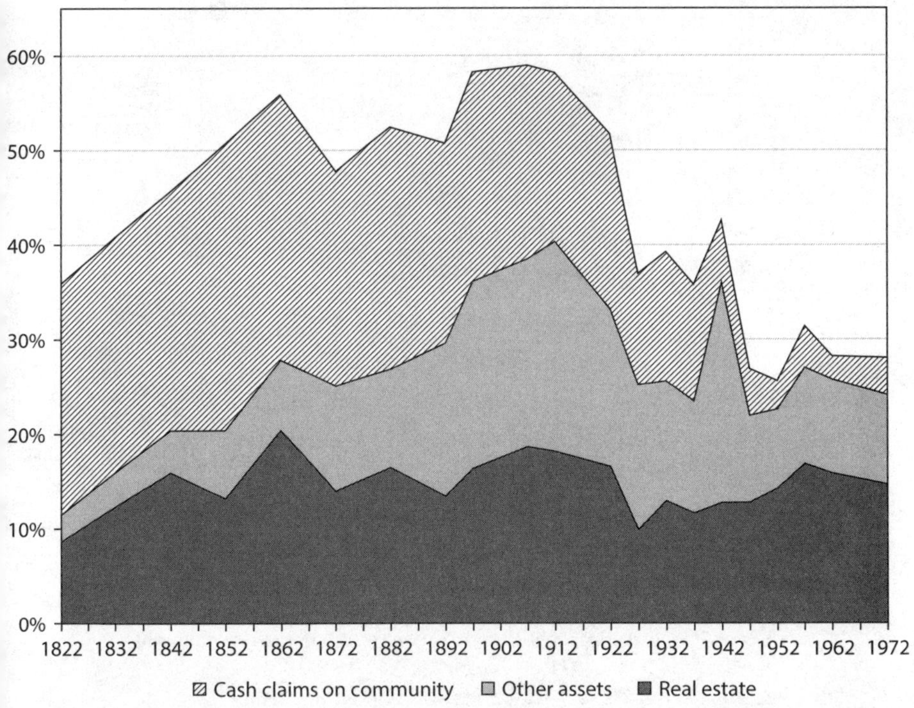

FIGURE 7.2. Personal assets in total wealth (%) in Paris, 1822–1972. *Note*: The sample is restricted to individuals who died married with community assets. *Source:* Estates dataset (see appendix 1).

and cash claims on the community. Overall, personal wealth is about 50% of married individuals' estates until WWI, when it falls abruptly and remains below 30% for the rest of our sample time. Personal real estate is variable but mostly bounces around 10%–15% of the value of the average estate. This decline in inherited wealth has two sources. First, a collapse of net inherited assets in cash. They had represented more than 30% of individual estates until the 1890s. There was some decline in the last years before WWI followed by a collapse in the 1920s with no recovery in that share as late as 1972. After 1937, inherited assets in cash never represented more than 5% of the average estate. After 1937, dowries and other cash receipts were severely eroded by inflation. A second pattern is that other inherited assets in kind grew to a peak between 15% and 20% in the three decades before 1914, then fell back to about 10% of the value of the estate. Over the longer term, cash claims declined most likely because large dowries became less and less frequent after WWI.

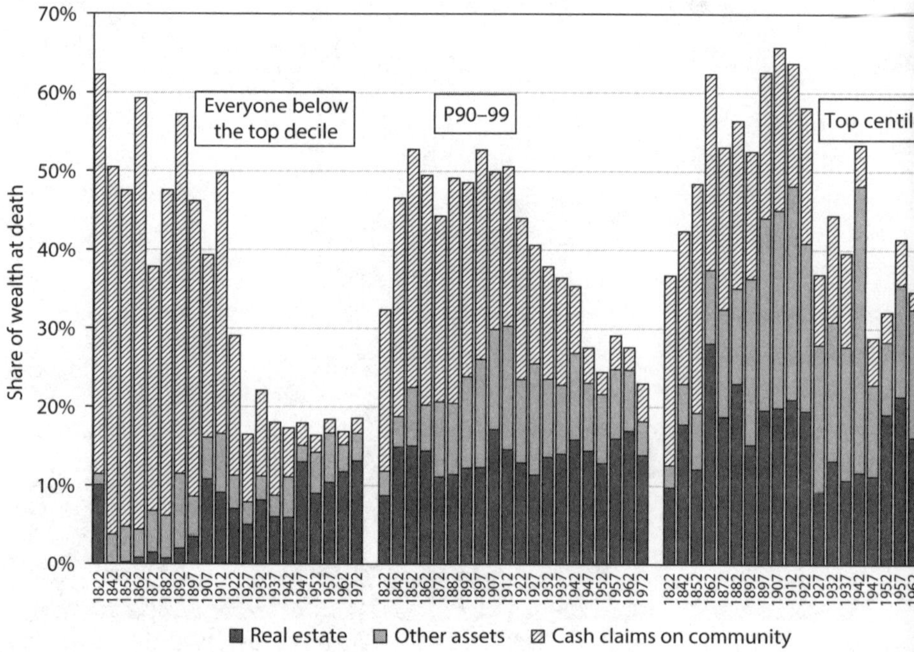

FIGURE 7.3. Personal assets in total wealth (%) by fractile in Paris, 1822–1972. *Note*: The sample is restricted to individuals who died married with community assets. *Source:* Estates dataset (see appendix 1).

Figure 7.3 displays the same data for three groups. The first is made up of everyone below the top 10%. This group represents most individuals who died with wealth, but only a tiny fraction of the wealth was transmitted from generation to generation. Here, the shock of WWI is dramatic possibly because couples provided resources to the community with their dowries or by selling inherited assets. With WWI, a long-term decline reduced personal wealth's share of estates to less than 30%, and all of the collapse can be accounted for by the decrease in inherited assets in cash that had been very large in the past. The middle panel shows the experience of the top decile net of the top 1%. It replicates, in more moderate terms, the experience of the individuals below the top decile. For the top decile, inherited assets in cash never represent more than half of all personal wealth; thus, the post-WWI decline is smaller. The top 1%'s personal assets' share is shown in the rightmost panel. For the top centile, personal assets do not drop as much as for the other groups because its members received and held more of their wealth in forms other than cash.

The reported inherited wealth share in figures 7.2 and 7.3 was above 40% before WWI in the aggregate and above 25% after WWII. There was clearly a large drop in inherited wealth during the great shocks of the twentieth century. Inherited wealth falls from about 40% before WWI to less than 20% for those below the top decile and from 50% or so to 25% or so for the members of the top decile. For the 1% group, the reported inherited wealth share grew through the nineteenth century, reaching 60%, and then it fell by almost half. However, as discussed in the section on Cousin and the Mortiers, the reported inherited wealth is massively downward biased because it omits forgone capital gains for the inherited assets in cash and capital income flows for the whole of inherited assets.

7.5. Capitalized Inheritances

As discussed above, the Mortier de Trévise family members are good examples of individuals who inherited vast sums and bequeathed substantial estates. After duke no. 2, all the title holders inherited assets near or larger than their bequests. When we include flow returns, they could all constitute their bequests entirely from inherited wealth. The last three dukes consumed both the entirety of the flow return and some of the nominal capital gains. Our dataset allows us to go beyond this one, or even a few, families. Using a statistical approach, we make similar computations for every individual who died, was married, and was sampled. We then use the individual data to compute inherited wealth shares. We start with the "raw" data, namely, the share of personal assets in estates. We take a statistical approach because we do not know when their parents died, so we do not know their intergenerational interval (the time during which they enjoyed their inheritances). We only know their age and when they married. Our statistical approach is to draw from a uniform distribution between 20 and 40 years before they died (to have an average intergenerational interval of 20 years) to date when they received their bequests in kind. We draw from a second uniform distribution in the ten-year interval following marriage to date the arrival of the inherited assets in cash. In both cases, we revised the dates so that individuals do not inherit before birth or receive some money after death.

To measure rates of return, we modify the calculations we used for the Mortiers. As discussed previously, we divided assets into four categories: equity, real estate, bonds, and all others. We create a set of capital gains based on the share of each category of assets for each fractile and apply these to all individuals in the fractile. There are two reasons for this. First, the share of assets that are not publicly traded is very large, and for those, we never have individual returns. Second, the investment decisions of any individual reflect various unobserved motivations. To simplify we borrow a page from finance where fund managers are measured against a benchmark. Here, the benchmarks are average portfolios for each fractile. As before, capital gains and capital income are based on a series of financial returns. All other assets have a zero return. Each fractile has an average return that reflects the weights of these four asset classes. It turns out, again, that the choice of rates of return and capital gains is not very important in most estates. For many, there is no inherited wealth (so their inherited wealth share is zero no matter what we use). For a smaller number of estates, the reported inherited wealth is greater than 50% of the total estate, and even a 3% cumulated return will bring inherited wealth to 100% after 25 years (see appendix 7 for more details).

In sum, our statistical approach produces a counterfactual inherited wealth share that is based on guessing three important variables: time of receipt of inherited assets in kind, time of receipt or sale of inherited assets in cash, and portfolio returns. These three guesses imply that the measured level of inherited wealth for an individual is very likely wrong. Take Antoine Adrien Cousin: it is very unlikely our random draw will give us an arrival date of 1832, 1847, or 1849 for inherited assets in kind or the appropriate average (about 1846). It is also unlikely to produce inheritance intervals for the first three Mortiers, 24, 23, and 24 years. Nevertheless, even in the 1%, there are enough observations to ensure that the impact of the draws washes out. The method also has the advantage of being easy to apply.

We start by simply adding the forgone capital gains to cash claims. Figure 7.4 shows both the share of cash claims to net estate and, on top of that, the imputed capital gains separately for men and women. Cash claims are large initially (25% of women's estates). They declined over time for both men and women, but until after WWII, they were always a bigger share of women's estates than men's. After WWII, cash claims were trivial since they were received sometime between 1920 and 1950, and thus, they suffered massively from inflation. The differences between men and women are small with respect to capital gains and reflect the systematic differences in wealth between men and women. More important is the variation in capital gains over time. Capital gains were important early on, reflecting the long-term decline in yields for government bonds (our source for capital gains). This process echoes the more recent experience in bond markets from 1980 to 2017. Capital gains were also important after WWI, but those reflect the

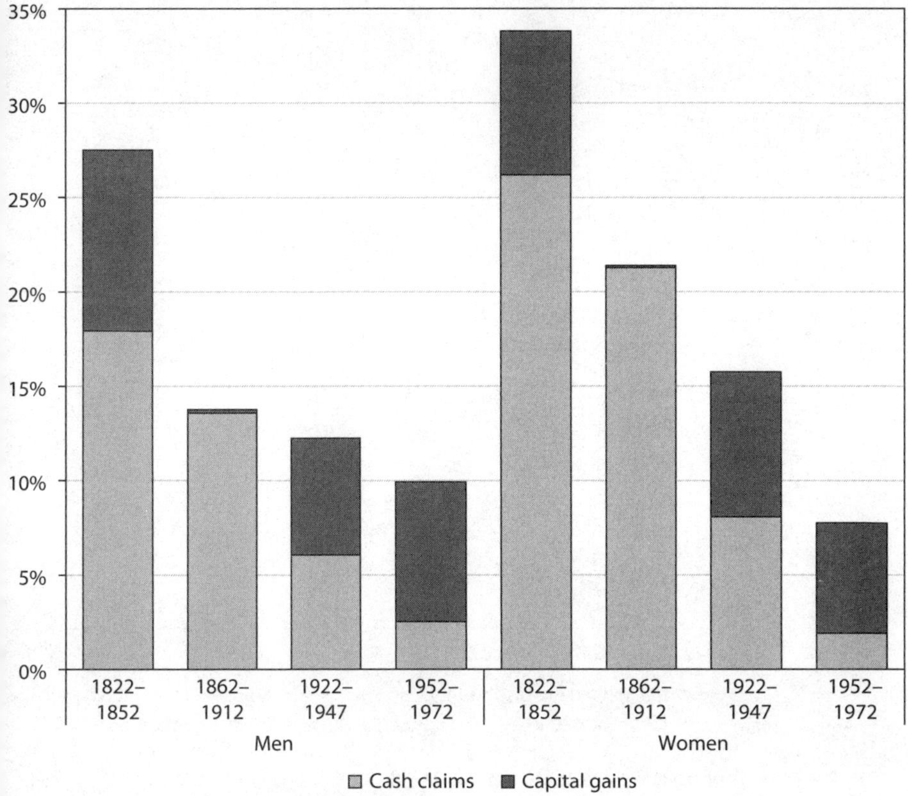

FIGURE 7.4. Cash claims and their imputed capital gains for men and women in Paris, 1822–1972. *Note*: The sample is restricted to individuals who died married with community assets. *Source:* Estates dataset (see appendix 1).

change in nominal prices of real estate and equities during the high inflation period that followed 1914. In both cases, however, valuing cash claims without their attendant capital gains would lead to a severe understatement of the value of this component of inherited wealth. For those who died between 1862 and 1912, capital gains are unimportant. Figure 7.4 shows that series with a long time span are useful because the importance of capital gains (and their source) varies massively over time.

The next step is to include the flow returns generated by the inherited wealth between the time it was received and the individual's death. These flow returns accrue in nearly all cases to the community and are either invested and grow the community assets, or they are consumed. In either case, they are not counted in the personal property. Yet the answer to the question "Could the deceased have constituted his or her bequests out of the wealth he or she

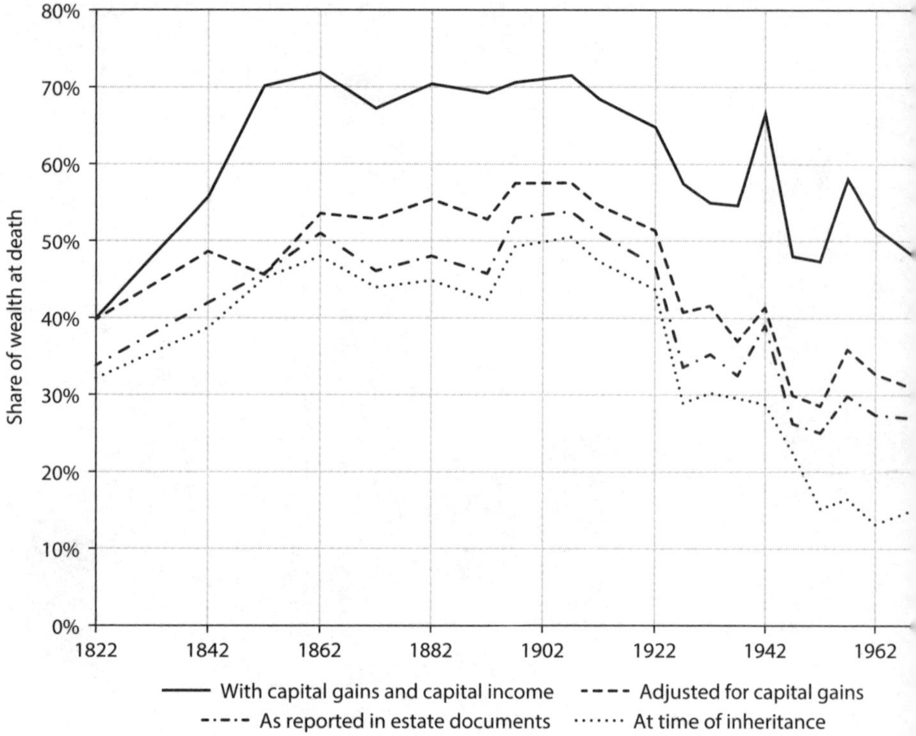

FIGURE 7.5. Inherited wealth shares in Paris, 1822–1972. *Note*: The sample is restricted to individuals who died married with community assets. Capital gains and capital income are computed using individual rates of return. *Source:* Estates dataset (see appendix 1).

inherited?" requires us to take these income flows into account. As noted above, we use market returns based on the observed part of each individual's portfolio at death to impute flow returns to each estate.

Figure 7.5 reports the inherited wealth share based on several measures. At the bottom of the figure lies the inherited wealth share computed from the nominal value of inheritances when received about 25 years before we observe the individual's estate. This corresponds most closely to an approach that matches parents' wealth and children's wealth two or three decades apart. The value of inherited wealth at time of receipt is net of flow returns and capital gains on all assets. The next (dashed) curve is the inherited wealth share based on inherited wealth as reported in estate documents when the individual died: it is always about 10% higher and more after the 1930s when inflation had been high over the deceased's lifetime. The third curve adds the capital gain to inherited assets in cash. This has a small but noticeable effect.

The top curve contains the data capitalizing capital gains and income flows. In each case, we constrain each individual's inherited wealth share to be no more than 100%. We do not display the unconstrained series because it is greater than 100% every year except 1822 (with a high of 213% in 1882 and a low of 114% in 1947). Thus, as we argued in Piketty et al. (2014), it is important to take into account that an individual's inherited share can be greater than his or her wealth. Over the whole period, inherited wealth rises from a low level of 40% in 1822 to 70% in 1850, remains at that level until 1912, then falls back to about 50% by the 1940s, and remains at about that level for the remainder of the period.

Given the number of steps that we took, the reader should be reassured that more than half of the inherited wealth is directly reported in the estate tax filings. Most of the decline in inherited wealth share we observe after WWI comes from a fall in the reported value of inheritances, not from a dramatic change in capital gains or income flows.

One unobserved element leads us to think our reported share of inherited wealth at death is likely lower than it is for the living. We only measure inherited wealth when it is reported. Still, such a report never occurs when an individual dies with nothing and has neglected or lost his or her inheritance. If we were to measure inherited wealth among the living, we would likely find individuals who, in the future, will die with nothing but have not completed the process of spending down their inheritances (and their inherited wealth shares would all be 100%).

The recovery of inherited wealth shares in the first half of the nineteenth century is hardly surprising. As we saw in chapter 1, the French Revolution and the Napoleonic periods were times when wealth was massively redistributed through gifts to key individuals in the First Empire, such as the first Duke of Trévise, and through real estate speculation, like the parents of the second duke's wife. It was also, for many, a period of dramatic drops in asset values, particularly for anyone who had invested in debt contracts before the revolution. The high levels of inherited wealth between 1852 and 1912 are also not surprising, as the development of financial markets and the decline in interest rates provided many opportunities for investors to increase their wealth. In the absence of any significant taxation, constituting a bequest when one had received some was comparatively easy. Finally, the decline of inherited wealth share in the twentieth century is also not surprising given the massive headwinds faced by wealthy families from 1914 to 1947. Yet even then, and despite major fiscal innovations, inherited wealth was still close to half of all wealth at the end of our period. Since 1972, the

FIGURE 7.6. Wealth, inherited wealth, and inequality in Paris, 1807–1972.
Source: Estates dataset (see appendix 1).

continued prosperity of the country and Paris makes it extremely likely that the inherited wealth share has rebounded from its lows.

Figure 7.6 puts together four dimensions of wealth and its distribution: average wealth at death in 1912 francs, the top 1% share, the share of individuals who die with an estate, and the share of inherited wealth. Each has its own independent history. The 1% share rose slowly and steadily to 1922 and then fell rapidly to a nadir in 1952. Average wealth grew rapidly to 1902, then fell unevenly to bottom out in 1947 before dramatically recovering through the 1960s. In contrast, the inherited wealth share jumped early, before the 1% share, and then dropped after 1907, also before the 1% share began to fall. Finally, the share of individuals with positive wealth at death was flat until 1922, when it started to grow rapidly.

Here, we observe only that these series move almost independently over time. Chapters 9 and 11 will consider why average wealth and inequality do not move in tandem. Here, it seems useful to point out that the first half of the nineteenth century was a period where inherited wealth was being rebuilt after the turmoil of the French Revolution. Once it reached its customary 70% in the 1850s, the wealth system was largely stable into the 1910s.

After WWI, it entered a long downward trend due partly to punishing inheritance taxes on collateral estates. Inequality, represented by the 1% share of wealth, has a later peak (1922) and a proportionally smaller decline, likely because some of the inequality is associated with the scale of enterprises in Paris, an issue we deal with directly in chapter 8.

7.6. The Consumption Value of Wealth

Wealth may have intrinsic value for dragons who like to sleep on piles of gold. For humans, though, wealth provides control rights over assets, including the right to use the assets and their income to sustain consumption. Economic historians have long focused on the consumption value of labor income, with research on the evolution of real wages and the distributions of skills and wages. Capital income has been relatively neglected even though there are entire social categories like aristocrats and rentiers whose labor earnings are a negligible part of their income. In this section, we focus on the value of capital income to show that relative to labor income, a massive transition took place between 1914 and 1947.

We will discuss the consumption value of wealth from the perspective of a dynast who limits his or her consumption to the flow income (rents, interest, dividends) and reinvests the capital gains. This dynast acted exactly like a rentier, as defined in the nineteenth century—an individual living off capital income (the classic rentier). The next section will consider a different definition, the one we used in our 2014 article (Piketty et al. 2014). A rentier is an individual whose capitalized inherited wealth is larger than what he or she bequeathed and, thus, who constituted his or her bequest without saving from labor income (the inheritor rentier). There are several distinctions between the two types of rentiers. The first definition assumes that rentiers are idle, the second makes no such assumption. Conversely, the classic rentiers may leave no bequest (if they invest the entirety of their wealth in life annuities). The inheritor rentier always receives and leaves a bequest. The classic rentiers include individuals who had careers and retired, but they also include a variety of younger individuals who have never worked. The inheritor rentiers may well have worked for most of their lives yet drew capital income from inheritances. The size of the classic rentier population is hard to apprehend in our data because occupation at death is closely correlated with age and secondarily to wealth. Instead, we turn to information in some of the censuses for Paris. The censuses report a category of individuals living off of their revenues (distinct from state pensioners). In 1881, 47,191 men and 53,805 women were reported as heads of households living from their revenues under the heading "rentiers" (Résultats Statistiques 1881, 217). Including their households brings these rentiers to 160,605 individuals or 7% of the Paris population.[3]

We use flow rates of capital incomes as defined by the portfolios from our samples and compute the consumption value of wealth for all estates both before and after the payment of estate taxes and before and after paying income taxes (we make labor income the marginal income). We then compare it to average labor income. Given the importance of inherited wealth, we ask: Is there a savings program that allows an estate to be transmitted over time and remain in the same fractile of the distribution? As we shall see, the answer to this question is always yes.

To measure the consumption value of wealth and the number of inheritor rentiers, we return to the population of all individuals who died with wealth. We first compute the flow rate of income based on the portfolio shares observed for each individual (for the period between 1807 and 1822, we use the 1822 weights). When there are missing data in the later years, we interpolate linearly. The key importance of using individual portfolios is that we measure the share of wealth held in zero-income assets. As we have pointed out in earlier chapters, the richer people are, the less of their wealth is held in assets that produce no income. As a result, the pretax consumption value of a franc of wealth is always higher for a member of the top 1% than for a member of the seventh decile. The differences across wealth groups are large, particularly before 1890, when income flows are almost twice as large as for those below the top decile.

Let's look at capital income itself. Below the top decile, relative to the CPI, capital income grew relatively more rapidly after WWII than before WWI, consistent with the trajectory of the real wealth of the seventh and eigth deciles. Relative to average French adult labor income, however, capital income for this group is small, rarely amounting to more than 20%, and that ratio falls over time. These results are not affected by taxes because these estates do not pay much in death duties. Before WWI, the only tax that mattered was the estate tax, which reduced the heirs' income by an average of 5%. After WWI, the estate tax bite was about 11%. If we include income taxes, the consumption of the heirs was 14% less than the pretax income of the bequeather. In short, the story below the top decile is that consumption possibilities depend far more on wealth trends than on tax rate changes.

The top decile tells an entirely different story. To begin with, capital income is massively larger than for the lower deciles and falls more dramatically after 1914. In the period before WWI, the top decile's capital income was about 10 times average labor income, or roughly 100 times the capital income of the next two deciles. As we discussed in chapter 6, wealth at the top did not recover from the great shocks. The top decile's average income before tax after WWI, deflated by the CPI, is about half what it was before. If deflated by average labor income, the value of wealth's fall is even steeper, and the top decile's capital income completely collapses. Before WWI, on average, capital income was 13 times the average adult labor income; it plummets to 3.5 after WWI and then drops again to about 1 after WWII. For rich individuals, there is another complication: deflating by the CPI is inappropriate because, in the nineteenth century, at least part of their consumption involved full-time personal services. As an alternative, we deflated by a measure of the labor cost. Before WWI, the average member of the top decile could afford about 10 full-time staff (maids and valets, cooks and butlers, grooms, or coachmen) from capital income; by the 1920s, that was only 2.5, and after WWII, less than 1.

The striking differences over time in pretax income are further magnified when we consider taxes. There are important differences not just between the top decile and everyone else. We must also account for differences between the top centile and the rest of the top decile. The estate tax reduced capital income by 5% before WWI, or about the same as for the lower part of the distribution, but after WWII, the estate tax was 20% for the rest of the top decile, a larger reduction in wealth. Income taxes were irrelevant before WWI, and the tax on interest and dividends (IRVM) is assessed on a small part of the portfolio. As a result, the capital income of the heir after he or she has paid all taxes is 95% of the pretax income of the bequeather for everyone in the top decile. For estates in P90–99, after WWI, the estate tax takes a 25% bite out of an estate when it is transmitted to direct heirs and income taxes another 30%, leaving the heir's disposable income at half of the bequeather's pretax income. One should bear in mind that the top decile has massively higher capital income than the rest of the wealthy. The ratio of average capital income of the top decile to the average of the next two deciles is more than 80 over the whole period. So, taxes only dampen the differences between the very rich and the rest of those with wealth a little bit.

Within the top decile, the top 1%'s capital income is 10 times that of the rest of the top decile. Rates of return are close enough within the top decile that the differences in pretax income directly reflect differences in total wealth. Estate taxes take nearly 30% of wealth for the top centile (because of much higher tax rates on indirect transmission) and are, thus, more and more severe for the top centile than for the rest of the top decile. As a result, the net income of the heir in the top centile is only 40% of the pretax income of the bequeather.

From the low-tax regime of the nineteenth century to the third quarter of the twentieth century, the consumption possibilities of the top centile fell by 90%, and the rest of the top decile saw a decline of 85%. This is in sharp contrast with the 50% total drop for the lower deciles. While the top 1% had enjoyed capital income greater than 80 times the average labor income, the increases in taxes and labor income brought those numbers down to single digits after WWII.

The evolution of the consumption value of capital income is easy to track. Relative to just evaluating wealth alone, measuring consumption makes two important adjustments to understand wealth's social evolution. First, large differences in rates of return between the top decile and everyone else in the nineteenth century indicate that wealth inequality understates consumption inequality. Second, taking income taxes into account gives us a sense of the extent of the reversal. After WWII the after-tax return for the top decile was less than the after-tax return for the eighth and ninth deciles; before WWI, it had always been higher. Our analysis of capital income demonstrates how the nineteenth century differed radically from the twentieth century in the relationship between top capital incomes and labor costs. Yet these consumptions are hypothetical, and there is no a priori good reason to say that top wealth holders in any period consumed less, more, or exactly the entirety of the flow incomes of their wealth. For this reason, we designed a method that sheds light on actual consumption patterns. It allows us to show that there was a sizable minority among the wealthy group for whom realized consumption was more than just the flow income from wealth.

We use our inherited wealth calculations and split wealthy individuals who died married into two groups. On one side are individuals whose estate is larger than the capitalized wealth they inherited. These "savers" added to their capitalized inherited wealth by saving from labor income broadly construed (from salary or by multiplying one's wealth through effort and talent). We leave them aside because the data say they invested rather than consumed the incomes from their inherited wealth. The other group comprises those individuals whose capitalized inherited wealth is larger than what they bequeathed—the inheritor rentiers. Even if we do not know their labor/talent income, we do know that they consumed some amount of their capitalized inheritance. While that group is never more than 12% of the population of decedents, it includes 40% of the top decile and nearly 60%

of the top centile. Overall, they alone account for about half of all the wealth bequeathed in Paris (60% in the nineteenth century and 43% after WWI). As noted above, their consumption from inherited wealth is just the difference between capitalized inherited wealth and what they leave behind. This measure of capitalized consumption is about 3.1 times their bequests over the whole period. We compare capitalized consumption with labor income capitalized appropriately.[4]

On average, for inheritor rentiers, capitalized inherited wealth consumption is 10 times capitalized labor income before WWI and 3.5 after WWI. If we divide this group of capital consumers by fractiles, the results are quite straightforward. Because capitalized labor income does not vary across the fractiles, consumption from inherited wealth is much higher at the top. These differences in consumption are easiest to measure as differences in flow income. On average, below the top decile, consumption from inherited wealth affords a bit less than one extra labor income per year before WWI, half a labor income in the interwar period, and a third from 1947 on. Although the consumption value of inherited wealth falls over time, it still significantly increases consumption relative to individuals without it. The consumption of those below the top centile is dwarfed by what the top 1% can afford. For these happy few, consumption from inherited wealth came to 75 years of labor income a year before WWI and 40 years in the interwar period. It was still 20 years from 1947 on.

In the paragraph prior, we measure the consumption possibilities afforded by the inheritances of the rentiers after they had paid whatever estate taxes were due but without consideration of income taxes. A quick way to evaluate income taxes' impact is to divide the consumption possibilities by a third at the top during the interwar period and by a half after WWII. Doing so brings down the consumption of the top 1% to about 10 times average labor income. Yet that remains far more than the 4.4 years of labor income produced by the dividends, rents, and interest that this group earned. Two additional factors explain this difference. First, the inherited wealth measure is backward-looking, so its implicit labor cost is, on average, about 15 years before death rather than the year before death for the capital income estimates. Given that labor income had a far more positive experience than wealth after WWI, the larger consumption from inherited wealth is not so surprising. The other reason is that the consumption of individuals who left estates smaller than their capitalized inherited wealth was not limited to income flows. It also included some of the capital itself, as was the case with the last Duke of Trévise.

Capital income has always been important to the consumption possibilities of the elites. In the nineteenth century, large fortunes enabled individuals to afford small armies of servants and purchase a whole variety of bespoke items while still leaving large estates to their heirs. In the twentieth century, these possibilities vanished. The massive decline in the real value of large fortunes from 1912 to 1947, accompanied by a heavy burden of taxation, dramatically reduced inflation-adjusted capital income. The rapid growth of labor costs meant that even the very rich could only afford a few domestics. For the eighth and ninth deciles, the value of capital income, while never large (less than 20% of labor income), was higher after WWII than it had been in the nineteenth century. For those below the top decile, taxes always remained limited, and they declined in the 1950s, confirming the state's interest in the constitution of middle-class estates. The same pattern holds for those below the top decile who consumed much of their inherited wealth: inherited wealth was a valuable but limited addition to labor income.

7.7. The Transformation of the Rentiers

The radical decline in the consumption value of wealth leads us to reevaluate the fate of the rentiers. While information on the number of rentiers disappeared from French censuses before WWI, we can ask how the population of inheritor rentiers evolved in our data. As noted before, the size of this population is anchored, at a minimum, by the share of individuals whose estates are smaller than their reported inherited assets (about 4% of the adult population). It hits an upper bound when we use a 5% capital income flow (about 10%). So, it is a small population (as reported in the late-nineteenth-century censuses). If we consider only the individuals who leave an estate, the rentier share is low at 18% in 1822, peaks at 34% in 1882, and then falls back to 18% in 1972. The rentier share is initially the same across all the fractiles, but by the 1850s the rentier share is increasing in wealth. It is generally three times higher for the top 1% than for those below the top decile, including in the 1960s and 1970s.

The low rentier share in the 1820s can be partly attributed to the French Revolution's redistribution. Beyond the extra churning of an unsettled period, three phenomena devalued inheritance. Families who chose exile when the monarchy fell lost much of their property. Private debts and government offices were reimbursed in quasi-worthless currency during the assignat inflation. Finally, in 1797, the restructuring of the public debt reduced bondholders' wealth by two-thirds. Other forces created new large wealth holders. Two are especially noteworthy. First, as discussed in chapter 1, and as we saw with the Mortier family, Napoleon richly rewarded his military and administrative elite (and many accumulated more wealth through other means). Second, if the assignat ruined many families with large financial positions, it delivered vast wealth to anyone who bought church and other property at public auction during the French Revolution and paid off those debts with devalued assignat. From then on, these astute or lucky speculators enjoyed rents that contributed to their lifestyle and the continued growth of their wealth. These buyers included other individuals we encountered earlier, like the Comte Roy and the parents of Anne Marie Lecomte, wife of the second duke of Trévise. It is thus not surprising that, in 1822, only 18% of the top decile were rentiers, and it was the only year in which the top centile had a smaller share of rentiers than other groups. By the 1850s, the rentier share in the top decile had reached 50%, and it remained above 40% into the 1930s. When the rentier share reached its nineteenth-century plateau, the first generation of new wealth holders had passed on, and the heirs of the revolutionary entrepreneurs dominated the data. For the whole population, it is striking that the share of decedents who were rentiers is very stable, between 9% and 12%, from 1852 to 1962. What changes is the extent to which members of this group could have lived exclusively from assets income.

Among the top 1% of wealthiest Parisians, rentiers predominate; by the time they die, they are extremely likely to be retired from whatever occupation they may have had. For them, age at death increased by nine years in the nineteenth century and another nine between 1902 and 1977. They are dying old and rich because they are in the top centile. By 1900, less than 25% of the top wealth holders die younger than 60, and the majority die past 70. As we discussed previously, there is no doubt that their capital income dwarfed their labor income. This is particularly true after 1922 as age-at-death continued to rise and individuals, whether inheritors or not, spent more time retired. Take, for example, the notary we saw in chapter 6, who died among the top wealth holders at age 89 in the late 1970s. He had sold his office some twenty years before and lived off his capital income from then on. The difference between the inheritor rentiers and the top 1% as a whole likely has to do with how long they enjoyed a large capital income. Those who started with little and constituted most of their wealth from savings had to set aside a significant fraction of their labor income and early capital returns to leave millions behind. Those who inherited vast sums could enjoy high consumption over their lifetimes. Luxurious lifestyles were particularly true before WWI. Then, being very rich afforded a standard of living that no one else could approximate. Although, in accounting terms, this consumption came at the expense of larger bequests, in practice, it did not prevent a large intergenerational wealth transmission at the top: from 1852 to 1912, the share of rentiers in the top 1% was higher than 60%, and on the eve of WWI it was 64% and overall, their inherited wealth share was 75%.

With the shock of WWI and its inflation, the value of financial assets fell, and labor costs rose. By the 1920s, consumption possibilities from capital income had shrunk. The fraction of rentiers in the top 1% had also declined from 64% to 55% after WWI, but it was still about 53% into the 1970s. The major changes thus came from the value of inherited wealth. As we saw in the previous sections, the income of the rentiers in the top 1% was divided by ten between 1912 and the 1950s. Rentiers in no way disappeared; they could no longer afford their extravagant lifestyles. For the rest of the top 1% (that only consumes labor income), the trade-off between leaving bequests and consumption was stark. If they had spent their entire capital income in the last years of life, their consumption would have been about five years' average labor income but then their bequests would have suffered.

The rest of the rentiers in the top decile tell a similar story of declining consumption possibilities, though they are a smaller share of that group and fall more—from 46% before WWI to about 32% after WWII. We will thus leave them aside. By contrast, the next two deciles tell an important and different story. There, the gains in average age did not begin until the 1920s, but then they rose quite steeply from an average of 53 years in 1912 to 74 in 1972, closing most of the gap with the top 1%. Before 1912, few in this group likely retired; 60% of all those whose wealth fell in the eighth and ninth deciles died before 60, and only about 20% died older than 70. The changes that followed WWI were dramatic: by 1960, only 22% of this group died younger than 60, and half were older than 70. Neither capital income nor inherited wealth consumption was particularly large (less than average labor income after 1900), and their inherited wealth share peaked in the 1880s at 24% and then oscillated around 18%. While there is no smoking gun, we speculate that some individuals in this group appear as rentiers because negative shocks to labor income (e.g., bouts of unemployment, ill health) led them to consume part of their inheritances. Others were created mechanically because the individuals in this group were poorly diversified and suffered large losses relative to the average portfolio. Hence, savings plans were likely very disturbed by the shocks of the first half of the twentieth century. For example, consider someone who inherited a pile of Russian bonds in 1907 and died in 1932. These bonds lost 92% of their value between these two dates. That person would be a rentier by most measures of inherited wealth unless they saved massively out of labor income to offset their losses from 1917. Such narrow portfolios abound in these deciles. It is likely for this group that the rentier/saver boundary lies with returns rather than consumption. For the top 1%, whose assets were much better diversified, efforts to maintain prewar consumption rather than increase savings were likely partly responsible for the continued importance of rentiers.

After WWII, the share of rentiers below the top decile continued to grow, but the value of inherited wealth remained smaller than before WWI. The rise in the rentier share, in fact, started around 1922, precisely as wealth per person in real terms was starting to collapse, and the fraction of individuals who died with wealth was starting to increase. The share of rentiers among those who died wealthy continued to hover around 20%. Instead, the rise in the rentier share overall is entirely attributable to the increase in the fraction of individuals who died wealthy. Between 1922 and 1957, the share of adult decedents who died wealthy and the fraction who were rentiers doubled. The three decades that followed 1945 are often viewed as a period of unusual equality. Still, the rise in the share of individuals leaving an estate also created more individuals who inherited wealth. Those inheritors could continue accumulating or "eat" part of the wealth they had received. The evidence for Paris suggests that some of the "new" inheritors continued to accumulate out-of-labor income while others did not.

7.8. Conclusion

Most of this chapter has focused on the magnitude of inherited wealth and the implications for the consumption of those who had the good fortune to be heirs. Inherited wealth provided many advantages. Top wealth holders generally inherited estates large enough that their consumption options were pretty much unimaginable to most of the population then and now. Given that, at its peak, inherited wealth constituted 70% or so of total wealth, a natural question arises about the extent of mobility in a society so weighed down by inherited wealth.

The answer for Paris is twofold, as with many things in economic history. First, it was difficult to die as a top 1% wealth holder and be a rentier without having received a large bequest. For 17 of the 20 sample years, more than half (60% on average) of the top 1% had received large enough bequests that they were rentiers. They also left a bequest that, if matched by their spouse, would have put their children in the same fractile of the wealth distribution. When we add the inherited wealth of the nonrentiers, the top 1% inherited wealth share is nearly 70%. So, inheriting a lot of money is a massive help in producing large bequests. Inheriting something is, in fact, almost necessary to be a top wealth holder: in an average year, 90% of the decedents in the top 1% had separate assets, and for the whole of the top decile, three-quarters had inherited something. Given that for most of our period, the top decile concentrates more than 90% of the wealth, the evidence is overwhelming that there is tremendous persistence in wealth over the whole period from 1807 to WWII. As our example suggests, the dukes of Trévise did not just transmit titles; they were very wealthy because wealth flowed through the generations.

At the same time, Parisian society was never a closed system. It would be wrong to consider societies as unequal as Paris as frozen with two or three groups that, as often envisioned, are self-reproducing classes. This is wrong because Paris, through the nineteenth century, could not sustain its growth without large-scale infusions of newcomers. In the later decades of the nineteenth century, more than half the individuals who died in the top 1% were not born in Paris, and the proportion is only slightly smaller than the share of the overall population of adult decedents. Migration to Paris clearly involved both poor individuals, like Balzac's Rastignac, coming to the city to make their fortunes and rich ones, like Léon Bel. It was possible to go from very little to nearly the very top of the wealth distribution. Aristide and Marguerite Boucicaut, whom we encountered earlier, are good examples of such trajectories. They started with little, and he was the tenth-wealthiest person to die in 1877; a decade later, Marguerite was the second-richest. Individuals like the Boucicauts who make it to the very top of the wealth distribution are rare, and, as a rule, the self-made individual did not enter the top 1% with much frequency. At the apex of the rentier society in 1897, 74% of the top 1% were rentiers, and the inherited wealth share was 80%. That means that a quarter of the top 1% saved out of labor income to produce its inheritance. In the same year, more than half of the top decile also did so. In other words, each generation saw the wealthy elite renewed because a quarter of its members had accumulated wealth well beyond what a passive dynastic strategy would have allowed.

8

Wealth at the Top

8.1. Motivation

One of the most striking patterns in our data is that, despite all the shocks and the documented turnover of the Parisian population, the wealth share of the super rich (top thousandth) was both very high, more than 20%, and almost constant between 1807 and 1972. Given the high Parisian share of French wealth, the Parisian super rich likely controlled about 5% of the country's wealth between 1890 and 1930. The threshold is arbitrary, but we believe it is meaningful. The super rich at death ranged from a dozen around 1807 to a peak of 37 in 1907 and then declined to 23 in 1977. The censuses tell us that these decedents corresponded to a population of living adults that numbered about 450 in 1807 and grew to 2,200 in 1932 before dropping back to 1,830 in the late 1970s. As we saw through chapters 1 to 6, the super rich were always massively wealthier than even the rest of the 1% and hundreds of times richer than the average Parisian. This chapter is devoted to investigating two related yet distinct questions. First, why was the wealth of the super rich so stable? A second question is, how persistent was membership in this group? In other words, were the super rich a closed class?

Others might use terms like aristocrats, capitalists, or plutocrats to describe the powerful and the wealthy. Because these terms are connected with the sociopolitical context, their use varies depending on context. The advantage of our definition is its precision; the advantage of other definitions is that they take in the interaction between wealth and power. From the left to the right, in academia and beyond, beliefs about the close connection between wealth and power run deep. Some of those beliefs come from the fact that the rich influence policy even if they have no formal role in the policymaking process. Such informal influence is difficult for us to quantify. Yet we document the frequency with which the super rich took on political roles at the national level (deputy, senator, minister) or local level (mayor, city, or departmental councilor). We also document the frequency with which the super rich had executive roles in businesses. We show that the formal political influence of the super rich remained significant after 1848, when property qualifications for voting ended. It only fell after 1870, when elections became competitive.

Whatever happened to the formal power of the super rich after 1870, they continued to exert massive influence over businesses and policy. One often presumes that this group is both self-contained and cohesive because the social network is sufficiently dense that the group's boundaries are well defined and understood by the members. In extreme cases, some consider the group as a closed and self-reproducing class. The most famous of such claims was made by radical politician Édouard Daladier. In a speech to the Radical Party in 1934, he asserted that 200 families controlled the Banque de France and the country itself. That claim was made as part of an attack on the governance structure of the Banque de France, where only the 200 top male shareholders could vote at the annual meeting. From prior chapters, we already know that in Paris, wealth tended to persist. In particular, even among the top 1%, a group ten times larger than the group we consider here, nearly everyone was married with a contract, thus with a dowry or the expectation of an inheritance. Moreover, as we saw in chapter 7, almost 60% of the top 1% had received a substantial inheritance. Nevertheless, a lot would be lost by assuming that the super rich or even the very rich formed a self-reproducing wealth class.

The extent to which the super-rich group is closed needs evaluation. First, there are demographic, political, and economic reasons to suspect that the super rich were constantly acquiring new members. Between a quarter and a third of the members of the 1% died without direct heirs. The fraction that died without sons was larger still; in this male-dominated society, demography alone imposed renewal. Second, the First Empire created a new group of wealthy individuals through speculation (like Comte Roy) or service (like Mortier de Trévise). These individuals entered the ranks of the rich elites and were accepted as such (at least if we examine whom their children married). Finally, economic change created new enterprises. Successful entrepreneurs like Henri Germain, who founded Crédit Lyonnais, joined the Parisian super rich. He died in 1905 with a fortune of 10 million francs because of the bank's success.[1]

Germain was born in Lyon, and the Crédit Lyonnais was founded there in 1863. The bank's headquarters were maintained in Lyon, but a central office was opened in Paris. Germain moved to the capital and died there. Indeed, Paris was open to immigrants of all kinds, and many of the most famous plutocrats established their fortunes by moving to the city. The Hottinguer family, whom we saw in chapter 5, for instance, came from German-speaking Switzerland in the 1790s. Indeed, rich migrants came from other parts of France, the colonies, and abroad. Some came to Paris to make their fortunes, while others had amassed substantial wealth before they came. Those who arrived rich came to access the capital's financial market, to take advantage of the city's wholesale markets, or, more simply, to enjoy the capital's amenities.

When considering wealth and power, an additional dimension is inescapable: the difference between men and women (Bessière and Gollac 2022). Formally, French women's assets were separate from their husbands. Practically, however, women married under the default regime or the modal contract surrendered their assets to the tutelage of their husbands. Women might have control of their assets if they were either single (and their parents had passed on) or widowed. For women, the link between wealth and power was, at best, indirect. Moreover, despite laws mandating inheritance equality, there were loopholes that privileged sons and men. When we look closely at how estates were divided, the *part de réserve*, the piece of the estate that a decedent could freely bequeath, almost always went to sons. Moreover, as we shall see, the state made little effort to force privately held firms to value their equity accurately. Keeping the capital value fixed advantaged sons because they received the equity, and the daughters were compensated with other assets of equal accounting value. It is not surprising, then, that between 1807 and 1870, men were 60% of the top 1% and slightly more of the top 0.1%. After 1870, men remained a constant 56% of the top wealth holders.

This chapter proceeds in four steps that correspond to different aspects of the extremely wealthy. We start with James Rothschild and his sons. James founded the Paris bank Rothschild Frères, which, along with its London analog, would dominate international finance until WWI. James and his sons' share of the inheritance flow in the years they died towers over any other we have measured. We then examine the top wealth holders in each cross section over time (35 individuals). This set does not include James or his sons because they did not die in a year that ends in a 2 or 7, but it includes two Rothschild women. We ask: Who were they? How did they change over time? What did they own? We then consider a third, much larger group, the top 0.1% wealth holders in every cross section (the top wealth holders included two additional Rothschild women), and ask the same questions. Finally, we examine the top 200 shareholders of the Banque de France sampled every five years from 1801 to 1936 and compare the social characteristics of the super-rich Parisians with those of the leading shareholders of the Banque de France.[2] We pay close attention to the fact that the individuals who comprise the top 0.1% are the wealth-race winners each year. This could be because they inherited massive amounts of wealth that they did not dilapidate or because the assets they owned performed exceptionally well in a short time around their death. They could also be their generation's most successful entrepreneurs and investors. Finally, like the Mortier de Trévise, they could be adept at securing the state's munificence. In each case, the top group emerges from the broader population because it had the greatest financial success.

Our analysis allows us to refine Daladier's claims significantly. We find that to close the "controlling group," we would have to make it much larger than 200 families, in fact, several thousand families instead of just 200. Second, even in finance, the powerful families extended well beyond the top shareholders of the Banque de France, and many families with the wealth to attend the meetings were not involved in banking. As we move from the Rothschilds to the top shareholders of the Banque de France, the size of the group we consider progressively increases from 3 to 35 to 920 to 2,100, and what we can learn about each person declines. Nevertheless, for the larger groups, we move beyond anecdotal description to a more illuminating statistical understanding of the diversity of wealth trajectories.

The analysis of our microdata confirms that family background counted a great deal more than individual genius or effort when it came to attaining great wealth. At the same time, we can show that the top 0.1% was definitely not a closed or self-reproducing class. Every generation contained many children of the super rich. In addition to these, there were a few "self-made" Parisians, and a larger group of migrants who, having established themselves in the provinces or abroad, moved to the city and joined its super rich. They then either continued their careers as newly minted Parisians or simply enjoyed its amenities as rentiers. The top 0.1% was far more like a club in this way. Entry was facilitated by being a member of a wealthy family, but that was not an exclusive criterion. Finally, many social rewards of wealth were very much clublike, from public distinctions like the Legion of Honor or political office, to sitting on boards of charitable institutions, business corporations, and public commissions.

Of course, being a member of the top 0.1% only requires meeting an arbitrary quantitative standard—much like running a hundred meters in under ten seconds. In either case, current members have little say about who is in and who is out. In the case of wealth at death, the standard is both arbitrary and unknown until the cohort dies. Moreover, those who have been "verified" by dying super rich are no longer around to exert influence, save in how they directed their bequests. Among the living, however, few can make it to the top without interacting with others, particularly with capital providers. Let us now turn to our analysis of the family that stood at the apex of the wealth pyramid for at least eight decades.

8.2. The Apex

James Rothschild died in 1868, leaving an estate of 100.6 million francs in Paris. We can add another 11.7 million francs of real estate in the Seine-et-Marne (where he had his country estate Ferrières), 4 million for an estate in Boulogne, and 4.1 million for Château Lafite, the wine estate near Bordeaux. The total comes to 120 million francs. James and his wife Betty had four sons and one daughter. Alphonse Rothschild, their eldest son, died in 1905, and his estate totaled 251 million francs. Gustave, the second son, left 242 million in 1911.[3] The Rothschild women were also extremely wealthy. When Betty Rothschild died in 1886, she left a Parisian estate of 74 million francs. Charlotte, James's daughter, left 85 million in 1899. Alphonse's wife, Leonora Rothschild, left 17 million in 1911. Cécile Anspach, Gustave's wife, left 29 million in 1912. James and his sons tower above all the hundreds of thousands of individual estates we have collected. Their wealth amounted to 15% of the inheritance flow in Paris in the year they died. The top estate in our sample years averages 4% of the total inheritance flow (with a wide dispersion as the standard deviation is 2%). We have found only two non-Rothschild estates greater than 7% of the inheritance flow (Louise d'Orléans in 1822 and Jules Lebaudy, a sugar magnate, in 1892). Today, the Rothschilds are no longer at the apex of Parisian wealth, but they remain very rich. The family exemplifies the secular persistence of families whose founders amassed massive fortunes.

James arrived in Paris in 1812 and quickly built the premier bank in Paris. Five and a half decades later, he was still at the top of the financial pecking order when he died. Nearly all of his estate consisted of three types of assets.

First and foremost was his stake in Rothschild Frères, which was reported at 40 million francs, followed by 35 million francs in Parisian real estate, and far behind was 13 million in the Chemins de Fer de Nord. His estate had no trace of the bank's international finance dealings. Betty Rothschild's estate of 74 million francs in 1886 also had its most prominent position in an account at Rothschild Frères (19 million). Still, she had 46.5 million francs of foreign assets. Three-quarters of these were in government bonds, the family bank's core business. We recovered the details of the estates of all their children.[4] Salomon James died young in 1864 with a "negligible" estate of 463,000 francs.[5] The others lived long and died between 1899 (Charlotte) and 1934 (Edmond). If we use 1912 francs to compare all these estates, James and his wife left 211 million while their four long-lived children (leaving aside their spouses) bequeathed 783 million francs, almost four times as much as their parents. The comparison is imperfect in several ways. First, James's foreign assets do not appear in his estate filings. Although there was a notarial document relating to these assets, it, along with many other family wealth documents, has disappeared from the notarial archives. Second, the brothers' valuation of their equity position in Rothschild Frères is idiosyncratic. Alphonse's estate reported a total stake in Rothschild Frères of 123.2 million francs (or about half his wealth). That includes 16.6 million for his third of the equity in the firm and another 106 million in a deposit account. The equity is valued at nominal capital (the firm maintained a nominal capital of 50 million francs from James's death in 1869 to at least 1925). Gustave's estate reports an equity stake of 55.3 million francs, amounting to 31.3% of the firm, and a deposit account of 103 million. Gustave's equity would put the value of Rothschild Frères at 180 million francs. We think the 55.3 million valuation of Gustave's equity combines 16.6 million of nominal capital with 38.7 million undistributed profits, which likely accrued between 1905 and 1911 and were later distributed.

When Edmond died in 1934, his estate reported a deposit account of 40.5 million and an equity stake of 3.8 million in nominal francs in Rothschild Frères. The last document we have of the nominal capital stock of Rothschild Frères is an indefinite prorogation (extension of the partnership term) of 1925: it maintains the nominal capital at 50 million. Relative to Alphonse and Gustave, Edmond is poorer by 100 million 1912 francs, and this gap is entirely accounted for by the decline in the value of his positions in Rothschild Frères. The real value of other assets is comparable to his male siblings. Perhaps his reduced stake was a form of tax avoidance or simply the reflection that in his eighties, he had already relinquished most of his stake in the firm to the next generation. In any case, the second generation of Rothschilds was extremely successful in maintaining its position at the apex of the wealth distribution.

James and his siblings in other European capitals exemplify the innovative entrepreneur who creates and retains control of an economically important sphere of activity. In Rothschild's case, this sphere was international finance. More importantly, James remained at the helm of the Paris firm even after 1860, when it was restructured so that the only equity holders were himself and his sons. Alphonse, Gustave, and Edmond exemplify heirs who are themselves successful entrepreneurs and modern investors. They inherited an important enterprise, ran it profitably, and retained effective control until they died. Alphonse and Gustave were, if anything, more invested in Rothschild Frères than their father had been.

We do not observe the distribution of James's estate because the fiscal authorities did not need to distinguish what went to Charlotte and her brothers. When Alphonse and Gustave died, the progressive inheritance tax regime required that the estate documents report how their wealth was distributed. Each only had one son (Édouard and Robert). In both cases, they directed the portion of their estate over which they had discretion (a quarter for Alphonse and a fifth for Gustave) to their sons. So, the sons, in practice, received twice as much wealth as their sisters or their heirs. Finally, the sons retained sole ownership of the firm's equity. Alphonse's equity was transferred by dissolving the existing partnership between Alphonse, Gustave, and Edmond. A successor firm, including Gustave, Edmond, and Édouard, was immediately created. In Robert's case, the 1905 partnership contract and 1909 amendments allowed Robert to exercise a right of first refusal to purchase the rest of his father's equity interest at a nominal value. Thus, the equity of Rothschild Frères was transmitted to the male heirs, who were all superlatively wealthy from their inheritance. This is not a particularity of the Rothschilds; we see the same patterns through the four generations of Hottinguer bankers we followed in chapter 5.

Extraordinary wealth was not confined to Rothschild men. As we saw above, it extended to their wives (themselves often other Rothschilds and thus cousins of their husbands) and daughters. When Charlotte, the daughter of James and Betty, left an estate of 98 million in 1900, she was almost surely the wealthiest person to die that year. The estate included 42 million in claims on Rothschild Frères but no equity. One is left to wonder what her wealth might have been if she had been deeded an equity stake in the firm. She also had substantial holdings in firms closely associated with the family bank (e.g., Chemins de Fer du Nord, Madrid-Saragosse, and the Banque de France). Julie, a Neapolitan Rothschild, moved to Paris and died in 1907. Her estate of 83 million francs made her the richest person in Paris to die that year. Her account at Rothschild Frères ran 13 million, and she also had a very large position in sovereign debt. Leonora Rothschild (Alphonse's wife) left an estate of 17 million in 1911. Three firms comprised 77% of her wealth (Rio Tinto, De Beers, and Suez). Cécile Anspach (Gustave's wife) died in 1912, the second-richest person in Paris. Her estate was just shy of 30 million francs, nearly half of it in shares of foreign mines and petroleum (De Beers, Rio Tinto, Boléo, and Royal Dutch). Adèle (Gustave's daughter) died the richest person in 1922. She had 5% of her 104 million francs in a deposit account at the firm. Her primary investments were in the same firms as her mother Cécile's, but with 139 different assets she was much more diversified. The extent of her portfolio's diversification is best illustrated by comparing it to her grandfather's. James concentrated the same nominal portfolio in less than 80 assets.[6]

The Rothschild women were all married under the separate property regime and exercised no profession. Unlike their husbands, they were not equity holders or managers of Rothschild Frères. Instead, the Rothschild women held accounts, at times massive, at the bank. Thus, their bequests were created from their inheritances. Throughout their lives, they were significant owners of assets. Dowries for Gustave's daughters ran 2.5 million francs, enough to immediately put them in or very near the top 0.1% of the wealth distribution. These women, however, did not exercise control over all their assets. Scrolling through the notarial archives,[7] one can see they had some control over their assets as they regularly provided powers of attorney to their money managers. However, the gender distinction is stark between men who have both ownership and control and women who own but do not control.

This exceptional family might be representative of the general strategies of the highly successful people who composed the top part of the wealth distribution. They might also be unique because of their German-Jewish origins, activity as bankers, propensity to intermarry, or maintain an international network of financiers. Our next step is to examine every top wealth holder, starting with Louise Jeanne Lory, who, in 1807, left an estate of 1.5 million francs or 1.25% of James Rothschild's nominal wealth.

8.3. The Richest Parisian to Die Each Year

From the Rothschilds, let's turn to the wealthiest Parisians who died in each of our cross sections, 36 in total from 1807 to 1977 (the full list is in appendix 8). When trying to determine occupation, place of birth, and other details, as discussed in prior chapters, the estate records are not entirely satisfactory. So we searched a genealogical database (Geneanet) and civil records to recover an occupation for each individual, their father, and, for those who were married, the occupations of their spouses and their spouses' fathers. One advantage of this group is that it is small, and wealth is highly correlated with having piqued the interest of a decedent or a historian to the extent that some records exist online. Except for Frédéric Vogel, who was single and died in Nuremberg, we could recover demographic details, occupations, and parental information for all the top wealth holders. They were heterogeneous in their places of birth and death: 15 of the 36 were born in Paris, the same number in the provinces, 5 abroad, and 1 we could not ascertain (for more detail, see table A6). Of the 32 that married, 14 did so in Paris, while 12 wed in the provinces and 5 abroad (1 could not be determined). They died mainly in Paris (28 of 36). As such, they were more Parisian than most. Taking the top wealth holders and their spouses, they, unsurprisingly, died old (68 for men and 71 for women). The men married later than average (31 versus 28), and they married much younger women (22 versus 25).

Regarding sectors of activity, we attribute the occupation of their husbands to women who do not report one. The primary sectors of activity were vast: 17 in industry, 10 in finance, and 9 in politics. When classifying occupations, we cannot ignore that individuals can have important activities in different sectors. We use a simple algorithm to assign an occupation to individuals for whom we found more than one. We regroup the occupation into three broad categories: finance (bankers, stockbrokers, notaries), industry (industrialists, merchants and traders, and the professions), and public service (government officials and the military). We privilege finance over industry and finance or industry over politics. Politics, thus, refers to individuals in the military or government service who were not financiers or entrepreneurs. Politics and finance dominated before 1870, and industry (manufacturing and trade) was the top branch afterward. Many of the top wealth holders had run the same firm as their parents (Stern in banking, Schneider with Le Creusot, and Roussel in pharmaceuticals). Still, a significant fraction of the men founded their businesses (often with some inherited wealth). Even this small group provides examples of how the very rich followed a wide variety of wealth accumulation strategies.

As discussed in chapter 1, the top wealth holders can have quite different origins. Like the Mortier de Trévise family, they can owe their wealth to the largesse of the sovereign or, more broadly, to political and military success. In more democratic periods, their wealth may come from government policies particularly favorable to their investments (e.g., Jules Lebaudy, for whom tariff protection allowed sugar beets to be a very profitable venture). Another path involves building wealth from financial activity, particularly seizing arbitrage possibilities like the Rothschilds did famously during the Napoleonic wars (Ferguson 1998, vol. 1). A final path to very high wealth involves investment in rapidly growing firms in industry, broadly defined, including manufacturing (Schneider), alcohol (Double), or petroleum (Boutheroue-Desmarais). Because wealth is durable, the individual we observe may be the descendent of the person who chose a particular sector and made a success of it. For instance, Olry-Roederer was not the first of the family to be heavily involved in champagne; Jacques Stern was the second generation of bankers in Paris. None of these different paths are exclusive. The Schneiders were financiers before they took over Le Creusot, for instance. Over time, the success of each strategy seems to have varied. Early on, we find many individuals whose wealth came from politics. For instance, Armand de Caulaincourt, a general and politician who was Napoleon's ambassador to Russia and later his last foreign minister, died the wealthiest person in Paris in 1827. His wealth cannot be attributed to economic entrepreneurship. His vast fortune was, in fact, a combination of inherited wealth and the bounty of being a close ally of the monarch. His father was a marquis who successfully navigated the French Revolution. Both his parents left Parisian estates in excess of 250,000 francs in movables, and he was also richly rewarded by Napoleon. Nevertheless, the way wealth and politics mixed changed over time.

A century after Caulaincourt, Jean Fougerat, whom we met in chapter 4, was the founder and sole owner of the Etablissements Industriels du Sirop Rami. When he died, that enterprise alone represented a third of his estate and was the primary source of his wealth. Rather than expand into other pharmaceutical products, Fougerat took the income from his cough syrup and reinvested it in various other assets. The data we assembled are consistent with some broad increasing tendency for individuals to maintain large equity stakes in their own businesses until they die, yet what is striking is the great diversity of wealth management strategy even at the top of the distribution.

Finally, entrepreneurs were not required to own their businesses at death.[8] For instance, the principal wealth holdings of Marc Michel (1852 top wealth holder) were in French and Spanish public debt, and these two assets accounted for 60% of his holdings. Except for some significant (18% of his estate) accounts with Noël, Place et Cie bankers, there is nothing in his estate that links him to his career as a banker with a dubious reputation. The lack of evidence of his financial career is not surprising since he died at 81. By then, he had clearly retired. Seven years before his death, he appears in the 1845 electoral roll; his only tax contribution is on his capital income (*cote personnelle et mobilière*).[9] He continued to be wealthy but did not seek to use his millions to influence any particular business.

Wealth, it seems, offered options rather than directives. Individuals could hold on to the businesses they inherited or sell them. They could stay in the same field as their parents or strike out on their own. Fougerat, who seems to be an excellent example of a founder who dies in the business saddle, was far from a single-minded investor. He also invested in the Cognac region from where he hailed and became quite involved in cognac quality issues in his later years. He also built himself a manor house with a park. However, his entire real estate holdings only came to 10% of the value of the syrup business. His core endeavors comprised 35% of his wealth, leaving him extremely well diversified.

Initially, the top wealth holders were poorly diversified and held large positions in either public debt (Michel, d'Andigné de la Blanchaye) or real estate (Baguenault, Lecomte, du Bouchet de Sourches, d'Aligre), but starting in the late 1860s the top asset share of wealth seems to have stabilized at about 25% to 30% of all wealth for women and men who were not in business. When a family business was important (e.g., Stern or Lebaudy) it could represent up to a third of the estate. The timing of portfolio concentration change in the last third of the nineteenth century suggests that capital markets were critical in allowing this diversification at the top.

The 35 top wealth holders are just too few to say much about change over time, but the date of death strongly correlates with industry and primary assets in the business. These correlations are primarily driven by the general disappearance of aristocrats and military officers from the top sample after 1860. Exploring how these patterns changed over time requires a larger sample: the entire top thousandth.

8.4. The Top Thousandth Wealth Holders

The top 0.1% is never a large group but it is always extremely wealthy. On average, their estates were 200 times richer than the average estate in Paris.[10] Even the poorest of the super rich died about one hundred times wealthier than the average Parisian. The threshold of the top thousandth represents a level of wealth commensurate with a great desire to bequeath and the ability to either accumulate or preserve vast amounts of wealth. From 1807 to 1977 our cross sections contain 921 members of this group. The oversample for the 1970s we discussed in chapter 6 adds another 127 individuals. We include each decedent's last spouse, so the top 0.1% sample contains 1,955 individuals. We repeated the same procedure as for the top wealth holders to seek additional data. Because of severe selection issues into what goes into genealogies, and due to the absence of civil registration records in Paris before 1860 (these were burned during the Commune), we were less successful than we were with the 35 top wealth holders of the prior section. We retrieved occupations for 77% of the 1,007 men who were part of the 0.1% sample themselves or the husbands of women who were members of the 0.1%.

When considering the entirety of the super rich, we find a similar breakdown between sectors of activity as for the 35 top wealth holders over the whole period (see table 8.1). Thirty-eight percent were in manufacturing and trade, 19% were in finance (e.g., bankers, stockbrokers, or notaries), 17% were in public service (half in the military and half in civil service), and the rest were in a variety of occupations, including law and art. Nobles here refers to individuals with a title for whom we could discern no occupation (Gustave Rothschild, baron and banker, is classified as finance. Auguste Louis de Stael-Holstein, son of the famous writer Madame de Staël, had no occupation and is a noble). Later, we will examine the share of individuals with titles irrespective of occupation, as we did in chapters 1 and 2.

Over time, three changes in occupational patterns are noteworthy. First, the share of individuals with aristocratic titles for whom we could decipher no occupation falls over time among the super rich, with a sharp break at WWI (from 10% to less than three after WWII). This break is noteworthy because it follows the rise of progressivity in income and wealth taxation, which means we observe rural real estate better after WWI (in the last three columns of table 8.1) than before, something that should have increased the presence of pure aristocrats among the top wealth holders in Paris. Real estate was the aristocratic asset par excellence, so their decline was not likely

TABLE 8.1. The occupations of the top 0.1% of men by period of death

	Occupations by period of death				
	1807–1867	1872–1912	1922–1942	1947–1962	1970–1977
Finance	20.4%	27.7%	21.5%	21.2%	7.6%
Manufacturing and trade	31.9%	33.0%	43.7%	51.8%	63.6%
Civil service	19.9%	9.0%	8.1%	8.2%	3.0%
Military	10.2%	10.1%	5.2%	7.1%	4.5%
Art	0.0%	0.0%	0.0%	0.0%	6.1%
Noble	10.2%	12.4%	5.9%	2.4%	3.0%
All others	7.5%	7.9%	15.6%	9.4%	12.1%
Number of observations	302	325	148	91	151
With occupation	226	267	135	85	132

Source: Estates dataset.

due to measurement error. It coincides with our measure of the fall in the inherited wealth share. To put it simply, aristocrats could no longer maintain their fortunes simply by managing the bequests they received. We will return to the broader decline in titles later in this chapter.

The second striking pattern involves a drop in the share of public officials, whether military or civil service. In the first seven decades of the nineteenth century, 30% of men in the top 0.1% were public officials. By 1922, their share was half as large. Initially, two-thirds of the public officials had military careers. By the 1870s, the military and civilian officials had almost equalized. It is important to note that this change is both large and complex. The Third Republic could draw its officials from a much broader range of society, and it did not have to provide the kind of largesse that Napoleon or the Bourbon kings bestowed on their key officials.

The final pattern involves an inverted U-shaped share for occupations in finance. At the beginning of the period, 20% of men with known occupations were in finance. That share rose in the golden age of capitalism (1872–1912) to 28%; from there, it fell to 20% for the interwar and immediate postwar periods and dropped further to 7% in the 1970s. Over time, the residual group, manufacturing and trade, grew to comprise more than half of all members of the top 0.1% as it replaced individuals in public service and finance.

To some extent, this detail by year of death obscures the transformation in occupations among the super rich because individuals died old. At the same time, they likely made their occupational choices decades earlier when they were in their twenties and thirties. For the last group that died in the 1970s, occupational choices were largely set before the crash of October 1929. The relatively high share of top wealth holders in finance in the 1930s does not reflect high returns to bankers or stockbrokers during the Great Depression but occupational choices made before WWI.

Table 8.2 displays the same occupations by cohorts of individuals based on the period when they reached 30. These data also reproduce the decline in the share of civilian and military officials and pure aristocrats; if anything, the patterns are stronger. These three groups represent the majority of the individuals who started their careers before the French Revolution. Financial careers, which were important at the start, grew to a peak of nearly 30% for those whose careers began between the Restoration and the Second Empire. The share of occupations in finance then fell by half after WWI, so for those who started their careers in the interwar period, only 14% were in finance. Consider an ambitious, well-educated young veteran of WWI. In the 1920s, a finance career might have been appealing; it was clear that governments would need the services of bankers and the like for a long time to come as they grappled with the debts incurred during the war. Working in a large bank like the Crédit Lyonnais might have seemed even better because the banker would get the chance to participate in large deals. Even if our ambitious banker were very successful and had gained some equity in the bank where he worked, he would have had a bad surprise in 1945. Then, the bank was nationalized, and his equity was exchanged for government debt. At best, he could have ended his career as an executive at the bank and received a nice pension. Bank nationalizations meant he was unlikely to repeat Henri Germain's performance and die with a top 0.1% estate. The rise of manufacturing and trade occupations was even more dramatic, from 25% for the cohort of the Old Régime to 63% for those of the interwar period. Over the long term, the Parisian experience privileged industrial over financial capitalism as the driver and maintainer of inequality.

The super rich provide a last set of lessons: the Rothschilds' experience was atypical, and one should be wary of the conclusion drawn from case studies that are so narrowly selected. Paris was the financial center of France, but in general, more considerable fortunes were made in manufacturing or trade. In that sense, the Hottinguer family we encountered in chapter 5 is more representative of finance than the Rothschild family. With fortunes in the neighborhood of 10 million 1912 francs, the Hottinguers were comfortably in the super rich from 1841 to 1943. Their wealth was between 5% and 10% of a Rothschild's because they were in a sector

BLE 8.2. The occupations of the top 0.1% of men by cohort

	Occupations by period when reached age 30					
	Before 1789	1789–1815	1816–1869	1870–1914	1915–1939	1940–1970
ance	16.4%	24.5%	28.8%	17.8%	13.9%	0.0%
nufacturing and trade	25.5%	26.5%	34.6%	46.2%	63.1%	63.2%
il service	16.4%	9.2%	8.8%	7.6%	2.5%	10.5%
litary	25.5%	23.5%	9.2%	8.5%	4.1%	5.3%
t	0.0%	0.0%	0.0%	0.8%	4.9%	0.0%
ble	9.1%	10.2%	11.5%	7.6%	2.5%	0.0%
others	7.3%	6.1%	7.1%	11.4%	9.0%	21.1%
mber of observations	63	123	350	265	274	83
th occupation	55	98	296	236	131	40

rce: Estates dataset.

(short-term credit) that was far more competitive and did not scale in the same way as sovereign debt.

A second lesson we can learn from examining the super rich relates to the stability of their share of Parisian wealth (roughly 23% from 1807 to 1972). Recall from chapter 7 that the share of individuals who could constitute their bequests from inherited assets (rentiers) and the attendant capital income increases with wealth. For the super rich, that share was 75% before 1870 and remained high at 70% in the next half-century. The share of rentiers fell after WWI but remained close to half even in the 1950s and 1960s. Thus, one explanation for the persistence of the high wealth share of the super rich is that it was easy for them to grow their wealth in periods of low inflation and taxes. It appeals to anyone who favors a class approach to wealth and power. Unfortunately, it runs into problems. First, self-reproduction might have worked in the nineteenth century because returns and taxes were such that individuals could constitute large enough bequests out of inherited wealth to provide their children with a similar wealth ranking to their own. This obviously was not feasible in the twentieth century when the rentier share fell severely and more of the top group came from self-made wealth. Second, it flies in the face of all the sectoral and geographic mobility we have described. It is a partial explanation that must be complemented by the fact that each generation created new businesses that could grow quickly and thus reach ever-wider markets. Indeed, what characterizes large fortunes is often the ability to scale, whether it is cough syrup (Fougerat) or steel (Schneider).

Scale interacts with the distribution of ownership. An individual like Fougerat, who owned 100% of his enterprise, could end up much wealthier than someone who owned a small fraction of a larger business. The experiences of Renault and Citroën automobile manufacturers, mostly focused on France, were good examples of this, although both founders lie outside our study. As we shall see in the conclusion of this book, the ability to sell on a global scale is a key source of the top French fortunes of the twenty-first century.

8.5. The Top Shareholders of the Banque de France

For comparison, we also compiled a sample of top shareholders and officers of the Banque de France for 1801 and every subsequent fifth year until 1936.[11] The Banque de France's charter limited access to the annual meeting to the top 200 (male and French) shareholders and that list was published annually. That list was the basis for Daladier's cry that 200 families controlled France. There are three important distinctions relative to the top wealth holder sample we just analyzed. First, one only dies once, but one can be a principal shareholder for many years. Gustave Rothschild was on the list for 60 years, and his brother Edmond James for 58. This is also a population that was almost entirely male. The exceptions are Joséphine Tascher de La Pagerie (Napoleon's first wife) and Hortense de Beauharnais (her daughter and the mother of Napoleon III), who appear early on along with two other women. After 1810, no women appeared on the list (by contrast, the 0.1% was 30% female before 1870 and more than 40% female afterward). Later, both the widows Callaghan and Clicquot, whose firms were principal shareholders of the Banque de France, had to be represented by men. Of course, everyone in our wealth sample is Parisian, while the principal shareholders could reside anywhere in France.[12] Our bank sample includes 2,362 shareholders, of whom 202 were organizations (e.g., business firms, cities, and charitable trusts), leaving 2,160 individuals, of whom only four were women. We completed the list with the 84 members of the board (*régents, censeurs, sous-gouverneurs, gouverneurs*) who were not principal shareholders. In the same way we did for top wealth holders, we searched genealogical databases for additional information because the shareholder lists are just lists of names. At the very least, we wanted each member's date and place of birth and death, and their occupation. Finally, the list was last published in 1936, the last year when the general meeting was restricted to the top 200 shareholders. The principal shareholders are an excellent group to compare to the top 0.1% since it is of the same magnitude.[13]

Being on the shareholders' list involved a significant investment. Between 1839 and 1936, we know the minimum number of shares needed to be included on the list. For most of this period, 60 shares were enough to "buy" entry into the general meeting. There was a sudden jump to 100 shares in 1858 following an imperial edict that doubled the number of shares outstanding. The new shares were distributed to current shareholders in return for a tax of 1,100 francs. These shares, it seems, were sold out slowly as the minimum number of shares to join the general meeting fell regularly (to below 80 in the 1870s and then stabilized around 60 until 1910, after which it was mainly in the upper fifties). However, the median number of shares held can be inferred to be about 100, except for a few years after 1857. If we take the minimum number of shares required to be on the list, it was always sizable, between a tenth and a fifth of the average wealth of the top 1% of Parisians, until the 1930s when the value of the bank's shares stopped keeping up with asset price inflation. For the top 0.1%, however, being a top shareholder involved placing about 5% of one's fortune in Banque de France shares before 1890. Most years after 1890, that threshold was less than 4%. Had there been more competition for control, many individuals in and out of Paris could have afforded to buy more shares—that they did not suggests that no one saw much value in contesting the status quo. As Manas points out (2019, 135), bank shares were not lucrative investments. They were safe, low-return assets. When the Restoration governments required entails (*majorats*) for the peerage that composed the legislature's upper house, they allowed these to be funded by land, public debt, and Banque de France shares.

The names alone help us reassess Daladier's assertion. The 2,156 distinct male shareholders represent 1,601 distinct names, which is a much larger number than the 200 families he said controlled France. Adding in the few women on the lists and directors leads to 1,652 names. Daladier was clearly wrong from a quantitative point of view. The number of families engaged in large-scale share ownership of the Banque de France was, in fact, eight times larger than what Daladier suggested. One reason for this was the relatively short tenure of most individual shareholders, with about half appearing only once (thus with an expected tenure membership of five years or less). There is also a group of about 400 whose average tenure was 30 years or so and who could represent a stable bloc.

What, then, were the principal shareholders' occupations? As table 8.3 shows, they were more likely to be bankers and a bit less likely to be in manufacturing and trade than super-rich Parisians. Beyond bankers, the occupational distribution of the shareholder list for the Banque de France is

TABLE 8.3. The occupations of Banque de France shareholders by cohort

	Occupations by period when reached age 30					
	Before 1789	1789–1815	1816–1869	1870–1914	1915–1939	1940–19
Finance	29.0%	35.1%	24.4%	17.4%	23.3%	29.0
Manufacturing and trade	19.2%	14.4%	23.8%	30.6%	33.3%	19.2
Civil service	22.8%	17.5%	13.1%	10.6%	3.3%	22.8
Military	10.9%	18.9%	12.1%	16.6%	10.0%	10.9
Noble	5.7%	7.4%	16.7%	10.6%	20.0%	5.7
All others	12.4%	6.7%	9.9%	14.3%	10.0%	12.4
Number of observations	244	351	627	403	46	244
With occupation	193	285	496	350	30	193

Source: Banque de France dataset.

rather broad and reflects the wide range of individual investors with wealth. It includes industrialists, lawyers, and aristocrats for whom we could find no occupation. Moreover, nobles and government officials always represent a larger share of the shareholders of the Banque de France than of top wealth holders. As others have noted (Szramkiewicz 1974; Plessis 1982, 1985), until the 1890s the bank was largely run by investment bankers in Paris, though the governor and his deputies were appointed by the minister of finance. After 1890, the bank came under closer government supervision (Michener and Monnet 2023). The leadership came directly from important posts in the finance ministry. However, the core elite bankers Hottinguer, Mallet, Rothschild, Vernes, and others continued to dominate the board of directors.

8.6. Wealth and Power

Having described both the top thousandth wealth holders and the top shareholders of the Banque de France, we now turn to one aspect that makes wealth more than just material resources. There is no doubt that wealth is a form of power, as Daladier pointed out. We examine both the extent to which wealth meant that individuals had direct control over specific assets and the extent to which wealth allowed its owners to have political power. This is not an issue only for the super rich because, on one level, the connection between wealth and economic power applies to everyone, from the butcher and the blacksmith who own their shops to the industrial and financial barons whose decisions influence the fate of thousands, if not millions, of individuals. On a second, broader level, the butcher and the baker have less political influence than the bankers and sugar beet magnates, and their businesses are not individually important. The connection between wealth and power, thus, has private and public dimensions.

On the private side, the Chicago school and Harold Demsetz (1972), in particular, have long emphasized the importance of principal-agent problems in large business entities and their connection to wealth inequality. Agency problems are multifold, but the most important is shirking by employees. This includes managers who must share the fruits of their labor with all the other owners. When they own a small fraction, they will neglect tasks because the personal return to their effort is less than the return to the firm. Shareholders must thus monitor and, if necessary, discipline management. Each shareholder has to share the fruits of his or her monitoring with all the owners and will not monitor if his or her stake is small enough. The solution, Demsetz argued, was that someone had to have an ownership stake sufficient to warrant paying the monitoring costs and implementing change when necessary. It follows that an increase in industrial concentration driven by technology will make the wealth distribution more unequal.[14] This would be true whether the monitors are owner-managers or simply owners, which is an issue of happenstance.

Another literature in corporate governance emphasizes an opposite motivation for holding on to a significant stake in a firm. That is to extract private benefits via self-dealing, excessive in-kind monetary compensation, and what other means the legal system allows. In these scenarios, the wealthy use their control of firms to divert returns from small investors to themselves (Rajan and Zingales 2004). In this light, the concentration of wealth reduces the extraction of private benefits by managers, but it is likely to be replaced by extraction by the dominant owners. In both cases, there is a deep connection between wealth and control of some enterprises. We examine this question by tracing the extent to which top wealth holders maintained large positions (our proxy for controlling stakes) in particular enterprises.

We already saw that at the very top, many men were in direct control of important enterprises. We traced the enterprises that the top 0.1% controlled in all the cross sections from 1902 to 1977. A few patterns emerge. First, among the firms the top 0.1% likely controlled, luxury trades were very important, including high-end restaurants (Bignon 1902), champagne (Roederer 1887 and 1932; Perrier-Jouët 1887; Mumm 1942), jewelry (Bruhl 1927), and leather goods (Hermès 1970). Second, sole proprietorships

TABLE 8.4. Elite status for top wealth holders and Banque de France shareholders

	Elite status		
	Noble	Political career	Legion of Honor
TOP 0.1% WOMEN			
To 1789	35.50%	.	0.00%
1789–1815	30.10%	.	0.00%
1816–1869	15.00%	.	0.00%
1870–1914	12.20%	.	2.40%
1915–1939	12.40%	.	3.40%
1940–1970	3.40%	.	0.80%
TOP 0.1% MEN			
To 1789	41.30%	14.30%	3.20%
1789–1815	30.90%	17.90%	19.50%
1816–1869	20.60%	15.70%	24.00%
1870–1914	16.20%	10.60%	38.50%
1915–1939	19.40%	3.10%	41.90%
1940–1970	8.60%	2.50%	6.20%
SHAREHOLDERS OF THE BANQUE DE FRANCE (MEN)			
To 1789	30.37%	8.90%	21.30%
1789–1815	28.57%	10.36%	23.21%
1816–1869	27.36%	8.14%	24.43%
1870–1914	22.30%	3.38%	19.59%

Source: See tables 8.2 and 8.3.

and partnerships dominate, in particular, when we include finance and asset management. These entrepreneurs included bankers (Mallet 1902; Stern 1902), notaries (20 occurrences from 1807 to 1977), construction (Daval 1902; Rontaix 1937; de Bancel 1952). They also included owners of large companies as well (e.g., iron and steel, Schneider 1942; sugar refiners, Say 1872 and Lebaudy 1887; petroleum, Fenaille 1937 and André 1962; textiles, Dollfus 1907 and Davey 1937).

The other important connection between wealth and power comes from the value of wealth in politics. Table 8.4 illustrates this second dimension of wealth and power by considering how often the members of our groups had aristocratic titles (an Old Régime marker of power), entered political careers, or received the Legion of Honor, which is awarded for service to the country.[15] Women did not have the right to vote, and very few held political office in France until after WWII. So, no women attained this kind of public power. However, women, as men, could hold aristocratic titles, and a few women did receive the Legion of Honor. Both men and women were very likely to have titles if they died in the 0.1% or were married to a member of that group. During the First Empire, more than a third of the top wealth holders had titles—this includes both the nobles without occupations from table 8.1 and all the individuals who had a title, as table 8.4 shows. Over the whole period, 28% of the super rich had titles, and though the proportion varies over time, there is little trend. Clearly, aristocrats, by this definition, were always massively overrepresented among the wealthy and (even more) among the top shareholders of the Banque de France.

Over time, the share of aristocrats declined among the super rich just as it did for Parisians as a whole (see chapter 2). Titled individuals comprised 28% of the top wealth holders before 1870 and only about 14% after WWII. Again, we measure membership in the aristocracy for men as having a title (prince, duke, marquess, count, or baron), whatever their occupation, and we measure membership for women as either having a title of their own (which was, in fact, rare) or because their fathers or husbands had a title. The decline in the density of aristocrats over time was not simply due to the end of the Old Régime in 1789. Starting with Napoleon and ending with his nephew, the regimes that followed the French Revolution continued to create titles. This process only ended in 1870 with the advent of the Third Republic. Titles were the highest honorific rewards that these regimes could bestow. The boundaries between the aristocracy and the working rich were very porous: elite bankers like the Rothschilds, the Pillet-Wills, Hottinguers, and Mallets were all barons or counts of something somewhere, as were industrialists like Worms, Seillière, and Oberkampf. Thus replacing the aristocrats in the Banque de France list with bankers and industrialists does not reduce the number of titles so much. There were also "true" aristocrats, dukes, marquesses, and counts for whom we could find no occupation, representing only about a third of the total. These included lines from the Old Régime like Colbert de Chabanais and more recent ones like Suchet d'Albufera. By the interwar period, the aggregate share of aristocrats had fallen by half, and the decline continued into the 1970s. Still, even after WWII, the descendants of aristocrats remained overrepresented in the top 0.1%.

As mentioned earlier, very few women had political careers until after WWII, and only a few received the Legion of Honor. Rich men, whether in the top 0.1% or among the shareholders of the Banque de France, were quite likely to have been appointed minister or to the upper chamber ("peer of the realm" before 1848 or "senator" under the Second Empire) or elected to either national office ("deputy" after 1815 or "senator" after 1870), or local office (general or municipal councilor, or mayor). Before 1870, shareholders of the Banque de France had an 8% chance of having a political career, and the members of the 0.1% were nearly twice as likely to hold elected office. After 1870, when universal male suffrage and competitive elections prevailed, both groups' rates fell to below 4%. Only after WWI did political careers become rare.

The Legion of Honor has been awarded for a variety of reasons, but receiving it implies some feat of merit and the recognition of that feat. Under Napoleon, the recipients were mostly in the military. The banker Benjamin Delessert did receive his medal for funding the development of sugar beet refining during the continental blockade. Delessert was a top shareholder and régent of the Banque de France from 1810 to his death in 1847, when he died with a fortune of more than 6 million francs, the third-richest person in the city. He was also a deputy to the National Assembly from 1815 to 1842. More bankers and industrialists would become legionnaires as governments increasingly emphasized economic growth. This explains the steady rise in the share of members of the 0.1% who received such a distinction, peaking with the cohort that reached their thirties in the interwar period, where 41% of the men received the award. The shareholders again had less success in attracting the honor, though 40% of the board members received the distinction.

Overall, economic and political power are interconnected. This was easy to see for the cohorts that came of age under the Old Régime and between 1815 and 1848, when wealth was a precondition for voting and standing for election. It was also true during the French Revolution and First Empire, except that the connection ran strongly from politics to wealth because the key officials of the empire were showered with wealth by Napoleon. However, the connection between wealth and political power endured long after. It can still be seen through the cohorts who reached adulthood after WWI. As we discussed earlier, wealth is an advantage in most endeavors, and it is not surprising to see individuals using their wealth to pursue noneconomic goals such as political careers. Conversely, wealthy individuals can enter politics or finance to further their financial goals. Returning for a moment to the banker Delessert, he funded the perfection of sugar beet refining at the behest of Napoleon to reduce France's dependence on Caribbean sugar. The successful production of sugar from beets set off a process of trade protection that endures to this day, for beet sugar is not competitive relative to cane sugar. The increasing demand for sugar in nineteenth-century France created and furthered some of the fortunes we see in top Parisian estates and the shareholders of the Banque de France, like Say or Lebaudy. At the same time, wealthy sugar refiners had to ensure that the sugar tariffs endured. Members of these top families represented the sugar beet lobby in politics and finance. The lessons from sugar beets extend widely.

A separate question involves the overlap between the top 0.1% and the top shareholders of the Banque de France. We focus on the period between 1807 and 1870. For that period, it is easy to compare the top shareholders of the Banque de France and the top wealth holders in Paris. To do so, we relied on information on the dates and places of death we extracted from the genealogies. We then sought out the principal shareholders in the TSA. For this period, the TSA report wealth, and they are often online. For 93% of those who resided in Paris and died there, we were able to find the TSA estimate of their wealth level. We were less successful for those who died in the city's orbit (Seine, Seine-et-Marne, Seine-et-Oise, Oise) and even less for those in the provinces simply because only about half of those départements have digitized their TSA, and in several cases, the records have been lost. Nevertheless, an 82% overall recovery rate (589 of 711) seems more than adequate. We focus on movable wealth because that sort of wealth was reported at the legal residence and transcribed in the TSA registers up to 1870. Figure 8.1 displays their movable wealth on a log scale and, for comparison, the lower threshold of the top 1% and top 0.1% in movable wealth for Parisians, where we linearly interpolate between the sample years. To improve legibility, the few values below 1,000 francs are reported as 1,000, and those above 10 million francs are reported as 10 million. On average, the top shareholders were twice as wealthy as the poorest member of the top 0.1% in Paris. The locality of residence does seem to matter. Shareholders who were residents in Paris had the highest movable wealth, 50% more than those who lived in the provinces and more than twice as much as those who resided in the Paris orbit.[16] The log figure shows that being a top shareholder of the Banque de France is highly predictive of dying wealthy: more than half died as members of the top 0.1%, and 73% are above the threshold to be in the top 1% in movable wealth. At the same time, being a top shareholder does not guarantee dying a plutocrat. Three top shareholders died with a negative estate, and two dozen estates were less than the value of a single share of the Banque de France—by the time of their death, they were no longer on the list of principal shareholders. If we set the threshold at 60 shares, a quarter

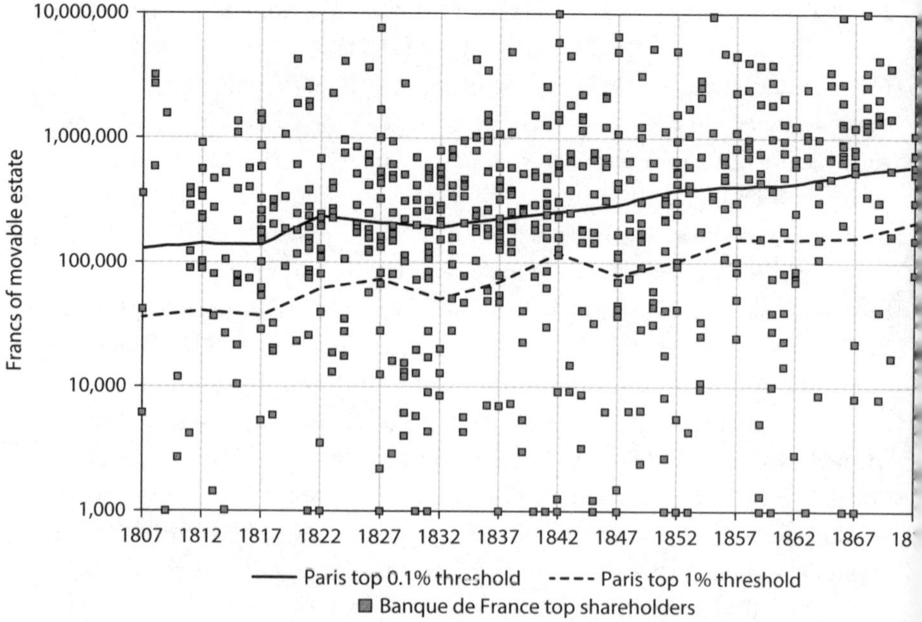

FIGURE 8.1. The movable wealth of top shareholders of the Banque de France versus the top 1% and top 0.1% thresholds for Parisians, 1807–1872. *Note*: Movable estates larger than 10 million are coded at 10 million francs. Movable estates smaller than 1,000 francs are coded at 1,000 francs. The top 0.1% or 1% threshold is the movable wealth at death of the poorest member of that group. *Source:* Banque de France sample.

of the top shareholders died with movable wealth less than the value of their stake in the Banque de France when they were on the list. Of course, they could have sold that position to buy land, which overstates their downward mobility. What is sure, however, is that they no longer owned a significant position in the bank at the time of their death.

We can also use the wealth of top shareholders of the Banque de France to evaluate whether the super rich are a club. Most members of the top 0.1% were not members of the top shareholders of the Banque de France. We can validate this claim in two ways. The first starts with the relative sizes of the cohorts of top wealth holders and principal shareholders. It is mathematically possible for all the top shareholders to belong to the top 0.1% in Paris from the 1870s to the 1930s. The number of top shareholders who died in any given year bounced from 10 to 24 from 1817 to 1947 with no discernible trend. The number of men in the top 0.1% of men rose from about 5 in 1807 to about 20 in 1880 and then declined after 1910 to about 15 on the eve of WWII. Although the two groups could coincide, they did not. When we looked at the wealth of the top shareholders before 1870, only about half died with sufficient wealth to be part of the super rich (and some 20% were not Parisian residents). If we return to our cross sections and ask how many of the top 0.1% ever appeared on the top shareholder list or were officers of the Banque de France, the answer is 71. It is better to restrict the data to the individuals who died from 1807 to 1932 to avoid any censoring problems due to the end of the limitation on who could attend the general meeting. In that interval, only 66 out of 411 top wealth holders ever appeared on the top shareholder list, and 33 out of the 300 men married to members of the top 0.1% appeared on the top shareholder lists. While there is also no doubt that entering the ranks of the super rich was difficult, only a small number of them were shareholders of the Banque de France.

As noted above, some 10% of the husbands of the women in the top 0.1% were significant shareholders of the Banque de France. This select group included Gustave Rothschild, who owned 66 shares upon his death and was a member of the 0.1%. His wife, Cécile Anspach, did not own any herself, even though most of the members of the Rothschild family did. Charlotte Rothschild owned 108 shares at the time of her death in 1899, placing her above the median holding of men entitled to attend the annual general meeting. Her husband Nathaniel, who died in 1870, owned no shares then and never appeared on the list of shareholders. While Charlotte could not vote her shares, the Rothschilds were abundantly represented at the general meeting. (From 1855 to 1905, they had three representatives—Gustave, his brother Adolphe, and a delegate for the firm.)

For women with large stakes in the Banque de France, disenfranchisement was complete. We trace their exclusion in different ways. Let us start with women who were wives of members of the general assembly: Antoinette Victorine Berthier and Louise Amélie Greffulhe. While neither woman was ever on the list, their husbands were. Antoinette Victorine Berthier (d. 1922) was married to Antoine Ernest Dervaux (d. 1965). She owned 61 shares of the bank, and in these years, the threshold for admission of men to the assembly was never above 56 shares. Dervaux himself first appeared on the list in 1916 and remained on it until 1936, but he was voting for his own shares, not his wife's. His position came from the shares he inherited from his father, Ernest Omer, who was a Banque de France assembly member from 1901 to 1916. Louise Amélie Greffulhe (d. 1932) married Arthur Robert des Acres de l'Aigle (d. 1931). She owned 180 shares. Des Acres de l'Aigle was a member of the general assembly from 1891 to 1926; thus he exited before his wife's death and is unlikely to have voted his wife's shares.

More frequently, women owned enough shares in the Banque de France to be listed and neither they nor their husbands appear on the lists. Therèse Stéphanie Feuillant (d. 1912) had married Pierre Gaspard Napoléon Cassagne de Beaufort de Miramon (d. 1914) in 1860. In her estate filing, we find 121 shares of the bank, putting her well above the median top shareholder. Marie Marguerite Ybry (d. 1927), the widow of Pierre Etienne Carraby, who had died in 1911, owned 81 shares. No individuals named Feuillant, Cassagne, Ybry, or Carraby are on the major shareholders' list. The Banque de France's governance simply excluded women from the general meetings. It also is clear that women were prevented from having their position represented by proxy. Their ownership was without power.

8.7. Conclusion

Starting with the Rothschild family, we have steadily broadened the definition of the super rich. The different layers of evidence we have brought lead to four substantive and methodological conclusions. First, unsurprisingly, one should be wary of examples like James Rothschild and his family. Examples are no more than that. They can illustrate a particular pattern of a financial dynasty among the super rich, but they can also lead us astray. The rest of the group of the super wealthy shows a much broader range of behavior. For those who are tempted to reduce the super rich to a group of individuals who persist through internal reproduction, even the Rothschilds, should give pause. To be sure, their affinity for intermarriage was unparalleled. However, if the super rich were a self-reproducing class, how could a German-Jewish immigrant not only join the list of the super rich of Paris but place himself and his direct descendants at its apex?

Second, as it bloomed in the second half of the nineteenth century, capitalism produced a raft of massive fortunes. Scholars have emphasized how financial capitalism provided new ways for individuals to secure capital for their enterprises. Here, we observe the converse effect: financial capitalism allowed the rich many new options to diversify their portfolios. Our sample is too small (one-fifth of Parisian decedents) to link generations. However, the evidence from portfolios in chapter 2 and from inherited wealth in chapter 7 highlights that the wealthy could adapt to the transformations of the economy. They moved large fractions of their wealth from public debt and rural real estate to equities and urban real estate. Over time, thus, as capital markets developed, being a rentier no longer led to low-return investments.

Third, for each individual, the portfolio of assets we observe at death results from many decisions over their adult life. Many never had an entrepreneurial career. Others did, but by the time they died, they had sold their interest and looked like any other rentier. Until WWI, their heirs often identified them as such. Distinguishing between these two groups required us to go beyond our fiscal sources to genealogical records. Using these additional records allowed us to trace out sectors of activities for nearly all of the super rich. While this does not always resolve the issue of sources of wealth neatly (e.g., were the Schneiders financiers or industrialists?), it does highlight a three-step transition from politics to finance to industry. Politics was a massive direct source of wealth during the upheaval of the French Revolution and First Empire, and it continued to be important through the Second Empire. From 1815, however, a reverse flow became visible: from wealth to politics. That flow led many rich people to exercise political power long after property qualifications were abolished.

Finally, we started with two questions. The first was about the persistence of the group's wealth share, and the other was about its permeability. The answer to the second is unambiguous: the top 0.1% was in constant flux. The French Revolution and First Empire created a whole new set of super rich that merged with the Old Régime aristocracy and the Parisian business elite before 1870. It is also likely that going back to the Middle Ages, a constant inflow of migrants and new industries kept adding new fortunes and new entrepreneurs to Paris (Sussman 2006). It is undoubtedly a constant in the composition of the top 0.1% in Paris. The super rich is a club rather than a closed class; new members join this club every generation as other families fade. Families fade for a variety of reasons: lack of heirs, poor business or investment decisions (for every Suez Canal, there is a Panama; for every Crédit Lyonnais, there is a Union Générale), and excessive consumption from capital income. Inherited wealth produced many of the top fortunes we see. At the same time, other important forces—namely technological change and migration—were renewing Parisian fortunes.

9

The Myth of the Grasshopper and the Ant

From 1807 to 1912, we collected 21 full populations of wealth at death. In each year, at least two-thirds of Parisians left no estate at all when they died. Moreover, the top decile of decedents never held less than 96% of Parisian wealth. Meanwhile, from trough (1812) to peak (1902), the average estate rose more than ninefold in real terms. Prosperity lifted some boats far more than others. After 1922, the fraction of individuals who left an estate began to rise, and by 1977, the fraction of Parisians with an estate was two-thirds. In a half-century, the proportion of rich and poor at death had flipped.

The very long period when the fraction who died with nothing was stable and high makes it tempting to conclude that Paris was composed of two groups. The first population, like the ant of the fable, were savers. They accumulated wealth over time and passed it on to their children. Like the grasshopper, the second population lived for the day, and any windfall from work or luck would be consumed. In recent years, scholars interested in long-term development have observed the radical difference in income across continents, countries, or even cities and have also noticed that these differences are correlated with household wealth and personal capital. Galor and coauthors (1993, 2004) and Clark (2007) have argued for either a genetic or cultural basis for wealth accumulation, with some lucky groups adopting the ants' attitudes and values while others remain grasshoppers. Given the secular length of the split between rich and poor in Paris, it is tempting to consider this evidence as confirming these theses.

The fable presents the ant and the grasshopper as having the same opportunities and very different savings behaviors. Thrift alone explains why one is rich at the beginning of winter and the other is not. The moral of the tale is that the profligate grasshopper does not deserve our sympathy. In this vein, a long-standing literature emphasizes the importance of teaching people to save. It has provided support for a host of policies designed to incentivize the savings of the poor (Boyer 2018; Gaillard 1998). Success in financial education and take-up of savings vehicles should turn grasshoppers into ants. At first blush, the Parisian wealth-at-death data provide credence to these views: the data support the notion that individuals are heterogeneous regarding their attitudes toward savings and bequests. However, that heterogeneity is unlikely to have a genetic or cultural basis. Moreover, the behavior of Parisians cannot be reduced to that of agents with common preferences that include a desire to make bequests to their children.

The first reason to suspect that parsimonious approaches that rely on fixed preferences are inadequate is internal to the period before WWI. As we saw in chapter 8, despite the stark difference between the wealthy and the rest of the population, this was a world of persistent mobility. The aristocrats and politically favored individuals who dominated the first set of cross sections had to accept the arrival of financiers and industrialists. Parisian-born elites had to make room for migrants from France and abroad. On the one hand, as we saw in chapter 7, being of poor parents (not receiving a bequest) almost closed the door to the very top of the wealth distribution. However, it was not completely shut, and every year a few of the very rich came from modest backgrounds. On the other hand, the majority of the individuals who died wealthy had not received significant bequests, while some of those who had gotten large inheritances left little behind.

If chapters 7 and 8 raise questions about the value of a two-population model, there are deeper reasons to doubt it. The fundamental problem is that slowly but surely, over the half-century following WWI, half the grasshoppers became ants. This transformation followed a long process of institutional innovation where ever more workers in France, particularly in Paris, were covered by mandatory retirement programs. As a result, after WWI, the stable system of the long nineteenth century was replaced by a steady growth in the share of the individuals who died with some wealth. By 1977, when our data end, we estimate that nearly two-thirds of Parisians died either with a pension (*retraités*), assets, or both. The decline in wealth concentration was much less dramatic than the drop in the share of people with no recorded estate. As late as the 1970s, the top two deciles of wealth holders still controlled 95% of all Parisian wealth at death. The growth of the share of individuals dying with estates started in the early 1920s when asset prices and wealth accumulation were generally falling. It continued after 1947 when asset prices and measures of wealth per capita were rising over the long term. The share of individuals who left a bequest was not connected to average wealth. Instead, the share of individuals with an estate rose along with those with pension assets or pension income. Before 1912, only 10% of estates contained either life-contingent or pension assets. In the interwar period, 19% had such assets, and after WWII, more than 22% did. Clearly, that proportion increased more and more with time. By 1972, 32% of Parisian estates had some pension income. If we add the retirees dying without reported wealth, more than 61% of all decedents had a retirement pension that year.

In this chapter, we take in both the periods before and after WWI to consider what the long history of wealth and inequality implies about the transmission of wealth over the generations. Our goal is to understand which Parisians decided to leave bequests to their heirs. This chapter explores five key findings about wealth at death that are surprising and at variance with simple life-cycle rationale approaches to savings and bequests. The first is that a far larger proportion of the population saved than died with wealth. Second, the importance of life-cycle savings was considerable but not as a substitute for bequests. Third, the accumulation practices of most Parisians were unusual because of the threshold wealth needed to acquire real estate in the city and the transaction costs of holding most financial assets. Fourth, government policy was critical in creating a large population of retirees and in increasing the share of the population with bequests. Fifth, the second half of the twentieth century saw a normalization of the portfolios of Parisians below the top 1%, with real estate now becoming the modal bequeathed asset. In each case, the perspective of behavioral economics helps inform our analysis. To start, however, we begin with the world of the living and the development of savings banks in Paris and France.

9.1. A World of Savers

Before the revolution, Paris had a variety of municipal, religious, and occupational welfare institutions. The revolution abruptly halted the church's and corporations' welfare activities, and the state (through the city) was left to shoulder the burden. With the end of Napoleon I's empire, the amount of poor relief fell abruptly despite concerns about political instability. A group of elite men, led by the Duke of La Rochefoucauld-Liancourt and Benjamin Delessert, sought to alleviate this problem by starting the Caisse d'épargne et de prévoyance de Paris (Christen-Lécuyer 2012). A royal authorization was granted in 1818. The plan was to turn the Parisian population into savers by subsidizing savings accounts at the bank. The bank would place the deposited funds in short-term government securities and pay the interest to depositors' accounts. It was supposed to cover its costs from the income of an endowment set up by its philanthropist founders. When an account reached a balance equal to the price of a 50-franc bond (capital about 1,000 francs), the bank would purchase it and hand the certificate to the deserving saver. The plan, from the founders' perspective, had three advantages. First, their holders could use their accounts to meet short-term financial needs. In economists' parlance, the bank provided a locus for precautionary savings. Its success would reduce the need for public investment in poverty assistance. Second, those savers who persisted would become holders of government bonds—wealthy! It is interesting that while this political stance was widely shared by European reformists, ranging from French *libéraux* like Delessert to Victorian advocates of an ethic of responsibility and self-help, it has long echoes in the political platforms of conservative parties, from Margaret Thatcher's privatization of public housing to George W. Bush's ownership society. The third goal was to create more demand for government bonds.

By 1835, the caisse was a massive success, so much so that it could not cover its costs from the foundation's income, and the government stepped in. It offered an annual subsidy of 0.25% of the value of accounts to cover administrative costs and provided a general mechanism for other localities to open caisses d'épargne. In Paris, the number of accounts raced to more than 100,000 by 1838, half a million by 1885, and nearly 700,000 on the eve of WWI. This growth led to an account for every seven inhabitants of Paris by the end of the Second Empire, and one for every five or so after 1900.

By 1900, there were over 1,900 savings bank offices in France and almost 7 million accounts. However, that was not all. In 1881, the government had also created a postal savings caisse. By 1900, there were 3.5 million postal savings accounts, of which 770,000 were in the Seine (at that time, Paris had three-quarters of the Seine's population). Taking postal and standard accounts together, there was about one account for every other person in the Seine, and it is unlikely that the proportion was less in the city than in the surrounding areas (RCE 1900). The Parisian caisse published data on the profession of individuals opening accounts. By the mid-1820s, 65% of all accounts were opened by workers (*ouvriers*) and domestic servants. A third category, employees in the service sector, represented less than 10% of all individuals opening accounts before 1870, and that share increased until it reached about 40% in the 1950s. Delessert would have been pleased. Moreover, the reports Delessert and later his son signed as presidents of the caisse's board of directors highlighted the connection between account growth and savings for the working classes.

This plethora of accounts may seem inconsistent with the fact that so few Parisians died with assets. The returns on caisse accounts, set at 4% or so, led to a doubling time of 20 years for those who let the interest accumulate. The caisse was a get-rich-slowly scheme. Since workers' wages were close to 12 francs per week, a worker who saved 5% of his or her income week in, week out would have accumulated a stock of savings equal to half a year's income after a decade and three times their annual income after thirty years. This kind of slow accumulation could be useful in building a buffer stock of wealth, but the returns were too small to provide for old-age security. In this sense, the caisse was a success, and Delessert would have been happy to know that by the beginning of the twentieth century, about 5% of the total value of accounts was used to purchase government bonds each year. That would have been enough to provide about 10,000 depositors each year with a bond. However, that was only 2% of active accounts. The caisse postale showed similar numbers as well.

The shiny picture of workers' long-term savings gets tarnished further when we look more closely at the distribution of accounts. This distribution was printed in the reports from 1883 into the 1930s. We find that before WWI, more than half the accounts had a balance under 100 francs, thus making it very unlikely that they would ever reach the threshold for the purchase of the smallest government bond (roughly 1,000 francs in 1818). Indeed, this sea of small accounts represented a tiny fraction of all the monies held by the caisse (5% to 10%). Moreover, later, when the government lowered the minimum investment needed to purchase a bond (Hautcoeur 2007, 127), the caisse did not follow through and buy more bonds for depositors whose accounts reached the threshold. The depositors showed little desire to become bondholders. Instead, they liked the liquidity of caisse accounts. More importantly, accounts with less than 100 francs do not seem to be part of any long-term wealth accumulation scheme. Instead, these accounts reflect the need for many Parisians to build up precautionary savings. These would likely get depleted near death. The argument that the caisse and its sister institution of postal savings were used to create precautionary savings is supported by the fact that more than 10% of all caisse accounts were closed each year through the end of the Second Empire. After 1880, that number fell to about 4%, in part because the caisse held on to accounts with no balance for a while to avoid the cost of creating new accounts when the same clients returned.

Taking it all together, it is clear that the caisse was, in fact, two conjoined operations. The first operation, which mattered to the caisse's working-class clients, involved precautionary savings. The more than 80% of the accounts that held less than 500 francs had an average balance of less than 80 francs through 1926. Those with less than 100 francs had an average balance of less than 23 francs. The other operation targeted the middle class and was associated with accounts above 500 francs. These larger accounts always amounted to about 90% of deposits at the caisse. These large accounts were those from which the bond purchases were funded. Had the caisse followed the vision of its founders, these accounts would never have reached much more than 1,000 francs. The structure of the government's subsidy that funded the caisse's operations implied that larger accounts were not just attractive but, in fact, necessary to the solvency of the institution. Data from the caisses d'épargne reports to the finance ministry demonstrate this. Between 1889 and 1892, the caisses lost money on all accounts of less than 200 francs. Later (1908–1911), they lost money on all accounts of less than 500 francs (RCE 1889–1911). The caisses balanced their books from the "profits" they made on the large accounts. This explains why the caisse barely implemented its plan to convert accounts that reached 500 francs into *rentes*. Instead, the transfer of balances from savings to bonds was mostly voluntary (among the purchases of bonds, those made at the request of the holders of the account were 20 times more frequent than those forced on the holders). Dealing with accounts in the paper ledger-passbook world that prevailed until the late twentieth century was costly and did not depend on the balance. The cost of opening or closing accounts was fixed, and the cost of recording deposits and withdrawals did not depend on the transaction size. Larger accounts probably had more transactions, but smaller accounts were opened and closed more often. The caisses evaluated the average cost of accounts as fixed at about 1 franc. An account with a balance of 100 francs received a subsidy of 0.25 francs, which was insufficient to pay the caisse's costs (Christen-Lécuyer 2012; de Coninck 2012).

The Parisian working class, including its large contingent of domestic workers, was thus neither ignorant of savings nor uninterested in them. On the contrary, once savings institutions, like the caisse d'épargne, became available, they opened accounts in great numbers and set aside money. However, their having only limited income required them to focus on precautionary savings rather than building up bequests or life-cycle savings.

9.2. Elusive Life-Cycle Savings

In earlier work (Piketty et al. 2006) that we published before collecting the microdata for the twentieth century, we noted that average wealth was strongly increasing in age before WWI, and then after WWII, the wealth-age profiles flattened. We revisit these findings with our full dataset here because the second motivation for individual savings is life-cycle concerns (distinct from precautionary or bequest motives). One simple reason that there were so few successful savers in our data could be that many Parisians had little motivation to build dynastic wealth but were concerned about building their businesses during their active years and planning retirement. As they got older, they could have sold their bonds and stocks and bought life-contingent assets. Parisians could also have moved out of Paris as they retired, and before 1901, they may have saved in ways we cannot see by purchasing real estate outside the city. The successful life-cycle savers would then either appear poor or disappear from the data, leaving Paris to look like a poverty trap for the elderly. While there is no doubt that the Parisian population was in constant flux with the rest of the country and the world, there is no clear evidence of this kind of selection in the wealth data. Regression analysis shows that, prior to the great shocks, the number of men with a positive estate was strictly increasing in age (the predicted peak is past 100). After 1912, the fraction of men who died wealthy peaked at around age 80. For women, the share of wealthy individuals peaked around age 70, earlier than for men, in both periods. The difference between men and women is not surprising, given the significantly lower incomes of women and their more limited access to retirement programs. More importantly, these ages are far from the 55–60 range we might have expected before 1920, or the 60–65 range afterward, if Parisians had been stalwart life-cycle savers or if Paris retained only the poor and downtrodden.

Here, we will discuss the relationship between age at death and wealth at death (age-wealth profiles) in two ways: first, by year of death, and second, by year of birth. If life-cycle savings are an important fraction of wealth, we should expect wealth to rise during productive years and then decline after retirement for some fraction of the population. If individuals annuitize some of their savings at retirement, their age-wealth profiles will have a large break

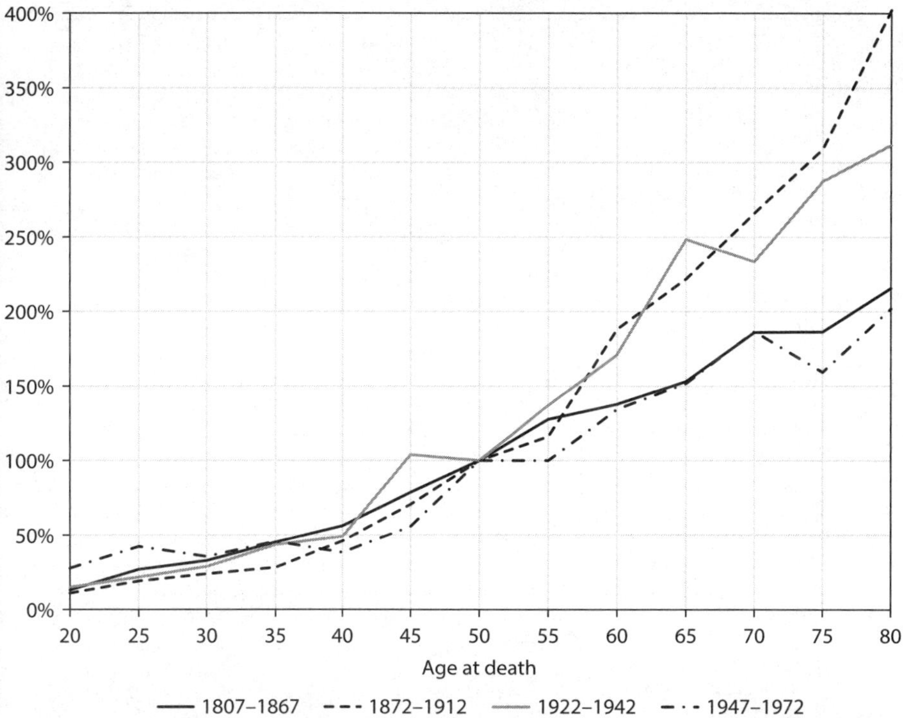

FIGURE 9.1. Age-at-death wealth profiles in Paris, 1807–1972 (all adult decedents). *Source:* Estates dataset (see appendix 1).

at retirement (since we do not observe the capitalized value of pensions). If retirees do not annuitize, their wealth should still fall after retirement as they sell assets to pay for their expenses. To look for a possible peak in the age-wealth profile around age 60, we deflate all estates to 1912 francs and aggregate the different cross sections into four periods. As figure 9.1 highlights, wealth at death rises with age in all periods, and age-wealth profiles never show the hump shape that a life-cycle saving pattern would have predicted. In the early years (1807–1867) and in the later ones (1947–1972), the slope of the age-wealth profile after 50 is about half as steep as in the two middle periods (1872–1912 and 1922–1942). Suppose we choose shorter time intervals (say, 20 years). In that case, the results remain the same: wealth profiles increase with age in all periods. Parisians who died between 1807 and 1817 have relatively flat profiles, but the slope increases steadily through the cohorts who died in 1902–1912, then falls steadily to the 1960s and 1970s. Relative to our 2006 article, where data we used for the twentieth century came from published sources, here we find much steeper age-wealth profiles

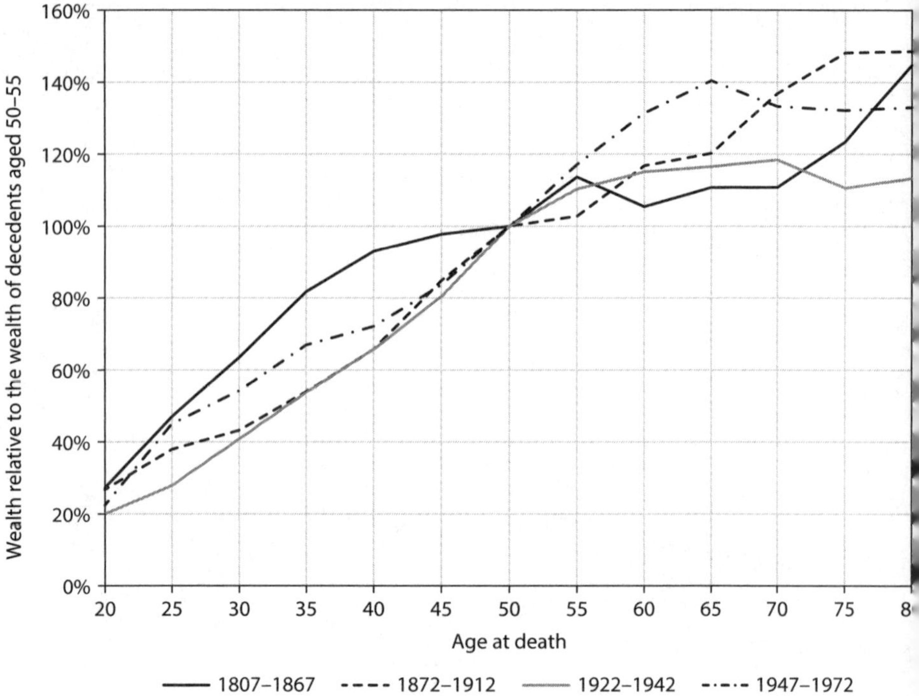

FIGURE 9.2. Age-at-death wealth profiles in Paris, 1807–1972 (top 5% excluded). *Source:* Estates dataset (see appendix 1).

after WWII. As discussed in chapter 3, the differences come from two forms of measurement error in the published data. In the 1930s and 1940s, the primary source bias in the published sources comes from the delayed filing of the richest estates. Starting in 1957, when inflation moderated, our higher measurement of inequality was driven by the high share of tax-avoiding assets in the very rich's portfolios.

Because we are looking for signs of life-cycle savings, we drop the richest 1% (and thus 50% or so of wealth at death). We do this because the top 1% of estates are hundreds of times the average annual income, so there is no reason to think that life-cycle motivations would be important in shaping the accumulation strategies of their owners. The steepness of the age-wealth profile is reduced when we focus on the bottom 99%, but the slope remains positive in all periods. This is no surprise as we know that the richest individuals tended to die particularly old. Moreover, the strongest effect arises for those who died after WWII, when the 80-year-olds are only 40% richer than the 50-year-olds (instead of 80%). If we remove the top 5% of the richest estates (figure 9.2), we at last see a wealth peak around age 60 for the

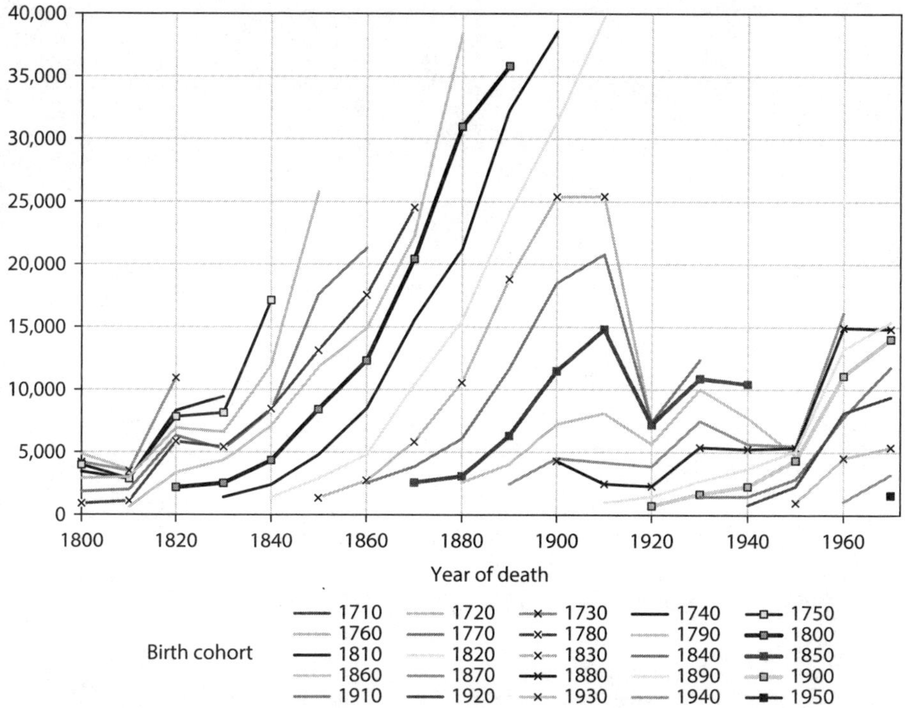

FIGURE 9.3. Wealth by birth cohort in 1912 francs (no top 1%). *Source:* Estates dataset (see appendix 1).

two cohorts that died after WWI. For the cohorts that died in the interwar period, the wealth of the bottom 95% of estates is essentially flat after 60 at 10% above those who died in their fifties. For the cohorts that died after WWII, wealth peaks at 65 and then declines to 90% of the wealth of 50-year-olds. Since these cohorts experienced the diffusion of retirement programs, one could easily conclude that this is evidence of life-cycle savings. In this view, the estates we saw in the nineteenth century were inferior substitutes to life-contingent assets like pensions, and when the latter became widespread, wealth at death fell. This attractive conclusion, however, must be discarded. When we examine the data from the perspective of birth cohorts, the timing of the change in behavior does not hold up.

Rather than death cohorts, birth cohorts are likely a better way to seek a pattern of life-cycle saving. This is precisely what our data allow us to do. Because we have an observation every five years, we can take the same data and estimate cohort wealth without the richest 1%. Scaled to 1912 francs, these profiles are displayed in figure 9.3 for every birth cohort. In each cohort, the youngest members to die are on the left of the line, and the

FIGURE 9.4. Wealth by birth cohort for those who faced WWI, in 1912 francs (no top 1%). *Source:* Estates dataset (see appendix 1).

oldest, at age 90, are on the right. Through the cohort born in the 1820s, the story of the graph is simple: over time, each cohort has higher and higher wealth by age, and there is no bend down past some hypothetical life-cycle peak wealth. The cohort born in the 1830s sees a sharp break between the wealth of those who died in 1912 and those who died in the 1920s. Beyond 1920, the data are massed in the lower part of the graph and do not have the clean structure of the earlier period. Age no longer matters much.

Figure 9.4 blows up the key period and now we can easily observe the break represented by WWI. At whatever age the cohorts hit the 1910s, their wealth peaked, followed by a sharp drop in the 1920s. Further, the drop is largest for the oldest group, whose members were the most likely to put a heavy portfolio weight on fixed-income securities. In the 1910s, the 1830 cohort was 5,000 francs richer than the 1840 cohort, and the 1840 cohort was 5,000 francs richer than the 1850 cohort. By the 1920s, all had about 10,000 francs on average. This break was not caused by the adoption of life-cycle savings strategies but by losses from the war and the financial and fiscal policies that followed.

The data on wealth by age lead to one further conclusion concerning return migration to the countryside at retirement. There was considerable migration between Paris and its labor market hinterland. In certain occupations, these migrations were annual (Magnac and Postel-Vinay 1997), in other they were long-term (Chevalier 1949; Farcy and Faure 2003). However, we see no sharp break at 50, 60, or 65 in the number of dead or mortality rates (Kesztenbaum and Rosenthal 2011, 2017). Thus, return migration to the countryside at retirement does not drive the age-wealth profile.[1]

More importantly, as we shall see, life-cycle savings mattered to growing numbers of Parisians over the twentieth century. However, that does not seem to have reduced their desire to accumulate wealth to transmit through inheritance. Despite pensions' increasing importance, most wealth accumulation was dynastic (driven by bequests). Let us now return to the problem of Parisian savings in the nineteenth century in order to understand the particular place of real estate in Parisian portfolios.

9.3. The Size of the Ante

Some investments are indivisible. Even if one can borrow some fraction of what is required, these indivisibilities create a correlation between portfolios and wealth. In the case of Paris, we must consider two different areas where indivisibilities might have mattered: real estate (where they did) and financial instruments (where they were less important). To be sure, the extent of indivisibility is endogenous. One of the most dramatic changes of the past two hundred years has resulted from a flood of innovations in finance and real estate that have allowed for a lower value of the minimal purchase of an asset. So, it is important to cast ourselves back to the beginning of the nineteenth century when the duke and the banker were dreaming up the Caisse d'épargne de Paris. That a large fraction of Parisians left no bequest between 1807 and 1817 is unsurprising. After all, the individuals who died in these years had their savings plans upended by the turmoil of the revolution and the First Empire. Parisian bondholders, in particular, suffered very large losses—90% for private bonds and 67% for public ones (see Hoffman et al. 2000). What is more surprising is that an equally large fraction of individuals left no bequest between 1827 and 1897. This was true even though the intervening seven decades had seen real GDP per adult rise by 110%, average adult labor income increase by 134% (Piketty and Zucman 2014, tables 1 and 3) and annual worker income by 52% (Bayet 1997, 29–30). These figures are for France as a whole, and there is every reason to think that Parisian incomes rose even faster.

As Delessert and other reformers argued, it was hard to save in Paris. After the inflation of the revolution, Parisians who found financial assets risky could, in principle, still build their wealth with real assets, particularly real estate. Unlike many other cities, however, Paris was bounded by its fortifications, and even though its size doubled in 1860, it has always been a dense city of apartments. In 1876, for instance, 76,000 buildings were divided into just under 700,000 dwelling units. The number of buildings grew slowly, and it was still under 100,000 on the eve of WWII. The number of dwellings grew to nearly 1 million by 1910. One million dwelling units are a lot of assets in a city of 2.9 million or so inhabitants. The problem was that until the 1960s, there was no simple legal structure to make the different dwellings in a building into legally separate entities. Buildings could be owned in common (as undivided inheritances). Buildings could also be purchased collectively and held in corporations or civil law companies. These vehicles were good for someone who wanted to invest in a building or set of buildings expecting to earn rents and capital gains. However, they did not directly confer the exclusive right to a particular part of the building for occupation. The law of June 28, 1938, provided mechanisms that accomplished joint ownership and exclusive use but were convoluted and not used frequently. Paris remained a city of renters until reform in the mid-1960s opened the door to individual apartment ownership on a large scale.

Parisian buildings were expensive, at least before WWII. We measure the distribution of real estate values reported in the estate filings at two points: the median and second decile threshold of real estate values reported in individual declarations. We use this second measure to ensure that we are not measuring housing units that, following some prior inheritance, were held in common by groups of siblings.

The median value of a building was larger than the wealth of the individuals in the 93rd centile of the wealth distribution. More striking still, before WWI, the value of a single real estate holding at the threshold of the second decile was larger than the wealth at the threshold of the top decile (P90). In other words, even if they had been willing to put their entire wealth into purchasing Parisian real estate, more than 90% of the individuals in Paris could not afford the cheapest unit in the second decile of the value distribution. Then, some improvement set in, and the lower limit of real estate value was closer to the 80th percentile in wealth. Unsurprisingly, before

WWI, only members of the top decile owned any significant Parisian real estate. Ninety-four percent of the ninth decile owned no real estate before WWII, and less rich decedents were even less likely to own real estate. The result is one of the more dramatic differences between Parisians and the rest of the French population. In France, real estate was the most broadly distributed asset category. One could buy a plot of land in the 1880s for under 500 francs and a house for under 1,000 francs outside Paris, while real estate in the capital was out of reach for nearly all its inhabitants (see chapter 10, figure 10.6).

Parisians were not, of course, precluded from owning real estate outside the city. Indeed, rural property ownership was commonplace for the very rich we have encountered, including the Mortiers de Trévise, the Rothschilds, Benjamin Delessert, and the Duke of La Rochefoucauld. For the very wealthy, a country estate was as much a necessity for one's social standing as a residence in a tony neighborhood of Paris. But what of the rest of the population? Given the small price of such assets, particularly in the poorer départements that drew lots of migrants to Paris, one can easily imagine that this would have been a desirable wealth accumulation strategy. Unfortunately, before the consolidated filings of the twentieth century, we only observe provincial real estate through rental payments due to the estate (see chapter 2). So we start with the period 1902–1912 when we know that a single declaration was filed in the bureau of legal residence. Then, the likelihood of holding provincial real estate for individuals in the eighth and ninth deciles was ten times that of holding Parisian real estate. Still, even then, the probability of owning provincial real estate was only 10%. Over time, however, the likelihood that members of the eighth and ninth deciles held Parisian real estate increased much faster than that of rural real estate. After WWII, Parisians were only twice as likely to own real estate outside Paris than in the city. Very few Parisians below the 95th percentile owned any real estate before WWI. Rural real estate was relatively unattractive because of its illiquidity and the transaction costs of supervising its use by tenants (Hoffman 1984, 1986). For the rich, who spent considerable time at their country estates, purchasing a cluster of farms nearby was a viable option since they could check on them as necessary. It was also economical for the very rich to hire a steward to oversee their landholdings. For the less wealthy, neither strategy made much sense. Entrusting one's fields to a family member was risky since the trustee could easily take advantage of the saver's absence to appropriate part of the income. Nevertheless, after 1945, individuals below the 95th percentile were attracted to real estate as a complement to other forms of wealth; more and more chose to own in Paris, but others chose to purchase a second home, including many who were renters for their primary residence.

Indivisibilities and other transaction costs meant that real estate was not a path to start building wealth. This brings up the question of whether financial assets in the capital could be a substitute for land and houses. Individuals with no wealth could not borrow, but perhaps they could be lenders or investors. For peer-to-peer loans, Paris was no friend to individuals of limited means. The second-decile loan size threshold was 1,000 francs in the 1820s and 2,000 francs by the 1850s (Hoffman et al. 2000). By all accounts, those who entered peer-to-peer markets as lenders were already wealthy and held a portfolio of loans. Besides savings accounts, the most divisible financial assets were the equities and bonds traded on and near the Paris Bourse. For a few years (1872, 1882, 1992, 1912), we reconstituted the nominal (par or reimbursement) and market value of equities and bonds traded on the exchange. There are always some very cheap assets available on the market. Indeed, people traded not just the financial assets of firms and governments with solid performance but also the dregs of the investment process whose market price was only a fraction of its par value (e.g., bonds of the Panama company after it failed in 1892 and the financial scandal that accompanied it, or Russian assets for two decades following 1917). We leave these aside, as long-term speculations on a miraculous discovery of a mechanism to dig through the isthmus or a Soviet accord on Russian bonds was not prudent for building wealth.

Let us start with the market value of bonds. Government bonds were largely reported as percent of par, and many issues had minimum bonds with face values of 100 francs; the threshold to the second decile of bond values is less than 100 francs. Even if an investor focused only on the safest securities, there were ample opportunities for individuals who wanted to place a few hundred francs in the market, and that is exactly what we see in the second half of the nineteenth century in the portfolios of the members of the ninth decile. For corporate bonds, the standard issue had a nominal value of 500 francs, with little variation around that number. When interest rose in the early twentieth century from the secular low, the 3% coupon corporate bonds lost some value, so it was feasible around 1912 to purchase such bonds for between 400 and 450 francs. Equities tell a similar story. Most of those issued between 1850 and 1912 had a par value of 500 francs. They also often only required an initial fractional payment (anywhere between one-half and three-quarters callable by the firm at a later date). Once on the market, like bonds, they traded at a wide range of values. A few very successful firms' shares traded at multiples of the issue price and some at fractions. Leaving out failed or nearly failing firms, buying a single share required several hundred francs, a reach for all the account holders with less than a hundred francs at the caisse d'épargne. As noted earlier, Parisians were late adopters of mutual funds, and the financial system was focused on individuals with means well beyond a hundred francs. This is no surprise given the cost of operating the ledgers to record ownership of assets. Firms that might have considered offering some sort of investment fund with, say, a minimum balance of 50 francs would surely have realized that they would have to charge such high fees that no one would want to avail themselves of the opportunity.

So far, we have described an equilibrium where assets are relatively indivisible and, thus, where individuals with tens or hundreds of thousands of francs of wealth have far more investment opportunities than those with just tens or hundreds of francs. Returning to the fable, this equilibrium could have just been because the grasshoppers had no desire for further savings, as the genetic or behavioral explanations would have it. Hence, securities issuers had no reason to issue a claim with a face value of 100 francs or less. Instead, we think this equilibrium was a structural feature of a financial market with high transaction costs. For the firms, these costs included the ledger problems we just alluded to. For savers, the additional issue was that the market was poorly regulated, and financial information was not entirely reliable (Bignon and Flandreau 2011). Finally, there were the issues of liquidity and risk. The attractiveness of the accounts at the caisse d'épargne was that it was costless to deposit or withdraw small sums of money. Instead, someone who had built a portfolio of 2,000 francs of corporate bonds and equities was seriously constrained in adjusting or diversifying his or her portfolio.

9.4. Learning to Retire

Prior to WWI, the poor died young in Paris, starting with high death rates in childhood and continuing with high death rates at all stages of adulthood (Kesztenbaum and Rosenthal 2011). Half of those with no estate died before age 50, and only 30% reached the age of 65. The mean age at death of adults without estates bottomed out at 43 in 1842 and grew rapidly to 1972 when it reached 66. From the perspective of saving for retirement, what is important is that the average age at death for the poor did not reach 60 until after WWII. Even taking in rich and poor, the median age at death remained below 60 until the interwar period for women and the late 1940s for men.

The rest of France was quite different. People in the provinces did not pay Paris's severe urban penalty; consequently, the average age at death among adults was substantially higher (between 7 and 12 years) before WWI. Starting in the 1920s, the differential narrowed until it vanished in the 1960s. Thus, one reason for building savings, the need for resources to live while retired, came late in Paris. In the countryside, the demand for life-cycle assets was small. For rural folks, saving in land was an efficient way to grow their farms and provide for old age. In Paris, the demand for life-cycle assets was also small, but this was because people worked until they died. As we shall see, it is ironic that retirement programs were more important in Paris than anywhere else in the country.

Providing for retirement, however, was a concern of public authorities going back to the seventeenth century. Our summary draws extensively from Bozio (2006, chapter 2).[2] After the French Revolution, retirement programs diffused slowly, starting with the civil service, as was also the case in many countries. A more general regime was initially voluntary (Thiveaud 1997) and began with the creation of the Caisse Nationale des Retraites pour la Vieillesse (CNRV) in 1850.[3] Over time, employers in key industries progressively agreed to make contributions for their employees. In 1853, the law extended retirement pensions beyond the military to all public employees reaching age 60 and having at least 30 years of service (for a subset of physically demanding occupations, the threshold was 55 and 25 years of service). In 1855, a retirement program was put in place for railroad workers, and in 1894, it was implemented for the mining industry. This structure, which endured through WWI with little change, produced two distinct retirement systems. One was run by the state and paid out of the national budget for most state and military employees. The other was run by the CNRV and paid for by employers. Its beneficiaries were primarily local public employees and railroad and mine workers. They also included employees of firms whose owners decided to participate even though they were not required to do so. In 1888, France had 220,000 public pensioners and 165,700 recipients of annuities from the CNRV. That number tells us that about 8.4% of the French aged 60 or older had pension access. In the same year, the CNRV reported 26,994 pensions were paid out to residents of the Seine, enough to cover 6.6% of the capital's 181,000 individuals older than 60. We must add public pensions to the pensions from the CNRV. Unfortunately, those are not broken down by département. So, we develop a bracket of estimates. The lower bound estimate for Parisian retirees with pensions assumes that the proportion of public pensions going to Parisians is equal to their share of the French elderly population, leading to more than 35,000 pensions covering nearly 20% of the city's population older than 60. A reasonable upper bound estimate considers the share of public pensions to be the same as that of private pensions (16%). That would imply about 63,000 total pensions covering nearly 35% of the Parisian population over 60. This latter estimate seems high, but many public pensioners (e.g., the military) could retire at age 50.

Beyond these precious indications of the Parisian share of pensions, we are on our own. The retirement system was national, and firms contributed at their headquarters—not where their employees resided. Not surprisingly, as late as 1900, 45% of all contributions were made in Paris. These were not all intended for Parisians. We used various information on employers to estimate the share of all contributions that benefited Parisian workers.[4] This leads to the conclusion that Parisians received 14% of all contributions in France in 1900, a share that did not seem to change much before WWI. If Parisians were the recipients of 14% of all CNRV pensions, that would be enough to cover 22% of the population aged 60 and above in 1914. If we take the lower bound estimate of public pensions, it implies that almost 30% of the Parisian population older than 60 was covered by a retirement program. In each case, this is a roughly 50% increase in the proportion of Parisians covered by pensions relative to the moderate estimate we generated for 1888. Moreover, the Parisian pension rate would be about twice that of the nation as a whole on the eve of WWI.

Understanding the development of retirement pensions from 1920 to 1940 is complicated by a scarcity of detailed data on the number of public pensions. The CNRV's annual payout doubled from 1914 to 1924, when it reached 111 million francs, only to jump to almost 900 million on the eve of WWII. This growth was largely due to inflation, as prices were multiplied by almost seven between 1914 and 1938. There was also an increase in annuitants (it rose by 166% from 1914 to 1938). That still leaves room for a 50% increase in the real value of pensions from 1914 to 1938. All this pales when we look at the public side of pensions. The state's annuitized debt (*dette viagère*) payments rose from 320 million francs in 1914 to 3.8 billion in 1924 and then to 13 billion in 1938. The huge jump from 1914 to 1924 was the result of war-related pensions after 1914. From 1924 to 1938, the annuitized debt rose more slowly than the CNRV payout. This annuitized debt can be broken down into three components: military pensions, civil pensions, and other annuitized debts. The first two correspond to the public pensions we saw earlier. The military pensions jumped from 183 million in 1914 to 9.5 billion in 1938. Annual numbers of pension recipients are not readily available. Most of these pensions went to soldiers who suffered traumatic injuries and to the spouses, children, or parents of those who died in the war. Around 2.9 million such pensions were paid in 1934 (Carrière 1935). Beyond the war pensions, in 1938, a half-million retired military personnel likely received pensions. Civil pensions would have covered another half million when the CNRV funded almost a million annuitants.

Paris's share of these pensions matters both for the argument that it led the country in the diffusion of life-cycle savings and for the importance of life-cycle savings in the city. The capital was most likely overrepresented in the CNRV and civil pensions. Indeed, large firms and state employment were a large share of the capital's employment base. The military pensions were, however, far more likely to have been evenly distributed across France after the war. This leads us to favor an estimate in which pensions covered almost 40% of the Parisian population over age 60 and about a quarter of those over 50 in the interwar period. After WWII, new legislation meant that all employees in Paris were essentially covered by some retirement scheme. Most of the difference with France as a whole came from agriculture, a sector for which retirement coverage spread slowly after 1955.

Our review of the history of retirement pensions leads us to highlight a few key findings. First, in the early decades of retirement programs, the number of contributors and pensioners grew slowly. The CNRV was put in place in 1850. Even in 1881, after the first cohort had fully retired after making 30 years of contributions, the CNRV only provided pensions to 2.2% of the Paris population. Adding the military and civil pension brings that share to 6.5% for France and 14% for Paris in 1881. It would take a century before a retirement system covered most people 65 or older. Second, WWI was a huge shock to the pension system because the state was committed to providing some welfare insurance to the injured and to the families of those who died. By 1934, WWI pensions were 4.3 times more numerous than all the active retirement pensions in 1914 and far more numerous than all other pensions in 1934. However, because individuals were unlikely to be called up past age 40 in WWI, most beneficiaries were young in 1920. Even by 1938, few were 60. They would reach "retirement" age in the 1940s and 1950s. The real sea change in retirement pensions came in the 1950s. Lastly, unlike the caisse d'épargne, a voluntary and very popular program, CNRV was a collective, scarce, and mandatory program. Although it had a voluntary component, where individuals could contribute independently, such individual contributions were rare. Indeed, only 46,000 individuals made voluntary contributions in 1911 when collective (employer-paid) deposits funded 1,036,000 accounts.[5] The same remained true in the interwar period. The working class, it seems, wanted retirement savings as part of collective bargaining, but its members would not participate individually.

The evidence of a regular (if not rapid) rise in the number of participants in retirement programs and the nearly complete failure of the voluntary part of the program to attract small deposits (less than 100 francs) raises the question of the demand for life-cycle assets. In one view, most forcefully expressed by Ransom and Sutch (1986), life-cycle capital is a modern and superior form of financial management. In an extreme form of life-cycle savings, parents have no further obligations to the younger generation after incurring the cost of raising their children. Consequently, bequests become a vanishing part of the intergenerational wealth flow because life-cycle savings allow parents to take care of themselves without the help of children, who, in turn, do the same. Even if we take a more moderate position, the life-cycle savings approach generally assumes bequests are an inefficient mechanism to deal with life-cycle problems. Thus, bequests and inherited wealth will fall as life-cycle savings take off. This last prediction was what spurred Franco Modigliani to try to measure the share of inherited wealth many decades ago. As we discuss later, there is little evidence in the Parisian data for this particular view of modernization.

The very structure of the CNRV implies that as early as 1850, policymakers did not consider the issue as trivial or settled. Indeed, the CNRV was devised to provide retirement pensions for workers. When it was set up, the state had already been providing pensions to many civilian and military employees. However, it did not follow the model put in place centuries before for soldiers. Instead, the CNRV provided two options to its account holders. In the first, individuals saved into accounts whose capitalized value would be converted into an annuity at retirement. In the other scheme, the pension would only involve the interest income from the government bonds purchased at the time of retirement. The capital would be reserved for the holder's heirs in this second case. If Modigliani had been right, everyone would have opted for pure annuities. However, persistently, from 1853 to 1910, about half the deposits by number and value were made into accounts that preserved the capital for bequests. In other words, while half the workers covered by the retirement programs might be "modern" life-cycle savers, half the others definitely were not. These mixed savers saw their accounts at the CNRV as an opportunity to contribute to both pensions and bequests.

From the CNRV data alone, we cannot conclude whether individuals interested or not interested in bequests increased their savings because it was not the only bucket into which individuals could deposit funds. One problem is that mandatory programs like the French retirement programs can be partially subverted by individuals who change their behavior. In the case of savings, individuals whose employers were making contributions for them could have stopped building up other savings once they were locked into annual contributions of 5% of their salaries. In extreme cases, one might see a decline in the number of individuals who leave bequests due to the rise of retirement programs. Indeed, in a world where life-cycle assets (annuities) are costly, scarce, or risky, individuals may invest in real or financial assets, intending to sell them as they age. The bequest we see would then come from those who died both old enough to have accumulated and young enough or rich enough not to have consumed their whole wealth. The Parisian data allow us to confirm that retirement programs were globally complementary to bequests savings. This is because the fraction of individuals with estates in Paris rose over time as more of the population became part of the retirement system and because, most often, individuals with pensions had other real or financial assets.

While the number of pensioners in Paris is important in itself, it also matters to know how they were divided among wealth classes. We remove the top decile because that group was so rich that its saving behavior was unlikely to be influenced by retirement assets, even though rich individuals with long public service, like the Maréchal de Trévise, often had pensions. Within the rest of the population, we sort the individuals with estates along two dimensions: those with or without retirement assets (pension and reversionary capital from the CNRV) and those with or without investment assets (shares, bonds, real estate). This gives us four categories, which are shown in figure 9.5. These groups involve individuals whose estates are composed of claims on family, movables, and bank accounts; individuals with retirement assets but no other investment assets; individuals with investment assets and no retirement assets; and, finally, individuals with both. The figure also displays the share of individuals who died without assets among those below the top decile.

The figure shows a long-term increase in the share of individuals who owned investment assets among those without pensions. This trend is quite remarkable and continues to the end of the period. This rise in investment assets started slowly in the nineteenth century and accelerated after 1907. Before 1907, the share of individuals with retirement assets was about 10% of

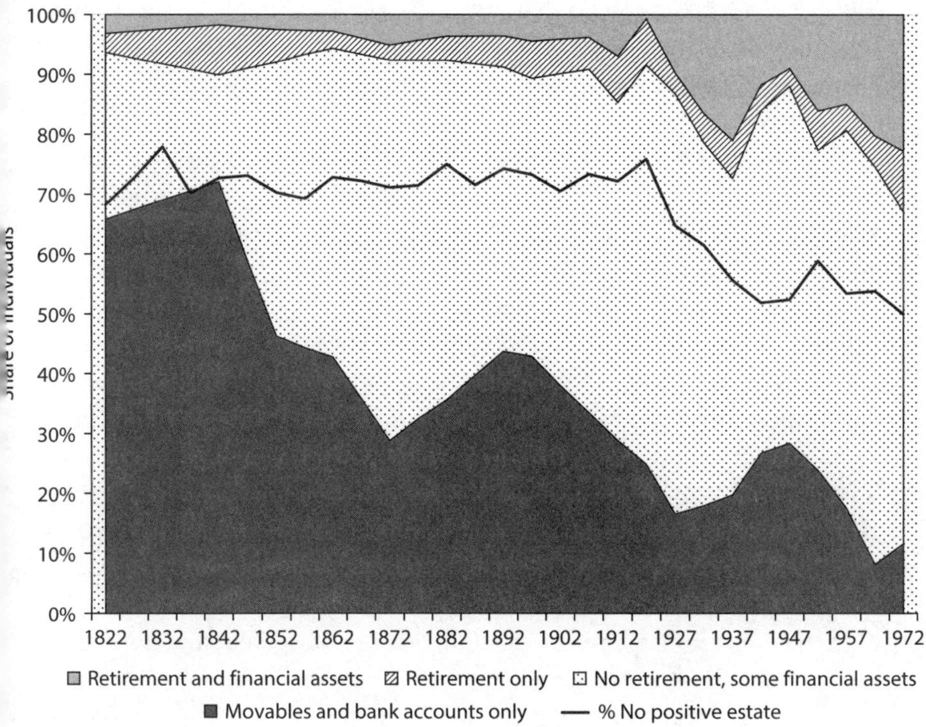

FIGURE 9.5. Fraction of individuals who died below the top 10% with different portfolios. *Source:* Estates dataset, sampled individuals (see appendix 1).

those who died with some wealth and unchanging over time. Among those with pensions, only about half had investment assets (a surprising echo of the CNRV before WWI, when half the depositors constituted a reversionary capital). After 1907, the share of individuals with pensions rose to 30% of those who left an estate below the top decile (more than in the general population of decedents, where that share topped out at 26%). More important, the share of pure life-cycle savers is essentially constant, ranging from 3% to 6%. The group that grew rapidly had both retirement and other assets. Before 1907, it was at most 4.5% of those who died wealthy, after WWI, it grew to a peak at 23% in 1972.

Clearly, then, in Paris, retirement assets were not substitutes for bequests. Instead, the share of the population that died wealthy increased because of conjoined life cycle and bequest savings efforts. The pensions themselves were almost negligible in the value of estates, representing 2.56% of the value of estates at their peak in 1912, while investment assets were 65% of the estates below the top decile on average (at peak in 1922, it was 80%).

This is not surprising since we observe two kinds of pension assets: first, the residual payment owed to the estate, often less than a month and very rarely more than three months of the annuity; second, the capitalized value of the reserved portion of the deceased account at the CNRV. One might be surprised that these capital accounts did not amount to more. Recall, however, that through the interwar period, civil and military pensions, which were strict annuities, represented at least half the pensions and were more generous than those of the CNRV. Only roughly half of the CNRV pensioners (representing only 6% of the pension flow) reserved the capital for their heirs. Most Parisians who wanted to pass on a bequest did not have access to a CNRV mechanism, and they used other financial strategies to do so.

The post-WWII period saw the expansion of retirement savings through a mix of negotiations and legislative decisions. Over time, an ever-greater fraction of the population was covered through a system where each individual was promised a fraction of their terminal wage times their years of participation in the pension program. As discussed in chapter 6, through the 1960s, the process of recording individuals in the TSA means that we did not collect information on the individuals whose heirs did not file an estate declaration, so we do not know the fraction of individuals identified as retirees and who also had no estate filings. In the 1970s, when the tax authorities established more documentation for every Parisian, we collected the taxable wealth and occupation for those who died rich and those who did not. The occupations include information on whether the deceased was retired. In part because of the general lengthening of life and in part because of the spread of retirement, the share of all adults who died in Paris in the 1970s who were retired reached an all-time high. By 1972, about two-thirds of all individuals who died were either noted as retired or had a retirement pension. As we shall see in the conclusion, this share continued to grow afterward. By 1972, there is evidence that the aggregate French inheritance flow was also increasing rapidly (Piketty 2010; Piketty and Zucman 2014), further proof that pensions are not substitutes but complements to bequests.

To close this section, we return to the rise of life-cycle savings. To begin with, in Paris in particular and Western Europe in general, life annuities were not the result of some breakthrough in financial engineering in the nineteenth century. They had their origins several centuries before our data begin (Hoffman et al. 2000). French kings and other monarchs issued many life-contingent annuities in the seventeenth century. There were already key policy debates surrounding the cost to the monarchy of using life-contingent schemes to raise money (Alter and Riley 1986; Weir 1989). The French Revolution's inflation demonstrated the risk inherent to long-term nominal investments, which were relatively rare in the first half of the nineteenth century. The reforms of the 1850s that gave birth to the CNRV draw from the same nexus of issues: encouraging life-cycle savings to ensure income in old age (and reducing pauperism) and creating a new flow of coin to fund the state's debt issues. A mechanism for life-cycle savings was put in place in 1850, but there was, initially, little demand for it. As we saw earlier, a succession of reforms was needed before large numbers of employers began to make contributions. Economic principles suggest that employers should be indifferent to whether worker expenses go to immediate or deferred compensation; the only thing that matters is the total cost of labor. So, the willingness of employers to agree to the government's entreaties to begin contributing would depend on either the ability of ministers to put pressure on certain branches of the economy or the perceived desire of workers for such benefits. These demands will be closely related to expectations about the length of life after retirement.

The irony of the high fraction of pensions in Paris is that if life was possibly less brutish or nasty than elsewhere in France, it was certainly shorter. In the 1850s, when the CNRV was founded, most people expected to work until they died. Only half of all 40-year-olds would reach 60. That probability grew slowly as age at death started to rise. It was only with growing life expectancy that more individuals would have found the retirement question relevant. That said, it is unclear how individuals, then or now, learn that their circumstances will be different from their parents' when a secular process of improvements in nutrition, infrastructure, and medicine rather than large shocks causes life expectancy to rise. Equally important, the estate data, like the CRNV data, point to some fundamental heterogeneity in behavior, with some caring more about bequests and others caring more about lifetime income. One explanation for these differences involves heterogeneity in direct descendants, an issue we turn to next.

9.5. Bequests Without Heirs

If bequests are an important motivation, then it is of some relevance to ask who received the intergenerational flow. The standard approach for economists is to focus on parents' altruistic or transactional motivations. In these models, the ebb and flow of bequests is related to the relative income of parents and children. In altruism models, parents seek to equalize their and their children's well-being. Thus, bequests are procyclical. When the cost of human capital acquisition for children increases (because its return increases), bequests fall because, early on, parents invest more in their children than later. In transactional models, parents hold back their wealth so children will care for them. The second type of model makes little sense for the wealthiest centile of Parisians because they never had a real need for their children's care as they could afford to purchase the services they needed. More generally, the transactional model makes the most sense when markets for services are thin; it is, therefore, less relevant for Paris than elsewhere in France.

Altruism involves care for others. Conceptually, it should not be limited to any particular category of individuals. In practice, economists have documented time and again that social distance measures profoundly affect an individual's willingness to devote resources to the welfare of others (Goere et al. 2013). The cohort of philanthropists who provided the initial endowment for the caisse d'épargne did so for Paris, the city they lived in, not for some random or general endeavor. It is, thus, not surprising that economists have focused heavily on the motivation for generosity to direct descendants and spouses, the closest possible individuals to the people leaving bequests.

French law limits the discretion with which individuals can distribute their estate to ensure that each legitimate child receives a share no less than 1/(n+1), where n is the number of heir lines (children and deceased children with living descendants).[6] Despite the counterexample of the Rothschilds in chapter 8, many children inherit equal amounts of wealth, particularly when there is no firm to transmit (cf. Bessière and Gollac 2022). When dowries differ across children, the estate distribution tends to equalize the children's portions. Measuring variation in altruism or transactions among individuals with children in this context is difficult. An individual with n children can freely distribute the "part de réserve" or 1/(n+1) of their wealth, and one should look at the frequency with which specific children (sons or the eldest son) were advantaged and the frequency and distribution of bequests beyond children. However, as we have constituted it, our dataset does not enable us to do so.[7]

However, we can examine the inheritance flow that goes beyond the direct line of descent by focusing on individuals who died without children—without direct descendants. Spouses are not considered direct descendants. Because the French estate tax is an inheritance tax, we have very good information on the relationship between the deceased and their principal heir type. These data were essential for the authorities to assess the tax, which is why they recorded such information in great detail. Their diligence allows us to uncover a startling set of facts.

Over the period from 1822 to 1972, more than 37% of the individuals who left estates had no direct descendants, and their bequests amounted to about 25% of the inheritance flow. The proportion of men and women without direct descendants was strikingly similar (37% and 38%), and men without direct descendants accounted for a slightly larger share of men's inheritance flow (26%) than was the case for women without descendants (23%). This difference was driven by differences in labor market options. The shares of men and women with no descendants were volatile with little trend over time: they tended to trend down to a low in 1912 of about 20%, rose in the interwar period, and then fell again toward 20%. Childlessness occurred at all fractiles of the distribution, but it was more frequent for those who left smaller estates. Only 23% of the top thousand richest died without a direct heir over the whole period. Meanwhile, 28% of those who comprised the rest of the top 1% died without direct heirs. For those who left an estate below the top two deciles, a full 45% had no direct heirs.

Across all estates, whether large or small, the recipients of indirect bequests were overwhelmingly family members. Before WWI, collateral branches (siblings, nieces and nephews, cousins, parents, and grandparents) received more than three-fourths of the flow, while spouses received 12% and nonrelatives received 11%. After WWI, the share of indirect bequests going to relatives dropped sharply, and that of spouses rose to about 40%, while the share of outsiders remained roughly stable. Parents and other ascendants were rarely principal heirs. The high rate of childlessness in Paris was consistent with the overall failure of the city's population to reproduce itself. Before WWI, dying without a direct heir could be attributed at least partly to the urban mortality penalty, which was particularly severe for the very young. This high rate of childlessness, however, persisted into the 1970s and hence points to more complex trade-offs between consumption and reproduction.

Of course, dying childless is the result of one of four life outcomes that combine either marrying or not and either having children survive to adulthood or not. Prior to the 1920s, we could not distinguish between marrying and having no children and marrying and having no children survive. However, those who died having never married (*célibataires*) are well identified, and by choice or circumstance, they could not rely on family for support in old age. Before WWII, among the never-married who left an estate, those with pensions were rare: no more than 12%. Then the proportion started to rise, reaching 20% for men in the 1960s and '70s and 40% for women. The fact that women who never married were so much more likely to die with a pension than men compensated for the lower rate for women who died after being married or widowed. On the whole, the proportions of men and women with pensions are similar. We leave to future research the task of untangling the intertwined influences of labor market forces, pension availability, and women's decision to marry (Goldin 1990, 2006). There does not seem to be any obvious adverse selection in terms of the average estate because, by the 1970s, single women with a pension were 20% or so richer than single women without (that was also true between 1870 and 1942, but not at other times).

We started this chapter with individuals who failed or were unable to save for posterity and who also largely failed or were unable to save for life's unforeseen events. The large share of individuals with no estate is certainly the main puzzle for the distribution of wealth in Paris prior to WWI. At the same time, there was always a sizable group, the childless, that saved "too much." While it is entirely possible that those in the eighth or even ninth decile could have left estates that reflected precautionary savings, it is impossible to make the same argument for individuals in the top decile. It seems more plausible that some individuals wanted to save for dynastic reasons and thus benefit siblings, nieces and nephews, and cousins when they had no children. Patrimony extends beyond children. This is not particular to Paris and is likely to be even stronger in societies where adoption is recognized (in France, adopted children were legally viewed as legitimated children rather than as heirs in direct line of descent).

Economists who have focused on cultural or genetic explanations for wealth accumulation, bequests, or its lack thereof should take heed. The childless are obviously from the same genetic base and cultural background as their brothers and sisters who do have children. However, in every generation, some individuals do not marry and do not have children, and some who marry and have children do not see their offspring reach adulthood. Whether these outcomes are driven by ill luck or preferences, it is obvious that this heterogeneity is important to family wealth trajectories because the childless do not all die poor. Similarly, some people moved to Paris while others stayed home, partly because of individual preferences and partly because of differences in opportunities. Given the high migration rate to Paris, family wealth dynamics extend back to the countryside where families may have "invested" in a migrant to Paris and then disadvantaged that child at inheritance.

9.6. The Supremacy of Real Estate

Modern finance's central tenets invite investors to trade off risk and return and, to that effect, diversify their portfolios. As we saw in chapters 1–6, Parisians outside the top centile did not follow this advice. We return to the issue of portfolio choice because risk-and-return trade-offs are very important to the persistence of inequality. We focus on real estate and in particular housing. Real estate has long had both good and bad press. On the good side, a long literature extols the virtues of owner-occupied housing, and home ownership is often a motivation for subsidies to young people. The argument that high levels of homeownership rates are associated with high-quality neighborhoods has driven many countries to provide subsidies to potential homeowners. These include reduced interest rates for first-time home buyers and exemptions from taxation, whether it is on interest payments as in the United States or on estate taxes in France, as we saw in chapter 6. All these policies are legitimated by arguments about the social value of high rates of homeownership. Real estate economists are well aware that nice neighborhoods may be nice in part because of the homeownership rate but also in part because of the high level of income of their inhabitants. The problem for Paris is slightly different because rich neighborhoods have a large number of renters and a higher-than-average level of amenities. Moreover, through the 1970s, homeownership likely diffused first in rich neighborhoods and then to other parts of Paris. Finally, many Parisians who had the means to purchase an apartment had to decide whether to buy in Paris or rent and purchase a country home.

Real estate also has its drawbacks as an investment. Its value tends to be closely correlated with local employment conditions. Thus, owner-occupied real estate creates an undesirable correlation between labor market outcomes and home value. Simply put, from a risk-return trade-off, it would be better to own real estate in a different labor market than one's own. Moreover, someone who has a lot of equity in a home may find it difficult to access that wealth in a downturn or may find it difficult to move to another location if their job disappears. A second problem with real estate subsidies is that in most countries, down payments are not trivial (in France, for most of the period we investigated, buyers were expected to pay half the value of their purchase up front). Thus, the subsidies benefit most of those with (inherited) wealth who can afford the downpayment.

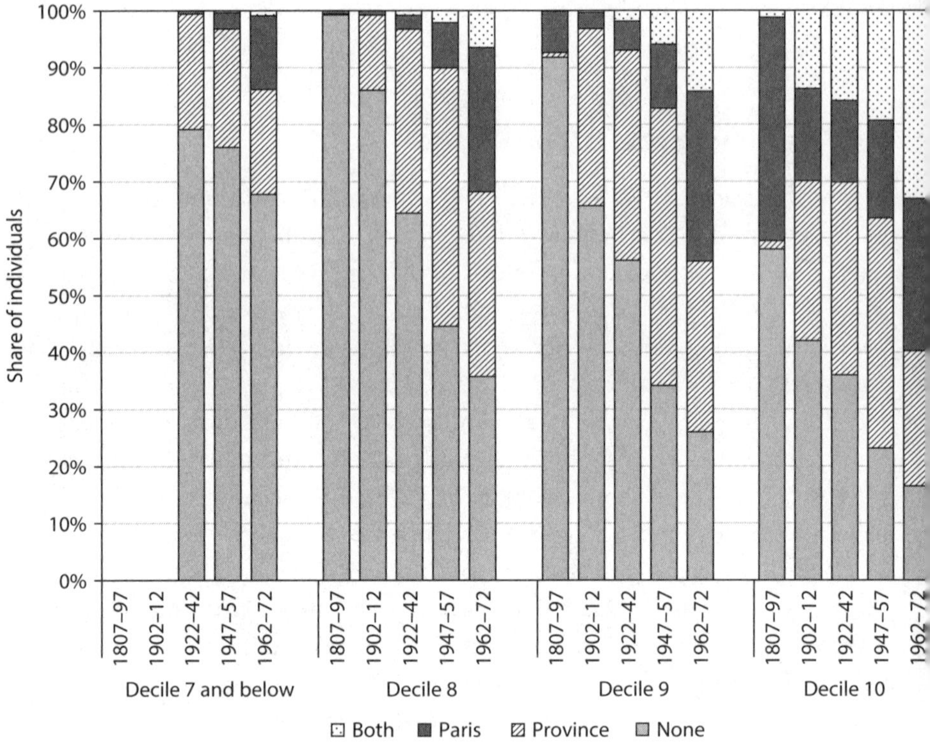

FIGURE 9.6. Share of estates that include real estate by fractile. *Source:* Estates dataset, sampled individuals (see appendix 1).

For Paris, however, real estate plays an unusual role because, before WWI, unlike in the rest of the country, it was not the first asset that individuals purchased, and the period after WWII marks a change that we do not see elsewhere (see chapter 10). While Paris is unique in the scale of the renter population (90% of the population around 1900), this was common in any city where buildings have more than two or three stories. The diffusion of multiowner housing units has a specific history for Paris, but it is not unique to the capital.

Nevertheless, in Paris, as elsewhere, when inequality declines, real estate is the asset that diffuses most rapidly. We examine the fraction of individuals who owned real estate and the real estate share in gross assets by decile. Because we do not accurately observe Parisians' ownership of provincial real estate before 1902, the whole nineteenth century is collapsed into one period. Figure 9.6 shows that real estate was increasingly popular within the eighth and ninth deciles. More importantly, when the fractions of individuals

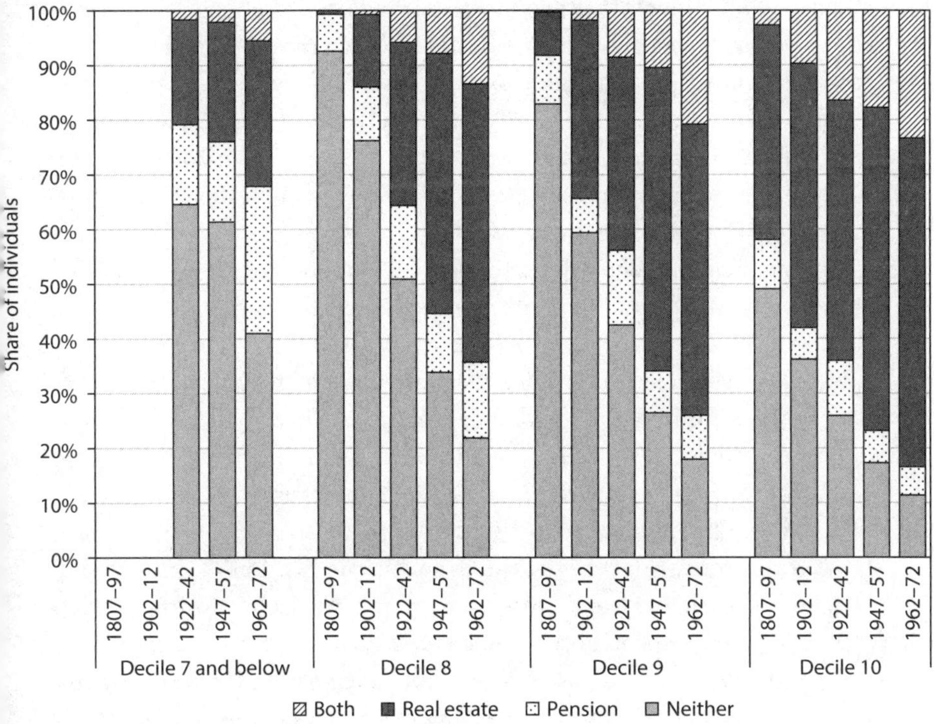

FIGURE 9.7. Share of estates with pension and/or real estate assets. *Source:* Estates dataset, sampled individuals (see appendix 1).

leaving an estate expanded beyond the top quarter of the distribution, so did the share of individuals holding real estate.

Although the dominant source of wealth for the lower deciles always involved items with no income flow (movables, bank accounts) the top investment asset was real estate in the last three periods (see figure 9.6). Moreover, real estate's share of total wealth per decile got larger as wealth increased. This is not surprising because although the value of movables, bank accounts, and other valuables with no return was increasing in wealth, its share of wealth was falling simultaneously.

To close, we examine the interaction between real estate and pensions. We break up the population of wealthy decedents into four categories: those with neither pension nor real estate, those with one or the other, and those with both (figure 9.7). At first glance, it seems that real estate or pensions were substitutes since, in every period, the vast majority of wealth holders in every wealth group had either neither or only one. However, the category of individuals that had both was the fastest growing. It tripled its share of individuals below the top decile between the 1900s and the 1970s. It should be noted that the share of individuals with pension assets in their estates was, at least until the 1970s, massively lower than the rate of the population that was covered by pensions. The difference in coverage rates arose because in each year those who died were far older than the population and covered at much lower rates.

Among the top decile, who were definitely not budget-constrained and therefore could afford both pensions and homeownership, the number of individuals with pensions and real estate increased by 62% from the 1920s to the 1960s. Moreover, after 1902, the individuals with only pensions were always as numerous as those with only real estate. Thus, dying wealthy was, over time, increasingly associated with having a pension. Dying wealthy was also closely related to real estate ownership, but that association came later and was more steeply related to wealth.

We interpret the data as providing another indication of complementarity in investments. Real estate ownership made sense for individuals who anticipated retiring because it allowed them to have a place to enjoy their pension without having to worry about paying rent. Interestingly, this complementarity strengthened over time, even though incumbent renters in Paris were the beneficiaries of major rent-control regulations.

9.7. The Life Cycle of Wealth Before the Bogle Rule

Our last examination of the preferences of Parisians is related to prudential wealth management, often known as the Bogle rule for John Bogle, the founder of Vanguard mutual funds. This rule advises investors to make their age the share of fixed-income securities in their portfolios (i.e., 50% bonds for a person aged 50), thereby lowering risk in later life. In her work, Daumard found some evidence for such a change when she detailed the differences in portfolios between retired and active businesspeople. However, she did not look at the impact of age on portfolios (Daumard 1973, 130–131). To examine the extent to which Parisians increased their portfolio share of bonds as they aged, we focus on the bond share of investment assets for the ninth decile of the distribution, the one that, in cross section, has the steepest age-to-bond-share curves. We will look at the data through three different projections, starting with cross sections by period of death. In figure 9.8, three early cohorts show a bond share that increases with age. After WWII,

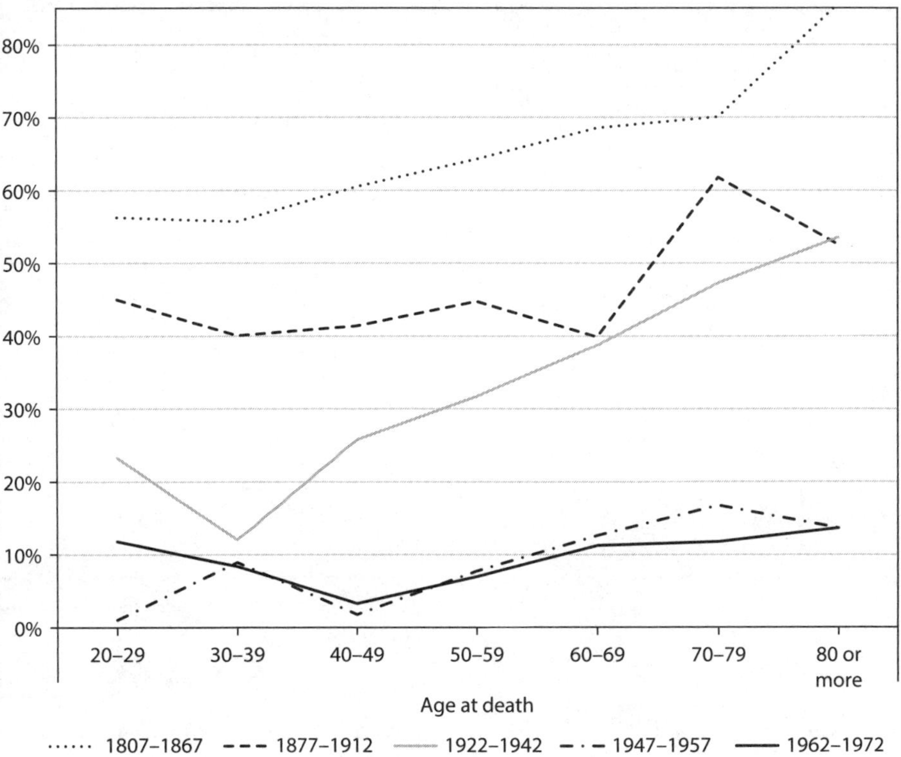

FIGURE 9.8. Age at death and bonds' share of income-producing assets in cross section (ninth decile of decedents). *Source:* Estates dataset, sampled individuals (see appendix 1).

however, the effect is small. The problem with the cross-sectional perspective is that it mixes people who set their investment strategies at different points in history.

As an alternative, in figure 9.9, we look at the same data but aggregate individuals by the 30-year birth cohort. The earliest cohort (born before 1800) has a flat age-bond profile. The next cohort, born before 1830 and died by 1910, is the only one that shows even a glimmer of an increase in bond share with age. The most one can say about the other later three cohorts is that they trend downward. Bond shares were high among those who died young, perhaps because parents used bonds to fulfill dowry obligations and the like. In any case, bond holdings tended to fall as individuals aged up to about 50 and then the holdings stabilized.

FIGURE 9.9. Age at death and bonds' share of income-producing assets by birth cohort (ninth decile of decedents). *Source:* Estates dataset, sampled individuals (see appendix 1).

One more graph helps explain the data. This time, we display the bond share so that each line is an age at death, and the horizontal axis is a period of birth (figure 9.10). Over the long run, there is very little difference in the bond share by age. Instead, there is a massive decline in the share of bonds over time. The explanations are twofold. Early on, there was little to invest in besides bonds and real estate, and for the ninth decile (and those below), real estate in Paris was too indivisible, so the bond share early on was high. The bond share fell when the stock market started to make other investments feasible. A second impetus for a declining share of bonds was the very sizable losses on domestic bonds relative to domestic stocks imposed by the inflation that occurred from 1914 to 1947. Those who held bonds lost dearly, a process that would tend to reduce their share of any portfolio. The younger generations inherited fewer bonds, and they seemed to have rebalanced their wealth in other directions. It is interesting to consider how historically sensitive investment advice can be. What made sense in the 1820s or the 1990s may have been downright unwise in the 1910s or the 2010s.

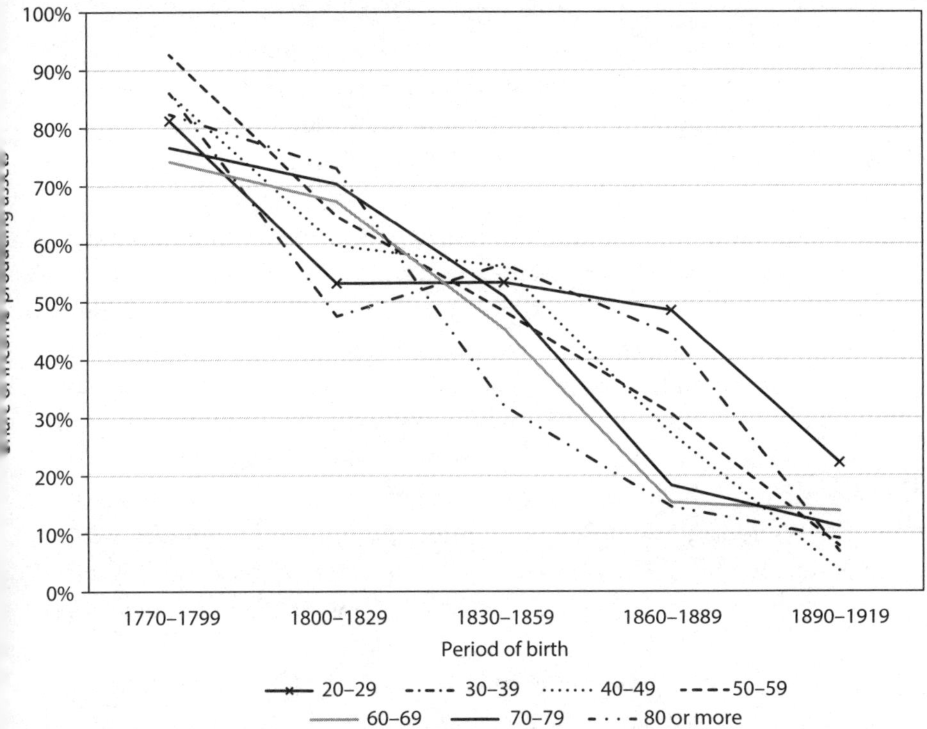

FIGURE 9.10. Age at death and bonds' share of income-producing assets by age at death (ninth decile of decedents). *Source:* Estates dataset, sampled individuals (see appendix 1).

9.8. Conclusion

The decline in the fraction of individuals who died with no estate is one of the more dramatic changes that we see in the twentieth century. The constitution of stable wealth holdings for a large fraction of the urban population was a political project of the nineteenth century. Absent mandatory retirement plans, this project largely failed before 1914. This is not because Parisians were entirely uninterested in saving. The ebb and flow of savings accounts and the very large number of such accounts we observed in the last quarter of the nineteenth century clearly show that the average Parisian was a saver. However, these savings were not durable, nor were they large. Three-quarters of all accounts had a balance smaller than a month's average wage. The Paris savings bank (Caisse d'Épargne) was popular as a place to hold precautionary savings. It was never an important vehicle for accumulating retirement savings (or significant wealth, even as Parisian income grew over the long nineteenth century). The doubling of the fraction of individuals who died wealthy was the result of several forces. Some were long-run forces, like the rise in incomes and life expectancy. Some were more medium-term, like retirement programs. Finally, some were quite short-term, like tax policy. What is clear is that change occurred over at least two generations in the long run.

Economic analysis is best suited to measuring the impact of short-term policy changes because a short interval avoids a host of thorny problems of endogeneity (say, in the desired level of wealth, number of children, or even residence in the capital). However, the evidence leads us to conclude that these short-term changes, like tax policy, had limited effects on inequality. Paris was very unequal in 1807, and it remains very unequal today. As we saw in earlier chapters, tax policy likely had more influence on where the very rich held their wealth and how they held it than on how much they held. The longer-term changes resulted in a doubling of the fraction of individuals who died with wealth and an increase in the wealth share of those leaving an estate below the top decile from less than 3% before WWI to 14% in 1972.

An important goal of this chapter has been to ask: What do these long-term changes tell us about the preferences of Parisians and the dynamics of wealth? Using representative agent models with average wealth is not very fruitful. The evidence we have presented here argues that the individuals who died in Paris exhibited vastly different wealth dynamics. The underlying diversity of experience that produces the distribution of wealth at death is very wide. It includes differences in inheritance, opportunity, talent, and the randomness of returns. Indeed, although average wealth at death is procyclical, the fraction of individuals who died wealthy evolved independently of the state of the economy. That fraction seems strongly influenced by the development of mandatory retirement programs and lengthening adult life expectancy.

As we shall see in chapter 10, Parisian wealth dynamics are radically different from the rest of France. Because a vast proportion of the individuals who died in Paris were either born outside the city or were the children of migrants, we can easily set aside the genetic theory of savings and consumption. Recall that in the nineteenth century, the migrants overwhelmingly came from rural areas in France, where 70% the population had a bequest when they died. To get the Parisian equilibrium of two-thirds dying with nothing from a genetic theory, we have to assume that migrants were heavily selected among nonsavers. But then how do we explain that migrants who decided to move to Paris before WWI from the same locations as before the Commune were the vanguard of the process that halved the share of individuals without a bequest? Similarly, if Parisians were selected to be massively nonsavers, then what explains the fact that after the diffusion of mandatory pension programs, the same individuals started to invest in real estate?

Among the theories that can explain the data, we favor those that combine the positive and negative shocks of life's circumstances with heterogeneous individual preferences over consumption versus bequests. In migrant families, some siblings stayed in the countryside and, by investing in land, were much more likely to leave a bequest. In contrast, the migrants played a lottery that most often led them to have no assets at death or die massively wealthier than their country brothers and sisters. Appealing to genetic differences within sibling groups is not a path contemplated in the literature. The history of Paris, however, emphasizes the centrality of other individual elements like learning, incentives, and simply luck.

Family matters, but in different ways. After all, we observe estates because someone inherits them. Beyond the nuclear family, the extended family often matters, given the large flow of inheritances outside the direct line of descent. The many estates bequeathed outside the direct line of descent are not the result of intergenerational conflicts that caused some parents to dispose of their wealth away from their children. Instead, the flows occur because of the lack of children. While one might attribute estates in the eighth decile that go to nieces, nephews, or cousins to one form or another of precautionary savings, this makes no sense for the large estates that flow to relatives from members of the 1%. Thus, a part of the population has dynastic motivations for savings. Altruism, for some, extends well beyond the direct line of descent.

Finally, France's path to retirement savings affords another glimpse of departures from savings as defined by a simple rational consumption bequest trade-off. Indeed, the rise of wealth at death for deciles below the top three was heavily influenced by life expectancy but with a large delay. Indeed, during the last third of the nineteenth century, individuals could only assume that they would die young, as most of their forebears had done. The rise in wealth was also influenced by the creation of mandatory retirement savings. In other words, some Parisians learned that they were increasingly likely to live into retirement and were induced to consider the wealth level they wanted in old age partly by the retirement programs themselves (Feller 2017). It is important to note that many of these policies diffused over time in ways that persistently advantaged men over women simply because male-dominated occupations (industry) were much more likely to be covered by a retirement scheme than female-dominated ones (e.g., domestic service and retail). Widows enjoyed the reversion of part of their husbands' pensions, but single men were at a serious advantage over single women. We return to the questions of differences between men and women when we consider other forms of inequality in chapter 11.

10

Inequality in Space

WITHIN PARIS AND BEYOND

For the past nine chapters, we have studied wealth and its distribution in Paris as a whole, with periodic brief nods to inequality across neighborhoods and the connections between the capital city, its suburbs, the rest of the country, and the globe. In doing so, we were fortunate that the city's formal boundaries changed only once, in 1860, even though Paris's relationship to the wider environment was in ceaseless flux. We know, for example, that Paris has attracted migrants from the rest of the country and beyond. For instance, in 1832, the prime minister, Casimir Perier, and the decipherer of hieroglyphs, Jean-François Champollion, both died in Paris, and both came from Grenoble. James Rothschild, the banker, and Max Ernst, the painter, died a century apart, and both came from Germany. Additionally, while labor income tends to be earned primarily near an individual's residence (and all the more so before the invention of commuting), wealth knows no such boundaries. Non-Parisians owned buildings in Paris, while Parisians certainly owned assets whose income came from well beyond the city's fortifications. This chapter examines the spatial patterns of inequality. We argue that although Paris was far more unequal than the rest of the country, in many ways, it was a forerunner of wealth transformations that spread to France as a whole a generation later.

Measuring wealth and wealth inequality requires defining a spatial unit over which the data can be collected. From a policy perspective, the unit that commands the most attention is the national one because that is how important taxes are set. Whether one considers inequality directly as the outcome of interest (taxes are simply redistributive) or whether one considers funding the provision of public goods, the pivotal decision-maker compares his or her wealth to some distribution. That distribution aggregates the wealth of individuals in some spatial unit. Even in the baseline proportional tax models (e.g., Meltzer and Richard 1981), higher inequality leads to higher taxes because it increases the gap between the pivotal decider (median voter) and the average level of wealth. The pivotal decider pays lower-than-average taxes and receives average benefits, and that gap increases as inequality worsens, leading to a higher willingness to pay taxes. Other models (Epple and Romano 1996; Lizzeri and Persico 2004; Acemoglu and Robinson 2006) embed this logic in a more complex institutional environment that is no longer a priori national. These endeavors are particularly fruitful (e.g., Chapman 2018, 2024) when local entities, say, Paris, have enough discretionary power over taxation to fund local services. However, analyses of this kind can only be carried out if one can measure the distribution of wealth at some subnational level—like Paris. A final complication is that the national distribution of wealth, as inferred from the national estate tax tabulation, is not the level of inequality that individuals care about or experience. What is likely to matter more to each individual is some local experience rather than the national level of inequality. At best, the measured national distribution of wealth is a statistic for what matters.

Adding a complication to the political economy of taxation and services is that, in highly centralized states, the main city (in our case, Paris) is not just the political center; it is also the apex of the financial market and houses many of the most important firms in the country. The largest city is also the center of the transportation network and the social and cultural dynamo of the country. For writers like Honoré de Balzac or Émile Zola (and for most of their characters), Paris was like a flame to a moth. Thus, Paris was a gravity well for various ambitions, not just for the "huddled masses." In periods of increasing urbanization (and more generally of increasing agglomeration economies), centers like Paris grow disproportionately (Wrigley 1967) and come to house the owners of an ever-larger fraction of the wealth in the country. Urbanization and the growth of Paris mean that over time, inequality in France increasingly reflects inequality in urban areas, particularly that of Paris. The reader can easily infer that the wealth share of the top 1% in France can increase, even if the wealth shares of different fractiles in Paris and the provinces do not change, simply because the weight of Paris in the national total rises. What are the consequences of such variations in wealth distribution on institutional change?

We examine these political economy issues in three steps. We start by looking within Paris. It will be no surprise to the reader that even in Paris, there is spatial variation in the distribution of wealth; there are rich and poor neighborhoods, richer and poorer streets within them, and nicer and worse buildings along many streets. Second, taking Paris as an isolated entity, we compare the wealth distribution of the city to that of the rest of the country. As with any other dominant city, inequality is more severe in Paris than elsewhere. However, we should be wary of measurements at the aggregate level because a 20% national wealth share for the top 1% may translate to higher levels of local inequality in company towns or villages dominated by one or two large landowners. The third section of the chapter revisits the same statistics to ask questions about how the units within a larger territory interact. One would like this analysis to encompass France's colonies and the rest of Europe, but the wealth data is lacking. Thus, we focus on France and include both départements within France and arrondissements within Paris. The question is important for understanding both what our data measure and the political economy of inequality. One example will suffice to flesh out these issues. In 1902, Édouard de Moracin, aged 70, died at his home in Le Vésinet, a village 20 kilometers west of Paris. This is not surprising since de Moracin was rich, a member of the 1%, and Le Vésinet was a fashionable place to have a country home. However, de Moracin's heirs filed his estate at his legal residence in Paris: 56 Boulevard Haussmann. Had de Moracin fully moved out of Paris when he retired (he is noted everywhere as a *propriétaire* and a baron), and had his estate been filed in Le Vésinet, that would have raised wealth outside the city and lowered it in Paris. It is also worth noting that de Moracin was born in Belgium and was thus an immigrant to the city. As this example shows, Paris was clearly not an isolated locality. Our data point to the fundamental nature of the flow of people and capital. The final section considers the implications of the first three sections for the political economy of inequality.

10.1. Inequality in Space, Looking In

Paris has never been a homogeneous city. Its western half, upwind from its own pollution, has long been better off than its eastern (downwind) half. This is nothing new, and it has also long been quantified. For instance, the real estate survey of 1876 (ASVP 1880, 392–400) enumerates rental

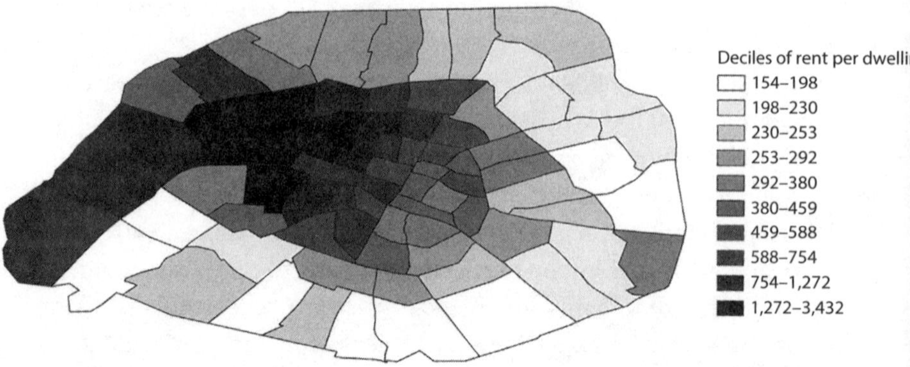

FIGURE 10.1. Average annual rental value in francs per dwelling by quartier in 1876. *Source:* ASVP 1880, 392–400.

values for just shy of 700,000 dwelling units. The average value per neighborhood is displayed in figure 10.1. Those in the top 1% paid a rent of 3,000 francs or more. Moreover, 60% of these homes were in the 7th, 8th, and 16th arrondissements, even though those three arrondissements accounted for less than 10% of all dwellings. Even within the 8th arrondissement, the average rent was 33% higher in the Champs Élysées neighborhood than in the still luxurious Chaussée d'Antin and 75% higher in the Europe neighborhood. The same census in 1910 provides very similar answers: 72% of dwellings with a rent of 4,000 or more francs (the top 1%) were in the same three arrondissements. Within the 8th and 16th arrondissements, the best neighborhoods had average rents twice that of the less desirable parts of their arrondissement. The median rent in the 8th arrondissement exceeded 1,000 francs, while the median rent in Paris was less than 400 francs. The average rent in the Champs Élysées was eight times that of the rent in the median Parisian neighborhood and seventeen times that in Charonne, the city's poorest neighborhood.

Jacques Bertillon, the head of the Paris statistical office in the late nineteenth century, was well aware of this inequality. In 1887, he noted a close connection between income and longevity: "We can compute for each neighborhood the share of male domestics (to evaluate the fraction of families living in luxury) and the share of female domestics (to assess the fraction of families living with ease)" (ASVP 1887, 124). He was in no way surprised that the ratio of male domestics to households in the Champs Élysées was .76, and the ratio of female domestics to households was .87, so that, on average, households had 1.6 domestic servants. For Paris as a whole,

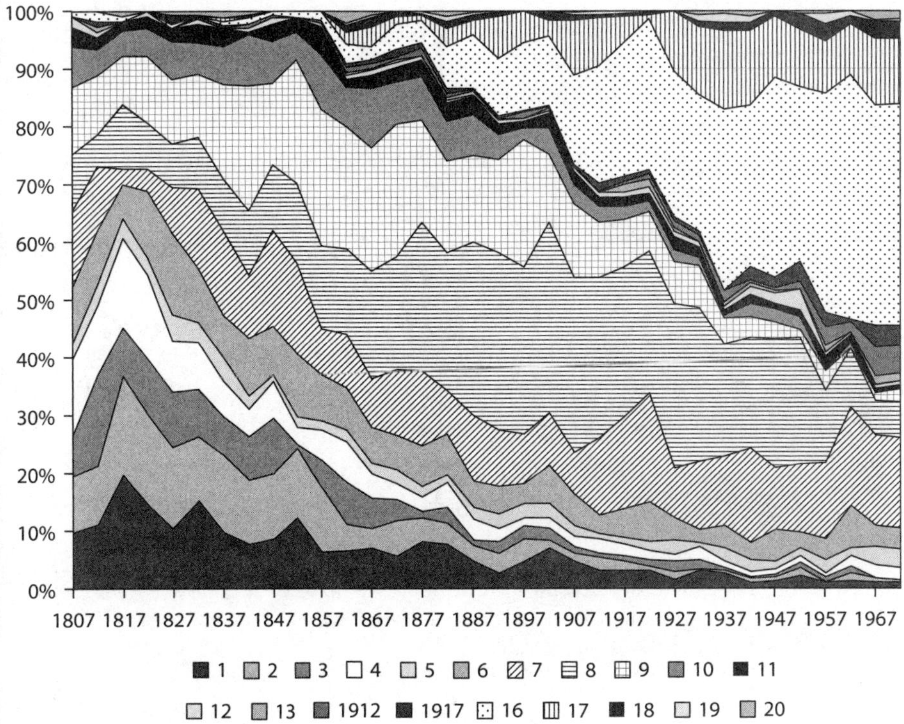

FIGURE 10.2. Arrondissement of residence of the top 1% wealthiest decedents in Paris, 1807–1972. *Source:* Estates dataset (see appendix 1).

these ratios were .05 and .17. At the other extreme of the distribution in the much poorer 18th arrondissement, there was one male domestic for every 250 households and one female domestic for every 24. Inhabitants were distributed unequally, with the wealthy ensconced in a few neighborhoods that most Parisians never visited. Bertillon then correlated the number of domestics to the mortality rate from different diseases and found that more domestics were associated with lower mortality rates. With Lionel Kesztenbaum, we verified this was also true with aggregate mortality by age (2017).

The dramatic spatial inequality is present in the data on wealth at death. Indeed, every estate declaration begins with the name of the deceased and his or her address. From 1807 to 1972, there are 7,820 members of the 1%. Because we collected their details, we have an address for nearly all of them and can allocate them to the post-1860 arrondissements (we could do so for lower fractiles, but we did not have enough addresses, particularly before 1870). Figure 10.2 displays the arrondissements where members of the top 1% had their legal residence. The core arrondissements (1st–4th)

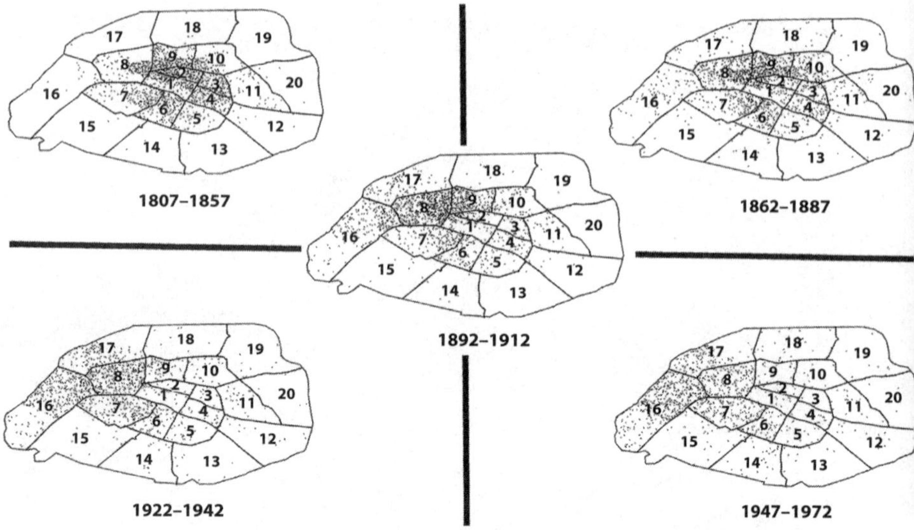

FIGURE 10.3. Maps of the residences of the top 1% decedents in Paris, 1807–1972. *Note*: The residences of the top 1% were identified up to their quartier, so the map provides a visualization of local density rather than pinpointing where each lived. *Source:* Estates dataset (see appendix 1).

were important in the First Empire and the Restoration years. Overtime, however, the core lost share to the 7th, 8th, and 9th arrondissements, where the rich died in very large numbers until 1900. After 1900, the rise of the 16th becomes the central spatial phenomenon.

The arrondissements were not alike, and as the city grew, it was inevitable that the central arrondissements' (1st to 4th in particular) share of the very wealthy would decline. They were small in surface area and cramped relative to the newer arrondissements. We provide further details and proceed in two steps. First, we divide the data in figure 10.1 into five distinct periods. Figure 10.3 has five maps that cover the five periods that we study. Before 1857, the very rich lived broadly across the city, with a concentration within a few hundred yards of the Bourse. Over time, their preference for the 8th, 7th, and 16th arrondissements became far more visible. They were always extremely scarce in the outer arrondissements (13th–20th) except for the 16th and the south of the 17th.

Their residential concentration is even more striking if we consider their share in total deaths. We divide the number of decedents who were part of the 1% by the number of adult individuals who died in the same arrondissement in the same year. We can only do so from 1872 to 1972 because deaths

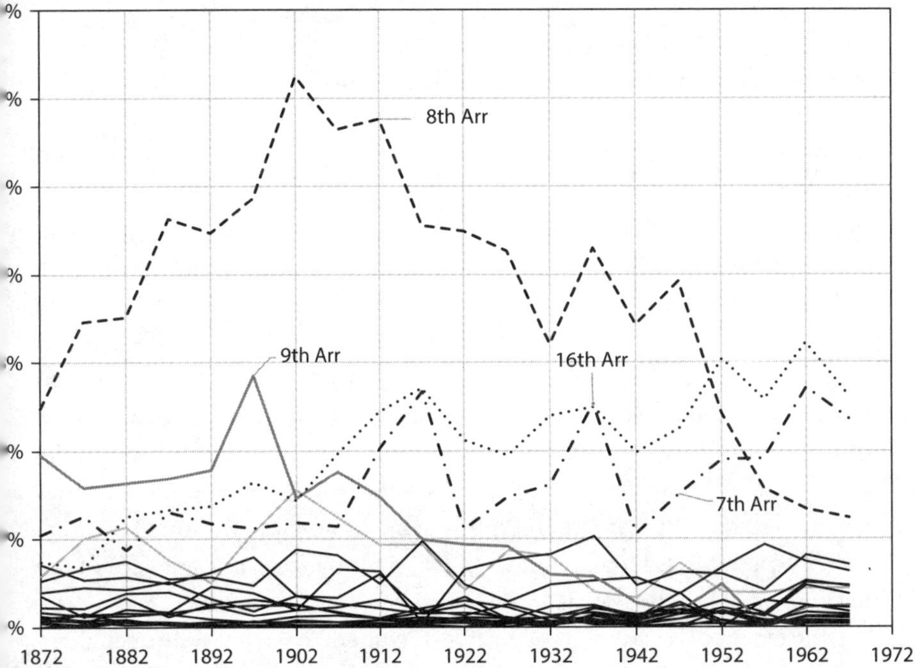

FIGURE 10.4. Fraction of adult decedents with top 1% estates by arrondissement.
Source: Estates dataset (see appendix 1).

by arrondissement were not available before 1872. The calculation produces the share of decedents who are members of the 1%. Figure 10.4 displays the results. For Paris as a whole, this share is, obviously, 1%, but it varies a lot within the city. The maps highlight the rich's predilection for the western parts of the city and the growing importance of the 16th arrondissement after 1900. At the lower end of figure 10.4, we find twelve arrondissements that are always below one-tenth of the average. They are virtually indistinguishable from one another. At the other end, four arrondissements are generally above the 2% mark. The 9th was dense early on, but, as seen in figure 10.4, it faded in the twentieth century. The 7th has an unstable share of the very rich, but they are most often three to four times as dense there than on average. The 16th has a growing share over time, and its top 1% represents about 5% of decedents after WWII. The 8th is entirely different: in its heyday before WWI, 10% of decedents were part of the 1%. For the 8th, figure 10.4 tells a more dramatic story than figure 10.2 because it is the arrondissement with the fewest deaths. More importantly, this figure shows the extent of the wealth fracture; except for the 16th that joined the wealthy western corner of the city, none of the new arrondissements managed to attract significant numbers of very rich people even though their populations were growing very rapidly (the 18th added 100,000 inhabitants between 1872 and 1932).

It is important to remember that Paris was a city of renters. To be sure, a few very wealthy individuals could afford a house and even fewer a mansion (*hôtel*). However, there were not enough parcels to indulge everyone. For instance, the 8th arrondissement only had 3,200 buildings and 23,000 dwelling units. Residential choice in Paris through the 1950s was not a rush to home ownership. Most of the renters in the 8th arrondissement could have afforded to own a building elsewhere in Paris, and many did. However, they did not live in the Parisian buildings they owned; instead, they preferred to rent in a richer neighborhood. Even those who could not afford a *hôtel*, or even a house in the city, could, after retirement, have moved out to the suburbs. Many wealthy Parisians owned real estate outside the city, and we often find them dying there. At the same time, they continued to enjoy the city's amenities by maintaining a home (as their legal residence) in the best neighborhoods in Paris.

The extent of residential sorting we document in Paris is only peculiar in that individuals chose to rent in rich neighborhoods rather than live in their properties. In other respects, it reproduces the same pattern we see in the west and east ends of London or in Manhattan and the outer boroughs of New York City. This pattern is, of course, found even in smaller cities, which also tend to have better and worse sides of town. This geography should have been irrelevant to the extent that elections were held in Paris to determine taxation and spending, as the poor were so much more numerous than the rich. However, landlords often prevailed in important debates about infrastructure (Kesztenbaum and Rosenthal 2017). One wonders whether their spatial congregation gave them coordination advantages.

10.2. Inequality When Locations Are Isolated Islands

The distribution of wealth in France is tautologically the aggregation of the distributions of wealth of its subcomponents. The subcomponents could be départements, arrondissements, municipalities, or even economic sectors. In his studies on inequality, Simon Kuznets emphasized the importance of structural change in reweighting the aggregate distribution of income. For Kuznets, there were two effects. First and most importantly, the new sectors (industry) had higher wealth per capita than the old ones (agriculture and handicrafts). That alone tended to add variance to the aggregate distribution as the weight of the new sector grew. The second, which Kuznets largely set aside, was that the new sectors could be more unequal than the old ones. Kuznets focused on income rather than wealth, so it is important to note that two additional factors matter to the impact of structural change on the distribution of wealth. The first is that the wealth distribution is not bounded in the same way as the labor income distribution. The latter is generally more equal because of the finite quantity of labor any individual can deploy and because households can live without wealth but need some income. Second, sectors differ by their economies of scale, so agriculture is generally characterized by a very large number of firms. In contrast, certain modern sectors are dominated by a small number of very large firms (e.g., railroads).

The economies of scale of modern industry, combined with the fact that ownership of modern sectors in France was far more concentrated than the traditional sectors of agriculture and handicrafts, can explain a rise of inequality with industrialization. This was definitely Marx's view since the new capitalist sector is one where the workers do not own the means of production, while in the old one, they do. The same argument can be made to explain urban versus rural differences in the last two centuries. The modern sectors, presumed to be more unequal, were largely urban, while the traditional sectors, presumed more equal, were to be found in small towns and the countryside. Kuznets's primary effect comes from increased urbanization—because nominal urban wages are higher than rural ones. The secondary effect comes from the fact that inequality is higher in cities. It is important to note that this approach considers each location as an island that can be treated as an independent observation. Workers and their families largely flow one way, from rural to urban locations, and the distribution of wealth in a location is mostly driven by endowments. In our case, Paris sits at the apex of the urban network. It has plausibly the highest average wealth and inequality, and its share of the French population is growing through the interwar period. We will revisit the island assumption in the next section but now turn to efforts to measure inequality in different locations in France.

For France as a whole, we have access to three key sources. These have been abundantly used in several articles and books by Thomas Piketty, us, and other coauthors. We revisit these sources and extend and confirm the early work presented in our 2006 article when we had collected less than a quarter of our current dataset. The first source is the aggregate inheritance flow (*annuité successorale*) available from 1826 to 1949. Second, estate tax filings by estate size are available from 1902 to 1964. Finally, we have the results of the TRA project (again, it collects the estates of all individuals

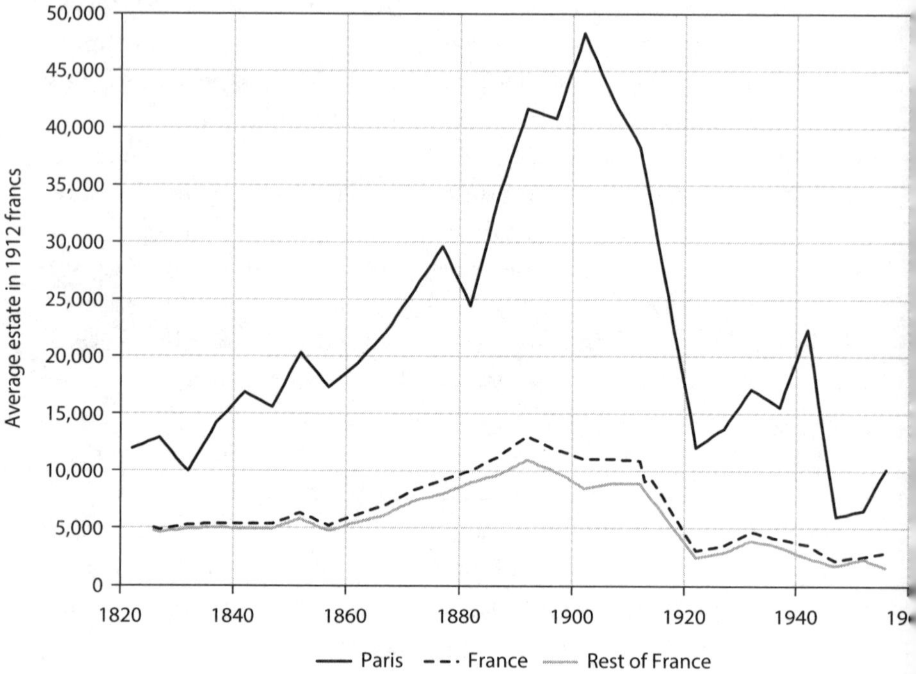

FIGURE 10.5. Average estate for adult decedents in 1912 francs. *Source:* Estates dataset for Paris. ASF for France.

whose last name starts with the letters TRA in the whole of France). Each source provides a different vantage point to evaluate Paris's role in shaping French wealth growth and inequality.

The most basic series is the total inheritance flow. From 1826 through 1964, the government published the total value of inheritances. We supplement this data with the total from our estate tabulations and arrive at a series from 1826 to 1964, with a gap from 1914 to 1925. Dividing the flow by the number of adult deaths produces average wealth. We can net out the value of inheritances in Paris to produce average wealth in the rest of France; this creates a Kuznetian primary effect. These data show that the average estate value in Paris was much higher than that of the rest of France (see figure 10.5). Starting at about two and a half times the French average, Parisian estates grew much faster and peaked around 1900, just above four times the national average. Then, the sharp drop in the value of estates that started in WWI brought Paris back to about 3.8 times the national average (evidence that the shocks were general across the country). The data after 1957 are missing the bottom of the distribution because the government exempted small bequests. Some of the divergence at the end is likely due to this censoring, but the rapid recovery of Paris set in right after WWII, a decade before the exemption was enacted.

The Kuznets two-sector effect is present, though small. It is present because figure 10.5 shows a growing gap between "the rest of the country" and the French average from 1827 to 1900. This was a period when the city steadily gained in population share. However, it is small because Paris did not grow to be a big enough share of the French population. To the extent Paris drove French inequality, it is because of the massive wealth accumulation and dis-accumulation the city has witnessed. Paris's share of the inheritance flow began in the 1820s and 1830s, ranging between 10% and 15%; it grew to between 25% and 28% in the two decades before WWI. Then, a decline set in. After WWII, Paris hit a nadir at 14% of the national inheritance flow before rebounding rapidly after that.

Because there is so much wealth in Paris, we must ascertain how it affects the national distribution of wealth. Following our 2006 article, we estimate the share of the richest decile, centile, and thousandth in France who died legally residing in Paris. To do so in the twentieth century, we compute the threshold value of each fractile using the published national tabulations of estates. These are available for nearly all the years for which we collected the Parisian cross sections. We then count the number of Parisians with an estate above that threshold. For instance, in 1912, 545,000 adults died in France, so 5,450 were in the top percentile. We compute the threshold to be 128,659 francs. Our cross section tells us that 1,459 Parisians had an estate larger than 128,659 francs. These Parisians thus composed 26.8% of all the top 1% wealth holders in France.[1] For the nineteenth century, we have no tabulations, so we assume that the share of each fractile followed change with the Parisian share of decedents. Except for 1832, when the cholera epidemic created a very large mortality spike in Paris, using adult decedent death shares rather than population shares does not matter much. We have not attempted to "fix" 1832. We were tempted to do so because the evidence internal to Paris is that the epidemic hit the poor much harder than the rich. However, some very rich and powerful individuals like Casimir Perier and the comte Chaptal did perish. Absent further information, it seems best to let the data lie where they fall. Overall, the evidence is consistent with an increasing concentration of the very wealthy (top 1% wealth holders) in Paris over the nineteenth century, followed by a slow decline in the twentieth century. Even then, fully 40% of the top 0.1% wealth holders in France died in Paris (see figure 10.6).

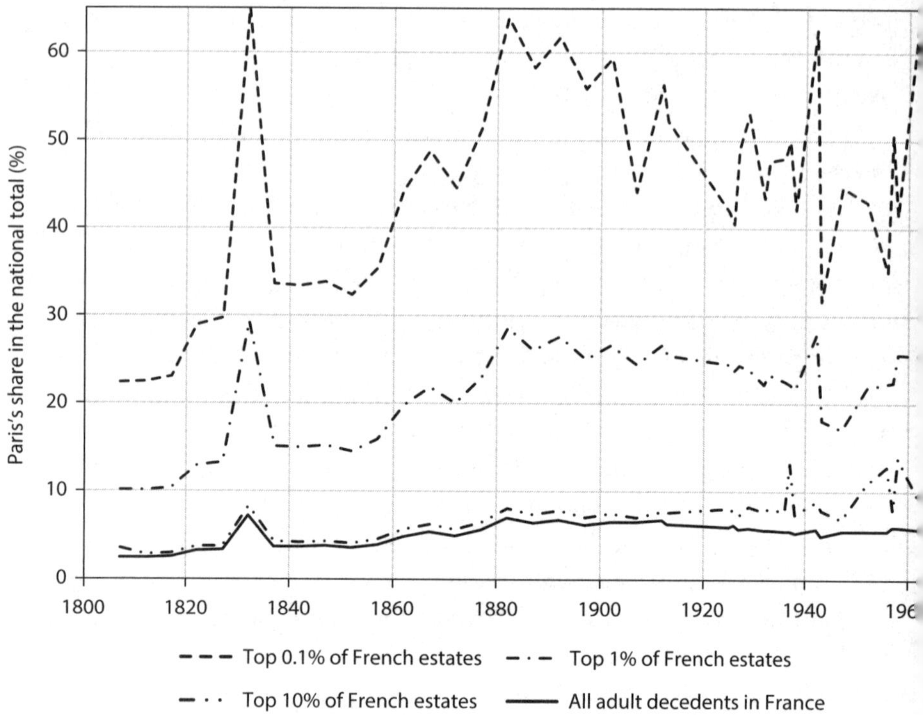

FIGURE 10.6. Share of Paris in French estates by fractile. *Source:* Estates dataset.

Like our data for Paris, the last source involves individual estates culled from the estate tax administration. In the 1980s, to improve on Louis Henry's estimates of French historical demography, which focused only on highly persistent families, Jacques Dupâquier launched the TRA project. The goal was to follow approximately 3,000 French families whose last names started with TRA over time. To do so, all vital records involving individuals with a last name beginning with these three letters would be collected starting in 1820. Because of France's wide name variation, Dupâquier correctly thought that individuals could be matched even if they migrated. A few years later, Denis Kessler realized that the TRA sample could yield additional insight by collecting data about wealth transmission (Dupâquier and Kessler 1992). Since the 1990s, Jérôme Bourdieu, Lionel Kesztenbaum, Gilles Postel-Vinay, and Akiko Suwa-Eisenman have pushed the project forward (Bourdieu et al. 2014). Dupâquier intended to create a representative sample of the French population. In many dimensions, the TRA sample answered just that ambition. Unfortunately, it is not representative of the very wealthy because it includes too few of the top 1% wealth holders.

In principle, one could combine the two datasets to produce a representative sample for France as a whole and compute national wealth shares by decade over the nineteenth century. Unfortunately, the TRA sample is too small to allow any such effort, even if the data are collapsed by decades. Nevertheless, the analyses of Bourdieu et al. (2003, 2011, 2019) help contrast Paris with the rest of France and delve further into the dynamics of wealth accumulation and inequality over time.

To begin, the proportion of adults leaving some wealth at death varied 60 and 70% in France as a whole and between 70 to 80% in the countryside. By contrast, it was only 50% in cities (not including Paris). Paris was a clear outlier here, with a proportion of individuals leaving an estate that was, at times, only half that of other cities and a third of the countryside. Also striking is the long-term decline of the share of individuals leaving an estate in France. That proportion was 70% early on but fell to about 55% at its nadir on the eve of WWI before recovering after the war. This decline was driven by the fall in the share of individuals with an estate both in rural and urban areas outside Paris and by the increasing weight of urban areas in the population. Interestingly, urban areas outside Paris were far more similar to that of the countryside than the capital. This is most likely due to the high price of real estate in Paris, which precluded a savings flow into real assets for most people. Equally important is that the fraction of wealthy individuals was always lowest in Paris and highest in rural areas (see figure 10.7).

As figure 10.8 shows, the data for France strongly support the hypothesis that real estate is the primary driver of the share of individuals leaving an estate. However, there is an important caveat. At all times except in the 1950s, the share of individuals with real estate falls as we move from small to large settlements, both among decedents and those with wealth.

When we examine financial assets, the reverse relationship holds. Figure 10.9's left panel shows that the relationship is compressed among all decedents; quantitatively, the differences are small. Nevertheless, picking out the Paris line that sits on top is easy. The right panel focuses on those with bequests, and we see that Paris's early lead continued to grow until the 1930s. Then, a convergence across all localities takes place, even if the order is unchanged.

To close this section, we return to the twentieth century and the tabulations of estates published by the administration using standard Pareto interpolation techniques. To reduce the impact of the random arrivals of very large estates on the distribution, we aggregate the data for 1902, 1903, 1904,

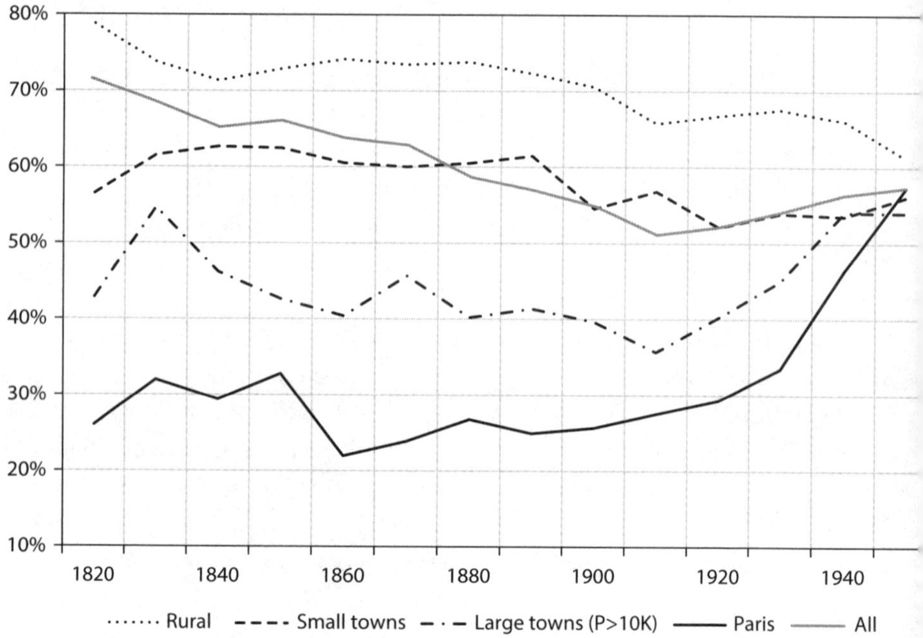

FIGURE 10.7. Fraction of decedents who died with positive wealth. *Source:* TRA sample (Bourdieu et al. 2014).

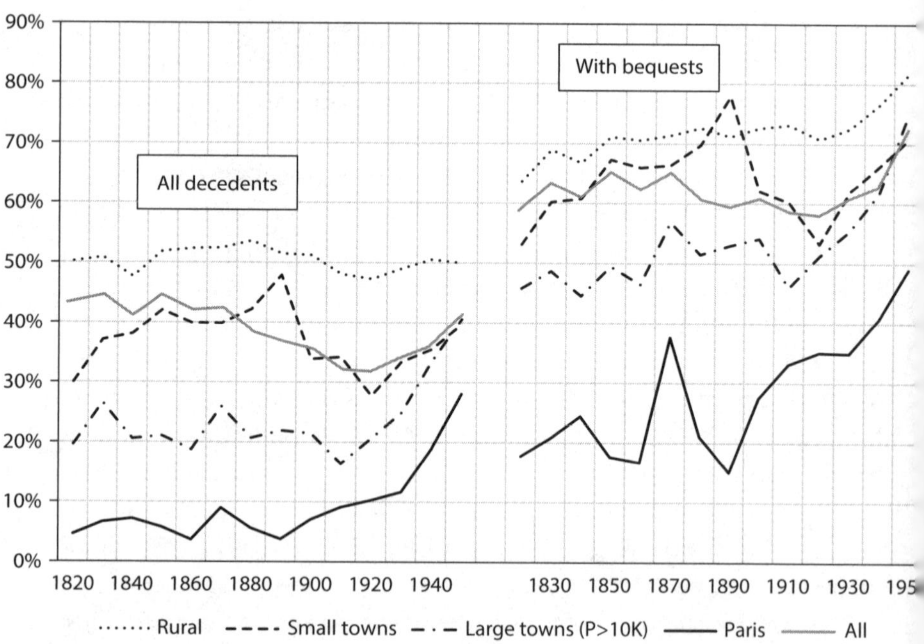

FIGURE 10.8. Fraction of decedents who owned real estate. *Note*: The left panel gives the proportion of decedents whose estate contained real estate, and the right panel provides the same proportion, looking only at those individuals who had an estate. *Source:* TRA sample (Bourdieu et al. 2014).

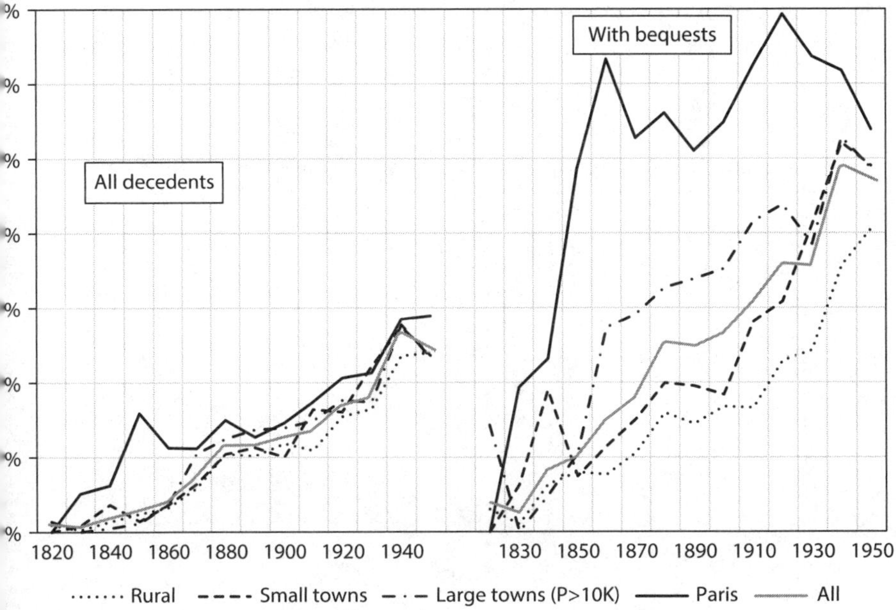

FIGURE 10.9. Fraction of decedents with financial assets. *Note*: The left panel gives the proportion of decedents whose estate contained financial assets, and the right panel provides the same proportion, looking only at those individuals who had an estate. *Source:* TRA sample (Bourdieu et al. 2014).

and 1905 and compute the distribution of estates by département, binned by average wealth. We set aside the Seine—where Paris is located—and three neighboring départements (Seine-et-Marne, Seine-et-Oise, Oise—henceforth the orbit), leaving eight départements per bin. Doing so reduces the impact of outliers that arrive randomly in any département. We also include our estimate of the distribution of wealth in Paris. It is composed of the 1902 distribution we collected and replicated four times. We augment that data with the six estates above 80 million francs reported in 1903, 1904, and 1905, at least three of which were worth more than 100 million francs. Figure 10.10 displays the distribution along with the fraction of positive estates (left axis) and the logarithm of both the value of the average estate and the average positive estate (right axis). The average estate was just under 3,000 francs for the poorest decile of départements and twelve times more, or 36,000 francs, for Paris. The richest decile départements had average estates of just above 11,000 francs. They were thus poorer than the départements in the Paris orbit, whose average estate was about 27,000 francs.

Figure 10.10 holds a second important lesson. The share of individuals with a positive estate (68% for France as a whole) has some volatility but

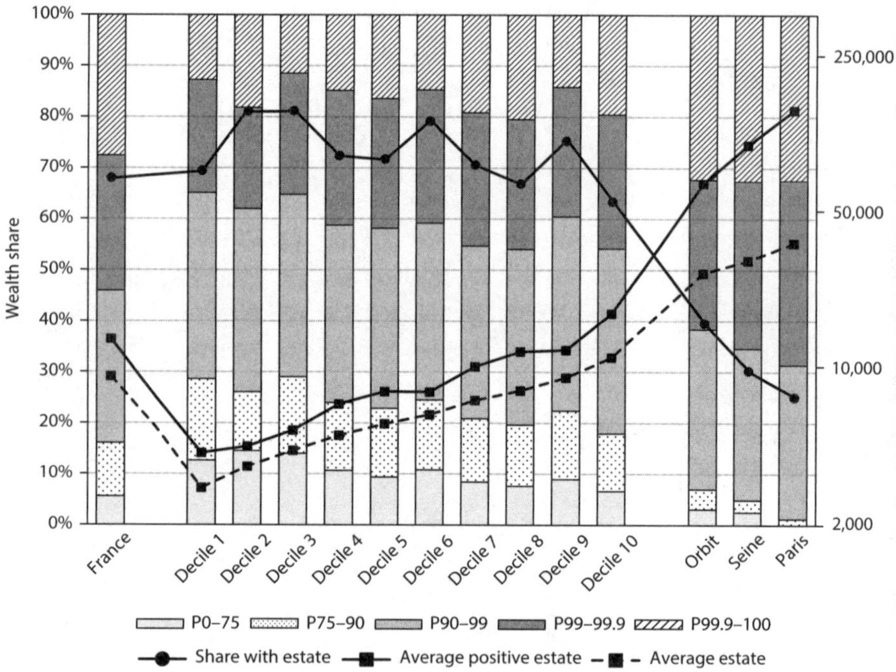

FIGURE 10.10. Wealth distribution across French départements, 1902–1905. *Note*: Départements are binned by average positive estate. *Source:* BSLC reports on estates by département for 1902, 1903, 1904, 1905. For references, see Piketty 2001, 749.

trends down as wealth increases. The big difference lies between the deciles and three locations on the far right of the graph: the Paris orbit, the Seine (which includes Paris), and the city itself. The deciles of départements have a share of individuals leaving a bequest that is at least 63%, while the Paris region tops out at 40% and Paris even further down with just about 26% of positive estates. The same pattern holds when looking at wealth shares. The poorest three deciles among départements have a 1% wealth share of about 36%, and the richest three have a 1% wealth share of about 43%. Again, the Paris orbit and the Seine are much more unequal, with a top 1% wealth share higher than 60% and a bottom 90% wealth share less than 10%. Paris was, and still is, a universe of its own. The top 1% wealth share comes in at almost 70%, the bottom 90% at 1.2%, and no measurable estate is below the top quartile. The stark differences from Paris come from concentration at the top. Paris was unequal because, as noted earlier, a huge fraction of the wealthiest French men and women lived and died there.[2]

We find the same positive correlation between inequality measures and average estate by département through the interwar period and after WWII.

We repeated the procedure that led to figure 10.10 for two additional sets of years (1929–1931 and 1952–1955). The results are strikingly similar. The most important implication of these departmental comparisons is that places that were quite poor in France were not particularly unequal. This stands in stark contrast to locations across the world where, for historical reasons, property was very concentrated and incomes and wealth were low. Although institutions in France were not homogeneous, there was little variation across space. It is hard to think of that variation coming from some local political processes designed to enshrine the existing distribution of wealth and power. Instead, variations in average wealth levels and inequality seem to owe mostly to economic forces, particularly market access. These forces were, until 1914, allowed to run largely unchecked in ways that benefited the wealthy in Paris.

The richer départements (including the Seine), which include all those with a large city and many of the most industrial ones, had the fastest population growth. Thus, they were the prime recipients of internal migration due to urbanization and industrialization. The poorer ones were, of course, the place of origin of these migrations. It is important to note that when choosing migration destinations, individuals had to choose between two opposed dimensions. On the one hand, they could decide to move to départements where access to property was easier (as defined by a larger proportion of positive estates) but where these estates were quite small. On the other, they could move to places with high average wealth. These were likely places where capital was relatively abundant and thus made labor more valuable. This choice, however, came at a cost: a lower likelihood of constituting a patrimony. The starkest of those choices was moving to Paris. It seems that before WWI, migrating individuals were not inequality-averse and were looking for high wages. It is also true that in doing so, they were moving into occupations where they would "rent" their labor to capitalists and that the path of owning a business (farm or other) was largely closed off.

10.3. Paris and France as Part of a System

Measuring the distribution of estates across départements has implications that require us to go beyond the very useful island model. Indeed, as we just saw in the last section, labor flowed across France, which is even more true about capital. As we saw in the first seven chapters, Parisians invested in the city, the rest of France, and abroad. In all cases, these flows were mediated by institutions and organizations. Organizations like financial markets were mostly centered in Paris, and formal institutions were overwhelmingly chosen by the national government and legislature in Paris. Around 1900, Paris accounted for 8% of French employment (6.5% of the population) and around 45% of jobs in financial firms and insurance companies. The revolution had bequeathed a political structure that privileged institutional uniformity, with taxes and laws applying in the same way everywhere and a judiciary controlled by Paris in charge of ensuring that uniformity. To be sure, some institutional variations did exist on the edges of the country, starting with Corsica, continuing with later acquisitions (Savoie, Haute-Savoie, and Alpes-Maritimes), and culminating with the départements recovered from Germany in 1918 (Moselle, Bas-Rhin, and Haut-Rhin). Yet these variations are small enough that we set them aside.

The economies of the capital and the rest of the country are also quite different. Paris had, of course, no meaningful agricultural sector, and it had an outsized share of employment in finance and legal services, as well as government employees. Within its diversified manufacturing sector, the city was home to a large luxury sector that produced or organized the production of goods for export (*articles de Paris*). The city was thus home to a higher-skilled population than the rest of France.

More important, the labor force was not homegrown. As we have seen before, a substantial fraction of the population of Paris was born elsewhere. This fact is displayed starkly in the birth, marriage, and death geography of individuals who left an estate. The data are sparser for the huge number of those who did not leave an estate. Nevertheless, the ASVP published death totals by place of birth, sex, and age from 1880 to 1885. Overall, roughly 55% of all individuals who died in Paris had not been born there. From 1881 to 1885, it is easy to remove all deaths of children (less than 15) who were almost all born in the city. The share of adults who died in Paris but were not born there was just under 80% for men and nearly the same for women. These very high shares of migrants in adult deaths are not due to negative selection among migrants who chose to move to Paris because, as discussed in chapter 9, in the same years, male migrants were, in fact, taller and lived longer than their native Parisian counterparts (Kesztenbaum and Rosenthal 2011). Instead, the huge share of migrants among the deceased in those years comes from the high growth rate of the Parisian population, a high mortality rate, and the out-migration of some native Parisians.

To better understand the extent of differences across the wealth distribution, we split the sample between the top 1% of wealth holders (who own half of all the wealth) and the rest of the population with positive estates.

For those at the top, we have tried to complete this information as much as possible using digitized records (including genealogical websites).[3] For births, we also leave out the period before 1860 when the destruction of civil registration records made the recovery of birthplaces extremely difficult. To parallel the last section, we break up the locations into four groups: Paris, the Parisian orbit (the rest of the Seine, Seine-et-Marne, Seine-et-Oise, Oise), the rest of contiguous France (including Alsace Lorraine and Corsica), and foreign locations (this category includes France's outlying départements and territories).

The bulk of wealthy Parisian decedents were always more likely to be born, get married, and to die in the city than its poor, but not by much, in the period corresponding to the data published in the ASVP (1862–1912). They were almost as likely to be born outside Paris (75%) as the population as a whole. They were less likely to have been born outside the city as time passed, but that share was always high (still 71% in 1962 and 1972). Strikingly, they were very often married in Paris. In the period before WWI, two-thirds of all the rich were married in Paris, and they died in Paris at an even higher rate (close to 90%). Even in the last period (1962 and 1972), when the health system of the city had expanded into its orbit, rich Parisians mostly died in the capital and, in fact, at home.

Consistent with the fact that a large fraction of the top 1% had inherited substantial estates themselves, these very rich wealth holders were more Parisian in birth and marriage than the other groups. They were less Parisian in death, consistent with their ability to afford one or more country estates while maintaining a legal residence in the city. They were also much more likely to be born in Paris or the orbit départements. Between 1862 and 1912, the majority of the top 1% was born in Paris, a rate that was twice that of the average decedent. Then, the rate declined but was never less than 40%. In marriage, the top 1% share is even higher—more than 70% were married in Paris. They were also much more likely to die outside the capital. Because the information about places of death for the sample of rich individuals is taken from the estate records, it is not subject to selection problems except for the fact that it was only recorded in some years. The relatively higher rate at which the richest 1% died outside the city is easily explained by the records themselves. For a few of those who died outside Paris, the records say they died while visiting a spa or the home of some other person. Nearly all, however, indicate that the person died in their summer residence, their property, or even their château.

Paris was not an isolated island. Even though its wealth distribution was always remarkably distinct from that of the rest of the country, it was and remains in constant exchange with France and the world more broadly. Individuals migrated to the city for a wide range of reasons, and they left the city for yet another wide range of reasons.

10.4. Political Economy and Inequality

The difference in wealth and inequality we have just documented also had consequences for the political economy of taxation and public services. All the regimes that followed the French Revolution have treated Paris differently from other municipalities, both because it is the capital of the country and because it is rich and unequal. At all times, the tax regime that applies to Paris and the resources that are devoted to public services have depended on the national government and, thus, on a negotiation between the capital and the provinces. This section investigates how this negotiation evolved.

For most of the Third Republic (1870–1940), legislative politics had a strong local flavor, as each arrondissement elected one deputy to the National Assembly by universal male suffrage. Returning to the Meltzer-Richards model, we can ask: What kind of fiscal regime might one expect given these circumstances? The model's equilibrium emphasizes that tax rates rise with inequality. It stands to reason that the differences in average wealth and the distribution of estates that we saw by département would only increase if we had data by arrondissement (this is certainly the case for Paris, as we saw at the start of the chapter).

At the national level, given the national distribution of wealth, proportional or progressive taxes on wealth should have garnered far more support than the proportional taxes on consumption, which were a major element of public revenue. Raising tax rates on wealth would have had no direct effect on the half of the population who owned little or nothing, and who could surely have benefited from additional public goods. Yet explaining the persistently low level of direct taxation before 1870 is relatively easy. Suffrage was massively restricted to the wealthiest men before 1848. After 1851, Napoleon III's ministers effectively chose candidates, thereby reducing the chances that advocates for more taxation would gain office. During the first three decades of the Third Republic (1872–1900), neither explanation holds because all adult men could vote, elections were competitive, and there were active leftist parties. The Meltzer-Richards model does not hold: no coalition formed to significantly raise taxes on the rich to increase social spending. So, we must look for other mechanisms that were at work.

Instead, and going back to the data from estate taxes in the early twentieth century, the explanation lies with the conservative goal of protecting private property. The demand for expanded taxation came largely from urban dwellers who wanted more infrastructure and possibly more education. This was blocked by an alliance that included those who lived in the countryside (the agricultural sector) and the very rich who lived in cities. Farmers had low demand for public goods (and all the more so because they had access to primary education). They were far more interested in protecting their property from taxation. In this respect, their priorities aligned with those of the very rich in cities (our top 1%), whose demand for public goods was low because they could privately procure many of these local services (Cagé and Piketty 2023) and certainly did not want to pay taxes. In short, we speculate that the arrondissement suffrage system enabled an Epple and Romano (1996) "ends against the middle" equilibrium. In this model, the rich, who can procure services privately, form an alliance with the poor, for whom private consumption is more valuable than public goods. Before WWI, more than 40% of the population remained engaged in agriculture (Marchand and Thélot 1991). This high agriculture share helped ensure the continuation of the low tax equilibrium. Across France, most arrondissements had a small urban population and a large agricultural sector. Under the Third Republic, the rule of one deputy per arrondissement remained unchanged, and the arrondissement map was barely altered before 1926. Thus, the rural vote remained extremely important in French elections even when the urban share of the population was growing.

Progressive taxation was initiated in France in the decade or so before WWI, partly in response to domestic fiscal needs (setting up the public pension system in 1910). To a large extent, however, the change in regime was needed to deal with increased military spending.[4] National defense, then as now, had a uniquely dramatic impact on fiscal policy, and taxes rose from 1900 to the 1930s. Returning to some of the discussion from chapter 3, it is important to highlight how much of the increased tax burden due to progressivity was borne by rich Parisians. In other words, when deputies in the 1920s or 1930s voted to raise tax rates at the top end of the scale, most knew full well how little their constituents would be affected. Take, for instance, the distribution of estates in the period from 1929 to 1931. We applied the 1929 tax law for those years and assumed that every estate paid the tax rate associated with two children. First, consider the richest individuals' estates, meaning those larger than 500,000 francs (roughly the top centile of the wealth distribution). In our calculations, these large fortunes paid almost two-thirds of all the taxes (the system was quite progressive). However, these estates were concentrated in Paris. Because the city was so rich, more than half the estates worth more than 500,000 francs were Parisian, and it is no surprise that the heirs to those Parisian estates paid more than half of all the taxes in that group. Because Paris had a higher share of individuals who died without descendants, its actual share of the tax burden was likely higher than we estimated. Over the whole distribution, our computation yields a share for Paris of 37% when its share of the value of estates was 25%. We can look at the estates below 50,000 francs for the rest of France. These account for 86% of all the estates; thus, upward movements in the rates were politically more meaningful because there were no exemptions. However, the rates in this range were, on average, less than 4%, or a fifth of what the richest estates paid. Consequently, the vast number of smaller estates brought in little revenue (15% of the total). The analysis above is another example that national politics is never simply macropolitics because its local implications are also very important.

The geography of wealth has one other likely consequence. Because wealth was so heavily concentrated in Paris, those who peddled tax avoidance strategies and tax dodging schemes found it most profitable to offer their services in the capital. Interestingly, French (Farquet 2018) and foreign banks (Guex 2006, 2022) participated in providing services to move money offshore as early as the 1930s. Thus, it is no surprise that the wealthiest and oldest Parisians held a larger share of their wealth in assets legally exempt from estate taxes than anyone else.

10.5. Conclusion

Wealth is dramatically unequally distributed over space and time. To begin with, it is important to have fine-grain data at a reasonable frequency simply because the distribution of wealth evolves. We find it remarkable that although the raw data exist in the archives, the French statistical apparatus became disengaged from collecting wealth-at-death statistics by 1964. This is particularly surprising since the estate tax is the only fine-grain source of data on the evolution of private wealth. For the past 60 years, French officials have been making wealth policy with a self-imposed blindfold. To be fair, France is not alone; many countries have either never had estate taxation or have limited the population at risk of taxation in ways that make their data relatively uninformative.

To understand wealth dynamics, it is also important to measure inequality over different units of aggregation from the national to the local. Even if one is interested in macroeconomic policy and, as in France, all important taxes are set in the capital, members of representative bodies (deputies and senators) are elected locally. They will be sensitive to inequality over space because their constituents are, ultimately, local. The voters' experience of inequality, including fears of the revolutionary tendencies of the poor or a sense of unfairness at the very wealthy's consumption, is partly at least local. When asked to support increased progressive taxation, the voters will want to know whether those revenues will be spent on local schools or hospitals on the one hand, or on elite services in the capital on the other.

This chapter has tried to put Paris in its French context. While both the country and the capital have their specific histories, the same kind of disaggregation would be useful in studying other places. Unless artificially created in strange places like Washington, DC, or Brasília, capitals generally concentrate not just the apex of political activity but also the most important economic and cultural organizations. Other large cities are also unequal simply because of scale effects. In 1800, the Parisian experience was exceptional because nearly everyone in nearly every country lived in the countryside (Davis 1955). Since 1800, the world's urban population has grown rapidly, and a majority of all people live in cities, especially in larger cities. The experiences of Parisians from 1807 to 1977, in many ways anticipated, by half a century or more, those of the inhabitants of capitals and other very large cities across the globe. As we shall see in chapter 11, those anticipations extend across a wide variety of private savings strategies and demand for public services.

Considering only the national aggregate simplifies things a great deal, particularly for countries and periods like France before WWI, where international migration is limited. Then, each country's experience can be taken as independent of the others. More recently, population and capital flows have made approaches that treat countries as independent observations untenable. Paris here offers an early and extreme version of an open economy. Even before the nineteenth century, Paris's population reflected a constant flux between the city, the rest of the country, and beyond (Sussman 2006; Slivinsk and Sussman 2019). The internal migration that fueled Paris's growth was from poor and relatively more equal rural areas to the capital's richer and more unequal economic environment. We find little evidence of inequality aversion and, instead, a tremendous pull from the higher wages Paris offered. Once in Paris, the working classes were active participants in the political conflicts that led to all the major changes in taxation and public expenditures. The rich, of course, understood that threat, so keeping control of the city and their wealth were constant concerns for political regimes from 1789 onward, well past the Commune (Chevalier 1958). Yet, the fact remains that Paris's growth was made possible by a large and persistent flow of individuals coming from more equal parts of the country.

For wealth, it is obvious that Paris's boundaries are essentially academic. For those Parisians who died wealthy, the bounds of the city were also not that important. Very large fractions of the wealth of its inhabitants involved claims on assets outside the city, and outsiders held some of its wealth. In 1898, for example, about 11% of the privately owned buildings in Paris were the property of non-Parisians (Brando et al. 2023). Returning to earlier chapters, the investment patterns in the capital diffused to the rest of the country. This was particularly true for life-cycle savings programs (like retirement) and precautionary savings organizations (like the caisses d'épargne). More broadly, the diversity of choices between saving in one's enterprise and real estate or financial assets, as well as choices between life-cycle and bequest motives, occurred throughout the country. Paris was different, however. It was different because the large share of individuals with wage or piece-rate employment went back to at least the early nineteenth century and anticipated the rest of the country by decades, if not a century. It was also different because, through the 1970s, most of the top fortunes were Parisian. The capital's capital remains to this day very unequally distributed.

11

Inequality Beyond Wealth

For ten chapters, our focus has been on private wealth. When other variables (sex, age, education) have been brought into the discussion, they have been deployed to help explain the processes that lead to wealth accumulation and distribution. At the outset, we emphasized a key motivation: understanding why some economies are rich and others poor. However, we also know that other elements matter to any notion of welfare. On the biological side, longevity and health are obviously important. For individual productivity, different measures of human capital like strength, education, and skills are important to making something of one's innate talent or ambitions. While these individual measures are partly the product of personal and parental investments, they are also the result of the distribution of services like education, safety net institutions, and health care. Thus, welfare also depends on access to public services. This chapter takes on the evolution of human capital in Paris as well as the structure of public services. Because the very rich can nearly always privately procure education, health care, and safety services, the diffusion of such services as public goods is of limited importance to them. Even though they may consume more such services than the average, the bulk of the expense goes to the "poor" or the middle class. This chapter thus focuses far more than any of the previous ones on the whole of the Parisian population, including those who left no estate.

Methodologically, it is useful to bear in mind that wealth and its distribution are not independent of the other activities we enumerated in the preceding paragraph. Societies where private wealth is high, for instance, have the resources to produce a high level of public services and sufficient protection of private property so that individuals are willing to accumulate resources. They remain rich when there is a flow of new ideas and enough are implemented. The flow of new ideas, in turn, depends on the level and distribution of human capital. It, in turn, is partly a function of public education expenditures. Similarly, as we saw earlier, both bequests and life-cycle savings are likely to increase when individuals have longer productive lives, which depends on social investments in health care. The endogeneity of private wealth, human capital, and public services is fundamental in the long run. Over a couple of centuries, Paris provides an important laboratory because different types of human capital change at different times, and different public services are deployed in different periods.

One can imagine two extreme cases over the long run and across space. At one end of the spectrum, the wealth distribution could be unrelated to measures of human capital, and public services could be either available to all or not at all. In this case, intervening in wealth inequality would be justified because we are morally averse to high levels of wealth inequality as prevailed in Paris before 1914, or because we desire to equalize consumption. At the other extreme, wealth distribution could be perfectly correlated with human capital, and public services are only available to a few. This would be most likely when the inherited wealth share is very high and the distribution of wealth is concentrated in two polar modes. In this second case, the arguments for intervention are more numerous because the lack of education for most of the population means a tremendous amount of talent is wasted. As we will see, Paris always lies somewhere between these two extremes. There is no doubt, however, that the decline in wealth inequality was at least partly driven by changes in access to public services.

To make this case, we rely mostly on spatial evidence, contrasting the rich arrondissements (7th, 8th, 16th) where the top 1% were concentrated, to the rest of the interior (1st–11th) and the other exterior arrondissements (12th–20th) and the near suburbs (arrondissements of Saint-Denis and Sceaux). Most of the data is derived from the ASVP and the BSM, whose publishers were keen to document the spatial heterogeneity of the city and its surroundings. We are fortunate that data on educational services (for boys and girls) and educational attainment (for men) are available for Paris for more than a century, often by arrondissement. We can also rely on prior research documenting the differential diffusion of water and sewer services. Our data allow us to establish age at death by arrondissement and top wealth fractiles over the entire period. We start with the most mundane perhaps of our indicators: burials.

11.1. Consumption in Death

That the very wealthy might have more lavish burials than the middle class makes intuitive sense. However, for most individuals, burial expenses are a small fraction of wealth, so they can choose how much to spend as part of a broader set of choices. These choices are likely influenced by religious attitudes and social expectations. Equally important is that some make their burial wishes known precisely in their wills, while others leave this to the discretion of their heirs, executors, or the city budget. We can check whether consumption at death is highly correlated with wealth at death. To do so, we rely on sources produced by the city.

The French Revolution upset many parts of the social order, not least the complex dance between religious and lay authorities in terms of assistance to the poor. After the revolution, the responsibility for ensuring burials was transferred to municipalities. They had to provide cemeteries and find ways to pay for the burial of the masses of lonely and destitute persons (Kselman 1993; Lassère 1997). Rich individuals' estates could be encumbered to pay for burial costs, and when individuals had families, so could parents or children, yet that left very large numbers of individuals without the means of a proper burial. Under the First Empire, and continuing for two centuries, the city established a concession regime whereby an entrepreneur leased the right to organize burials in the city's cemeteries. The leaseholder had to bury the poor, but in return, he could collect a burial tax on those who could afford a burial service. The leaseholder proposed a menu of burial services and accoutrements divided into nine classes. The cheapest in 1872 cost under 20 francs, while top-class services started at 4,000 francs—at the time, an average income was about 1,000 francs, and the average wealth among Parisian decedents was 24,000 francs. Additionally, the deceased or his or her family could choose a religious or civil ceremony.

Figure 11.1 displays the data reported in the ASVP from 1872 to 1967. We can clearly distinguish two periods. The first ends with WWI; it features a majority of all burials being either free or in the cheapest possible class (9). For those burials where families made an expense of 20 francs or more, most were less than 1,000 francs. Only about 5% of all burials involved more than 1,000 francs, and the share of expensive burials dropped steadily from 1872 to WWI. More episodic data for the 1820s and 1840s strongly suggest that the pattern that prevailed before 1900 was stable throughout the nineteenth century. After WWI, the rise of significant inflation during the war and continuing to the 1950s meant that prices for burials had to evolve. Not

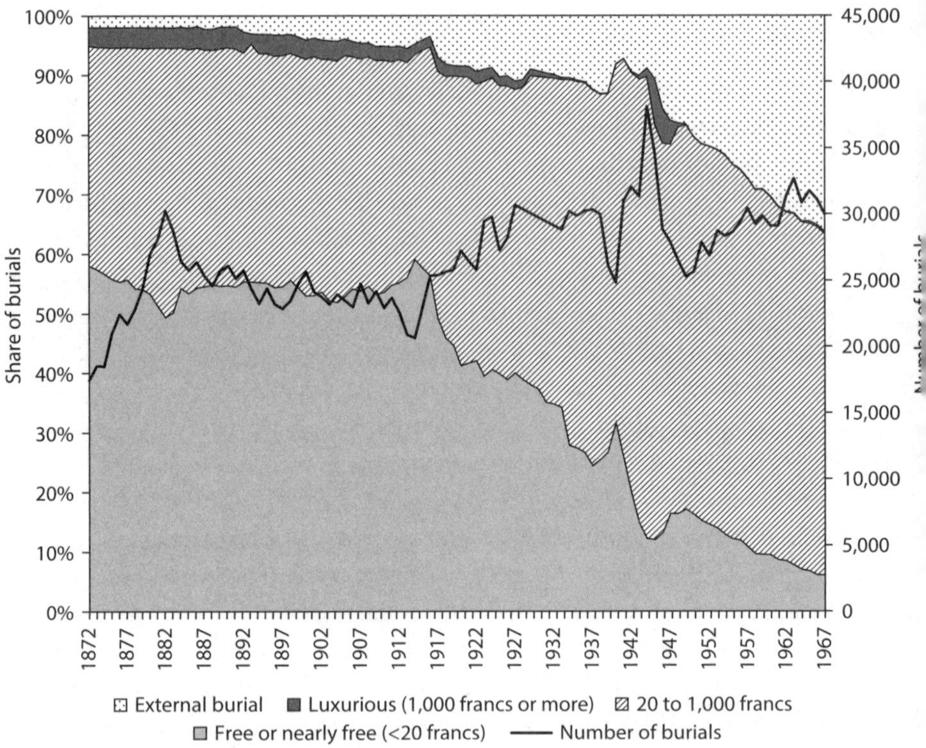

FIGURE 11.1. Burial classes in Paris. *Source:* ASVP.

surprisingly, tariffs, which until then had been published at decadal frequencies, were published at an annual frequency by the 1930s. The inflation also corresponded to a massive change whereby free burials went into a permanent and nearly continuous decline. The demand for cheap-class burials also declined. For those buried in Parisian cemeteries, the classes between 20 and 1,000 francs seemed to satisfy the majority. Another trend involves the acceleration of burials outside Parisian cemeteries, in part because those within the city limits were both expensive and full.

To show further patterns of inequality, we again divide Paris into three groups of arrondissements. The 7th, 8th, and 16th form the rich group because this is where more than half the top-centile wealth holders had a legal residence. The rest of the interior arrondissements (1st–11th) form a second group, while the exterior arrondissements form a third group. Figure 11.2 shows just how much more prevalent high-expense burials were in the rich arrondissements than elsewhere until 1927. In the rich arrondissements, high-expense burials amounted to 10% of the total, and most burials

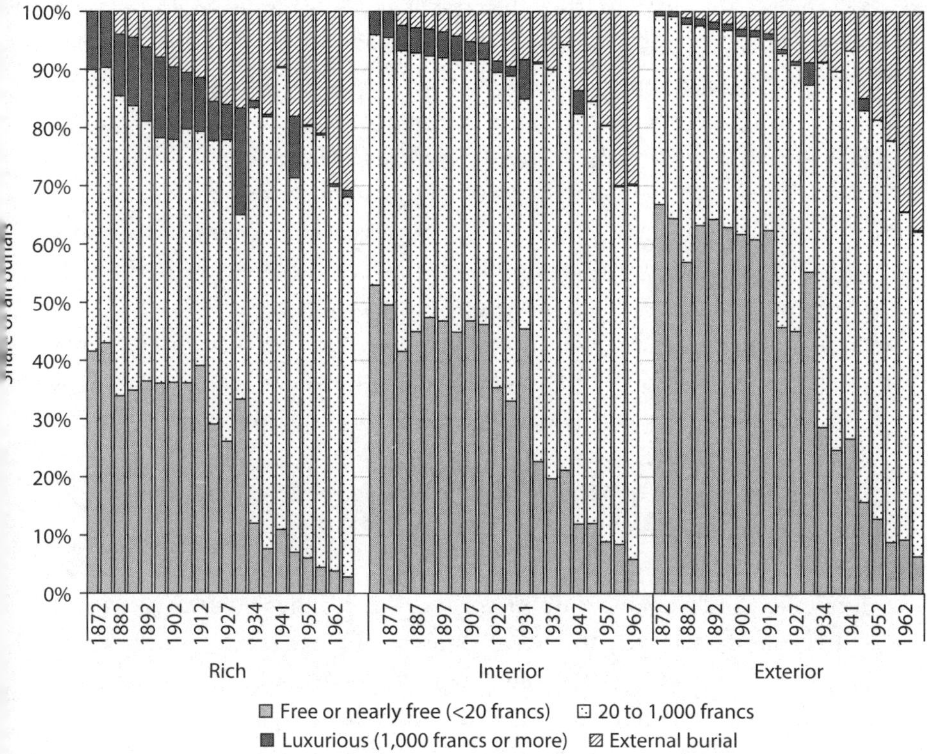

FIGURE 11.2. Types of burials by types of arrondissements. *Source:* ASVP.

involved expenses above 20 francs before WWI. Immediately after 1931, these areas essentially had no more free burials. The interior arrondissements show the same patterns but weaker: high-expense burials were less than 5% of the total, and a bare majority of burials involved expenses above 20 francs. After 1931, the fraction of free burials declined but did not disappear. Before WWI, there were few high-expense burials in the exterior arrondissements, and 70% of all burials were free. The spatial components of wealth inequality we saw earlier are repeated with burial expenses.

We have access to another pattern by arrondissement, namely the type of grave. The city provided concessions for individual graves or family plots, and the concessions ran for different lengths of time at various prices (those who could not afford a concession were buried in a common grave). The most prized concession is perpetual and lasts as long as a family member visits the cemetery and makes herself or himself known to the caretakers. In other cases, the grave is reopened after the lease expires, and the remains are moved to the ossuary in the catacombs to release the grave site. From

1882 to 1931, the city published the type of grave that individuals were buried in. It is important to note that the grave purchase is a separate expense from the burial ceremony. Individuals who belong to long-established families are likely to have access to a family concession and thus not bear that expense. Not surprisingly, the share of perpetual concessions was higher in the rich arrondissements than elsewhere. There, between a quarter and a third of all individuals were buried in perpetual concessions, a share that tended to increase over time, even as burial expenses fell. The share oscillated between 10% and 13% in the rest of Paris. Kselman (1993, 184), who examined an earlier period, points to an even smaller share of perpetual concessions, but he does not report the data by arrondissement. Overall, it is clear that the patterns of inequality in wealth were mirrored in the patterns of inequality in burial.

Interestingly, inequality in burial costs was almost as stark as wealth inequality before 1914 (the share of no-wealth individuals and no burial expenses were relatively close). The changes in the range of burial costs anticipated and were more dramatic than the decline in wealth inequality that began in Paris with the Great Depression. One reason for the fall in the share of free burials was the decline in infant mortality that started in the 1880s and accelerated in the twentieth century. Indeed, the share of free burials among children younger than 7 was about 80%, while the share among individuals older than 7 was just about 50%. For adults, the decline came from a greater ability of families to pay for burial expenses either because some insurance or pension schemes covered them or because children were more willing to contribute to the burial of their parents. For those burials above 20 francs, families chose how much to spend on burying their relatives, so private resources mattered. The data suggest that in the interwar period, lavish burials had come to an end. At the same time, fewer and fewer people were being buried using the redistributive scheme put in place under Napoleon.

11.2. Unequal Lifespan

The strong connection between wealth and lifespan is one of the most important elements that our data can address. If we take men and women together, we find massive differential mortality, with the poorer groups always dying at younger ages (the lines are strictly ordered by wealth). Figure 11.3 displays the data for men on the left and women on the right. The graph shows two periods: from 1807 to about 1860, the age-at-death differences between

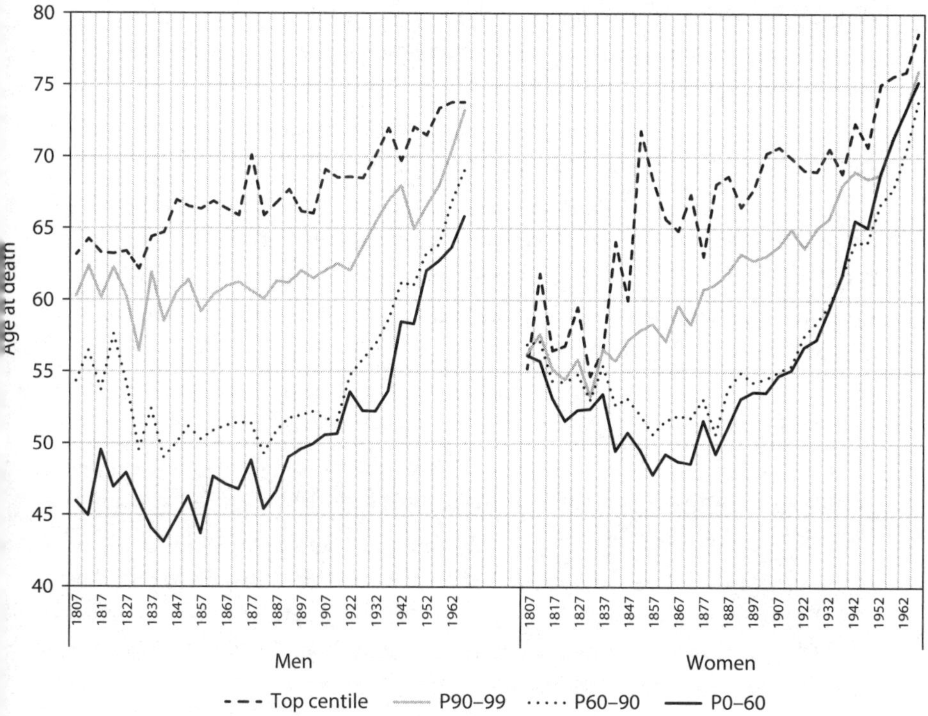

FIGURE 11.3. Age at death by wealth fractile for men and women in Paris, 1807–1972. *Source:* Estates dataset (see appendix 1).

groups widened, while between 1880 and 1972, the age-at-death differences shrank. At the same time, age at death increased for all groups. In general, individuals with sufficient wealth and strong bequest motives have positive savings rates at all ages. However, longer lifespans imply longer accumulation periods. Breaking the data down between men and women helps sort out the causality.

Figure 11.3 shows that men and women had radically different experiences, although they shared the same biological and disease environments. Between 1807 and about 1890, rich and poor men had huge differences in age at death of 15 or more years, and there was little change over time, though the members of the top decile (top two lines) saw some increase in age. For women, the picture is entirely different. Before 1832, there was little differential demography, and women's average age at death was falling. Then the top decile started to make large gains in age at death, while all other women continued to die younger and younger. Thus, by the 1870s, a differential mortality gap equal to that of men had opened up. After 1880, all groups saw their ages at death rise rapidly, but women made greater gains than men, and all four female groups died older than 70 after 1962. The picture thus tells a story of remarkable progress in life expectancy. It also shows that the gains were unequally distributed over time and sex. Given that men and women had similar estates overall, age-wealth patterns were likely driven by wealth, which helps preserve life in a variety of ways and which we will document in the section on health in this chapter. However, the differential patterns between men and women were at least partly biological since women alone face the mortality risk of childbirth. They were, consequently, the sole beneficiaries of the decline in childbirth, a slow and steady phenomenon that diffused down the social structure and from urban to rural areas.

Another pointer to the connection between wealth and other dimensions of inequality comes from military recruits. From the revolution onward, French men were subject to conscription. Before 1872, rich individuals had access to different forms of buying out, but at all times, a very large fraction of the cohort reaching 20 in a given year was measured, and their education levels were tallied. These data are available by arrondissement from 1890 to 1912, and we have supplemented them with data from the late 1820s and 1921 (the last year the archives were opened).

A few preliminary observations are necessary. Most obviously, the data concern only men. Second, the military conscripted soldiers based on the residence of their parents or legal guardians on January 1 of their conscription year. This means the data do not represent the cohort of 20-year-olds who lived in any one place or who were born there (though most were born and resided in Paris). Individuals who had migrated to Paris in their teens from other parts of France had to return to their "home" to be examined. While it is conceptually feasible to match the estate records to conscription records for men, it is complicated because the estate records do not tell us where individuals' parents lived when the deceased reached age 20.[1] The conscription records follow individuals into their forties so they could be called up in case of war, but no further. Finally, the last cohort for which we have this human capital information (1921) was born in 1901, and given that the average age at death was in the upper sixties, these individuals would appear principally in the estate data for the 1960s and 1970s.

To start, we split the data between the three rich arrondissements and the rest of Paris and examine three indicators: average height, the share of men shorter than 163 centimeters, and the share of men taller than 170 centimeters (see figure 11.4). We take the difference between the level in the rich arrondissements and the rest of the city each year. The rich

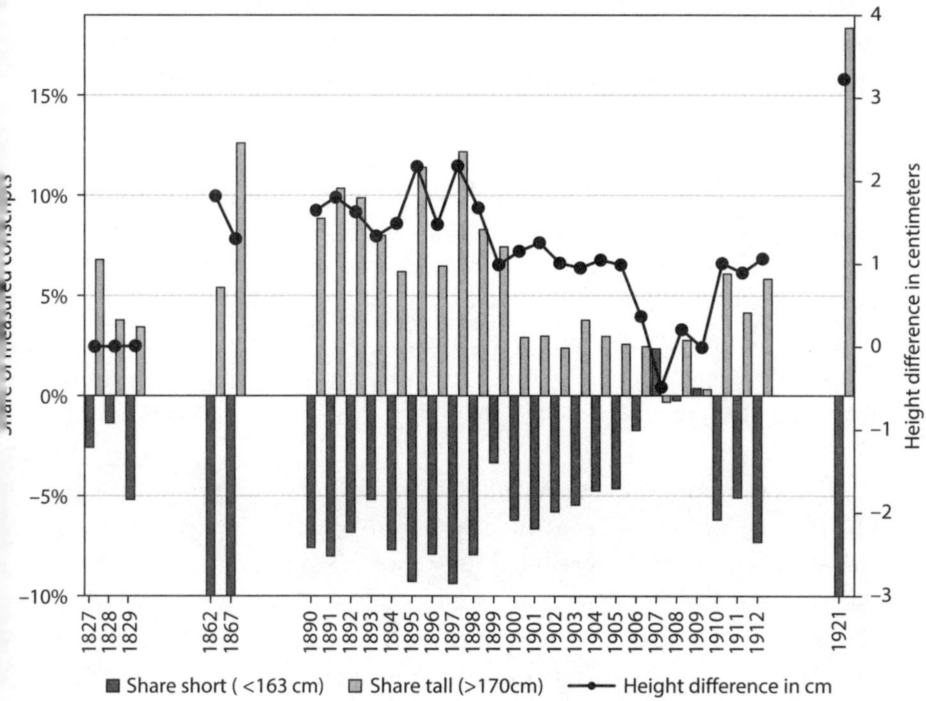

FIGURE 11.4. Differences in height among conscripts in Paris, 1827–1921. *Note*: Rich arrondissements relative to all others. *Sources:* 1827–1829: Military conscripts online dataset; 1862 and 1867: ADP D4R1:8–10, 318–323; 1890–1909: ASVP; 1921: ADP D3R1 270–73.

arrondissements generally had taller cohorts than the rest of Paris (the only exception is 1907). Because the height bins vary over time, we prefer a simpler approach: compute the share of short (less than 1.63 meters) and tall (more than 1.70 meters) people. Here, even in the 1820s, the rich arrondissements had a smaller share of short individuals (the bars are downward) and a larger share of tall individuals (those bars are positive). The rich arrondissements had cohorts that were taller than the rest.

The same differential pattern holds when we look at the fraction of the cohort that receives a degree (*brevet* or *baccalauréat*). In the 1820s, degree holders were exceedingly rare, and the samples were too small to establish any differential pattern. From 1882 to 1921, the rich arrondissements had a share of degree holders more than three times the share of the rest of Paris. Because the share of the population that lived in the three rich arrondissements was at most 10%, most degree holders lived in other parts of the city, and most of the cohort of the rich arrondissements were not degree holders by the eve of WWI. However, this larger difference in degree holders is exactly what we would expect, given the cost of education. The cohort of 1921 shows the same difference, except that it is magnified, as more than two out of five rich conscripts had a degree (and most of those were baccalaureate degrees).

In France, less than 2% of the cohorts who could have sat for the exams right after WWI received a baccalaureate. For Paris, the rate was about three times higher than for France as a whole, and for the rich arrondissements, more than ten times. Given that the state administered the exams to anyone who had completed the requisite schooling in a public or private school, these differences can only come from differences in family resources or from the supply of schools. We will examine the latter in section 11.5 of this chapter.

The state also asked individuals about their literacy and split them into up to four classes. Class 0 contained those who could neither read nor write. Class 1 tallied those who claimed they could read but not write. Class 2 involved those who claimed they could do both. Class 3 was added in the 1880s and involved those who had completed some primary education. Looking back at the 1820s, the lowest groups accounted for 20% to 30% of all conscripts. By the 1860s, nearly all Parisians reported that they could read and write. By 1900, the modal education was "can read and write" and "has completed primary education without a degree" (classes 2 or 3); these accounted for 90% or so of the population in Paris. So, there was progress over time. Class 0 (can neither read nor write) and class 1 (can only read) were at most 5% of a cohort (though somewhat more in the poorest exterior neighborhoods). This was also true for heights. Parisians were almost 15 centimeters shorter in the 1820s than today, but they had gained about 3 centimeters by 1900.

The information on conscripts' education changes over time. As a result, we cannot trace a fine chronology of educational inequality over as long a period as we would like. Before the 1870s, the military did not record degrees with any coherence, and it seems that only about one-third of each cohort was measured (including none of those excused for physical or medical reasons), so we face serious selection problems that vary over time. After 1921, the individual files are still closed to scholars, so we have no data, though the regime of measuring nearly everyone that was put in place after the defeat of 1870 prevailed well into the 1980s. To deepen our examination of the progress of human capital and its heterogeneity, we look next at health before returning to education in the last section.

11.3. Health and Wealth

When Emmanuel Le Roy Ladurie initiated the collection and analysis of heights, he sought a marker of health that would allow him to go back in time before there were censuses or health statistics (Le Roy Ladurie and Bernageau 1971; Le Roy Ladurie et al. 1969). Later, Robert Fogel (2004) used the same sources to document the decline in the physical sources of mortality in Europe. For Fogel, improved nutrition was the primary source of mortality decline before the twentieth century. Differences in income largely explain the variation in access to calories over space and time. To nutrition, we must also add the private provision of other services (domestic or medical) that improve health and lifespan, as well as the public provision of infrastructure and services that do the same. Importantly, the supply of privately purchased services can be tracked across Paris, as can the provision of several public services. In each case, the connection between wealth and access to these services is inescapable.

We start with two private services that improve health outcomes: domestic servants and doctors. The director of the statistical office, Bertillon, established the close connection between the wealth of a neighborhood and the number of female servants per household (see chapter 10). His tables show a strong positive association between rental values and female servants, and the same holds at the arrondissement level for wealth at death. He also documented a strong negative association between mortality measures and domestic servants' share. What Bertillon did not do was chart the evolution of domestic servants over time. The Parisian censuses show that from 1846 to 1901, domestics accounted for roughly 10% of adults (for comparison, the share of domestics in France in 1896 and 1901 is less than 3% of adults). This huge number is consistent with the extraordinary concentration of wealth in Paris. From 1901, the number of domestics began to decline, and the drop was particularly severe for men (who, by 1962, were at 10% of their absolute number in 1866). Thus, the women's share of domestics rose over time and passed 95%. Some of the rising share of women was due to a decline in the demand for coachmen or chauffeurs and to the relative scarcity of men after WWI. Had servants and nutrition been the only mechanism to improve lifespan, the city would have done poorly over time. Of course, other forces were at work since even though fewer and fewer households could afford live-in help, age at death kept rising.

The second factor in altering lifespan is the private provision of health services. To chart the market for such services, we used the city directory

(Annuaire du commerce) to apportion Paris's doctors to their arrondissements (of residence) in 1862, 1900, 1927, and 1951). Unlike domestic servants, the number of doctors in Paris rises over time and across every neighborhood.

The results show a massive concentration of health services in the richer arrondissements and nearby neighborhoods. In 1862, Paris had twice as many doctors per capita as the rest of the country (Léonard 1978, 87). Within the city, we find roughly twice as many doctors per capita (one for every 620 inhabitants) in the rich arrondissements than in the rest of Paris; the range is not perfectly correlated with the share of 1% wealth holders because the 6th has many, thanks to the medical school. The 1st and 9th (rich neighbors of the 8th) and the 16th, which remained relatively rural then, also had few doctors (less than one per 620 inhabitants). Across Paris, the range is astounding, with the poor eastern (12th, 13th, 18th, 19th, 20th) arrondissements all having fewer than one doctor for 3,000 inhabitants. The number of doctors in the city grew from roughly one per thousand in 1862 to one per 450 inhabitants by 1951. Over time, the growth in medical services seems quite smooth: there was a doctor per 900 persons in 1900 and one per 712 in 1927. However, most of the additional doctors were absorbed by either population growth or the needs of the rich arrondissements. In the 7th, 8th, and 16th, the number of doctors grew faster than the population (one doctor per 340 persons in 1900, one per 225 in 1927, and one per 161 persons in 1951). In the rest of Paris, there was essentially no progress before the Great Depression: there was still only one doctor per thousand persons in 1927. By 1951, the rest of Paris has finally passed the doctor density of the rich arrondissements of 1862. This market equilibrium was more a reflection of patients' ability to pay than their biological demand (that is, illness). That is exactly what basic economics would predict if doctors are sufficiently concerned about their incomes. To be sure, individuals were not restricted to seeing doctors in their arrondissements. However, the pattern implies that the poor had to do the most traveling and may have been discouraged from doing so because of the time and cost of getting to the doctors in the posh districts.

11.4. Disease and Wealth

Health is not entirely provided by private means. It also depends on public infrastructure and the extent of socialization of health care costs. The chronology of public investment in health is complicated and varies considerably across different forms of morbidity.

Here, we take two important causes of mortality and morbidity in the nineteenth century whose role in mortality decline has been well documented in the literature: waterborne diseases and tuberculosis. Consistent with the themes of this chapter, we document that the extent of these mortality risks was highly correlated with wealth and housing prices in the nineteenth century. We also document that following public investment in sewers (for waterborne diseases), tuberculosis detection centers, and later antibiotics, these sources of mortality essentially vanished. Given the scale of mortality from these sources, their reduction explains an important part of the reduction in mortality risk we documented earlier in the chapter. Equally important for this chapter's themes is that the interventions' chronology was radically different, with water a main focus of investment during the Second Empire, while the effort to rein in tuberculosis started in the 1920s. Consequently, we begin with water and borrow heavily from Kesztenbaum and Rosenthal (2016 and 2017).

In 1850, the city stank. Water infrastructure in Paris was quasi-medieval; water for urban uses came from many sources, including the Seine. The city had begun to provide piped water to buildings for a fee that depended on the pipe's diameter. There was also a large number of public fountains that those who did not have running water could use. For those who distrusted the public supply, there were also water carters. Dealing with waste and wastewater, in particular, was also antiquated. By 1880, the city had largely banned discharge into the street, so building owners had to procure the services of private night soil companies to take their waste away. The sewer system at the time collected filtered soiled water from some streets and discharged it partly in the Seine and partly in fields to the west of Paris. The private provision of waste removal services was profitable, both because building owners had to pay the companies and because night soil could be turned into fertilizer. This was far from the most healthful environment in general, but matters had to have been worse in the populous eastern neighborhoods, where housing was dense and many women worked outside the home and had less time for cleaning.

As part of the rebuilding of Paris that started under prefect Rambuteau (1833–1848) and accelerated under Haussmann (1853–1870), the city captured water supplies up to 100 kilometers from Paris from a variety of sources (Dhuis and Vannes rivers). It brought them to Paris in aqueducts that are still in use today. The initial effort was largely focused on increasing quantity rather than quality (though both occurred over time). It was not until the 1890s that the piped water supply was fully cleansed. Preston and van de Walle (1978) note the connection between cohort-level mortality declines and the diffusion of clean water. While it is likely that improved water supply reduced mortality, the process had too little geographic variation within the city to allow for statistical analysis across neighborhoods.

As part of Haussmann's plan, a sewer pipe was laid in every street in Paris over time, starting in the 1860s. Initially, the engineers were concerned with the system's ability to handle waste solids. Although buildings could connect their waste pipes to the sewer, they could only do so after passage through a filtration tank, which the night soil companies would periodically empty. A major innovation occurred in 1888 when direct connection to the sewers was made available for a fee. Take-up was left to the discretion of building owners. Over time, the schedule of fees was modified to encourage the owners of cheap or small buildings to connect. The city kept track of the number of buildings directly connected to the sewers by neighborhood from 1888 into the 1920s. Kesztenbaum and Rosenthal document that the sewer connection rate was highly correlated with rents, with the poorest neighborhoods achieving the same connection rate as the top quartile with a thirty-year delay. They also document that the time path of connection in a neighborhood directly correlated to the time path of mortality decline, particularly for the very young.

There is no doubt that these investments reduced mortality and morbidity. Whatever cultural indifference the rich may have had to the brief lifespans of the poor, the economics of human capital would suggest that, as capitalists, they had a serious interest in reducing ill health, as it would reduce lost days of work and time wasted training new workers. However, there is evidence that the rich were reluctant to engage in any social program to reduce waterborne diseases. The water system was the city's property, but the collection of fees from subscribers to the tap water was let out to a private company, the Générale des Eaux. The system's goal was that the fees would pay the interest on the bonds used to invest in clean water and for the water that the city made available freely. The same was true for sewers. In both cases, the connection was left up to the building owners. They, in turn, made a private benefit calculation: Would the connection pay for itself through increased rents? In the 1890s, building owners thought the answer to the question was yes in the rich neighborhoods and no in the poor ones. The result of a public intervention that provided a service with fees was twofold. First, individuals in rich neighborhoods who could afford the fees benefited soonest, even though they were the least exposed to waterborne diseases. Second, poorer neighborhoods did catch up with much delay. None of this is particularly surprising, yet it does point to another mechanism whereby inequality can persist because individuals living in rich neighborhoods with less exposure to waterborne diseases will suffer fewer interruptions to their lives and be more productive.

In the long run, the massive investments of the city in water infrastructure proved very successful. By the 1920s, the number of such causes of mortality had declined massively. The investment that started in the 1850s does point to a connection that is not often emphasized, namely that the wealth accumulation of the city in the decades before the launch of the Haussmann project was critical (cf. chapter 1). Without that, there would have been an insufficient tax base to pay for the public part of the infrastructure, little demand for water or sewer connections, and finally, little demand for all the new buildings that sprang up from 1860 to 1914. It is noteworthy that the bulk of the investment in water services was completed before WWI, when Paris's wealth inequality was at its highest. Clearly, this very unequal society was able to produce a public good, in particular, because it was not easy to provide a private alternative.

Tuberculosis has a different chronology and a strikingly similar spatial pattern. An airborne bacterium that is easily communicated from person to person, tuberculosis was a major killer in Paris throughout the 1930s. The ASVP and its predecessor allow us to trace the share of all deaths in Paris attributed to tuberculosis from 1865 to 1967. Tuberculosis was a cause of death for adults (it accounts for less than 5% of deaths of children younger than 14 in 1868 and in 1920), so the reduction in child mortality (35% of deaths in the 1860s and 10% in the 1920s) is independent of the course of tuberculosis.

In the last days of the Second Empire, tuberculosis was responsible for 15% to 20% of all deaths in the city. It was particularly devastating for adults between the ages of 20 and 45, and it killed more men than women. From 1872 to 1914, the numbers rose, and the share of all deaths attributable to tuberculosis was close to 21% in most years. Following the war, and without any real medical innovation, the share of death began to drop: it had fallen to about 10% by 1938. This reduction started with a municipal effort to detect individuals infected and send them to sanatoria or hospitals that the city operated in the provinces. Then, after WWII, the use of penicillin and other antibiotics brought the disease under control. Tuberculosis fatalities plummeted, falling to about 1% of all deaths when the series ended in 1967. By then, lung and throat cancer deaths were more than twice as numerous as tuberculosis deaths. In the aggregate, the history of tuberculosis is a good example of a successful social intervention (Guillaume 1986; Shapiro 1985, 157).

While there are famous tuberculosis fatalities, none is more perhaps than Alphonsine Plessis, the Dame aux Camélias made famous by Alexandre Dumas *fils* and further immortalized by Verdi's opera *La Traviata.* Alphonsine Plessis died in 1847, age 23, of tuberculosis, and she is remembered for the melodramatic tale of agony that took place in an unheated garret that features both in the play and the opera. In fact Alphonsine Plessis died in the swank neighborhood of La Madeleine and left a sizable estate of 43,000 francs. Deaths from tuberculosis were entirely anonymous because the individual cause-of-death certificates were not preserved. Using the same data as Kesztenbaum and Rosenthal to identify the wealth of neighborhoods, it is easy to show the massive social differentiation of tuberculosis. Starting in the late 1870s, it is possible to evaluate the impact of wealth by comparing it to the rental value in each neighborhood. In 1876, the strongest relationship was with the tuberculosis mortality rate (rental values explain 32% of the variance). However, the share of deaths reported as tuberculosis was identical in the poorest decile and richest deciles of neighborhoods. Hence, the mortality rate from tuberculosis is a finding on aggregate mortality (see figure 11.5). The richest neighborhoods had fewer tuberculosis deaths because of lower aggregate mortality.

In subsequent years, the explanatory power of wealth rose and reached 51% in 1912. In that year, the poorest decile neighborhoods had a share of deaths from tuberculosis that exceeded 27%, at the same time when the top decile was at half that ratio (13%). Combining this with the aggregate differential mortality leads the mortality rate for tuberculosis to be five times higher in the poorest deciles than in the richest ones. After WWI, the city put in place screenings for tuberculosis, and the connection between tuberculosis and wealth began to decline. By the late 1930s, the share of the variance of tuberculosis deaths explained by wealth had fallen by half, and by 1957, it had fallen by 90%.

The chronology of tuberculosis is entirely different from that of waterborne diseases; much of the improvement in water (and the attendant decline in mortality) preceded WWI, when tuberculosis was getting worse. Moreover, the share of deaths attributed to tuberculosis was either flat or declining in the rich arrondissement when it was increasing markedly in the rest of Paris (Shapiro 1985, 81–82, 144). The near elimination of tuberculosis by 1957 does not imply that social differences had been eliminated. The rich arrondissements had a tuberculosis rate almost 40% lower than the

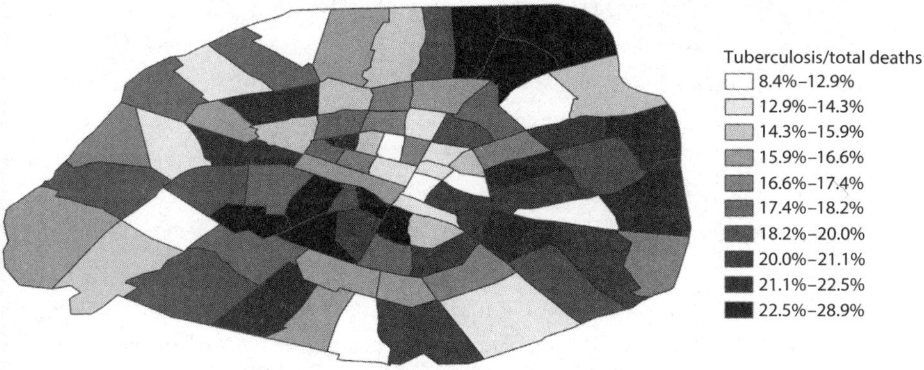

FIGURE 11.5. Tuberculosis deaths by neighborhood in 1878. *Source:* BSM 1878, 710.

rest of Paris. However, with an average rate of 2%, it was far less important than in 1912, when the average share of deaths from tuberculosis was 20%. The lesson here is not that social interventions banish inequality but rather, more realistically, that they can dampen differences and accelerate change.

11.5. Education

Education is the last element of human capital we can trace over a long period. We focus on three indicators: basic literacy, primary school attendance, and secondary school (high school) attendance.

As for literacy, like we saw previously, conscription data allow us to evaluate basic literacy for men. The conscript evidence is consistent with the view that Parisian men achieved basic literacy in the 1860s (Blum and Houdaille 1985). A different source, marriage records, allows us to parse both men's and women's ability to sign their contracts. About 20% of the population of Paris never married (ASVP sample years 1882–1967, see also Brée 2016), and celibacy was more likely for the poor than the rich. As a result, this source positively selects income and wealth and thus likely overreports this form of human capital. It does have the advantage over conscript data of covering women and a large fraction of migrants because, as we saw in chapter 10, they were more likely than not to marry in the city. The data start in 1865, when 10% of brides and 3.6% of grooms marked their marriage record with an X. Thirty-odd years later, in 1892, only 60 of 47,666 brides and grooms could not sign. The ability to sign was thus very broad even before the diffusion of free public primary education due to the Ferry reforms of the early 1880s.

The signature rate rose steadily, if slowly. By 1878, 98% of men and 94% of women signed; by 1887, 99.7% of men and 99.3% of women signed. At the same time, even this convergence did not eliminate inequality. While there had been improvement everywhere, the share of brides and grooms who could not sign was always lower in the rich arrondissements than in the poor ones, and always higher for women than men. However, for Paris at least, the Ferry reforms were not important to basic literacy, which came almost entirely before primary school became mandatory. The high rate we observe as early as 1865 came from both a high literacy for Parisian conscripts and a positive selection of migrants (Farcy and Faure 2003, 253). The spread of primary education in France was accelerated by the Guizot law of 1833. It mandated a school per municipality. Primary education grew further into the 1870s through a hodgepodge of municipal decisions, denominational investments, and private initiatives. Starting in the 1830s, Paris expanded public primary education on a two-track system (secular and religious) combined with a two-track private system (secular and religious). Using a 3% and then 7% addition to the basic direct tax levies, the city opened schools over time and collected tuition only from a subset of students who were attending primary school longer in order to prepare for the *brevet* diploma and gain admission to high school (*primaire supérieur*). As Jeanne Gaillard (1997) documents, during the Second Empire, the city was committed to a rapid expansion of primary education, but it faced two conjoined problems: in the capital's budget, building schools was expensive, and the expansion of Paris in 1860 meant that many schools had to be built. The deficit of schools in the new arrondissements was not resolved until the 1880s. There Gaillard documents that some classes were very large (in excess of 100 pupils). The other challenge involved the annual budget, including teacher salaries. In 1877, Octave Gréard published an extensive report on the state of schooling in Paris. It is the first that contains enough detail to ask questions about social differentiation.

In 1875, there were 275 primary schools, more than twice the number in 1831, and another 109 buildings that welcomed children younger than 6 (*salles d'asile*, precursors of kindergartens). Enrollment in public schools had grown from under 20,000 in 1831 to 115,000 (one would like to add attendance in private schools, but that was not tabulated until 1875). Gréard was an advocate of universal public schooling for all children aged 1 to 14. He pointed out that the school-age population had grown from 118,000 to 292,000. As a result, he argued, the city faced a substantial need for additional schools to allow universal enrollment from age 1 to 14. He strove to show that this need was particularly acute in the outer arrondissements (12th–20th), already underserved and growing rapidly. The city continued to build schools but, likely for budgetary reasons, did not take up Gréard's challenge of universal public schooling.

Instead, from 1881 to 1883, the government passed a series of laws, the Ferry laws, that made public primary schools free and, in principle, required attendance from age 6 through 13. The education ministry also assumed authority over the nonreligious component of instruction for both public and private schools. In return, the state paid teachers (including teachers in private schools). The capital cost of new schools was left to municipalities, but the cost was subsidized because the government provided low-interest loans. We can track the impact of the Ferry laws in Paris because, after a short interval, when private enrollments were not reported, the ASVP published the numbers of boys and girls enrolled by arrondissement for each type of school.

We review the sharp rise in enrollment from 1875 to 1915 and the differential take-up of education in the city in 1901.[2] In 1875, 157,000 students were enrolled in primary school, a tenfold increase from the Restoration period. This pre-Ferry growth was associated with a large increase in attendance since the population of school-age children increased by less than three times from 1817 to 1875, and as Gaillard points out, by the end of the Empire, 70% or so of school-age children attended school. The Ferry reforms also led to a substantial rise in enrollment. Indeed, between 1880 and 1905, the number of children in school increased by 60% when the number of children increased by 10%. Yet this growth continued the trajectory that had begun in the 1830s under Guizot. Equally important, the rise of free primary education changed the mix of public and private school attendance. Private school attendance by boys collapsed from 40% in the 1870s to 15% on the eve of WWI. Private school attendance by girls remained much stronger, falling only from 40% to 30%. After WWI, the same trends continued, and the enrollment rate seems to have been about 90%.

What of differential access? In 1880, on the eve of the reforms, the enrollment rate in the rich arrondissements was almost 12% higher than the rest of Paris managed. There is a caveat. Enrollment in public schools was higher in the poor arrondissements (32%) than in the rich ones (24%) because private schools were dense in the 7th and 8th arrondissements (in particular for girls). If we move to 1886, enrollments have passed 80%, with the rich having an average enrollment of 84% and the rest of Paris 80.6%. The gap between rich and poor had narrowed quite a bit, and nearly all the change came from public enrollment, which grew massively. The 1890 and 1895 data largely reproduce these results. At the end of the 1890s, the rich arrondissements had enrollment rates of 94% or so, and the rest of Paris was 10% lower, though that is 20% higher than the 60% of 1880. State intervention improved school attendance in the capital, even though it did not eliminate the large disparities in educational attainment.

The history of secondary education is entirely different. First and foremost, public high schools were rare and, for most potential students, expensive. Unlike public primary schools, which had a vocation of serving a neighborhood, high schools drew their pupils citywide, and we have no enrollment data by arrondissement. Like many other public services, the French Revolution and Napoleonic era rewrote all the rules. By 1820, public secondary education was rationed in a way that would remain almost unchanged until after WWI. In 1880, 6,037 male students were enrolled in one of seven public *lycées* and another 5,800 in private secondary schools. Thus, about 10% of the age cohort of boys attended high school and none of the girls (the first tally for both sexes was given in 1892 when 621 girls were enrolled in a lycée). To put the capital's numbers in context, that is about four times more than the rest of the country (Isambert-Jamati 1970, 376). In 1911, the ASVP (550–553) reported that 2,800 girls and 11,400 boys attended a lycée. The 1911 census reports 116,000 girls and 107,000 boys aged 15–19. Attendance was low for both sexes. Change did come, but only slowly. On the eve of WWII, the 12 lycées reserved for boys enrolled 22,750 students, while the 8 lycées for girls welcomed 11,001 students. This corresponds to more than 3,000 male and 1,500 female students for a given year of birth. A birth cohort is about 13,000 for men and 14,000 for women (little changed from 60 years before). At best, then, 25% of an age cohort of men in Paris attended high school and 11% of women. By the mid-1960s, another jump had occurred, and the number of students enrolled in lycées in Paris corresponded to about half the population at risk. The inequality between the sexes is stark. Through 1959, the last year when boys and girls were tallied separately, girls still made up only 40% of all secondary students (by the end of the 1960s, girls were the majority of high school students and baccalaureate recipients).

Nevertheless, from 1892 to 1921, the evidence points to a form of change that could increase rather than decrease educational inequality. Initially, public high schools were both fee-paying and selective. The elite could not hope to place all its boys in these establishments, creating a demand for private lycées (like Sainte Barbe or Stanislas in Paris). The slow and steady expansion of the supply of seats in public lycées and the admission of a measurable number of girls in the 1910s primarily benefited the elites early on because only they could afford the tuition. A different source on secondary education that comes from census data (Cagé and Piketty 2023) suggests the share of the adult Parisian population that earned a baccalaureate degree increased from less than 10% in the early 1960s to about 30% in the 1970s, 50% in the 1990s, and by 2022 more than 75%. During these decades, however, the frontier of education has shifted to higher education. While less than 5% of the Parisian population had a university degree in the 1960s, it is now almost two-thirds. However, although this growth was rapid, it was not progressing at a pace sufficient to lead to complete convergence. The difference between rich arrondissements, as we defined them in the 1890s, remained. In 1977, the share of adults with at least a baccalaureate was 43% in the rich arrondissements, 10% more than in the rest of Paris.

11.6. Conclusion

Half a century ago, Fogel (1967) argued for an economic history that made its case by the weight of diverse sources of evidence. He reasoned that economic history is not an experimental discipline, particularly regarding economic change. He also pointed out that we often lack the clean data we would like to have, so economic historians must proceed by accumulating a variety of evidence that bears on the issue. Paris's history of inequality is exactly in line with Fogel's eclectic empirical approach. We would prefer to have all of the inequality indicators at the individual level and tied to each individual's wealth at death. Lacking such data, we have tried to emulate Fogel's approach to understanding change over time. We focused on a set of human capital outcomes that were tabulated by the authorities and mostly published in the ASVP or the BSM. However, the underlying data about conscripts, burial costs, and baccalaureate holders survive at the individual level for many years. Future scholars will likely want to take advantage of major advances in digitization and matching to forge an even more detailed dataset than ours. Every one of our sources is in some way unsatisfactory. All highlight, in a different way, the extent to which Paris was unequal in dimensions other than wealth. Each also indicates that the heterogeneity in its outcome (say, sewer connection) is highly correlated spatially with wealth. It seems that the very high levels of wealth inequality are time and again associated with other dimensions of inequality.

It is important to note that some indicators bear immediately on the individuals whose wealth is recorded in our dataset while others do not.

For instance, when we measure tuberculosis mortality, burial costs, or age at death in 1952, the individuals in the data are the same as those for whom we collected wealth-at-death data for 1952. When we consider individuals married by contract or not, the inequality involves a population that will appear in the wealth data in the next thirty years only if they do not leave Paris and only if they remain in the arrondissement where they were married. The same is true for conscripts, except they have even longer to live. The time disconnect is even greater when we measure primary school attendance or infant mortality.

The cross-sectional variation is remarkably stable, something that limits the extent of inquiry. Serial correlation in spatial data is a general phenomenon in social sciences data, and it has given us a large body of literature that emphasizes persistence in many social and economic variables. Our advantage is that we can track this correlation over time and chart how our welfare indicators change. The value of the Fogel approach of using diverse indicators is that it allows us to show that each of our welfare indicators has a different chronology and that different forms of intervention lead to dramatic changes in the level of the observed variable (e.g., tuberculosis essentially disappears).

The levels of inequality in our different indicators evolve almost independently from each other and from wealth inequality. For instance, large-scale water infrastructure started and was largely completed before WWI, when inequality was at its highest. It was also a period when the cost of capital was low, which helped long-term investment. Though the tax regime at the time was strictly linear, the participants were keenly aware that infrastructure was largely financed by a tax on real estate rather than the octrois and thus weighted heavily on the very rich who owned nearly all of it (Guibert 1848). Most of that process was completed before the Great Depression, when wealth inequality began to fall. The rise and fall of tuberculosis also has connections to inequality because wealthier individuals are generally healthier and live in less dense environments. It is also true that individuals who are ill for long periods will likely see their wealth fall rather than rise. Primary education's long-term path is one of increasing schooling over time. In Paris, the reforms of Jules Ferry led to a break that advanced the fraction of children in school by a decade or two from the pace that the city on its own would have achieved. Secondary and tertiary education are at the opposite end of the spectrum from water infrastructure. For a century and a half or more, lycées welcomed a small fraction (less than 25%) of possible pupils, and it was not until the 1950s that a large break happened.

The extraordinary diversity of trajectories of these different elements of human capital allow us to conclude that the primary channel to explain cross-sectional or spatial inequality is from wealth to other outcomes rather than the other way around. For each of the other indicators of inequality, we find massive spatial differentiation where rich neighborhoods do better than poor ones. When the state or the city intervenes to reduce private costs, the spatial variation is massively reduced, even if it does not go away. The timing of intervention on sewers, tuberculosis, or secondary education is very different, and they all reduce inequality in their domain only. So, the only general pattern is wealth inequality. It is not surprising that being able to afford to live in the best neighborhoods in western Paris depends on income and wealth. If anything, this chapter solidifies our view that understanding the evolution of wealth and its distribution is essential.

The massive differences in the timing of improvement in social indicators also support our emphasis on public intervention. Fiscal innovation produced additional revenue through capital taxation. The initial reforms up to 1918 aimed to help with the military budget. Those following the armistice focused on debt service, rebuilding the damaged parts of France, and paying pensions to those who suffered in WWI. Most public sewer expenditures were undertaken before 1914 and paid for with 3% city of Paris bonds (and their service was getting cheaper by the day after 1914). These heavy investments were undertaken when capital was cheap after 1860. Subsequent expenditures to connect buildings were, in comparison, small and not paid for by the city. For housing, the city planned significant expansion of low-rent housing (*habitations à bon marché* and *habitations à loyers modérés*), but little was built during the interwar period. The same, as we saw, was also true for schooling. Fiscal stringency did not allow any rapid increase in spending on human capital, and the number of students grew quite slowly. As we can see, the real estate market and secondary education underwent a profound transformation after 1945.

The combination of a stable structure of spatial heterogeneity (the rich arrondissements stay richer than the poor ones) and completely different temporal trajectories of improvement of social indicators results in one dimension of inequality (wealth inequality) that is extremely persistent. In contrast, others are capable of considerable change. Spatial persistence is not fate. Looking back at the Parisian experience raises important questions for the literature on social capital and public spending. Social capital is often thought to be a necessary condition for increases in public spending because if individuals do not share the same goals and do not trust each other, they will prefer privately provided services to public ones (Alesina et al. 1999; Goldin and Katz 2000; Putnam 1994). Given the massive rates of migration, the radical difference in wealth across the city, and the persistent political conflict that produced a revolution roughly every generation from 1789 to 1870, it would seem that Paris would have low social capital and poorly functioning social services. Yet the history of the city argues otherwise; in fact, it suggests that social capital in successful economies is not found in culture alone, it is embedded in the infrastructure and in the labor force that provides social services.

Conclusion

CAPITAL AND CAPITALS IN THE LONG RUN

A Parisian looking back from 2025 can contemplate the result of the eight decades of economic growth and wealth accumulation that started in 1945 with the end of WWII. That perspective is thus similar to the view that people might have had around 1900, following the eight decades of growth that started with the end of the First Empire in 1815. Even the slow rate of accumulation of the nineteenth century (2.3%) per year had led to a sevenfold increase in private wealth, enough to bring about a radical transformation of society. In the last two decades, however, debates about wealth inequality have come to the forefront among academics and the public alike because the growth that has occurred since WWII has most certainly had uneven benefits. Beyond the long-run averages, historians are also aware of the importance of shocks to these dynamics. In particular, the shocks of WWI, the Great Depression, and WWII were so severe that nearly all the wealth gains from 1807 to 1912 were lost by 1947. Those shocks have an antecedent in the French Revolution (1789–1815), though we do not know how much wealth fell then. Our dataset of 780,295 Parisian estates over 170 years has the merit that it allows us to capture both the long-term structures of wealth change and the impact of important shocks. In this conclusion, we will use our findings for the period 1807–1977 to interpret the experience of the last half-century.

After a brief review of the key findings of the book, we turn to the evolution of wealth and inequality in Paris in the last 50 years. We point out that the spatial persistence that we identified in chapters 10 and 11 has continued to the present. The 7th, 8th, and 16th arrondissements continue to be richest. At the same time, wealth concentration has likely resumed an upward trajectory despite the fact that the fraction of Parisians who die with some wealth has also continued to increase. We then turn to the problems posed by studies that are multisecular—simply put, everything changes. Change affects what is recorded by tax officials, and thus, it impacts the very nature of the wealth we can measure. Change also alters wealth itself. For instance, there were no automobiles in 1807. Thus, no one owned them, and there were also no automobile factories, auto-repair shops, gas stations, and very few paved roads or parking structures. Change alters the population that holds that wealth in terms of skills, occupation, and age at death. This chapter's final section makes the case that for economic history to be relevant, it has to do more than either tell the story of economic development or do the statistical work of documenting the impact of specific policies. It has to merge these two efforts to consider the political economy of institutional change.

C.1. Wealth and Its Distribution in the Long Run

Our contribution to the analysis of the evolution of wealth and inequality is twofold. First, we offer a new dataset that is more detailed, larger, and covers a longer period of the wealth of a specific population than any dataset that has come before. In our case, that is the wealth at death of Parisians. The choice of Paris as a locus of investigation was partly practical since the records are all held at the city archive, while the rest of the French records are dispersed in 90 or so locations. It was also consequential. As we show for 1832 and even more so for the WWII period, shocks hit Paris particularly hard. This is part of a general phenomenon: shocks are spatially differentiated, and new institutions matter more in some places than in others. More importantly, a very large fraction of the wealthiest people in France maintained a legal address in the toniest neighborhoods of Paris throughout our period of study. So, the capital is a very good place to study the evolution of wealth and its distribution in Paris, in France as a whole, and between the capital and the country. Assembling the wealth at death of every adult decedent in nearly all the years starting in 2 or 7 from 1807 to 1977 allows us to offer scholars a new tool for research. As described in the introduction, the dataset involves 19 complete cross sections of wealth for the nineteenth century and another 15 for the twentieth century. Using this dataset, anyone interested in placing an individual or a group in a context can now do so.

Our effort's second contribution is an initial analysis of the data. This book has largely focused on the cross sections. We made a limited effort to match these data across the generations or to other sources like the censuses, as many scholars do today. We view our effort not as closing opportunities for understanding the evolution of inequalities but as a step toward systematic inquiry. The whole dataset is available for scholars who want to push the research in new directions. Readers will find additional details in the appendices. This book has five key findings. These concern (1) the time path of wealth, (2) the share of individuals who left an estate at death, (3) the share of wealth held by top estates, (4) the inherited wealth share, and (5) the structure of wealth itself. Each reflects the history of Paris and France in different ways. Although it is clear that the great political and economic shocks of the twentieth century created the most radical changes over time, a careful reading reveals other important forces at work.

We start with our first finding involving the most volatile series we have assembled: average wealth per decedent in 1912 francs. From a low in 1812 of 5,400 francs, wealth per adult decedent had multiplied more than eight times when it peaked at 48,400 francs in 1902. It then fell by more than 85% to a low of 6,900 francs by 1947; by 1972, it had returned to 30,600 francs. The 1807–1902 rise in wealth reflects a twenty-six-fold increase in total wealth at death in Paris, moderated by the threefold rise in the number of individuals who died each year. The 1902–1947 decline in aggregate wealth (90%) is more severe than the per capita drop because of the decline in the number of individuals dying each year, which fell by a third (a combination of increasing lifespan and declining city population). The post-1947 recovery occurred when the number of deaths was almost constant, so average and total wealth had the same time path. Our microdata show that the post-1947 recovery was understated because of increasing tax avoidance and tax evasion by the very rich. Unlike the published data on taxable wealth, our individual-level datasets cover all reported wealth, including taxable and exempt. It also accounts for the increasing top wealth share that is legally tax-sheltered.

The huge long-term variation in average wealth at death reflects a wide combination of phenomena discussed in the chapters, including economic growth rates, capital returns, taxation, aging, and financial market deepening, as well as destruction and nationalization of assets. At a very broad level, the history of Parisian wealth at death matches the growth process of the city, the country, and the globe. After the nineteenth century's long period of growth, WWI ushered in brutal change: average wealth fell almost 4.3% a year over 45 years. It reminds us that destroying wealth is far easier than building it. The nadir reached in 1947 reflects both a real decline in the wealth stock during WWII (plant, equipment, and buildings were not maintained, and investment was low), a price effect due to concerns about the future of private property, and another price effect due to high rates of taxation. The post-WWII recovery reflects a high investment rate and a decline in expropriation risk.

Our second major finding concerns the share of individuals who died poor—that is, with no estate. The share of adults who died in Paris and left no bequests of interest to the tax authorities was more than two-thirds through 1922. Then, an inflection set in, and the share of individuals who died poor fell. By 1977, we estimate that more than 60% or so of adult Parisians died with an estate that would have been recorded had there been no minimum threshold (almost 20% had an estate that owed no tax).

Taking both series together, it is clear that in the long nineteenth century (1807–1912), prosperity did not lift all boats. Instead, one-third of the population had all the wealth while two-thirds died with nothing. After WWI, wealth declined rapidly and then rose again, while the proportion of individuals dying with an estate increased steadily. Wealth's rise and fall reflected nearly entirely accumulation decisions and market valuation of estates in the top decile (which alone accounted for 90% of all the wealth). In contrast, the growth in the share of taxable estates reflected voluntary and involuntary savings over the previous half-century in the rest of the population. Even in a gentrifying Paris, a large group, more than one-third of the population, left no bequest and seems not to have been covered by retirement pensions. On the one hand, this group reminds us of the sharp divide that continues to exist in most societies between the haves and have-nots. On the other hand, we must admit that the data we use do not capture their behavior.

Our third finding involves the distribution of wealth. Like the share of those who died poor, concentration only changed course once during the Great Depression. The top decile accounted for at least 96% of all wealth until 1927. From then on, its wealth share slowly declined to about 80% by the 1970s. Note that the wealth measured here takes into account tax-exempt assets. Admittedly, the decline reported here is overstated because we do not see assets held in tax havens. There was, however, a significant reduction in inequality. Moreover, even though some assets are missing at the top, the fact that individuals below the top decile held at least 10% of all the wealth after WWI is a major change from before WWI when their wealth in most years accounted for only 2% of the total. One would have to think that undeclared foreign assets accounted for a huge fraction of the top wealth holders' assets to significantly reduce the importance of the rise of Parisian estates in the sixth to ninth decile, what, for lack of a better term, we can call middle-class wealth. At the same time, Paris has always been very unequal. This, we think, is not specific to the city itself but to capitals that tend to concentrate a country's core financial, entrepreneurial, political, and cultural functions and, more generally, to other very large cities.

Our fourth finding concerns the inherited wealth share. It dropped from above 65% in 1922 to about 60% in the interwar period and about 50% after WWII. Here, the underreporting of assets we mentioned above likely makes the second drop seem larger than it really is. Indeed, the inherited wealth share by fractile was relatively stable (25% or so below the top decile, 50% for those in P90–99, and 65% or more for the top 1%) but increasing with wealth. As a result, if we could observe all wealth, we would also find a higher inherited wealth share than we report. Given the dramatic decline in wealth observed from 1912 to 1947, one might have surmised that the next period would be a halcyon time for new wealth and new enterprises and a low point for inherited wealth. To be sure, 1947 and 1952 had the lowest inherited wealth share, but by the 1960s, inherited wealth was on the rise again. It would, however, take decades to return to a long-run equilibrium because most individuals who died in the 1960s and 1970s had received only "small" inheritances in the 1930s and 1940s when asset values were very low.

Over the longer term, what is striking is that wealth endures even when its value fluctuates dramatically. The persistence of inherited wealth should come as no surprise. Many individuals like to bequeath, and in affluent societies, a lot of durable capital has to move through the generations. Wealth embodied in objects and buildings endures and must be transmitted inter vivos or when individuals die. For Paris, at least, the large value and concentration of wealth at death tells us that individuals hold on to vast amounts of wealth until they die and that this basic fact does not change much over time. The desire of individuals to hold on to their wealth implies that the inherited wealth share will be substantial whenever societies like that of Paris and France are rich. Indeed, the younger generations would prefer to receive their bequests earlier (as we saw in chapters 2 and 7) because that is when they are most capital-constrained.

Taking the four series together, we come to a surprising fifth point: the rise of wealth in the sixth and seventh deciles was poorly connected to the decline in the absolute value of the wealth of the top 1% and its share in total wealth. While the rise of a wealth share of about 15% below the top decile reduced the wealth share of the top decile, it had almost no effect on the very top (0.1%). On the one hand, the dynamics of middle-class wealth have involved human capital, life expectancy, retirement, and real estate. Institutional innovation in these areas (e.g., public schooling, compulsory savings, and condominium legislation) played an essential role in the growth of middle-class wealth. On the other hand, the wealth dynamics of the wealthiest people depend entirely on the size of the companies they control and in which they hold significant stakes. Time and again (and often with the state's help), entrepreneurs have built firms that are much larger than their competitors, and when they have succeeded, their wealth has become very large indeed. In the nineteenth century, this was true of the Rothschilds and the Schneiders, for example. Other entrepreneurs followed them in petroleum, automobiles, pharmaceuticals, and perhaps more specifically, in the French case, consumer goods and luxury items. We will return to this point because the European Union and globalization have renewed the returns to scale.

Our large samples of portfolios lead to our last conclusion, namely that Parisian portfolios from the 1840s to the 1960s were very different from those in the rest of France and, in most ways, far more modern. For legal and practical reasons, Parisians overwhelmingly owned whole buildings rather than specific dwellings well past WWI. As a result, for those below the top 5%, the real estate portfolio share was very small, and for those below the top decile, it was not until after WWII that real estate diffused into their holdings. More generally, the real estate share remained fairly stable and low in Parisian estates compared with France as a whole. As a consequence, Parisians owned a lot of financial assets. A second and often overlooked aspect of Parisian portfolios was that the share of wealth that produced no income (movables, bank accounts, and the like) fell with total individual wealth, which meant that the rate of return on wealth was also increasing in total wealth.

Taking all these findings together, the history of wealth and inequality in Paris is very rich and involves multiple processes whose chronologies differ. These include the rise of modern financial markets starting in the 1850s and 1860s, demographic changes and reduction in life expectancy differentials in the following decades, the creation of a pension system over many decades starting in the 1850s, changes in taxes beginning around 1900, and, of course, technological change.

C.2. Wealth After 1977

At the time of this writing, the archives' holdings of estate tax documents end in 1977, so we cannot rely on our sources to detail the evolution of wealth in the city after that date. Nevertheless, the recent evolution of wealth and inequality is key to putting our findings in perspective. In the last fifty years, the city has become, in some ways, less important to France and its surrounding region, as it fell from 23% of the Ile-de-France population in 1975 to 17% in 2021. At the same time, French economic growth was quite sustained, and real per capita income in France almost doubled in real terms. Nevertheless, Paris's relative aggregate decline in population or wealth share says little about the relative incomes of Parisians. Cagé and Piketty (2023) suggest that Paris has increased its income advantage over France, with per capita income jumping from 2.5 times the national average to above 3. It is, therefore, likely that the wealth of Parisians has grown in recent decades, but what about inequality? To answer this question, we proceed in four steps. First, we review the economic transformations of France and Paris over the past five decades and examine two very imperfect sources (top wealth holders in France and real estate prices in Paris and France) to argue not only that the spatial patterns we highlighted in chapter 10 persist to this day but also that it is extremely likely that wealth inequality as defined by the top 0.1% or top 1% has indeed risen since 1977. Using a sample of death registrations for 1986, we also suggest that the continuing evolution of the Parisian population implies that an ever-increasing proportion of the population is dying with some assets. In this respect, the last five decades show great continuity with the 1945–1977 period.

France (and Paris) have undergone radical changes in the last fifty years. Indeed, 1977 was quickly followed by the 1981 victory of François Mitterrand and the last wave of nationalizations (including most banks). Five years later, as the right won the next elections, a wave of privatization began that continued for three decades. Given the centralization of the financial market, much of the income from these capital transformations accrued to Parisian bankers. Yet, at the same time, the state has continued to jump into the economy in various ways. The French have also maintained a much larger social safety net than Anglo-Saxon countries and, as Piketty (2013) has noted, escaped the rise in income inequality experienced by the United States and the United Kingdom. The French top 1% income share fell from about 20% before WWI to about 10% in 1945, and it has stayed at about that number since then. The same stability does not characterize the wealth distribution.

According to the World Inequality Lab's results, the top 1% held about 50% of France's wealth through 1922. This share then fell precipitously, reaching a nadir of 15% in 1982, and it has since rebounded to just shy of 27% (Alvaredo et al. 2018). This rise is smaller than what was experienced by the United States where the top share is now 35%.

The drivers of these changes are complex (Piketty 2011; Piketty and Zucman 2014). We are particularly interested in those that matter to the differences between Paris and France. The low point in inequality in the early 1980s is hardly surprising, as the value of large enterprises was depressed by the socialist electoral victory and the ensuing nationalizations. The socialists' 1981 program failed to bring about either the collapse of the French economy or a significant surge in employment. Instead, by the mid-1980s, France embarked on two decades of reform that liberalized trade and financial markets. This included increasing integration within the European Union (or its predecessor, the Common Market). More broadly, leading French firms became more dependent on international sales and production. In a way, the internationalization of the French economy after 1980 continued and possibly accelerated a process that had started in the 1960s. Change in financial markets, however, was more radical. The process began when Mitterrand nationalized all banks with capital over 1 billion francs. The goal was to push the financial sector to provide more resources to French industry. However, that goal ran into a stark fact. Nationalization stripped the public sector of investment resources, and the banks could provide no more resources than they would have had they remained private. The alternative was that the banks ignored investment rules, and the government (their owner) would have to absorb investment losses. Even the socialists mainly chose the first option. To be sure, there remained essential components of state finance, for instance, with the Caisse des Dépôts, but quasi-market rules prevailed. At the same time, the state needed access to financial markets for its own needs and faced pressures from the largest banks who wanted to compete in the European and global markets. The result was the reprivatization of the capital market and the adoption of open capital market rules in 1985. To be sure, some of this was due to the radical reforms pushed by Margaret Thatcher on British financial institutions. The last key element in economic policy was the global declining interest rate regime from the 1990s to 2021, something that echoes the nineteenth century. Low interest rates are a fantastic gift to equity holders because firms can finance their expansion with debt, leaving a greater residual to the shareholders. The logic of low interest rates applies to a wide variety of asset classes, including real estate, small-

and medium-sized businesses, and large firms. Because much of these gains will be reflected in asset prices and thus are passive, they cannot be offset by changes in taxation unless the assets are sold or inherited.

We expect these policy changes to have increased private wealth. Their impact on wealth inequality would be nil save that richer individuals have more flexibility and invest more of their resources in high-return assets. Beyond the financiers who benefited directly from the rebirth of financial markets after 1985, the market also favored firms whose products were in high global demand. Those firms grew much faster than others, and their principal owners saw their wealth increase faster than anyone else. When considering how these changes might have affected the distribution of wealth in France relative to Paris, we have to answer two questions: Were the portfolios of Parisians more likely to experience gains than those of individuals in the rest of France? Did Parisian real estate and Parisian firms experience greater rates of capital appreciation? Simple economics would suggest the answer should be no to the first and yes to the second. By 1980, buying a share of Peugeot or Siemens was as easy in Aurillac, a small town in the center of France, as in Paris. The reason that wealth and inequality may have charted a different path in Paris than elsewhere lies in the extent to which the city and the firms headquartered there have benefited from increasing returns as the capital of France, or, more generally, as a metropolis.

As noted by many other scholars, detailed data on wealth distribution by locality after 1977 are scarce. The periodic samples of estates (10,000) produced by the statistical arm of the finance ministry are too small to establish the distribution of wealth in Paris with any reliability (and, in fact, for France as a whole). Moreover, the wealth tax imposed by the socialists in 1981 covers an insufficient range of wealth and range of assets. So, we cannot answer this question directly. Nevertheless, we can access two sources that speak to our hypothesis that wealth inequality is on the rebound. The first, and the more suspect, involves the distribution of top fortunes in France as reported by the magazine *Challenges*, which we have digitized for 1999, 2000, 2005, 2010, 2015, 2020, and 2023. These lists detail the 500 top French fortunes as uncovered by the magazine's staff. We have analyzed these lists, which seem to comprise the equity positions of controlling shareholders in publicly traded and private firms. We made adjustments to them so they better reflect the wealth of individuals living in France. For reasons never explained, the data usually refer to families rather than individuals. In recent years, the lists include individuals whose businesses are not in France. We, therefore, have removed people like Evan

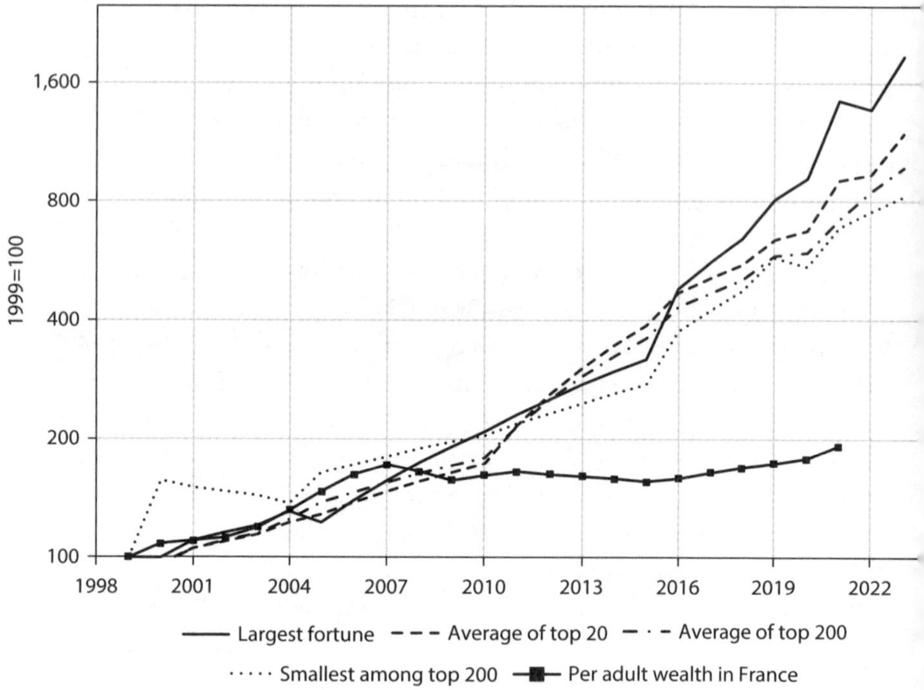

FIGURE C.1. Top fortunes in France, 1999–2023 (1999 = 100 for each rank). *Source: Challenges*, "Les 500 plus grandes fortunes de France," 1999, 2004, 2005, 2010, 2015, 2019–2023.

Spiegel, the founder of Snapchat, from the list because he is not a French resident and Snapchat is not a French company. We have also traced the headquarters (*siège social*) of nearly all the companies and removed those firms whose headquarters are abroad, even if French citizens run them. On the one hand, this data source underestimates wealth in two ways. First, our data show that individuals do not concentrate all their wealth on their business and, as they get older, they tend to diversify. Second, individuals with large, diversified fortunes do not appear in the dataset even though there are, in all likelihood, diversified individual fortunes of 150 million euros or more (enough to be on the list). On the other hand, the magazine's decision to focus on families rather than individuals overstates wealth, particularly when the current owners are several generations away from the founder. Figure C.1 reports these findings and compares the evolution of these top equity positions with average-per-adult wealth in France from 1999 to 2023, as reported by Garbinti et al. (2018). We focus on the most significant fortune (rank 1), the average among the top 20, the average among the top

200, and the smallest of that group, and scale everything by their values in our first year, 1999. We find that the top 200 fortunes have grown 3.7 times faster than average wealth—undoubtedly a sign that inequality in France has been increasing in recent decades. The top 20 equity positions have grown almost five times faster than average French wealth, so concentration at the very top is also likely growing.

The 200 equity positions that we have culled from the *Challenges* lists represent a tiny fraction of the 50 million or so adults in France, even allowing for the fact that there may be more than one claimant to that fortune through marriage and inheritance. Even if we took in all the 500 fortunes in the magazine's list and assumed each belonged to two individuals, this would amount to knowing the principal wealth holding of only 2 out of every 5,000 French people. The data are thus suggestive, rather than conclusive, of a rise in inequality for France.

Returning to the hypotheses about how economic change affects the relationship between Paris and France, we coded the location of the top wealthholders' firms' headquarters. We identified the top fortunes associated with firms headquartered in Paris and its orbit. The metropolis's share of top fortunes seems to be declining (in 2010, 18 of 20 top fortunes came from firms in greater Paris; in recent years, that number has fallen to 14). Among the top 200 fortunes, Paris's share is steady, between 60% and 65%.

Beyond just the increasing inequality it documents, the figures on the valuation of fortunes reveal another, more specific lesson about the relationship between Paris and France in our time. What is unique about the recent experience is that many of the top firms are old in name and young in structure. They are old because many are enterprises with names that hark back to the nineteenth century or before. They are young and modern because they have a global reach, and most seek to reach a much larger fraction of the world's population than a half-century ago. Luxury and beverage firms are prominent examples. In 1999, 6 of the 20 biggest fortunes were in these sectors; by 2023, it was 11, and throughout the quarter century that the data span, almost a quarter of all top 200 fortunes are in these sectors. These firms are dependent on an international market and easy access to capital. Given their customer base, had they remained focused either on French customers or customers who visited their stores in France, they would never have reached capitalizations in the tens of billions of euros. The international market is what makes the difference between a major fashion house of the first three decades of the twentieth century, Soeurs Callot, and a leading firm in post-WWII Paris, Christian Dior. The first depended entirely on demand for one-of-a-kind clothing and operated a pure haute couture enterprise. It could not survive the combination of the Great Depression and the death of its principal designer. Dior also started out as a pure haute couture house; it now retails a wide variety of items to a global clientele. It has also continued to grow long past Christian Dior's 1957 sudden passing. Dior and many other elite luxury brands are now part of luxury conglomerates. The construction of the luxury megafirms depended on favorable financial markets that facilitated debt issuance, acquisitions, and mergers. While we did see the rise of large-scale retailing with the Bon Marché in the late nineteenth century, a department store, however magnificent, cannot compare in scale to the twenty-first-century global luxury firms. Luxury is but a poster child of the forces at work in the modern economy. And among the French firms that succeeded in meeting these global challenges, being in Paris proved valuable.

Because the data on fortunes are inaccurate, we complement it with a second source to evaluate the path of wealth and inequality in Paris after 1977. The second set of data involved real estate prices, which have been collected by Julia Cagé and Thomas Piketty (2023). For our analysis here, we combine the transactions of the rich arrondissements (7th, 8th, 16th), and we split the remaining transactions between those that involved dwellings in interior (1st–11th) and exterior (12th–20th) arrondissements. We also tally the transactions in the two départements most closely associated with the old suburban areas around Paris, the Seine-Saint-Denis and the Hauts-de-Seine, which we break up into the six priciest municipalities and the rest of the département. We normalize all the series by the average price of real estate transactions in France, so when a line hits four, it implies that the locality has transactions that are four times as expensive as the average for that year in France. Figure C.2 shows that all the Parisian arrondissement groups have much higher real estate values than France. The figure also shows that prices have risen faster in the Paris region, except for the Seine-Saint-Denis, than in France as a whole. These prices support the thesis that wealth in Paris has been growing and has remained higher than in the rest of the country.

There is also a clear demarcation between the rich arrondissements (7th, 8th, 16th) and the rest of the localities in and around Paris. The pattern repeats the rankings that we have seen time and again. Indicative that inequality within Paris has likely risen since 1977, prices have risen slower for the exterior arrondissements (12th–20th) than for the rest. The interior arrondissements (1st–11th) have seen substantial appreciation, but that is

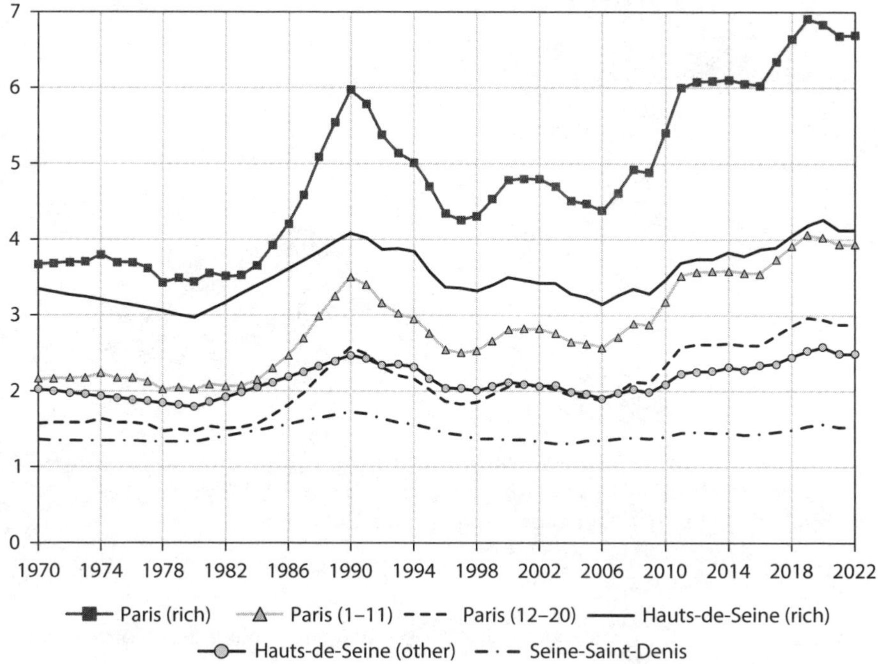

FIGURE C.2. Real estate prices in and around Paris (relative to the national average). *Source:* Cagé and Piketty 2022.

dwarfed by the price increases for the wealthiest three (7th, 8th, 16th), as defined by the share of top 1% estates around 1900. To be sure, there has been some change. The 7th arrondissement is now more desirable than the 16th or the 8th, but overall, recent generations have not changed the geography of wealth in Paris. Similarly, there has been quite a lot of gentrification in many neighborhoods of Paris. However, gentrification has not changed the order of things within the city; it has just made the northeastern arrondissements less like the lower-rent communities on the other side of the city's boundary.

The real estate market offers one last lesson when we compare real estate values by neighborhood. Figure C.3 includes the map of rental values by dwelling for 1876, which we saw in chapter 10. The other map displays transaction prices for Paris apartments between 2019 and 2024 in thousands of 2024 euros (Paris real estate values). The two maps are strikingly similar. The correlation between housing values 150 years apart at the quartier level is a massive .86. Spatial inequality is vast and persists amid extensive changes in wealth and population. The range across quartiers in 1876 runs from 1 to 20,

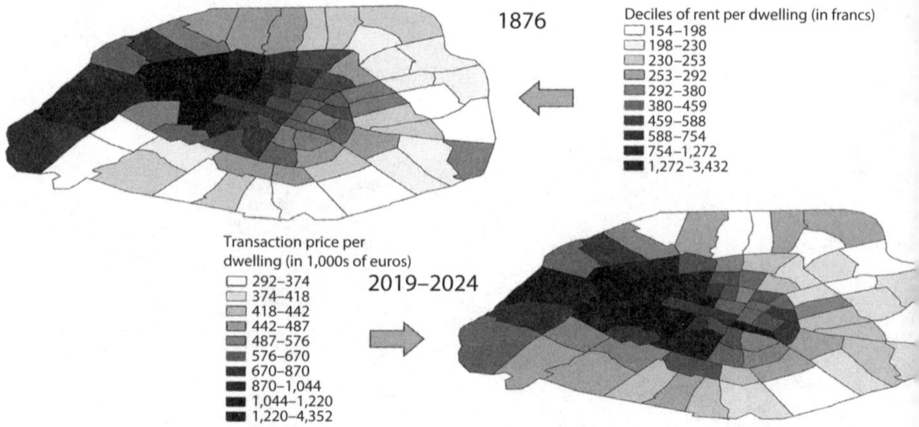

FIGURE C.3. Maps of real estate values by quartiers in 1876 and 2019–2024. *Source:* BSM 1878, 710; Explorateur de données de valeurs foncières (https://explore.data.gouv.fr/fr/immobilier?onglet=carte&filtre=tous).

while in the 2020s, the range is 1 to 14; thus, there has been a bit of compression in real estate values across neighborhoods. Focusing on the more recent map, we compare the neighborhoods of the rich arrondissements to those in the exterior arrondissements. An apartment in a rich neighborhood is, on average, four times more expensive than the average exterior dwelling. This is in part because apartments in rich neighborhoods are, on average, 60% larger than in poorer ones, and in part because prices per square meter are 2.23 times higher.

As noted earlier, one of the striking changes of the twentieth century for Paris involved the rise of pensions. In 1977, our last year of observations, old-age pensions covered a majority of the population. It seemed likely that this share would continue to rise for some time. To check whether pensions continued to diffuse, we collected a sample of death records for Paris for 1986, which is the last year that the records are easily available. We collected all the registrations filed in January 1986 for all the rich arrondissements (7th, 8th, 16th) and more than half of the registrations for the rest of Paris. Of the 1,336 Parisians in the sample, 70% were reported as retirees (for those above 60 it nears 80%). Their estates would have included at least their last pension payments and some furniture. Moreover, men and women are recorded as retirees at almost the same rate (69.6% vs. 70.8%). Given that the share of retirees was 62% for the rich core (arrondissements 7th, 8th, 16th) and 71% in the rest of Paris, it is likely that the fraction of individuals who died with assets in 1986 was more than 75% since we would have to add to the retirees those who died younger with some wealth and those in rich arrondissements who died rich and were reported without an occupation. The high share of retirees reflects the success of policies put in place after WWII that aimed to bring nearly everyone within the public retirement system. Agriculture was the last sector to be covered, so the progress of retirement in the provinces may well have been slower than in the capital. Obviously, given the very large share of the metropolitan population that lived in the suburbs rather than in Paris, an important question is whether the urban ring around Paris had the same high proportion of retirees.

We bring this section to a close with the remark that the more recent period also shows that increasing wealth inequality is not incompatible with a reduction in the number of individuals who die poor. The dynamic at the top reflects success for some in European and global markets. At the same time, more average individuals seem to have benefited from policies put in place thirty to forty years before.

C.3. Secular Change in Wealth and Inequality

The most salient arguments about the evolution of wealth and its distribution are all either explicitly or implicitly secular. Simple growth models like Solow's (1956) describe processes that occur at the generational level or longer. Even economies that have had stunning rates of capital accumulation for several decades, like China, would have to continue on their course for several more decades before catching up to high-income countries. More complex models are usually deployed to explain growth that is slower than expected as opposed to growth that is faster than expected. When we consider variations of wealth inequality absent wholesale nationalization (like what occurred in Russia in 1917), change is slow. This is why datasets that span a few decades are of limited value. We need secular datasets to document secular change. More importantly, the high degree of spatial persistence in wealth within Paris hides an unceasing process of change. In the case of Paris, the structure of the city, the characteristics of its population, and the structure of private wealth today would have been difficult for someone in 1807 to recognize. The same would have been true for someone in 1860 when the boundaries of the city changed.

C.4. Change in the City

As noted earlier, between 1800 and the present, Paris's formal boundaries changed only once, in 1860. It then doubled its territory, increasing its population by about 20%. At first glance, this stability is an advantage: we collect wealth at death for those who died as legal residents of a constant space. Moreover, prior to WWI, the neighboring areas, the two suburban arrondissements (Sceaux and Saint-Denis), were only partly urbanized. As late as 1915, some parts of Paris remained sufficiently empty that Citroën could build a massive munition factory in the 15th arrondissement and later replace it with France's first automobile assembly line. Citroën was not the only automaker within city limits (e.g., Panhard et Levassor). Beyond these large plants, Paris remained an industrial city (a fact confirmed by the occupations reported by the military conscripts of 1921) well into the post-WWII period. While the workers and the owners of Parisian factories did not live in the same neighborhoods, they lived in the same city. Investigating the wealth of Parisians also involves investigating the capital of its enterprises.

Yet even before 1930, much of the production of Parisian firms had begun to migrate beyond the city's boundaries. This was true in heavy industry (e.g., Schneider and Le Creusot) and for the *fabrique de Paris*, whose textiles were primarily produced to the north and east of the city. One location was heavily favored: the suburban zone just to the north and east of Paris in what is now the Seine-Saint-Denis département. In this area, rural villages evolved slowly into industrial towns. The growth of the suburbs reflected radical changes in transportation, not just in the rise of mass transport like railroads or trams that made it possible to move goods and people far more easily than horses and carriages, but also the automobile and the telephone that allowed individuals in Parisian offices to participate in the management of economic activities at a greater distance.

The growth of the suburbs and their growing industrial share into the 1980s allowed Paris to export much of its industrial labor force, most of whom had died poor in the city in the nineteenth century and now died elsewhere. So, the positive story about the fall of inequality in Paris after 1922 may well have to be tempered. We would likely find less progress if we extended the investigation to include the Parisian suburbs. Very partial investigations of neighboring towns reveal a wide range of situations, from Ivry, east of Paris, where the part of the population that died without a taxable estate reached up to 80% into the 1960s, and Nanterre, in the west, that looked far more provincial with a large proportion of positive estates among decedents. While the initial investigations suggest that the most striking aspect of inequality in the towns near Paris is their heterogeneity, future investigations could elucidate whether the forces that reduced inequality in Paris, like the spread of pensions and changes in the real estate market, also reduced inequality in the suburbs.

The foregoing takes the boundaries of Paris as exogenous, largely because they have remained frozen since 1860. Yet that stasis reflects a political equilibrium, which implies that the boundary is endogenous. The administrative entity that contains Paris is the Ile-de-France region. Consistent with the view that stasis is not exogenous, in the past four decades, there have been many discussions among regional and national politicians about the creation of institutions that recognize that the metropolis extends well beyond the ring road that has replaced the fortifications of Paris (Subra 2012). Everyone recognizes that official Paris is, in many ways, "too" small and that the region of Ile-de-France is, in many different ways, too large to coordinate public services. However, the discussions about expanding Paris to include the dense ring of suburban municipalities have never been acted upon. Faced with the imperative to rationalize a variety of services, in particular transport and health, the system has evolved. Instead of a wholesale merger of municipalities, new organizations for specific services have been put in place. For instance, the Paris transport authority, the RATP, has not disappeared. Instead, it has been subsumed into a larger entity, the Société des Transports d'Ile-de-France (STIF). The obvious redundancies of these multiple layers of administration notwithstanding, the region has developed its infrastructure well enough. Indeed, the Ile-de-France region has continued to enjoy the highest per capita income in France and rapid population growth. This is the case even though Paris itself, with its falling population, is becoming progressively less important. Nevertheless, municipalities remain closely associated with the delivery of local public goods and other infrastructure-dependent services. As we argued in chapter 11, understanding the spatial distribution of such services brings us back to wealth and its distribution. The stasis of Paris's limits over time is a direct outcome of large differences in wealth over space. Indeed, the affluent municipalities of the Hauts-de-Seine have never seen much advantage in joining Paris. Meanwhile, they and most Parisians would oppose having to share common resources with the much poorer towns of the Seine-Saint-Denis.

C.5. Capital and Capital

Paris, the rich capital of France, has also seen a massive transformation of its infrastructure and building stock since 1807. Scholars usually track these changes by documenting the spread of certain services, like schooling, tramways, or sewers. Another approach looks at the private and public organizations that make the investments and tracks how much they spend. Our data offer a different perspective by asking who owns the shares and bonds that these organizations issue. Paris's water supply began to improve in the 1850s with the building of an aqueduct to bring clean water to the city. Further investments would follow into the 1890s. The city owned the infrastructure, and we see a rise in the city of Paris's bond holdings in the portfolios of Parisians. Similarly, private companies provided natural gas for street lighting and private use; they issued stocks and bonds, and these financial assets turned up in the portfolios of Parisians. Prior to WWII, inequality meant that absent national directives, most local services remained fee-based and partly private. The rich favored the expansion of fee-based services, partly for welfare reasons but also likely because such expansion offered new investment opportunities. After WWII, when these companies were nationalized, the state issued Gaz de France bonds, and Parisians bought them. Our data are thus a very good source for tracking the transformations of the city.

Studying wealth over the long run in Paris has led us to document tremendous changes in productive capital, human capital, and the infrastructure of the city. Because Paris is relatively small, the extent of change is dramatic. Returning to Solow, growth over the long term implies adopting new technologies that demand new capital and make some old capital obsolete. More generally, most of the capital stock of 1807 would be of very little value in 2024. This is obvious for business capital. For instance, in Paris, the number of horses fell after 1900 as they were replaced by other means of transport like electric tramways and automobiles. It is also true for real assets, including the huge residences of the super rich of the Napoleonic era, given the decline in the number of live-in servants. Over time, increases in income have led to the replacement of slums by improved housing. It is also true for infrastructure: for instance, the six-foot-wide streets with an open sewer of central Paris before Haussmann would be very impractical for today's Parisians. This lesson applies to all rich economies.

Paris, thus, reflects one of the imperatives faced by societies that want to remain rich: if technology and tastes change, the structure of wealth must evolve. By the mid-nineteenth century, the dynamics of wealth required a financial market because the scale of important enterprises meant that they had to be financed partly by outside investors. During and after WWI, the state became an important funder of private enterprise. For housing and infrastructure, as we have seen in chapter 11, the distribution of activities between public and private varied considerably over time. The need for change then interacts with inequality because the distribution of gains to innovation is different from the existing distribution of wealth and because innovators' rewards partly depend on the wealth they can invest in their ventures. The extent of those rewards also depends on circumstances. The history of Paris vividly demonstrates that private wealth accumulation is profoundly dependent on whether one lives in a period of relative stability, before WWI or after WWII, or in a period of severe and persistent adverse shocks such as 1914–1945 (and to a lesser degree in recent years).

Parisians have long been a very important source of financial capital in France. The city's inhabitants were wealthy, and into the 1950s, the share of real estate in their portfolios was relatively low. Thus, the changes we described above are reflected in the structure of Parisians' wealth. Parisian portfolios reflect change beyond simply the type of capital that is being deployed. Over the past two hundred years, legal and financial innovation has altered the way individuals hold their wealth dramatically. Going back to chapter 2, the assets that were available in the 1810s are far less important to wealth nowadays. To give an example of innovation, in the 1850s, the rise of the Bourse, coinciding with a flood of share and bond issues, brought important changes to the portfolios of Parisians. Another innovation came much later in France: the rise of mutual funds (OPC in France) in the 1970s. This innovation came so late that we observed very few such funds in the portfolios of Parisians in 1972 when our data ended, but they have diffused rapidly since then. The absence of mutual funds had implications for inequality because it meant that the individuals below the top centile could not replicate the risk-return profiles of the very rich.

C.6. Structure and Change in Economic History

We borrow the title of this section from Douglass North's 1981 book. He focused on the institutions (the rules of the political and economic process) that guaranteed the security of private property. Among the themes that animated North was the complicated political economy that led the state to commit not to expropriate its citizens or subjects. These issues are central to any study of private wealth. It is, after all, reasonable to assume that if the state or other social actors are likely to be rapacious, individuals will either refuse to accumulate or do their best to hide their assets. Another strand of research sees the same institutions as the key drivers of inequality because they either favor or hinder political and economic competition (Engerman and Sokoloff 2012; Acemoglu and Robinson 2006). We return to these themes with two sets of observations on the structures of Parisian wealth and on what drives change.

For Paris, some institutions (e.g., family law or women's authority over their property) changed at very low frequency, and in some cases, the changes came so late that the impact of the reforms does not appear in our data. Other institutions changed infrequently, including the real estate market, the financial market, and pensions, but still all underwent profound transformations over the past two centuries. Other institutional changes were one-time events, like the annexation of territory in 1860, which created a new landscape but was followed by a century or more of stability. Still, other institutions remained in place only briefly; these include the contracts that governed the private firms of Paris. They were renegotiated every few years. Finally, the inheritance tax system itself had a century of stasis followed by a half-century when the rates were changed every few years. Institutional change is thus a very broad and complex process.

The same richness characterizes the patterns in the wealth data itself. Some findings are extraordinarily persistent, including a high top 1% wealth share, a high inherited wealth share, and a low real estate share for everyone below the top 5% up to 1922. One could add the rough parity of wealth between men and women. These are clearly not institutions but outcomes. Taken as such, they indicate the presence of deeper social or economic structures that produce such outcomes. We also have important dimensions of wealth that sometimes change after decades of stasis, like the share of inherited wealth, the real estate share below the top 5% after 1947, or the share of bonds in Parisian portfolios after 1922. There were also much more abrupt transformations, like the real level of wealth per adult after 1914 or the rise of railroads in Parisian portfolios after 1850. For some dimensions of the distribution of wealth, change is unidirectional; for others, there are reversals. Finally, there is the ceaseless flow of population into Paris (and likely out of Paris). Thus, different institutions and different facets of the distribution of wealth evolve at quite varying paces.

More importantly, perhaps, prosperity is not simply an issue of respecting property rights. In fact, the Paris that exists in 2024 is the result of a century and a half of periodic expropriations in which the state used a process for acquiring property and setting compensation that did not require the owners' consent. The legality of the state's expropriations has been central to the building and rebuilding of Paris. Paris's history shows us that North's proposition that failing to uphold property rights leads to poverty is too extreme. From 1940 to 1946, French authorities produced the greatest number of violations of private property since the French Revolution. The attacks on private property include the Vichy regime's expropriations of the Jews, German exactions during the war, and the large-scale nationalization of industries in the immediate post-WWII period. Nevertheless, from 1947 to 1962 private Parisian wealth boomed. One reason, we would argue, is that the state made important investments in services that were essential to the private sector. The growth of Parisian wealth since 1807 is the story of a division of labor between private enterprise and the public sector. Public spending was central to the reduction of many dimensions of inequality beyond wealth. Inequality in health and secondary education remained largely unchanged from 1807 to 1914 despite a massive increase in private wealth per capita and an even larger increase in total private wealth. Until a break in policy at the beginning of the twentieth century, very rich Parisians had succeeded in minimizing their expenses on public services even though these proved valuable to the health and skills of their workers. Then, after a half-century of change in public education and sanitation, the workers who entered the labor force in the mid-1950s were indeed much better educated and much healthier than those who had entered around 1850 or 1900. Such spending does not come for free, and thus, the returns to private wealth must be shared with the state in ways that were unthinkable in the nineteenth century.

Radical changes to fiscal policy do not come simply from a logic of domestic politics whereby the elites, fearing the growing masses of urban workers, consent to sharing power and to redistributive politics. The key reforms come largely as consequences of international conflict and from the pressures of international competition. When French tax revenues were squeezed at the end of the nineteenth century, a moderately progressive estate tax proved efficacious. When the threat of war with Germany loomed, it proved easy to raise the rates and collect more revenue, because, at the time, private wealth was large. The transformation of the inheritance tax produced a twofold regime from 1922 on (see figure 3.1). For estates going to children, the tax was moderate for everyone except the top decile of wealth in Paris, and even then, it was quickly capped at 25% for a single child. For estates going to other family members or third parties, the state chose to take a very large chunk, as the political cost of interfering with testamentary freedom was far less than that of interfering with patrimony. Most other countries chose an estate rather than an inheritance tax and their politics had to be different. By the 1920s, French population growth was slow, and the number and value of indirect inheritances were large. These issues mattered to the political economy of France, and they mattered to the fiscal burden of Paris versus the rest of France.

On the one hand, in a progressive tax system, individuals with very large absolute amounts of wealth will pay very large absolute amounts of taxes. On the other hand, the cost of reducing one's tax burden will likely also decline with wealth (as in the cost of financial intermediation) and, because there are increasing returns to the intermediaries that provide such services, tax avoidance will be practiced most by the super wealthy and found spatially where they are most dense. Parisians were thus both heavily taxed and the richest among them had access to the most advanced services to reduce that tax burden—this issue remains with us to this day. The financial liberalization of the 1980s combined with the increasing international exposure of many French firms only makes the problem of monitoring wealth more difficult than it was a hundred or even fifty years ago. That, in turn, makes it harder for the state to intervene to alter the wealth distribution, even if the resources are to be spent on improving French productivity, let alone reducing welfare disparities.

The challenge of finding the resources to fund social services is not new. In fact, as we saw in chapter 11, one of the constraints of such expansion in Paris was the unequal distribution of wealth in the city. Since not all cities are as unequal as Paris, the problem brings us back to what of the preceding chapters is specific to Paris. Let us consider Western Europe as a whole. In 1807, Paris was the second-largest city in Europe, where more than 80% of the population lived in settlements smaller than 5,000 inhabitants; the rest of the world was similarly rural. So, the dynamics of wealth in Paris were very different from those in most other places. Combining its very high population with a high population density (Paris, unlike London, was bounded by its municipal limits) meant that its wealth dynamics were also different from most other cities of Europe. Over time, Western Europe urbanized, and today, half the population lives in cities of half a million or more, making it more like Paris than it was in 1807. The rest of the world has followed at different rates, but the movement of populations to very large and unequal metropolitan areas is a distinguishing feature of the last century. On this dimension, Paris is thus now more representative of the lived experience of French people than it was in 1807.

The reader who wishes to extrapolate from our work should not expect that other places will have the exact same history. Nevertheless, most of the forces behind large shifts in wealth and its distribution are similar across space; these include asset returns, migration and labor market returns, demography, taxation, and public services. Because the urbanization rate varied across countries and time, the chronology of change varies across space for economic reasons such as differences in the demand for sanitation and education, or the rate of growth of manufacturing versus services, and for political economy reasons like variation in the timing of universal suffrage and the political weight of rural and urban dwellers.

Because the evolution of wealth is well documented in Paris, we can answer some important questions that are hard to pin down in other places. In particular, the spatial patterns of wealth inequality and various forms of human capital are strongly correlated over space but much less correlated over time. The different time paths of wealth and social indicators allow us to argue that public intervention is critical to reducing disparities in lifespan, health, and education because these changed far less before 1914 than the trajectory of wealth would have led us to expect. Social disparities also fell markedly after 1914, both in the period when average real wealth was falling and when it was growing. Moreover, public intervention in areas like pensions prompted large swaths of the Parisian population to accumulate bequests and, for the first time, reduce the fraction of individuals who died with no wealth to less than half. In the long run, sewers, electric lighting, and free secondary education have diffused across the rich cities of the globe. The pace at which these problems are tackled is deeply contingent on each country's political economy and its inequality.

In the first third of the nineteenth century, Paris was very unequal, and, compared to what would come later, poor. It was also cruel. Individuals without wealth lived brief lives with a very small and shrinking safety net; for two-thirds of the population, incomes were too low and erratic to allow for any sustained accumulation of wealth to transmit to the next generation. Much of the wealth in the city came from individuals who benefited from the state's largesse in one way or another. This book is about the transformations of wealth and inequality over nearly two centuries, from 1807 to 1977. By the time our data series end, wealth had been radically transformed, and the patterns of inequality irrevocably altered. All this change might seem to indicate that this history has little use for policy. Our views and the arguments we have deployed suggest otherwise. On the one hand, the economic forces at play in the third decade of the twenty-first century continue to make Paris, like the other major cities of the globe, rich and unequal. On the other hand, the differences between today and two hundred years ago are partly technological. Still, they also reflect a radical change in the social willingness to provide the infrastructure needed to dampen the impact of wealth inequality on social outcomes. Looking forward, funding social infrastructure will be easier for the biggest cities that have large locational advantages, and harder if it is feasible for the very wealthy to enjoy these advantages while living in tax havens or in locations whose real estate values exclude almost everyone. The world's biggest cities will remain dynamic, rich, and unequal, but they will be challenged to maintain their commitment to public urban amenities and human welfare.

APPENDIX 1

The Datasets

The analysis in the book is based on two sets of data and one set of command files that are all being made available for anyone who wants to use them. The first group of datasets was collected from the estate tax documents (DQ7 and DQ8 series) at the Archives de Paris. The second group contains a wide variety of data sources that help put wealth at death in Paris in its context. Both these two groups are archived as CSV files. Finally, scholars have access to a group of command files for STATA to go from the raw data to the dataset we analyze. These are archived as text files.

Wealth in Paris, 1807–1977

PART 1: DATA FROM ESTATE TAX FILINGS IN PARIS

1) The TSA dataset, 1807–1867
2) The RMD dataset, 1822–1972
3) Annual files
4) The Estates dataset, 1807–1972
5) The Assets dataset
6) Estate mutiplier dataset
7) Estate tax dataset
8) Variable definitions for files 1–7

These are all accessible at: https://data.progedo.fr/studies/doi/10.13144/lil-1702, La richesse à Paris 1807–1977. Données issues des déclarations de successions après décès. Base individus - Gilles Postel-Vinay, Jean-Laurent Rosenthal (producteurs), PROGEDO-ADISP (diffuseur).

PART 2: SOCIAL AND DEMOGRAPHIC DATASETS

1. Asset return and prices
2. Estate tax rate schedules, 1790–1960
3. Banque de France shareholders
4. Caisses d'épargne
5. Retraites
6. *Challenges* list
7. Estate tax data by Département
   a. All years
   b. 1902–1905
   c. 1929–1931
   d. 1951–1954
8. Education in Paris
9. Death records, 1862
10. Burials
11. Marriages
12. Military recruits
    a. 1862
    b. 1867
    c. 1892
    d. 1921
    e. ASVP
13. Doctors by arrondissement
14. Demographic data
    a. Censuses
    b. Deaths by age
15. Deportation
    a. Samples from MdM, MdlC, MdlS
    b. Matched sample

These are all accessible at: https://nakala.fr/collection/10.34847/nkl.dd74950v. La richesse à Paris 1807–1977. Données socio-démographiques, by Gilles Postel-Vinay, Jean-Laurent Rosenthal.

APPENDIX 2

The Data from Estate Tax Filings

A2.1. The TSA Data

Each tax bureau maintained two sets of volumes: the Registres de Mutations par Décès (RMD) and the Tables de Successions et Absences (TSA).

The RMD contain estate declaration filings in chronological order by filing date. Prior to 1901, the clerks transcribed the declarations on their registers. After 1901, the records are the declarations submitted by the heirs. There are more than 50,000 such volumes in the Paris archives for the period 1800–1977.

The TSA, the source we use for chapter 1, focused on individuals, and its goal was to keep track of all decedents. The pages were initially filled by transcribing the death records of all decedents who died in the bureau, the death records of residents who died elsewhere, and those who owned real property in the bureau but did not reside or die there. On the left side of the volumes, the scribes entered each individual's first and last name, address, occupation, age, location of death if not at home, and marital status. For some bureaus (e.g., B8 from the 1810s to the 1830s), clerks noted marriage dates and recorded notaries for marriage contracts and probate inventories. Finally, on the far right, they transcribed the list of estate tax filings in the bureau (date, number, and type). These lists of filing dates allow us to find all the filings that relate to the same individual in the bureau.

A decedent may appear in multiple locations (where he or she died, where he or she had a legal residence, and any bureau where he or she owned real estate). Thus, collating lists of deaths from the TSA would greatly overstate the number of individuals who died. For the local officials, the list had the merit of detailing who was at risk for paying estate taxes.

As noted above, before 1870, the registers also reported the value of movable wealth and real estate wealth in the bureau (with the address and type of real estate).

A2.1.1. THE COLLECTION METHOD, 1807–1867

The primary challenge was to link real estate assets not located in the bureau of the deceased's legal residence to his or her main filing. In Paris, one has to own whole buildings. Buildings were expensive (the average value of a real asset in Paris declared in a bureau was more than twice the average wealth of an individual with a positive estate). As it turns out, real estate was overwhelmingly concentrated in the estates of the wealthiest 5% of Parisians (they were two-thirds of the property owners, with more than 70% of the properties and 94% of the real estate wealth). We decided to follow a simple procedure.

We opened the TSA register, and for any individual with positive wealth, we noted age, gender, and marital status. For anyone whose wealth was above a threshold (16,000 francs to 1822, 30,000 for 1827 and 1832, and 40,000 from 1837 to 1867), we also included first name, last name, address, occupation, and type of heir. When an estate had real assets, we also noted the property's address. We took down the same information for estates that only included real estate (to link it with the primary declaration in another bureau that included the movables). The procedure allowed us to attribute nearly all real assets to an individual for whom we also observed their movable wealth.

A2.2. The RMD Data

A2.2.1. THE COLLECTION PROCESS FOR RMD DATA

On the mundane side, the TSA volumes follow death dates: individuals who died in a given month and year appear essentially in one volume per initial letter of last name. However, RMD volumes place declarations by the date and time when they are filed. By law, estate declaration had to be filed within six months of death, and in 1822, some 90% of all estates had filed by the end of 1823, but some would come in 1824 and a few into the 1830s (mainly because it took time to find the heirs). To recover all the filings of individuals who died in a particular year requires opening more than one year of RMD volumes.

Our initial foray into the RMD was a limited effort to bridge the gap between the TSA records and the published distributions of estates that began in 1902 (Piketty 2001). We thus collected the TSA dataset for 1877 and 1887 (that gave us dates of filing for individuals) and then opened the RMD registers to recover movable and real estate wealth. We also collected 1902, but because individuals only appeared in one bureau after the estate tax became progressive, we did not collect the data in the TSA dataset before opening the RMD register.[1] These three years have the same limited information as the TSA dataset. We then decided to launch a deeper investigation to provide much more detail about wealth and the individuals who held it.

With a limited budget, we had to make decisions that traded off accuracy for breadth. A complete transcription of each estate would have maximized the details of the dataset. Recovering every available estate every year would have provided perfect coverage but was not feasible. Challenges inherent to the source implied that we had to sample.

The first challenge, as noted above, was that the records that pertain to the individuals who died in a given year were dispersed. Retrieving volumes from the archives takes time, so our goal was to recover 90% of the estates for each year. For 1912, the first year that we collected in detail, in order to recover 90% of all the estates, we had to open the registers from January 1912 through June 1914. From then on, we examined the volumes of the RMD for the 30 months that follow January of the sample year. This rule means we do not quite have the entire population of estates. For the cross sections from 1872 to 1912, we have recovered 91% of the Parisians who filed an estate tax and were recorded by the TSA volumes as having paid the tax. In other words, we missed 9% of these individuals either because they paid the tax with delay or because we failed to find them due to some transcription error on our part or the part of the clerks of the estate tax. In the end, we opened half of all the volumes of estate tax declarations between 1872 and 1964. We also collected data from similar sources for 1972 and 1977.

A second set of decisions involved how much information we wanted to collect for each decedent. Initially, we decided to follow the model of the data collection for the TSA datasets (age, gender, marital status, and aggregate wealth) for everyone and much more detail for a stratified sample (100% of the top 2%, 50% of P94–98, 25% of P86–94, 12.5% for all individuals with positive wealth). Over time, we realized that all names, addresses, and heir types should be collected. The dataset thus reflects the evolution of our thinking on these matters.

A third set of decisions involved what to record regarding information about assets. We decided to compromise because the number of assets recorded in the declarations is vast. We aimed to transcribe enough information to understand what entity issued each financial claim (specific firm, individual, or agency) and if it was equity, debt, or a deposit of some kind. We decided not to worry about the specificity of bonds (e.g., issued in 1876 with a 5% coupon or 1875 with a 3% coupon, and so forth) or the type of equity (e.g., nominative or bearer shares). As we shall see, even this coarse coding of the data is amply sufficient to demonstrate the astounding diversity of Parisian wealth holdings in the aggregate and the utter lack of diversification of most portfolios.

Finally, there was an issue of whether the RMD dataset was consistent with that of the TSA. We built RMD datasets for 1822, 1842, 1852, and 1862 that can be matched directly with the TSA data.[2] We collected the full asset details for the stratified sample for those years. The TSA provides sufficient information to compile reliable wealth inequality statistics.[3]

A2.2.2. CREATING THE POPULATIONS OF RICH AND POOR

Having collected data about the rich, we also needed to include the poor. We start with the total number of adult deaths, N, and a total number of adult rich Parisians $N^R = N_0^R + N_A^R + N_{NT}^R$ where N_0^R is the number of rich for whom we do not have age, N_A^R is the number of rich whose age at death we know, and N_{NT}^R is the number of individuals in the TSA but for whom we did not find a declaration. Denote $N_B^R = N_0^R + N_A^R$ these are the rich for which we have a wealth observation

By construction, the total number of poor $N^P = N - N^R$ and $N^P = N_0^P + N_A^P + N_{NT}^p$

Now we assume $N_B^p = N_0^P + N_A^P = (N - N^R)\dfrac{N_{NT}^R}{N^R}$.

We further assume that $\dfrac{N_0^P}{N_B^p} = \dfrac{N_0^R}{N_B^R}$

Thus $N_0^P = (N - N^R)\dfrac{N_{NT}^R}{N^R}\dfrac{N_0^R}{N_B^R}$ and $N_A^P = (N - N^R)\dfrac{N_{NT}^R}{N^R}\dfrac{N_A^R}{N_B^R}$

Generally, we find about 90% of the individuals, so we include about 90% of the individuals (85% of the records of those we find provide an age, and nearly all are adults).

To assess public debt holdings, our task was relatively simple because the estate reported both taxable interest amounts (the 838.35 francs in Madame Aujolet's estate) and the coupon value of the annuity (1,784 francs in her case) and going from the coupon value to the nominal value of the asset requires multiplying it by 20. One can then deflate it by the market price since that is always expressed as a percent of par. For real estate, things are not so easy. The rental arrears reported do not list the time covered, so it is unclear what the proper capitalization factor should be. For 1822, we compared the total value of reported real estate (25.2 million) to the total value of reported rents (188,000 francs), which gives an implicit capitalization.

To create the Estates dataset we merged the RMD data described above with the TSA dataset to include TSA observations for years when we did not collect the RMD data.

To create the Estates multiplier dataset we used the age distribution drawn from censuses and the age distribution drawn from death records to reweight the Estates dataset so it is representative of the living rather than the dead.

A2.3. Portfolios

The data for the portfolios were recovered through a series of interlinked processes. The data comprise 693,905 distinct items, of which 555,934 pertain to individuals we sampled to recover the details of their wealth at death, and 137,971 pertain to individuals we did not sample.

As noted in the text, continuing the effort from the TSA, we tried to collect information on all real estate assets (including their location). Thus, the assets file includes 48,072 real estate assets (16,900 of which belong to individuals we did not sample).

Second, to ensure that we had examined each declaration, we collected information about tax payments made in advance or after the primary declaration. There are 30,279 such "assets" in the file, 14,250 of which concern individuals we did not sample.

Third, we collected full asset details for a stratified sample of individuals. These are the data we use for portfolios, and they involve 494,962 assets. Associated with these assets are 32,128 items that involve accounting between spouses (e.g., claims on the community).

Fourth, because we only decided whether an individual would or would not be sampled when we encountered his or her principal filing, additional data were collected. For instance, we entered partial declarations when we encountered them to avoid having to reopen these filings when the principal filing indicated that the individual should be in the detailed sample. When producing the portfolios we took care to only include the partial filings that had not been repeated in the principal one.

For individuals who were not married when they died, there is only one list, so the total corresponds to the value of the estate. For individuals who died married, the estate includes both a personal and a community asset list. Conceptually, the value of an individual estate is the value of his or her personal estate plus half the community net of claims. Because these claims are quite variable and net communities can be negative, the share of personal property can be less than 1 and the share of the gross community ranges

TABLE A2.1. Investment by sector for international financial assets (% of total)

	Industry	Mines and land	Utilities	Finance	Rail-roads	Trans-port	Services	Public de
1842		0.8%	0.0%	0.1%	0.4%	0.0%	0.2%	98.5%
1852		3.4%	0.6%	1.6%	0.7%	0.6%	0.0%	92.1%
1862		9.6%	3.6%	5.3%	39.8%	1.1%	0.0%	36.0%
1872		9.2%	0.2%	5.0%	43.1%	2.5%	0.0%	39.3%
1882	0.4%	3.6%	0.2%	8.7%	33.1%	9.1%	0.2%	44.6%
1892	0.8%	3.5%	0.7%	8.1%	34.6%	9.0%	0.6%	42.7%
1897	1.4%	6.8%	1.7%	3.3%	27.9%	9.4%	0.0%	49.5%
1907	1.4%	10.6%	0.6%	5.3%	21.5%	7.7%	1.2%	51.7%
1912	2.1%	15.0%	1.5%	9.6%	19.8%	7.1%	0.3%	44.5%
1922	6.1%	26.2%	1.9%	11.2%	10.1%	11.4%	0.6%	32.5%
1927	9.5%	22.0%	4.4%	14.6%	12.2%	13.1%	3.2%	21.0%
1932	7.8%	17.7%	3.9%	11.1%	19.9%	17.2%	0.9%	21.6%
1937	8.0%	41.4%	2.1%	11.1%	6.5%	14.2%	0.6%	16.1%
1942	13.6%	39.6%	3.1%	11.0%	4.1%	12.1%	2.2%	14.3%
1947	18.8%	36.9%	2.5%	11.0%	8.3%	12.3%	2.3%	7.9%
1952	15.5%	50.1%	2.5%	9.8%	4.7%	9.2%	2.8%	5.5%
1957	15.5%	57.4%	2.2%	6.2%	2.6%	11.2%	0.5%	4.5%
1962	26.2%	47.8%	2.9%	11.6%	1.3%	6.4%	1.2%	2.5%
1972	35.7%	44.7%	1.2%	12.8%	0.9%	2.8%	0.8%	1.1%

Source: Assets dataset.

TABLE A2.2. *Patente* payers as a function of age in the 1845 electoral roll

	Age					
	45 or less	46–50	51–55	56–60	61–70	70+
% patente payers	72.1	58.9	44.5	34.7	20.3	7.0

Source: ADP VD4 4149-50-51.

from 0 to 1. For practical reasons, we produce portfolios that have weights of 1 for personal assets and one-half for community assets. For the top decile, this procedure is on average right. Below the top decile, the average community share is closer to 44%, likely because widow's portions are a more important share of the estate.

The Assets dataset includes all assets for which details were collected and an indicator for whether the individual was sampled or not.

APPENDIX 3

Progressive Taxation

A3.1. Estate Tax Laws

This list gives the publication dates in the *Journal Officiel*.

1798 December 12 (2 Frimaire An VII)
1801 March 18 (27 Ventose An IX)
1873 December 31 (8273–83)
1901 February 26 (2131–6)
1902 March 30 (2274)
1910 April 10 (3159–61)
1917 June 12 (4580), December 21 (3423)
1920 June 26 (8994)
1926 August 4 (8788)
1929 December 30 (13986)
1934 July 6 (not found), July 12 (7028–30)
1937 January 1 (115–6), November 7 (not found)
1939 July 30 (9623–4)
1940 December 10 (6041–2)
1941 September 5 (3767–8)
1942 March 15 (1027–8)
1946 October 29 (9197–8)
1947 March 29 (2689)
1949 January 1 (75–6)
1950 April 30 (4538–40)
1951 January 8 (338)
1952 April 15 (3929–30)
1959 December 29 (12467–8)
1968 December 29 (12340)

Note: The law of 1956 was not recovered. Details are in SEF (Supplément 127, July 1959).

A3.2. Responses to High Taxes: Filing Delays

The creation of progressivity did not just increase the share of estates flowing to state coffers; it also changed the frequency and timing of tax filings. Although estate taxes still had to be paid within six months of death, heirs seem to have found it increasingly difficult or undesirable to complete a consolidated balance sheet. This explains the rise in preliminary filings (*partielles*) and partial payments (*acomptes*). At the same time, both filers and the state had stronger incentives to establish the "correct" valuation for assets when the marginal rate was 20% than when it was 2%. For publicly traded assets, valuation was never much of an issue since the rule was that the value would be determined by the price closest to the day of death. Nevertheless, 60% of all assets were not publicly traded, and their value was established from a variety of sources (e.g., rent, balance sheet multiples, and insured valuations). Establishing correct valuations became more complicated after 1914 because government regulation of income flows became more complex (e.g., rent freezes) and because macroeconomic conditions made prices much more unstable. These issues increased the number of corrective filings (e.g., *insuffisances*). Finally, the wide range of policy interventions of the interwar period combined with taxpayers' desire to avoid taxes drove up the number of supplementary filings (*complémentaires* and *omissions*) where heirs reported assets that had not been tabulated in the principal filing (*définitive*). Lastly, the tax rates from the nineteenth century could largely be paid out of a fraction of capital income or by selling a few assets. When the average rate hit 20%, the state had to be more flexible in receiving its due, particularly since it wanted to allow heirs to retain patrimonial assets and avoid the fire sale of the assets of liquidity-constrained businesses. Thus, the period that followed progressivity saw the invention of installment payments (*payments fractionnés*).

These different changes in the interaction between the tax authorities and heirs inevitably introduced biases in the tabulations of estate tax filings that we discuss in section A3.3. The fiscal authorities did not have the patience to tabulate the estates' final value since it was unclear when this total would be established. They could have tabulated all asset filings separately, including the *partielles, omissions,* and *insuffisances,* that would have produced much lower inequality measures by creating many small filings. Instead, it seems the tax authorities chose to tabulate only principal filings. This shortcut lowered wealth totals because it omitted the value of assets in corrective filings. The shortcut also reduced measured inequality because corrective filings were more likely for large estates than small ones. Before WWI, 95% of all estates involved less than two filings. The average number of filings peaked in the first decade following WWII. The distribution of filings also shifts further to the right as we go from a lower to a higher fractile. By 1952, the top 1% of estates required ten filings or more.

A different set of forces also pushed individuals to file later. As figure A3.1 shows, more and more principal filings were delayed past the six-month deadline, leading to increases in the average time between death and the recording of the principal filing. Before WWI, the average filing time was six months, which was exactly the delay allowed by the estate tax. Further, nearly 94% of all filings were completed within our sampling interval (30 months starting on January 1 of the sample year). Then, as the dotted line shows, filing delays increased, reaching a year by 1922 and remaining above that level for the rest of our period. On average, heirs filed within a year of death except for 1942, which we will examine later. Through 1932, the top 1% of estates were filed at roughly the same time as the rest of the rich (half a year before WWI and a year afterward). Then the two series diverge, and while the rest of the population continues to file within a year of death, filings by the top 1% become ever more delayed, reaching a full two years in 1947 and stabilizing at that level.

Figure A3.1 also shows that the top 0.1% has a story of its own. Despite the few observations that compose the very top, it is clear they filed particularly late in the period after WWI and continued to do so over time. Interestingly, the extent of delay at the very top is closely related to the average inflation rate over the prior five years. This connection is not a coincidence. Indeed, delay was one mechanism to reduce the tax burden in high inflation periods since the nominal tax burden was calculated based on the estate's value at the time of death. In contrast, the real cost of paying that burden depended on how prices changed between the date of death and the time of payment. In periods of inflation, delay was profitable, while in periods of deflation, it was not. Consistent with much of the tax literature that points out that the very rich are the most able to take advantage of tax avoidance strategies, it is the very top of the distribution that best mirrors the evolution of the price level. When inflation became acute in the 1940s, the super rich filed on average three years late (two years later than most of the population) and received massive discounts on their liabilities. This is no mere anecdote because, as we have seen before, the top thousandth estates represented 20% of all the wealth left behind and would have had to pay a tax rate of no less than 20% and, on average, closer to 30%.

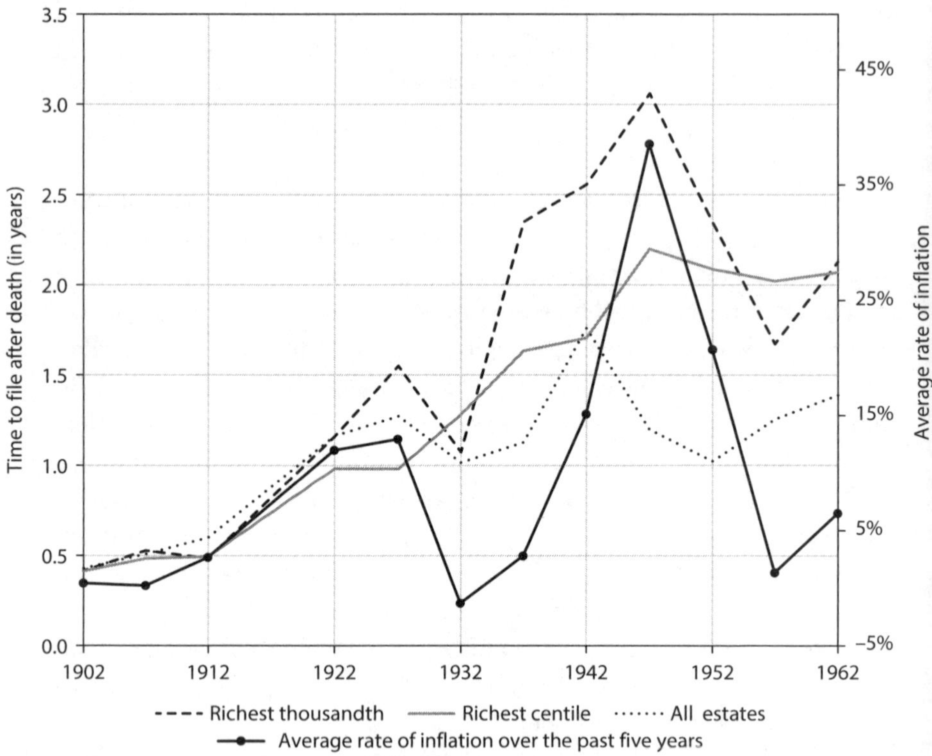

FIGURE A3.1. Average time to file per decedent in Paris, 1902–1962. *Source:* Our samples.

A3.3. Delay as Tax Avoidance

As noted above, the government published tabulations for estates whose principal filing occurred in a given year without regard to when the individual had died. Heirs were supposed to file within six months of the date of death. Supposing they used the maximum legal delay, we can contrast the public tabulations with our efforts using the 1907 cross section. As instructed, civil servants would have collected data from filings received between January 1 and December 31, 1907, which would have recorded the estates of individuals who died between July 1, 1906, and June 30, 1907. Our tabulations focus on those who died in 1907, and their estates were filed from January 1, 1907, through June 30, 1909. So, had everyone filed at the legal maximum and had there been no inflation, our two series should be different but statistically identical except for the richest individuals' random date of death. This we call the Gustave Rothschild problem (see chapter 8). With an estate of 242 million francs, he was one of the richest persons to ever die in Paris, accounting for 13.1% of all the wealth in that year. He died on November 28, 1911, and is not in our 1912 sample. His heirs filed in 1912, and his estate appears alone in the top category of the tabulation published for 1912. So, our estimate of the 1% share for 1912 differs from what the tabulations report. Of course, had we collected every year, Gustave Rothschild would have appeared in 1911, and the spike in the published data would also appear in our data. Our one-in-every-five-year sampling strategy means that we miss some of these spikes.

Late filings influence the measurement of inequality only when there is a high degree of time variation in nominal asset prices (either high inflation or deflation) and a systematic difference in time to file by estate size. Suppose asset price inflation is 1% or 2% per year. In that case, the fact that some individuals' estates are filed a year later than everyone else only underestimates their wealth a tiny bit, and it will not affect the trend. Similarly, if there is a lot of asset price inflation but no differential filing time depending on the size of the estate, there would also be few effects on the trend. If the 1942 published tabulation referred to individuals who all died in 1940 instead of 1942, we would measure wealth inequality correctly for that population. Such tabulations would, however, produce downward biased estimates of the wealth individuals left behind in 1942 because it refers to an earlier year when the price level was lower.

Consider now a world with significant asset price inflation (10% or more) and where the very rich delay a year more than everyone else. In this case, compiling inequality from published sources will lead to comparing the assets of the middle class from one year to the assets of the very rich from the year before, inducing a downward bias in the measurement of inequality. The higher the rate of asset price inflation, the greater the downward bias in the published sources. In the quarter century before 1912, the average inflation rate was 2.5%; in the quarter century that preceded 1937, it was 7%, but in the quarter century that preceded 1962, it was 15%, with some peaks (1947) at 50%. As we saw earlier, before 1937, there was little differential delay; everyone filed more or less in the same interval. Until then, the published series for Paris and the our own samples matched well. After 1937, the bias from differential delay increased over time, contributing to the fall in inequality, as measured from the published tabulations. It also eroded the value of the estate tax as more and more tax payments were made late. A 25% tax paid two years late with 15% inflation is a 19% tax; after three years, it is only 16.5%. It is no surprise that heirs delayed payments. By 1947, the treasury applied penalties for late payments, but they were ineffective.

TABLE A3.1. Estates filed in 1947 by year of death

	Filings		Value in 1947 adjusted for	
	Number	Value (million francs)	CPI inflation	Asset price inflation
1942 or earlier	16,785	2,825	14,498	5,057
1943	5,927	1,783	7,367	2,961
1944	16,192	4,987	16,849	7,861
1945	30,305	9,059	20,653	13,331
1946	117,122	28,183	42,105	34,402
1947	122,195	23,539	23,539	23,539
Total	308,526	70,376	125,013	87,153
Share filed in 1947	39.6%	33.4%	18.8%	27.0%

Source: SEF, no. 4, 1949, 671–742.
Note: The average value of estates of 1943 decedents filed in 1947, once adjusted for asset price inflation, is 2.6 times larger than the average value of 1947 decedents filed in 1947.

The tax authorities had more than an inkling that delay was a problem. They, however, mostly avoided calculating how this affected estate tax revenues. In 1947, they published estate tax filings by year of death in terms of numbers and value for France (see table A3.1). Only 40% of filings and 33% of the value of estates concerned individuals who died in 1947. The evidence of differential delay is striking. The average estate for individuals who died in 1943 and was filed in 1947 was 300,000 francs, or 50% larger than the average filings of individuals who both died and filed in 1947. The average value of estates of individuals who died in 1944 and 1945 is also 50% larger than those who died and filed in 1947. We can take that data and perform two adjustments. The first is to adjust the value of the estates to reflect consumer price inflation (CPI) between the year of death and 1947. This is the relevant calculation for the public purse because the government is a consumer. Had the state imposed an inflation adjustment to tax payments, the state's income would have been 77% higher, and the average estate from 1943 would have had a CPI-adjusted value of more than a million francs. For wealth calculations, adjusting by asset price inflation makes more sense. The goal is to produce a wealth estimate that approximates what the different estate vintages would have been worth if everyone had died in 1947. Asset prices rose more slowly than the consumer price level (partly because bond prices do not respond upward to inflation), so our revisions are smaller than with the CPI. Had the state required that wealth be adjusted for asset price inflation, the wealth subject to the estate tax in 1947 would have been at least 24% higher than reported. The authorities did not report the delay by estate size, so we cannot compute the impact of delays on reduced progressivity. In any case, the tabulations underestimate both wealth and inequality.

A3.4. Measures of Inequality

Filing delays also matter in how inequality is measured. From 1902 to 1964, the French finance ministry regularly published tabulations of estate filings for a particular year. The same tabulations exist for every département, including the Seine, for many years up to 1958, and we used them in our early article (Piketty et al. 2006). We have improved the series by adding eight tabulations for years that we had not found a decade and a half ago.[1] We also make one substantive change to the 2006 data.[2] Government officials produced tabulations for the Seine from 1902 to 1913. Then, from 1925 to 1942, they divided the Seine département into two lines, one with roughly 60% of the estate filings and 85% of the wealth reported for the Seine—the totals for Paris—and the other line containing about 40% of the estate filings and 15% of the wealth—the totals for the suburban arrondissements of Saint-Denis and Sceaux.[3] Then, from 1943 to 1958, there was again a single line for the Seine. In 2006, we erroneously assumed that the post-1943 data reported only Paris. We have corrected this mistake and computed estimates for both the Seine and Paris from 1925 to 1942 and the Seine from 1943 to 1958.

Figure A3.2 presents four estimates of the 1% share of wealth. First, we show the estimate from published data for the Seine (1902 to 1958) and for Paris (1925 to 1942). The third estimate follows our 2006 procedure (1902 to 1958). Fourth, we display our microsample estimates as dots. Comparing our inequality estimates to those from published tabulations produces two surprising findings. First, before WWII, all the estimates roughly agree (keeping in mind that the 2006 series follows the Seine before 1914 and the Paris series to 1942). Any differences between the series can be attributed to the volatility of top estates. By chance, measuring inequality at the Paris or the Seine level does not matter as much as one might think for the 1% share. When one computes wealth inequality for the Seine relative to Paris, two opposing effects tend to neutralize each other. The suburban arrondissements are poorer and more equal than Paris. These characteristics of the suburbs bring down inequality. Conversely, the suburb's large demographic weight massively increases the number of individuals who comprise the top

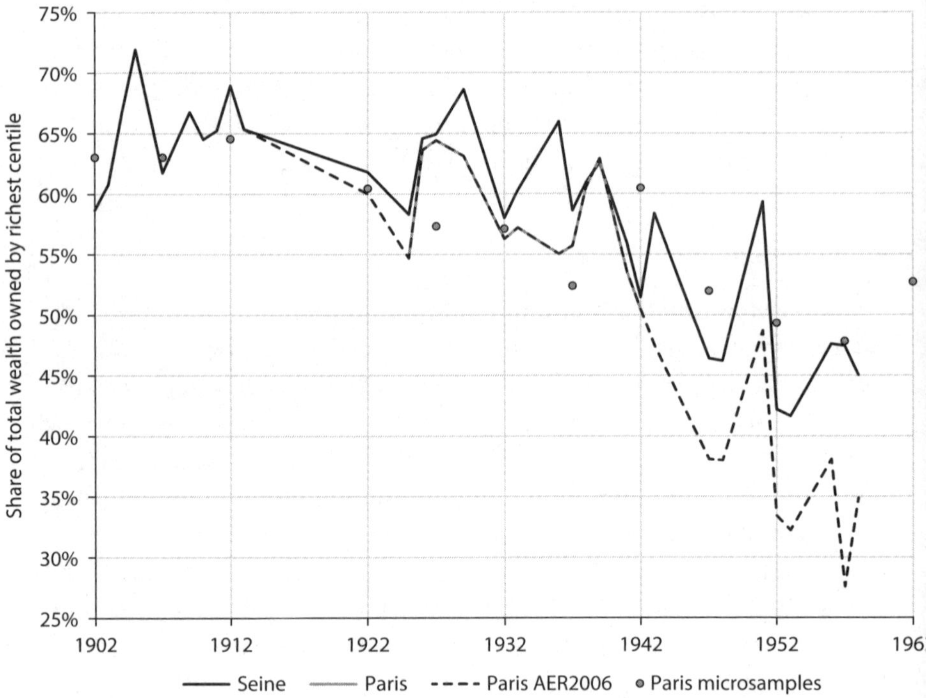

FIGURE A3.2. Estimates of the top 1% share of wealth at death. *Note*: The Seine and Paris top 1% shares are extracted from the official tabulations of estates (for details, see chapter 3). The AER 2006 reproduces the data reported in Piketty et al. 2006. The Paris top 1% microsamples come from the Estates dataset (see appendix 1).

1%. Because the outlying arrondissements are relatively poor, the Seine's 1% is disproportionately composed of richer Parisians.[4] This second effect tends to drive inequality up. In general, the second effect dominates, and Parisian inequality is estimated to be less than that of the Seine. Starting in 1937, the Seine and Paris were indistinguishable. After 1942, however, the 2006 estimate diverges from the others, which is unsurprising since, as we saw above, it has the wrong denominator. The Seine series roughly tracks the microsamples.

We view the concordance of the microsamples and the Seine series as a chance event. Indeed, when we consider the transformations of the inheritance tax system from the timing of filings through the rise of a minimum threshold and tax-exempt assets, the distribution of taxable wealth should generally be less unequal than that of all wealth. The distribution of estates filed in a year is less unequal than that of estates in that year, save for years of nominal asset declines like the Great Depression. In fact, before 1952, our higher estimates for the 1% share are largely due to systematic differences in delays in filing in the context of high inflation. After 1952, the introduction of tax-exempt assets further reduced the value of estates in the tabulations but not in our microsamples.

A3.5. Tax-Exempt Assets

After 1947, the most important wedge between the published tabulations and our series comes from tax-exempt assets. Between 1952 and 1973, part of the national debt was held in Pinay bonds. These bonds, issued in 1952 and 1958, had a variety of advantages that compensated for a low nominal yield (3.5%). In particular, they were not subject to the estate tax when transmitted to heirs. As early as 1952 (see table A3.2), the propensity to hold these bonds increased with wealth. In subsequent years, the share of Pinay bonds in every fractile's portfolio grew. More importantly, the top 1% of wealth holders placed at least twice as much of their wealth in Pinay bonds as did other groups.

The wedge between taxable and economic wealth grew further in the mid-1960s when the government enacted a law that made new construction free of inheritance tax at first transmission. By 1972, estates included large amounts of Pinay bonds and two forms of new construction: buildings owned entirely by the deceased or jointly with his or her spouse and investment funds issued by France's largest banks deployed in new real estate. Unsurprisingly, in 1972, more than 18% of the net wealth of the top wealth holders (0.1%) was in such assets. In contrast, the share of wealth held in tax-exempt assets was systematically lower for lower fractiles. For estates below the top decile, who paid little estate tax, the share of wealth held in exempt assets was trivial (less than 1%). Such assets were unimportant to tax officials focused on taxable wealth and tax revenue, but since they form a part of a decedent's overall wealth, they matter to us. Our microsamples have the advantage that such assets are often reported and almost always valued. They allow us to compute a measure closer to economic wealth than its taxable counterpart. Both of these measures reflect one of the tensions of high tax regimes. When taxes are low, providing tax exemptions has limited effects on incentives. Recall that until 1850, the public debt was exempt from the estate tax precisely to encourage the public to hold these bonds. When taxes are meaningfully high (20% or 25% in direct line of descent), the government is tempted to use tax exemptions to achieve policy

TABLE A3.2. Share of tax-exempt assets in total net estate

	1952	1957	1962	1972	1972
	Pinay bonds				New real estate
Top 0.1%	1.7%	10.3%	6.40%	17.60%	0.86%
Rest of top 1%	0.7%	5.2%	10.40%	8.74%	2.08%
P90–99	0.3%	1.4%	3.50%	1.51%	2.45%
P50–89	0.05%	0.1%	0.20%	0.01%	0.01%

Source: Estates dataset.

goals. Because taxes are high, wealthy individuals are motivated to direct part of their wealth into areas the government wishes to encourage. This behavior has a severe cost with progressive taxation: it undermines the tax's progressivity. The richest are the most likely to take advantage of the exemption because they have the highest marginal rate of taxation. In chapters 5 and 6, we show that the richest decedents used these exemptions to markedly reduce the progressivity of the estate tax system.

A3.6. Donations

Up to 1942, tax avoidance could take yet another form. Inter vivos gifts (donations) might not appear in an individual's consolidated estate. When tax rates were linear, the state did not care. It taxed donations in the same way it taxed estates. Under the progressive tax system in 1901, the incentive to make inter vivos gifts rose. Although inter vivos gifts were taxed at the same progressive rate as estates, someone who divided half her estate among her heirs during her lifetime would certainly reduce the bite of taxes because the whole estate was not exposed to the higher brackets. To mitigate such practices, starting in 1942, donations (inter vivos gifts) had to be integrated into the estate. This introduces thorny value problems for assets whose prices changed markedly over time. Nevertheless, since such donations tended to increase in frequency with wealth, their differential treatment should have reduced measured inequality before the reform (because they were omitted) and increased it afterward. The extent to which this effect is large or small depends on what the heirs reported. When the state produced statistics on inter vivos gifts, it distinguished three kinds: dowries, distributions to children, and all others. In the lead-up to the establishment

ABLE A3.3. Inter vivos transfers

		Total		Dowries		Distribution to children		Other	
		Number	Value	N	V	N	V	N	V
898	Seine	2,185	182,862,945	86.7%	76.3%	5.6%	20.4%	7.7%	3.4%
898	France	8,275	987,936,561	52.9%	53.7%	33.5%	39.4%	13.6%	6.9%
hare Seine		2.6%	18.5%						
954	Seine	2,277	3,474,500	48.0%	23.5%	11.4%	38.5%	40.7%	38.1%
954	France	92,837	6,6248,400	26.6%	12.6%	26.8%	52.2%	46.6%	35.2%
hare Seine		2.5%	5.2%						
962	Seine	2,000	125,366	34.1%	9.3%	1.0%	47.9%	53.9%	42.9%
962	France	80,246	1,636,869	18.5%	6.9%	29.3%	53.4%	52.2%	39.7%
hare Seine		2.5%	7.7%						

urces: BSLC 1899, 343–69 (Les donations en 1898). SEF, Supplément no. 91, July 1956, p. 823 and . 204, December 1965, p. 1705

of progressive taxation, the state collected and reported donations by these three types and by département for 1898, 1954, and 1962 (see table A3.3).

Table A3.3 shows that in 1898, most donations were dowries by number or value. This was particularly true for the Seine, where 86% of all donations were dowries (to brides and grooms). This category accounted for 76% of all the value transferred in registered inter vivos transfers. Dowries appeared quite systematically in estates of different kinds. They appeared in the estate of the parent who died first, either because parents insisted on equal division, even if dowries had been unequal among the children, or because the couple's marriage contract required that the estate of the first to die reimburse the community for the children's dowries. The other two categories of donations were less likely to be reported in estate tax documents before 1942. If an individual distributed her wealth to her children as a *donation-partage* before her death, there is no reason for us to see this estate. Fortunately, this kind of King Lear behavior was rare in the Seine: as late as 1898, it concerned less than 200 individuals a year (when the number of positive estates was more than 15,000). In Paris, at least, the residual category was tiny. France, as a whole, tells a different story; dowries comprised about half of the total, and estate divisions were nearly a third of the total. If we jump to 1962, the share of donations from marriage contracts plummets everywhere, particularly in Paris, and is largely replaced by "other donations."

One might well worry, then, that the interwar period featured lots of donations to reduce the tax burden of transmitting wealth to children. This hypothesis we leave to others to investigate. Since these transfers involved significant amounts of wealth (the average donation was close to the average estate) and were likely made by wealthy individuals who also left an estate, we underestimate inequality. However, the extent of the bias is likely to be small. In the Seine, at least, the number of donations was relatively small, about two thousand, or a tenth of the number of estates filed in any given year, and about the same fraction of the wealth transmitted. Given that through the mid-1950s, at least half the donations were dowries in the Seine, we may miss at most 5% of the wealth transmitted through families.

A3.7. Inventing Tax-Exempt Estates

In 1956, legislators broke with a century-and-a-half-long commitment to collect inheritance taxes on the first franc. They introduced a threshold such that estates of 1 million francs or less were not liable for the estate tax. From then on, the published tabulations reported no estates below the million-franc threshold. As table A3.4 reports, for France as a whole, the number of filings fell by about 75% between 1955 and 1956. The reform thus had a dramatic effect on the number of reported estates. In the Seine, tabulated estates fell by two-thirds. The threshold was kept at the same level through 1977. Asset price inflation led the number of estates liable for the tax to grow rapidly, and by 1959, they had recovered to about half what they had been before the reform. The impact of the reform on the wealth subject to the tax was much more muted. Estates smaller than a million represented about 20% of French estates and less than 10% of the value of estates in the Seine, and even less in Paris. As we discussed in chapter 10, because of the high levels of Parisian wealth and inequality, the politics of taxation played out entirely differently in the capital and the provinces.

One might have expected the number of estates we could collect for 1957 to fall by the same amount, but that is not the case. The number of estates for which we have wealth information in 1957 and 1962 is over 13,000 or 30% more than in 1952. There are also more estates than what was reported for the Seine as a whole in the same years. The discrepancy between the Paris microsamples and the Seine published data comes from two conjoined forces. First, because the published data included the suburban arrondissements

TABLE A3.4. Number of estates tabulated by the government and in our cross sections, 1952–1962

	Number of estates			% below the threshold			Value below the threshold		
	Published		Microsamples	Published		Microsamples	Published		Microsamples
	France	Seine	Paris	France	Seine	Paris	France	Seine	Paris
1952	285,768	24,921	10,719	83.40%	75.80%	66.60%	25.80%	12.70%	6.50%
1953	259,223	20,023		80.40%	73.90%		24.30%	13.20%	
1954	285,932	25,220		77.70%	69.00%		21.00%	9.80%	
1955	248,960	21,500		75.40%	65.40%		19.10%	8.50%	
1956	64,735	7,351							
1957	68,727	8,519	13,546			56.70%			3.70%
1958	84,038	10,053							
1959	88,560	12307							
1960	89,654								
1961									
1962	104,652	12,218	13,355			33.30%			0.80%

Sources: Reported tabulations for France (SEF 1955–1965; see Piketty 2001, appendix J). Microsample values come from the Estates dataset.

Note: No data were published for 1961; in 1960, only the French totals were published.

with many small filings, the aggregate number of estates in the published tabulations fell steeply—more steeply than they would have for Paris alone. The different geography of published tabulations and microsamples can only explain a smaller decline in Paris than in the Seine, but it cannot explain the rise in the number of filings we observe.

The increase comes from the fact that individuals below the threshold in Paris continued to file at very high rates. These estates accounted for about half of all estates filed in 1957 *after* the reform. Whether the tax officials tabulated them or not, we will never know because the archives of the finance ministry's statistical bureau seem to have disappeared. In any case, they did not publish them. The relative plethora of small filings matters for computations of the fraction of decedents who left something behind, but it is largely irrelevant to the measurement of inequality. Half of all estates were less than the threshold but accounted for less than 4% of the wealth in 1957. Omitting these small estates increases the share of the top 0.1% by less than 0.6% and the top 1% by less than 2%.

3.8. Tax Evasion

The devices we have discussed, such as delay, donations, and tax-exempt assets, were legal stratagems that allowed individuals to reduce or eliminate the tax burden of their inheritances when the state did not eliminate them via the threshold of 1956. Another important element is tax evasion. There are two significant mechanisms for tax evasion: underreporting assets and holding assets abroad. We observe traces of attempts at underreporting, but there is essentially no trace of assets held abroad that the heirs were able not to report.

We observe underreporting through the additional filings that follow the principal estate declaration. We have "insuffisances" where the state forced a revaluation of assets undervalued by the heirs—most often, this concerned real estate. We also have "omissions" where additional assets were brought into the estate. The published totals underestimate wealth since these additional values were not counted in the principal declaration. The impact on inequality depends on how much corrections correlated with wealth. A simple test of this is to take the ratio of the principal declaration to the total estate, which we report in table A3.5 from 1907 to 1957. In most cases, the share of the estate included in the principal declaration declines with wealth. After 1937, for the top 0.1%, additional filings account for 5% to 10% of all wealth; for other fractiles, 5% generally represents an upper bound.

TABLE A3.5. Ratio of the value of principal filings to total net estate

	Fractiles of the wealth distribution				
	P0–70	P80–89	P90–98	P99–99.9	Top 0.1%
1907	100%	98%	97%	95%	95%
1912	100%	100%	100%	99%	99%
1922	100%	99%	99%	99%	97%
1927	99%	99%	99%	99%	100%
1932	99%	99%	99%	99%	100%
1937	99%	99%	99%	98%	93%
1942	97%	96%	96%	92%	90%
1947	98%	97%	94%	95%	93%
1952	100%	95%	96%	96%	93%
1957	99%	97%	94%	96%	95%

Source: Estates dataset.

Beyond the underreporting caught by the tax authorities, fiscal evasion has received considerable attention. Gabriel Zucman and coauthors have estimated the share of wealth parked in tax havens. Their work can be used to evaluate the extent to which the decline in observed wealth inequality that followed WWI was due to tax evasion rather than a more even distribution of assets within France.

We know from the work of Zucman that, in the last half-century, substantial amounts of wealth have been hidden in Switzerland and elsewhere—anywhere between 2% and 4% of all European wealth (Zucman 2013, 2015). It proves useful to define onshore wealth as wealth held in France or abroad that appears in estate tax filings. It is important to note that onshore wealth includes any asset held in the home country (in our case, France), including foreign assets. We define foreign assets as the part of onshore wealth that involves stocks and bonds of foreign entities (e.g., the Russian government or US Steel). Like Zucman, we define offshore wealth as wealth held abroad and not reported to fiscal authorities (this could include French stocks and bonds). Zucman's work raises issues about how much of the observed decline in inequality after WWI is due to the rise of offshore wealth. The top part of the wealth distribution that faces the highest tax liabilities has, of course, the greatest incentives and abilities to move its money around. Zucman's initial work, however, produced no estimate of the extent of evasion by income or wealth. Offshore wealth could include French assets, but that detail is not reported.

Since then, Zucman and his coauthors Annette Alstadsæter and Niels Johannesen (2018–2019) have shown that offshore tax evasion is largely reserved for the top 1% and even more so for the top 0.1%. Below the top 1%, tax evasion occurs at a low rate (the top centile holds 93% of all offshore wealth). In the appendix to their paper, Alstadsæter et al. also report the offshore wealth of top (0.1% and 0.01%) wealth holders from various countries, including France, over the twentieth century. For France, offshore wealth was trivial before WWI, when the foreign share of wealth in top portfolios in Paris was about 20%. According to Alstadsæter et al., the importance of offshore wealth rose steadily over time, starting in the 1920s. This is also the view of Christophe Farquet (2018) and Sébastien Guex (2022). They focus on the rise of Switzerland as a fiscal haven. By 1972, the French top 0.1% held about 13% of its wealth offshore; at the same time, we estimate that about 5% of top Parisian wealth holders' onshore portfolio was composed of foreign assets. Interestingly, combining offshore wealth with onshore foreign assets leads to a relative constancy of the foreign portfolio share at 20%.

Alstadsæter et al. report some key statistics, including the share of offshore wealth held by different fractiles for the distribution of onshore wealth alone and including offshore wealth. These statistics allow us to compute the offshore wealth change over time. Those data are enough to compute two estimates of the offshore wealth share of top fractiles in Paris for 1957 and 1962. To do so, we must choose a path that takes us from Alstadsæter et al.'s estimates of offshore wealth held by the top 0.1% at the national level to the wealth of the dead in Paris. Given our general results that the inequality of the dead is similar to that of the living, we assume that the dead are an unselected sample for the living. Below, we detail the method we used for the top 0.1% in 1957. Table A3.6 reports the computations using the same methods for 1957 and 1962 for the top 0.1% and 0.01%.

The tabulation published by the government reports that total onshore wealth at death in France in 1957 was 286.7 billion francs. We measured Parisian onshore wealth at death to be 59 billion or 20% of the French total. Alstadsæter et al. tell us to increment the French onshore total by 0.76% of 2.2 billion to account for offshore wealth, for a total of just under 288.9 billion. This may seem like a small issue, but distribution here is everything. To allocate the share of those 2.2 billion to the top 0.1%, Alstadsæter et al. report that the top 0.1% held 77% of all offshore wealth (1.7 billion francs). The next question is, how much of this 1.7 billion belonged to Parisians?

Our first estimate assumes that very wealthy Parisians' propensity to evade taxes is the same as that of other French people of the same wealth

TABLE A3.6. Offshore wealth held by Parisians in 1957 and 1962

		1957 (million francs)		1962 (francs)	
		Top 0.1%	Top 0.01%	Top 0.1%	Top 0.01%
France	Wealth thresholds	44.13	146,00	798,372	2,694,351
	Evasion rate	4.1%	11.6%	5.2%	12.0%
France	# estates in top bracket	465.31	46.53	510.01	51.00
Paris	# estates in top bracket	243	44	340	64
DOMESTIC WEALTH					
Paris	Total	58,697	58,697	1,596,014	1,596,014
	Top fractile	26,481	12,100	957,623	583,999
	Top share	45.1%	20.6%	60.0%	36.6%
France	Total	286,688	286,688	6,287,051	6,287,051
	Top fractile	42,747	14,140	863,111	291,284
	Top share	14.9%	4.9%	13.7%	4.6%
COMPUTING OFFSHORE AND TOTAL WEALTH					
France	(Offshore/private) wealth	0.76%	0.76%	0.97%	0.97%
	Offshore wealth	2,207	2,207	61,406	61,406
	Private wealth	288,895	288,895	6,348,457	6,348,457
Paris	Domestic wealth share	20.5%	20.5%	25.4%	25.4%
COMPUTING PRIVATE WEALTH TOP SHARE IF EVASION IS NOT CORRELATED TO PARIS RESIDENCE					
Paris	Top Paris/France	61.9%	85.6%	111.0%	200.5%
	Other Paris/France	13.2%	17.1%	11.8%	16.9%
	Offshore wealth total	1,119	1,156	54,059	68,503
	Offshore wealth top	1,051	974	52,389	63,482
	Private wealth total	59,816	59,854	1,650,072	1,664,516
	Private wealth top	27,533	13,073	1,010,012	647,481
	Private wealth top share	46.0%	21.8%	61.2%	38.9%
HIGHEST CASE: ALL EVASION IS PARISIAN					
Paris	Total wealth	60,904	60,904	1,657,420	1,657,420
Paris	Total wealth top group	28,178	13,238	1,004,841	615,662
	Top group wealth share	46.3%	21.7%	60.6%	37.1%

Note: The wealth thresholds are computed from reported tabulations for France in 1957 and 1962. The evasion rate, offshore wealth share, and top offshore wealth share were computed by Alstadsæter et al. (2018, online appendix). Paris wealth levels were computed from our 1957 and 1962 microsamples. The rest is arithmetic.

levels. So, we only need to find out how many Parisians are in the top 0.1% of France. We start by computing wealth thresholds for France based on the published reports of estate tax filings. In 1957, the published statistics indicated that the 0.1% wealth-at-death threshold was 44 million francs and that there were 465 such estates in France. Our Parisian sample set tells us there were 243 estates that size filed in Paris in 1957. Parisians thus accounted for 52% of the top estates in France, but even within that group, they were richer than average and accounted for 62% of the wealth in top French estates that year. Given the assumption that all top wealth holders evaded at the same rate, Parisians owned 62% of the 1.7 billion francs held offshore by the top 0.1% of wealth holders. This neatly comes to 1 billion francs (evasion below the top group adds another 100 million to Parisian wealth). This allows us to compute the total wealth share for that top group. The share of onshore French wealth at death in Paris for those with at least 44 million is 45.1% (26.5/58.7). To consider total wealth, we add 1.1 billion to the denominator (total offshore Parisian wealth) and 1 billion to the numerator (27.5/59.8) for a revised top wealth share of 46%.

Parisians living in France's most sophisticated financial market might have had a greater propensity to hold foreign assets than other rich French people simply because it was easier. If true, we must increase the foreign wealth share to reflect this tendency. In the extreme, assume that Parisians hold all of the offshore assets. As the bottom panel of table A3.6 shows, we must add 2.2 billion to the denominator and 1.6 billion to the numerator (28.1/60.9) for a top wealth share of 46.3%, not much different from the onshore ratio we computed. The French 0.1% wealth threshold is close to the 1% threshold for Paris (38.9 million), so the computation gives a good idea of how much we need to revise the top wealth share. The top 0.01% threshold for France (146 million) is lower than the 0.1% threshold for Paris (193 million), so the 0.01% computation is illustrative of the impact of offshore wealth at the very top. Offshore wealth increases their share of Parisian wealth from 20.6% to 21.8%. Overall, as table A3.6 shows, the effect of offshore wealth is rather small up to 1962. It does matter to total wealth calculation, but its distribution, while far more unequal than the distribution of wealth in France, is not very different from the massively unequal distribution of onshore wealth in Paris.

One of the dramatic findings of Alstadsæter et al. is that the move toward owning offshore holdings accelerated at the end of the 1960s. Unfortunately, we lack the sources to track that change (namely, an accurate distribution of wealth at death in France in 1972). Before 1957, Alstadsæter et al. reported only small shares of offshore wealth, so we do not worry much about the issue for the earlier years. Before WWII, our data at least suggest that wealthy Parisians held a lot of their foreign assets onshore. After 1962, we must acknowledge that the estimates of offshore wealth rose and that systematically biases our estimates of wealth inequality and the share of Parisians in French wealth.

APPENDIX 4

Deportation and Wealth Distribution

Critical to measuring inequality is the total population to be considered. In our case, that is the total number of adult Parisian residents who died in a given year. In most years, it is sensible to take the death by sex and age tables from the ASVP as the base population for our dataset. For 1942, this approach could not work. The ASVP reported 36,125 deaths in Paris in 1942. Of those, 18,089 were male (16,524 adults) and 18,036 female (16,684 adults). The 33,208 adult deaths of 1942 are very similar to the totals for 1941 and 1943 and about 10% higher than the numbers before the war.

The TSA records yield 15,315 rich decedents for 1942. We found estates for 14,548. The 771 we missed could have been filed outside our data collection period from January 1, 1942, to June 30, 1954. Because the war years were so complicated, we knew many estates were filed late or very late, so we opened all the estate tax volumes over eleven and a half years. We also found a few estates in later periods. We may have just missed some estates. If we follow the same procedure as we do for other years, given that we have found 95.5% (14,548 out of 15,315) of the rich and we want a similar proportion of the poor, we reduce the total population of adult deaths to be considered from 33,208 to 31,699.[1]

This number total of death has a huge problem: as a result of the agreements between the Vichy government and the Nazis, the Paris police used lists of Jews that had been compiled after the defeat of 1940 to go door to door and arrest 13,152 Jews, mostly foreign-born Jews, on July 16 and 17, 1942. The police brought their captives to the Paris cycling track: the Vélodrome d'Hiver. This tragedy has gone down in history as the Vél d'Hiv roundup and is emblematic of the collaboration between the Parisian authorities and the Nazis. From Paris, the arrested Jews were sent to concentration camps in France like Drancy and then deported to Auschwitz. Less than 100 survived (for more details, see Joly 2023).

While the Vél d'Hiv was the most dramatic episode, it was neither the start of mass deportation in France (the first convoy left in March 1942) nor its end. In 1942, nearly 42,000 Jews were sent to Auschwitz from France (more than half of all deportations of Jews during the war); only 700 survived. Since very few Jews were deported from France before 1942, fixing the 1942 population at risk requires us to learn what proportion of the 1942 deported Jews were adults, lived in Paris, and died in 1942.

No one has produced such an estimate. While the total number of Jews deported from France between 1940 and 1944 who died in deportation is known, the number of Parisian Jews who died in deportation in 1942 is not.

A4.1. Parisian Deportees Who Died in 1942

There are, however, sources that allow us to calculate this number. The defense ministry and the site Mémoire des Hommes (MDH) have posted a dataset of 111,858 individuals who died in deportation; this includes many individuals who were not Jewish or did not die in death camps. The site Morts dans les Camps (MdlC) contains 80,909 mentions of individuals who died in death camps and whose death records have retrospectively been inscribed as died in deportation. Neither of these datasets provides a place of residence at the time of death. Finally, the Mémorial de la Shoah (MdlS) website provides a list of the 76,240 Jews captured in France and deported in one of 77 convoys (overwhelmingly to Auschwitz). This last list can be used to consult the individual files of deportees that provide residence information.

We first drew all individuals whose last names started with A or T in the MdH, MdlC, and MdlS datasets and matched them. Table A4.1 provides the details. Each dataset contributes individuals who are not in the other datasets. Some of this is due to how different datasets handle married versus maiden names, while some is due to how the sources create their lists. For instance, the MdlS list includes survivors. Matching reduces the dataset to 10,401 individuals. Table A4.1 provides the totals and the share of individuals whose last names started with T or A. The MdlS sample is exclusively Jewish, and nearly everyone died in a concentration camp.

Because we want to recover individuals who died in 1942, we remove all individuals who died in 1941 or after December 31, 1942 (all individuals in convoys numbered 46 and higher because they left France in 1943 or later); all Spanish prisoners deported from France and all children (individuals

TABLE A4.1. The samples of deportees

	MdlS match		
Data sources	Convoy found	Convoy not found	Total
Matched MdlC and MdH	2,608	2,336	4,944
Only MdH	99	2,543	2,642
Only MdlC	547	824	1,371
Only MdlS	1,444	0	1,444
Total	4,698	5,703	10,401

Source: MdlC, Morts dans les camps, https://lesmortsdanslescamps.com; MdH, Mémoire des Hommes, https://www.memoiredeshommes.sga.defense.gouv.fr/; MdlS, Mémorial de la Shoah, https://www.memorialdelashoah.org.

TABLE A4.2. The residences of Jews deported from France

	Year of Death			
	1942	Unknown	Unknown	
Residence	Age 20+	Age 20+	Age unknown	Total
Paris	1137	269	1	1407
Other	312	109	3	424
No address	177	233	6	416
Total	1626	611	10	2247

Source: See table A4.1.

born after 1922). Lastly, we removed the 62 individuals who survived. These we could all remove safely. Another 615 individuals were not matched to a convoy (216 died in 1942, and 399 had no death date). We also left them aside because these individuals could not be characterized precisely (in particular, we cannot hope to get a residence).

The resulting dataset totals 2,247 individuals matched to a convoy that left France in 1942 (see table A4.2). It comprises 1,626 adult individuals that MdlC states died in 1942, 611 for whom no death date is known, and 10 individuals for whom we have no age and leave aside. For the 2,237 remaining individuals, we consulted the individual files at the Mémorial de la Shoah and established their last known address.

As table A4.2 shows, nearly half of the sample's adults known to have died in 1942 have a Parisian address (1137/2237 = 50.8%). Another 269 adults have a Parisian address, but we do not know when they died. Thus, 62.9%

TABLE A4.3. Adult Jews deported from France and killed in 1942

	Total deaths		
Adult Jews deported	34,420	34,420	34,420
Scenario	Low	Medium	High
Share Parisian	50.8%	62.4%	75.8%
Total Parisian adults	17,495	21,493	26,095

Source: See table A4.1.

(1,406/2,237) of all adult deported Jews in 1942 had a reported address in Paris. Over the whole dataset, 96.6% of those deported in 1942 died in the same year. If we assume that death date reporting is random, 259 of the 269 deportees without a known death date died in 1942. This implies that 62.4% of Jews deported in 1942 were Parisians.

Finally, another 416 had no address. Again, if we assume they were Parisians in the same proportion as those for whom we have an address, we would have to add 300 individuals to the Parisian total; the share of deported Jews who died in 1942 and resided in the capital would then be 75.8%.

The convoy statistics report that 41,959 Jews were deported in convoys that left France before the end of 1942. Of these, 6,839 were children, and 700 survived. That leaves 34,420 adult Jews deported who died. The three estimates of the proportion of adult Parisian Jews killed in deportation in 1942 discussed just above can be applied to that total to form estimates of the total number of adult Parisian Jews killed in deportation (see table A4.3).

If we take the least restrictive set of assumptions (both reported death and reported residence are random), then 26,095 adult Parisian Jews were deported and killed in 1942. That is a high estimate; it is quite likely too high because many Jews who arrived before the defeat of 1940 were interned and thus never established a residence. The lower bound, which assumes that all Jews without a death date died in 1943 or later, implies that at least 17,495 adult Parisian Jews were deported and killed in 1942. We think this is likely too low because the primary reason for not having a death date is a lack of a death registration in official French documents. Such official records were more likely when other family members survived and wanted to establish claims on private assets or public indemnities. It should be noted that 53% of adults deported in 1942, and almost all the children, were sent to the gas chambers directly upon arrival at Auschwitz; finding their heirs would be extremely difficult. The middle estimate—where we assume that unreported deaths are random but unreported residence implies something other than Paris—suggests that 21,493 adult Parisian Jews were deported and killed in 1942.

A4.2. The Distribution of Wealth Among the Dead in 1942

Taking in the 21,493 adults who died in deportation, the correct number of Parisians at risk of filing an estate tax was not 33,208 but 54,701. Considering that we found 95.5% of those who had a filing, the reference population is 52,239. That, in turn, implies the top 1% is the richest 522 individuals rather than the richest 332. Given that we have been exhaustive in our wealth collection, taking into account those who died in deportation changes the number of individuals who died poor and the other measures of the distribution of wealth.

The results are displayed in table A4.4. Inequality is obviously much higher if we take the deportees into account, and, as was the case for nearly all of them, they had been despoiled of any wealth. Here, we assume the 21,493 deported Jews, save for the few hundred already in the tax records, were all poor. Doing so nearly doubles the total population of decedents and does not change the number of estate tax filings. As a result, the fraction of Parisian residents who died in 1942 and for whom no estate was ever filed jumps from 55% to 62%. Other inequality measures also reflect this change, with the top 1% share moving from 53% to 63%.

A4.3. WEALTHY AND POOR DEPORTED JEWS

The assumption that only deported Jews for whom an estate tax filing was recovered were wealthy is untenable because tax filing requires heirs, and they know how to find the deceased's wealth. The Vichy government confiscated most Jews' assets, the Holocaust destroyed entire families, and financial institutions' secrecy, in fact, made it difficult for distant heirs to learn about relatives' wealth. There is some evidence, however, that suggests that many deported Jews had wealth, and some had quite large amounts before Vichy and the Holocaust made them poor. The most systematic source involves lists of bank account holders.

As part of the Commission Mattéoli's effort to quantify the financial losses suffered by Jews in France during WWII, Claire Andrieu and her

TABLE A4.4. The distribution of wealth in Paris, taking the number of deported Jews into account

	Wealth shares per intermediate fractile								
Deported Jews	P0–50	P50–60	P60–70	P70–80	P80–90	P90–95	P95–99	P99–99.9	P99.9–100
No	0%	0%	1%	2%	6%	9%	29%	32%	21%
Yes	0%	0%	0%	1%	4%	7%	27%	36%	27%
	Wealth shares per top fractile								
	P0–50	P50–60	P60–70	P70–80	P80–90	P90–95	P95–99	P99–99.9	P99.9–100
No	100%	100%	99%	99%	97%	91%	82%	53%	21%
Yes	100%	100%	100%	100%	100%	96%	90%	63%	27%
	Percentiles thresholds (current francs)								
	P0–50	P50–60	P60–70	P70–80	P80–90	P90–95	P95–99	P99–99.9	P99.9–100
No	0	0	7,402	23,104	63,129	225,920	659,132	3,718,598	20,600,000
Yes	0	0	0	0	16,805	92,037	315,546	2,362,234	14,500,000
	Average net estate per intermediate fractile (current francs)								
	P0–50	P50–60	P60–70	P70–80	P80–90	P90–95	P95–99	P99–99.9	P99.9–100
No	0	2,421	14,184	39,712	120,303	387,279	1,500,978	7,291,774	43,900,000
Yes	0	0	0	6,449	43,946	169,833	840,618	4,981,248	32,986,792
	Average net estate per top fractile (current francs)								
	P0–50	P50–60	P60–70	P70–80	P80–90	P90–95	P95–99	P99–99.9	P99.9–100
No	206,944	413,875	516,738	684,256	1,006,529	1,892,754	3,398,230	10,987,242	43,900,000
Yes	125,429	250,848	313,556	418,095	623,859	1,203,884	2,237,935	7,813,870	32,986,792

Source: See table A4.1.

N[illegible]: The "no" lines report values of the wealth distribution not taking the deportees into account. The "yes" lines report values of the wealth distribution team collected a sample of deported Jews (those whose last names started with BA) (Andrieu 2000, 177–180). That work produced an estimate that 16% of the individuals who had had their bank accounts seized during the war were deported if we take both certain and likely matches. For our purposes, the share of seized bank account holders is only a first step because we are interested in the share of deported Jews who had bank accounts and use it to estimate the share of Jews who were wealthy.

We merged our dataset of 1,407 Parisian deported Jews with the list of seized-bank-account holders whose last names started with T or A established by the Commission Mattéoli. The merge produced 162 individuals with the same first and last names and address in the two sources and another 42 for whom last name and address match (for extra confidence, we leave these 42 aside). The 162 fully matched individuals represent 14.2% of the 1,137 adult deportees who lived in Paris and died in 1942. In the 1942 Parisian estate dataset, 52% of wealthy individuals have financial accounts.[2] Assuming the same proportion for deportees leads to 27.4% of deportees, or 5,889, who likely had an estate.

The next question is how many deported Jews came to the attention of the estate tax authorities. In principle, the TSA listed all legal residents of Paris when they died and were thus potentially liable for the estate tax. The registers should, therefore, have included all Parisian deported Jews who died in 1942, whether or not there was wealth to transmit. We sought out who among the deportees in our sample was inscribed in the TSA by examining all the volumes for individuals whose last names started in A and T and whether or not they had a tax filing. We collected all the names that appear at the end of each month's list from March to December. Some entries mention that the person died in Auschwitz or Drancy, but most do not. In 1942, the TSA included only a small fraction of those who died in a concentration camp (in France or abroad).

The procedure yielded 308 individuals. Our list included 43 children, one person who was deported in convoy 68 and thus could not have died in 1942, and one person who appears in the MdlS as not deported. That leaves 263 adult Parisian Jews in the TSA, and 246 were matched to the deported dataset. We left the 17 that we could not find aside as they could have died in France rather than being deported. The 246 adults with a TSA represent 17.5% of the 1,406 deported Jews with an age and a Paris address, leading us to estimate that out of the total 21,493 deported Jews, some 3,757 could have been inscribed in the TSA. Of the 246 we found in the TSA, only 30 (2.1%) ever had a filing of any kind, a stark contrast with the nearly 45%

filing rate for the nondeported population. This also implies that about 458 deported Jews (21,493*2.1%) should appear in the full dataset we collected for Paris in 1942.

In any case, the share of deported Jews for whom an estate was filed seems to have been on the order of a tenth (2.8%) of what the bank sample would lead us to expect (24.8%). Thus, and unsurprisingly, the tabulations of estates, like the death totals, are inaccurate because both omit nearly all Jews killed as part of the Holocaust.

APPENDIX 5

Reported and Capitalized Inheritance

Reported inherited wealth is a mix of cash claims and assets in kind. We rely on rates of capital gains and flow returns to reach either the value of inherited wealth at the time of receipt (about 25 years before death) or capitalized inherited wealth. Our computations of the fraction of individuals whose inherited wealth is larger than their estate, or the fraction of inherited wealth, both rise if we increase the rates of capital gains or flow returns but by less than would be expected because a substantial part of the distribution lies at extremes of the distribution, as table A5 reports.

At one end of the distribution, nearly half of all individuals and at least 20% of the wealth are associated with individuals whose reported inherited wealth share requires a 10% rate of return or more before their capitalized inherited wealth would reach the size of the estate. These are the estates in the first two rows of both panels of table A5. At the other end of the distribution, 30% before WWI and 10% afterward, are individuals whose capitalized inherited wealth will reach 100% if the nominal return to their assets is 2%. The yield on government bonds is always higher than that, so they are rentiers for any flow rate of return we might use. Moreover, before WWII, these estates amounted to about one-third of all the wealth, and those where inherited wealth reached bequeathed wealth with a 3% flow return were a majority of the wealth from 1852 to 1922.

TABLE A5. The distribution of reported inherited wealth shares

	Reported inherited wealth shares			
	1822–1852	1862–1912	1922–1947	1952–1972
	Share of estates			
None	44.2%	55.0%	58.1%	37.6%
Less than 10%	3.6%	7.1%	12.9%	35.6%
10% to 25%	2.9%	6.3%	6.4%	6.7%
25% to 60%	5.9%	10.7%	10.3%	9.7%
60% to 100%	7.2%	11.9%	8.4%	7.5%
100% or more	36.2%	9.1%	3.8%	2.9%
	Share of wealth			
	1822–1852	1862–1912	1922–1947	1952–1972
None	12.2%	8.9%	14.6%	13.6%
Less than 10%	17.8%	13.1%	20.6%	28.7%
10% to 25%	12.5%	11.9%	13.1%	17.2%
25% to 60%	25.9%	18.7%	19.0%	17.7%
60% to 100%	18.4%	24.3%	23.0%	18.9%
100% or more	13.2%	23.1%	9.6%	3.9%

Source: Estates dataset.

APPENDIX 6

The Top Wealth Holders Each Year

TABLE A6. The top wealth holders

Year	Last Name	Sex	Marital status	Age: Death	Age: Marriage	Wealth/mean wealth	Branch or career	Family business	Principal asset: Personal	Principal asset: Community
1807	Hulot	F	M	28	17	855	Industry	Continues	Not sampled	
1812	D'espinay de St Luc	F	W	62	24	245	Politics	Continues	PtoP loans	
1817	Baguenault	M	M	57	32	507	Finance	Founder	Business	
1822	D'Orléans	F	M	66	20	1031	Politics	Continues	Family debt	
1827	De Caulaincourt	M	M	53	40	478	Politics	Continues	Pub debt	
1832	Foucard	F	W	60	16	413	Service	Founder	Pub debt	
1837	du Bouchet de S.	F	W	71	17	332	Politics	Continues	Real Estate	
1842	Leroux	M	W	70	29	898	Finance	Founder	Fr-Shares	
1847	d'Aligre	M	W	77	40	939	Politics	Continues	Real Estate	
1852	Michel	M	W	81	32	1024	Finance	.	Pub debt	
1857	D'andigné	M	W	93	22	296	Politics	Continues	Pub debt	
1862	Dossier	M	W	72	43	1273	Finance	Continues	Pub debt	
1867	Maynaud de P.	F	W	76	20	519	Politics	Continues	Pub debt	
1872	Wey	F	W	47	19	801	Industry	Founder	Real Estate	Business
1877	Blanc	M	M	71	48	1672	Industry	Founder	Business	
1882	Vogel	M	C	.	.	510	Finance	.	Business	
1887	Roederer	F	M	48	20	1006	Industry	Continues	Fr-Bonds	Business
1892	Lebaudy	M	W	65	38	3260	Industry	Continues	Business	Fr-Bonds

1897	Leleu	F	W	78	17	860	Industry	Founder	Pub debt
1902	Stern	M	W	63	46	1205	Finance	Continues	Business
1907	De Rothschild	F	W	77	20	2244	Finance	Continues	Business
1912	Javal	M	M	68	44	866	Industry	Founder	Brokerage
1922	De Rothschild	F	W	79	19	2390	Finance	Continues	Fo-Shares
1927	Goldet-G.	M	M	55	26	183	Industry	Continues	Fo-Shares
1932	Fougerat	M	C	68	.	520	Industry	Founder	Business
1937	Deutsch De La M.	F	C	45	.	1520	.	Continues	Fr-Shares
1942	Schneider	M	M	74	26	1232	Industry	Continues	Business
1947	Boutheroue-D.	M	W	82	27	1144	Industry	Continues	Business
1952	Letellier	F	D	82	31	898	Industry	Founder	Fo-Shares
1957	Fantauzzi	M	M	77	29	918	Industry	Continues	Business
1962	Worms	M	M	72	22	348	Finance	Founder	Business
1970	Double	F	W	101	18	1301	Industry	Continues	Not Sampled
1972	Roussel	M	M	50	21	.	Industry	Continues	Business
1974	Delore	F	W	78	24	846	Industry	Continues	Not Sampled
1976	Richard	M	M	67	33	.	Industry	Founder	Not Sampled
1977	Devaye	F	M	68	19	.	Industry	Founder	Not Sampled

Source: Estates dataset.

Note: The wealth column is the ratio of the individual estate to the average estate (including zeros). In the principal asset columns, PtoP loans are peer-to-peer loans, and Fr-Share are shares in French joint stock companies. Fo-Shares are shares in foreign companies. A period signifies that we have no information for that entry in the table.

NOTES

Introduction: Wealth and Inequality in the Long Run

1. Adeline Daumard (1962) collected a sample of probates for eighteenth-century Paris. Alice Hanson Jones (1980) did so for the United States on the eve of the French Revolution.

2. We skipped 1917 because many Parisian men died at the front during WWI and were not recorded in the Parisian death totals. We skipped 1967 because the records were incomplete.

3. ASF (1914, 243) Seine 67,318 loans for 3.96 billion francs; France 194,591 loans for 6.87 billion francs. Communal loans Seine 901 loans for 1.055 billion francs and France 51,958 loans for 4.762 billion francs.

4. The IRS reports income data by zip code, but it is top-coded at $200,000. For New York City, that top group is 7% of the filings and has 58% of the gross income and an average income of $930,000; the same 7% in the United States have half the gross income and an average income of $609,000. The city is both richer and more unequal than the state and country that contains it.

5. The annual value of the pension is given in a fraction of cases, but it is not reported in many other cases. Further, most retirement programs do not contain a capital component, so we cannot observe retirement benefits for working individuals. This is an obvious way that our source understates wealth.

1. Dukes and Indigents: Inequality in the Shadow of the Old Régime, 1807–1867

1. The city published data on the cause of death for all decedents starting in the 1820s. Unfortunately, those records were not preserved.

2. The civil registration records of births, deaths, and marriages begin in 1860; the earlier records burned during the Commune.

3. As appendix 1 shows, the estate declarations are more accurate, but the TSA provides an excellent picture of the distribution of wealth at death in Paris.

4. The number of bureaus grew to 7 in 1846, 9 in 1860, 14 in 1921, and 20 in 1955.

5. To be precise, Daumard collected the declarations filed in a given year, not the declarations of individuals who died in a given year.

6. In the paper we published with Thomas Piketty (Piketty et al. 2006), we only had the years starting at 7, so the dataset we analyze here is twice as large. Our results are, however, closely consistent with what we produced then.

7. The published data run to 1967, when the ASVP ceased publication, but can easily be extended forward from the INED databases. Lionel Kesztenbaum kindly provided that data.

8. From 1797 to 1872, the French franc was worth 4.5 grams of silver and about 0.29 grams of gold. After 1873, the franc was pegged only to gold (Flandreau 2004).

9. ADP, DQ8 #485, p. 8.

10. Burials were tallied for 1821–1826, then 1839–1848, and annually from 1872 to 1967.

11. Duma reports these values in *livres,* the Old Regime unit of account. The livre's gold and silver content was essentially identical to that of the franc.

12. The Duc de Penthièvre's fortune breaks down into 9.2 million francs in movables, 3.4 million in urban real estate, 10 million in bonds, 8 million in government offices, and 72 million in rural real estate (Duma 1995, 57). Had that estate been evaluated according to the estate tax rules of the nineteenth century, the Parisian declaration would have 28.6 million plus the current year's prorated rural property income; the 72 million in rural wealth would have been declared outside Paris; and at least 2 million of the bonds were public bonds and free of the estate tax, only the prorated interest would have appeared. Leaving aside price changes in real estate, the French Revolution would have amputated about 18 million from this massive fortune, but that would have been almost two-thirds of his Parisian estate. We do not observe fortunes in the range of 30 million francs until the 1860s.

13. A variety of scholars have shown that inheritance favors sons (cf. Bessière and Gollac 2022). Our data are silent on the matter because we did not trace how wealth was divided. The high share of men's wealth between 1807 and 1972 (see figure 4) suggests it's not so much men's estates that are undervalued as the shares and composition of what goes to sons (see chapter 8). We also emphasize that the relative equality of the estates of women and men does not offset the fact that men had extensive control over, and use of, their wives' wealth.

14. This finding comes from extending the dataset past 1862; see chapter 2.

15. These inverses can be obtained by taking the ratio of the population in an age group to the number of dead in that age group. By chance, the French ran censuses frequently, starting with 1821 in years ending in 1 and 6, which proves convenient given that our death totals are for the following year. For years without a census, we average the two adjoining years, and for the early years, we use the weights obtained for 1817.

16. This approach is simple to compute. Let M be the mortality of the whole population, M_p the mortality of the poor, and M_r that of the rich, and α the share of the poor in the population. We assume $M_p=2M_r$, by definition $M=\alpha M_p+(1-\alpha)M_r$, that implies $M_r=M(1+\alpha)$.

2. Belle Époque Capitalism, 1852–1912

1. Also see Archives du Ministère des Finances, Etats des montants des rôles généraux des contributions directes, B-0034393/11.

2. In 1970, the TSA system changed from volumes with a line per person to a binder system with a page per person that only listed individuals who died with some wealth.

3. Before 1901, the clerks transcribed the estate filings onto bound registers; after 1901, the registers were, in effect, binders that contained the filings individuals brought to the estate tax authorities.

4. The absence of civil registration records for those years makes it impossible to allocate the errors between TSA and RMD.

5. We add individuals in the RMD from the TSA either because they were entered in the wrong year in the TSA register or because that register was unavailable when we collected the TSA. We remove individuals we cannot find in the RMD (the year and number of the filing refer to a different individual) because they turn out not to be Parisians or because that filing belongs to another person from a different bureau.

6. APD DQ7 3931, p. 46 (declaration 709, August 7, 1822).

7. These values are consistent with those Daumard reported for 1820 and 1847.

8. We deflate wealth by the cost of hiring a worker for 2,750 hours a year, a figure that we take from Bayet (1997). The average number of hours worked per year is 2,750 over the full range of time (1840–1977) our study covers.

9. Dherbécourt's finding is that the assessed value of each child's inheritance is almost always the same. It remains that sons were more likely to inherit control in family enterprises. Although such control has value, it is never priced. Thus, equal sharing is a first approximation. See chapter 8.

10. *Communauté réduite aux acquets.*

11. Nobles do own more rural real estate. It represents 13% of their gross wealth in 1902, 9.6% in 1907, and 23% in 1912, while commoners have 6.7%, 8.5%, and 9.7% in rural real estate in the same years.

12. The distribution of international investments by year from 1842 to 1972 is given in table A2.1

3. The Laboratory Evolves: Progressive Taxation, 1901–1972

1. 1902, 1910, 1917, 1920, 1926, 1929, 1934, 1937, 1939, 1940, 1942, 1947, 1948, 1949, 1951, 1952, 1957, 1959, 1965, 1968.

4. The Collapse of Wealth, 1912–1947

1. Interestingly, the share of officers among the dead was almost two and half times larger for residents of Paris than for the rest of France. This is not surprising since there were many officers in various military administrations in the capital, and many such officers were then sent to combat positions. It is telling, however, that among those born in Paris, the proportion of officers was double that of the rest of France, testifying to yet another connection between wealth and social status.

2. We do not differentiate between individuals with savings or bank accounts (what the commission measured) versus those with accounts at brokers, notaries, or other agents. The bias, therefore, leads us to understate the fraction with wealth.

3. The other candidate households, given Parisians' low fertility, would be families with two parents and only one child—in which case our measures of inherited wealth are too low, and estate taxes and income taxes will be higher.

4. These tabulations are based on the inventories for ADP series D2R7.

5. ADP PEROTIN/1011/44/1/17 1. The same document also tabulates death from bombardments at 587, the number of wounded at 1,020, and the number of displaced persons at 21,000. Given a population of 2.6 million, the casualty rate was small relative to nearly all other major cities of Northern Europe.

6. ADP series 50W and 1131W.

5. Inequality when Wealth Collapses, 1912–1947

1. See Pastore (2024). In work not yet published, Angelo Riva and Amaury de Vicq are producing more detailed individual returns for the publicly traded part of individuals' estates for the first third of the twentieth century.

2. One thing is clear: portfolios are inconsistent with clear foresight, as they are far too dispersed to take advantage of the asset class with the highest realized return.

3. We use a log-log specification (setting zeros to 1) to estimate an elasticity. This specification also has the advantage that wealth at the very top does not drive the regression results. The regressions using datasets without the top 0.1% or the top 1% produce similar results.

4. Almost all asset values discussed in this chapter were culled from this database (and for each year, we use the closest price to June 30). Where we use other data, they are referenced in the text.

5. The Herfindhal index is the sum of the squared ratio of each asset to total wealth.

6. ADP DQ7 32161 #806.

7. ADP DQ7 31250 #299.

8. ADP DQ7 36081 #1872.

9. ADP DQ7 33972 #2026.

10. Jean Conrad: Boissy Saint Léger AD94 3Q1746, p. 95; Jean Henri ADP DQ7 10682 #374, Rodolphe DQ7 31951 864 31983#637 31989 #895, Jean Henri DQ7 32310 #204 DQ7 32388 #1991.

11. The estate details the assets included in the 1937 and 1942 donations but not those in 1934. The 7.5 million of this first donation was revalued to 12.3 million using the asset price inflation index of Piketty and Zucman (2014). We found prices for nearly all the financial assets for the other two donations. The real estate appears in the estates of Jean Henri's father, and we used real inflation indexes to value it.

12. In 1943, there was no official or market rate. The pound-to-dollar exchange rate was near 4 for the whole war. The last quotes for francs in the *New York Times* are in January 1941, and they are 50 francs per dollar. The first quotes at the end of August 1945 are again at 50 francs to the dollar. In January 1946, the franc was massively devalued to 120 to the dollar. We chose the lower values.

13. ADP DQ7 32082 #1141.

14. ADP DQ7 32071 #1637.

15. The pension value reported in the declaration involves the payment due to the deceased (from when he or she received his or her last payment to his or her day of death). To estimate the capital value of a pension, one would first have to compute the annual value of the pension (it is often reported, but we rarely note it) and then capitalize that value. To the extent that pensions were capitalized ten times the annual value and paid half yearly, it is not unreasonable to think that the capital values of retirement assets could be twenty times the reported liability. That would suggest they were the most important asset for retirees in the P70–89.

16. ADP, DQ7 28162 #2162 (Dujat); DQ7 35164 #148 (Leclère); DQ 7 30298 #360 (Ménard); DQ7 28137 #1058 (Jamet); DQ7 28923 #2260 (Gamon).

17. ADP DQ7 35176 #2063 and #2064.

18. Annuaire Bottin 1905, 1906, 1924, 1926, Archives commerciales de La France 02-22-1905, p. 224.

6. Recovery, 1947–1977

1. For this group, taxable wealth has to be amended to include tax-exempt assets (for details on Pinay bonds and real estate, see chapter 3 and appendix table A3.5). Tax-exempt assets are reported in the filings from 1952 to 1972, and we imputed them for estates in 1977 because we do not observe them in the summary files we used for that year.

2. ADP DQ7 37388 #594.

3. The currency reform of 1960 created a new franc worth 100 old francs, thus the 10,000 franc exemption in 1972 corresponds to the 1 million old franc exemption of the 1950s.

4. With the end of the publication of the ASVP in 1967, we lost easy access to death totals by arrondissement. The total number of adult deaths in Paris, however, fell from above 27,000 in the 1960s to 23,500 in 1977. This decline in adult death was steeper than before or afterward but is part of a long-term trend.

5. We sought the super rich and their spouses out in genealogy records to get marriage information. This is lacking from both the estate tax one-page summaries (*fiches*) and death records, as well as information on the places of death of the spouses. We consulted the files of those who were members of the Legion of Honor. We recovered the date and place of birth (289/305) and death for nearly all individuals (287/305) and less for place and date of marriage (219/305). We consulted the civil registration records for deaths and marriages when they were available online and otherwise trusted online genealogies recovered from Geneaweb (there are errors, but not many). The most glaring gap comes from marriages, where we only verified the data for 136 individuals out of 224 possible because civil records are not online.

6. The overall dataset is balanced with 153 women and 152 men. The simple reason for the difference in recovered death records is that male spouses were older at marriage and thus died in earlier years (on average, around 1970 relative to 1983 for women) in time and are more likely to be found in online records.

7. Capital and the Family: Inherited Wealth, 1822–1972

1. Before 1901, inheritance taxes on rural properties were assessed at their location (see chapter 1). This implies that we generally do not observe their value in Paris.

2. Seeking these transfers out in the Parisian notarial archives would be very time-consuming and we would only recover them if they had occurred in Paris.

3. Numbers for 1886 (p. 99) and 1891 (p. 174) are similar. From 1896 forward, the census tabulations on occupations focus on the active population and the rentier tallies disappear.

4. Wealth is a stock variable, while labor income is a flow variable. The appropriate stock version of labor income is its capitalized value over the same number of years as we estimated individuals enjoyed their inheritances (about 20 years).

8. Wealth at the Top

1. ADP DQ7 31682 #1064.

2. These include the firm (Rothschild Frères) and Rothschild men, none of whom are in our cross sections.

3. James's estate is filed ADP DQ7 10702 #455. TSA volumes for Lagny, Tournant, and Rosay en Brie accessed online. Betty's estate is found ADP DQ7, 12185 #27.

4. Alphonse ADP DQ7 26796 #1760 26798 #476; 26818 #1132; 26824 #1943; 26879 #1077; Gustave ADP DQ7 31813 #902; Edmond ADP DQ7 32194 #1225; 32250 #305.

5. Salomon James's estate: ADP DQ7 10579 # 867; 10624 #591.

6. Charlotte's estate: ADP DQ7 31550 #27; 31558 #817; 31603 #835. Her husband's Parisian wealth is ADP DQ7 10624 #725 (most of his estate was likely in London). The TSA of Pauillac for Château Lafite and the TSA for Neuilly are online. Leonora's estate ADP DQ7 26879 #1991; Cécile's estate: ADP DQ7 31829 #613.

7. See AN /MC/ET/CXVIII, passim.

8. The connection between age and activity is particularly striking in the Paris electoral list of 1845 (ADP VD4 4149-50-51). See table A2.2

9. ADP VD4 4149–51.

10. One could draw other comparisons, for instance, to the median or other means. The median taxable estate is zero even in 1972 so we leave it aside. The average of the top thousandths estate is roughly one thousand times the size of the median among positive estates and 70 times the average positive estate, and 90 times the average positive estate below the top thousandth.

11. Arnaud Manas, historian at the Banque de France, kindly provided copies of the shareholders' lists.

12. In the first few years, there are principal shareholders from Amsterdam and Switzerland. But by 1806, all those whom we can identify reside in France.

13. Fully sampling the top 0.1% from 1801 to 1936 would have led us to collect 3,400 individuals in the top 1% of which 1,700 were men—almost the same number as we have found in the shareholders' list.

14. Other models have far less stark predictions (e.g., Baker and Hall 2004)

15. For the top 0.1% group and the stockholders of the Banque de France, because we sought out their genealogical information, we have more direct information about their titles than what

is reported in the estate tax records. So the definition here is more demanding than in chapters 1 and 2, but our coverage is more extensive.

16. James Rothschild, an obvious outlier, was excluded from the computations.

9. The Myth of the Grasshopper and the Ant

1. Electoral rolls were regularly updated. They record voters who had died as well as those who had left Paris, in which case they often provided their destinations. However, as this information is not systematic, and concerns only men, it is not possible to determine the extent of these return migrations.

2. This summary is partial and leaves out mutual retirement societies and other reforms, including the law of 1910.

3. The CNRV was a government institution that managed retirement contributions by investing them in public debt.

4. We use census information for the share of employment in a sector that is Parisian and the reports of the CNRV for contributions by sector.

5. Voluntary contributions were also large (four times the average collective deposit), suggesting they came from well-compensated individuals rather than typical workers.

6. Children born out of wedlock who were not recognized at marriage had no claim on their parents' wealth. Children who were legitimated or adopted also had no claims on the estate if there were children from a marriage.

7. Moreover, before 1901, information about what each child received was lacking because the tax rate was the same for all children. After 1901, progressivity on inheritances means we do observe the value of each child's part of the estate.

10. Inequality in Space: Within Paris and Beyond

1. We also compute the Parisian shares of top wealth holders using the published tabulations for Paris as a robustness check. Before WWII, the alternative shares are indistinguishable. After 1942, our shares were generally higher because our totals (which include omissions and tax-exempt assets) were generally larger (see appendix 3.4).

2. As a counterfactual, we added to the Paris distribution enough estates to give it a positive estate share equal to France's (68%), and we gave each of these additional estates a value of 2,000 francs (a value larger than the largest estate below the top quartiles for the poorest departments). That leads to a tiny decline in the wealth share of the 1% and an increase in the wealth share below the top decile from 1.6% to 2%.

3. We did so because we were concerned with various forms of selection due to the inherent mobility of the top 1% (between their primary residence in Paris, their country estates, spas, and seaside resorts).

4. Defense spending accounted for a growing share of public spending: 26% in 1872, 38% in 1900, and 41% in 1912. Between the wars, it hovered around 30% (Delorme and André 1983, 727).

11. Inequality Beyond Wealth

1. A linkage would be possible through marriage records, but that path presents obvious selection problems.

2. We computed enrollment rates as the reported number of children enrolled in schools divided by 80% of the total number of children counted in the census between ages of 5 and 14 (to get an estimate of the population aged 6 to 13).

Conclusion: Capital and Capitals in the Long Run

1. We opened the TSA volumes much later to recover the ages and addresses.

2. The years 1852 and 1862 were assembled on the same principles as other RMD samples; 1822 and 1842 were completed as we wrote the book and are not as extensive. For details, see appendix 2.

3. The TSA biases wealth upward because some clerks seem to have recorded total community assets rather than the value of the estate.

Appendix 3. Progressive Taxation

1. The complete list of years for which tabulations were published is (those in bold are the ones that we added to the 2006 dataset): 1902, 1903, 1904, 1905, 1907, 1909, 1910, 1911, 1912, 1913, 1925, 1926, **1927**, 1929, **1932**, 1933, 1936, **1937**, 1938, **1939**, **1941**, **1942**, 1943, 1947, **1948**, 1951, 1952, 1953, 1956, **1957**, 1958.

2. We also restored the very large estates that created spikes in 1938 and 1951. In 2006, we had no microsamples for the twentieth century and focused the analysis on the long-term trend, but with the higher frequency of data we now have, it seems best to include them.

3. The archives of the finance ministry hold no records that would illuminate how these statistics were produced.

4. More than 90% of all individuals above the 1% threshold for the Seine are Parisians. For the interwar period, the ratio of Seine adult deaths to Parisian adult deaths is 1.7. The ratio of wealth at death in the Seine to wealth at death in Paris is about 1.2. Because Parisians dominate the top of the wealth distribution for the Seine, a good approximation of the 1% Seine share is the ratio of the wealth of the top 1.7% richest Parisians divided by 1.2% of Parisian wealth.

Appendix 4. Deportation and Wealth Distribution

1. These calculations, in turn, would imply that there were 31 individuals in the top 0.1% of decedents in 1942, 317 in the top 1%, and 3,176 in the top decile.

2. We do not differentiate between individuals with savings accounts, or bank accounts (what the commission measured), versus those with accounts at brokers, notaries, or other agents. The bias therefore leads us to understate the fraction with wealth.

REFERENCES

Archival Sources

ADP Archives de Paris

TSA Tables des successions et absences, 1807–1953 (DQ8) (online)

RMD Registres des mutations par décès (DQ7)

État civil (online)

Conscription ADP D4R1:18–10, 318–323; D3R1 270–73

Listes électorales, année 1845 (ADP VD4 4149-50-51)

AN Archives Nationales

Notarial contracts of the Rothschilds (AN/MC/ET/ CXVIII)

Notarial contracts of the Cousins (AN MC ET LXIV 588)

SAEF Service des archives économiques et financières

Etat du montant des rôles généraux. Seine: B 34393/11

Serial Publications

Annuaire Desfossés: valeurs cotées en banque à la Bourse de Paris, Paris, E. Desfossés et Fabre frères 1907–1949.

Annuaire et almanach du commerce, de l'industrie, de la magistrature et de l'administration; Annuaire du commerce Didot-Bottin.

ASF *Annuaire statistique de la France,* 1878–99; 1953–1966.

ASVP *Annuaire statistique de la ville de Paris, Seine* (France), 1880–1967. Bureau de la statistique départementale et communale.

RCE *Rapport au Président de la République sur les opérations des caisses d'épargne et de prévoyance,* 1836–1910. Paris, Imprimerie Nationale.

AGAP *Administration générale de l'assistance publique à Paris,* 1851–1864.

BSLC *Bulletin de statistique et législation comparée,* 1899–1930.

BSM *Bulletin de statistique municipale—ville de Paris,* 1865–1879.

Challenge, Les 500 plus grandes fortunes de France, 1999–2024.

Mémento-guide du propriétaire. Annuaire officiel de la chambre syndicale des propriétés immobilières de la ville de Paris. 1951–1952. Imp. commerciales du Maine libre (non daté)

Résultats statistiques du recensement général de la population en. . . . Statistique générale de la France & Insee. Données historiques de la statistique générale de France https://www.insee.fr/fr/statistiques/2653233?sommaire=2591397.

SEF *Statistiques et études financières, supplément statistiques,* 1949–1952; *Statistiques et études financières,* supplément, 1953–1968. Paris, Imprimerie Nationale.

Online Databases

Base Cassini, population by municipality for France, http://cassini.ehess.fr/cassini/fr/html/6_index.htm

Base de données des Morts pour la France de la Première Guerre mondiale, http://www.memoiredeshommes.sga.defense.gouv.fr/fr/

Data for Financial History, https://dfih.fr/

Geneanet, https://en.geneanet.org/

Military conscripts, doi:10.13144/lil-1572

Mission d'étude sur la spoliation des Juifs de France: rapport d'étape: janvier–décembre 1998 La doumentation francaise, https://www.vie-publique.fr/rapport/24212-mission-detude-sur-la-spoliation-des-juifs-de-france-rapport-general

Mémorial de la Shoah, https://www.memorialdelashoah.org

Morts dans les Camps, https://lesmortsdanslescamps.com

Offshore wealth, https://gabriel-zucman.eu/offshore/

Paris real estate values (Explorateur de données de valeurs foncières), https://explore.data.gouv.fr/fr/immobilier?onglet=carte&filtre=tous (Note: we downloaded the transactions from 2019 to 2024).

World Bank Economic Indicators, Gini, https://data.worldbank.org/indicator/SI.POV.GINI

Secondary Sources

Acemoglu, Daron, and James A. Robinson. 2006. *Economic Origins of Dictatorship and Democracy*. Cambridge University Press.

Aicardi, M. 1988. *La Fiscalité du patrimoine: Rapport au ministre d'état ministre de l'Economie, des finances et de la privatisation et au ministre chargé du budget*. La Documentation française.

Alesina, Alberto, Reza Baqir, and William Easterly. 1999. "Public Goods and Ethnic Divisions." *Quarterly Journal of Economics* 114, no. 4: 1243–1284.

Alfani, Guido. 2019. "Wealth and Income Inequality in the Long Run of History." In *Handbook of Cliometrics*, edited by Claude Diebolt and Michael Haupert, 1–30. Springer.

———. 2023. *As Gods Among Men: A History of the Rich in the West*. Princeton University Press.

Alstadsæter, Annette, Niels Johannesen, and Gabriel Zucman. 2018. "Who Owns the Wealth in Tax Havens? Macro Evidence and Implications for Global Inequality." *Journal of Public Economics* 162: 89–100.

———. 2019. "Tax Evasion and Inequality." *American Economic Review* 109, no. 6: 2073–2103.

Alter, George, and James C. Riley. 1986. "How to Bet on Lives: A Guide to Life Contingent Contracts in Early Modern Europe." *Research in Economic History* 10: 1–53.

Alvaredo, Facundo, Lucas Chancel, Thomas Piketty, Emmanuel Saez, and Gabriel Zucman, eds. 2018. *World Inequality Report*. Harvard University Press.

Ando, Albert, and Franco Modigliani. 1963. "The 'Life Cycle' Hypothesis of Saving: Aggregate Implications and Tests." *American Economic Review* 53, no. 1: 55–84.

Andrieu, Claire. 1993. "Les nationalisations bancaires en France de 1945 à nos jours." In *Les banques en Europe de l'Ouest de 1920 à nos jours*, edited by Maurice Lévy-Leboyer, 101–136. Comité pour l'histoire économique et financière de la France.

———. 2000. *La Spoliation financière: Rapport de la Mission d'étude sur la spoliation des Juifs de France*. La Documentation Française.

Andrieu, Claire, Lucette Le Van, and Antoine Prost, eds. 1987. *Les Nationalisations de la Libération: De l'utopie au compromis*. Presses de la fondation nationale des sciences politiques.

Aron, Jean-Paul, Paul Dumont, and Emmanuel Le Roy Ladurie. 1972. *Anthropologie du conscrit français d'après les comptes numériques et sommaires du recrutement de l'armée (1819–1826): Présentation cartographique*. Mouton.

Atkinson, Anthony. 2015. *Inequality: What Can Be Done?* Harvard University Press.

Atkinson, Anthony, and Alan Harrison. 1978. *Distribution of Personal Wealth in Britain, 1923–1972*. Cambridge University Press.

Atkinson, Anthony, and Thomas Piketty, eds. 2007. *Top Incomes over the Twentieth Century: A Contrast Between Continental European and English-Speaking Countries*. Oxford University Press.

Audiganne, Armand, et al. 1861. *Paris dans sa splendeur: Monuments, vues, scènes historiques, descriptions et histoire*. 3 vols. Henri Charpentier.

Baker, George P., and Brian J. Hall. 2004. "CEO Incentives and Firm Size." *Journal of Labor Economics* 22, no. 4: 767–798.

Bayet, Alain. 1997. "Deux siècles d'évolution des salaires en France (Document de travail INSEE)."

Benjamin, Walter. 1982. *Das Passagen-Werk*. Vol. 5. In *Gesammelte Schriften*, edited by Rolf Tiedemann and Hermann Schweppenhäuser, with Theodor W. Adorno and Gershom Scholem.

Benjamin, Walter. 1989. *Paris, capitale du XIXe siècle. Livre des passages*. Le Cerf.

Berlanstein, Lenard R. 1991. *Big Business and Industrial Conflict in Nineteenth-Century France: A Social History of the Parisian Gas Company*. University of California Press.

Bessière, Céline, and Sibylle Gollac. 2022. *"Le genre du capital." Comment la famille reproduit les inégalités*. La Découverte. English translation: *The Gender of Capital: How Families Perpetuate Wealth Inequality*. Harvard University Press, 2023.

Bignon, Vincent, and Mark Flandreau. 2011. "The Economics of Badmouthing: Libel Law and the Underworld of the Financial Press in France Before World War II." *Journal of Economic History* 71, no. 3: 616–653.

Blum, Alain, and Jacques Houdaille. 1985 "L'alphabétisation aux XVIIIe et XIXe siècles: L'illusion parisienne?" *Population* 40, no. 6: 944–951.

Bourdieu, Jérôme, Lionel Kesztenbaum, and Gilles Postel-Vinay. 2011. "Thrifty Pensioners: Pensions and Savings in France at the Turn of the Twentieth Century." *Journal of Economic History* 71, no. 2: 383–412.

———. 2014. *L'enquête TRA, histoire d'un outil, outil pour l'histoire*. Vol. 1. 1793–1902. Édition de l'INED.

Bourdieu, Jérôme, Lionel Kesztenbaum, Gilles Postel-Vinay, and Akiko Suwa-Eisenmann. 2019. "Intergenerational Wealth Mobility in France, 19th and 20th Century." *Review of Income and Wealth* 65, no. 1: 21–47.

Bourdieu, Jérôme, Gilles Postel-Vinay, and Akiko Suwa-Eisenmann. 2003. "Pourquoi la richesse ne s' est pas diffusée avec la croissance? Le degré zéro de l'inégalité et son évolution en France, 1800–1940." *Histoire & mesure* 18, no. 1/2: 147–198.

Bourdieu, Pierre. 1984. "L'espace social et la genèse des 'classes.'" *Actes de la recherche en sciences sociales*. 52–53: 3–14. https://www.persee.fr/doc/arss_0335-5322_1984_num_52_1_3327.

———. 1989. *La noblesse d'État: Grandes écoles et esprit de corps*. Editions de Minuit.

Bourdieu, Pierre, and Jean-Claude Passeron. 1964. *Les Héritiers: Les étudiants et la culture*. Editions de Minuit.

Bourdieu, Pierre, and Monique de Saint-Martin. 1978. "Le patronat." *Actes de la recherche en sciences sociales* 20–21: 3–82.

Boyer, George R. 2018. *The Winding Road to the Welfare State: Economic Insecurity and Social Welfare Policy in Britain*. Princeton University Press.

Bozio, Antoine. 2006. *Réforme des retraites: Estimations sur données françaises*. PhD diss., Paris, EHESS.

Brando, Carmen, Gabriela Elgarrista, Alena Lapatniova, Frédérique Mélanie-Becquet, and Gilles Postel-Vinay. 2023. "Les propriétaires de Paris et du département de la Seine—1898." https://data.progedo.fr/studies/doi/10.13144/lil-1578.

Brée, Sandra. 2016. *Paris l'inféconde: La limitation des naissances en région parisienne au XIXe siècle*. INED éditions.

Broustail, Joël. 2020. *Citroën et le citroënisme*. Au Pont 9.

Cagé, Julia, and Thomas Piketty. 2023. *Une histoire du conflit politique: Élections et inégalités sociales en France, 1789–2022*. Seuil.

Carrière, Jean. 1935. *La dette viagère française depuis la guerre: Thèse pour le doctorat*. Sirey.

Chapman, Jonathan. 2018. "Democratic Reform and Opposition to Government Expenditure: Evidence from Nineteenth-Century Britain." *Quarterly Journal of Political Science* 13, no. 4: 363–404.

———. 2024. "Gradual Franchise Extensions and Government Spending in Nineteenth-Century England." *Journal of Politics* 86, no. 1: 369–374.

Charle, Cristophe. 2006. *Les élites de la République, 1880–1900*. Fayard.

———. 2021. *Paris "capitale" des XIXe siècles*. Points.

Chetty, Raj, Nathaniel Hendren, Patrick Kline, and Emmanuel Saez. 2014. "Where Is the Land of Opportunity? The Geography of Intergenerational Mobility in the United States." *Quarterly Journal of Economics* 129, no. 4: 1553–1623.

Chetty, Raj, Nathaniel Hendren, Patrick Kline, Emmanuel Saez, and Nicholas Turner. 2014. "Is the United States Still a Land of Opportunity? Recent Trends in Intergenerational Mobility." *American Economic Review* 104, no. 5: 141–147.

Chevalier, Louis. 1949. *La Formation de la population parisienne au XIXe siècle*. PUF, collection Travaux et documents, Institut national d'études démographique.

———. 1958. *Classes laborieuses et classes dangereuses à Paris pendant la première moitié du XIXe siècle*. Plon.

Christen-Lécuyer, Carole. 2004. *Histoire sociale et culturelle des Caisses d'Épargne en France 1818–1881*. Economica. https://books.openedition.org/ined/1546#text.

Clark, Catherine E. 2018. *Paris and the Cliché of History: The City and Photographs, 1860–1970*. Oxford University Press.

Clark, Gregory. 2007. *A Farewell to Alms: A Brief Economic History of the World*. Princeton University Press.

Crouch, Gregory. 2018. *The Bonanza King: John Mackay and the Battle over the Greatest Riches in the American West*. Scribner.

Cummins, Neil. 2022. "The Hidden Wealth of English Dynasties, 1892–2016." *Economic History Review* 75, no. 3: 667–702.

Daumard, Adeline. 1962. "Structures sociales et classement socio-professionnel: L'apport des archives notariales au XVIIIe et au XIXe siècle." *Revue historique* 227, no. 1: 139–154.

———. 1963. *La bourgeoisie parisienne de 1815 à 1848*. Albin Michel.

———. 1973. *Les fortunes françaises au XIXe siècle: Enquête sur la répartition et la composition des capitaux privés à Paris, Lyon, Lille, Bordeaux et Toulouse d'après l'enregistrement des déclarations de succession*. Mouton.

Davis, Kingsley. 1955. "The Origin and Growth of Urbanization in the World." *American Journal of Sociology* 60, no. 5: 429–437.

Davis, Lance E., and Robert A. Huttenback. 1986. *Mammon and the Pursuit of Empire: The Political Economy of British Imperialism, 1860–1912*. Cambridge University Press.

Davis, Lance E., Larry D. Neal, and Eugene N. White. 2003. "How It All Began: The Rise of Listing Requirements on the London, Berlin, Paris, and New York Stock Exchanges." *International Journal of Accounting* 38, no. 2: 117–143.

De Coninck, Séverine. 2012. *Le livret de Caisse d'Épargne: Histoire et anthropologie d'un instrument financier (1818–2008)*. Economica.

Delorme, Robert, and Christine André. 1983. *L'Etat et l'économie: Un essai d'explication de l'évolution des dépenses publiques en France 1870–1980*. Le Seuil.

Demsetz, Harold. 1972. "Wealth Distribution and the Ownership of Rights." *Journal of Legal Studies* 1, no. 2: 223–232.

De Saint-Jacob, Pierre. 1960. *Les paysans de la Bourgogne du nord au dernier siècle de l'Ancien Régime*. Les Belles Lettres.

Dherbécourt, Clément. 2013. *La fortune des héritiers: Essais sur la transmission du capital en France du Second Empire à la Seconde Guerre mondiale*. PhD diss., Paris, EHESS.

Duma, Jean. 1995. *Les Bourbon-Penthièvre (1678–1793): Une nébuleuse aristocratique au XVIIIe siècle*. Les Éditions de la Sorbonne.
Dupâquier, Jacques, and Denis Kessler, eds. 1992. *La société française au XIXe siècle*. Fayard.
Eichholtz, Piet, Matthijs Korevaar, Thies Lindenthal, and Ronan Tallec. 2021. "The Total Return and Risk to Residential Real Estate." *Review of Financial Studies* 34, no. 8: 3608–3646.
Emigh, Rebecca Jean. 2003. "Property Devolution in Tuscany." *Journal of Interdisciplinary History* 33, no. 3: 385–420.
Engerman, Stanley L., and Kenneth L. Sokoloff. 2012. *Economic Development in the Americas Since 1500: Endowments and Institutions*. Cambridge University Press.
Epple, Dennis, and Richard E. Romano. 1996. "Public Provision of Private Goods." *Journal of Political Economy* 104, no. 1: 57–84.
Fack, Gabrielle, and Julien Grenet. 2009. "Sectorisation des collèges et prix des logements à Paris." *Actes de la recherche en sciences sociales* 180, no. 5: 44–62.
Farcy, Jean-Claude, and Alain Faure. 2003. *La mobilité d'une génération de Français: Recherche sur les migrations dans et vers Paris à la fin du XIXe siècle*. INED.
Farquet, Christophe. 2018. *Histoire du paradis fiscal suisse: Expansion et relations internationales du centre offshore suisse au XXe siècle*. Presses de Sciences Po.
Feller, Elise. 2017. *Du vieillard au retraité: La construction de la vieillesse dans la France du XXe siècle*. L'Harmattan.
Ferguson, Nial. 1998. *The House of Rothschild: Money's Prophets, 1798–1848*. Vol. 1. Viking.
Flandreau, Marc. 2004. *The Glitter of Gold: France, Bimetallism, and the Emergence of the International Gold Standard, 1848–1873*. Oxford University Press.
Fogel, Robert W. 1967. "The Specification Problem in Economic History." *Journal of Economic History* 27, no. 3: 283–308.
———. 1984. *The Escape from Hunger and Premature Death, 1700–2100: Europe, America, and the Third World*. Cambridge University Press.
Frain de La Gaulayrie, Pierre. 1909. *Les majorats depuis le premier Empire jusqu'à nos jours: Thèse pour le doctorat*. Rennes. http://catalogue.bnf.fr/ark:/12148/cb321232502.
Friggit, Jacques. 2007. "Long Term (1800–2005) Investment in Gold, Bonds, Stocks and Housing in France with Insights into the USA and the UK: A Few Regularities." *Note de Travail, CGPC*.
Gaillard, Jeanne. 1998. *Paris, la ville (1852–1870)*. L'Harmattan.
Gale, William G., James R. Hines, and Joel Slemrod. 2001. *Rethinking Estate and Gift Taxation*. Brookings Institution Press.
Galor, Oded, and Omer Moav. 2004. "From Physical to Human Capital Accumulation: Inequality and the Process of Development." *Review of Economic Studies* 71, no. 4: 1001–1026.
Galor, Oded, and Joseph Zeira. 1993. "Income Distribution and Macroeconomics." *Review of Economic Studies* 60, no. 1: 35–52.
Garbinti, Bertrand, Jonathan Goupille-Lebret, and Thomas Piketty. 2018. "Income Inequality in France, 1900–2014: Evidence from Distributional National Accounts (DINA)." *Journal of Public Economics* 162: 63–77.
Gerschenkron, Alexander. 1951. "Economic Backwardness in Historical Perspective." Reprinted in *The Political Economy Reader: Markets as Institutions*, edited by Alexander Gershenkron, 211–228. Belknap Press, 1962.
Goeree, Jacob K., Margaret A. McConnell, Tiffany Mitchell, Tracey Tromp, and Leeat Yariv. 2010. "The 1/d Law of Giving." *American Economic Journal: Microeconomics* 2, no. 1: 183–203.
Goldin, Claudia. 1990. *Understanding the Gender Gap: An Economic History of American Women*. Oxford University Press.
———. 2006. "The Quiet Revolution That Transformed Women's Employment, Education, and Family." *American Economic Review* 96, no. 2: 1–21.

Goy, Joseph. 1988. "La Révolution française et la famille." In *Histoire de la population française*, edited by Jacques Dupâquier et al., vol. 3, 84–115. PUF.

Gréard, Octave. 1877. *L'instruction primaire à Paris et dans le département de la Seine, en 1875: Mémoire adressé à M. le préfet de la Seine*. Imprimerie Paul Dupont.

Gribaudi, Maurizio. 2014. *Paris ville ouvrière: Une histoire occultée*. La Découverte.

Guex, Sébastien. 2006. "Les scandales des fraudes fiscales de la Banque commerciale de Bâle." In *Des économies et des hommes: Mélanges offerts à Albert Broder*, edited by F. Bourillon et al. Editions Bière.

———. 2022. "The Emergence of the Swiss Tax Haven, 1816–1914." *Business History Review* 96, no. 2: 353–372.

Guibert, Adrien. 1848. *Dictionnaire géographique et statistique*. Jules Renouard et cie.

Guillaume, Pierre. 1986. *Du désespoir au salut: Les tuberculeux aux XIXe et XXe siècles*. Aubier.

Hautcoeur, Pierre-Cyrille. 2007. *Le marché financier français au XIXe siècle: Récit*. Vol. 1. Publications de la Sorbonne.

Hautcoeur, Pierre-Cyrille, and Sandrine Grotard. 2001. "Taxation of Corporate Profits, Inflation and Income Distribution in France, 1914–1926." *Communication au congrès d'Oxford de l'European historical economics society*.

Herlihy, David, and Christiane Klapisch-Zuber. 1985. *Tuscans and Their Families: A Study of the Florentine Catasto of 1427*. Yale University Press.

Higgs, David. 2019. *Nobles in Nineteenth-Century France: The Practice of Inegalitarianism*. Johns Hopkins University Press.

Hoffman, Philip T. 1984. "The Economic Theory of Sharecropping in Early Modern France." *Journal of Economic History* 44, no. 2: 309–319.

———. 1986. "Taxes and Agrarian Life in Early Modern France: Land Sales, 1550–1730." *Journal of Economic History* 46, no. 1: 37–55.

Hoffman, Philip T., Gilles Postel-Vinay, and Jean-Laurent Rosenthal. 2000. *Priceless Markets: The Political Economy of Credit in Paris, 1662–1869*. University of Chicago Press.

———. 2019. *Dark Matter Credit: The Development of Peer-to-Peer Lending and Banking in France*. Princeton University Press.

Homer, Sidney, and Richard Eugene Sylla. 1996. *A History of Interest Rates*. Rutgers University Press.

Hugo, Victor. 1867. *Paris-Guide par les principaux écrivains et artistes de la France*, introduction by Victor Hugo. 2 vols. Librairie internationale.

Isambert-Jamati, Viviane. 1970. *Crises de la société, crise de l'enseignement: Sociologie de l'enseignement secondaire français*. PUF.

Joly, Laurent. 2022. *La rafle du Vel d'Hiv Paris, juillet 1942*. Grasset.

Jones, Alice H. 1980. *Wealth of a Nation to Be: The American Colonies on the Eve of the Revolution*. Columbia University Press.

Jones, Joseph. 1984. *The Politics of Transport in Twentieth-Century France*. McGill-Queen's University Press.

Kesztenbaum, Lionel. 2008. "Places of Life Events as Bequestable Wealth: Family Territory and Migration in France, 19th and 20th Centuries." In *Kinship and Demographic Behavior in the Past*, edited by Tommy Bengtsson and Geraldine P. Mineau, 155–184. Springer.

Kesztenbaum, Lionel, and Jean-Laurent Rosenthal. 2011. "The Health Cost of Living in a City: The Case of France at the End of the 19th Century." *Explorations in Economic History* 48, no. 2: 207–225.

———. 2016. "The Democratization of Longevity: How the Poor Became Old in Paris, 1880–1913." In *New Approaches to Death in Cities During the Health Transition*, edited by Diego Ramiro Fariñas and Michel Oris, 137–154. Springer.

———. 2017. "Sewers' Diffusion and the Decline of Mortality: The Case of Paris, 1880–1914." *Journal of Urban Economics* 98: 174–186.

Kopczuk, Wojciech, and Emmanuel Saez. 2004. "Top Wealth Shares in the United States, 1916–2000: Evidence from Estate Tax Returns." *National Tax Journal* 57, no. 2: 445–487.

Kselman, Thomas A. 1993. *Death and Afterlife in Modern France*. Princeton University Press.

Kuznets, Simon. 1955. "Economic Growth and Income Inequality." *American Economic Review* 45, no. 1: 1–28.

———. 1971. *Economic Growth of Nations*. Belknap Press. Reprinted 2013.

Lagneau-Ymonet, Paul, and Angelo Riva. 2012. *Histoire de la Bourse*. La Découverte.

Lamoreaux, Naomi R., and Jean-Laurent Rosenthal. 2005. "Legal Regime and Contractual Flexibility: A Comparison of Business's Organizational Choices in France and the United States During the Era of Industrialization." *American Law and Economics Review* 7, no. 1: 28–61.

Lampman, Robert. 1962. *The Share of Top Wealth-Holders in National Wealth, 1922–1956*. Princeton University Press.

La Porta, Rafael, Florencio Lopez-de-Silanes, Andrei Shleifer, and Robert W. Vishny. 1997. "Legal Determinants of External Finance." *Journal of Finance* 52, no. 3: 1131–1150.

———. 1998. "Law and Finance." *Journal of Political Economy* 106, no. 6: 1113–1155.

Lassère, Madeleine. 1997. *Villes et cimetières en France. De l'Ancien Régime à nos jours. Le territoire des morts*. L'Harmattan.

Le Bris, David, and Pierre-Cyrille Hautcoeur. 2010. "A Challenge to Triumphant Optimists? A Blue Chips Index for the Paris Stock Exchange, 1854–2007." *Financial History Review* 17, no. 2: 141–183.

Lemercier, Claire. 2003. *Un si discret pouvoir: Aux origines de la Chambre de commerce de Paris 1803–1853*. La Découverte.

———. 2004. "Classer l'industrie parisienne au XIXe siècle." In *Nomenclatures et classifications: Approches historiques, enjeux économiques*, edited by Jérôme Bourdieu, Martin Breugel, and Alessandro Stanziani, 237–271. Actes et communications de l'INRA, no. 21.

Léonard, Jacques. 1978. *La France médicale: Médecins et malades au XIXe siècle*. Gallimard.

Le Roy Ladurie, Emmanuel, and Nicole Bernageau. 1971. "Étude sur un contingent militaire (1868): Mobilité géographique délinquance et stature, mises en rapport avec d'autres aspects de la situation des conscrits." *Annales de démographie historique*: 11–337.

Le Roy Ladurie, Emmanuel, Nicole Bernageau, and Yvonne Pasquet. 1969. "Le conscrit et l'ordinateur: Perspectives de recherche sur les archives militaires du XIXe siècle français." *Studi Storici* 10, no. 2: 260–308.

Levine, Ross. 1997. "Financial Development and Economic Growth: Views and Agenda." *Journal of Economic Literature* 35, no. 2: 688–726.

Lévy, Claude, and Henry Louis. 1960. "Ducs et pairs sous l'Ancien Régime. Caractéristiques démographiques d'une caste." *Population* 15, no. 5: 807–830.

Lévy-Leboyer, Maurice, 1977. *La position internationale de la France. Aspects économiques et financiers. XIXe-XXe siècles*. Editions de l'EHESS.

———. 1979. Le patronat de la seconde industrialisation. *Cahiers du "Mouvement Social."* Editions ouvrières.

Lévy-Leboyer, Maurice, and François Bourguignon. 1990. *A Macroeconomic Model of France During the 19th Century: An Essay in Econometric Analysis*. Cambridge University Press.

Lindert, Peter H. 1986. "Unequal English Wealth Since 1670." *Journal of Political Economy* 94, no. 6: 1127–1162.

———. 2004. *Growing Public: Volume 1, The Story; Social Spending and Economic Growth Since the Eighteenth Century*. Vol. 1. Cambridge University Press.

Lizzeri, Alessandro, and Nicola Persico. 2004. "Why Did the Elites Extend the Suffrage? Democracy and the Scope of Government, with an Application to Britain's 'Age of Reform.'" *Quarterly Journal of Economics* 119, no. 2: 707–765.

Lucas, Robert E. Jr. 1990. "Why Doesn't Capital Flow from Rich to Poor Countries?" *American Economic Review* 80: 92–96.

Lyon-Caen, Nicolas. 2018. "Combien vaut Paris? La monarchie, les loyers et les boues de la capitale à l'époque moderne." In *Annuaire-Bulletin de la société de l'histoire de France 2014*, 179–199. https://hal.science/hal-02013079/document.

Magnac, Thierry, and Gilles Postel-Vinay. 1997. "Wage Competition Between Agriculture and Industry in Mid-Nineteenth Century France." *Explorations in Economic History* 34, no. 1: 1–26.

Mallet, Bernard. 1908. "A Method of Estimating Capital Wealth from the Estate Duty Statistics." *Journal of the Royal Statistical Society* 71, no. 1: 65–101.

Mallet, Bernard, and H. C. Strutt. 1915. "The Multiplier and Capital Wealth." *Journal of the Royal Statistical Society* 78, no. 4: 555–599.

Manas, Arnaud. 2019. "Banque de France's Shareholders (1800–1945): Passive Petit-Rentiers." *Research in Economic History* 35: 133–163.

Marchand, Olivier, and Claude Thélot. 1991. *Deux siècles de travail en France: Population active et structure sociale, durée et productivité du travail*. Editions de l'INSEE.

Marnata, Francoise. 1961. *Les loyers des bourgeois de Paris: 1860–1958*. Armand Colin.

Masson, André. 2025. *L' Héritage au XXIe siècle*. Odile Jacob.

McGarry, Kathleen, and Robert F. Schoeni. 2000. "Social Security, Economic Growth, and the Rise in Elderly Widows' Independence in the Twentieth Century." *Demography* 37, no. 2: 221–236.

Meltzer, Allan H., and Scott F. Richard. 1981. "A Rational Theory of the Size of Government." *Journal of Political Economy* 89, no. 5: 914–927.

Mitchener, Kris James, and Eric Monnet. 2023. "Connected Lending of Last Resort." No. w30869. National Bureau of Economic Research.

Modigliani, Franco, 1986. "Life Cycle, Individual Thrift and the Wealth of Nations." *American Economic Review* 76, no. 3: 97–313.

Monnet, Eric. 2018. *Controlling Credit: Central Banking and the Planned Economy in Postwar France (1948–1973)*. Cambridge University Press.

Moriceau, Jean-Marc, and Gilles Postel-Vinay. 1992. *Ferme, entreprise, famille: Grande exploitation et changements agricoles: Les Chartier, XVIIe-XIXe sièces*. Éditions de l'ÉHESS.

Moulinier, Pierre. 2012. *Les Étudiants étrangers à Paris au XIXe siècle: Migrations et formation des élites*. Presses universitaires de Rennes.

———. 2014. "Les étudiants étrangers à Paris en 1900." *Hommes & migrations* 1308: 119–127.

Neal, Larry D. 2000. "How It All Began: The Monetary and Financial Architecture of Europe During the First Global Capital Markets, 1648–1815." *Financial History Review* 7, no. 2: 117–140.

Neymarck, Alfred. 1903. "Statistique nouvelle sur le morcellement des valeurs mobilières (chemins de fer, rentes, Banque de France, Crédit foncier, etc.)." *Journal de la société statistique de Paris* 44: 127–142.

———. 1908. *Les émissions et remboursements d'obligations des six grandes compagnies de chemins de fer en 1907*. Editions Félix Alcan.

North, Douglass C. 1981. *Structure and Change in Economic History*. W. W. Norton.

Oosterlinck, Kim, Loredana Ureche-Rangau, and Jacques-Marie Vaslin, 2019, "Aristocratic Privilege. Exploiting 'Good' Institutions." CEPR Discussion Paper No. DP14071. https://ssrn.com/abstract=3474443.

Parent, Antoine, and Christophe Rault. 2004. "The Influences Affecting French Assets Abroad Prior to 1914." *Journal of Economic History* 64, no. 2: 328–362.

Passy, Louis. 1874. *Frochot préfet de la Seine*. Imprimerie Auguste Herrissey.

Pastore, Thomas. 2024. "The Belle-Epoque of Portfolios? How Returns, Risk, and Diversification Correlated with the Wealth Distribution in Paris in 1912." *Journal of Economic History* 84, no. 3: 767–806.

Piette, Christine, and Barrie M. Ratcliffe. 1993. "Les migrants et la ville: Un nouveau regard sur le Paris de la première moitié du XIXe siècle." *Annales de démographie historique. Société de Démographie Historique*.

Piketty, Thomas. 2001. *Les hauts revenus en France au XXᵉ siècle*. Grasset.
———. 2013. *Le capital au XXIᵉ siècle*. Éditions du Seuil. English translation: *Capital in the Twenty-First Century*. Harvard University Press, 2014.
Piketty, Thomas, and Anthony Atkinson, eds. 2010. *Top Incomes: A Global Perspective*. Oxford University Press.
Piketty, Thomas, Gilles Postel-Vinay, and Jean-Laurent Rosenthal. 2006. "Wealth Concentration in a Developing Economy: Paris and France, 1807–1994." *American Economic Review* 96, no. 1: 236–256.
———. 2014. "Inherited vs. Self-Made Wealth: Theory & Evidence from a Rentier Society (Paris 1872–1927)." *Explorations in Economic History* 51: 21–40.
Piketty, Thomas, and Gabriel Zucman. 2014. "Capital Is Back: Wealth-Income Ratios in Rich Countries 1700–2010." *Quarterly Journal of Economics* 129, no. 3: 1255–1310.
Plessis, Alain. 1982. *La Banque de France et ses deux cents actionnaires sous le Second Empire*. Vol. 40. Librairie Droz.
———. 1985. *Régents et gouverneurs de la Banque de France sous le Second Empire*. Vol. 44. Librairie Droz.
Postel-Vinay, Gilles. 2002. "The Dis-integration of Traditional Labour Markets in France: From Agriculture and Industry to Agriculture or Industry." In *Labour Market Evolution: The Economic History of Market Integration, Wage Flexibility and the Employment Relation*, edited by George Grantham and Mary MacKinnon, 78–97. Routledge.
Preston, Samuel H., and Etienne van de Walle. 1978. "Urban French Mortality in the Nineteenth Century." *Population Studies* 32, no. 2: 275–297.
Rajan, Raghuram G., and Luigi Zingales. 2004. *Saving Capitalism from the Capitalists: Unleashing the Power of Financial Markets to Create Wealth and Spread Opportunity*. Princeton University Press.
Ransom, Roger L., and Richard Sutch. 1986. "The Labor of Older Americans: Retirement of Men On and Off the Job, 1870–1937." *Journal of Economic History* 46, no. 1: 1–30.
Roehl, Richard. 1976. "French Industrialization: A Reconsideration." *Explorations in Economic History* 13, no. 3: 233–281.
Rosenthal, Jean-Laurent, and R. Bin Wong. 2011. *Before and Beyond Divergence: The Politics of Economic Change in China and Europe*. Harvard University Press.
Rydqvist, Kristian, and Rong Guo. 2021. "Performance and Development of a Thin Stock Market: The Stockholm Stock Exchange, 1912–2017." *Financial History Review* 28, no. 1: 26–44.
Saez, Emmanuel, and Gabriel Zucman. 2016. "Wealth Inequality in the United States Since 1913: Evidence from Capitalized Income Tax Data." *Quarterly Journal of Economics* 131, no. 2: 519–578.
———. 2019. *The Triumph of Injustice: How the Rich Dodge Taxes and How to Make Them Pay*. W. W. Norton.
Sagot, Mariette. 2019. "Gentrification et paupérisation au cœur de l'Île-de-France—Évolutions 2001–2015." Institut d'aménagement et d'urbanisme d'Île-de-France.
Scheidel, Walter. 2018. *The Great Leveler: Violence and the History of Inequality from the Stone Age to the Twenty-First Century*. Princeton University Press.
Scheve, Kenneth, and David Stasavage. 2012. "Democracy, War, and Wealth: Lessons from Two Centuries of Inheritance Taxation." *American Political Science Review* 106, no. 1: 81–102.
Schumpeter, Joseph A. 1950. *Capitalism, Socialism and Democracy*. 3rd ed. Harper-Collins.
Séailles, J. 1910. *La répartition des fortunes en France*. Editions Félix Alcan.
Shapiro, Anne-Louise. 1985. *Housing the Poor of Paris, 1850–1902*. University of Wisconsin Press.
Sicsic, Pierre. 1992. "City-Farm Wage Gaps in Late Nineteenth-Century France." *Journal of Economic History* 52, no. 3: 675–695.
Slivinski, Al, and Nathan Sussman. 2019. "Tax Administration and Compliance: Evidence from Medieval Paris." CEPR discussion paper, no. DP13512.

Smith, Matthew, Owen Zidar, and Eric Zwick. 2023. "Top Wealth in America: New Estimates under Heterogeneous Returns." *Quarterly Journal of Economics* 138, no. 1: 515–573.

Solow, Robert M. 1956. "A Contribution to the Theory of Economic Growth." *Quarterly Journal of Economics* 70, no. 1: 65–94.

Soltow, Lee, and Jan Luiten Van Zanden. 1998. *Income and Wealth Inequality in the Netherlands, 16th–20th Century*. Het Spinhuis.

Stamp, Josiah C. 1919. "The Wealth and Income of the Chief Powers." *Journal of the Royal Statistical Society* 82, no. 4: 441–507.

Stourm, René. 1905. *Systèmes généraux d'impôts*. Guillaumin et Cie.

Strutt, H. C. 1910. "Notes on the Distribution of Estates in France and the United Kingdom." *Journal of the Royal Statistical Society* 73, no. 6/7: 634–644.

Subra, Philippe. 2012. *Le grand Paris: Géopolitique d'une ville mondiale*. Armand Colin.

Sussman, Nathan. 2006. "Income Inequality in Paris in the Heyday of the Commercial Revolution." Working paper.

Sutch, Richard. 1991. "All Things Reconsidered: The Life-Cycle Perspective and the Third Task of Economic History." *Journal of Economic History* 51, no. 2: 271–288.

Szramkiewicz, Romuald. 1974. *Les régents et censeurs de la Banque de France nommés sous le Consulat et l'Empire*. Librairie Droz.

Thiveaud, Jean-Marie. 1997. "La lente construction des systèmes de retraite en France de 1750 à 1945." *Revue d'économie financière*: 21–54.

Vuillermot, Catherine. 2003. "La nationalisation de l'électricité en France en 1946: Le problème de l'indemnisation." *Annales historiques de l'électricité* no. 1: 53–69.

Waldenström, Daniel. 2024. *Richer and More Equal: A New History of Wealth in the West*. John Wiley & Son.

Waldman, Michael. 1994. "Systematic Errors and the Theory of Natural Selection." *American Economic Review* 84, no. 3: 482–497.

Weir, David R. 1989. "Tontines, Public Finances, and Revolution in France and England, 1688–1789." *Journal of Economic History* 49, no. 1: 95–124.

Woloch, Isser. 1994. *The New Regime: Transformations of the French Civic Order, 1789–1820s*. Norton.

Wrigley, E. Anthony. 1967. "A Simple Model of London's Importance in Changing English Society and Economy, 1650–1750." *Past & Present* 37: 44–70.

Yates, Alexia. 2015. *Selling Paris: Property and Commercial Culture in the Fin-de-Siècle Capital*. Harvard University Press.

Zucman, Gabriel. 2013. *La richesse cachée des nations: Enquête sur les paradis fiscaux*. Le Seuil. English translation: *The Hidden Wealth of Nations: The Scourge of Tax Havens*, translated by Teresa Lavender Fagan. University of Chicago Press, 2015.

———. 2020. *Le Triomphe de l'injustice: Richesse, évasion fiscale et démocratie*. Seuil.

INDEX

Page numbers in *italics* refer to illustrations and text boxes.

The Princeton Economic History of the Western World
Joel Mokyr, Series Editor

Recent titles

Raised to Obey: The Rise and Spread of Mass Education, Agustina Paglayan

The Laissez-Faire Experiment: Why Britain Embraced and Then Abandoned Small Government, 1800–1914, W. Walker Hanlon

The Hidden Victims: Civilian Casualties of the Two World Wars, Cormac Ó Gráda

The Wealth of a Nation: Institutional Foundations of English Capitalism, Geoffrey M. Hodgson

The Corporation and the Twentieth Century: The History of American Business Enterprise, Richard N. Langlois

Pioneers of Capitalism: The Netherlands 1000–1800, Maarten Prak and Jan Luiten van Zanden

Pawned States: State Building in the Era of International Finance, Didac Queralt

Pliny's Roman Economy: Natural History, Innovation, and Growth, Richard P. Saller

Plague's upon the Earth: Disease and the Course of Human History, Kyle Harper

The Israeli Economy: A Story of Success and Costs, Joseph Zeira

Credit Nation: Property Laws and Institutions in Early America, Claire Priest

The Decline and Rise of Democracy: A Global History from Antiquity to Today, David Stasavage

Going the Distance: Eurasian Trade and the Rise of the Business Corporation 1400–1700, Ron Harris

Escape from Rome: The Failure of Empire and the Road to Prosperity, Walter Scheidel

Trade in the Ancient Mediterranean: Private Order and Public Institutions, Taco Terpstra

Dark Matter Credit: The Development of Peer-to-Peer Lending and Banking in France, Philip T. Hoffman, Gilles Postel-Vinay, and Jean-Laurent Rosenthal

The European Guilds: An Economic Analysis, Sheilagh Ogilvie

The Winding Road to the Welfare State: Economic Insecurity and Social Welfare Policy in Britain, George R. Boyer

The Mystery of the Kibbutz: Egalitarian Principles in a Capitalist World, Ran Abramitzky

Unequal Gains: American Growth and Inequality since 1700, Peter H. Lindert and Jeffrey G. Williamson

Uneven Centuries: Economic Development of Turkey since 1820, Şevket Pamuk

The Great Leveler: Violence and the History of Inequality from the Stone Age to the Twenty-First Century, Walter Scheidel

The Rise and Fall of American Growth: The U.S. Standard of Living since the Civil War, Robert J. Gordon

Brazil in Transition: Beliefs, Leadership, and Institutional Change, Lee J. Alston, Marcus André Melo, Bernardo Mueller, and Carlos Pereira

The Vanishing Irish: Households, Migration, and the Rural Economy in Ireland, 1850–1914, Timothy W. Guinnane

Fragile by Design: The Political Origins of Banking Crises and Scarce Credit, Charles W. Calomiris and Stephen H. Haber

The Son Also Rises: Surnames and the History of Social Mobility, Gregory Clark

Why Did Europe Conquer the World?, Philip T. Hoffman

Lending to the Borrower from Hell: Debt, Taxes, and Default in the Age of Philip II, Mauricio Drelichman and Hans-Joachim Voth

Power to the People: Energy in Europe over the Last Five Centuries, Astrid Kander, Paolo Malanima, and Paul Warde

Cities of Commerce: The Institutional Foundations of International Trade in the Low Countries, 1250–1650, Oscar Gelderblom

Why Australia Prospered: The Shifting Sources of Economic Growth, Ian W. McLean

The Chosen Few: How Education Shaped Jewish History, 70–1492, Maristella Botticini and Zvi Eckstein

Creating Wine: The Emergence of a World Industry, 1840–1914, James Simpson

Distant Tyranny: Markets, Power, and Backwardness in Spain, 1650–1800, Regina Grafe

The Evolution of a Nation: How Geography and Law Shaped the American States, Daniel Berkowitz and Karen B. Clay

The Roman Market Economy, Peter Temin

States of Credit: Size, Power, and the Development of European Polities, David Stasavage

Power over Peoples: Technology, Environments, and Western Imperialism, 1400 to the Present, Daniel R. Headrick

Unsettled Account: The Evolution of Banking in the Industrialized World since 1800, Richard S. Grossman

A Farewell to Alms: A Brief Economic History of the World, Gregory Clark

Power and Plenty: Trade, War, and the World Economy in the Second Millennium, Ronald Findlay and Kevin H. O'Rourke

War, Wine, and Taxes: The Political Economy of Anglo-French Trade, 1689–1900, John V. C. Nye

Cultures Merging: A Historical and Economic Critique of Culture, Eric L. Jones

The European Economy since 1945: Coordinated Capitalism and Beyond, Barry Eichengreen

Feeding the World: An Economic History of Agriculture, 1800–2000, Giovanni Federico

Understanding the Process of Economic Change, Douglass C. North

The Strictures of Inheritance: The Dutch Economy in the Nineteenth Century, Jan Luiten van Zanden and Arthur van Riel

Farm to Factory: A Reinterpretation of the Soviet Industrial Revolution, Robert C. Allen

Quarter Notes and Bank Notes: The Economics of Music Composition in the Eighteenth and Nineteenth Centuries, F. M. Scherer

The Big Problem of Small Change, Thomas J. Sargent and François R. Velde

The Great Divergence: China, Europe, and the Making of the Modern World Economy, Kenneth Pomeranz

Black '47 and Beyond: The Great Irish Famine in History, Economy, and Memory, Cormac Ó Gráda

Growth in a Traditional Society: The French Countryside, 1450–1815, Philip T. Hoffman

A NOTE ON THE TYPE

This book has been composed in Adobe Text and Gotham. Adobe Text, designed by Robert Slimbach for Adobe, bridges the gap between fifteenth- and sixteenth-century calligraphic and eighteenth-century Modern styles. Gotham, inspired by New York street signs, was designed by Tobias Frere-Jones for Hoefler & Co.